Advances in Medical Imaging

From Behavioral Analysis to Disease Prevention and Rehabilitation

Advances in Medical Imaging

From Behavioral Analysis to Disease Prevention and Rehabilitation

Edited by

Dilber Uzun Ozsahin
Medical Diagnostic Imaging Department, College of Health Sciences,
Research Institute for Medical and Health Sciences,
University of Sharjah, Sharjah, United Arab Emirates;
Irfan Suat Gunsel Operational Research Institute,
Near East University, Nicosia,
Turkish Republic of Northern Cyprus (TRNC), Turkey

Ilker Ozsahin
Irfan Suat Gunsel Operational Research
Institute, Near East University, Nicosia,
Turkish Republic of Northern Cyprus (TRNC), Turkey

Leena R. David
Medical Diagnostic Imaging Department,
College of Health Sciences,
University of Sharjah, Sharjah, United Arab Emirates;
Department of Medical Imaging Technology,
Manipal College of Health Professions,
Manipal Academy of Higher Education,
Manipal, Karnataka, India

Suresh Sukumar
Department of Medical Imaging Technology,
Manipal College of Health Professions,
Manipal Academy of Higher Education,
Manipal, Karnataka, India

Berna Uzun
Irfan Suat Gunsel Operational Research Institute, Near East University, Nicosia,
Turkish Republic of Northern Cyprus (TRNC), Turkey; Medical Diagnostic Imaging Department,
College of Health Sciences, University of Sharjah, Sharjah, United Arab Emirates

ELSEVIER

ACADEMIC PRESS
An imprint of Elsevier

Academic Press is an imprint of Elsevier
125 London Wall, London EC2Y 5AS, United Kingdom
50 Hampshire Street, 5th Floor, Cambridge, MA 02139, United States

ISBN: 978-0-443-28967-5

For information on all Academic Press publications visit our website at https://www.elsevier.com/books-and-journals

Publisher: Mara Conner
Acquisitions Editor: Sonnini Yura
Editorial Project Manager: Sudip Saha
Production Project Manager: Maria Bernard
Cover Designer: Bakyalakshmi S

Typeset by Integra Software Services

Contents

3 Medical imaging, behavioural changes, and their effects on memory and learning

Venu Madhav Nelluri, Sudarshan Surendran, Suresh Sukumar, Leena R. David, Sneha Ravichandran, Berna Uzun, Dilber Uzun Ozsahin, M.G. Ramesh Babu, Rajagopal Kadavigere, Dilip Shettigar, and Ilker Ozsahin

Part II
Sedentary behavior, lifestyle, and brain health

4 Medical imaging in the assessment of sedentary behavior

Dilip Shettigar, Rajagopal Kadavigere, Leena R. David, Sneha Ravichandran, Baskaran Chandrasekaran, Berna Uzun, Dilber Uzun Ozsahin, Suresh Sukumar, Sathya Sabina Muthu, Ilker Ozsahin, Priya Seagal V.K., and K.R. Banumathe

5 Impact of sedentary behaviour and nutritional intake on structural brain network among young adults

Sathya Sabina Muthu, Rajagopal Kadavigere, K. Vaishali, Hari Prakash Palaniswamy, Sneha Ravichandran, Dilip Shettigar, Leena R. David, Dilber Uzun Ozsahin, Berna Uzun, Ilker Ozsahin,and Suresh Sukumar

Part IV Frailty, musculoskeletal health, and imaging

17 The role of ultrasound in assessing repetitive motion injuries in musculofascial system

Kalyan Chakravarthi Kosuri, Venu Madhav Nelluri, Rajagopal Kadavigere, Suresh Sukumar, Sneha Ravichandran, Leena R. David, Berna Uzun, Dilber Uzun Ozsahin, and Ilker Ozsahin

Part V
Chronic disease, yoga, and cognitive well-being

18 Muscle imaging in chronic obstructive pulmonary disease

Anup Bhat, K. Vaishali, Mukesh Kumar Sinha, Leena R. David, Sneha Ravichandran, Suresh Sukumar, Rajagopal Kadavigere, Dilber Uzun Ozsahin, Ilker Ozsahin, and Berna Uzun

Part VII
Traumatic brain injury and auditory neuroscience

24 Rehabilitation and medical imaging strategies for coma recovery following traumatic brain injuries

Guruprasad Vijayasarathi, Shovan Saha, Leena R. David, Dilip Shettigar, Sneha Ravichandran, Dilber Uzun Ozsahin, Berna Uzun, Suresh Sukumar, Rajagopal Kadavigere, Ilker Ozsahin, and Banumathe Karuppaya

25 Role of magnetic resonance imaging in observing the structure of amygdala and hippocampus to assess the stress related behavioral changes and the impact on learning and memory

Sudarshan Surendran, Venu Madhav Nelluri, Suresh Sukumar, Leena R. David, Sneha Ravichandran, Berna Uzun, Dilber Uzun Ozsahin, M.G. Ramesh Babu, Rajagopal Kadavigere, Dilip Shettigar, and Ilker Ozsahin

26 Imaging in auditory neuroscience

Pradeep Yuvaraj, Aravind Kumar Rajasekaran, Rajagopal Kadavigere, Leena R. David, Sneha Ravichandran, Dilip Shettigar, Suresh Sukumar, Berna Uzun, Dilber Uzun Ozsahin, and Ilker Ozsahin

27 Inner ear: Imaging and functional correlation

Pradeep Yuvaraj, Aravind Kumar Rajasekaran, Sneha Ravichandran, Leena R. David, Rajagopal Kadavigere, Dilber Uzun Ozsahin, Dilip Shettigar, Suresh Sukumar, Ilker Ozsahin, and Berna Uzun

Part VIII
Imaging in oncology and systemic diseases

28 Retrocochlear condition: Imaging and clinical correlates

Pradeep Yuvaraj, Aravind Kumar Rajasekaran, Rajagopal Kadavigere, Leena R. David, Berna Uzun, Dilber Uzun Ozsahin, Sneha Ravichandran, Suresh Sukumar, Ilker Ozsahin, and Dilip Shettigar

29 The role of imaging in oncology

Winniecia Dkhar, Rajagopal Kadavigere, S. Sharath, Abhimanyu Pradhan, Leena R. David, Berna Uzun, Dilber Uzun Ozsahin, Ilker Ozsahin, and Neil Abraham Barnes

30 Recent trends and techniques in breast imaging

S. Sharath, Rajagopal Kadavigere, Veena Mayya, Leena R. David, Berna Uzun, Dilber Uzun Ozsahin, Abhimanyu Pradhan, Winniecia Dkhar, Ilker Ozsahin, and Neil Abraham Barnes

Contributors

Meshari Almeshari Department of Diagnostic Radiology, College of Applied Medical Science, University of Hail, Hail, Saudi Arabia

Revati Amin Department of Physiotherapy, Kasturba Medical College Mangalore, Manipal Academy of Higher Education, Manipal, India

K. Annapoorna Division of Yoga, Manipal Academy of Higher Education, Centre for Integrative Medicine and Research, Manipal, India

M.G. Ramesh Babu Division of Physiology, Department of Basic Medical Sciences, Manipal Academy of Higher Education, Manipal, India

K.R. Banumathe Department of Occupational Therapy, Manipal University College Malaysia, Melaka, Malaysia

Neil Abraham Barnes Department of Medical Imaging Technology, Manipal College of Health Professions, Manipal Academy of Higher Education, Manipal, Karnataka, India

Anup Bhat Department of Physiotherapy, Manipal College of Health Professions, Manipal Academy of Higher Education, Manipal, India

Baskaran Chandrasekaran Symbiosis School of Sports Sciences, Symbiosis International (Deemed University), Symbiosis Knowledge Village, Pune, India

Leena R. David Department of Medical Imaging Technology, Manipal College of Health Professions, Manipal Academy of Higher Education, Manipal, Karnataka, India; Medical Diagnostic Imaging Department, College of Health Sciences, University of Sharjah, Sharjah, United Arab Emirates

Elsa Sana Tombi Devi Department of Medical Surgical Nursing, Manipal Academy of Higher Education, Manipal, India

Winniecia Dkhar Department of Medical Imaging Technology, Manipal College of Health Professions, Manipal Academy of Higher Education, Manipal, Karnataka, India

Tina Agnes Dsouza Department of Physiotherapy, Manipal Academy of Higher Education, Manipal College of Health Professions, Manipal, India

Rajagopal Kadavigere Department of Radiodiagnosis and Imaging, Kasturba Medical College, Manipal, India

Koustubh Kamath Department of Medical Imaging Technology, Manipal College of Health Professions, Manipal Academy of Higher Education, Manipal, Karnataka, India

Banumathe Karuppaya Occupational Therapy Program, School of Rehabilitation and Medical Sciences, College of Health Sciences, University of Nizwa, Nizwa, Sultanate of Oman; Department of Occupational Therapy, Manipal University College Malaysia, Melaka, Malaysia

Kalyan Chakravarthi Kosuri Department of Anatomy, Manipal Univeristy College Malaysia, Melaka, Malaysia

Veena Mayya Department of Information & Communication Technology; Manipal Institute of Technology, Manipal Academy of Higher Education, Manipal, India

Sidhiprada Mohapatra Department of Physiotherapy, Manipal College of Health Professions, Manipal Academy of Higher Education, Manipal, India

Sathya Sabina Muthu Department of Medical Imaging Technology, Manipal College of Health Professions, Manipal Academy of Higher Education, Manipal, Karnataka, India

Rajesh Navada Department of Physiotherapy, Manipal Academy of Higher Education, Manipal College of Health Professions, Manipal, India

Kaushik Nayak Department of Medical Imaging Technology, Manipal College of Health Professions, Manipal Academy of Higher Education, Manipal, Karnataka, India

Venu Madhav Nelluri Department of Anatomy, Manipal Univeristy College Malaysia, Melaka, Malaysia

Dilber Uzun Ozsahin Medical Diagnostic Imaging Department, College of Health Sciences, University of Sharjah, Sharjah, United Arab Emirates; Irfan Suat Gunsel Operational Research Institute, Near East University, Nicosia, Turkish Republic of Northern Cyprus (TRNC), Turkey

Ilker Ozsahin Irfan Suat Gunsel Operational Research Institute, Near East University, Nicosia, Turkish Republic of Northern Cyprus (TRNC), Turkey

Hari Prakash Palaniswamy Department of Speech and Hearing, Manipal Academy of Higher Education, Manipal College of Health Professions, Manipal, India

Nitika C. Panakkal Department of Medical Imaging Technology, Manipal College of Health Professions, Manipal Academy of Higher Education, Manipal, Karnataka, India

Poovitha Shruthi Paramashiva Division of Yoga, Manipal Academy of Higher Education, Centre for Integrative Medicine and Research, Manipal, India

Abhimanyu Pradhan Department of Medical Imaging Technology, Manipal College of Health Professions, Manipal Academy of Higher Education, Manipal, Karnataka, India

N. Praveen Department of Radiodiagnosis & Imaging, Kasturba Hospital, Manipal, India

Priya Seagal V.K. Adichunchanagiri University, Bellur, Karnataka, India

Nikhil Raj Department of Medical Imaging Technology, Manipal College of Health Professions, Manipal Academy of Higher Education, Manipal, Karnataka, India

Deepika Raja Department of Medical Imaging Technology, Manipal College of Health Professions, Manipal Academy of Higher Education, Manipal, Karnataka, India

Aravind Kumar Rajasekaran Department of Speech Pathology and Audiology, National Institute of Mental Health and Neurosciences, Bengaluru, India

Preetha Ramachandra Department of Physiotherapy, Manipal Academy of Higher Education, Manipal College of Health Professions, Manipal, India

Anuradha Rao Manipal Institute of Technology, Manipal Academy of Higher Education, Manipal, India

Ashok Rao Manipal Institute of Technology, Manipal Academy of Higher Education, Manipal, India

Sneha Ravichandran Department of Medical Imaging Technology, Manipal College of Health Professions, Manipal Academy of Higher Education, Manipal, Karnataka, India

Shovan Saha Department of Occupational Therapy, Manipal Academy of Higher Education, Manipal College of Health Professions, Manipal, India

M. Sanjana Manipal Academy of Higher Education, Department of Speech and Hearing, Manipal College of Health Professions, Manipal, India

S. Sharath Department of Medical Imaging Technology, Manipal College of Health Professions, Manipal Academy of Higher Education, Manipal, Karnataka, India

Dilip Shettigar Department of Medical Imaging Technology, Manipal College of Health Professions, Manipal Academy of Higher Education, Manipal, Karnataka, India

Aishwarya Shetty Department of Medical Imaging Technology, Manipal College of Health Professions, Manipal Academy of Higher Education, Manipal, Karnataka, India

K.B. Shishira Department of Medical Imaging Technology, Manipal College of Health Professions, Manipal Academy of Higher Education, Manipal, Karnataka, India

Shwetha Tumkur Shivakumar Department of Clinical Psychology, Manipal Academy of Higher Education, Manipal College of Health Professions, Manipal, India

K.N. Shivshankar Department of Medicine, Manipal Academy of Higher Education, Kasturba Medical College, Manipal, India

Kirti Singhal Department of Clinical Psychology, Manipal Academy of Higher Education, Manipal College of Health Professions, Manipal, India

Mukesh Kumar Sinha Department of Physiotherapy, Manipal College of Health Professions, Manipal Academy of Higher Education, Manipal, India

Prateek Srivastav Department of Physiotherapy, Manipal Academy of Higher Education, Manipal College of Health Professions, Manipal, India

Suresh Sukumar Department of Medical Imaging Technology, Manipal College of Health Professions, Manipal Academy of Higher Education, Manipal, Karnataka, India

Sudarshan Surendran Department of Anatomy, American University of Antigua College of Medicine, Antigua, Antigua and Barbuda

Nayak Ullas U. Division of Anatomy, Manipal Academy of Higher Education, Manipal, India

Berna Uzun Irfan Suat Gunsel Operational Research Institute, Near East University, Nicosia, Turkish Republic of Northern Cyprus (TRNC), Turkey; Medical Diagnostic Imaging Department, College of Health Sciences, University of Sharjah, Sharjah, United Arab Emirates

K. Vaishali Department of Medical Imaging Technology, Manipal College of Health Professions, Manipal Academy of Higher Education, Manipal, Karnataka, India

Anand Venugopal Manipal Academy of Higher Education, Manipal, India

Guruprasad Vijayasarathi Occupational Therapy Program, School of Rehabilitation and Medical Sciences, College of Health Sciences, University of Nizwa, Nizwa, Sultanate of Oman; School of Rehabilitation and Medical Sciences, College of Health Sciences, University of Nizwa, Nizwa, Sultanate of Oman

T. Visakh Department of Medical Imaging Technology, Manipal College of Health Professions, Manipal Academy of Higher Education, Manipal, Karnataka, India

Meliz Yuvalı Near East University Faculty of Medicine, Department of Biostatistics, Nicosia, Cyprus

Pradeep Yuvaraj Department of Speech Pathology and Audiology, National Institute of Mental Health and Neurosciences, Bengaluru, India

Preface

Medical imaging has become an essential tool for understanding, diagnosing, and managing a wide range of health conditions, while behavioral science provides insight into the psychological, cognitive, and social factors that influence health outcomes. This book brings together these two fields, offering a comprehensive resource for researchers, clinicians, and students across medicine, radiology, physiotherapy, occupational therapy, psychology, psychiatry, and related disciplines. It is also intended to inform policymakers seeking evidence-based guidance for healthcare strategies and resource allocation.

The book addresses foundational principles of medical imaging, software-assisted diagnostic tools, and behavioral assessment, alongside applications in neurology, musculoskeletal health, vascular and cardiovascular studies, oncology, and rehabilitation. Special emphasis is placed on the effects of lifestyle factors—such as sedentary behavior, physical activity, yoga, and pranayama—on brain structure, cognitive function, musculoskeletal health, and overall well-being. The text also explores advanced imaging techniques—including MRI, ultrasound, DEXA, fMRI, and EEG—and demonstrates how they can be integrated with behavioral and physiological measures to improve diagnosis, treatment, and patient outcomes.

Chapters have been carefully curated and authored by experts from multiple disciplines, combining theoretical foundations with clinical and practical perspectives. This multidisciplinary approach ensures that readers gain a comprehensive understanding of how imaging and behavioral science can work together to advance research, improve patient care, and guide evidence-based interventions. Emerging topics such as computational imaging in oncology, rehabilitation strategies for stroke and traumatic brain injury, and workplace interventions are also discussed, providing insight into current challenges and future directions in the field.

By presenting these topics in a cohesive and accessible format, this book provides a reference that supports both academic learning and professional practice. It aims to foster interdisciplinary collaboration, encourage innovation, and ultimately improve health outcomes across diverse populations.

Acknowledgements

The editors express their heartfelt gratitude to their families for their unwavering support, patience, and encouragement during the preparation of this work. They respectfully acknowledge their parents—N. S. Rajan (late), Lalitha Bai Rajan (late), R. Balaraj (late), Elisi Balaraj, Sukumar S., and Saraswati S.—whose values, sacrifices, and commitment to education laid the foundations for their academic and professional journeys.

Dr. Ilker Ozsahin and Dr. Dilber Uzun Ozsahin would like to acknowledge their children, Mehmet Ada Ozsahin and Mavi Su Ozsahin, for their support and patience throughout the phases of this book.

Dr. Leena R. David and Dr. Suresh Sukumar extend their deepest appreciation to their spouses, David Raj, B.E., and Dr. Poovitha Shruthi Paramashiva, for their understanding, support, and encouragement during the most demanding phases of this book. They also warmly recognize their children, Dayana David, Leyana David, and Saindhavi Suresh, whose love, resilience, and inspiration motivated them throughout the process.

Dr. Leena R. David additionally expresses her sincere gratitude to Prof. Rajani Kanth Rao Vedula, Prof. Peter Hogg, Dr. Leslie Robinson, and Dr. Michael Heinrich Commichau (late) for their mentorship and belief in her abilities, which have significantly shaped her academic and professional journey.

Chapter 1

Introduction to advances in medical imaging

Leena R. David[a,b], Ilker Ozsahin[c], Sneha Ravichandran[b], Dilip Shettigar[b], Suresh Sukumar[b], Meliz Yuvalı[d], Koustubh Kamath[b], Berna Uzun[c], Dilber Uzun Ozsahin[a,c,e], and Sathya Sabina Muthu[b]

[a]*Medical Diagnostic Imaging Department, College of Health Sciences, University of Sharjah, Sharjah, United Arab Emirates*

[b]*Department of Medical Imaging Technology, Manipal College of Health Professions, Manipal Academy of Higher Education, Manipal, Karnataka, India*

[c]*Irfan Suat Gunsel Operational Research Institute, Near East University, Nicosia, Turkish Republic of Northern Cyprus (TRNC), Turkey*

[d]*Near East University Faculty of Medicine, Department of Biostatistics, Nicosia, Cyprus*

[e]*Research Institute for Medical and Health Sciences, University of Sharjah, Sharjah, United Arab Emirates*

1 The evolution of medical imaging

Over the past several decades, medical imaging has undergone a profound transformation from a predominantly diagnostic instrument into a sophisticated, multidisciplinary field that integrates the expertise of radiologists, clinicians, neuroscientists, engineers, and public health specialists. Early imaging technologies, such as conventional radiography and computed tomography (CT), were primarily employed to visualize gross anatomical structures and detect overt pathologies. However, the scope of medical imaging has significantly expanded. It now encompasses highly advanced modalities such as magnetic resonance imaging (MRI), positron emission tomography (PET), and three dimensional ultrasound, which not only provide structural detail but also yield quantitative information on physiological, metabolic, and molecular processes (de Souza et al., 2022; Kantarcı et al., 2025).

This transition towards higher resolution, multimodal imaging has enabled the investigation of dynamic biological functions in vivo. Techniques such as functional MRI (fMRI) enable real time examination of neural activity, thereby accelerating progress in cognitive neuroscience and behavioral medicine (Zhao, 2025). Diffusion tensor imaging (DTI) permits the mapping of white matter architecture and connectivity, advancing our understanding of neurodegenerative diseases, traumatic brain injuries, and developmental conditions (Owens-Walton et al., 2024). PET imaging, with its capability to trace metabolic pathways and receptor binding, has emerged as a pivotal tool in oncology, cardiology, and psychiatry (Nerella et al., 2022; Cervenka et al., 2022). Collectively, these advancements have positioned medical imaging as an indispensable component of both clinical practice and biomedical research.

Furthermore, the integration of computational methods and artificial intelligence (AI) into imaging science has led to a significant transformation of the imaging field. Machine learning algorithms, particularly in radiomics and image recognition, now assist clinicians in identifying subtle radiological patterns, predicting disease progression, and stratifying patient risk (van Timmeren et al., 2020; Li et al., 2021; Zhang et al., 2024). Large scale imaging datasets, supported by cloud computing and federated learning frameworks, facilitate population-level research and enable the development of predictive models that inform preventive health strategies (da Silva et al., 2023; Ur Rehman et al., 2023). These developments indicate a paradigm shift from reactive disease management to proactive, data driven, and personalized medicine.

One of the emerging frontiers in medical imaging lies in its capacity to examine the interplay between behavior, lifestyle, and health outcomes. Imaging studies have provided empirical evidence linking sedentary behavior, stress, nutrition, and physical activity to structural and functional changes in the brain, vascular system, and musculoskeletal tissues (Staempfli et al., 2024; Sinha and Sinha, 2024; Townsend et al., 2022; Gogniat et al., 2025). This interdisciplinary convergence has opened new avenues for research into behavioral determinants of health and the role of imaging in lifestyle interventions, rehabilitation, and public health policy.

Advances in Medical Imaging. https://doi.org/10.1016/B978-0-443-28967-5.00002-2

This book, *Advances in Medical Imaging: From Behavioral Analysis to Disease Prevention and Rehabilitation*, synthesizes diverse perspectives and research contributions that highlight the expanding role of imaging in contemporary healthcare sciences. The chapters in this book collectively demonstrate how imaging technologies are being employed to elucidate behavioral patterns, monitor physiological adaptations, assess disease risk, and evaluate therapeutic efficacy. By bridging diagnostic imaging with behavioral science, rehabilitation, and preventive medicine, this volume reflects an evolving scientific ethos in which health is understood not merely as the absence of disease but as a dynamic interaction between biology, behavior, and environment.

2 Aim and scope of the book

This book volume aims to present the breadth of medical imaging research from its technical foundations to its applications in behavioral neuroscience, chronic disease assessment, and rehabilitation. The book highlights how imaging can decode brain behavior, assess lifestyle impacts, and guide interventions that promote physical and mental wellbeing. It integrates perspectives from neurology, radiology, physiology, public health, and rehabilitation sciences, showcasing imaging not as a stand-alone discipline but as a unifying force across medical domains.

3 Overview of the chapters

3.1 Chapters 2–3: Imaging and behavioral science

The opening chapters of this volume examine the integration of neuroimaging techniques with advanced software tools to deepen our understanding of brain function and behavior. They highlight how modalities such as MRI, PET, and electroencephalogram (EEG), when paired with platforms like PsychoPy and EPrime, have enabled precise experimental design, stimulus control, and the realtime recording of neural responses. This fusion of imaging and cognitive software has transformed neuroscience research, allowing investigators to correlate specific behavioral tasks with dynamic changes in brain activation and connectivity.

Building on this foundation, the subsequent chapter explores the role of imaging in assessing behavioral modulation of memory and learning. It demonstrates how changes in attention, emotion, or motivation can be visualized at the neural level, providing direct insights into neuroplasticity and cognitive flexibility. Together, these chapters underscore the growing capacity of imaging technologies to map the brain's functional adaptation in response to experience, training, and behavior.

3.2 Chapters 4–8: Sedentary behavior, lifestyle, and brain health

This section addresses sedentary behavior, a growing public health concern with wide ranging effects on physical and neurological health. The chapters investigate how prolonged physical inactivity leads to measurable alterations in brain structure, including reduced gray matter volume, impaired white matter integrity, and compromised vascular function. Using advanced imaging techniques such as high resolution MRI, diffusion imaging, and ultrasound based vascular assessments, researchers document how sedentary behavior and poor nutritional habits contribute to cognitive decline, inflammation, and obesity related brain changes.

Chapters in this section reveal that lifestyle factors, particularly physical activity and balanced nutrition, can significantly influence neural connectivity, metabolic efficiency, and cognitive performance. These findings not only quantify the biological impact of inactivity but also emphasize the reversibility of certain neural changes through targeted lifestyle interventions. Collectively, the insights highlight the indispensable role of movement, a healthy diet, and active living in safeguarding brain health across the lifespan.

3.3 Chapters 9–12: Vascular function and ultrasound applications

The middle chapters address the interplay between vascular health, cognitive function, and physical activity. Using tools like Doppler ultrasound and MRI, these chapters assess how microbreaks, exercise, and sitting patterns influence vascular elasticity and cognition in office workers. Subsequent chapters extend these findings to overweight and obese populations, emphasizing the value of ultrasound-based vascular assessments for the early detection and prevention of cardiovascular risks.

3.4 Chapters 13–17: Frailty, musculoskeletal health, and imaging

This section of the book focuses on aging and musculoskeletal health, addressing the growing importance of imaging in evaluating frailty and functional decline in older adults. These chapters demonstrate how tools such as ultrasound, MRI, and radiographic analysis provide valuable biomarkers for assessing muscle mass, tissue composition, joint integrity, and degenerative changes. Particular emphasis is placed on identifying early signs of sarcopenia, osteoporosis, and repetitive motion injuries, which are increasingly prevalent in both aging populations and occupational settings.

In addition to diagnosis, these chapters highlight the utility of imaging in monitoring the effectiveness of exercise-based interventions, physiotherapy, and ergonomic modifications. Imaging modalities enable clinicians to track tissue regeneration, muscular adaptation, and recovery from work-related musculoskeletal disorders. Collectively, this section underscores the critical role of imaging in supporting preventive care, personalized rehabilitation, and the promotion of lifelong musculoskeletal resilience, bridging clinical management with public health and occupational medicine.

3.5 Chapters 18–20: Chronic disease, yoga, and cognitive well-being

This section addresses the intersection of respiratory, vascular, and cognitive health through the lens of lifestyle-based interventions. The chapters examine chronic obstructive pulmonary disease alongside the effects of yoga practices on muscle structure, vascular function, and cognitive performance in sedentary individuals. Imaging techniques, such as ultrasound and MRI, are used to quantify improvements in blood flow, endothelial function, and muscle quality following yoga-based protocols. Novel findings from interventions like Y Break Yoga and pranayama reveal measurable central and peripheral vascular enhancements, demonstrating how traditional practices, when evaluated through modern imaging, offer promising therapeutic benefits for both physical and mental health.

3.6 Chapters 21–23: Bone health, meditation, and stroke recovery

These chapters examine diverse applications of medical imaging in understanding physical and psychological healing. Topics include bone density evaluation through dual energy X-ray absorptiometry (DEXA), yoga based physiological assessment, neuroplastic changes during meditation, and imaging-supported stroke rehabilitation. By integrating modalities such as MRI, EEG, and DTI, researchers are able to visualize structural and functional changes in the brain and body in response to physical activity and mental practices. Findings demonstrate measurable effects on memory, emotional regulation, neural connectivity, and motor recovery, contributing to the evidence based validation of interventions like yoga and mindfulness.

The importance of this work lies in its ability to objectively quantify therapeutic outcomes, offering clinicians and researchers reliable biomarkers for monitoring progress and tailoring treatments. These studies reinforce the growing role of imaging not just in disease diagnosis but also in promoting holistic recovery, bridging the gap between behavioral science, rehabilitation, and personalized healthcare.

3.7 Chapters 24–27: Traumatic brain injury and auditory neuroscience

This section focuses on rehabilitation imaging in complex neurological conditions, including coma recovery, stress induced neural alterations, and auditory system disorders. The chapters illustrate how MRI, functional imaging, and diffusion techniques provide critical insights into brain activation patterns, structural recovery, and neural connectivity following traumatic brain injury. In addition, specialized imaging of the auditory and vestibular systems is explored to evaluate inner ear and retrocochlear pathologies. By integrating multiple imaging modalities, these studies demonstrate the value of precise visualization in diagnosis, treatment planning, and monitoring therapeutic progress, reinforcing the indispensable role of imaging in neurorehabilitation and sensory system evaluation.

3.8 Chapters 28–35: Imaging in oncology and systemic diseases

The final section of the volume focuses on the expanding role of medical imaging in oncology and systemic health, highlighting its evolution beyond traditional diagnosis to becoming a critical tool in precision medicine and patient-centered care. The chapters examine how advanced imaging modalities including multiparametric MRI for prostate cancer, contrast enhanced techniques in breast and thoracic oncology, and hybrid PET/CT for abdominal malignancies enhance early detection, staging, and treatment monitoring.

A significant emphasis is placed on the accelerating integration of computational approaches in cancer imaging. Machine learning and radiomic analyses are explored for their potential in tumor characterization, treatment response prediction, and clinical decision support. This reflects a shift toward data-driven oncology, where imaging contributes to personalized treatment strategies through high dimensional feature extraction and predictive modeling. Beyond structural and molecular imaging, the section also engages with emerging areas such as psycho oncology, where neuroimaging provides insights into the cognitive and emotional impacts of cancer and its treatment. These contributions underscore the importance of understanding not only tumor biology but also the psychological dimensions of survivorship and patient quality of life.

In addition, one of the chapters examines imaging in exercise physiology and metabolic disease, showing how vascular and cardiac imaging can quantify the benefits of physical activity in conditions like dyslipidemia. This reinforces a central theme of the volume: imaging as a tool not only for diagnosing disease but also for visualizing and supporting prevention and rehabilitation. Together, these chapters illustrate the broadening scope of medical imaging in oncology and systemic conditions, reflecting its growing relevance to both the clinical and behavioral dimensions of healthcare.

4 Integrating technology, behavior, and health

Across all 35 chapters, a common theme emerges that states imaging is not only a diagnostic tool but also a bridge between **technology and human behavior**. Whether applied to studying brain connectivity, assessing frailty, or guiding yoga based interventions, imaging provides measurable, objective insights into the body's response to internal and external factors.

As technology advances, imaging continues to move from the hospital to the community, supporting preventive health, workplace wellness, and personalized medicine. The convergence of imaging, AI, and behavioral science promises a future where diseases can be anticipated, lifestyles optimized, and recovery accelerated.

5 Conclusion

This book offers a comprehensive overview of the evolving role of medical imaging in shaping modern healthcare and biomedical research. From detecting microscopic neuronal changes to the macroscopic structures of organs, imaging has become an indispensable tool for visualizing physiological processes that were previously inaccessible. Its applications now extend beyond traditional diagnostic settings, permeating the domains of public health, cognitive science, and preventive medicine.

By integrating technological advancements with clinical and behavioral inquiry, *Advances in Medical Imaging: From Behavioral Analysis to Disease Prevention and Rehabilitation* demonstrates how imaging contributes to a multidimensional understanding of human health. The work presented herein underscores the potential of imaging not only to identify pathology but also to inform therapeutic interventions, monitor recovery, and support health promoting behaviors.

Collectively, the contributions in this book encourage a forward looking perspective, wherein medical imaging serves not merely as a record of disease but as a catalyst for innovation, resilience, and restorative care.

References

Cervenka, S., Frick, A., Bodén, R., Lubberink, M., 2022. Application of positron emission tomography in psychiatry—methodological developments and future directions. Transl. Psychiatry 12, 248. doi:10.1038/s41398-022-01990-2.

da Silva, F.R., Camacho, R., Tavares, J.M.R.S., 2023. Federated learning in medical image analysis: a systematic survey. Electron. (Basel 13, 47). doi:10.3390/electronics13010047.

de Souza, N.M., Oprea-Lager, D.E., Fournier, L.S., 2022. Editorial: quantitative imaging for clinical decisions. Front. Oncol. 12. doi:10.3389/fonc.2022.858372.

Gogniat, M.A., Won, J., Cruz, C., Aranda, A., Verma, A., Gujral, S., et al, 2025. Sedentary behavior, cognition, and brain health in older adults: a systematic review. Front. Aging Neurosci., 17. doi: 10.3389/fnagi.2025.1622049.

Kantarcı, M., Aydın, S., Oğul, H., Kızılgöz, V., 2025. New imaging techniques and trends in radiology. Diagn. Interv. Radiol. doi:10.4274/dir.2024.242926.

Li, S., Deng, Y.-Q., Zhu, Z.-L., Hua, H.-L., Tao, -Z.-Z., 2021. A comprehensive review on radiomics and deep learning for nasopharyngeal carcinoma imaging. Diagnostics 11, 1523. doi:10.3390/diagnostics11091523.

Nerella, S.G., Singh, P., Sanam, T., Digwal, C.S., 2022. PET molecular imaging in drug development: the imaging and chemistry perspective. Front. Med. (Lausanne 9. doi:10.3389/fmed.2022.812270.

Owens-Walton, C., Nir, T.M., Al-Bachari, S., Ambrogi, S., Anderson, T.J., Aventurato, Í.K., et al, 2024. A worldwide study of white matter microstructural alterations in people living with Parkinson's disease. NPJ Parkinsons Dis 10, 151. doi:10.1038/s41531-024-00758-3.

Sinha, U., Sinha, S., 2024. Magnetic resonance imaging biomarkers of muscle. Tomography 10, 1411–1438. doi:10.3390/tomography10090106.

Staempfli, J.S., Kistler-Fischbacher, M., Gewiess, J., Bastian, J., Eggimann, A.K., 2024. The validity of muscle ultrasound in the diagnostic workup of sarcopenia among older adults: a scoping review. Clin. Interv. Aging 19, 993–1003. doi:10.2147/CIA.S463917.

Townsend, R.F., V., W.J., Prinelli, F., O'Neill, R.F., McEvoy, C.T., 2022. Associations between dietary patterns and neuroimaging markers: a systematic review. Front. Nutr. 9. doi:10.3389/fnut.2022.806006.

Ur Rehman, M.H., W, H.L.P., Nachev, P., Teo, J.T., Ourselin, S., Cardoso, M.J., 2023. Federated learning for medical imaging radiology. Br. J. Radiol. 96. doi:10.1259/bjr.20220890.

van Timmeren, J.E., Cester, D., Tanadini-Lang, S., Alkadhi, H., Baessler, B., 2020. Radiomics in medical imaging—"how-to" guide and critical reflection. Insights Imaging 11, 91. doi:10.1186/s13244-020-00887-2.

Zhang, C., Zhong, M., Liang, Z., Zhou, J., Wang, K., Bu, J., 2024. MRI-based radiomic and machine learning for prediction of lymphovascular invasion status in breast cancer. BMC Med. Imaging 24, 322. doi:10.1186/s12880-024-01501-3.

Zhao, L., 2025. Advances in functional magnetic resonance imaging-based brain function mapping: a deep learning perspective. Psychoradiology 5. doi:10.1093/psyrad/kkaf007.

Part I

Imaging and behavioral science

Chapter 2

Application of software and neurological diagnostic tools in medical imaging & behavioral science

M. Sanjana[a], Leena R. David[b,c], Suresh Sukumar[c], Dilber Uzun Ozsahin[b,d,e], Berna Uzun[e], Sneha Ravichandran[c], Dilip Shettigar[c], Hari Prakash Palaniswamy[a], Shovan Saha[f], Ilker Ozsahin[e], Rajagopal Kadavigere[g], Guruprasad Vijayasarathi[h], K.R. Banumathe[i], and Baskaran Chandrasekaran[j]

[a]*Manipal Academy of Higher Education, Department of Speech and Hearing, Manipal College of Health Professions, Manipal, India*

[b]*Medical Diagnostic Imaging Department, College of Health Sciences, University of Sharjah, Sharjah, United Arab Emirates*

[c]*Department of Medical Imaging Technology, Manipal College of Health Professions, Manipal Academy of Higher Education, Manipal, Karnataka, India*

[d]*Research Institute for Medical and Health Sciences, University of Sharjah, Sharjah, United Arab Emirates*

[e]*Irfan Suat Gunsel Operational Research Institute, Near East University, Nicosia, Turkish Republic of Northern Cyprus (TRNC), Turkey*

[f]*Manipal Academy of Higher Education, Department of Occupational Therapy, Manipal College of Health Professions, Manipal, India*

[g]*Department of Radiodiagnosis and Imaging, Kasturba Medical College, Manipal, India*

[h]*School of Rehabilitation and Medical Sciences, College of Health Sciences, University of Nizwa, Nizwa, Oman*

[i]*Department of Occupational Therapy, Manipal University College Malaysia, Melaka, Malaysia*

[j]*Symbiosis School of Sports Sciences, Symbiosis International (Deemed University), Pune, India*

1 Introduction

The convergence of software tools and neurological diagnostics represents one of the most significant advances in modern medical science, fundamentally transforming our ability to understand, diagnose, and treat neurological conditions. This integration has created unprecedented opportunities for both researchers and clinicians to explore the complexities of the human brain with remarkable precision and detail (Bajwa et al., 2021; Surianarayanan et al., 2023). In recent decades, the development of sophisticated imaging technologies, coupled with advanced software tools, has revolutionized our approach to studying brain structure and function. These technological advances have enabled researchers to observe neural activity in real time, track disease progression, and evaluate treatment efficacy with unprecedented accuracy (Xue et al., 2010; Yen et al., 2023). The impact of these developments extends beyond pure research applications, significantly influencing clinical practice and patient care.

The field of neurological diagnostics has undergone a remarkable transformation since the introduction of computerized imaging techniques in the 1970s. What began with basic computerized tomography has evolved into a sophisticated array of imaging modalities, each providing unique insights into brain structure and function (Schulz et al., 2021). This evolution has been particularly accelerated by:

1. Increased computational power, enabling complex image processing
2. Development of machine learning algorithms for image analysis
3. Advanced software tools for data interpretation and visualization
4. Integration of multiple imaging modalities for comprehensive analysis

Recent years have witnessed several groundbreaking developments in neurological imaging and diagnostics. Artificial intelligence and machine learning algorithms have become increasingly integral to image analysis and interpretation, leading to more accurate diagnoses and improved patient outcomes (Lui et al., 2020). Key trends include:

Advances in Medical Imaging. https://doi.org/10.1016/B978-0-443-28967-5.00017-4

AI-enhanced image analysis: Implementation of deep learning algorithms for automated lesion detection and classification
Multimodal imaging integration: Combining different imaging techniques for comprehensive brain mapping
Real-time processing: Development of software tools enabling immediate image analysis during scanning
Cloud-based solutions: Implementation of distributed computing for complex image processing tasks

2 Different types of modalities of assessment

There are several imaging modalities for the standard assessments of neurological diagnostics and also to provide real-time functioning of the brain. They are as follows:

2.1 Magnetic resonance imaging

Magnetic resonance imaging (MRI) represents a cornerstone of modern neurological diagnostics, providing detailed anatomical images of the brain without ionizing radiation. The technology utilizes strong magnetic fields and radio waves to generate high-resolution images of brain structure. The key applications include structural brain mapping, tumor detection and characterization, assessment of neurodegenerative disorders, and evaluation of vascular issues.

Recent advances in MRI technology have led to improved resolution and faster scanning times, with particular progress in higher field strength magnets (7T and above), advanced pulse sequences for better contrast, motion correction algorithms, and quantitative imaging techniques (Westbrook and Talbot, 2018).

2.2 Functional MRI

Functional MRI (fMRI) has revolutionized our understanding of brain activity by enabling the visualization of neural activation patterns in real time. This technique measures changes in blood oxygen level-dependent signals as an indirect marker of neural activity. The applications include diagnosing neurological disorders, investigating brain connectivity patterns, cognitive function assessment, preoperative measures, etc. The current developments include improved temporal resolution, advanced statistical analysis methods, real-time fMRI feedback, and integration with other modalities (Orringer et al., 2012).

2.3 Electroencephalography

Electroencephalography (EEG) provides unique insights into brain function by measuring electrical activity directly from the scalp. Modern EEG systems incorporate advanced digital signal processing and analysis tools, enabling increased temporal resolution of brain activity, resting-state activity, evoked response potentials, brain–computer interface applications, etc. The recent advancements in this modality of assessment are immense and include the following: improved signal processing, portable and wireless recordings, integration with other modalities, high-density electrode arrays, etc. (Louis et al., 2016; Chegondi et al., 2022).

2.4 Positron emission tomography

Positron emission tomography (PET) imaging offers unique capabilities in visualizing metabolic processes and neurotransmitter activity in the brain. This technique utilizes radioactive tracers to track biological processes at the molecular level (Kapoor et al., 2025 Belcari et al., 2023). The main applications of this modality include amyloid imaging for Alzheimer's disease, metabolic studies, cancer detection, and monitoring disease progression. The current focus is on improved detection and hybrid imaging processes, combining fMRI and quantitative analysis methods (Berger, 2003).

3 Integration of imaging modalities through software tools

The trend toward multimodal imaging approaches has led to a more comprehensive understanding of brain structure and function. Integration of different imaging modalities provides complementary information, enabling better spatial and temporal resolution, improved diagnostic accuracy, effective treatment planning, and comprehensive structural and functional analysis.

Modern neuroscience research demands precise control over stimulus presentation, response recording, and synchronization with various imaging modalities. The evolution of specialized software tools has revolutionized how researchers conduct experiments, analyze data, and integrate different methodological approaches. This section explores the major software platforms that have become indispensable in contemporary behavioral neuroscience research.

3.1 Psychopy: An open-source solution for psychological research

PsychoPy is a significant advancement in psychological research software, offering the flexibility of Python programming along with an intuitive graphical user interface. Developed by Jonathan Peirce at the University of Nottingham, this open-source platform has gained popularity among researchers conducting neuroimaging studies. It is widely used in laboratories around the world for psychophysics, cognitive neuroscience, and experimental psychology.

3.1.1 Technical framework

PsychoPy is built on Python and utilizes powerful libraries such as NumPy and Matplotlib. This foundation offers researchers exceptional precision in stimulus timing and presentation, which is crucial for neuroimaging experiments that require millisecond accuracy. The software has two main interfaces:

Builder interface: This is a graphical interface that enables researchers to create experiments without needing coding knowledge. Users can easily drag and drop components onto a timeline, adjust parameters using dialog boxes, and visualize the flow of the experiment.

Coder interface: This is a Python-based programming environment designed for those who require more complex experimental designs. This interface offers complete control over stimulus properties, timing, and data collection.

3.1.2 Key features for neuroimaging research

Precise timing capabilities with microsecond accuracy
Support for a wide range of stimulus types (visual, auditory, video)
Integration with various response devices and eye-tracking systems
Built-in synchronization with MRI triggers
Real-time data collection and online preprocessing
Export capabilities compatible with major analysis packages

3.1.3 Applications in neuroimaging

PsychoPy has become an essential tool in neuroimaging research due to its precise timing capabilities and flexible integration with various neuroimaging modalities. Here's how PsychoPy is applied across different neuroimaging techniques:

Functional magnetic resonance imaging

- Scanner synchronization: PsychoPy can receive MRI trigger pulses (transistor–transistor logic, TTL signals) to precisely time stimulus presentation with image acquisition. This is critical for event-related designs, where stimuli must be presented at specific points in the scanning sequence (Peirce et al., 2019).
- Repetition time (TR) monitoring: PsychoPy's built-in functions allow for monitoring TR and adjusting stimulus delivery based on the scanner's timing, ensuring that experimental protocols match the hemodynamic response being measured.
- Response collection: The platform enables the recording of participant responses during scanning using MRI-compatible input devices, allowing researchers to correlate behavioral data with neural activity (Bridges et al., 2020).

EEG

- Marker generation: The software can send event markers to EEG recording systems to tag the precise timing of stimulus presentation, allowing for accurate event-related potential (ERP) analysis.
- Parallel port communication: PsychoPy supports direct communication through parallel ports, which many EEG systems use for high-precision timing markers (Mathôt et al., 2012)
- Lab streaming layer (LSL) integration: Support for the LSL allows for synchronized data collection across multiple devices, ideal for combined EEG-behavioral studies (Bigdely-Shamlo et al., 2016).

Magnetoencephalography

- Precise visual timing: PsychoPy's frame-based timing ensures accurate presentation of visual stimuli, which is critical for magnetoencephalography (MEG) studies examining early sensory processing (Gramfort et al., 2014).
- Trigger output: Similar to EEG, PsychoPy can output trigger signals that are recorded alongside MEG data for event synchronization.

Near-infrared spectroscopy

- Block design implementation: The software allows researchers to create optimized block designs that account for the slower hemodynamic response measured in near-infrared spectroscopy (NIRS) (Pinti et al., 2020).
- Multimodal integration: PsychoPy facilitates synchronized recording of NIRS data with behavioral responses, enabling a more comprehensive analysis of cognitive processes.

Eye-tracking integration

- Native support: PsychoPy includes built-in support for several eye-tracking systems (like SR Research EyeLink, Tobii, and GazePoint), allowing gaze data to be collected simultaneously with neuroimaging data.
- Real-time processing: The platform enables online processing of eye movement data, allowing for gaze-contingent stimulus presentation in neuroimaging studies (Mathôt et al., 2012).

Practical applications

- Cognitive neuroscience: Studies examining attention, memory, and decision-making use PsychoPy to present complex stimulus sequences while recording brain activity (Zaki et al., 2016).
- Clinical neuroscience: Researchers investigating psychiatric and neurological disorders use PsychoPy to deliver standardized cognitive tasks during neuroimaging (Thomas et al., 2013).
- Developmental neuroscience: PsychoPy's engaging visual interface helps create child-friendly paradigms for developmental neuroimaging studies (Monks et al., n.d.) (Figs. 2.1 and 2.2).

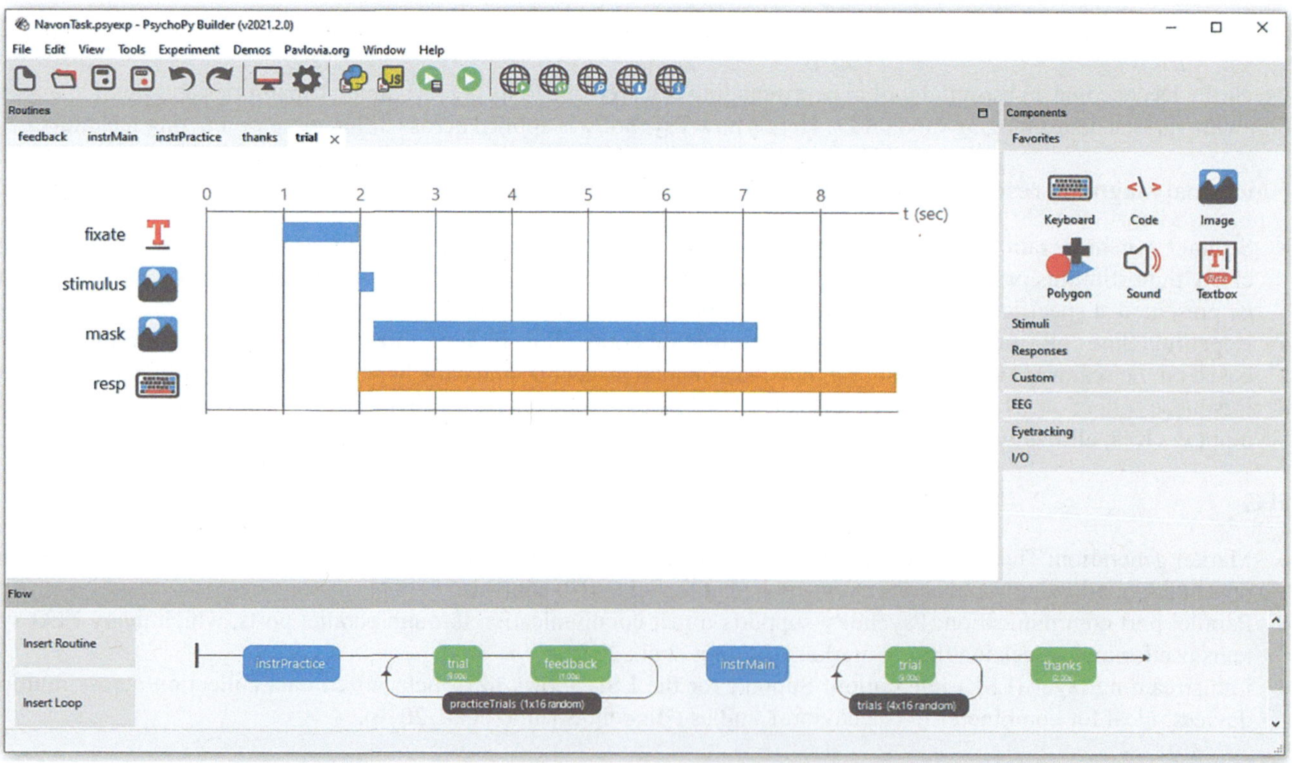

FIG. 2.1 Screenshot of the experiment builder in PsychoPy. PsychoPy experiment builder interface.

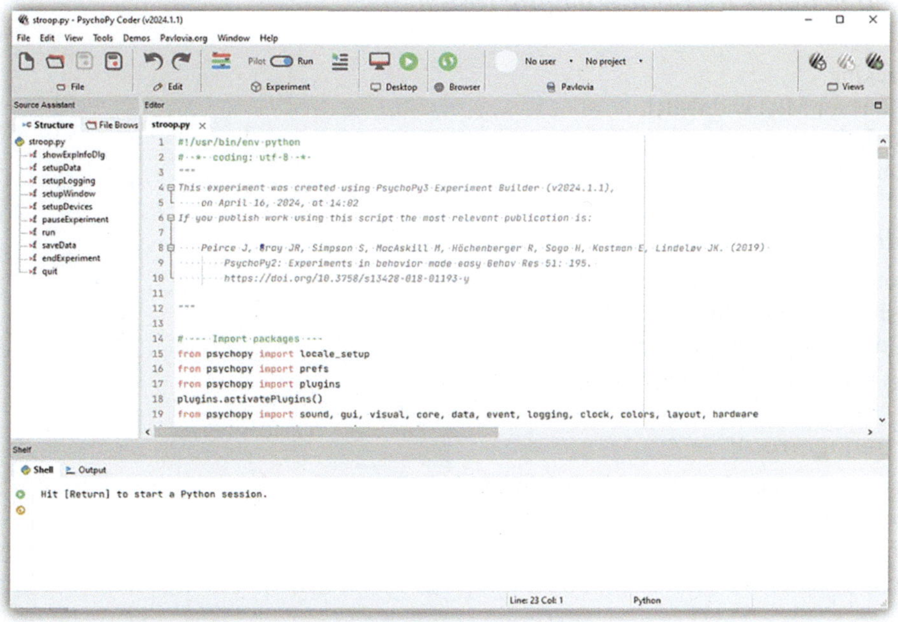

FIG. 2.2 Image of the coder view in PsychoPy. Screenshot of the coder view in PsychoPy.

3.2 E-prime—Psychology software tools

E-Prime, developed by Psychology Software Tools (PST), was introduced in the early 2000s as the successor to the disk operating system (DOS)-based Micro Experimental Laboratory (MEL) software. Since then, it has become one of the most comprehensive and widely used platforms for experimental design in psychological and neuroimaging research. The software is well-regarded for its precise timing capabilities, user-friendly interface, and strong support for various research hardware configurations.

E-Studio serves as the central hub for experimental design within E-Prime. It features a graphical user interface that visualizes the experimental structure through a hierarchical timeline representation. This interface includes:

Experiment explorer: A tree-view structure displaying all experimental objects
Workspace: The main design area where experimenters arrange stimuli and responses
Properties window: Detailed parameter settings for each experimental object
Output window: Real-time debugging and validation information
Toolbar: Quick access to commonly used functions and tools

3.2.1 E-basic scripting language

While E-Prime's graphical interface suffices for many experiments, its proprietary E-Basic scripting language provides additional flexibility for complex designs. Based on visual basic for applications, E-Basic offers:

Full programmatic control over experimental parameters
Custom calculation and data manipulation capabilities
Inline script insertion at any point in the experiment
Access to system-level functions and hardware control
Extended mathematical and string-processing functions

3.2.2 Specialized features for neuroimaging research

MRI integration
Head motion tracking
E-Prime extensions for MRI and fMRI
MRI sound stimulation
Vision correction lenses

3.2.3 Sync timing system

Direct integration with scanner trigger pulses
Automated synchronization with TR
Support for multiple trigger-detection methods
Precision timing adjustment for scanner delays

3.2.4 Display management

Dual-display support for experimenter and participant views
Resolution matching for MRI-compatible displays
Refresh rate synchronization
Gamma correction capabilities

3.2.5 Stimulus presentation

E-Prime offers researchers a powerful and versatile system for designing psychological and neuroimaging experiments using various integrated methods. The software features a graphical, timeline-based interface that allows users to easily drag and drop experimental components to create their study designs. These components include slides for presenting stimuli, input forms for collecting responses, and feedback objects for engaging participants.

The software excels in managing complex experimental designs through its advanced list management system. This system facilitates both simple and intricate randomization schemes, counterbalancing, and nested designs. Researchers can utilize conditional branching to create adaptive experiments that adjust their flow based on participants' responses or performance. E-Prime's precision in timing is especially impressive, offering millisecond-level accuracy for both stimulus presentation and response collection. The platform accommodates various stimulus types, including text, images, videos, and audio, with extensive customization options for each.

For advanced users, there is the possibility of enhancing these features through E-Basic scripting. This allows for custom calculations, dynamic stimulus generation, and complex trial logic. This combination of visual design tools and programming flexibility makes E-Prime suitable for a broad range of studies—from straightforward reaction time tests to complex fMRI paradigms that require precise timing and synchronization with scanners.

3.2.6 Software compatibility

The platform maintains broad compatibility with:

Analysis packages (MATLAB, R, SPSS)
Data collection systems
Third-party extensions
Custom software through API access

3.2.7 Applications of E-prime in neuroimaging

E-Prime has established itself as a cornerstone technology in neuroimaging research, offering robust integration with various neuroimaging modalities through its specialized features and precise timing capabilities. Here's how E-Prime is applied across different neuroimaging techniques:

fMRI

E-Prime extensions for fMRI: These dedicated extensions facilitate seamless integration with scanner hardware, allowing researchers to synchronize stimulus presentation with volume acquisition through TR monitoring (Schneider et al., 2002).

SyncBox integration: E-Prime's compatibility with the PST SyncBox hardware enables precise timing between scanner pulses and stimulus presentation, which is critical for event-related designs requiring millisecond accuracy (Fishburn et al., 2019).

Multidisplay support: The software's dual-display capability allows experimenters to monitor participants' performance while presenting stimuli on MRI-compatible displays, maintaining experimental control during scanning (MacWhinney et al., 2001).

EEG

Serial/parallel port communication: The software provides robust support for sending event markers through serial or parallel ports, which is essential for time-locking EEG data to stimulus events (Dien et al., 2004).

Dedicated EEG extensions: E-Prime offers specialized extensions for major EEG systems, facilitating synchronized data collection without requiring custom programming (Moeller et al., 2008).

E-Basic scripting for complex paradigms: The scripting capabilities allow for the implementation of adaptive paradigms that respond to real-time EEG features, enabling neurofeedback and brain-computer interface applications (Ries et al., 2014).

MEG

Precise timing controls: E-Prime's millisecond-level timing precision is crucial for MEG studies examining early sensory and cognitive processes (Wiesman and Wilson, 2019).

Configurable trigger options: The software supports various trigger protocols used by MEG systems, allowing for accurate event synchronization without specialized hardware (Dumas et al., 2013).

NIRS

Block design implementation: The timeline-based interface facilitates the creation of optimal block designs for measuring the hemodynamic response in NIRS studies (McKendrick et al., 2015).

Integration with NIRS systems: E-Prime can communicate with NIRS acquisition software to synchronize experimental events with hemodynamic measurements (Pinti et al., 2015).

Eye-tracking integration

Direct interface with major systems: Built-in support for eye-tracking hardware from manufacturers like SensoMotoric Instruments (SMI), Tobii, and SR Research (SR) enables synchronized data collection without additional programming (Ryan et al., 2007).

Gaze-contingent paradigms: E-Basic scripting allows for the implementation of real-time, gaze-contingent experimental designs, where stimuli change based on participants' eye movements (Thigpen et al., 2019).

Practical applications

Cognitive neuroscience: The software is widely used in studies of attention, working memory, and executive function, where precise timing of complex stimulus sequences is critical (Owen et al., 2005).

Clinical neuroimaging: E-Prime facilitates the standardized administration of cognitive tasks in clinical populations during neuroimaging, supporting research on psychiatric and neurological disorders (Siegle et al., 2007).

Developmental neuroscience: The user-friendly interface and robust timing controls make E-Prime suitable for developmental neuroimaging studies across the lifespan (Luna et al., 2010).

Multimodal imaging: E-Prime's flexibility enables simultaneous data collection across multiple imaging modalities, such as combined EEG-fMRI studies (Mullinger and Bowtell, 2011).

3.3 Inquisit: Bridging laboratory and web-based research

Inquisit, developed by Millisecond Software, represents a significant advancement in psychological research software by uniquely bridging the gap between traditional laboratory-based research and web-based data collection. This distinctive characteristic has positioned Inquisit as a valuable tool in the modern researcher's arsenal, particularly as the field increasingly embraces online methodologies while maintaining rigorous scientific standards (Figs. 2.3 and 2.4).

At its core, Inquisit operates through two complementary platforms: Inquisit Lab and Inquisit Web. The laboratory version serves as a robust desktop application, providing researchers with the full spectrum of hardware integration capabilities and maximum timing precision essential for controlled experimental environments. Complementing this, Inquisit

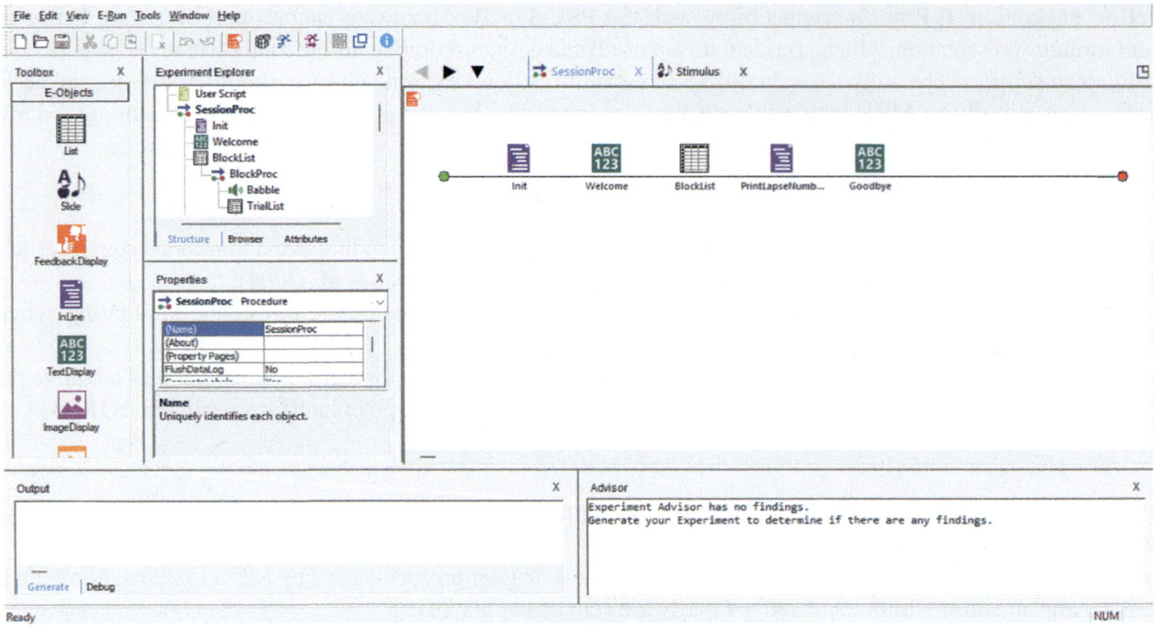

FIG. 2.3 Screenshot of the E-Prime software paradigm builder. Paradigm builder in E-Prime software.

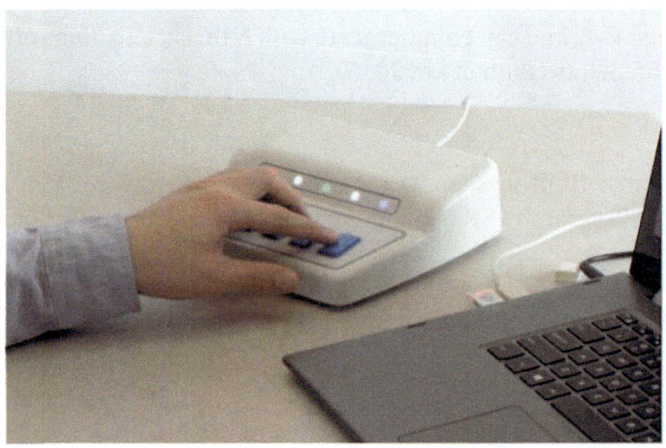

FIG. 2.4 External response pad—Chronos. Chronos is an external response pad used for precise stimulus-response timing in experiments.

Web extends these capabilities to browser-based implementation, offering unprecedented flexibility in data collection while maintaining the platform's commitment to timing accuracy.

3.3.1 Experimental design and implementation

Inquisit's approach to experimental design centers on a scripting language specifically crafted for psychological research. While this may present an initial learning curve for new users, it provides researchers with extraordinary flexibility in creating and implementing experimental paradigms. The scripting system offers intuitive syntax for experimental design, built-in timing controls, and access to extensive stimulus libraries, allowing researchers to create everything from simple reaction time studies to complex adaptive experiments.

The platform's stimulus presentation capabilities encompass a wide range of modalities, including visual, auditory, and multimedia content. Researchers can implement precise timing control, dynamic content generation, and adaptive display management, all while maintaining resolution-independent scaling across different display configurations. This flexibility proves particularly valuable in web-based studies, where participant hardware varies considerably.

3.3.2 Laboratory research and hardware integration

In laboratory settings, Inquisit demonstrates its versatility through extensive hardware integration capabilities. The software supports various research equipment, including response boxes, eye trackers, and physiological monitors, while maintaining precise timing synchronization. Performance optimization features include display refresh synchronization, processing thread management, and hardware acceleration, ensuring maximum precision in controlled research environments.

3.3.3 Research applications and practical considerations

The platform's versatility makes it particularly well-suited for a wide range of research domains, including cognitive psychology, clinical assessment, neuropsychological testing, and social psychology research. Its ability to seamlessly combine laboratory and web-based data collection makes it especially valuable for mixed-method studies, multisite research, and longitudinal investigations. This flexibility has proven particularly beneficial for researchers conducting cross-cultural studies or large-scale data collection efforts, where traditional methods may fall short. The software's unique position in bridging laboratory and web-based research makes it an invaluable tool for researchers adapting to the changing landscape of psychological research. Its ability to maintain timing precision across different implementation contexts, combined with comprehensive data management capabilities, positions Inquisit as a forward-looking solution for researchers seeking to combine traditional experimental rigor with modern data collection methodologies (Fig. 2.5).

3.3.4 Applications of Inquisit in neuroimaging

fMRI

Scanner synchronization: Inquisit Lab's precise timing mechanisms allow researchers to synchronize stimulus presentation with TR pulses from MRI scanners, which is critical for event-related designs (Forman et al., 1995).

Script-based experimental control: The scripting language facilitates complex, adaptive paradigms that can adjust based on participant responses or neural activity patterns during scanning (De Houwer et al., 2001).

Response collection: Inquisit supports MRI-compatible response devices, allowing researchers to collect behavioral data concurrently with neural measurements (Phan et al., 2002).

EEG

Event marker generation: The software can send precisely timed event markers to EEG acquisition systems, which are essential for time-locking neural responses to stimuli (Kappenman and Luck, 2012).

Millisecond precision: Inquisit's focus on timing accuracy ensures that stimulus onset latencies are recorded with sufficient precision for ERP studies (Whelan, 2008).

Complex paradigm implementation: The scripting language supports sophisticated protocols that can adapt based on ongoing EEG data, enabling neurofeedback applications (Zotev et al., 2013).

MEG

High-precision timing controls: The software's millisecond-level timing accuracy is crucial for MEG studies examining rapid neural dynamics (Garrido et al., 2015).

Visual stimulus optimization: Inquisit's display management features ensure consistent presentation across the specialized visual projection systems used in MEG environments (Hämäläinen et al., 1993).

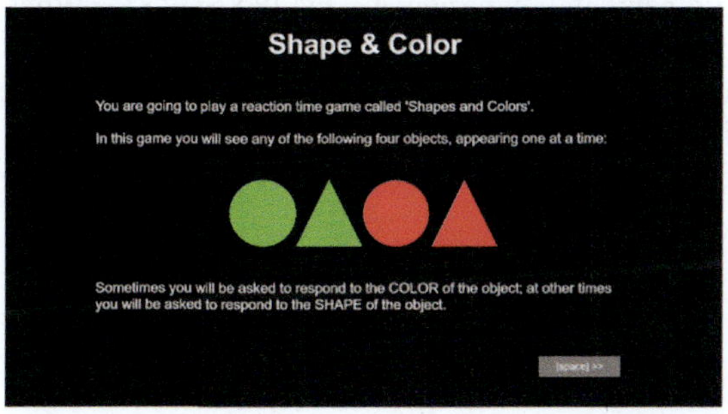

FIG. 2.5 Screenshot of a task running in Inquisit. Inquisit is used for running psychological experiments and cognitive tasks with precision.

NIRS

Block design implementation: The scripting language facilitates the creation of optimized block designs suitable for measuring the slower hemodynamic response in NIRS (Tak and Ye, 2014). Portable testing solutions: Inquisit's ability to run on standard hardware complements the portability advantages of modern NIRS systems, enabling field research (Piper et al., 2014).

Eye-tracking integration

Direct interface with eye-tracking hardware: The software supports leading eye-tracking systems, allowing synchronized collection of gaze data with neuroimaging measures (Peterburs et al., 2012).

Gaze-contingent stimulus presentation: Advanced scripting enables stimuli to adapt based on participants' gaze patterns during neuroimaging (Wilms et al., 2010).

Multimodal imaging applications

Synchronized data collection: The platform facilitates concurrent stimulus presentation across multiple recording systems, which is essential for combined EEG-fMRI or NIRS-EEG studies (Liu et al., 2016).

Consistent paradigm implementation: Researchers can implement identical paradigms across different imaging modalities, which is crucial for multimethod validation studies (Bridwell et al., 2018).

Practical applications in neuroimaging research

Cognitive neuroscience: The software supports complex cognitive paradigms, examining attention, memory, and executive function during neuroimaging (MacLeod and MacDonald, 2000).

Affective neuroscience: Inquisit's precise timing and multimedia support enable emotion induction and regulation paradigms in neuroimaging contexts (Etkin and Wager, 2007).

Clinical neuroimaging: The platform facilitates the standardized administration of neurocognitive assessments during clinical brain imaging, supporting research on psychiatric and neurological disorders (Insel et al., 2010).

Developmental neuroscience: Inquisit's engaging interface and adaptive capabilities make it suitable for developmental neuroimaging studies across the lifespan (Casey et al., 2005).

Distributed and remote neuroimaging research

Prescan/postscan behavioral testing: Researchers can deploy identical cognitive tasks before and after scanning sessions, even when participants are remote (Lumsden et al., 2016).

Multisite standardization: The platform ensures consistent task implementation across different neuroimaging facilities in collaborative research networks (Van Horn and Toga, 2014).

Longitudinal protocol management: Inquisit facilitates consistent paradigm delivery across multiple time points in longitudinal neuroimaging studies (Siegle et al., 2007).

3.4 SuperLab

SuperLab, crafted by Cedrus Corporation, stands as a testament to the principle that sophisticated research tools need not be complicated. In the landscape of experimental psychology software, SuperLab has carved out a unique position by combining straightforward usability with the precision demanded by scientific research.

At the core of SuperLab's appeal lies its visual experiment builder—a refreshing departure from text-heavy, script-based alternatives. Imagine building your experiment like arranging blocks in a visual timeline, where what you see during design is exactly what your participants will experience. This WYSIWYG (What You See Is What You Get) approach transforms complex experimental design into an intuitive, visual process.

One of SuperLab's standout features is its real-time preview functionality. Researchers can immediately visualize their experimental flow, catching potential issues before they become problems during data collection. This immediate feedback loop significantly reduces development time and minimizes experimental design errors.

3.4.1 Multimodal stimulus support

SuperLab handles an impressive array of stimulus types with remarkable efficiency:

Visual stimuli (images, shapes, text)
Audio presentations (tones, speech, music)

Video content (digital video files, animations)
Combined multimodal presentations

SuperLab is widely used to design and execute task-based fMRI paradigms, where participants perform specific cognitive, motor, or perceptual tasks inside the scanner:

Cognitive tasks: Memory recall, language comprehension, or decision-making studies
Perceptual tasks: Visual or auditory discrimination tasks
Motor tasks: Finger-tapping or other motor coordination tasks
SuperLab ensures precise control over task timing, stimulus sequence, and response collection, all synchronized with the scanner's trigger signals

3.4.2 Research domains

SuperLab proves particularly valuable in:

Cognitive psychology studies
Perception research
Attention experiments
Psycholinguistic investigations
Educational research
Clinical assessments

SuperLab stands out as a versatile and reliable tool for behavioral and psychological research. Its robust feature set, ease of use, and compatibility with external devices make it a preferred choice for researchers across disciplines. While it does come with certain limitations, such as cost and hardware dependence, its advantages far outweigh these challenges.

By streamlining the process of experiment design and data collection, SuperLab empowers researchers to focus on what truly matters: advancing our understanding of human behavior and cognition. Whether you are a seasoned researcher or a student embarking on your first project, SuperLab provides the tools needed to bring your experiments to life.

3.4.3 Applications of SuperLab in neuroimaging

fMRI

Scanner synchronization: SuperLab interfaces seamlessly with MRI trigger systems to synchronize stimulus presentation with volume acquisition, essential for event-related designs (Gazzaley et al., 2005).

Block and event-related designs: The visual timeline interface facilitates the creation of both block-design paradigms and complex event-related experiments with precise onset timing (D'esposito et al., 2000).

Response collection: SuperLab supports MRI-compatible response devices from Cedrus and other manufacturers, allowing researchers to record behavioral responses during scanning (Kanwisher and Yovel, 2006).

TR monitoring: The software can receive and process scanner pulses to maintain synchronization with the scanner's repetition time throughout experiments (Poldrack et al., 2011).

EEG

Event marker precision: The software can send precisely timed event markers to EEG acquisition systems via a parallel port or dedicated interfaces, which is crucial for ERP studies requiring millisecond accuracy (Luck, 2005).

Stimulus timing verification: SuperLab's real-time preview and timing verification features help ensure that stimulus presentation meets the precise requirements of EEG research (Woodman, 2010).

Multimodal stimulation: The platform's support for combined visual, auditory, and video stimuli enables diverse EEG paradigms, examining different sensory processing pathways (Hillyard and Anllo-Vento, 1998).

MEG

Precise visual timing: SuperLab's display synchronization features ensure accurate presentation timing, which is critical for MEG studies examining early visual processing (Bayle and Taylor, 2010).

Trigger output flexibility: The software supports various triggering protocols used by MEG systems, allowing for accurate event synchronization.

NIRS

Block design implementation: The visual timeline interface simplifies the creation of block designs optimized for the hemodynamic response measured in NIRS (Huppert et al., 2006).

Portable testing solutions: SuperLab's ability to run on standard hardware complements the portability advantages of NIRS systems, enabling more naturalistic testing environments (Cutini and Brigadoi, 2014).

Eye-tracking integration

Integration with major systems: The software interfaces with popular eye-tracking systems, allowing synchronized collection of gaze data with neuroimaging measures (Henderson and Ferreira, 2004).

Gaze-contingent designs: SuperLab's event system enables the creation of paradigms where stimulus presentation adapts based on participants' gaze behavior (Duchowski Andrew, 2007).

Practical applications in neuroimaging

Cognitive neuroscience: The software supports paradigms examining attention, memory, and executive function during neuroimaging, with its visual builder facilitating the rapid implementation of standard tasks like Stroop, n-back, and flanker tasks (Miller and Cohen, 2001).

Perceptual neuroscience: SuperLab's precise timing and multimodal stimulation capabilities make it suitable for investigating neural correlates of perceptual processes across sensory modalities (Calvert et al., 2004).

Clinical neuroimaging: The platform's user-friendly interface facilitates the implementation of clinical protocols for examining brain function in patient populations, where standardized administration is crucial (Etkin et al., 2013).

Developmental neuroscience: SuperLab's engaging interface and straightforward stimulus presentation make it appropriate for developmental neuroimaging studies across different age groups (Casey et al., 2008).

Multisite implementation

Standardized protocols: The visual builder ensures consistent implementation across different research sites, which is critical for large-scale collaborative studies (Van Horn and Toga, 2009).

Portable experiment files: SuperLab experiments can be easily shared between institutions, maintaining methodological consistency in multisite investigations (Biswal et al., 2010) (Fig. 2.6).

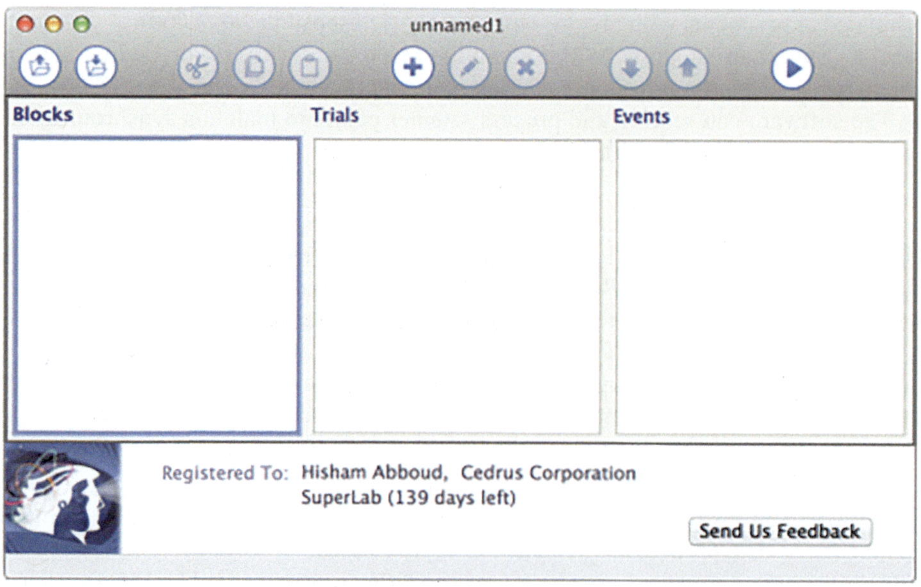

FIG. 2.6 Screenshot of the experimental window of the SuperLab software. Opening screen of SuperLab.

4 Discussion

The integration of software tools and neuroimaging technologies represents a transformative advancement in neurological diagnostics and research. This convergence has fundamentally altered how researchers and clinicians approach brain structure and function analysis, enabling unprecedented precision in investigating neural processes. The software platforms discussed—PsychoPy, E-Prime, Inquisit, and SuperLab—each offer unique capabilities that address specific research needs while sharing the common goal of facilitating precise experimental control in neuroimaging environments.

The evolution of these software tools parallels the development of neuroimaging technologies themselves. As imaging modalities like MRI, fMRI, EEG, and PET have become more sophisticated, the software required to interface with these systems has similarly advanced. This coevolution has created synergistic benefits, where improvements in one domain drive innovations in the other. For instance, the increasing temporal and spatial resolution of modern fMRI systems necessitates software capable of synchronizing stimulus presentation with millisecond precision—a capability now standard in platforms like E-Prime and SuperLab.

A notable trend emerging from this review is the growing emphasis on accessibility and flexibility. Open-source solutions like PsychoPy have democratized access to sophisticated research tools, allowing laboratories with limited resources to conduct cutting-edge neuroimaging research. Meanwhile, platforms like Inquisit have bridged the gap between laboratory and web-based research, enabling researchers to combine the precision of controlled experiments with the scale and diversity of online participant pools.

The multimodal integration capabilities of these software platforms highlight another significant advancement. Researchers can now seamlessly combine data from different imaging techniques, allowing for complementary insights into brain function that would be impossible with any single modality. This integration has led to a more comprehensive understanding of neural processes, combining the spatial precision of techniques like MRI with the temporal resolution of EEG.

5 Future directions

Advanced AI integration: While machine learning algorithms are already being applied to image analysis, deeper integration of AI into experimental design and real-time data processing represents a frontier with significant potential. Future software may incorporate adaptive experimental paradigms that evolve based on neural responses, optimizing data collection in real-time.

Enhanced cloud-based solutions: The migration toward cloud-based platforms will likely continue, enabling more powerful distributed computing capabilities for complex analyses and facilitating multisite collaborations with standardized protocols and centralized data storage.

Augmented and virtual reality integration: As AR/VR technologies mature, their integration with neuroimaging software will enable more naturalistic experimental paradigms while maintaining experimental control. This could revolutionize how we study social cognition, spatial navigation, and other complex behaviors in neuroimaging environments.

Mobile and wearable integration: The development of software specifically designed to interface with mobile neuroimaging technologies, such as portable EEG and fNIRS systems, will expand opportunities for ecological validity in research and potential clinical applications outside traditional laboratory settings.

Open science infrastructure: Further development of open-source platforms and standardized data formats will promote transparency, reproducibility, and collaborative advancement in the field, potentially accelerating discoveries through broader access to tools and data.

6 Challenges

Technical complexity: The learning curve associated with specialized software platforms can be steep, particularly for researchers without programming backgrounds. This complexity can limit adoption and create barriers to entry for new researchers.

Standardization issues: The diversity of software platforms, data formats, and analysis approaches can complicate data sharing and study replication. While this variety offers flexibility, it also creates challenges for establishing field-wide standards.

Hardware dependencies: Many software solutions remain tied to specific hardware configurations, limiting their flexibility and potentially increasing research costs through the requirement of proprietary equipment.

Balancing precision and usability: There often exists a trade-off between software that offers maximum experimental control and software that prioritizes user-friendly interfaces. Striking the optimal balance remains challenging.

Data security and privacy: As neuroimaging research increasingly collects sensitive data, particularly in clinical applications, ensuring appropriate security and privacy protections becomes increasingly important, yet technically demanding.

7 Recommendations

Invest in training: Institutions should prioritize comprehensive training in neuroimaging software tools, incorporating these skills into graduate curricula and offering continuing education for established researchers.

Promote open standards: The field should continue to develop and adopt open standards for data formats, experimental protocols, and analysis procedures to enhance reproducibility and collaboration.

Cross-platform validation: Researchers should validate experimental paradigms across different software platforms to identify platform-specific effects and ensure the robustness of findings.

User-centered design: Software developers should incorporate user experience research into the development process, creating tools that balance power and usability for diverse user groups.

Interdisciplinary collaboration: Strengthening collaborations between neuroscientists, computer scientists, and software engineers will accelerate innovation in neuroimaging software tools and ensure they meet the evolving needs of researchers.

8 Limitations

Timing precision variability: While all the software platforms reviewed emphasize timing precision, actual performance can vary based on hardware configurations, operating systems, and experimental complexity. This variability can introduce confounds, particularly in studies of rapid neural processes.

Ecological validity constraints: Even with advances in naturalistic paradigms, laboratory-based neuroimaging experiments often sacrifice ecological validity for experimental control. Software solutions addressing this tension remain limited.

Resource requirements: Many advanced neuroimaging software solutions require substantial computational resources, potentially limiting access for researchers in resource-constrained settings.

Clinical translation challenges: Despite the potential for clinical applications, many software tools remain optimized for research rather than clinical environments, creating barriers to translational implementation.

Compatibility issues: Not all software solutions work seamlessly across different operating systems or with all neuroimaging hardware, potentially restricting researchers' choices based on existing infrastructure.

9 Conclusion

The integration of software tools and neurological diagnostics has fundamentally transformed our ability to investigate brain structure and function. Platforms such as PsychoPy, E-Prime, Inquisit, and SuperLab have each contributed unique capabilities to the neuroimaging ecosystem, enabling increasingly sophisticated experimental paradigms with precise timing and control. The continued evolution of these tools, particularly in their ability to integrate multiple imaging modalities and adapt to diverse research questions, promises to further advance our understanding of the human brain.

As the field moves forward, addressing the identified challenges while building upon existing strengths will be crucial. The trend toward greater accessibility through open-source solutions and web-based implementations suggests a future where advanced neuroimaging research becomes increasingly democratized. Similarly, the growing incorporation of artificial intelligence and machine learning approaches points toward more automated, efficient, and potentially more insightful analysis methods.

The ultimate value of these technological advances lies in their application to fundamental questions in neuroscience and clinical challenges in neurological and psychiatric disorders. By continuing to refine the tools that allow us to observe and interact with the brain, we expand our capacity to understand its complexities and develop more effective interventions for brain-related conditions. The synergistic relationship between software development and neuroimaging technology promises continued innovation in this vital field of research.

References

Andrew, D., 2007, Eye Tracking Methodology, second ed. Springer London, London.

Bajwa, J., Munir, U., Nori, A., Williams, B., 2021. Artificial intelligence in healthcare: transforming the practice of medicine. Future Healthc. J. 25146645. 8 (2), e188. doi:10.7861/fhj.2021-0095.

Bayle, D.J., Taylor, M.J., 2010. Attention inhibition of early cortical activation to fearful faces. Brain Res. 00068993. 1313, 113–123. doi:10.1016/j.brainres.2009.11.060.

Belcari, N., Bisogni, M.G., Del Guerra, A., 2023. Positron emission tomography: its 65 years and beyond. Riv. Nuovo Cim. 46, 693–785. doi:10.1007/s40766-024-00050-3.

Berger, 2003. How does it work? Positron emission tomography. BMJ 326 (7404), 1449–1451. doi:10.1136/bmj.326.7404.1449.

Bigdely-Shamlo, N., Makeig, S., Robbins, K.A., 2016. Preparing laboratory and real-world EEG data for large-scale analysis: a containerized approach. Front. Neuroinform. 16625196. 10. Available from: http://journal.frontiersin.org/article/10.3389/fninf.2016.00007/pdf. doi:10.3389/fninf.2016.00007.

Biswal, B.B., Mennes, M., Zuo, X.-N., Gohel, S., Kelly, C., Smith, S.M., Beckmann, C.F., Adelstein, J.S., Buckner, R.L., Colcombe, S., Dogonowski, A.-M., Ernst, M., Fair, D., Hampson, M., Hoptman, M.J., Hyde, J.S., Kiviniemi, V.J., Kötter, R., Shi-Jiang, L., Lin, C.-P., Lowe, M.J., Mackay, C., Madden, D.J., Madsen, K.H., Margulies, D.S., Mayberg, H.S., McMahon, K., Monk, C.S., Mostofsky, S.H., Nagel, B.J., Pekar, J.J., Peltier, S.J., Petersen, S.E., Riedl, V., Rombouts, S.A.R.B., Rypma, B., Schlaggar, B.L., Schmidt, S., Seidler, R.D., Siegle, G.J., Sorg, C., Teng, G.-J., Veijola, J., Villringer, A., Walter, M., Wang, L., Weng, X.-C., Whitfield-Gabrieli, S., Williamson, P., Windischberger, C., Zang, Y.-F., Hong-Ying Zhang, F.X.C., Milham, M.P., 2010. Toward discovery science of human brain function. Proc. Natl. Acad. Sci. USA. 0027-8424. 107 (10), 4734–4739. doi:10.1073/pnas.0911855107.

Bridges, D., Pitiot, A., MacAskill, M.R., Peirce, J.W., 2020. The timing mega-study: comparing a range of experiment generators, both lab-based and online. PeerJ. 21678359. 8. Available from: https://peerj.com/archives/?journal=peerj. doi:10.7717/peerj.9414.

Bridwell, D.A., Rachakonda, S., Silva, R.F., Pearlson, G.D., Calhoun, V.D., 2018. Spatiospectral decomposition of multi-subject EEG: evaluating blind source separation algorithms on real and realistic simulated data. Brain Topogr. 15736792. 31 (1), 47–61. Available from: www.wkap.nl/journalhome.htm/0896-0267. doi:10.1007/s10548-016-0479-1.

Calvert, G.A., Spence, C., Stein, B.E., 2004. The Handbook of Multisensory Processes. The MIT Press. doi:10.7551/mitpress/3422.001.0001.

Casey, B.J., Jones, R.M., Hare, T.A., 2008. The adolescent brain. Ann. N.Y. Acad. Sci. 17496632. 1124, 111–126. Available from: http://onlinelibrary.wiley.com/journal/10.1111/(ISSN)1749-6632. doi:10.1196/annals.1440.010.

Casey, B., Tottenham, N., Liston, C., Durston, S., 2005. Imaging the developing brain: what have we learned about cognitive development? Trends Cogn. Sci. 13646613. 9 (3), 104–110. doi:10.1016/j.tics.2005.01.011.

Chegondi, Lin, Naqvi, Sendi, Totapally, 2022. The effect of electroencephalography abnormalities on cerebral autoregulation in sedated ventilated children. Pediatr. Rep. 15 (1), 9–15.

Cutini, S., Brigadoi, S., 2014. Unleashing the future potential of functional near-infrared spectroscopy in brain sciences. J. Neurosci. Methods. 1872678X. 232, 152–156. Available from: www.elsevier.com/locate/jneumeth. doi:10.1016/j.jneumeth.2014.05.024.

D'esposito, M., Ballard, D., Zarahn, E., Aguirre, G.K., 2000. The role of prefrontal cortex in sensory memory and motor preparation: an event-related fMRI study. NeuroImage. 10538119. 11 (5), 400–408. doi:10.1006/nimg.2000.0571.

De Houwer, J., Crombez, G., Baeyens, F., Hermans, D., 2001. On the generality of the affective Simon effect. Cogn. Emot. 14640600. 15 (2), 189–206. doi:10.1080/0269993004200051.

Dien, J., Spencer, K.M., Donchin, E., 2004. Parsing the late positive complex: mental chronometry and the ERP components that inhabit the neighborhood of the P300. Psychophysiology. 00485772. 41 (5), 665–678. Available from: http://www.blackwellpublishing.com/aims.asp?ref=0048-5772. doi:10.1111/j.1469-8986.2004.00193.x.

Dumas, T., Dubal, S., Attal, Y., Chupin, M., Jouvent, R., Morel, S., George, N., Keil, A., 2013. MEG evidence for dynamic amygdala modulations by gaze and facial emotions. PLoS ONE. 1932-6203. 8 (9), e74145. doi:10.1371/journal.pone.0074145.

Etkin, A., Gyurak, A., O'Hara, R., 2013. A neurobiological approach to the cognitive deficits of psychiatric disorders. Dialogues Clin. Neurosci. 12948322. 15 (4), 419–429. Available from: http://www.ncbi.nlm.nih.gov/pmc/articles/PMC3898680/pdf/DialoguesClinNeurosci-15-419.pdf.

Etkin, A., Wager, T.D., 2007. Functional neuroimaging of anxiety: a meta-ana lysis of emotional processing in PTSD, social anxiety disorder, and specific phobia. Am. J. Psychiatry. 0002953X. 164 (10), 1476–1488. doi:10.1176/appi.ajp.2007.07030504.

Fishburn, F.A., Ludlum, R.S., Vaidya, C.J., Medvedev, A.V., 2019. Temporal Derivative Distribution Repair (TDDR): a motion correction method for fNIRS. NeuroImage. 10959572. 184, 171–179. Available from: http://www.elsevier.com/inca/publications/store/6/2/2/9/2/5/index.htt. doi:10.1016/j.neuroimage.2018.09.025.

Forman, S.D., Cohen, J.D., Fitzgerald, M., Eddy, W.F., Mintun, M.A., Noll, D.C., 1995. Improved assessment of significant activation in functional magnetic resonance imaging (fMRI): use of a cluster-size threshold. Magn. Reson. Med. 15222594. 33 (5), 636–647. doi:10.1002/mrm.1910330508.

Garrido, M.I., Barnes, G.R., Kumaran, D., Maguire, E.A., Dolan, R.J., 2015. Ventromedial prefrontal cortex drives hippocampal theta oscillations induced by mismatch computations. NeuroImage. 10959572. 120, 362–370. Available from: http://www.elsevier.com/inca/publications/store/6/2/2/9/2/5/index.htt. doi:10.1016/j.neuroimage.2015.07.016.

Gazzaley, A., Cooney, J.W., McEvoy, K., Knight, R.T., D'Esposito, M., 2005. Top-down enhancement and suppression of the magnitude and speed of neural activity. J. Cogn. Neurosci. 17 (3), 507–517. doi:10.1162/0898929053279522.

Gramfort, A., Luessi, M., Larson, E., Engemann, D.A., Strohmeier, D., Brodbeck, C., Parkkonen, L., Hämäläinen, M.S., 2014. MNE software for processing MEG and EEG data. NeuroImage. 10959572. 86, 446–460. Available from: http://www.elsevier.com/inca/publications/store/6/2/2/9/2/5/index.htt. doi:10.1016/j.neuroimage.2013.10.027.

Hämäläinen, M., Hari, R., Ilmoniemi, R.J., Knuutila, J., Lounasmaa, O.V., 1993. Magnetoencephalography theory, instrumentation, and applications to noninvasive studies of the working human brain. Rev. Mod. Phys. 00346861. 65 (2), 413–497. doi:10.1103/RevModPhys.65.413.

Henderson, J.M., Ferreira, F. (Eds.), 2004. The interface of language, vision, and action: Eye movements and the visual world. Psychology Press.

Hillyard, S.A., Anllo-Vento, L., 1998. Event-related brain potentials in the study of visual selective attention. Proc. Natl. Acad. Sci. USA. 0027-8424. 95 (3), 781–787. doi:10.1073/pnas.95.3.781.

Huppert, T.J., Hoge, R.D., Diamond, S.G., Franceschini, M.A., Boas, D.A., 2006. A temporal comparison of BOLD, ASL, and NIRS hemodynamic responses to motor stimuli in adult humans. NeuroImage. 10538119. 29 (2), 368–382. doi:10.1016/j.neuroimage.2005.08.065.

Insel, T., Cuthbert, B., Garvey, M., Heinssen, R., Pine, D.S., Quinn, K., Sanislow, C., Wang, P., 2010. Research domain criteria (RDoC): toward a new classification framework for research on mental disorders. Am. J. Psychiatry. 15357228. 167 (7), 748–751. Available from: http://ajp.psychiatryonline.org/cgi/reprint/167/7/748. doi:10.1176/appi.ajp.2010.09091379.

Kanwisher, N., Yovel, G., 2006. The fusiform face area: a cortical region specialized for the perception of faces. Philos. Trans. R. Soc. Lond. B Biol. Sci. 09628436. 361 (1476), 2109–2128. Available from: http://rstb.royalsocietypublishing.org/. doi:10.1098/rstb.2006.1934.

Kapoor, M., Sharma, A., Tripathi, M., 2025. Positron emission tomography (PET) scanning. In: StatPearls. StatPearls Publishing. Available from: https://www.ncbi.nlm.nih.gov/books/NBK559089/.

Kappenman, E.S., Luck, S.J., 2012. The Oxford Handbook of Event-Related Potential Components the Oxford Handbook of Event-Related Potential Components. Oxford University Press, United States, pp. 1–664. Available from: http://www.oxfordhandbooks.com/view/10.1093/oxfordhb/9780195374148.001.0001/oxfordhb-9780195374148. doi:10.1093/oxfordhb/9780195374148.001.0001.

Liu, Y., Bengson, J., Huang, H., Mangun, G.R., Ding, M., 2016. Top-down modulation of neural activity in anticipatory visual attention: control mechanisms revealed by simultaneous EEG-fMRI. Cereb. Cortex. 14602199. 26 (2), 517–529. Available from: http://cercor.oxfordjournals.org/. doi:10.1093/cercor/bhu204.

Louis, E.K.S., Frey, L.C., Britton, J.W., Hopp, J.L., Korb, P.J., Koubeissi, M.Z., et al., 2016. Electroencephalography (EEG): An Introductory Text and Atlas of Normal and Abnormal Findings in Adults, Children, and Infants. American Epilepsy Society. Available from: https://api.semanticscholar.org/CorpusID:151886401.

Luck, S.J., 2005. An Introduction to the Event-Related Potential Technique. MIT Press. Available from: https://mitpress.mit.edu/9780262525855/an-introduction-to-the-event-related-potential-technique/.

Lui, Y.W., Chang, P.D., Zaharchuk, G., Barboriak, D.P., Flanders, A.E., Wintermark, M., Hess, C.P., Filippi, C.G., 2020. Artificial intelligence in neuroradiology: current status and future directions. Am. J. Neuroradiol. 1936959X. 41 (8), E52. Available from: http://www.ajnr.org/content/ajnr/41/8/E52full.pdf. doi:10.3174/ajnr.A6681.

Lumsden, J., Edwards, E.A., Lawrence, N.S., Coyle, D., Munafò, M.R., 2016. Gamification of cognitive assessment and cognitive training: a systematic review of applications and efficacy. JMIR Serious Games. 22919279. 4 (2). Available from: https://games.jmir.org/2016/2/e11. doi:10.2196/games.5888.

Luna, B., Padmanabhan, A., O'Hearn, K., 2010. What has fMRI told us about the development of cognitive control through adolescence? Brain Cogn. 02782626. 72 (1), 101–113. doi:10.1016/j.bandc.2009.08.005.

MacLeod, C.M., MacDonald, P.A., 2000. Interdimensional interference in the Stroop effect: uncovering the cognitive and neural anatomy of attention. Trends Cogn. Sci. 13646613. 4 (10), 383–391. doi:10.1016/S1364-6613(00)01530-8.

MacWhinney, B., James, J.S., Schunn, C., Ping, L., Schneider, W., 2001. STEP—A system for teaching experimental psychology using E-prime. Behav. Res. Methods Instrum. Comput. 0743-3808. 33 (2), 287–296. doi:10.3758/bf03195379.

Mathôt, S., Schreij, D., Theeuwes, J., 2012. OpenSesame: an open-source, graphical experiment builder for the social sciences. Instrum. Comput. Behav. Res. Methods Instrum. Comput. 15543528. 44 (2), 314–324. doi:10.3758/s13428-011-0168-7.

McKendrick, R., Parasuraman, R., Ayaz, H., 2015. Wearable functional near infrared spectroscopy (fNIRS) and transcranial direct current stimulation (tDCS): expanding vistas for neurocognitive augmentation. Front. Syst. Neurosci. 16625137. 9. Available from: http://journal.frontiersin.org/article/10.3389/fnsys.2015.00027/pdf. doi:10.3389/fnsys.2015.00027.

Miller, E.K., Cohen, J.D., 2001. An integrative theory of prefrontal cortex function. Annu. Rev. Neurosci. 0147006X. 24, 167–202. doi:10.1146/annurev.neuro.24.1.167.

Moeller, F., Siebner, H.R., Wolff, S., Muhle, H., Granert, O., Jansen, O., Stephani, U., Siniatchkin, M., 2008. Simultaneous EEG-fMRI in drug-naive children with newly diagnosed absence epilepsy. Epilepsia. 15281167. 49 (9), 1510–1519. doi:10.1111/j.1528-1167.2008.01626.x.

Monks, C.P., Tsermentseli, S., Sarah, O.C.E., Executive function and theory of mind as predictors of aggressive and prosocial behavior and peer acceptance in early childhood.

Mullinger, K., Bowtell, R., 2011. Combining EEG and fMRI. Methods Mol. Biol. 711, 303–326. Available from: https://www.springer.com/series/7651. doi:10.1007/978-1-61737-992-5_15.

Orringer, D.A., Vago, D.R., Golby, A.J., 2012. Clinical applications and future directions of functional MRI. Semin. Neurol. 10989021. 32 (4), 466–475. doi:10.1055/s-0032-1331816.

Owen, A.M., McMillan, K.M., Laird, A.R., Bullmore, E., 2005. N-back working memory paradigm: a meta-analysis of normative functional neuroimaging studies. Hum. Brain Mapp. 25 (1), 46–59. doi:10.1002/hbm.20131.

Peirce, J., Gray, J.R., Simpson, S., MacAskill, M., Höchenberger, R., Sogo, H., Kastman, E., Lindeløv, J.K., 2019. PsychoPy2: experiments in behavior made easy. Instrum. Comput. Behav. Res. Methods Instrum. Comput. 15543528. 51 (1), 195–203. Available from: http://www.springerlink.com/content/1554-351x/. doi:10.3758/s13428-018-01193-y.

Peterburs, J., Gajda, K., Koch, B., Schwarz, M., Hoffmann, K.P., Daum, I., Bellebaum, C., 2012. Cerebellar lesions alter performance monitoring on the antisaccade task-An event-related potentials study. Neuropsychologia. 18733514. 50 (3), 379–389. doi:10.1016/j.neuropsychologia.2011.12.009.

Phan, K.L., Wager, T., Taylor, S.F., Liberzon, I., 2002. Functional neuroanatomy of emotion: a meta-analysis of emotion activation studies in PET and fMRI. NeuroImage. 10538119. 16 (2), 331–348. doi:10.1006/nimg.2002.1087.

Pinti, P., Aichelburg, C., Lind, F., Power, S., Swingler, E., Merla, A., Hamilton, A., Gilber, S., Burgess, P., Tachtsidis, I., 2015. Using fiberless, wearable fnirs to monitor brain activity in real-world cognitive tasks. J. Vis. Exp. 1940087X. 2015 (106). Available from: http://www.jove.com/video/53336. doi:10.3791/53336.

Pinti, P., Tachtsidis, I., Hamilton, A., Hirsch, J., Aichelburg, C., Gilbert, S., Burgess, P.W., 2020. The present and future use of functional near-infrared spectroscopy (fNIRS) for cognitive neuroscience. Ann. N.Y. Acad. Sci. 0077-8923. 1464 (1), 5–29. doi:10.1111/nyas.13948.

Piper, S.K., Krueger, A., Koch, S.P., Mehnert, J., Habermehl, C., Steinbrink, J., Obrig, H., Schmitz, C.H., 2014. A wearable multi-channel fNIRS system for brain imaging in freely moving subjects. NeuroImage. 10959572. 85, 64–71. doi:10.1016/j.neuroimage.2013.06.062.

Poldrack, R.A., Mumford, J.A., Nichols, T.E., 2011. Handbook of Functional MRI Data Analysis. Cambridge University Press. doi:10.1017/CBO9780511895029.

Ries, A.J., Touryan, J., Vettel, J., McDowell, K., David Hairston, W., 2014. A comparison of electroencephalography signals acquired from conventional and mobile systems. J. Neurosci. Neuroeng. 21682011. 3 (1), 10–20. doi:10.1166/jnsne.2014.1092.

Ryan, J.D., Hannula, D.E., Cohen, N.J., 2007. The obligatory effects of memory on eye movements. Memory. 14640686. 15 (5), 508–525. doi:10.1080/09658210701391022.

Schneider, W., Eschman, Zuccolotto, A., 2002. E-prime User's Guide. Available from: https://psy.swan.ac.uk/staff/lucignoli/e-prime/ReferenceGuideV2_2012.pdf.

Schulz, R.A., Stein, J.A., Pelc, N.J., 2021. How CT happened: the early development of medical computed tomography. J. Med. Imaging. 2329-4302. 8 (05). doi:10.1117/1.jmi.8.5.052110.

Siegle, G.J., Thompson, W., Carter, C.S., Steinhauer, S.R., Thase, M.E., 2007. Increased amygdala and decreased dorsolateral prefrontal BOLD responses in unipolar depression: related and independent features. Biol. Psychiatry. 00063223. 61 (2), 198–209. doi:10.1016/j.biopsych.2006.05.048.

Surianarayanan, C., Lawrence, J.J., Chelliah, P.R., Prakash, E., Hewage, C., 2023. Convergence of artificial intelligence and neuroscience towards the diagnosis of neurological disorders—a scoping review. Sensors. 14248220. 23 (6). Available from: http://www.mdpi.com/journal/sensors. doi:10.3390/s23063062.

Tak, S., Ye, J.C., 2014. Statistical analysis of fNIRS data: a comprehensive review. NeuroImage. 10959572. 85, 72–91. doi:10.1016/j.neuroimage.2013.06.016.

Thigpen, N., Petro, N.M., Oschwald, J., Oberauer, K., Keil, A., 2019. Selection of visual objects in perception and working memory one at a time. Psychol. Sci. 14679280. 30 (9), 1259–1272. Available from: http://www.sagepub.com/home.nav. doi:10.1177/0956797619854067.

Thomas, L.A., Kim, P., Bones, B.L., Hinton, K.E., Milch, H.S., Reynolds, R.C., Adleman, N.E., Marsh, A.A., Blair, R.J.R., Pine, D.S., Leibenluft, E., 2013. Elevated amygdala responses to emotional faces in youths with chronic irritability or bipolar disorder. Neuroimage Clin. 22131582. 2 (1), 637–645. doi:10.1016/j.nicl.2013.04.007.

Van Horn, J.D., Toga, A.W., 2009. Multisite neuroimaging trials. Curr. Opin. Neurol. 13507540. 22 (4), 370–378. doi:10.1097/WCO.0b013e32832d92de.

Van Horn, J.D., Toga, A.W., 2014. Human neuroimaging as a "Big Data" science. Brain Imaging Behav. 19317565. 8 (2), 323–331. doi:10.1007/s11682-013-9255-y.

Westbrook, C., Talbot, J., 2018. MRI in Practice, fifth ed. Wiley-Blackwell, pp. 01–416.

Whelan, R., 2008. Effective analysis of reaction time data. Psychol. Rec. 0033-2933. 58 (3), 475–482. doi:10.1007/BF03395630.

Wiesman, A.I., Wilson, T.W., 2019. The impact of age and sex on the oscillatory dynamics of visuospatial processing. NeuroImage. 10959572. 185, 513–520. Available from: http://www.elsevier.com/inca/publications/store/6/2/2/9/2/5/index.htt. doi:10.1016/j.neuroimage.2018.10.036.

Wilms, M., Schilbach, L., Pfeiffer, U., Bente, G., Fink, G.R., Vogeley, K., 2010. It's in your eyes-using gaze-contingent stimuli to create truly interactive paradigms for social cognitive and affective neuroscience. Soc. Cogn. Affect. Neurosci. 17495024. 5 (1), 98–107. doi:10.1093/scan/nsq024.

Woodman, G.F., 2010. A brief introduction to the use of event-related potentials in studies of perception and attention. Attent. Percep. Psychophys. 1943-3921. 72 (8), 2031–2046. doi:10.3758/app.72.8.2031.

Xue, G., Chen, C., Lu, Z.-L., Dong, Q., 2010. Brain imaging techniques and their applications in decision-making research. Acta Psychol. Sin. 0439-755X. 42 (1), 120–137. doi:10.3724/SP.J.1041.2010.00120.

Yen, C., Lin, C.-L., Chiang, M.-C., 2023. Exploring the frontiers of neuroimaging: a review of recent advances in understanding brain functioning and disorders. Life. 2075-1729. 13 (7), 1472. doi:10.3390/life13071472.

Zaki, J., Wager, T.D., Singer, T., Keysers, C., Gazzola, V., 2016. The anatomy of suffering: understanding the relationship between nociceptive and empathic pain. Trends Cogn. Sci. 1879307X. 20 (4), 249–259. Available from: www.elsevier.com/locate/tics. doi:10.1016/j.tics.2016.02.003.

Zotev, V., Phillips, R., Young, K.D., Drevets, W.C., Bodurka, J., 2013. Prefrontal control of the amygdala during real-time fMRI neurofeedback training of emotion regulation. PLoS ONE. 19326203. 8 (11). Available from: http://www.plosone.org/article/fetchObject.action?uri=info%3Adoi%2F10.1371%2Fjournal.pone.0079184&representation=PDF. doi:10.1371/journal.pone.0079184.

Chapter 3

Medical imaging, behavioural changes, and their effects on memory and learning

Venu Madhav Nelluri[a], Sudarshan Surendran[b], Suresh Sukumar[c], Leena R. David[c,d], Sneha Ravichandran[c], Berna Uzun[e], Dilber Uzun Ozsahin[d,e,f], M.G. Ramesh Babu[g], Rajagopal Kadavigere[h], Dilip Shettigar[c], and Ilker Ozsahin[e]

[a]*Department of Anatomy, Manipal University College Malaysia, Melaka, Malaysia*

[b]*Department of Anatomy, American University of Antigua College of Medicine, St. John's Antigua, Antigua and Barbuda*

[c]*Department of Medical Imaging Technology, Manipal College of Health Professions, Manipal Academy of Higher Education, Manipal, Karnataka, India*

[d]*Medical Diagnostic Imaging Department, College of Health Sciences, University of Sharjah, Sharjah, United Arab Emirates*

[e]*Irfan Suat Gunsel Operational Research Institute, Near East University, Nicosia, Turkish Republic of Northern Cyprus (TRNC), Turkey*

[f]*Research Institute for Medical and Health Sciences, University of Sharjah, Sharjah, United Arab Emirates*

[g]*Division of Physiology, Department of Basic Medical Sciences, Manipal Academy of Higher Education, Manipal, India*

[h]*Department of Radiodiagnosis and Imaging, Kasturba Medical College, Manipal, India*

1 Introduction

1.1 History of medical imaging

The discovery of X-rays by Wilhelm Conrad Roentgen in the 19th century played a pivotal role in the field of medical imaging. This discovery helped physicians visualize the internal structures of the human body without invasive procedures. From then on, the development and advancement of medical imaging technologies progressed at a very high pace, leading to the creation of fluoroscopes and radiographs. By the mid-20th century, the advent of computed tomography (CT scans) advanced this field by providing cross-sectional details of the body. Following this, magnetic resonance imaging (MRI) emerged in the 1970s, utilizing powerful magnets and radio waves to create highly detailed images compared to CT scans. After these technologies designed for structural visualization, positron emission tomography (PET scans) and single-photon emission CT (SPECT) were introduced as highly valuable tools for studying physiological processes.

The evolution of medical imaging has been nothing less than spectacular and continues to evolve with innovations like 3D printing, functional MRI (fMRI), and molecular imaging, assisting in precise diagnosis and the treatment of various medical conditions. The trajectory of development has been particularly remarkable in the last two decades, with the integration of computational advances enabling unprecedented clarity and precision. The digital revolution transformed medical imaging from analog technologies to sophisticated digital systems capable of processing, storing, and transmitting vast amounts of data. This digitization facilitated the emergence of teleradiology, allowing specialists to examine images remotely, thereby democratizing access to expert diagnoses across geographical boundaries (Hricak et al., 2011).

1.2 Medical imaging techniques

Medical imaging has played a major role in most clinical setups, where diagnoses often begin with these techniques prescribed by physicians in appropriate cases. These technologies help in diagnosis, treatment planning, and continuous monitoring of certain medical conditions.

Brain medical imaging plays an important role in medicine and healthcare for reasons such as diagnosis, treatment of certain conditions, and monitoring various neurological and psychiatric conditions. The diagnosis of brain disorders has seen major improvements with advances in medical imaging technology. Techniques such as MRI and CT contribute

Advances in Medical Imaging. https://doi.org/10.1016/B978-0-443-28967-5.00009-5

significantly to the diagnosis of tumors, malformations in the vascular architecture, and structural abnormalities of the brain. In conditions such as stroke, the extent of brain damage can be evaluated, allowing for appropriate treatment and prognosis. In cases of head trauma, intracranial bleeding, fractures, or any other type of structural damage can be assessed using medical imaging technologies.

All over the world, there is a rise in patients with epilepsy, and imaging technologies can assist in planning effective surgical interventions and providing appropriate treatment. In cases involving tumor, monitoring the condition plays a major role in adjusting treatments for the progression of the condition based on its growth, shrinkage, or stability during the treatment period. Another major neurological condition is Alzheimer's disease, where PET scans play a crucial role in the evaluation of the condition, along with other forms of dementia. In psychiatric disorders such as depression, bipolar disorder, or schizophrenia, brain imaging technologies can help in planning custom-tailored treatment options.

When considering neurosurgery, planning to the finest detail possible helps neurosurgeons in performing their procedures effectively. This is an essential feature in such procedures to minimize damage by choosing the most effective approach and, in turn, maximize the probability of a successful outcome. Cerebrovascular diseases can be fatal in certain conditions, as the blood supply to every part of the brain is essential for its complete and integrated function with the body. Many related diseases can lead to stroke, resulting in neurological problems.

The clinical relevance of medical imaging extends beyond diagnostic applications to therapeutic interventions. For instance, in the treatment of brain tumors, stereotactic radiosurgery utilizes high-precision imaging to deliver targeted radiation therapy, minimizing damage to surrounding healthy tissue. This approach has revolutionized treatment paradigms, reducing the need for invasive surgical procedures and improving patient outcomes. Similarly, deep brain stimulation for Parkinson's disease relies on precise imaging to locate specific brain regions for electrode placement, highlighting the crucial role of imaging in guiding interventions (Okun, 2012).

Radiation therapy has been used appropriately in certain conditions, and monitoring the response to specific treatments has been effective with the use of medical imaging technologies. The brain, being a complicated organ to study structurally, requires functional understanding to provide a greater ability to assess certain conditions. Techniques such as fMRI and SPECT provide physicians with the ability to study the activity patterns of a functioning brain, aiding in the evaluation of cognitive and behavioral functions. In addition to advancements in understanding the structure, connections, and functioning of the brain, research remains an essential core need. For the development of new therapies based on this understanding of the brain, researchers and scientists utilize medical imaging technologies to create advanced treatment modalities for neurological conditions. Overall, the contribution of medical imaging technologies has played a pivotal role in medicine and healthcare, enabling professionals to excel in their fields.

2 Behavioral changes and brain-behavior interactions

An individual's thoughts, actions, or emotions, if altered, lead to behavioral changes, and these could be linked to the underlying changes that occur in the brain itself. Behavior could be considered the central core of the outcomes resulting from changes within the brain. The dynamic nature of the brain provides it with the capacity to undergo changes, which is termed neuroplasticity. These changes are linked to the response mechanisms triggered by inputs from the environment and the experiences it generates. This is connected to the behavior of the individual and their cognitive abilities. The connections within the brain shed light on the comprehension of the mechanisms responsible for behavioral changes, particularly in the realms of memory and learning.

The intricate web of connections within the structure of the brain, the blanket of medical imaging over this structure, behavior, memory, and learning together form the basis for a better understanding of this chapter. By improving the understanding of these connections, their potential for use in diagnosis, treatment, and the enhancement of cognitive performance could be realized. In the following pages, we will probe into the deeper aspects of these areas and make an attempt to relate to real-world examples along with scientific findings to exemplify the intricacies of this complex interdisciplinary field.

The neurobehavioral interface is particularly evident in conditions like posttraumatic stress disorder (PTSD), where imaging studies have revealed alterations in the amygdala and prefrontal cortex that correspond to behavioral manifestations such as hypervigilance and emotional dysregulation. These findings demonstrate how specific neural circuits mediate behavioral responses to traumatic experiences, providing a neurobiological framework for understanding the condition. Furthermore, studies on substance addiction have elucidated the roles of the brain's reward pathways in driving compulsive behaviors, illustrating how dysregulation in these circuits can manifest as maladaptive behavioral patterns (Volkow et al., 2016).

The overall aim of this chapter is to explore the intricate relationship between medical imaging techniques, behavioral changes, and their profound implications for understanding memory and learning. Shifts in behaviors could indicate changes in brain function, which could also be influential in the context of learning and memory.

3 Types of medical imaging and their application in neuroscience

Medical imaging plays a crucial role in studying the human brain and its relationship with memory, learning, and behavioral changes. With recent advances in imaging technologies, various intricate structural and complex functional aspects of the brain can be studied in an intriguing manner.

Of the various technologies, MRI, using its strong magnetic fields and radio waves, provides detailed images of the brain's structural and functional components. The images produced by MRI are of very high resolution, and the advantage of not exposing patients to ionizing radiation makes it a safer option in cases requiring repeated usage. It is used in understanding or studying the anatomy of the brain, detecting any abnormalities, and this advantage could be utilized for relating alterations in brain structure to memory and learning (Smith et al., 2004).

Recent technological advances in MRI have elevated its capabilities significantly. High-field MRI systems operating at 7T and above have enabled visualization of brain structures at unprecedented levels of detail, revealing microstructural features previously inaccessible to neuroimaging. These advanced systems can detect subtle alterations in hippocampal subfields—critical regions for memory processing—allowing researchers to correlate specific structural changes with memory function. Moreover, the development of multiparametric MRI, which combines different contrast mechanisms in a single examination, provides comprehensive characterization of brain tissue properties, enhancing our understanding of the neurobiological underpinnings of learning processes (Kraff et al., 2015).

CT is a scanning technology that combines X-rays taken from different angles to create cross-sectional images of the brain. The use of CT has the advantage of providing quick and detailed visualization of the brain's structure. CT is often used in cases involving brain injuries, tumors, or other structural abnormalities (Brenner and Hall, 2007).

Modern CT technology has progressed remarkably with the introduction of multidetector CT scanners capable of acquiring multiple image slices simultaneously, significantly reducing scan time while improving spatial resolution. This advancement has been particularly valuable in emergency settings, where rapid assessment of brain trauma can be lifesaving. In addition, dual-energy CT, which utilizes two different energy levels to distinguish between tissues with similar attenuation values, has enhanced the differentiation of various brain pathologies, providing more precise diagnostic information in cases of stroke, hemorrhage, and other acute neurological conditions (Patino et al., 2016).

PET is performed with the injection of a radioactive tracer into the body, which emits positrons to be detected by the scanner. This study is mainly used for the measurement of metabolic activity. This technology offers insights into the functioning of the brain and its activity. In studies involving the understanding of memory and learning tasks, the metabolism of glucose or the activity of neurotransmitters is used as the basis for this technique (Herholz and Heiss, 2004).

The development of novel radiotracers has revolutionized PET's contribution to neuroscience. Beyond glucose metabolism, modern tracers can now visualize specific neurotransmitter systems, neuroinflammation, and protein aggregates associated with neurodegenerative diseases. For instance, amyloid and tau PET tracers have transformed our understanding of Alzheimer's disease progression, revealing that pathological changes begin decades before clinical symptoms emerge. This knowledge has profound implications for early intervention strategies. Similarly, dopamine system tracers have provided insights into reward learning mechanisms, while serotonin system imaging has enhanced our understanding of mood regulation and its disorders (Grimm et al., 2015).

fMRI is a specialized technique combined with MRI that measures the blood flow and oxygenation patterns associated with brain activity. This technique has the added advantage of studying the brain in real time. This advantage is used to investigate the regions of the brain involved in tasks related to memory and learning, as well as to study the dynamics of neural networks (Logothetis, 2008).

In educational neuroscience, fMRI has illuminated the neural bases of different learning strategies, demonstrating that techniques like spaced repetition engage memory consolidation networks more effectively than massed practice. Such findings have direct applications in optimizing educational approaches. In addition, real-time fMRI neurofeedback has emerged as a promising tool for enhancing cognitive function, allowing individuals to modulate their own brain activity through visual feedback, with demonstrated benefits for attention, working memory, and emotional regulation. These applications showcase how fMRI has evolved from a purely research tool to having practical implications for cognitive enhancement and rehabilitation (DeBettencourt et al., 2015).

Diffusion tensor imaging (DTI) is another specialized form of MRI that uses the diffusion of water molecules within brain tissue. This technology is used for mapping the brain's white matter, which is known as the tracts. As it is used for this purpose, the major role of this technology lies in understanding the connectivity in the brain and the various pathways that influence behavior patterns along with cognitive abilities (Assaf and Pasternak, 2008).

Advanced diffusion techniques have expanded beyond the tensor model to include high angular resolution diffusion imaging, diffusion spectrum imaging, and neurite orientation dispersion and density imaging. These sophisticated approaches provide a more comprehensive characterization of white matter microstructure and connectivity. Longitudinal DTI studies have revealed how white matter pathways develop during skill acquisition, showing that learning to juggle or master a musical instrument leads to measurable changes in white matter architecture. These findings highlight the structural basis of experience-dependent plasticity, demonstrating how learning physically reshapes the brain's connectome (Zatorre et al., 2012).

Electroencephalography (EEG) is used for the measurement of electrical activity in the brain by the placement of electrodes on the scalp. The major advantage of this technique is the production of high temporal resolution, which is critical in the understanding of instant brain responses requiring rapid capturing of signals. The investigation of brainwaves associated with memory encoding, retrieval of stored memory, and other learning processes is done using this technology (Niedermeyer and da Silva, 2005).

Contemporary EEG research has benefited from high-density electrode arrays and sophisticated signal processing algorithms, allowing for improved source localization and functional connectivity analyses. These advancements have enabled researchers to identify specific oscillatory patterns associated with different memory processes—theta oscillations during encoding, gamma oscillations during retrieval, and sleep spindles during memory consolidation. Mobile EEG systems now permit the study of brain activity during natural learning situations outside laboratory environments, providing ecological validity to neurocognitive research. Furthermore, machine learning approaches applied to EEG data have demonstrated the ability to predict individual learning outcomes based on neural signatures, offering potential for personalized educational interventions (Krigolson et al., 2017).

Magnetoencephalography (MEG) represents another valuable imaging modality that detects the magnetic fields produced by neuronal activity. With temporal resolution comparable to EEG but superior spatial precision, MEG has provided unique insights into the rapid dynamics of neural processes underlying memory formation and retrieval. Studies using MEG have revealed that successful memory encoding involves precisely timed synchronization between the prefrontal cortex and medial temporal lobe within milliseconds, highlighting the exquisite temporal coordination required for effective learning. When combined with structural imaging techniques, MEG offers a comprehensive view of both the "when" and "where" of cognitive processing (Hari et al., 2018).

Near-infrared spectroscopy (NIRS) has emerged as a noninvasive optical imaging technique particularly suited for studying brain function in naturalistic settings. By measuring hemodynamic responses through the intact skull, NIRS allows researchers to monitor cortical activity during everyday learning activities, including classroom instruction or while children engage with educational materials. This accessibility has made NIRS especially valuable for developmental cognitive neuroscience, revealing how brain networks supporting attention and working memory mature during critical learning periods. The relative affordability and portability of NIRS systems have democratized access to neuroimaging research, expanding the study of learning mechanisms across diverse populations and contexts (Ferrari and Quaresima, 2012).

All of these medical imaging techniques play their own crucial role in uncovering the relationship between the structure of the brain, its functioning, behavioral changes, and the overall effect of these on memory and learning. Researchers often use a combination of these techniques or methods to gain insight into the structure and functioning of the brain (Bandettini, 2012; Damoiseaux and Greicius, 2009). With the combination of appropriate technologies, understanding can be extended to gain insights into the complex interplay between the structure of the brain and cognitive processes (Filippi and Agosta, 2011; Bullmore and Sporns, 2009).

4 Artificial intelligence and deep learning in medical imaging analysis

The integration of artificial intelligence (AI) and deep learning into medical imaging analysis has revolutionized how we process, interpret, and extract information from neuroimaging data. These computational approaches have transformed the field by enabling automated feature detection, pattern recognition, and prediction at scales and levels of complexity beyond human capabilities.

Convolutional Neural Networks (CNNs), a class of deep learning algorithms particularly suited for image analysis, have demonstrated remarkable success in automatically identifying brain structures and abnormalities from MRI and CT scans. These networks learn hierarchical features from raw imaging data, progressively extracting higher-level representations that can distinguish subtle patterns associated with different cognitive states or pathological conditions. For instance, CNNs trained on structural MRI data can now classify Alzheimer's disease with accuracy comparable to expert neuroradiologists, often detecting changes before they become clinically apparent (Lu et al., 2018).

Deep learning techniques have proven particularly valuable for analyzing functional connectivity in the brain. By applying recurrent neural networks (RNNs) and transformer models to fMRI time series data, researchers can now identify complex temporal patterns in brain activity associated with specific cognitive processes. These algorithms have revealed previously unrecognized brain states during learning tasks, demonstrating how different neural networks dynamically reconfigure as individuals acquire new skills or consolidate memories (Thomas et al., 2019).

The application of generative adversarial networks (GANs) has addressed critical limitations in neuroimaging research, such as small sample sizes and data quality issues. These models can synthesize realistic brain images with specific characteristics, augmenting training datasets and enabling more robust machine learning pipelines. Furthermore, GANs have shown promise in cross-modality synthesis, where one imaging modality (e.g., MRI) can be used to predict the appearance of another (e.g., PET), potentially reducing the need for multiple scans and minimizing patient exposure to radiation (Wolterink et al., 2017).

Multimodal deep learning approaches that integrate data from different imaging techniques (e.g., MRI, PET, EEG) have pushed the boundaries of what can be learned about brain-behavior relationships. These models leverage the complementary strengths of each modality—structural details from MRI, metabolic activity from PET, and temporal dynamics from EEG—to construct comprehensive representations of brain function during learning and memory processes. Such integrated analyses have identified biomarkers that predict learning outcomes with greater accuracy than any single imaging approach alone (Dafflon et al., 2020).

In clinical applications, AI-assisted image analysis has enhanced diagnostic accuracy while reducing interpretation time. Automated lesion detection systems can now flag subtle abnormalities that might be missed during visual inspection, while AI-driven volumetric analysis can track changes in brain structures over time with precision beyond manual measurement capabilities. These tools are particularly valuable for monitoring disease progression or treatment response in conditions affecting learning and memory, such as multiple sclerosis or traumatic brain injury (Litjens et al., 2017).

Looking ahead, explainable AI represents a critical frontier in neuroimaging analysis. As these algorithms become increasingly complex, understanding how they arrive at specific conclusions becomes essential for both scientific validity and clinical trust. Techniques that visualize feature importance and decision pathways are being developed to make AI "black boxes" more transparent, ensuring that insights gained from these powerful tools can be meaningfully integrated into our understanding of brain-behavior relationships (Yang and Wang, 2020).

5 Role of neuroplasticity in memory and learning

Neuroplasticity, also known as brain plasticity or neural plasticity, is a fundamental concept in neuroscience that refers to the ability of the brain to reorganize and adapt throughout a person's life. It is the capacity of the brain to adapt or change in response to experiences, learning, and environmental influences. Neuroplasticity plays a crucial role in how the brain processes information, learns new things, and extends to the understanding of recovery mechanisms from injuries. Neuroplasticity can be either structural or functional, relating to physical changes in brain structure or the ability of the brain to redistribute its functions across different areas, respectively. The physical changes include the formation of new neurons (neurogenesis) or connections (synaptogenesis), or the pruning of unused connections (synaptic pruning). In the latter case, if there is damage to one of the regions of the brain, other regions might compensate for the lost function.

Recent advances in our understanding of neuroplasticity have revealed sophisticated mechanisms that underlie the brain's remarkable adaptability. At the molecular level, experience-dependent plasticity involves complex cascades of gene expression and protein synthesis that ultimately reconfigure synaptic connections. For example, learning-induced changes activate immediate early genes like c-fos and zif268, initiating molecular processes that strengthen specific neural circuits. Long-term potentiation (LTP), a cellular mechanism underlying memory formation, involves the insertion of AMPA receptors into postsynaptic membranes, enhancing synaptic efficacy. Conversely, long-term depression (LTD) weakens connections through receptor internalization, enabling the selective refinement of neural networks based on experience (Kandel et al., 2014).

Critical periods represent windows of heightened plasticity during development when specific brain circuits are exceptionally receptive to environmental input. Modern research has identified molecular "brakes" on plasticity that close these periods, including perineuronal nets—specialized extracellular matrix structures that stabilize mature neural connections. Fascinatingly, recent studies demonstrate that these brakes can be temporarily released in adulthood through specific interventions, reopening periods of enhanced plasticity and potentially facilitating learning or recovery from injury. These findings have profound implications for educational approaches across the lifespan and rehabilitation strategies after brain damage ()Hensch and Bilimoria, 2012.

Neuroplasticity manifests differently across various learning domains. Procedural learning, such as motor skill acquisition, induces structural changes primarily in the basal ganglia, cerebellum, and motor cortex, with decreased activation as skills become automatic—reflecting neural efficiency. In contrast, declarative memory formation engages hippocampal-cortical networks, with information initially processed by the hippocampus before gradual transfer to distributed cortical regions for long-term storage through systems consolidation. These distinct plasticity mechanisms explain why different learning strategies are optimal for different types of knowledge acquisition (Doyon and Benali, 2005).

Sleep plays a crucial role in consolidating neuroplastic changes initiated during waking hours. During slow-wave sleep, the hippocampus "replays" neural activity patterns from recent learning experiences, strengthening specific synaptic connections while generally downscaling others—a process that enhances memory specificity and prevents network saturation. REM sleep appears particularly important for procedural learning and emotional memory processing. Medical imaging studies have visualized these consolidation processes, revealing coordinated activity between the hippocampus and neocortex during sleep spindles, which predict subsequent memory performance (Rasch and Born, 2013).

Environmental enrichment dramatically enhances neuroplasticity through multiple mechanisms. Enriched environments increase neurotrophic factors like BDNF (brain-derived neurotrophic factor), promote angiogenesis (the formation of new blood vessels), enhance dendritic branching, and stimulate adult neurogenesis in the hippocampus. Conversely, chronic stress and glucocorticoid exposure can impair these same processes, particularly affecting the hippocampus and prefrontal cortex—regions critical for learning and executive function. This biological understanding helps explain how educational environments and stress management can significantly impact learning outcomes (Voss et al., 2017).

Exercise represents one of the most potent enhancers of neuroplasticity. Aerobic activity increases cerebral blood flow, upregulates neurotrophic factors, and stimulates hippocampal neurogenesis through mechanisms involving BDNF and IGF-1 (insulin-like growth factor 1). Neuroimaging studies have documented increased hippocampal and prefrontal cortex volumes following regular exercise programs, with corresponding improvements in memory and executive function. These findings have led to the integration of physical activity into cognitive enhancement programs and rehabilitation protocols for various neurological conditions (Cotman et al., 2007).

Aging affects neuroplasticity through multiple pathways, including decreased neurogenesis, reduced neurotrophic factor production, increased inflammation, and the accumulation of oxidative stress. However, cognitive reserve—the brain's resilience to age-related changes—can be enhanced through lifelong learning, social engagement, and physical activity. Longitudinal imaging studies demonstrate that individuals with greater cognitive reserve show more efficient neural network utilization and alternative network recruitment when facing cognitive challenges or pathological changes, explaining why some people maintain cognitive function despite significant brain pathology (Stern, 2012).

6 Role of medical imaging in memory and learning

Medical imaging plays a crucial role in studying the structure of the brain and its functional aspects. This capability of the technology is instrumental in advancing our understanding of memory and learning.

Medical imaging techniques like MRI and CT have the capability of providing us with very detailed images of the brain's anatomical structures. It is a boon for researchers and healthcare professionals in the identification of the characteristics of various brain regions in normal and disease conditions. In conditions such as tumors, lesions, or vascular malformations that can impact the normal functioning of the brain, these imaging techniques can help us identify their location. The brain, being a very compact structure with thousands of neurons concentrated in very small regions, means that specific lesions even in a small part of the brain can have profound effects on learning and memory. The overall size and shape of the brain can be studied in comparative studies between individuals. Medical imaging can help us understand the details of variations in such studies, which enable us to explore the potential correlations between the regions involved in cognitive functions.

The clinical relevance of medical imaging in memory disorders is exemplified by its application in early Alzheimer's disease detection. Volumetric MRI analyses can identify hippocampal atrophy years before clinical symptoms appear, potentially allowing for earlier intervention. In amnestic mild cognitive impairment (MCI), a precursor to Alzheimer's, medical imaging can differentiate individuals likely to progress to dementia from those with stable conditions based on patterns of brain atrophy and functional connectivity alterations. This prognostic capability has significant implications for patient care planning and clinical trial enrollment for disease-modifying therapies (Jack et al., 2018).

Medical imaging has been instrumental in differentiating various types of memory systems within the brain. Studies combining fMRI with carefully designed cognitive tasks have dissociated the neural substrates of explicit (conscious) versus implicit (unconscious) memory, declarative versus procedural learning, and episodic versus semantic memory. For instance, case studies of patients with focal hippocampal damage, who retain procedural learning abilities while losing declarative memory, have been validated through structural and functional imaging, confirming the anatomical separation of these memory systems. These distinctions have practical implications for rehabilitation approaches, suggesting that different strategies may be optimal for the remediation of different types of memory impairments (Squire and Wixted, 2011).

In many neurological conditions and neurodegenerative diseases, monitoring the progress of the condition is essential. Medical imaging technologies help us achieve the required information for diagnosis and prognosis. With the advancement of functional imaging technologies such as fMRI, PET, or SPECT, the observation of the functioning of the brain in real time is now possible. Another important aspect of functional imaging is seen in cognitive task activations. Researchers use fMRI for such studies to understand the effect of various cognitive tasks on specific brain regions. This paves the way for understanding the neural mechanisms behind the processes of learning and memory.

Educational neuroscience has benefited tremendously from medical imaging by elucidating the neural bases of different instructional approaches. Comparative imaging studies have revealed that active learning strategies engage broader neural networks and lead to more robust memory traces than passive approaches. Similarly, research on multimodal learning (combining visual and auditory information) has demonstrated enhanced recruitment of integration areas like the superior temporal sulcus, correlating with improved retention. These findings provide neurobiological validation for specific pedagogical techniques and offer guidance for developing evidence-based educational practices that align with the brain's natural learning mechanisms (Ansari and Coch, 2006).

Medical imaging has been instrumental in revealing the concept of neuroplasticity, where regions of the brain reorganize their structural and functional components based on learning and memory. The structure of the brain, being highly plastic in nature, allows frequent modifications in its structure and connectivity. With the help of imaging studies, monitoring, understanding, and making sense of various adaptations is possible. In a clinical setting, these techniques aid in diagnosing neurological and psychiatric conditions linked to memory and learning. This is especially useful in conditions such as Alzheimer's disease, traumatic brain injuries, and epilepsy. In a study by Shu et al. (Shu et al., 1988) a new area of the brain called the marginal division (MrD) was identified for the unique structure of its neurons and the distribution of neurotransmitters. This study demonstrated its involvement in auditory working memory, which was confirmed using fMRI. The use of medical imaging to confirm the functions of newly recognized vital regions implicated in diseased conditions is a boon to neuroscience. The major functional capacity of the brain depends on its connectivity networks within its structure. Techniques such as fMRI are used to understand the functional connectivity networks of the brain. This sheds light on the various communications and collaborations between different regions of the brain during the processes of learning and memory.

As the structural plasticity of the brain is highly dynamic, keen attention is needed for its understanding, along with continuous monitoring. This monitoring is possible by conducting longitudinal studies, where the brain is studied over longer periods of time. Such studies are crucial for apprehending the alterations undergone by the brain in learning and memory, as well as the evolution of these abilities with age or in response to interventions (Lövdén et al., 2013).

The most important advantage of medical imaging in neuroscience is that the technique is noninvasive. This is ethically advantageous when studying vulnerable populations, such as children or individuals with cognitive impairments, who participate in such studies. Since these techniques are noninvasive, they can be used in longitudinal studies in cases requiring frequent monitoring over longer periods of time.

Advanced medical imaging has revealed sex differences in memory-related brain function that have important implications for understanding learning variability. Functional imaging studies have shown that women and men sometimes utilize different neural strategies for the same memory tasks, with women typically showing more bilateral activation patterns compared to men's more lateralized responses. These differences are particularly pronounced in spatial and verbal memory tasks. In addition, hormonal influences on memory-related brain activity have been visualized through imaging, demonstrating how estrogen and testosterone levels modulate hippocampal function and connectivity. These findings highlight the importance of considering biological variability in educational approaches and memory disorder treatments (Cahill, 2006).

Medical imaging techniques have been crucial in documenting how memory systems develop across the lifespan, from childhood through aging. Developmental neuroscience, using structural and functional imaging, has traced the maturation of memory circuits, showing that different memory systems develop along distinct timelines—procedural memory systems maturing earlier than episodic memory networks, which depend on late-developing prefrontal-hippocampal connections. At the other end of the age spectrum, longitudinal imaging studies have characterized the normal aging trajectory of memory-related brain structures and identified protective factors that preserve cognitive function despite age-related changes. These insights inform age-appropriate learning strategies and interventions to support memory function throughout life (Ofen and Shing, 2013).

Overall, medical imaging is invaluable for understanding the structure and functional composition of the brain in relation to learning and memory. Such techniques provide a comfortable and convenient bridge between the physical and cognitive aspects of the brain. Studies involving these technologies help us understand the relationships between various brain regions, their neural pathways, and the cognitive processes related to these structures and their networks. Our understanding of the alterations undergone by the brain in learning and memory is improved by these technologies, and they also provide appropriate information related to the diagnosis and treatment of cognitive disorders. On the whole, medical imaging has advanced our knowledge about the structure and function of the brain in various learning and memory situations over different stages of life, both under normal conditions and in the presence of impediments.

7 Technological advances in image processing for behavioral analysis

Recent technological advances in image processing have dramatically enhanced our ability to analyze behavioral correlates of brain activity, offering unprecedented insights into the neural mechanisms of learning and memory. These innovations span from improved data acquisition methods to sophisticated analytical techniques that extract meaningful patterns from complex neuroimaging datasets.

Machine learning algorithms have revolutionized the analysis of functional brain images, moving beyond traditional univariate approaches to multivariate pattern analysis (MVPA). These methods can identify distributed neural representations associated with specific cognitive states or memories, revealing how information is encoded across brain networks rather than in isolated regions. For example, MVPA applied to fMRI data can decode what specific content a person is remembering based solely on activation patterns, demonstrating that memories maintain distinct neural signatures. This "mind-reading" capability has profound implications for understanding how memories are represented and accessed in the brain (Norman et al., 2006).

Dynamic causal modeling (DCM) represents another significant advance, allowing researchers to test hypotheses about directional influences between brain regions during cognitive processing. By modeling how neural activity in one area affects activity in connected regions, DCM can reveal the causal architecture of memory networks—whether prefrontal regions are driving hippocampal activity during encoding, for instance, or vice versa during retrieval. These analyses have transformed our understanding of memory from static mapping of activated regions to dynamic models of information flow through neural circuits (Friston et al., 2003).

Graph theoretical approaches have provided new mathematical frameworks for analyzing brain connectivity patterns. By representing brain regions as nodes and connections as edges, these methods quantify network properties like clustering, path length, and centrality. Applied to memory research, graph theory has identified hub regions that play disproportionately important roles in information integration during learning. Furthermore, changes in network efficiency metrics have been shown to predict successful memory formation and retrieval, suggesting that optimal cognitive performance depends on specific patterns of brain network organization (Bassett and Sporns, 2017).

Computational modeling combined with neuroimaging has enabled researchers to test mechanistic hypotheses about how learning and memory operate. By implementing formal computational models (e.g., reinforcement learning algorithms) and relating their parameters to brain activity, researchers can identify neural correlates of specific computational processes, such as prediction error calculation or value updating. This approach has been particularly valuable for understanding how different brain regions contribute to distinct aspects of learning, such as exploration versus exploitation or model-based versus model-free strategies (O'Doherty et al., 2003).

High-performance computing has enabled more sophisticated analyses of neuroimaging data, including techniques that were previously computationally prohibitive. For instance, representational similarity analysis (RSA) compares patterns of brain activity to theoretical models of how information might be organized, revealing the organizational principles of memory representations. Similarly, hyperalignment techniques can align functional data across individuals in a common representational space, allowing for more powerful group analyses that preserve fine-grained information patterns while accounting for individual differences in brain organization (Haxby et al., 2014).

Real-time processing capabilities have opened new frontiers in neurofeedback and brain-computer interfaces. Closed-loop systems that analyze brain activity on the fly can provide immediate feedback or adjust experimental parameters based on an individual's neural state. These technologies have shown promise for enhancing memory through targeted reinforcement of specific brain patterns associated with successful encoding or by delivering memory reactivation cues during optimal brain states for consolidation (Sitaram et al., 2017).

Mobile and wearable neuroimaging technologies, such as ambulatory EEG and functional NIRS, have enabled the study of brain activity during natural behavior outside laboratory settings. These approaches bridge the gap between controlled experiments and real-world learning contexts, providing ecological validity to neuroimaging findings. Data from these technologies have revealed how attention fluctuates during everyday learning experiences and how environmental factors influence memory encoding in naturalistic settings (Reis et al., 2014).

8 Case studies and research findings

With medical imaging, the understanding of neuroscience has seen major breakthroughs. Though not related to medical imaging, the famous case of H.M. is well-known to many in the field of neuroscience. H.M. underwent surgery for the removal of his hippocampus and amygdala as a measure to control his seizures. Postsurgery, the result was a reduction in the seizures, but an additional observation was that he could not form any new memories or remember certain prior years of his life before the surgery. He had anterograde amnesia (the inability to form many new types of memories), where he could not form memories of new faces or facts. This study led to the discovery of the link between memory and the hippocampus. In a similar manner, there are various mysteries within the cranial cavity. Medical imaging, especially functional imaging, has helped researchers understand the characteristics and functional connections of the various regions of the brain related to learning and memory. In a healthcare setting, physicians often depend on such advanced imaging techniques to make decisions on cases related to the brain. They also use medical imaging techniques for the continuous monitoring of patients who require regular observation and attention.

While H.M.'s case predated modern neuroimaging, contemporary reimaging of his preserved brain using high-resolution MRI has provided extraordinary insights into the precise anatomical structures affected by his surgery. These detailed post-mortem analyses revealed that portions of his hippocampus remained intact, helping to refine our understanding of which hippocampal subregions are most critical for different memory functions. This case illustrates how medical imaging continues to extract new knowledge from historically significant cases, bridging classical neuropsychology with modern neuroscience methods (Annese et al., 2014).

The London Taxi Driver Study represents one of the most compelling examples of experience-dependent structural plasticity captured through neuroimaging. Eleanor Maguire and colleagues at University College London used structural MRI to examine the brains of London taxi drivers who had undergone extensive training to memorize the complex layout of London streets—a requirement for licensure. The study revealed significantly larger posterior hippocampi in taxi drivers compared to controls, with the size correlating with years of experience. Follow-up studies demonstrated that this enlargement reversed upon retirement, highlighting the dynamic nature of structural brain changes in response to cognitive demands. This research provides compelling evidence for the "use it or lose it" principle in adult neuroplasticity and has implications for cognitive enhancement strategies based on targeted mental exercise (Maguire et al., 2000).

A particularly illuminating application of neuroimaging in memory research comes from studies of "exceptional memorizers"—individuals who can perform extraordinary feats of memory, such as memorizing thousands of digits of pi or the order of multiple decks of cards. Contrary to expectations, structural imaging revealed that these individuals do not possess unusual brain anatomy. Instead, functional imaging during memory tasks showed that they recruit spatial navigation networks (including the hippocampus) to implement mnemonic strategies like the "method of loci," effectively repurposing brain systems evolved for navigation to support exceptional memory performance. These findings demonstrate how cognitive strategies can dramatically enhance memory by exploiting the brain's existing neural architecture, offering practical applications for memory training in educational and clinical contexts (Mallow et al., 2015).

Research has shown that stress can affect memory centers; in particular, the effects are seen in the amygdala and the prefrontal cortex (McEwen et al., 2016; Roozendaal et al., 2009). The effect of chronic stress is demonstrated by changes observed in the hippocampus that negatively impact memory function. In neurological diseases such as Alzheimer's, MRI and PET scans have shown brain atrophy in regions of the brain particularly related to learning and memory. This atrophy is associated with cognitive decline and memory impairment. The effects of neurodegenerative diseases can have a devastating impact on learning and memory. Neurofeedback has been studied in various cases, and it could provide real-time information related to brain activity for the enhancement of learning and memory.

Imaging studies have clarified the differential impact of acute versus chronic stress on memory systems. While acute stress can enhance memory consolidation through amygdala-mediated mechanisms involving stress hormones, chronic stress leads to structural remodeling of dendrites in the hippocampus and prefrontal cortex, with corresponding impairments in spatial and working memory. Longitudinal MRI studies in individuals experiencing prolonged stress, such as those with PTSD, have documented progressive volume reductions in these regions, corresponding with declining memory performance. These findings have significant implications for understanding how chronic life stressors, including poverty and trauma, may impact learning capacity through neurobiological pathways (Kim et al., 2015; Chakravarthi and Avadhani, 2013, 2014).

The neuroimaging of memory during sleep has yielded fascinating insights into offline memory consolidation processes. PET and fMRI studies during sleep have visualized "replay" activity, where neural patterns from learning experiences are reactivated, particularly during slow-wave sleep. This reactivation strengthens memory traces and facilitates their integration into existing knowledge networks. Remarkably, targeted memory reactivation studies have shown that presenting subtle cues associated with prior learning during deep sleep can selectively enhance the consolidation of specific memories. These findings have inspired novel interventions for memory enhancement, such as sleeping cap devices that detect deep sleep stages and deliver carefully timed sensory cues to reinforce learning (Rasch and Born, 2013).

Medical imaging in neuroscience also has the benefit of understanding intrinsic alterations in various regions of the brain, even in improved brain conditions. In individuals who are bilingual or those learning a second language, structural alterations in brain regions have been reported. Memory enhancement is another area of research that attracts many researchers. The neurofeedback mechanism in memory enhancement has recently gained attention as a topic of interest, where neural feedback is collected in real-time to understand brain activity (Sitaram et al., 2017). These studies provide deeper clarity on the functioning of various regions of the brain involved in learning and memory (Koush et al., 2017). Such studies could help improve the normal functioning of the brain or be used in disease conditions affecting memory. The use of fMRI in such studies is ideal as it provides real-time monitoring of brain activity during neurofeedback sessions. The results observed in such studies are promising because they reveal the mechanisms responsible for alterations in brain structure and the various connections that drive such structural changes. The plastic nature of the brain's structure makes studies capable of providing real-time feedback highly essential.

Neuroimaging studies of bilingualism have revealed fascinating structural and functional adaptations. Bilinguals show increased gray matter density in language-related regions and enhanced white matter integrity in tracts connecting these areas. However, the most striking finding is that bilingualism appears to enhance domain-general executive control networks beyond language processing, potentially contributing to the "bilingual advantage" in cognitive control tasks. Furthermore, longitudinal studies of second-language acquisition have documented progressive changes in brain structure and function that correlate with proficiency gains, demonstrating that adult language learning induces measurable neuroplasticity. These findings have educational implications, suggesting that language learning may serve as a form of cognitive exercise with benefits extending beyond linguistic competence (Abutalebi and Green, 2016).

Music training provides another compelling example of learning-induced brain changes captured through neuroimaging. Professional musicians show enlarged motor cortex representations for their playing digits, expanded auditory cortex regions responsive to musical tones, and enhanced white matter connectivity between auditory and motor areas. Remarkably, longitudinal studies in children beginning music lessons have documented the emergence of these changes within months of training initiation, with the magnitude of structural and functional alterations correlating with practice intensity. These findings illustrate how specialized skill learning reshapes relevant brain circuits and have inspired music-based interventions for various neurological conditions affecting learning and memory (Herholz and Zatorre, 2012).

Such studies can be extrapolated to understand the structure and functioning of the brain in larger populations. One such study is the London taxi driver study conducted by Eleanor Maguire and her team at University College London. In this study, MRI was used to examine the brains of taxi drivers from London. The study focused on the part of the brain involved in memory: the hippocampus. The hippocampus has been studied extensively and is known to be related to memory. In this taxi driver study, this was supported, as the hippocampus was found to be larger in those who had memorized the complex maps of London. The hippocampus is also involved in spatial navigation, and an increase in its size, as seen in this study, supports this fact. Such studies provide clarity on the learning and memory processes in the brain, and medical imaging plays a pivotal role in such cases.

Age-related changes leading to a decline in memory are another area of concern. In such cases, it becomes imperative that the structure and function of the brain are studied over a period of time, where MRI plays a major role (Fjell and Walhovd, 2010). The structural changes observed in such studies, when combined with testing cognitive abilities over a longer period of time, could provide a better interpretation of the link between the changes and the abilities (Nyberg et al., 2012).

Advanced neuroimaging has transformed our understanding of age-related memory changes by revealing that cognitive decline is not simply a result of generalized atrophy but involves specific patterns of brain alterations. Longitudinal studies have demonstrated differential vulnerability across brain regions, with medial temporal and prefrontal structures showing earlier changes than sensory cortices. Importantly, functional imaging has revealed compensatory mechanisms in healthy aging, where older adults recruit additional brain regions for tasks that engage more focal networks in younger individuals. This pattern, termed hemispheric asymmetry reduction in older adults, suggests that the aging brain adaptively reorganizes to maintain performance despite structural decline. Such insights guide the development of cognitive interventions designed to leverage preserved neural mechanisms while supporting vulnerable processes (Cabeza et al., 2018).

Medical imaging has played a crucial role in elucidating the neural basis of dyslexia and other learning disabilities. Structural and functional abnormalities in left hemisphere language networks have been consistently identified in individuals with dyslexia, particularly in regions involved in phonological processing. DTI has revealed alterations in white matter tracts connecting these regions, correlating with reading performance. Importantly, intervention studies using pretreatment and posttreatment imaging have documented neuroplastic changes following effective reading remediation, with neural activation patterns normalizing as reading skills improve. These findings have improved the identification of learning disabilities and validated specific intervention approaches based on their demonstrated effects on underlying neural systems (Vandermosten et al., 2012).

In the realm of mathematical cognition, neuroimaging has identified specialized brain circuits for numerical processing, including regions in the intraparietal sulcus that respond to quantity regardless of the format in which numbers are presented (e.g., Arabic numerals, dot arrays). Developmental imaging studies have traced how these numerical circuits mature and become increasingly specialized with age and education. Studies of individuals with dyscalculia (a math learning disability) have revealed structural and functional anomalies in these networks. This research has practical applications for mathematical education, suggesting that teaching approaches emphasizing the conceptual meaning of numbers rather than rote procedures may better engage these core numerical processing systems (Ansari, 2008).

On the whole, medical imaging has advanced the understanding of the link between structural changes in the brain and the functional alterations caused by these changes to the next level. Such studies are a minimal representation of the vast capabilities of medical imaging in neuroscience. With advancements in these techniques, there will be deeper clarity regarding one of the most complex organs of living beings: the brain.

9 Literature review of recent advances in medical imaging and behavior

A comprehensive review of recent literature reveals significant advances in understanding the relationship between brain structure, function, and behavior through medical imaging. This section synthesizes findings from key studies published in the last few years, highlighting emerging themes and their implications for learning and memory research.

Multimodal imaging approaches have gained prominence, combining the strengths of different imaging techniques to provide more comprehensive characterization of brain-behavior relationships. A metaanalysis of 37 studies using combined EEG-fMRI demonstrated that the integration of these modalities significantly enhanced the prediction of memory performance compared to either method alone. The temporal precision of EEG, coupled with the spatial resolution of fMRI, provided complementary information about memory processes, identifying both when and where critical neural events occur during encoding and retrieval. Similarly, studies combining PET with MRI have revealed the relationship between neurotransmitter activity and functional connectivity patterns during different learning states (Jorge et al., 2014).

Emerging research on the default mode network (DMN) has transformed our understanding of its role in memory. While initially characterized as a "task-negative" network that deactivates during cognitive challenges, recent high-resolution imaging studies have revealed that specific DMN subsystems actively contribute to memory processes. The posterior medial subsystem supports contextual representations, while the anterior temporal subsystem processes conceptual information. During successful memory encoding, these subsystems show distinct activity patterns and coordination with the hippocampus. These findings challenge simplistic views of the DMN and highlight how precise parcellation of brain networks can reveal functional specialization relevant to learning and memory (Andrews-Hanna et al., 2014).

Advances in characterizing hippocampal subfields have revolutionized memory research. Ultra-high-field MRI at 7T can now distinguish between hippocampal subregions (CA1, CA3, dentate gyrus) in living humans, allowing researchers to test theories about their differential roles in memory processing, which were previously derived from animal models. Recent studies have confirmed that pattern separation (the discrimination of similar experiences) relies primarily on the dentate gyrus, while pattern completion (the retrieval of memories from partial cues) depends more heavily on CA3. These distinctions have implications for understanding specific memory impairments in conditions like mild cognitive impairment, where particular subfields may show preferential vulnerability (Yassa and Stark, 2011).

Longitudinal neuroimaging studies tracking individuals over extended periods have provided unprecedented insights into how learning changes the brain over time. A five-year study following adults learning a complex cognitive skill revealed three distinct phases of structural and functional adaptation: initial expansion of gray matter volume in task-relevant regions, followed by pruning as performance became more efficient, and finally stabilization once expertise was achieved. These findings demonstrate that learning-induced neuroplasticity follows a nonlinear trajectory that would not be captured in cross-sectional designs, highlighting the importance of extended longitudinal approaches for understanding learning-related brain changes (Wenger et al., 2017).

Research on reconsolidation—the process by which memories become temporarily labile when reactivated—has benefited substantially from neuroimaging approaches. PET studies using radiotracers for protein synthesis have confirmed increased molecular activity in specific brain regions during memory reactivation, consistent with the reconsolidation hypothesis. fMRI studies have further demonstrated that interfering with reactivated memories produces measurable changes in their neural representations, with corresponding effects on behavioral performance. These findings have significant implications for educational practices, suggesting that the strategic reactivation of knowledge can create opportunities for strengthening or updating existing memories (Nader and Hardt, 2009).

Interoception—the perception of internal bodily signals—has emerged as an important influence on learning and memory through neuroimaging research. Studies using cardiac-gated fMRI have demonstrated that memory encoding is enhanced when information is presented during specific phases of the cardiac cycle, when baroreceptors are active. The anterior insula and anterior cingulate cortex, key nodes in the interoceptive network, show increased functional connectivity with the hippocampus during successful encoding, particularly for emotionally salient material. These findings highlight how the brain integrates bodily states with cognitive processing, with implications for understanding how physiological factors influence learning outcomes (Garfinkel and Critchley, 2013).

Neuroimaging studies of individual differences in learning strategies have revealed distinct neural signatures associated with various approaches to knowledge acquisition. Verbal learners show greater recruitment of left-lateralized language networks during encoding, while visual learners engage occipitotemporal visual processing regions more extensively. Interestingly, individuals who spontaneously organize information hierarchically show enhanced activity in the rostrolateral prefrontal cortex, a region associated with relational integration. These findings suggest that personalized learning approaches might be optimized by aligning instructional methods with individual neural propensities, although considerable work remains to translate these descriptive findings into prescriptive educational interventions (Sanfey et al., 2014).

Sleep-dependent memory consolidation has been extensively characterized through neuroimaging, with recent studies using innovative approaches to visualize consolidation processes in action. Simultaneous EEG-fMRI during sleep has captured coordinated oscillations between the hippocampus and neocortex during slow-wave sleep, with the strength of this synchronization predicting subsequent memory performance. Novel "decode-and-trigger" paradigms use real-time monitoring of sleep states to deliver sensory cues precisely when the brain is most receptive to consolidation enhancement. These techniques have demonstrated that targeted memory reactivation during optimal brain states can nearly double the retention of selected information, opening new avenues for augmenting learning through sleep-based interventions (Oudiette and Paller, 2013).

10 Summary

The role of medical imaging in the understanding of learning and memory processes, as well as the behavioral changes associated with them, is substantial. There are various roles of medical imaging that are useful in both normal and diseased conditions. To summarize the capabilities of this technology, it is used in identifying behavioral changes as indicators, conducting case studies and research, understanding neuroplasticity, studying the effects of learning and memory, diagnosing neurological conditions and neurodegenerative diseases affecting learning and memory, and exploring the practical implications of these findings. The future of medical imaging in research and healthcare settings looks very promising for advancing knowledge and providing a broader perspective on the processes of learning, memory, and behavior.

Medical imaging has transformed our understanding of the neural basis of learning and memory by allowing us to visualize brain structure and function with unprecedented detail. Structural techniques like MRI and CT have mapped the anatomical substrates of memory systems, while functional modalities like fMRI, PET, and EEG have revealed the dynamic patterns of brain activity associated with different learning processes. Advanced analytical approaches, including multivariate pattern analysis and network connectivity methods, have further enriched our understanding of how distributed neural systems coordinate to support cognitive functions.

The integration of artificial intelligence and deep learning with medical imaging has accelerated progress in this field, enabling automated feature detection, pattern recognition, and predictive modeling at scales beyond human capability. These computational approaches have revealed subtle patterns in brain structure and function associated with learning outcomes and memory performance, potentially paving the way for personalized educational interventions based on individual neural characteristics.

Neuroplasticity—the brain's remarkable capacity to reorganize in response to experience—has been directly visualized through medical imaging, confirming that learning induces measurable changes in brain structure and function across the lifespan. From the dramatic rewiring observed during early development to the more subtle but significant changes that occur during adult skill acquisition, imaging studies have documented how the brain physically adapts to accommodate new knowledge and abilities.

Case studies and research findings highlighted in this chapter illustrate the diverse ways in which medical imaging has illuminated the neural mechanisms of learning and memory. From studies of extraordinary memory performers to investigations of bilingual individuals, from monitoring age-related cognitive changes to documenting recovery from brain injury, medical imaging has provided crucial insights into how the brain supports cognitive functions in both typical and atypical conditions.

Recent technological advances in image processing have enhanced behavioral analysis capabilities, allowing researchers to extract increasingly nuanced information from neuroimaging data. These innovations enable the detection of subtle patterns in brain activity associated with specific cognitive states and behavioral outcomes, bridging the gap between neural events and observable behavior.

Looking ahead, emerging developments suggest an exciting future for medical imaging in learning and memory research. Ultra-high-field imaging systems will provide even more detailed views of brain microstructure, while real-time analysis capabilities will enable closed-loop systems for cognitive enhancement. Mobile and wearable technologies will extend neuroimaging beyond laboratory settings, capturing brain activity during natural learning experiences in everyday environments.

This chapter emphasizes the above by shedding light on the intricate connections between the structure of the brain and its functioning, as well as the effects on learning and memory. The convergence of advanced imaging techniques, computational methods, and cognitive neuroscience theory promises to yield an increasingly sophisticated understanding of how the brain acquires, stores, and retrieves information—knowledge that will inform educational practices, clinical interventions, and our fundamental understanding of human cognitive potential.

References

Abutalebi, J., Green, D.W., 2016. Neuroimaging of language control in bilinguals: neural adaptation and reserve. Bilingualism. 14691841. 19 (4), 689–698. Available from: http://www.cambridge.org/uk/journals/journal_catalogue.asp?mnemonic=BIL. doi:10.1017/S1366728916000225.

Andrews-Hanna, J.R., Smallwood, J., Spreng, R.N., 2014. The default network and self-generated thought: component processes, dynamic control, and clinical relevance. Ann. N.Y. Acad. Sci. 17496632. 1316 (1), 29–52. Available from: http://onlinelibrary.wiley.com/journal/10.1111/(ISSN)1749-6632. doi:10.1111/nyas.12360.

Annese, J., Schenker-Ahmed, N.M., Bartsch, H., Maechler, P., Sheh, C., Thomas, N., Kayano, J., Ghatan, A., Bresler, N., Frosch, M.P., Klaming, R., Corkin, S., 2014. Postmortem examination of patient H.M.'s brain based on histological sectioning and digital 3D reconstruction. Nat. Commun. 2041-1723. 5 (1). doi:10.1038/ncomms4122.

Ansari, D., 2008. Effects of development and enculturation on number representation in the brain. Nat. Rev. Neurosci. 1471-003X. 9 (4), 278–291. doi:10.1038/nrn2334.

Ansari, D., Coch, D., 2006. Bridges over troubled waters: education and cognitive neuroscience. Trends Cogn. Sci. 13646613. 10 (4), 146–151. doi:10.1016/j.tics.2006.02.007.

Assaf, Y., Pasternak, O., 2008. Diffusion tensor imaging (DTI)-based white matter mapping in brain research: a review. J. Mol. Neurosci. 0895-8696. 34 (1), 51–61. doi:10.1007/s12031-007-0029-0.

Bandettini, P.A., 2012. Twenty years of functional MRI: the science and the stories. NeuroImage. 10959572. 62 (2), 575–588. doi:10.1016/j.neuroimage.2012.04.026.

Bassett, D.S., Sporns, O., 2017. Network neuroscience. Nat. Neurosci. 15461726. 20 (3), 353–364. Available from: http://www.nature.com/neuro/index.html. doi:10.1038/nn.4502.

Brenner, D.J., Hall, E.J., 2007. Computed tomography—an increasing source of radiation exposure. N. Engl. J. Med. 0028-4793. 357 (22), 2277–2284. doi:10.1056/nejmra072149.

Bullmore, E., Sporns, O., 2009. Complex brain networks: graph theoretical analysis of structural and functional systems. Nat. Rev. Neurosci. 1471-003X. 10 (3), 186–198. doi:10.1038/nrn2575.

Cabeza, R., Albert, M., Belleville, S., Craik, F.I.M., Duarte, A., Grady, C.L., Lindenberger, U., Nyberg, L., Park, D.C., Reuter-Lorenz, P.A., Rugg, M.D., Jason Steffener, M.N.R., 2018. Maintenance, reserve and compensation: the cognitive neuroscience of healthy ageing. Nat. Rev. Neurosci. 1471-003X. 19 (11), 701–710. doi:10.1038/s41583-018-0068-2.

Cahill, L., 2006. Why sex matters for neuroscience. Nat. Rev. Neurosci. 1471-003X. 7 (6), 477–484. doi:10.1038/nrn1909.

Chakravarthi, K.K., Avadhani, R., 2013. Beneficial effect of aqueous root extract of Glycyrrhiza glabra on learning and memory using different behavioral models: an experimental study. J. Nat. Sci. Biol. Med. 4 (2), 420–425. doi:10.4103/0976-9668.117025.

Chakravarthi, K.K., Avadhani, R., 2014. Enhancement of hippocampal CA3 neuronal dendritic arborization by glycyrrhiza glabra root extract treatment in Wistar albino rRats J. Nat. Sci. Biol. Med. 5 (1), 25–29. doi:10.4103/0976-9668.127279.

Cotman, C.W., Berchtold, N.C., Christie, L.A., 2007. Exercise builds brain health: key roles of growth factor cascades and inflammation. Trends Neurosci. 01662236. 30 (9), 464–472. Available from: www.elsevier.com/locate/tins. doi:10.1016/j.tins.2007.06.011.

Dafflon, J., Pinaya, W.H.L., Turkheimer, F., Cole, J.H., Leech, R., Harris, M.A., Cox, S.R., Whalley, H.C., McIntosh, A.M., Hellyer, P.J., 2020. An automated machine learning approach to predict brain age from cortical anatomical measures. Hum. Brain Mapp. 10970193. 41 (13), 3555–3566. Available from: http://onlinelibrary.wiley.com/journal/10.1002/(ISSN)1097-0193. doi:10.1002/hbm.25028.

Damoiseaux, J.S., Greicius, M.D., 2009. Greater than the sum of its parts: a review of studies combining structural connectivity and resting-state functional connectivity. Brain Struct. Funct. 18632661. 213 (6), 525–533. doi:10.1007/s00429-009-0208-6.

DeBettencourt, M.T., Cohen, J.D., Lee, R.F., Norman, K.A., Turk-Browne, N.B., 2015. Closed-loop training of attention with real-time brain imaging. Nat. Neurosci. 15461726. 18 (3), 470–478. Available from: http://www.nature.com/neuro/index.html. doi:10.1038/nn.3940.

Doyon, J., Benali, H., 2005. Reorganization and plasticity in the adult brain during learning of motor skills. Curr. Opin. Neurobiol. 09594388. 15 (2), 161–167. doi:10.1016/j.conb.2005.03.004.

Ferrari, M., Quaresima, V., 2012. A brief review on the history of human functional near-infrared spectroscopy (fNIRS) development and fields of application. NeuroImage. 10538119. 63 (2), 921–935. doi:10.1016/j.neuroimage.2012.03.049.

Filippi, M., Agosta, F., 2011. Structural and functional network connectivity breakdown in Alzheimer's disease studied with magnetic resonance imaging techniques. J. Alzheimers Dis. 18758908. 24 (3), 455–474. doi:10.3233/jad-2011-101854.

Fjell, A.M., Walhovd, K.B., 2010. Structural brain changes in aging: courses, causes and cognitive consequences. Rev. Neurosci. 03341763. 21 (3), 187–221. Available from: http://www.reference-global.com/loi/revneuro. doi:10.1515/REVNEURO.2010.21.3.187.

Friston, K.J., Harrison, L., Penny, W., 2003. Dynamic causal modelling. NeuroImage. 10538119. 19 (4), 1273–1302. doi:10.1016/s1053-8119(03)00202-7.

Garfinkel, S.N., Critchley, H.D., 2013. Interoception, emotion and brain: new insights link internal physiology to social behaviour. Soc. Cogn. Affect. Neurosci. 1749-5024. 8 (3), 231–234. doi:10.1093/scan/nss140.

Grimm, O., Gass, N., Weber-Fahr, W., Sartorius, A., Schenker, E., Spedding, M., Risterucci, C., Schweiger, J.I., Böhringer, A., Zang, Z., Tost, H., Schwarz, A.J., Meyer-Lindenberg, A., 2015. Acute ketamine challenge increases resting state prefrontal-hippocampal connectivity in both humans and rats. Psychopharmacology. 0033-3158. 232 (21-22), 4231–4241. doi:10.1007/s00213-015-4022-y.

Hari, R., Baillet, S., Barnes, G., Burgess, R., Forss, N., Gross, J., Hämäläinen, M., Jensen, O., Kakigi, R., Mauguière, F., Nakasato, N., Puce, A., Romani, G.-L., Schnitzler, A., Taulu, S., 2018. IFCN-endorsed practical guidelines for clinical magnetoencephalography (MEG). Clin. Neurophysiol. 13882457. 129 (8), 1720–1747. doi:10.1016/j.clinph.2018.03.042.

Haxby, J.V., Connolly, A.C., Guntupalli, J.S., 2014. Decoding neural representational spaces using multivariate pattern analysis. Annu. Rev. Neurosci. 15454126. 37, 435–456. Available from: http://arjournals.annualreviews.org/loi/neuro. doi:10.1146/annurev-neuro-062012-170325.

Hensch, Bilimoria, 2012. Re-opening windows: manipulating critical periods for brain development. Cerebrum 2012, 1–18.

Herholz, K., Heiss, W., 2004. Positron emission tomography in clinical neurology. Mol. Imaging Biol. 15361632. 6 (4), 239–269. doi:10.1016/j.mibio.2004.05.002.

Herholz, S.C., Zatorre, R.J., 2012. Musical training as a framework for brain plasticity: behavior, function, and structure. Neuron. 10974199. 76 (3), 486–502. doi:10.1016/j.neuron.2012.10.011.

Hricak, H., Brenner, D.J., Adelstein, S.J., Frush, D.P., Hall, E.J., Howell, R.W., McCollough, C.H., Mettler, F.A., Pearce, M.S., Suleiman, O.H., Thrall, J.H., Wagner, L.K., 2011. Managing radiation use in medical imaging: a multifaceted challenge. Radiology. 15271315. 258 (3), 889–905. Available from: http://radiology.rsna.org/content/258/3/889.full.pdf+html. doi:10.1148/radiol.10101157.

Jack, C.R., Bennett, D.A., Blennow, K., Carrillo, M.C., Dunn, B., Haeberlein, S.B., Holtzman, D.M., Jagust, W., Jessen, F., Karlawish, J., Liu, E., Molinuevo, J.L., Montine, T., Phelps, C., Rankin, K.P., Rowe, C.C., Scheltens, P., Siemers, E., Snyder, H.M., Sperling, R., Elliott, C., Masliah, E., Ryan, L., Silverberg, N., 2018. NIA-AA research framework: toward a biological definition of Alzheimer's disease. Alzheimers Dement. 1552-5260. 14 (4), 535–562. doi:10.1016/j.jalz.2018.02.018.

Jorge, J., Van der Zwaag, W., Figueiredo, P., 2014. EEG-fMRI integration for the study of human brain function. NeuroImage. 10959572. 102 (1), 24–34. Available from: http://www.elsevier.com/inca/publications/store/6/2/2/9/2/5/index.htt. doi:10.1016/j.neuroimage.2013.05.114.

Kandel, E.R., Dudai, Y., Mayford, M.R., 2014. The molecular and systems biology of memory. Cell. 10974172. 157 (1), 163–186. Available from: https://www.sciencedirect.com/journal/cell. doi:10.1016/j.cell.2014.03.001.

Kim, E.J., Pellman, B., Kim, J.J., 2015. Stress effects on the hippocampus: a critical review. Learn. Mem. 15495485. 22 (9), 411–416. Available from: http://learnmem.cshlp.org/content/22/9/411.full.pdf+html. doi:10.1101/lm.037291.114.

Koush, Y., Meskaldji, D.E., Pichon, S., Rey, G., Rieger, S.W., Linden, D.E.J., De Ville, D.V., Vuilleumier, P., Scharnowski, F., 2017. Learning control over emotion networks through connectivity-based neurofeedback. Cereb. Cortex. 14602199. 27 (2), 1193–1202. Available from: http://cercor.oxfordjournals.org/. doi:10.1093/cercor/bhv311.

Kraff, O., Fischer, A., Nagel, A.M., Mönninghoff, C., Ladd, M.E., 2015. MRI at 7 Tesla and above: demonstrated and potential capabilities. J. Magn. Reson. Imaging. 15222586. 41 (1), 13–33. Available from: http://onlinelibrary.wiley.com/journal/10.1002/(ISSN)1522-2586. doi:10.1002/jmri.24573.

Krigolson, O.E., Williams, C.C., Norton, A., Hassall, C.D., Colino, F.L., MUSE, C., 2017. Validation of a low-cost, portable EEG system for ERP research. Front. Neurosci. 1662453X. 11. Available from: http://journal.frontiersin.org/article/10.3389/fnins.2017.00109/full. doi:10.3389/fnins.2017.00109.

Litjens, G., Kooi, T., Bejnordi, B.E., Setio, A.A.A., Ciompi, F., Ghafoorian, M., van der Laak, J.A.W.M., van Ginneken, B., Sánchez, C.I., 2017. A survey on deep learning in medical image analysis. Med. Image Anal. 13618423. 42, 60–88. Available from: http://www.elsevier.com/inca/publications/store/6/2/0/9/8/3/index.htt. doi:10.1016/j.media.2017.07.005.

Logothetis, N.K., 2008. What we can do and what we cannot do with fMRI. Nature. 14764687. 453 (7197), 869–878. Available from: http://www.nature.com/nature/index.html. doi:10.1038/nature06976.

Lövdén, M., Wenger, E., Mårtensson, J., Lindenberger, U., Bäckman, L., 2013. Structural brain plasticity in adult learning and development. Neurosci. Biobehav. Rev. 01497634. 37 (9), 2296–2310. doi:10.1016/j.neubiorev.2013.02.014.

Lu, D., Popuri, K., Ding, G.W., Balachandar, R., Beg, M.F., Weiner, M., Aisen, P., Petersen, R., Jack, C., Jagust, W., Trojanowki, J., Toga, A., Beckett, L., Green, R., Saykin, A., Morris, J.C., Shaw, L., Kaye, J., Quinn, J., Silbert, L., Lind, B., Carter, R., Dolen, S., Schneider, L., Pawluczyk, S., Beccera, M., Teodoro, L., Spann, B., Brewer, J., Vanderswag, H., Fleisher, A., Heidebrink, J., Lord, J., Mason, S., Albers, C., Knopman, D., Johnson, K., Doody, R., Villanueva-Meyer, J., Chowdhury, M., Rountree, S., Dang, M., Stern, Y., Honig, L., Bell, K., Ances, B., Carroll, M., Creech, M., Franklin, E., Mintun, M., Schneider, S., Oliver, A., Marson, D., Grifth, R., Clark, D., Geldmacher, D., Brockington, J., Roberson, E., Love, M.N., Grossman, H., Mitsis, E., Shah, R., deToledo-Morrell, L., Duara, R., Varon, D., Greig, M., Roberts, P., Albert, M., Onyike, C., D'Agostino, D., Kielb, S., Galvin, J., Cerbone, B., Michel, C., Pogorelec, D., Rusinek, H., de Leon, M., Glodzik, L., Santi, S.D., Doraiswamy, P., Petrella, J., Borges-Neto, S., Wong, T., Coleman, E., Smith, C., Jicha, G., Hardy, P., Sinha, P., Oates, E., Conrad, G., Porsteinsson, A., Goldstein, B., Martin, K., Makino, K., Ismail, M., Brand, C., Mulnard, R., Thai, G., Mc-Adams-Ortiz, C., Womack, K., Mathews, D., Quiceno, M., Levey, A., Lah, J., Cellar, J., Burns, J., Swerdlow, R., Brooks, W., Apostolova, L., Tingus, K., Woo, E., Silverman, D., Lu, P., Bartzokis, G., Graf-Radford, N., Parftt, F., Kendall, T., Johnson, H.,

Farlow, M., Hake, A.M., Matthews, B., Brosch, J., Herring, S., Hunt, C., Dyck, C., Carson, R., MacAvoy, M., Varma, P., Chertkow, H., Bergman, H., Hosein, C., Black, S., Stefanovic, B., Caldwell, C., Hsiung, G.Y.R., Feldman, H., Mudge, B., Assaly, M., Finger, E., Pasternack, S., Rachisky, I., Trost, D., Kertesz, A., Bernick, C., Munic, D., Mesulam, M.M., Lipowski, K., Weintraub, S., Bonakdarpour, B., Kerwin, D., Wu, C.K., Johnson, N., Sadowsky, C., Villena, T., Turner, R.S., Johnson, K., Reynolds, B., Sperling, R., Johnson, K., Marshall, G., Yesavage, J., Taylor, J., Lane, B., Rosen, A., Tinklenberg, J., Sabbagh, M., Belden, C., Jacobson, S., Sirrel, S., Kowall, N., Killiany, R., Budson, A., Norbash, A., Johnson, P.L., Obisesan, T., Wolday, S., Allard, J., Lerner, A., Ogrocki, P., Tatsuoka, C., Fatica, P., Fletcher, E., Maillard, P., Olichney, J., DeCarli, C., Carmichael, O., Kittur, S., Borrie, M., Lee, T.Y., Bartha, R., Johnson, S., Asthana, S., Carlsson, C., Potkin, S., Preda, A., Nguyen, D., Tariot, P., Burke, A., Trncic, N., Reeder, S., Bates, V., Capote, H., Rainka, M., Scharre, D., Kataki, M., Adeli, A., Zimmerman, E., Celmins, D., Brown, A., Pearlson, G., Blank, K., Anderson, K., Flashman, L., Seltzer, M., Hynes, M., Santulli, R., Sink, K., Gordineer, L., Williamson, J., Garg, P., Watkins, F., Ott, B., Querfurth, H., Tremont, G., Salloway, S., Malloy, P., Correia, S., Rosen, H., Miller, B., Perry, D., Mintzer, J., Spicer, K., Bachman, D., Pomara, N., Hernando, R., Sarrael, A., Relkin, N., Chaing, G., Lin, M., Ravdin, L., Smith, A., Raj, B.A., Fargher, K., 2018. Multimodal and multiscale deep neural networks for the early diagnosis of Alzheimer's disease using structural MR and FDG-PET images. Sci. Rep. 20452322. 8 (1). Available from: www.nature.com/srep/index.html. doi:10.1038/s41598-018-22871-z.

Maguire, E.A., Gadian, D.G., Johnsrude, I.S., Good, C.D., Ashburner, J., Frackowiak, R.S.J., Frith, C.D., 2000. Navigation-related structural change in the hippocampi of taxi drivers. Proc. Natl. Acad. Sci. USA 97 (8), 4398–4403. doi:10.1073/pnas.070039597.

Mallow, J., Bernarding, J., Luchtmann, M., Bethmann, A., Brechmann, A., 2015. Superior memorizers employ different neural networks for encoding and recall. Front. Syst. Neurosci. 1662-5137. 9 (September). doi:10.3389/fnsys.2015.00128.

McEwen, B.S., Nasca, C., Gray, J.D., 2016. Stress effects on neuronal structure: hippocampus, amygdala, and prefrontal cortex. Neuropsychopharmacology. 1740634X. 41 (1), 3–23. Available from: http://www.nature.com/npp/index.html. doi:10.1038/npp.2015.171.

Nader, K., Hardt, O., 2009. A single standard for memory: the case for reconsolidation. Nat. Rev. Neurosci. 1471-003X. 10 (3), 224–234. doi:10.1038/nrn2590.

Niedermeyer, E., da Silva, F.L., 2005. Electroencephalography: basic principles, clinical applications, and related fields, fifth ed. Lippincott Williams & Wilkins.

Norman, K.A., Polyn, S.M., Detre, G.J., Haxby, J.V., 2006. Beyond mind-reading: multi-voxel pattern analysis of fMRI data. Trends Cogn. Sci. 13646613. 10 (9), 424–430. doi:10.1016/j.tics.2006.07.005.

Nyberg, L., Lövdén, M., Riklund, K., Lindenberger, U., Bäckman, L., 2012. Memory aging and brain maintenance. Trends Cogn. Sci. 13646613. 16 (5), 292–305. doi:10.1016/j.tics.2012.04.005.

O'Doherty, J.P., Dayan, P., Friston, K., Critchley, H., Dolan, R.J., 2003. Temporal difference models and reward-related learning in the human brain. Neuron. 08966273. 38 (2), 329–337. Available from: www.neuron.org. doi:10.1016/S0896-6273(03)00169-7.

Ofen, N., Shing, Y.L., 2013. From perception to memory: changes in memory systems across the lifespan. Neurosci. Biobehav. Rev. 18737528. 37 (9), 2258–2267. doi:10.1016/j.neubiorev.2013.04.006.

Okun, M.S., 2012. Deep-brain stimulation for Parkinson's disease. N. Engl. J. Med. 15334406. 367 (16), 1529–1538. Available from: http://www.nejm.org/doi/pdf/10.1056/NEJMct1208070. doi:10.1056/NEJMct1208070.

Oudiette, D., Paller, K.A., 2013. Upgrading the sleeping brain with targeted memory reactivation. Trends Cogn. Sci. 1879307X. 17 (3), 142–149. doi:10.1016/j.tics.2013.01.006.

Patino, M., Prochowski, A., Agrawal, M.D., Simeone, F.J., Gupta, R., Hahn, P.F., Sahani, D.V., 2016. Material separation using dual-energy CT: current and emerging applications. Radiographics. 15271323. 36 (4), 1087–1105. Available from: http://pubs.rsna.org/doi/pdf/10.1148/rg.2016150220. doi:10.1148/rg.2016150220.

Rasch, B., Born, J., 2013. About sleep's role in memory. Physiol. Rev. 0031-9333. 93 (2), 681–766. doi:10.1152/physrev.00032.2012.

Reis, P.M.R., Hebenstreit, F., Gabsteiger, F., von Tscharner, V., Lochmann, M., 2014. Methodological aspects of EEG and body dynamics measurements during motion. Front. Hum. Neurosci. 1662-5161. 8. doi:10.3389/fnhum.2014.00156.

Roozendaal, B., McEwen, B.S., Chattarji, S., 2009. Stress, memory and the amygdala. Nat. Rev. Neurosci. 14710048. 10 (6), 423–433. doi:10.1038/nrn2651.

Sanfey, A.G., Stallen, M., Chang, L.J., 2014. Norms and expectations in social decision-making. Trends Cogn. Sci. 1879307X. 18 (4), 172–174. Available from: www.elsevier.com/locate/tics. doi:10.1016/j.tics.2014.01.011.

Shu, S.Y., Penny, G.R., Peterson, G.M., 1988. The 'marginal division': a new subdivision in the neostriatum of the rat. J. Chem. Neuroanat. 08910618. 1 (3), 147–163.

Sitaram, R., Ros, T., Stoeckel, L., Haller, S., Scharnowski, F., Lewis-Peacock, J., Weiskopf, N., Blefari, M.L., Rana, M., Oblak, E., Birbaumer, N., Sulzer, J., 2017. Closed-loop brain training: the science of neurofeedback. Nat. Rev. Neurosci. 1471-003X. 18 (2), 86–100. doi:10.1038/nrn.2016.164.

Smith, S.M., Jenkinson, M., Woolrich, M.W., Beckmann, C.F., Behrens, T.E.J., Johansen-Berg, H., Bannister, P.R., De Luca, M., Drobnjak, I., Flitney, D.E., Niazy, R.K., Saunders, J., Vickers, J., Zhang, Y., De Stefano, N., Brady, J.M., Matthews, P.M., 2004. Advances in functional and structural MR image analysis and implementation as FSL. NeuroImage. 10538119. 23 (1), S208–S219. doi:10.1016/j.neuroimage.2004.07.051.

Squire, L.R., Wixted, J.T., 2011. The cognitive neuroscience of human memory since H.M. Annu. Rev. Neurosci. 15454126. 34, 259–288. doi:10.1146/annurev-neuro-061010-113720.

Stern, Y., 2012. Cognitive reserve in ageing and Alzheimer's disease. Lancet Neurol. 14744422. 11 (11), 1006–1012. doi:10.1016/s1474-4422(12)70191-6.

Thomas, A.W., Heekeren, H.R., Müller, K.R., Samek, W., 2019. Analyzing neuroimaging data through recurrent deep learning models. Front. Neurosci. 1662453X. 13. Available from: https://www.frontiersin.org/journals/neuroscience#. doi:10.3389/fnins.2019.01321.

Vandermosten, M., Boets, B., Wouters, J., Ghesquière, P., 2012. A qualitative and quantitative review of diffusion tensor imaging studies in reading and dyslexia. Neurosci. Biobehav. Rev. 01497634. 36 (6), 1532–1552. doi:10.1016/j.neubiorev.2012.04.002.

Volkow, N.D., Koob, G.F., McLellan, A.T., 2016. Neurobiologic advances from the brain disease model of addiction. N. Engl. J. Med. 15334406. 374 (4), 363–371. Available from: http://www.nejm.org/doi/pdf/10.1056/NEJMra1511480. doi:10.1056/NEJMra1511480.

Voss, P., Thomas, M.E., Cisneros-Franco, J.M., de Villers-sidani, É., 2017. Dynamic brains and the changing rules of neuroplasticity: implications for learning and recovery. Front. Psychol. 16641078. 8. Available from: https://www.frontiersin.org/articles/10.3389/fpsyg.2017.01657/full. doi:10.3389/fpsyg.2017.01657.

Wenger, E., Brozzoli, C., Lindenberger, U., Lövdén, M., 2017. Expansion and renormalization of human brain structure during skill acquisition. Trends Cogn. Sci. 13646613. 21 (12), 930–939. doi:10.1016/j.tics.2017.09.008.

Wolterink, J.M., Dinkla, A.M., Savenije, M.H.F., Seevinck, P.R., van den Berg, C.A.T., Išgum, I., 2017. Deep MR to CT synthesis using unpaired data. In: Lecture Notes in Computer Science (Including Subseries Lecture Notes in Artificial Intelligence and Lecture Notes in Bioinformatics), vol. 10557. Springer Verlag Netherlands. doi:10.1007/978-3-319-68127-6_2. 16113349, pp. 14–23. http://springerlink.com/content/0302-9743/copyright/2005/.

Yang, G.R., Wang, X.J., 2020. Artificial neural networks for neuroscientists: a primer. Neuron. 10974199. 107 (6), 1048–1070. Available from: http://www.cell.com/neuron/home. doi:10.1016/j.neuron.2020.09.005.

Yassa, M.A., Stark, C.E.L., 2011. Pattern separation in the hippocampus. Trends Neurosci. 1878108X. 34 (10), 515–525. doi:10.1016/j.tins.2011.06.006.

Zatorre, R.J., Fields, R.D., Johansen-Berg, H., 2012. Plasticity in gray and white: neuroimaging changes in brain structure during learning. Nat. Neurosci. 15461726. 15 (4), 528–536. doi:10.1038/nn.3045.

Part II

Sedentary behavior, lifestyle, and brain health

Chapter 4

Medical imaging in the assessment of sedentary behavior

Dilip Shettigar[a], Rajagopal Kadavigere[b], Leena R. David[a,c], Sneha Ravichandran[a], Baskaran Chandrasekaran[d], Berna Uzun[e], Dilber Uzun Ozsahin[c,e,f], Suresh Sukumar[a], Sathya Sabina Muthu[a], Ilker Ozsahin[e], Priya Seagal V.K.[g], and K.R. Banumathe[h]

[a]*Department of Medical Imaging Technology, Manipal College of Health Professions, Manipal Academy of Higher Education, Manipal, Karnataka, India*

[b]*Department of Radiodiagnosis and Imaging, Kasturba Medical College, Kasturba Medical College, Manipal, India*

[c]*Medical Diagnostic Imaging Department, College of Health Sciences, University of Sharjah, Sharjah, United Arab Emirates*

[d]*Symbiosis School of Sports Sciences, Symbiosis International (Deemed University), Pune, Maharashtra, India*

[e]*Irfan Suat Gunsel Operational Research Institute, Near East University, Nicosia, Turkish Republic of Northern Cyprus (TRNC), Turkey*

[f]*Research Institute for Medical and Health Sciences, University of Sharjah, Sharjah, United Arab Emirates*

[g]*Adichunchanagiri University, Bellur, Karnataka, India*

[h]*Department of Occupational Therapy, Manipal University College Malaysia, Melaka, Malaysia*

1 Sedentary lifestyle

1.1 Introduction to sedentary behavior

There has been a drastic increase in sedentary behavior or physical inactivity globally over the past few years. Sedentary behavior encompasses any activity during waking hours that demands less than 1.5 metabolic equivalent tasks, including actions like sitting or leaning, indicating minimal energy expenditure (Tremblay et al., 2017). Sedentary behavior includes actions characterized by minimal energy expenditure, such as watching television, working on a computer, or extended periods of sitting or lying down with limited physical movement (Hamilton et al., 2008; González et al., 2017).

The distinction between sedentary behavior and physical inactivity is important to understand. While they are related concepts, they represent different aspects of movement patterns in daily life. Sedentary behavior specifically refers to low-energy activities performed in a sitting or reclining position, whereas physical inactivity refers to insufficient engagement in recommended levels of moderate to vigorous physical activity. Both contribute significantly to public health concerns but require different intervention approaches.

1.2 Technological advancements and changes in work culture

The advancements in technology have allowed for remote work access. The rapidly spreading pandemics have compelled the disruption in normal work and education systems, resulting in the increase of work-from-home and education-from-home cultures. This transition has fundamentally altered how people interact with their environments, creating conditions where individuals may spend the majority of their waking hours in sedentary positions.

The proliferation of digital devices has further exacerbated this trend. Smartphones, tablets, and laptops have made entertainment, communication, and work accessible without requiring physical movement. These devices often encourage prolonged sitting or lying positions, as users engage with screens for extended periods. The convenience of technology has inadvertently contributed to lifestyle patterns that minimize physical exertion and maximize sedentary time.

In addition, urban planning and transportation developments have reduced the necessity for physical movement in daily routines. Car-centric infrastructure, elevators, escalators, and automated systems have eliminated many opportunities for incidental physical activity that were once integrated into daily life. These conveniences, while improving efficiency, have collectively diminished the natural movement patterns that previously helped maintain basic fitness levels.

Advances in Medical Imaging. https://doi.org/10.1016/B978-0-443-28967-5.00021-6

1.3 Impact of the pandemic on activity levels

Studies consistently report a significant rise in sedentary behavior and a decrease in physical activity across various populations due to lockdowns and home confinement. For instance, a study examining Spanish older adults found that confinement increased sedentary lifestyles in addition to causing a reduction in physical activity, potentially leading to negative health consequences in this age group (Fernández-García et al., 2021). Similarly, research on cardiovascular disease patients in the Netherlands revealed a 25-minute daily increase in sedentary time after 12 weeks of lockdown, mainly due to more sitting during leisure activities (van Bakel et al., 2021).

The pandemic's impact extended beyond adults to significantly affect children's activity patterns. A review article revealed that the COVID-19 pandemic caused a substantial decrease in physical activity and a notable rise in sedentary behavior among children aged 3–13 years (Mansoor Rahman and Chandrasekaran, 2021). The review highlighted a 34% reduction in physical activity and an 82% surge in sedentary behavior during this period.

These changes in activity patterns during the pandemic reflect not just temporary adaptations but potentially lasting behavioral shifts. Many individuals who adopted more sedentary routines during lockdown periods have maintained these patterns even after restrictions were lifted. The habituation to sedentary behaviors and the restructuring of work and education around remote participation have created lasting changes in how people allocate their time between active and inactive pursuits.

1.4 Global prevalence of sedentary behavior

Sedentary lifestyles represent a significant contemporary public health challenge, yet they remain insufficiently addressed, with an estimated 60%–85% of individuals globally, spanning both industrialized and developing countries, engaging in sedentary behaviors. This high prevalence indicates a widespread public health concern that transcends geographical and economic boundaries.

The scale of this issue varies across regions and demographic groups, but the overall trend shows increasing sedentary time across populations worldwide. Industrialized nations typically show higher rates of sedentary behavior, often associated with service-oriented economies where desk-based work predominates. However, rapidly developing countries are quickly catching up as they adopt similar technological and occupational patterns.

Urban populations tend to exhibit more sedentary behaviors than rural populations, partly due to differences in occupational demands and transportation options. This urban-rural divide represents one dimension of the complex distribution of sedentary behavior patterns globally. The transition from agricultural and manual labor economies to service and information-based economies has accelerated this shift toward more sedentary lifestyles.

1.5 Physical inactivity defined and recommended guidelines

The term "physical inactivity" refers to an individual's inability to meet recommended levels of exercise or physical activity necessary for maintaining optimal health and well-being. World Health Organization (WHO) has advised that "adults aged 18–64 years should do at least 150 minutes of moderate-intensity aerobic physical activity throughout the week or do at least 75 minutes of vigorous-intensity aerobic physical activity throughout the week, or an equivalent combination of moderate- and vigorous-intensity activity" (World Health Oranization, 2010).

These guidelines are based on extensive research demonstrating the minimum activity levels required to maintain cardiovascular health, muscle strength, and metabolic function. They represent a threshold below which individuals face significantly increased health risks. The recommendations include not only aerobic activity but also strength training components, recognizing the importance of diverse forms of physical activity for overall health.

It's worth noting that these guidelines have evolved over time as research has provided more insights into the relationship between physical activity and health outcomes. Earlier recommendations focused primarily on moderate-to-vigorous physical activity (MVPA), but newer guidelines increasingly acknowledge the importance of reducing sedentary time, even among individuals who meet aerobic exercise recommendations.

1.6 Health consequences of physical inactivity

Global concern is rising as physical inactivity becomes the fourth leading cause of death, linked to an increase in noncommunicable diseases (NCDs) like heart disease, diabetes, and cancer, alongside risk factors like high blood pressure and obesity (World Health Organization, 2009). Insufficient physical activity among approximately 31% of individuals aged ≥15 years globally contributes to roughly 3.2 million deaths annually (World Health Organization, 2020). According to statistics from WHO, 25% of men and 44% of women in India over the age of 18 are physically inactive (World Health Organizations, 2022).

The health implications of sedentary behavior extend beyond these major disease categories. Research has shown associations between prolonged sitting and an increased risk of depression, anxiety, and cognitive decline. The musculoskeletal system also suffers from prolonged sedentary time, with an increased prevalence of back pain, neck discomfort, and joint problems among those who sit for extended periods.

Metabolically, sedentary behavior disrupts normal glucose metabolism and lipid processing, leading to insulin resistance and dyslipidemia, even in otherwise healthy individuals. These metabolic disruptions can occur independently of body weight, meaning even physically fit individuals who spend long periods sitting face increased health risks. The phrase "sitting is the new smoking" has emerged to capture the pervasive and serious health consequences of prolonged sedentary behavior.

1.7 Independent effects of sedentary behavior

Despite engaging in moderate-to-intense physical exercise, various epidemiological studies have consistently shown that prolonged periods of sedentary behavior can negatively impact several health outcomes. This finding highlights an important nuance in understanding movement patterns and health: the harmful effects of prolonged sitting cannot be entirely offset by periodic exercise sessions.

The concept of the "active couch potato" describes individuals who meet physical activity guidelines through structured exercise but spend the majority of their remaining time engaged in sedentary behaviors. Research suggests that these individuals still face elevated health risks compared to those who distribute movement more evenly throughout the day. This understanding has led to the recommendation to "break up" sitting time with brief movement periods, even if these movements do not qualify as moderate or vigorous exercise.

To promote general health and well-being, it is imperative to address both physical inactivity and sedentary behavior, as they are linked to detrimental health effects. This dual approach recognizes that these are related but distinct behavioral patterns, requiring different intervention strategies.

1.8 Factors contributing to sedentary lifestyles

Several factors are thought to contribute to low engagement in physical activity, with age-related declines being one of the most notable reasons. As individuals age, physiological changes, increased health concerns, and shifting priorities often lead to reduced physical activity. This age-related decline is not inevitable but reflects a combination of biological, psychological, and social factors that collectively influence activity patterns across the lifespan.

Some socioeconomic factors include income disparities, educational attainment, and access to recreational facilities and green spaces. Lower-income communities often have fewer safe outdoor spaces for recreation, limited access to fitness facilities, and higher demands on time due to multiple jobs or lengthy commutes. These structural barriers create environments where sedentary behavior becomes the default option rather than an active choice.

Technological advancements, such as the proliferation of electronic devices and the influence of social media and digital entertainment, are highly associated with increased sedentary lifestyles. The design of digital platforms often aims to maximize user engagement time, inadvertently promoting prolonged sitting. Screen-based entertainment has largely replaced more active leisure pursuits, particularly among younger generations who have grown up with ubiquitous digital technology.

Psychological factors also play a role in the lack of physical activity, as individuals might face a lack of motivation, stress, and mental health issues. Depression and anxiety can create cycles where reduced energy and motivation lead to increased sedentary time, which, in turn, exacerbates mental health symptoms. The perceived effort of physical activity often outweighs the perceived benefits, particularly when immediate rewards (such as those from digital entertainment) are readily available.

1.9 Cultural and environmental influences

Cultural attitudes toward physical activity and leisure time significantly influence movement patterns. In some cultures, structured exercise is valued and promoted, while in others, incidental physical activity through manual labor or active transportation has traditionally provided sufficient movement. As globalization shifts cultural practices and economic activities, many societies are losing traditional forms of physical activity without replacing them with new movement opportunities.

Built environments profoundly impact movement opportunities. Urban design features such as walkability, bicycle infrastructure, public transportation access, and mixed land use can either facilitate or hinder physical activity. Car-dependent suburban developments, for example, often create environments where walking or cycling for transportation is impractical or unsafe, effectively requiring sedentary transportation modes for daily activities.

Workplace environments have increasingly been recognized as critical settings influencing sedentary behavior. The typical office environment, with desk-based work performed in a seated position, often enforces prolonged sedentary time. Organizational cultures that value presenteeism and extended work hours can further reduce opportunities for movement during the workday.

1.10 Interventions and solutions

Addressing the sedentary lifestyle epidemic requires multilevel interventions targeting individual behaviors, social norms, and environmental contexts. Individual-level approaches include education about the risks of sedentary behavior and strategies for increasing physical activity. Behavioral techniques, such as goal-setting, self-monitoring, and habit formation, can help individuals develop more active routines.

At the environmental level, designing activity-friendly communities with safe walking paths, bicycle infrastructure, and accessible recreational facilities can make physical activity more convenient and appealing. Workplace interventions, such as standing desks, walking meetings, and activity breaks, can reduce occupational sitting time. School-based programs that incorporate movement into the academic day can help establish healthy habits among children.

Policy approaches include urban planning regulations that prioritize active transportation, workplace policies that support physical activity breaks, and healthcare initiatives that include physical activity counseling and prescription. Economic incentives, such as reduced insurance premiums for meeting activity goals or subsidies for active transportation, can also motivate behavior change.

Technology, while often contributing to sedentary behavior, can also be leveraged to promote physical activity. Mobile applications, wearable activity trackers, and exergames (video games requiring physical movement) can motivate and monitor physical activity. Online communities can provide social support and accountability for maintaining active lifestyles.

The global increase in sedentary behavior represents a significant public health challenge, with wide-ranging consequences for physical and mental health. Understanding the complex factors contributing to this trend is essential for developing effective interventions. By addressing sedentary behavior at multiple levels—from individual habits to built environments and policies—we can work toward creating societies where regular physical activity is the natural and enjoyable default rather than the exception.

To promote population health in an increasingly digital and automated world, finding ways to reintegrate movement into daily routines will be crucial. This may require rethinking fundamental aspects of how we structure work, education, transportation, and leisure time. The challenge of sedentary lifestyles calls for innovative approaches that make physical activity accessible, convenient, and rewarding for people of all ages and backgrounds.

2 Health risks arising from sedentary lifestyles

A sedentary lifestyle with a physical inactivity is associated with health risks. A sedentary lifestyle is closely linked to the emergence of NCDs. It poses significant risks to both physical and mental health. Prolonged periods of inactivity contribute to a multitude of health issues, such as obesity, cardiovascular disease, diabetes, musculoskeletal problems, and an increased risk of certain cancers. Sedentary behavior also negatively impacts mental well-being, increasing the likelihood of depression and anxiety. In addition, physical inactivity is associated with reduced life expectancy. The body is designed for movement, and when it remains inactive for extended periods, it disrupts essential physiological processes, leading to a cascade of health complications.

2.1 Cardiovascular disease

In a study involving older women without a prior history of heart attacks or strokes, researchers found that those who spent the most time being sedentary (approximately 11 hours or more per day) showed a notably increased hazard of cardiovascular disease ($HR = 1.62$; $P < .001$) (Bellettiere et al., 2019). Another investigation found that individuals who participated in extensive sedentary behavior while also engaging in minimal MVPA experienced an increased likelihood of developing

TABLE 4.1 Impact of sedentary behavior on cardiovascular disease risk.

Study	Population	Sedentary behavior	Outcome	Hazard ratio (HR)	*P*-value
(Bellettiere et al., 2019)	Older women without a prior history of heart attacks or strokes	~11 hours or more per day	Increased hazard of cardiovascular disease	1.62	<0.001
(Paudel et al., 2023)	Middle-aged adults	Extensive sedentary behavior with minimal moderate-to-vigorous physical activity	Increased likelihood of developing cardiovascular disease	1.22	<0.001

cardiovascular disease (HR = 1.22; $P < .001$) (Paudel et al., 2023). Table 4.1 provides a summary of the findings related to the impact of sedentary behavior on the risk of cardiovascular disease.

2.2 Diabetes

Sitting for extended periods and insufficient physical activity can hinder the body's capacity to regulate blood sugar levels, potentially resulting in insulin resistance and, eventually, type 2 diabetes. A study by Stephens et al. (2011) found that sitting for a day led to a 39% decrease in insulin's effectiveness in absorbing glucose in healthy adults. In addition, regularly interrupting sitting with short periods of upper body movement can significantly reduce postmeal blood sugar and insulin levels by around 57% and 20%, respectively, compared to just sitting for long periods (McCarthy et al., 2017). In a study conducted among women aged 65 and above, those in the second, third, and fourth quartiles of sedentary time had a higher risk of developing incident diabetes, with hazard ratios of 1.20, 1.33, and 1.21, respectively. This study also found a significant trend ($P_{trend} = 0.04$) indicating a correlation between increased periods of sedentary behavior and an increased risk of developing diabetes. Substituting sedentary behavior with 30 minutes of MVPA was linked to a 15% decrease in the likelihood of developing diabetes (Bellettiere et al., 2021). Multiple cross-sectional studies have demonstrated an inverse relationship between sedentary time and insulin sensitivity. Extended periods of sedentary behavior have consistently been linked to diminished insulin sensitivity, resulting in insulin resistance and an increased risk of type 2 diabetes, as evidenced by numerous cross-sectional studies (Balkau et al., 2008; Maher et al., 2014; Lahjibi et al., 2013; Healy et al., 2011; Buman et al., 2014).

2.3 Obesity

Physical inactivity and prolonged hours of sitting are closely associated with the high prevalence of obesity. According to reports, the chance of becoming overweight (13%) and gaining significant abdominal fat (26%) increases with every hour of sedentary behavior (Byun et al., 2012). A systematic review involving 638,000 adults revealed a strong correlation between obesity and both sedentary behavior [odds ratio (OR) = 1.45] and lack of physical activity (OR = 1.52) (Silveira et al., 2022). Another systematic review identified a consistent positive correlation solely between sedentary behavior and indicators of obesity, including heightened body fat, greater severity of obesity, and elevated body mass index (BMI) (Mabele et al., 2019; Schumacher et al., 2020; Baillot et al., 2020; Castro et al., 2017). Figure 4.1 illustrates how a magnetic resonance imaging (MRI) scan can be used to measure abdominal fat content.

2.4 Hypertension

A sedentary lifestyle characterized by physical inactivity is strongly correlated with the development of hypertension, as insufficient exercise contributes to weight gain, poor cardiovascular health, and increased levels of stress hormones, all of which can raise blood pressure levels and elevate the risk of hypertension. According to a study, physical inactivity contributes to approximately 1.6% of the proportion of NCDs attributable to hypertension, and the largest absolute burden of people impacted by physical inactivity is notable in middle-income countries (Katzmarzyk et al., 2022). A study conducted among government officials in Sri Lanka found a significant association between physical inactivity and an elevated risk of hypertension among both senior officers (OR = 2.08) and managerial assistants

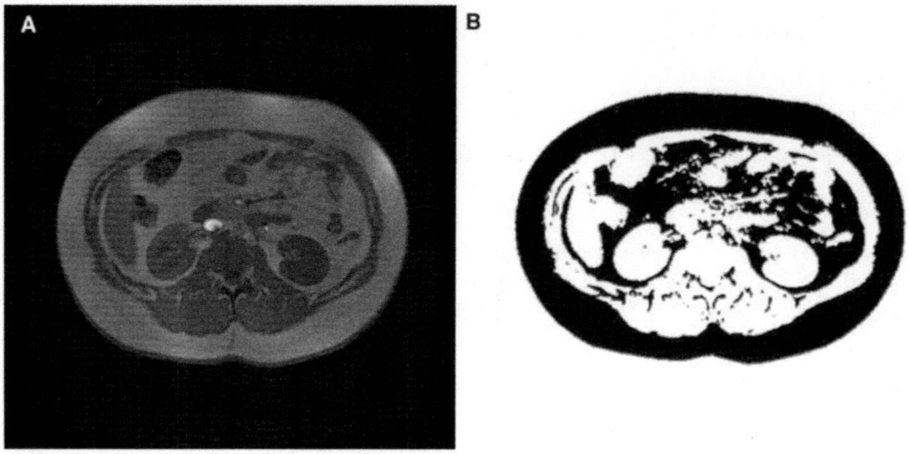

FIG. 4.1 MRI of an obese male at the section of L3–L4 section. (A) Shows MRI image. (B) Black color shows fat (visceral and subcutaneous) content in a binary image. *(Image courtesy: From Eloi, J.C., Epifanio, M., De Gonçalves, M.M., Pellicioli, A., Vieira, P.F.G., Dias, H.B., Bruscato, N., Soder, R.B., Santana, J.C.B., Mouzaki, M., Baldisserotto, M., 2017. Quantification of abdominal fat in obese and healthy adolescents using 3 T MRI and free software for image analysis. PLoS ONE, 12(1). https://doi.org/10.1371/journal.pone.0167625)*

(OR = 2.8) (Gamage and Seneviratne, 2021). Sustained sitting among individuals with hypertension leads to elevated blood pressure levels, alterations in flow-mediated dilation and nitroglycerin-mediated dilation, increased arterial stiffness, changes in plasma nitric oxide levels, variations in intima-to-media thickness, and pulse wave velocity (PWV) (Ravichandran et al., 2022). Figure 4.2 shows an example of the evaluation of pulmonary arterial hypertension with the help of a cardiac MRI scan.

2.5 Musculoskeletal problems

Sedentary behavior is associated with various musculoskeletal problems, including, but not limited to, back pain, neck pain, and poor posture. A meta-analysis involving 83,111 participants determined that engaging in sedentary behavior for more than 3 hours per day was linked to disability related to lower back discomfort (OR = 1.24) (Alzahrani et al., 2022). A cross-sectional study (Nawrocka et al., 2019) emphasized the importance of following the physical activity recommendations established by the World Health Organization. It demonstrated that adult females who engaged in higher levels of physical activity reported fewer musculoskeletal problems in both their upper and lower extremities.

2.6 Neurocognitive health

Extended periods of sedentary behavior have been linked to negative impacts on brain health, such as alterations in brain structure and cognitive functioning. Sitting for extended periods has been associated with cognitive decline and an increased risk of neurodegenerative conditions, such as Alzheimer's disease. While the direct link between sedentary behavior and neurodegenerative conditions is not fully established, some studies suggest a potential connection. For instance, a systematic review and meta-analysis found that physical activity interventions improved cognitive function in sedentary older adults (Zhao et al., 2022). This implies that sedentary behavior might be a risk factor for cognitive decline and neurodegenerative diseases. Brain atrophy is a marker of cognitive decline in elders. Possible causes for the reduction in brain volume and age-related cognitive decline include high sedentary time and poor physical activity (Raja et al., 2022). A cross-sectional study (Willey et al., 2016) has shown a decline in the cognitive performance of participants with low levels of physical activity. Another study conducted on the Japanese population demonstrated an improvement in cognitive functions and brain volume in the exercise group compared to the nonexercise group (Tamura et al., 2015). In addition, a cross-sectional study in the Indian population found a positive relationship between physical activity and hippocampal volume (Raja et al., 2022). A review article emphasized the importance of interrupting prolonged sitting with intermittent standing or engaging in low to moderate physical activity (Chandrasekaran et al., 2021). This approach may enhance cognitive functions by counteracting the vascular and hormonal changes linked to sedentary behavior.

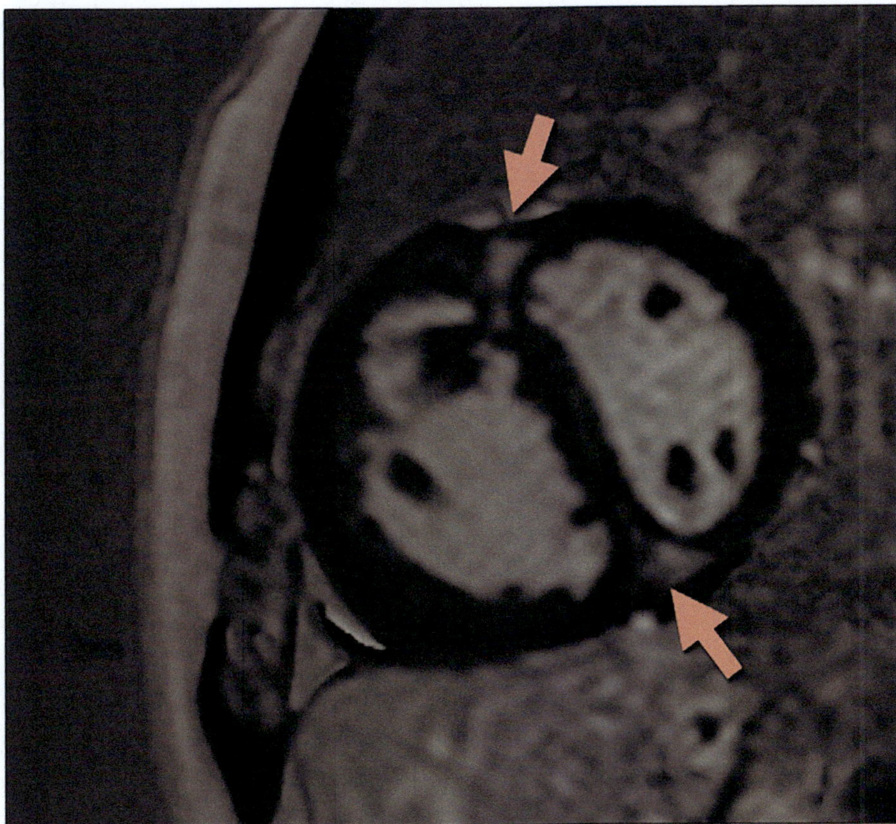

FIG. 4.2 A cardiac MRI scan showed late gadolinium enhancement in the areas where the right ventricle connects to the rest of the heart in a patient with pulmonary arterial hypertension. The arrows on the image point to late gadolinium contrast enhancement at right ventricular insertion points. *(Image courtesy: Freed, B.H., Collins, J.D., François, C.J., Barker, A.J., Cuttica, M.J., Chesler, N.C., Markl, M., Shah, S.J., 2016. MR and CT Imaging for the evaluation of pulmonary hypertension. JACC Cardiovasc. Imaging. 18767591. 9 (6), 715–732. Available from: http://www.elsevier.com. https://doi.org/10.1016/j.jcmg.2015.12.015).*

3 Medical imaging techniques in assessing sedentary behavior patterns

Medical imaging techniques, such as MRI, Computed Tomography, and ultrasound, play crucial roles in diagnosing various medical conditions and assessing the effects of sedentary behavior on human health. Medical imaging is essential for understanding the physiological consequences of sedentary behavior and guiding interventions to mitigate its adverse effects on health. By providing objective measures of tissue structure, function, and blood flow, imaging techniques help healthcare professionals assess the impact of sedentary lifestyles and develop targeted strategies for prevention and treatment.

3.1 Magnetic resonance imaging

MRI scans rely on meticulously chosen parameters to visualize specific tissues or physiological alterations. These parameters, including pulse sequence, echo time, repetition time, and flip angle, are tailored to the target tissue or process under investigation. In addition, the desired contrast between tissues influences parameter selection. Optimizing image quality and contrast is an iterative process. Initial parameters are chosen based on the research question or clinical need, then refined to best depict the tissue or physiological change of interest. There has been growing research interest in utilizing MRI scans to investigate the consequences of prolonged sitting and insufficient physical activity on the body. MRI is a valuable tool for assessing the influence of sedentary behavior on both brain structure and function. Studies have demonstrated associations between physical activity levels and differences in brain volume and connectivity, and MRI can help identify structural changes that may occur in individuals who engage in sedentary behavior (Raja et al., 2022; Siddarth et al., 2018; Yamamoto et al., 2017). A recent investigation delved into the impact of sedentary behavior on physically inactive individuals, examining MRI-derived fat distribution (Henson et al., 2018). The findings of this study revealed a

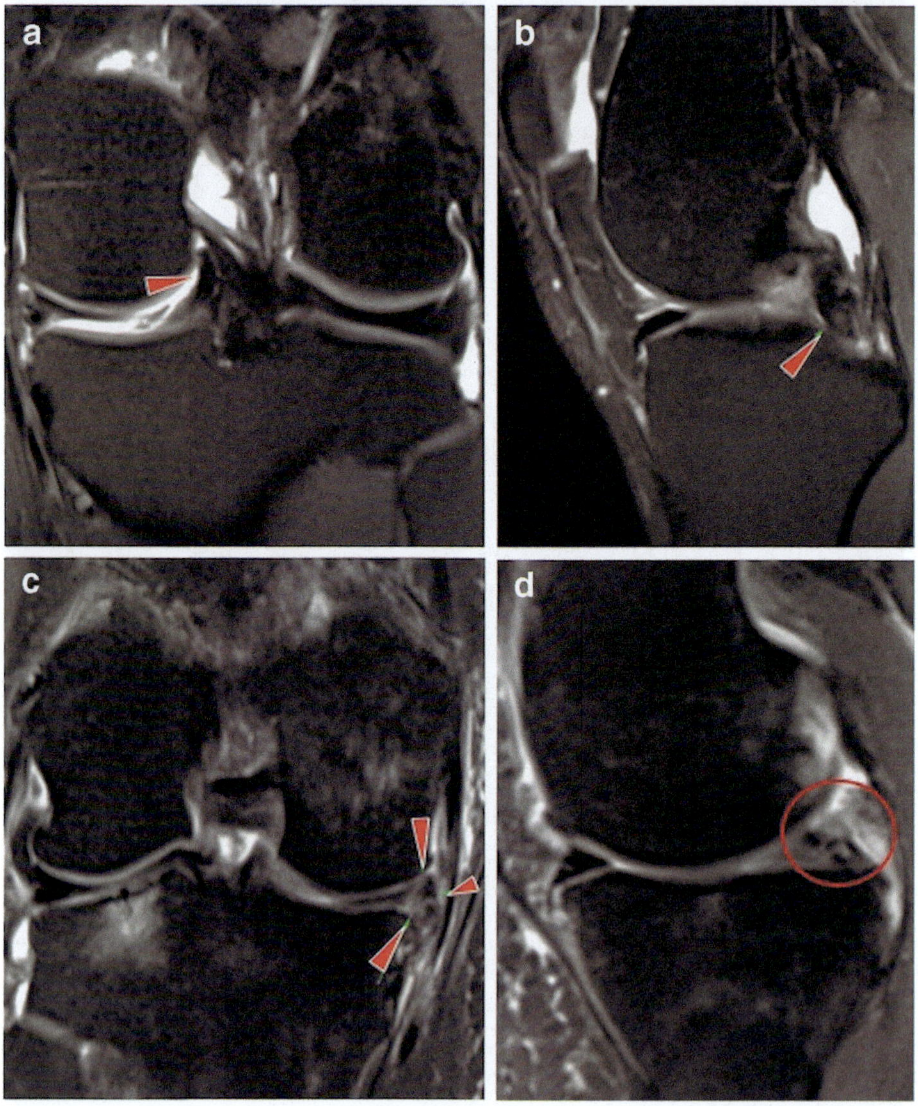

FIG. 4.3 Coronal proton-density fat-saturated MR images (A, C) and sagittal images (B, D) demonstrate a bucket-handle tear (A, B; arrowheads) in the left knee of a 54-year-old man, and a complex macerated (C, arrowheads; D, circle) meniscal tear in the right knee of a 57-year-old woman. *(Image courtesy: From Horga, L.M., Hirschmann, A.C., Henckel, J., Fotiadou, A., Di Laura, A., Torlasco, C., D'Silva, A., Sharma, S., Moon, J.C., Hart, A.J., 2020. Prevalence of abnormal findings in 230 knees of asymptomatic adults using 3.0 T MRI. Skelet. Radiol. 49(7), 1099–1107. https://doi.org/10.1007/s00256-020-03394-z).*

notable association: for every hour spent in sedentary activities, there was a discernible rise in liver fat (2.16%, $P < .001$), visceral fat (0.54 L, $P = .014$), and total abdominal fat (1.80 L, $P = .005$). The amount of abdominal fat in two men with similar amounts of sedentary time but different levels of MVPA. An additional investigation uncovered adverse correlations between the frequency of sedentary bouts per day lasting over 30 minutes and gray matter volume (−0.44, $P < .05$) in individuals with multiple sclerosis (Klaren et al., 2017). Research conducted in older adults unveiled a significant connection between prolonged uninterrupted sedentary time exceeding 10 hours per day and diminished cognitive functions. Structural MRI markers were identified as partial mediators for the observed decline in cognitive functions (Han et al., 2023). Examination of MRI knee scans from 115 sedentary individuals without prior injury unveiled several notable findings: 30% showed meniscal tears, 21% displayed moderate-intensity lesions in tendons, and 6% exhibited high-grade tendonitis. Notably, nearly all knees of asymptomatic adults revealed abnormalities in at least one knee structure upon MRI assessment (Horga et al., 2020). Fig. 4.3 shows meniscal tears identified by MRI scans in two asymptomatic sedentary adults. Overall, while MRI can provide valuable insights into the physiological effects of sedentary behavior, it is just one tool in a larger toolkit for studying this complex issue. Integrating MRI findings with other measures, such as physical activity monitoring and health assessments, can provide a more comprehensive understanding of the relationship between sedentary behavior and health outcomes.

3.2 Ultrasonography

Ultrasound imaging can be utilized to assess certain aspects of sedentary behavior and its effects on the body. While ultrasound may not directly measure sedentary behavior itself, it can provide valuable information about the physiological changes associated with prolonged sitting or inactivity. A review article summarizes the impact of sedentary behavior in hypertensive adults by assessing the changes in vascular functions through ultrasonography (Ravichandran et al., 2022). A recent study by Shruthi et al. (2024) found that yoga asana intervention improved vascular function in desk-based workers. Specifically, the yoga asana intervention showed a significant increase in the diameter of the carotid artery (CCA) and superficial femoral artery compared to prolonged sitting and pranayama intervention. The study also observed that the yoga asana intervention led to a decrease in CCA velocity and shear rate over time, indicating improved vascular function. Several studies have indicated that extended periods of sitting, exceeding three hours, are associated with a rise in carotid-ankle PWV by 0.27 m/s (Ravichandran et al., 2022), and similarly, an increase in carotid-femoral and carotid-radial PWV by 0.03 m/s has been observed (Feairheller et al., 2014; Morishima et al., 2020) along with a decrease in intima-to-media thickness by 6.4% and an increase in the ratio of flow-mediated dilation (FMD) to nitroglycerin-mediated dilation (NMD) by 36.2% (Feairheller et al., 2014). A study found that sitting continuously for 4 hours resulted in a reduction of the diameter of the common carotid artery by 5% (Carter et al., 2018). Overall, while MRI may offer more detailed anatomical information, ultrasound can provide a portable, cost-effective, and real-time imaging solution for studying the effects of sedentary behavior on the body. Like MRI, ultrasound can be used in combination with other measures to gain a more comprehensive understanding of the relationship between sedentary behavior and health outcomes.

3.3 Computed tomography

The health implications of sedentary behavior can sometimes be visualized through CT scans. A longitudinal cohort study carried out in the United States found significant links between sedentary behavior and the accumulation of abdominal fat depots, as assessed through CT scans. In this investigation, the total duration of sedentary behavior correlated with elevated levels of visceral adipose tissue (VAT), intermuscular adipose tissue (IMAT), and liver attenuation, particularly notable among White men. Television viewing was notably associated with higher VAT, SAT, and IMAT, especially in White men, suggesting a link between sedentary behavior, abdominal adiposity, and liver health. There is limited evidence for specific imaging biomarkers identified through CT that correlate with sedentary behavior. One relevant finding is that a recent study explored the connection between physical activity and a lung imaging marker referred to as the expiratory-to-inspiratory ratio of mean lung density (E/IMLD), which indicates resting lung hyperinflation in individuals with chronic obstructive pulmonary disease (COPD). This study found that COPD patients who were sedentary had higher E/IMLD scores (0.983) compared to those who were nonsedentary, with a score of 0.972. CT imaging biomarkers, such as the E/I ratio of mean lung density in COPD patients, show significant correlations with sedentary behavior. These biomarkers can potentially be used for early detection and monitoring of physical inactivity in these populations. Understanding the relationship between sedentary behavior and its potential health consequences, as revealed by CT scans, underscores the importance of promoting an active lifestyle for overall well-being. Table 4.2 highlights the key strengths and limitations of the imaging modalities in assessing the physiological and anatomical impacts of sedentary behavior.

4 Conclusion

The global trend toward increased sedentary time represents one of the most significant shifts in human behavior patterns in recent history. This transition has been accelerated by technological advancements, urbanization, and changes in occupational demands, creating environments where sitting has become the default posture for many daily activities. The concerning prevalence rates of sedentary behavior, estimated to be between 60% and 85% globally, indicate that this is not merely an individual behavioral issue but a systemic public health challenge that spans industrialized and developing nations alike.

The health implications of sedentary behavior extend far beyond simple physical inactivity. The evidence presented throughout this article demonstrates that prolonged sitting triggers a cascade of physiological responses that negatively impact multiple body systems. Cardiovascular health is particularly vulnerable, with studies showing clear associations

TABLE 4.2 Comparison of imaging techniques for assessing the impacts of sedentary behavior.

Imaging modalities	Strengths	Limitations	Case studies/Examples
Magnetic resonance imaging	• High-resolution images of soft tissues • Can assess brain structure and function • Noninvasive	• Expensive • Time-consuming • Limited availability in some regions	• Study showing increased liver, visceral, and total abdominal fat associated with sedentary behavior (Henson et al., 2018) • Research links prolonged sedentary time with decreased gray matter volume in MS patients (Klaren et al., 2017) • Examination of knee scans revealing meniscal tears and tendon lesions in sedentary individuals (Horga et al., 2020)
Ultrasonography	• Real-time imaging • Portable and cost-effective • Good for assessing vascular changes	• Limited in imaging deep tissues • Operator-dependent • Lower resolution compared to MRI and CT	• Study showing improved carotid and femoral artery function after a yoga intervention for sedentary desk workers (Shruthi P et al., 2024) • Research indicates increased carotid-ankle pulse wave velocity with prolonged sitting (Ravichandran et al., 2022)
Computed tomography	• Excellent for imaging bones and dense tissues • Fast and widely available • Useful for measuring abdominal fat deposits	• Exposure to ionizing radiation • Less effective for soft tissue contrast compared to MRI	• Study linking sedentary behavior with increased visceral and intermuscular adipose tissue (Whitaker et al., 2017) • Research on sedentary COPD patients shows higher lung density ratios (Murata et al., 2023)

between sedentary time and an increased risk of cardiovascular disease. The research by Bellettiere et al. demonstrated a 62% increased hazard of cardiovascular disease among older women who spent approximately 11 hours or more per day being sedentary.

The metabolic consequences of sedentary behavior are equally concerning. The disruption of normal glucose regulation can lead to insulin resistance and, eventually, type 2 diabetes. As highlighted in the findings of Stephens et al., a single day of sitting led to a remarkable 39% decrease in insulin's effectiveness in absorbing glucose in otherwise healthy adults. This acute response to sedentary behavior underscores the immediate physiological impact of prolonged sitting, even before chronic patterns are established.

The relationship between sedentary behavior and obesity presents another critical health concern. The data indicating a 13% increased chance of becoming overweight and a 26% rise in significant abdominal fat with every hour of sedentary behavior highlight the direct correlation between sitting time and adiposity. The systematic reviews showing strong correlations between obesity and both sedentary behavior (OR = 1.45) and lack of physical activity (OR = 1.52) further reinforce this connection. These findings are particularly alarming given the global rise in obesity rates and the associated health complications.

Musculoskeletal health also suffers under the influence of prolonged sedentary behavior. The meta-analysis involving 83,111 participants, which determined that engaging in sedentary behavior for more than 3 hours per day was linked to disability related to lower back discomfort (OR = 1.24), provides compelling evidence of this relationship. The cross-sectional study emphasizing the importance of following the WHO's physical activity recommendations further supports the connection between physical activity levels and musculoskeletal health.

Perhaps most concerning is the emerging evidence regarding the impact of sedentary behavior on neurocognitive health. The findings linking sedentary time to alterations in brain structure, cognitive functioning, and potentially increased risk of neurodegenerative conditions like Alzheimer's disease suggest that the consequences of sedentary behavior extend beyond physical health to cognitive wellbeing. The systematic review and meta-analysis showing improved cognitive function in sedentary older adults following physical activity interventions point to the potential reversibility of these effects, offering a glimmer of hope in addressing this aspect of sedentary behavior.

Medical imaging techniques have been instrumental in advancing our understanding of the physiological consequences of sedentary behavior. These technologies provide objective measures of tissue structure and function that might otherwise remain hidden, offering insights into the mechanisms through which sedentary behavior impacts health.

MRI stands out for its ability to provide detailed anatomical information about structural changes associated with sedentary behavior. The study revealing a discernible rise in liver fat (2.16%), visceral fat (0.54 L), and total abdominal fat (1.80 L) for every hour spent in sedentary activities demonstrates the precision with which MRI can quantify the metabolic consequences of sitting. Similarly, the research uncovering adverse correlations between the frequency of sedentary bouts lasting over 30 minutes and gray matter volume in individuals with multiple sclerosis highlights the potential neurological implications of sedentary behavior. The examination of knee scans from sedentary individuals without prior injury, revealing meniscal tears (30%), moderate-intensity lesions in tendons (21%), and high-grade tendonitis (6%), provides compelling evidence of the musculoskeletal impact of inactivity.

Ultrasonography offers complementary insights, particularly in assessing vascular changes associated with sedentary behavior. The findings that extended periods of sitting are associated with increases in carotid-ankle PWV, carotid-femoral and carotid-radial PWV, as well as changes in intima-to-media thickness and flow-mediated dilation, highlight the vascular consequences of sedentary behavior. The observation that sitting continuously for 4 hours resulted in a 5% reduction in the diameter of the common carotid artery demonstrates the acute vascular response to prolonged sitting. The study showing improved vascular function through yoga asana intervention in desk-based workers offers promising evidence for potential interventions to mitigate these effects.

CT provides additional perspectives, particularly in visualizing abdominal fat deposition and lung function changes associated with sedentary behavior. The longitudinal cohort study, finding significant links between sedentary behavior and the accumulation of visceral adipose tissue, intermuscular adipose tissue, and liver attenuation, provides valuable insights into the distribution patterns of adipose tissue related to sedentary behavior. The research on sedentary COPD patients, showing a higher E/IMLD density scores compared to nonsedentary patients, highlights the potential respiratory implications of sedentary behavior in vulnerable populations.

While each imaging modality offers unique advantages, the integration of findings from multiple techniques provides the most comprehensive understanding of the relationship between sedentary behavior and health outcomes. MRI excels in providing high-resolution images of soft tissues and assessing brain structure and function but is limited by its cost, time requirements, and availability in some regions. Ultrasonography offers real-time imaging capabilities and is both portable and cost-effective, making it particularly useful for assessing vascular changes, although it is limited in imaging deep tissues and is operator-dependent. CT is excellent for imaging bones and dense tissues and is useful for measuring abdominal fat depots but involves exposure to ionizing radiation and offers less effective soft tissue contrast compared to MRI.

The complementary nature of these imaging techniques suggests that a multimodal approach to studying sedentary behavior would yield the most comprehensive insights. For example, using MRI to assess structural changes in the brain and abdominal fat distribution, ultrasonography to monitor vascular function, and CT to evaluate bone density and lung function could provide a more complete picture of the systemic effects of sedentary behavior.

Future research should focus on longitudinal studies that track changes in imaging biomarkers over time in relation to sedentary behavior patterns. Such studies would help establish causal relationships and identify early indicators of adverse health effects before clinical symptoms develop. In addition, research exploring the differential impacts of various patterns of sedentary behavior (e.g., prolonged, uninterrupted sitting versus frequent breaks, sedentary time distribution throughout the day) would provide more nuanced insights into the most harmful aspects of sedentary behavior.

The findings from imaging studies have significant implications for public health interventions targeting sedentary behavior. The evidence of increased abdominal fat deposition, reduced gray matter volume, vascular dysfunction, and musculoskeletal abnormalities associated with sedentary behavior provides powerful motivation for individuals to reduce sitting time. Visual representations of these physiological changes could be incorporated into public health messaging to increase awareness and motivate behavior change.

Moreover, the findings regarding the benefits of interrupting sedentary time with brief periods of activity or standing suggest that even modest changes in behavior could yield meaningful health benefits. Public health interventions should emphasize not only reducing total sedentary time but also breaking up prolonged sitting periods with short bouts of movement or postural changes. The study demonstrating improved vascular function through a yoga asana intervention in desk-based workers offers an example of a targeted intervention that might be effective in mitigating the vascular consequences of occupational sitting.

The differential impacts of sedentary behavior across various demographic groups, as highlighted in some of the studies, suggest that interventions may need to be tailored to specific populations. For instance, the finding that sedentary behavior was notably associated with higher visceral adipose tissue, subcutaneous adipose tissue, and intermuscular adipose tissue, especially in white men, suggests that this demographic group might benefit from targeted interventions addressing television-viewing habits.

Despite the growing evidence regarding the adverse health effects of sedentary behavior, addressing this issue at a population level presents significant challenges. The integrated nature of sitting in many aspects of modern life, including work, transportation, and leisure activities, makes behavior change difficult. In addition, environmental and social factors often reinforce sedentary patterns, creating barriers to increased physical activity.

However, these challenges also present opportunities for innovative approaches to promoting active lifestyles. Workplace interventions, such as standing desks, walking meetings, and structured activity breaks, could help reduce occupational sitting time. Urban planning initiatives that prioritize walkability, active transportation infrastructure, and accessible recreational facilities could create environments that encourage physical activity. Educational programs that emphasize the health benefits of reducing sedentary time could help shift social norms around sitting and physical activity.

The role of technology in both contributing to and potentially mitigating sedentary behavior presents another interesting paradox. While digital devices and screen-based entertainment have contributed significantly to increased sitting time, technology could also be leveraged to promote active behaviors. Mobile applications that monitor sedentary time and prompt movement breaks, wearable devices that track physical activity, and virtual platforms that facilitate remote participation in exercise programs represent potential technological solutions to the sedentary behavior challenge.

In conclusion, the extensive evidence regarding the adverse health effects of sedentary behavior, as visualized through various imaging modalities, underscores the urgent need to address this pervasive public health issue. The multisystem impact of prolonged sitting, affecting cardiovascular, metabolic, musculoskeletal, and neurocognitive health, highlights the wide-ranging consequences of this seemingly benign behavior. By integrating findings from multiple imaging techniques with other measures of health and behavior, researchers can continue to deepen our understanding of the complex relationship between sedentary behavior and health outcomes.

Ultimately, addressing the sedentary behavior epidemic will require coordinated efforts across multiple sectors, including healthcare, education, urban planning, and workplace design. By creating environments and cultures that encourage movement and limit prolonged sitting, we can work toward a future where physical activity is integrated naturally into daily life rather than being an additional burden requiring conscious effort. The health benefits of such a shift would be substantial, potentially reducing the prevalence of numerous NCDs and improving the quality of life across the population. The challenge is significant, but the potential rewards in terms of improved public health make it a worthwhile endeavor.

References

Alzahrani, H., Alshehri, M.A., Alzhrani, M., Alshehri, Y.S., Al Attar, W.S.A., 2022. The association between sedentary behavior and low back pain in adults: a systematic review and meta-analysis of longitudinal studies. PeerJ. 21678359. 10. Available from: https://peerj.com/articles/13127/. doi:10.7717/peerj.13127.

Baillot, A., Black, M., Brunet, J., Romain, A.J., 2020. Biopsychosocial correlates of physical activity and sedentary time in adults with severe obesity. Clin. Obes. 17588111. 10 (3). Available from: https://onlinelibrary.wiley.com/loi/17588111. doi:10.1111/cob.12355.

Balkau, B., Mhamdi, L., Oppert, J.M., Nolan, J., Golay, A., Porcellati, F., Laakso, M., Ferrannini, E., 2008. Physical activity and insulin sensitivity the RISC study. Diabetes. 1939327X. 57 (10), 2613–2618. Available from: http://diabetes.diabetesjournals.org/cgi/reprint/57/10/2613. doi:10.2337/db07-1605France.

Bellettiere, J., Lamonte, M.J., Evenson, K.R., Rillamas-Sun, E., Kerr, J., Lee, I.M., Di, C., Rosenberg, D.E., Stefanick, M.L., Buchner, D.M., Hovell, M.F., Lacroix, A.Z., 2019. Sedentary behavior and cardiovascular disease in older women: the OPACH study. Circulation. 15244539. 139 (8), 1036–1046. Available from: http://circ.ahajournals.org. doi:10.1161/CIRCULATIONAHA.118.035312.

Bellettiere, J., Lamonte, M.J., Healy, G.N., Liles, S., Evenson, K.R., Di, C., Kerr, J., Lee, I.M., Rillamas-Sun, E., Buchner, D., Hovell, M.F., Lacroix, A.Z., 2021. Sedentary behavior and diabetes risk among women over the age of 65 years: the OPACH study. Diabetes Care. 19355548. 44 (2), 563–570. Available from: https://diabetesjournals.org/care/article-pdf/44/2/563/531987/dc200709.pdf. doi:10.2337/dc20-0709.

Buman, M.P., Winkler, E.A.H., Kurka, J.M., Hekler, E.B., Baldwin, C.M., Owen, N., Ainsworth, B.E., Healy, G.N., Gardiner, P.A., 2014. Reallocating time to sleep, sedentary behaviors, or active behaviors: associations with cardiovascular disease risk biomarkers, NHANES 2005-2006. Am. J. Epidemiol. 14766256. 179 (3), 323–334. doi:10.1093/aje/kwt292.

Byun, W., Dowda, M., Pate, R.R., 2012. Associations between screen-based sedentary behavior and cardiovascular disease risk factors in Korean youth. J. Korean Med. Sci. 15986357. 27(4), 388–394. Available from: http://www.ncbi.nlm.nih.gov/pmc/articles/PMC3314851/. doi: 10.3346/jkms.2012.27.4.388United.

Carter, S.E., Draijer, R., Holder, S.M., Brown, L., Thijssen, D.H.J., Hopkins, N.D., 2018. Regular walking breaks prevent the decline in cerebral blood flow associated with prolonged sitting. J. Appl. Physiol. 15221601. 125 (3), 790–798. Available from: https://www.physiology.org/doi/pdf/10.1152/japplphysiol.00310.2018. doi:10.1152/japplphysiol.00310.2018.

Castro, E.A., Júdice, P.B., Silva, A.M., Teixeira, P.J., Benito, P.J., 2017. Sedentary behavior and compensatory mechanisms in response to different doses of exercise - a randomized controlled trial in overweight and obese adults. Eur. J. Clin. Nutr. 14765640. 71 (12), 1393–1398. Available from: http://www.nature.com/ejcn/index.html. doi:10.1038/ejcn.2017.84.

Chandrasekaran, B., Pesola, A.J., Rao, C.R., Arumugam, A., 2021. Does breaking up prolonged sitting improve cognitive functions in sedentary adults? A mapping review and hypothesis formulation on the potential physiological mechanisms. BMC Musculoskelet. Disord. 14712474. 22 (1). Available from: http://www.biomedcentral.com/bmcmusculoskeletdisord/. doi:10.1186/s12891-021-04136-5.

Eloi, J.C., Epifanio, M., De Gonçalves, M.M., Pellicioli, A., Vieira, P.F.G., Dias, H.B., Bruscato, N., Soder, R.B., Santana, J.C.B., Mouzaki, M., Baldisserotto, M., 2017. Quantification of abdominal fat in obese and healthy adolescents using 3 tesla magnetic resonance imaging and free software for image analysis. PLoS ONE. 19326203. 12 (1). Available from: http://journals.plos.org/plosone/article/file?id=10.1371/journal.pone.0167625&type=printable. doi:10.1371/journal.pone.0167625.

Feairheller, D.L., Diaz, K.M., Kashem, M.A., Thakkar, S.R., Veerabhadrappa, P., Sturgeon, K.M., Ling, C., Williamson, S.T., Kretzschmar, J., Lee, H., Grimm, H., Babbitt, D.M., Vin, C., Fan, X., Crabbe, D.L., Brown, M.D., 2014. Effects of moderate aerobic exercise training on vascular health and blood pressure in African Americans. J. Clin. Hypertens. 17517176. 16 (7), 504–510. Available from: http://onlinelibrary.wiley.com/journal/10.1111/(ISSN)1751-7176. doi:10.1111/jch.12328.

Fernández-García, Á.I., Marin-Puyalto, J., Gómez-Cabello, A., Matute-Llorente, Á., Subías-Perié, J., Pérez-Gómez, J., Lozano-Berges, G., Mañas, A., Guadalupe-Grau, A., González-Gross, M., Ara, I., Casajús, J.A., Vicente-Rodríguez, G., 2021. Impact of the home confinement related to COVID-19 on the device-assessed physical activity and sedentary patterns of Spanish older adults. Biomed Res. Int. 23146141. 2021. Available from: http://www.hindawi.com/journals/biomed/. doi:10.1155/2021/5528866.

Gamage, A.U., Seneviratne, R.D.A., 2021. Physical inactivity, and its association with hypertension among employees in the district of Colombo. BMC Public Health. 1471-2458. 21 (1). doi:10.1186/s12889-021-12013-y.

González, K., Fuentes, J., Márquez, J.L., 2017. Physical inactivity, sedentary behavior and chronic diseases. Korean J. Fam. Med. 20926715. 38 (3), 111–115. Available from: https://synapse.koreamed.org/Synapse/Data/PDFData/0001KJFM/kjfm-38-111.pdf. doi:10.4082/kjfm.2017.38.3.111.

Hamilton, M.T., Healy, G.N., Dunstan, D.W., Zderic, T.W., Owen, N., 2008. Too little exercise and too much sitting: inactivity physiology and the need for new recommendations on sedentary behavior. Curr. Cardiovasc. Risk Rep. 19329563. 2 (4), 292–298. doi:10.1007/s12170-008-0054-8.

Han, X., Song, L., Li, Y., Dong, Y., Liu, R., Han, Q., Wang, X., Mao, M., Cong, L., Tang, S., Hou, T., Zhang, Q., Liu, C., Han, X., Shi, L., Nyberg, L., Launer, L.J., Wang, Y., Du, Y., Qiu, C., 2023. Accelerometer-measured sedentary behavior patterns, brain structure, and cognitive function in dementia-free older adults: a population-based study. J. Alzheimers Dis. 13872877. 96 (2), 657–668. doi:10.3233/jad-230575.

Healy, G.N., Matthews, C.E., Dunstan, D.W., Winkler, E.A.H., Owen, N., 2011. Sedentary time and cardio-metabolic biomarkers in US adults: NHANES 200306. Eur. Heart J. 15229645. 32 (5), 590–597. doi:10.1093/eurheartj/ehq451.

Henson, J., Edwardson, C.L., Morgan, B., Horsfield, M.A., Khunti, K., Davies, M.J., Yates, T., 2018. Sedentary time and MRI-derived measures of adiposity in active versus inactive individuals. Obesity. 1930739X. 26 (1), 29–36. Available from: http://onlinelibrary.wiley.com/journal/10.1002/(ISSN)1930-739X. doi:10.1002/oby.22034.

Horga, L.M., Hirschmann, A.C., Henckel, J., Fotiadou, A., Di Laura, A., Torlasco, C., D'Silva, A., Sharma, S., Moon, J.C., Hart, A.J., 2020. Prevalence of abnormal findings in 230 knees of asymptomatic adults using 3.0 T MRI. Skeletal Radiol. 14322161. 49 (7), 1099–1107. Available from: https://www.link.springer.de/link/service/journals/00256/index.htm. doi:10.1007/s00256-020-03394-z.

Katzmarzyk, P.T., Friedenreich, C., Shiroma, E.J., Lee, I.M., 2022. Physical inactivity and non-communicable disease burden in low-income, middle-income and high-income countries. Br. J. Sports Med. 14730480. 56 (2), 101–106. Available from: http://bjsm.bmjjournals.com/. doi:10.1136/bjsports-2020-103640.

Klaren, R.E., Hubbard, E.A., Wetter, N.C., Sutton, B.P., Motl, R.W., 2017. Objectively measured sedentary behavior and brain volumetric measurements in multiple sclerosis. Neurodegener. Dis. Manag. 17582032. 7 (1), 31–37. doi:10.2217/nmt-2016-0036.

Lahjibi, E., Heude, B., Dekker, J.M., Højlund, K., Laville, M., Nolan, J., Oppert, J.M., Balkau, B., 2013. Impact of objectively measured sedentary behaviour on changes in insulin resistance and secretion over 3years in the RISC study: interaction with weight gain. Diabetes Metab. 18781780. 39 (3), 217–225. doi:10.1016/j.diabet.2012.12.006.

Mabele, G.K., Ekisawa, C.N., Delecluse, C., Bompeka, F.L., 2019. Level of physical activity and eating behavior: risk factors associated with sedentariness among obese employees of a company in the city of Kinshasa province, Democratic Republic of Congo. Sci. Sports. 07651597. 34 (3), 156–164. doi:10.1016/j.scispo.2018.11.009.

Maher, C., Olds, T., Mire, E., Katzmarzyk, P.T., 2014. Reconsidering the sedentary behaviour paradigm. PLoS ONE. 19326203. 9 (1). Available from: http://www.plosone.org/article/fetchObject.action?uri=info%3Adoi%2F10.1371%2Fjournal.pone.0086403&representation=PDF. doi:10.1371/journal.pone.0086403.

Mansoor Rahman, A., Chandrasekaran, B., 2021. Estimating the impact of the pandemic on Children's physical health: a scoping review. J. Sch. Health. 0022-4391. 91 (11), 936–947. doi:10.1111/josh.13079.

McCarthy, M., Edwardson, C.L., Davies, M.J., Henson, J., Rowlands, A., King, J.A., Bodicoat, D.H., Khunti, K., Yates, T., 2017. Breaking up sedentary time with seated upper body activity can regulate metabolic health in obese high-risk adults: a randomized crossover trial. Diabetes, Obesity and Metabolism. 14631326. 19 (12), 1732–1739. Available from: http://onlinelibrary.wiley.com/journal/10.1111/(ISSN)1463-1326. doi:10.1111/dom.13016.

Morishima, T., Tsuchiya, Y., Ueda, H., Tsuji, K., Ochi, E., 2020. Sitting-induced endothelial dysfunction is prevented in endurance-trained individuals. Med. Sci. Sports Exerc. 1530-0315. 52 (8), 1770–1775. doi:10.1249/mss.0000000000002302.

Murata, Y., Hirano, T., Doi, K., Fukatsu-Chikumoto, A., Hamada, K., Oishi, K., Kakugawa, T., Yano, M., Matsunaga, K., 2023. Computed tomography lung density analysis: an imaging biomarker predicting physical inactivity in chronic obstructive pulmonary disease: a pilot study. J. Clin. Med. 20770383. 12 (8). Available from: https://www.mdpi.com/journal/jcm. doi:10.3390/jcm12082959.

Nawrocka, A., Niestrój-Jaworska, M., Mynarski, A., Polechoński, J., 2019. Association between objectively measured physical activity and musculoskeletal disorders, and perceived work ability among adult, middle-aged and older women. Clin. Interv. Aging. 11781998. 14, 1975–1983. Available from: https://www.dovepress.com/getfile.php?fileID=53866. doi:10.2147/CIA.S204196.

Paudel, S., Ahmadi, M., Phongsavan, P., Hamer, M., Stamatakis, E., 2023. Do associations of physical activity and sedentary behaviour with cardiovascular disease and mortality differ across socioeconomic groups? A prospective analysis of device-measured and self-reported UK Biobank data. Br. J. Sports Med. 0306-3674. 57 (14), 921–929. doi:10.1136/bjsports-2022-105435.

Poovitha Shruthi, P., Kamath, K., Vaishali, K., Shivashankar, K.N., Sugumar, S., Ravichandran, S., David, L.R., Hogg, P., Guruprasad, V., Banumathe, K.R., Saha, S., Kadavigere, R., 2024. Impact of yoga on the central and peripheral vascular function among desk-based workers: a single-centred trial study. F1000Research. 2046-1402. 13, 277. doi:10.12688/f1000research.135239.1.

Raja, D., Ravichandran, S., Chandrasekaran, B., Kadavigere, R., Babu, M.G.R., Almeshari, M., Alyahyawi, A.R., Alzamil, Y., Abanomy, A., Sukumar, S., 2022. Association between physical activity levels and brain volumes in adults visiting radio-imaging center of tertiary care hospital. Int. J. Environ. Res. Public Health. 16604601. 19 (24). Available from: http://www.mdpi.com/journal/ijerph. doi:10.3390/ijerph192417079.

Ravichandran, S., Sukumar, S., Chandrasekaran, B., Kadavigere, R., Shivshankar, K.N., Palaniswamy, H.P., Uppoor, R., Ravichandran, K., Almeshari, M., Alzamil, Y., Abanomy, A., 2022. Influence of sedentary behaviour interventions on vascular functions and cognitive functions in hypertensive adults—a scoping review on potential mechanisms and recommendations. Int. J. Environ. Res. Public Health. 16604601. 19 (22). Available from: http://www.mdpi.com/journal/ijerph. doi:10.3390/ijerph192215120.

Schumacher, L.M., Thomas, J.G., Vithiananthan, S., Webster, J., Jones, D.B., Bond, D.S., 2020. Prolonged sedentary time adversely relates to physical activity and obesity among preoperative bariatric surgery patients. Surg. Obes. Relat. Dis. 18787533. 16 (4), 562–567. Available from: http://www.elsevier.com/wps/find/journaldescription.cws_home/703772/description#description. doi:10.1016/j.soard.2019.12.016.

Siddarth, P., Burggren, A.C., Eyre, H.A., Small, G.W., Merrill, D.A., 2018. Sedentary behavior associated with reduced medial temporal lobe thickness in middle-aged and older adults. PLoS ONE. 19326203. 13 (4). Available from: http://journals.plos.org/plosone/article/file?id=10.1371/journal.pone.0195549&type=printable. doi:10.1371/journal.pone.0195549.

Silveira, E.A., Mendonça, C.R., Delpino, F.M., Elias Souza, G.V., de Souza Rosa, L.P., de Oliveira, C., Noll, M., 2022. Sedentary behavior, physical inactivity, abdominal obesity and obesity in adults and older adults: a systematic review and meta-analysis. Clin. Nutr. ESPEN. 24054577. 50, 63–73. Available from: http://www.journals.elsevier.com/clinical-nutrition-espen. doi:10.1016/j.clnesp.2022.06.001.

Stephens, B.R., Granados, K., Zderic, T.W., Hamilton, M.T., Braun, B., 2011. Effects of 1 day of inactivity on insulin action in healthy men and women: interaction with energy intake. Metabolism. 00260495. 60 (7), 941–949. doi:10.1016/j.metabol.2010.08.014.

Tamura, M., Nemoto, K., Kawaguchi, A., Kato, M., Arai, T., Kakuma, T., Mizukami, K., Matsuda, H., Soya, H., Asada, T., 2015. Long-term mild-intensity exercise regimen preserves prefrontal cortical volume against aging. Int. J. Geriatr. Psychiatry. 0885-6230. 30 (7), 686–694. doi:10.1002/gps.4205.

Tremblay, M.S., Aubert, S., Barnes, J.D., Saunders, T.J., Carson, V., Latimer-Cheung, A.E., Chastin, S.F.M., Altenburg, T.M., Chinapaw, M.J.M., Aminian, S., Arundell, L., Hinkley, T., Hnatiuk, J., Atkin, A.J., Belanger, K., Chaput, J.P., Gunnell, K., Larouche, R., Manyanga, T., Gibbs, B.B., Bassett-Gunter, R., Biddle, S., Biswas, A., Chau, J., Colley, R., Coppinger, T., Craven, C., Cristi-Montero, C., de Assis Teles Santos, D., Del Pozo Cruz, B., Del Pozo-cruz, J., Dempsey, P., Do Carmo Santos Gonçalves, R.F., Ekelund, U., Ellingson, L., Ezeugwu, V., Fitzsimons, C., Florez-Pregonero, A., Friel, C.P., Fröberg, A., Giangregorio, L., Godin, L., Halloway, S., Husu, P., Kadir, M., Karagounis, L.G., Koster, A., Lakerveld, J., Lamb, M., LeBlanc, A.G., Lee, E.Y., Lee, P., Lopes, L., Manns, T., Ginis, K.M., McVeigh, J., Meneguci, J., Moreira, C., Murtagh, E., Patterson, F., da Silva, D.R.P., Pesola, A.J., Peterson, N., Pettitt, C., Pilutti, L., Pereira, S.P., Poitras, V., Prince, S., Rathod, A., Rivière, F., Rosenkranz, S., Routhier, F., Santos, R., Smith, B., Theou, O., Tomasone, J., Tucker, P., Meyer, R.U., van der Ploeg, H., Villalobos, T., Viren, T., Wallmann-Sperlich, B., Wijndaele, K., Wondergem, R., 2017. Sedentary behavior research network (SBRN) - terminology consensus project process and outcome. Int. J. Behav. Nutr. Phys. Act. 14795868. 14 (1). Available from: http://www.ijbnpa.org/home/. doi:10.1186/s12966-017-0525-8.

van Bakel, B.M.A., Bakker, E.A., de Vries, F., Thijssen, D.H.J., Eijsvogels, T.M.H., 2021. Changes in physical activity and sedentary behaviour in cardiovascular disease patients during the COVID-19 lockdown. Int. J. Environ. Res. Public Health. 16604601. 18 (22). Available from: https://www.mdpi.com/1660-4601/18/22/11929/pdf. doi:10.3390/ijerph182211929.

Whitaker, K.M., Pereira, M.A., Jacobs, D.R., Sidney, S., Odegaard, A.O., Behavior, S., 2017. Physical activity, and abdominal adipose tissue deposition. Med. Sci. Sports Exerc. 15300315. 49 (3), 450–458. Available from: http://www.lww.com/product/?0195-9131. doi:10.1249/MSS.0000000000001112.

Willey, J.Z., Gardener, H., Caunca, M.R., Moon, Y.P., Dong, C., Cheung, Y.K., Sacco, R.L., Elkind, M.S.V., Wright, C.B., 2016. Leisure-time physical activity associates with cognitive decline. Neurology. 1526632X. 86 (20), 1897–1903. Available from: http://www.neurology.org. doi:10.1212/WNL.0000000000002582.

World Health Oranization, 2010. Global recommendations on physical activity for health. PMID: 26180873. World Health Oranization, Geneva.

World Health Organization, 2009. Global Health Risks : Mortality and Burden of Disease Attributable to Selected Major Risks. World Health Organization.

World Health Organization, 2020. Physical Inactivity: A Global Public Health Problem, World Health Organization, Geneva.

World Health Organizations, 2022. Physical activity India 2022 country profile.

Yamamoto, M., Wada-Isoe, K., Yamashita, F., Nakashita, S., Kishi, M., Tanaka, K., Yamawaki, M., Nakashima, K., 2017. Association between exercise habits and subcortical gray matter volumes in healthy elderly people: a population-based study in Japan. eNeurologicalSci. 24056502. 7, 1–6. doi:10.1016/j.ensci.2017.03.002.

Zhao, Y., Li, Y., Wang, L., Song, Z., Di, T., Dong, X., Song, X., Han, X., Zhao, Y., Wang, B., Cui, H., Chen, H., Li, S., Wu, L.-Y., 2022. Physical activity and cognition in sedentary older adults: a systematic review and meta-analysis. J. Alzheimers Dis. 13872877. 87 (3), 957–968. doi:10.3233/jad-220073.

Chapter 5

Impact of sedentary behaviour and nutritional intake on structural brain network among young adults

Sathya Sabina Muthu[a], Rajagopal Kadavigere[b], K. Vaishali[c], Hari Prakash Palaniswamy[d], Sneha Ravichandran[a], Dilip Shettigar[a], Leena R. David[a,e], Dilber Uzun Ozsahin[e,f,g], Berna Uzun[g], Ilker Ozsahin[g], and Suresh Sukumar[a]

[a]*Department of Medical Imaging Technology, Manipal College of Health Professions, Manipal Academy of Higher Education, Manipal, Karnataka, India*
[b]*Department of Radiodiagnosis and Imaging, Kasturba Medical College, Manipal, India*
[c]*Department of Physiotherapy, Manipal Academy of Higher Education, Manipal college of Health Professions, Manipal, India*
[d]*Department of Speech and Hearing, Manipal Academy of Higher Education, Manipal College of Health Professions, Manipal, India*
[e]*Medical Diagnostic Imaging Department, College of Health Sciences, University of Sharjah, Sharjah, United Arab Emirates*
[f]*Research Institute for Medical and Health Sciences, University of Sharjah, Sharjah, United Arab Emirates*
[g]*Irfan Suat Gunsel Operational Research Institute, Near East University, Nicosia, Turkish Republic of Northern Cyprus (TRNC), Turkey*

1 Introduction

Modern day lifestyles, coupled with the rapid and ever evolving advancements in technology, have undeniably brought immense convenience and transformed the way we live and work. From automating household chores to digitalizing most professional and recreational activities, these developments have significantly reduced the need for physical exertion in our day to day lives. While these innovations have undoubtedly made life more comfortable and efficient, they have also inadvertently contributed to a growing public health concern: the increasing prevalence of lifestyle related diseases (Isaac et al., 2017). A key contributor to this trend is the marked rise in desk based occupations and computer related tasks, which often involve extended hours of sitting. This prolonged sedentary posture has become a defining feature of modern work life, giving rise to what is now recognized as a sedentary lifestyle.

A sedentary lifestyle is defined as a way of living that involves little to no physical activity and requires minimal metabolic energy expenditure. It typically involves engaging in activities such as sitting or lying down for extended periods while working, studying, watching television, or using digital devices (González et al., 2017). This lack of physical movement has been strongly associated with numerous health concerns. Not only does it impair physical wellbeing, but it also significantly affects mental health. Scientific studies and epidemiological data have consistently shown that sedentary behavior contributes to the rising rates of obesity, type 2 diabetes, hypertension, and cardiovascular diseases. In addition, it is increasingly recognized as a potential risk factor for mental health conditions, including anxiety, depression, and even cognitive decline. The detrimental effects of prolonged inactivity emphasize the urgent need to better understand how these behaviors impact overall health and to promote strategies for incorporating regular physical activity into daily routines.

Beyond physical inactivity, dietary patterns also play a fundamental role in influencing health outcomes. An unbalanced diet, characterized by a lack of essential nutrients and a high intake of processed foods, exacerbates the risks posed by sedentary behavior. Lifestyle related noncommunicable diseases (NCDs) are now among the leading causes of disability and mortality worldwide (Noce et al., 2021), accounting for a significant portion of the global disease burden. Preventive strategies must, therefore, focus not only on reducing sedentary behavior but also on improving dietary habits. A health promoting diet includes a diverse array of nutrient rich foods such as fresh fruits, green and colorful vegetables, whole grains, legumes, olive oil, and fatty fish. Such a diet should also include moderate amounts of dairy products and eggs while limiting the intake of red meat and foods rich in saturated fats, added sugars, and excessive salt (Noce et al., 2021). These

Advances in Medical Imaging. https://doi.org/10.1016/B978-0-443-28967-5.00032-0

healthier food choices are typically abundant in natural antioxidants, fiber, polyphenols, vitamins, and essential fatty acids, all of which are critical for maintaining metabolic and cognitive health.

Both sedentary behavior and poor dietary choices exert profound effects on the body's metabolic systems and the central nervous system, particularly the brain. The growing field of neuroimaging research suggests that a sedentary lifestyle may negatively impact cerebral blood flow, leading to compromised white matter integrity and reduced brain volume (Konwar et al., 2023). White matter, composed of bundles of myelinated axons, is essential for efficient communication between different regions of the brain. It ensures the smooth transfer of information across neural circuits, supporting various cognitive functions such as memory, attention, and decision making. Disruption in the organization or structure of white matter can hinder these processes and is associated with the development of several neurodegenerative conditions, including dementia, Alzheimer's disease, and multiple sclerosis (Bae et al., 2020).

On the other hand, dietary intake can offer a protective effect on the brain. Nutrients such as omega3 polyunsaturated fatty acids, polyphenols, and vitamins B and D have shown positive associations with brain health and neuroplasticity (Moore et al., 2018). Diets enriched with these elements, such as the Mediterranean diet, have been shown to enhance brain volume, improve cognitive performance, and protect against age related cognitive decline. The Mediterranean dietary pattern, in particular, has been associated with greater white matter integrity, as evidenced by higher fractional anisotropy (FA) values and lower diffusivity measures on neuroimaging studies, both of which indicate a well-preserved white matter microstructure (Rodrigues et al., 2020). By promoting vascular health, reducing oxidative stress, and supporting synaptic function, such diets contribute to a healthier and more resilient brain.

The importance of integrating physical activity and balanced nutrition becomes even more critical when considering their combined impact on cognitive function and brain aging. Although previous studies have explored the relationships between sedentary behavior, diet, and brain structure, much of the available data has focused on middle-aged or elderly populations. There remains a relative paucity of research on young adults, despite this being a formative period in terms of brain development and lifelong health habits. Young adult brains are still undergoing structural and functional changes and, as such, may be particularly sensitive to the effects of environmental and lifestyle factors.

In this context, the present study aims to investigate the potential alterations in white matter integrity, total brain volume, and cognitive performance in young adults, with a focus on their sedentary behavior and dietary habits. Utilizing advanced neuroimaging techniques such as diffusion tensor imaging (DTI) and amide proton transfer (APT) chemical exchange saturation transfer (CEST) magnetic resonance imaging (MRI) sequences, this research seeks to uncover subtle brain changes associated with these lifestyle factors. By shifting the lens to a younger demographic, the study hopes to capture early indicators of brain vulnerability and resilience, potentially offering valuable insights into how early-life interventions can support better brain health throughout adulthood.

In summary, adopting a healthy lifestyle early in life, one that integrates consistent physical activity and a balanced, nutrient rich dietis not only beneficial for general wellbeing but is also critically important for protecting brain structure and preserving cognitive function. Early lifestyle choices play a pivotal role in shaping brain development, particularly during young adulthood when the brain is still maturing and remains highly susceptible to both positive and negative environmental influences. By establishing health promoting habits at a young age, individuals can potentially reduce their vulnerability to a range of neurological and cognitive impairments that may arise later in life. The outcomes of this study may offer valuable insights into the intricate and multifaceted relationship between lifestyle behaviors such as sedentary time and dietary patterns and neural health indicators, including white matter integrity, brain volume, and cognitive performance. By leveraging advanced neuroimaging techniques and standardized cognitive assessments, the research aims to uncover early biomarkers that reflect how lifestyle factors influence the structural and functional aspects of the brain. These findings can serve as a foundation for developing targeted public health strategies, clinical interventions, and educational programs aimed at promoting healthier living from an early age.

Furthermore, the implications of this research extend beyond individual health; they resonate with broader societal goals, including reducing the long-term burden of neurodegenerative diseases such as Alzheimer's disease and other forms of dementia. By identifying and addressing modifiable risk factors early in life, it may be possible to slow or prevent the onset of such conditions, thus enhancing quality of life and reducing healthcare costs at a population level. Ultimately, this study reinforces the message that proactive, preventive approaches to health grounded in physical activity and dietary mindfulness are crucial not only for maintaining optimal cognitive function throughout the lifespan but also for building a healthier, more resilient society.

2 Study objectives

2.1 Primary objective

- To determine the changes in white matter integrity between sedentary and nonsedentary young adults based on dietary intake.

2.2 Secondary objective

- To determine the brain volume changes between sedentary and nonsedentary young adults based on dietary intake.
- To explore the changes in cognitive performance between sedentary and nonsedentary young adults based on dietary intake.
- To explore the changes in APT signal intensities between sedentary and nonsedentary young adults based on dietary intake.

3 Methodology

This observational study will be conducted using a 3.0 Tesla United Imaging UMR 780 MRI system located in the Department of Radiodiagnosis and Imaging at Kasturba Hospital, Manipal. The advanced imaging capabilities of this high-field MRI system will allow for the acquisition of high resolution structural and functional brain data, which are critical for assessing white matter integrity and brain volume in the context of this study.

Prior to the initiation of any research related procedures, the study protocol will be submitted for review and approval by the Institutional Ethics Committee (IEC) of Kasturba Hospital, Manipal. All participants enrolled in the study will receive comprehensive verbal and written explanations regarding the objectives, procedures, potential risks, and benefits of the research. They will be given adequate time to consider participation and ask questions. Written informed consent will be obtained from each participant before any data collection begins, ensuring that their participation is entirely voluntary.

In addition, the IEC will be responsible for the periodic monitoring and review of the study's progress, particularly with respect to data collection, participant safety, and adherence to ethical standards. This ongoing oversight will ensure the integrity and ethical conduct of the study throughout its duration.

3.1 Participants

Participants for this study will be individuals who are already scheduled to undergo MRI scans at the Department of Radiodiagnosis and Imaging, Kasturba Hospital, Manipal. These participants will be referred by their respective consulting physicians as part of their routine clinical evaluation or diagnostic workup. Only those patients who meet the study's predefined inclusion and exclusion criteria will be considered for participation. The selection will be based on clinical indications, referral notes, and the absence of contraindications for MRI. The recruitment process will ensure that participants are enrolled in an ethically compliant manner without interfering with their standard medical care.

Sample size: The difference between the two means was calculated considering a 95% confidence interval and 80% power. The total sample size calculated was 124, with 62 participants in each group, i.e., sedentary and nonsedentary.

Inclusion criteria: Individuals between the ages of 18 and 40 years, referred for MRI brain scans, who can comprehend the instructions.

Exclusion criteria: Individuals who have had head trauma, individuals clinically diagnosed with cognitive impairment or dementia, and patients who have been bedridden for a few months. Patients clinically diagnosed with hormonal disorders.

4 Procedure

This cross sectional observational study will be conducted following ethical standards and institutional guidelines. Prior to the commencement of the study, necessary approvals will be obtained from both the Institutional Research Committee of the Manipal College of Health Professions and the IEC of Kasturba Hospital, Manipal. These approvals ensure that the study adheres to ethical principles involving human research and complies with national and institutional standards for the protection of participants' rights and safety. The study population will include patients aged between 18 and 40 years who are undergoing MRI scans at the Department of Radiodiagnosis and Imaging, Kasturba Hospital, Manipal. This age group has been selected as it represents young adults, a population particularly vulnerable to the long term effects of a sedentary lifestyle and poor dietary habits on brain development and cognitive function. Participants will be recruited based on physician referrals, and only those who meet the study's inclusion and exclusion criteria will be considered eligible. Informed written consent will be obtained from all participants after providing them with a detailed explanation of the study objectives, procedures, and potential risks and benefits.

4.1 Baseline measurements

Demographic and clinical information: Basic demographic information, such as the participant's age, sex, occupation, and educational background, will be systematically documented at the time of enrollment. In addition to demographic data, comprehensive clinical details will be gathered to better understand the health profile and potential confounding variables of each participant. This will include the assessment of cardiovascular and lifestyle related risk factors, such as smoking status, alcohol consumption habits, level of physical activity, and sleep patterns. Information on these risk factors will be obtained through structured interviews and standardized questionnaires to ensure consistency and reliability. Furthermore, a thorough medical history will be recorded, encompassing any previous diagnoses, chronic conditions, or significant medical events. Current medication usage, including prescription drugs, over-the-counter medications, and dietary supplements, will also be documented through both patient selfreport and verification via available medical records. Collecting this comprehensive set of baseline data is crucial to contextualize the findings of the study, control for confounding variables, and ensure accurate interpretation of the associations between lifestyle behaviors, brain health, and cognitive function.

4.2 Physical activity measurement

To evaluate sedentary behavior and levels of physical activity, the Global Physical Activity Questionnaire (GPAQ), developed by the World Health Organization, will be utilized. The GPAQ assesses physical activity in three domains: activity at work, travel to and from places, and recreational activities. It also captures information on sedentary time spent during a typical day. After receiving a detailed verbal explanation and providing written informed consent, each participant will be asked to complete the questionnaire, which is designed to be completed in approximately 5 to 10 minutes. This validated tool provides a reliable estimate of an individual's physical activity and is especially suited for large scale epidemiological studies.

4.3 Dietary intake assessment

Nutritional intake will be assessed using a semiquantitative Food Frequency Questionnaire (FFQ), which has been validated for use in similar populations. The FFQ is designed to evaluate the frequency and portion size of a wide range of food items consumed over a defined reference period (typically the past month or three months). It will capture details about the intake of fruits, vegetables, grains, dairy, meat, fish, oils, processed foods, and beverages, as well as nutrient dense and high fat or sugary foods. This tool will help determine the dietary patterns and overall nutritional quality of the participants' diets, which will later be categorized as healthy or unhealthy based on established dietary guidelines.

Together, these baseline measurements will provide a comprehensive overview of the participants' physical activity levels, dietary habits, and relevant lifestyle factors, serving as the foundation for analyzing their association with MRI based brain parameters and cognitive outcomes.

5 Structural brain analysis

The neuroimaging procedures will be performed at the Department of Radiodiagnosis and Imaging, Kasturba Hospital, Manipal, using a 3 T United Imaging UMR 780 MRI system. Participants will be positioned supine within the MRI scanner, and standard safety protocols will be followed throughout the imaging process. The imaging protocol will include sequences for assessing white matterintegrity, brain volume, and APT signal intensities, which together offer a comprehensive overview of the structural and molecular state of the brain.

5.1 Brain white matter integrity

White matter integrity will be assessed using DTI, a noninvasive technique that enables the visualization and quantification of microstructural properties of white matter tracts. DTI scans will be acquired with participants in the supine position using standardized diffusion weighted sequences optimized for whole brain coverage. The raw DTI data will undergo postprocessing using the Functional Magnetic Resonance Imaging of the Brain (FMRIB) Software Library (FSL), specifically FMRIB's Diffusion Toolbox. Preprocessing steps will include motion correction, eddy current correction, and brain extraction. Subsequently, diffusion tensors will be fitted to the data to generate quantitative maps of diffusion metrics, including FA, mean diffusivity (MD), axial diffusivity, and radial diffusivity. These metrics will be analyzed to evaluate the integrity of major white matter tracts and to identify potential microstructural alterations associated with sedentary behavior and dietary habits.

5.2 Brain volume analysis

Structural MRI scans will be acquired using a T1weighted fast spoiled gradient echo (FSPGR) sequence to obtain high-resolution anatomical images necessary for volumetric analysis. Participants will remain in the supine position throughout the acquisition.

The acquired T1weighted images will undergo voxel-based morphometry (VBM) processing using standard neuroimaging software such as FSL or Statistical Parametric Mapping (SPM), including steps such as tissue segmentation, spatial normalization, and modulation. Total and regional brain volumes including gray matter, white matter, and cerebrospinal fluid will be quantified. These data will be used to explore potential structural differences in brain volume between the sedentary and nonsedentary groups and to assess any associations with dietary patterns.

5.3 Amide Proton Transfer Signal Intensities

To evaluate molecular changes in brain tissue, CEST imaging will be performed, focusing specifically on APT. This advanced MRI technique detects endogenous mobile proteins and peptides by measuring proton exchange between amide protons and water. Participants will undergo the CEST sequence in the supine position, and APT signal intensities will be quantified by calculating the Magnetization Transfer Ratio (MTR) asymmetry values at ±3.5 ppm offset from the water resonance. These values provide insights into tissue pH, protein content, and cellular metabolic activity, and may reflect early neurochemical alterations not detectable by conventional MRI techniques.

6 Cognitive function assessment

To evaluate cognitive performance among participants, two standardized neuropsychological tasks will be administered: the Eriksen Flanker Task and the N Back Memory Test. These assessments target key domains of cognitive functioning, including attention, inhibitory control, working memory, and processing speed. The tasks will be delivered using Inquisit 6 Lab software (developed by Millisecond Software, www.millisecond.com), a reliable platform for conducting computerized psychological experiments. All cognitive tests will be conducted in a quiet, distraction free room. Participants will be seated approximately two feet away from a computer monitor, with their feet flat on the ground and hands resting comfortably on the keyboard. Detailed instructions and practice trials will be provided prior to the actual assessment to ensure participant understanding and task familiarity.

6.1 Eriksen flanker test

The Eriksen Flanker Task is designed to assess selective attention, response inhibition, and executive control by requiring participants to focus on a central target stimulus while ignoring distracting flanker stimuli presented simultaneously. This task provides insight into the participant's cognitive control and attentional flexibility. Fig. 5.1 illustrates the Eriksen Flanker test.

Stimulus Presentation: A string of five letters will appear on the screen for a brief duration (1 s). The target letter will always be the center letter, flanked by either congruent or incongruent letters on both sides.

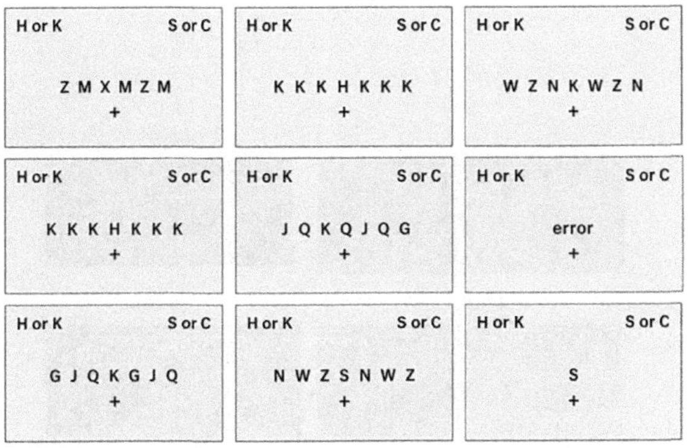

FIG. 5.1 Illustrates the Eriksen Flanker test.

Instructions: Participants will be instructed to respond only to the target letter in the center and ignore the flanking letters. They will press:

"Q" if the target letter is H or K
"P" if the target letter is S or C

Trial Design: Each participant will complete twelve test trials designed to evaluate both reaction time (RT) and response accuracy. The trials will include a mix of congruent and incongruent stimuli to test the participant's ability to suppress irrelevant information.

Data Recording: The software will log: Response time (in milliseconds) for each trial, accuracy (correct vs. incorrect responses), and the type of trial (congruent/incongruent).

6.2 N back memory test

The N Back Test is a widely used working memory task that measures cognitive load handling, updating of memory representations, and mental tracking. In this test, participants are presented with a sequential stream of letters on the screen and are required to respond when a stimulus matches one presented "N" steps earlier in the sequence. Fig. 5.2 illustrates the N Back memory test.

Four levels of the N-Back task will be used:

0 Back Test:

- Participants are required to press the "A" key every time the letter "M" appears on the screen.
- No other response is expected for any other letter.
- This serves as a baseline condition, requiring minimal working memory engagement.

1 Back Test:

- Participants press the "A" key if the current letter is identical to the one presented immediately before (i.e., one position back).
- No key press is required if there is no match.

2 Back Test:

- Participants press the "A" key if the current letter matches the letter shown two positions earlier in the sequence.
- Nonmatching stimuli require no response.

3 Back Test:

- Participants press the "A" key if the current letter matches the letter presented three steps earlier.
- Again, no response is made if there is no match.

Stimulus timing and format: Each letter will be presented for a fixed duration (typically 500 ms to 1 s), with a short interstimulus interval. The task gradually increases in difficulty from 0-back to 3-back, allowing for the assessment of working memory performance under varying cognitive loads.

Data collection: For each level of the N-back test, the following parameters will be recorded: Accuracy (hit rate and false alarm rate), RT for correct responses, missed responses, and incorrect hits.

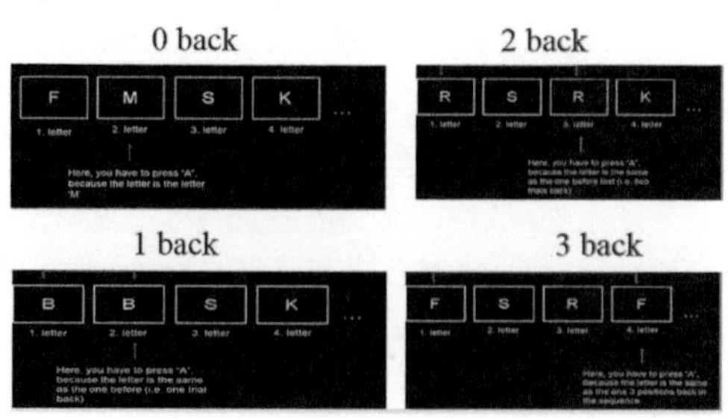

FIG. 5.2 Illustrates the N-Back memory.

7 Outcome measurement

The primary and secondary outcome measures of the present crosssectional study will be derived from MRI data and computerized cognitive assessments using the Inquisit 6 Lab software (Millisecond Software, www.millisecond.com). All participants will undergo MRI at Kasturba Hospital, Manipal, as part of their routine clinical referral. Prior to the imaging procedure, participants will complete the FFQ for dietary intake assessment and standardized cognitive function tests, including the N Back Memory Test and the Eriksen Flanker task.

7.1 Primary outcome measures: Brain white matter integrity

The primary outcome of interest is the assessment of white matter integrity using DTI, a quantitative MRI technique that evaluates the microstructural properties of white matter tracts.

7.2 Diffusion parameters

MD: This parameter reflects the overall average motion of water molecules within brain tissues. Elevated MD values may indicate axonal loss, increased extracellular space, or compromised cellular structure, potentially pointing to early white matter abnormalities or neurodegenerative changes.

FA: FA measures the directional coherence of water diffusion. High FA values are typically associated with **intact,** well-myelinated, and organized white matter fibers, whereas reduced FA values may suggest demyelination, fiber degeneration, or disruption of axonal architecture.

7.3 Secondary outcome measures: Brain volume, cognitive functions, and APT signal

Structural MRI scans will be acquired using a T1weighted FSPGR sequence to evaluate total and regional brain volume. VBM will be performed using software such as SPM or FSL to quantify gray matter and white matter volume. These volumetric measurements will be compared between groups and correlated with lifestyle parameters (sedentary behavior, dietary patterns) and cognitive performance. Alterations in brain volume may reflect the early impact of lifestyle-related changes on brain health in young adults.

Cognitive function will be assessed through two computerized tasks conducted via Inquisit 6 Lab. The Eriksen Flanker task measures selective attention and response inhibition by evaluating participants' ability to suppress interference from incongruent flanking stimuli. RT and accuracy will be recorded for each trial to assess cognitive control. The N Back Memory test assesses **working memory capacity** and processing speed through increasing levels of cognitive load (0-back to 3-back).

To explore metabolic and molecular changes in the brain, CEST imaging will be performed using an MRI based APT sequence. APT signal intensity will be measured by evaluating the MTR asymmetry values at ± 3.5ppm, indicative of endogenous protein and peptide concentrations. Alterations in APT signal intensities may reflect molecular changes related to cognitive function and white matter health.

8 Statistical analysis

The statistical analysis for this study will be conducted using SPSS version 16.0 (IBM Corp., Armonk, NY). The primary objective of the analysis is to explore the associations between brain structural integrity, brain volume, cognitive function, physical activity levels, and dietary intake patterns in young adults. Both descriptive and inferential statistical techniques will be employed to address the research objectives and test the formulated hypotheses.

8.1 Descriptive statistics

Descriptive statistics will be used to summarize demographic data (including age, gender, height, weight, and BMI), physical activity levels (as assessed using the GPAQ and expressed in terms of MET scores), and dietary intake patterns (measured using the FFQ). In addition, MRI-derived parameters, such as FA, MD, intracranial volume, and APT signal intensities, will be summarized. Cognitive performance metrics, including accuracy and RTs from the Eriksen Flanker and N Back memory tests, will also be included in the descriptive analysis. Normally distributed variables will be expressed as mean, standard deviation, and range, while nonnormally distributed variables will be reported as median and interquartile range. Data normality will be assessed using the Kolmogorov Smirnov and Shapiro Wilk tests.

8.2 Inferential statistics

Inferential statistics will be employed to determine significant differences and relationships between the sedentary and nonsedentary groups. An independent samples *t*-test will be used to compare continuous variables such as FA, MD, brain volumes, APT signal intensities, and cognitive scores between groups. If the assumptions of normality and equal variances are not met, the Mann-Whitney *U* test will be used as a nonparametric alternative. The chi-square test will be used to compare categorical variables such as gender distribution or the presence of lifestyle-related risk factors. A mixed linear model will be applied to assess how dietary intake, as quantified through FFQ scores, influences structural brain metrics, accounting for potential confounding variables such as age, sex, BMI, and physical activity levels.

9 Anticipated outcome

We hypothesize that young adults exhibiting sedentary behavior will demonstrate significant alterations in key brain regions when compared to their nonsedentary group. Specifically, we anticipate that prolonged physical inactivity will be associated with reduced white matter integrity, characterized by lower FA and higher MD values, suggesting compromised microstructural organization of white matter fibers. In addition, we expect observable reductions in total brain volume, particularly in regions sensitive to activity-related neuroplasticity. Cognitive function is also hypothesized to be negatively impacted in the sedentary group. We anticipate that these individuals will demonstrate slower RTs, decreased accuracy, and overall reduced cognitive performance, particularly in domains related to attention, executive control, and working memory. Furthermore, we propose that sedentary individuals will show altered APT signal intensities when compared to physically active individuals.

Collectively, we hypothesize that the neuroimaging and cognitive findings will indicate that a sedentary lifestyle, even in young adults, can adversely affect the brain's structural integrity, functional capacity, and metabolic environment. These anticipated outcomes support the critical need for early lifestyle interventions to preserve brain health and cognitive resilience throughout adulthood.

10 Conclusion

In conclusion, this study aims to comprehensively examine the impact of sedentary behavior and dietary intake on the structural and functional aspects of the brain, with a particular focus on white matter integrity, brain volume, cognitive performance, and APT signal intensities in young adults, to identify early quantifiable markers of neurobiological changes linked to lifestyle factors. The findings from this study have the potential to inform and guide preventive strategies by emphasizing the importance of regular physical activity and balanced nutritional intake for preserving brain health and cognitive function. In an era where sedentary lifestyles and poor dietary habits are becoming increasingly prevalent, particularly among youth and working-age populations, understanding the underlying neural mechanisms is essential for implementing effective health interventions. Early identification of lifestyle-related brain alterations may not only facilitate timely modifications in individual behavior but also influence public health policies and healthcare practices.

Moreover, this research aligns strongly with global health and development priorities, particularly the United Nations Sustainable Development Goals (SDGs). Specifically, it supports SDG 3: *Good Health and Well-being* by contributing to the understanding and prevention of NCDs, which are among the leading causes of morbidity and mortality worldwide. Promoting healthier behaviors can improve quality of life, reduce the burden on healthcare systems, and foster a more health-conscious society. In addition, this study supports SDG 8: *Decent Work and Economic Growth* by offering valuable insights into how sedentary behavior and dietary patterns can affect cognitive function and productivity in the workforce. As cognitive health is crucial for optimal performance and mental well-being in professional settings, this research can inform the design of workplace wellness programs, advocate for active work environments, and encourage healthier dietary habits among employees.

Ultimately, the study underscores the importance of a proactive, preventive approach to health that integrates neuroscience, lifestyle medicine, and public health policy. By bridging the gap between everyday behaviors and long-term brain health, it advocates for a shift toward sustainable, health-promoting environments that empower individuals and communities to make informed, conscious choices that benefit their cognitive and overall well-being throughout the lifespan.

References

Bae, H.G., Kim, T.K., Suk, H.Y., Jung, S., Jo, D.G., 2020. White matter and neurological disorders. Arch. Pharm. Res. 43 (9), 920–931.
González, K., Fuentes, J., Márquez, J.L., 2017. Physical inactivity, sedentary behavior, and chronic diseases. Korean J. Fam. Med. 38 (3), 111–115.

Isaac, A., Olanipekun, O., Adetunji, J., 2017. Sedentary life-style as inhibition to good quality of life and longevity. J. Educ. Pract. 8 (13), 35–38.

Konwar, S., Manca, R., De Marco, M., Soininen, H., Venneri, A., 2023. The effect of physical activity on white matter integrity in aging and prodromal to mild Alzheimer's disease with vascular comorbidity. Front. Aging Neurosci. 15, 1096798.

Moore, K., Hughes, C.F., Ward, M., Hoey, L., McNulty, H., 2018. Diet, nutrition, and the ageing brain: current evidence and new directions. Proc. Nutr. Soc. 77 (2), 152–163.

Noce, A., Romani, A., Bernini, R., 2021. Dietary intake and chronic disease prevention. Nutrients 13 (4), 1358.

Rodrigues, B., Coelho, A., Portugal-Nunes, C., Magalhães, R., Moreira, P.S., Castanho, T.C., et al., 2020. Higher adherence to the Mediterranean diet is associated with preserved white matter integrity and altered structural connectivity. Front. Neurosci. 14, 786.

Chapter 6

Magnetic resonance imaging assessment of cerebral grey matter volume: The role of physical activity and sedentary behavior in brain health

Deepika Raja[a], Rajagopal Kadavigere[b], Leena R. David[a,c], Sneha Ravichandran[a], Baskaran Chandrasekaran[d], Berna Uzun[e], Dilber Uzun Ozsahin[c,e,f], M.G. Ramesh Babu[g], Meshari Almeshari[h], Ilker Ozsahin[e], Suresh Sukumar[a], and Banumathe Karuppaya[i]

[a]*Department of Medical Imaging Technology, Manipal College of Health Professions, Manipal Academy of Higher Education, Manipal, Karnataka, India*

[b]*Department of Radiodiagnosis and Imaging, Kasturba Medical College, Manipal, India*

[c]*Medical Diagnostic Imaging Department, College of Health Sciences, University of Sharjah, Sharjah, United Arab Emirates*

[d]*Symbiosis School of Sports Sciences, Symbiosis International (Deemed University), Symbiosis Knowledge Village, Pune, India*

[e]*Irfan Suat Gunsel Operational Research Institute, Near East University, Nicosia, Turkish Republic of Northern Cyprus (TRNC), Turkey*

[f]*Research Institute for Medical and Health Sciences, University of Sharjah, Sharjah, United Arab Emirates*

[g]*Division of Physiology, Department of Basic Medical Sciences, Manipal Academy of Higher Education, Manipal, India*

[h]*Department of Diagnostic Radiology, College of Applied Medical Science, University of Hail, Hail, Saudi Arabia*

[i]*Department of Occupational Therapy, Manipal University College Malaysia, Melaka, Malaysia*

1 Understanding cerebral gray matter volume

Cerebral gray matter comprises synapses, glial cells, dendrites, and neuronal cell bodies. “Gray matter volume” refers to the quantity and distribution of gray matter in the brain’s cerebral hemispheres, categorized as subcortical and cortical structures. It plays crucial roles in memory, emotions, decision-making, motor control, and sensory perception. The complex mosaic of cognitive abilities throughout the human lifespan is influenced by gray matter changes across different brain areas. Gray matter volume measurement frequently evaluates regional brain integrity, with individual variations affected by age, genetics, and environment (Baxi et al., 2020; Sanchis-Segura et al., 2019).

Gray matter volume changes occur throughout life, peaking before age 10, with dynamic nonlinear changes continuing during adolescence that vary by brain region. Recent longitudinal studies provide insights into developmental trajectories, showing that prefrontal regions develop later than primary sensory areas, potentially explaining asynchronous cognitive ability development (Tamnes et al., 2017).

Understanding microstructural composition is essential for interpreting volume changes. Gray matter consists of neuronal cell bodies, unmyelinated axons, dendrites, glial cells, and synaptic connections. Modern neuroimaging can differentiate between changes in neuronal density versus alterations in synaptic density or glial cell populations, providing a nuanced understanding of volume changes (Zatorre et al., 2012). This distinction is particularly important when interpreting exercise-induced changes, as they may reflect different cellular processes.

Gray matter volume alterations have been linked to attention-deficit/hyperactivity disorder, schizophrenia, and autism spectrum disorders (Li et al., 2022; Zhang et al., 2018). These conditions often present region-specific patterns of volume changes that could serve as potential biomarkers. For instance, decreased prefrontal cortex volume appears in ADHD, while temporal lobe gray matter reductions characterize schizophrenia (Hoogman et al., 2017).

Advances in Medical Imaging. https://doi.org/10.1016/B978-0-443-28967-5.00006-X

Gray matter volume partly explains cognitive function diminishment in various diseases. Gray matter volume and processing speed are mediated via white matter hyperintensities connected with small vessel disease (Wang et al., 2022). The relationship between gray matter volume and cognitive function appears causally linked through neural efficiency and network integration mechanisms. Longitudinal studies show that volume changes often precede detectable cognitive decline, suggesting that volumetric measurements could serve as early predictors of cognitive impairment (Jack et al., 2015).

Through gray matter volume investigation, researchers and clinicians can develop intervention methods focusing on brain anomaly reduction. Neuroimaging advancements, including structural magnetic resonance imaging (MRI) and diffusion tensor imaging, enable higher-detail and precision investigation. Quantitative susceptibility mapping and magnetic resonance spectroscopy further enhance gray matter composition characterization, detecting subtle changes in iron content, myelin, and neurotransmitter concentrations that are not apparent in conventional analyses (Deistung et al., 2017).

2 Physical activity and brain health

2.1 Physical activity and increased gray matter volume

Physical activity positively impacts gray matter volume through multiple mechanisms. It stimulates hippocampal neurogenesis, improving spatial memory. Higher physical fitness correlates with improved executive function, faster information processing, and increased gray matter volume in regions including the accumbens nucleus, caudate nucleus, and hippocampus. Aerobic activity specifically increases gray matter volume in various brain regions (Paillard et al., 2015). In overweight and obese children, physical activity increases intelligence and cognitive flexibility (Ortega et al., 2022).

Exercise affects different brain regions distinctively. While the hippocampus has been extensively studied due to its role in memory and sensitivity to exercise, other regions show differential responses. The prefrontal cortex demonstrates significant volumetric increases with aerobic exercise, particularly in older adults (Colcombe et al., 2006). The anterior cingulate cortex, involved in attention and error detection, also responds positively to physical activity. In contrast, the primary visual cortex shows fewer exercise-related changes, suggesting selective rather than global effects (Stillman et al., 2016).

Sex differences exist in physical activity's effects on gray matter. Research shows increases in cortical gray matter, total gray matter, and right hippocampal volumes in males but not in females (Sanchis-Segura et al., 2019). This sex-specific response raises questions about underlying mechanisms. Hormonal differences may play a role, with estrogen potentially providing females with additional neuroprotective effects that might mask some exercise benefits. Testosterone in males may enhance certain exercise-induced neuroplasticity aspects, leading to more pronounced volumetric changes (Barha et al., 2017).

Independent of glucose management, regular physical activity positively impacts gray matter volumes in individuals with type 2 diabetes, potentially reducing adverse brain consequences. Among older adults with intact cognitive function, home-based physical activity correlates favorably with brain volume assessments, particularly gray matter volume, compared to recreational activities.

Physical activity timing across the lifespan appears critical. Evidence suggests sensitive periods when the brain is particularly responsive to exercise-induced changes. Early-life physical activity may establish "cognitive reserve," protecting against age-related decline, while mid-life exercise might maintain these reserves (Cheng, 2016). In older adults, even beginning exercise later in life yields significant benefits for gray matter volume and cognitive function, though the effect magnitude may differ from lifelong exercisers (Hörder et al., 2018).

Studies indicate that physical activity beneficially affects human myelin content, with lifelong myelin expansion resulting from intense aerobic exercise (Dunton, 2017). Incorporating physical activity into daily routines improves psychological and anatomical states in individuals with mild cognitive deficits and Alzheimer's disease, as demonstrated by various neuroimaging modalities (Wu et al., 2021).

Physical activity interacts with other lifestyle factors, including nutrition, sleep quality, cognitive stimulation, and social engagement. The Mediterranean diet, rich in antioxidants and antiinflammatory compounds, may enhance exercise-induced neuroplastic effects (Vauzour et al., 2017). Adequate sleep appears to be necessary for consolidating physical activity-induced structural changes. Cognitive training combined with exercise shows synergistic effects on gray matter volume, suggesting that multimodal interventions may surpass single-component approaches (Bamidis et al., 2014).

2.2 Mechanisms: Neuroplasticity and neurotrophic factors

Physical activities like exercise and gardening improve neuroplasticity and increase neurotrophic factor synthesis, including brain-derived neurotrophic factor (BDNF), glial cell-derived neurotrophic factor (GDNF), and nerve growth factor (NGF), which are essential for brain function improvement (van Praag and Wrann, 2022; Lentoor et al., 2023). These activities alter neurogenesis, growth factor production, and synaptic plasticity, thereby improving cognition in human and animal models (De Sousa Fernandes et al., 2020; Phillips, 2017).

The molecular cascade initiated by exercise involves multiple interacting pathways. Muscle contractions release myokines, including irisin, which cross the blood-brain barrier and directly influence neuronal health. Exercise increases cerebral blood flow, delivering oxygen and nutrients while removing waste. This enhanced perfusion stimulates growth factor release, particularly BDNF, which is central to neuroplasticity (Wrann et al., 2013). BDNF promotes the survival of existing neurons while encouraging the growth, differentiation, and formation of new neurons and synapses. In the hippocampus, BDNF activates the TrkB receptor, triggering signaling cascades that influence neuronal structure, function, and gene expression (Liu and Nusslock, 2018).

Physical activity and nutrition impact neuroplasticity substrates and are important for preserving brain health during aging (Arnardottir et al., 2016; Voss et al., 2013; Erickson et al., 2014; Aslanyan et al., 2023). Aerobic exercise positively influences brain volume, particularly in aging populations, inducing structural alterations through neurogenesis, synaptogenesis, and angiogenesis, which increase white and gray matter volume. Studies on elderly people with amnestic mild cognitive impairment (MCI) show a positive association between aerobic fitness and gray matter volume, especially in the superior frontal cortex, with higher aerobic fitness linked to enhanced white matter tract microstructural integrity and increased frontal area gray matter density (Teixeira et al., 2016).

Beyond BDNF, other neurotrophic factors play important roles in exercise-induced neuroplasticity. Insulin-like growth factor 1 (IGF-1) increases with physical activity, supporting neuronal growth and survival while protecting against oxidative stress. Vascular endothelial growth factor stimulates angiogenesis, creating new capillaries and improving cerebral blood flow. Exercise also modulates dopaminergic, serotonergic, and glutamatergic signaling, influencing mood, cognition, and neuroplasticity (Pedersen, 2019).

Animal studies provide valuable insights into the cellular mechanisms underlying exercise-induced gray matter volume changes. In rodent models, voluntary wheel running increases hippocampal neurogenesis, dendritic branching, and spine density. These structural changes correlate with improved spatial memory task performance (Van Praag et al., 2005). Transgenic mouse models further elucidate specific genes and proteins mediating these effects, showing that BDNF expression is necessary for many exercise-related cognitive benefits (Vaynman et al., 2004).

Aerobic exercise intervention duration plays a crucial role in inducing neuroplasticity, with 2.5 months being sufficient for structural changes and more than 3 months required for increased functional connectivity. High-intensity exercise leads to larger neurotrophic factor increases compared to lower intensities, improving motor skill retention. Increased cerebral blood flow sustains exercise-induced brain activity by providing vital nutrients to neurons. Physical activity is promising for promoting optimal brain health and potentially delaying neurodegenerative diseases. Studies show that habitual physical activity positively correlates with gray matter volume in the anterior hippocampal and parahippocampal regions, independent of age and gender (Demirakca et al., 2014).

Clinical studies elucidate the translational relevance of these mechanisms. In early-stage Alzheimer's disease patients, six months of aerobic exercise increased serum BDNF levels, correlating with preserved hippocampal volume and improved memory performance (Baker et al., 2010). Similarly, in individuals with Parkinson's disease, high-intensity interval training increased serum BDNF and IGF-1, with corresponding improvements in motor symptoms and executive function (Marusiak et al., 2015).

Exercise has epigenetic effects, modifying DNA methylation and histone acetylation patterns, which affect neuroplasticity and neuroprotection gene expression (Sølvsten et al., 2018). These epigenetic changes can persist beyond the acute exercise period, potentially explaining the long-term benefits of regular physical activity. Evidence suggests that these epigenetic modifications may be transmissible to offspring, indicating that parental exercise habits could influence brain development in subsequent generations (McGee and Hargreaves, 2019).

2.3 Exercise recommendations: Type, intensity, and duration

World Health Organization guidelines provide physical activity recommendations across three age ranges (World Health Organization, 2010):

While these guidelines provide a solid foundation for general health benefits, emerging research suggests that specific exercise parameters may be particularly beneficial for brain health and gray matter volume. The optimal "dose" of exercise for brain health appears to differ from cardiovascular benefits. Moderate-intensity aerobic exercise performed for at least 150 minutes weekly shows the most consistent benefits for gray matter volume, particularly in the hippocampus and prefrontal cortex (Gomes-Osman et al., 2018). The intensity threshold for inducing neuroplastic changes appears to be around 60–70% of maximum heart rate, though higher intensities may yield greater benefits in certain populations (Northey et al., 2018).

2.3.1 Age group: 5–17 years old

Various forms of physical activity include play, games, sports, transportation, leisure, physical education, and organized exercise within family, school, and community contexts.

Children and youth should engage in at least 60 minutes daily of moderate-to-intense physical activity.

Physical activity beyond 60 minutes provides additional health benefits.

Aerobic exercise should comprise most daily activity, with vigorous-intensity bone and muscle-strengthening exercises included at least three times weekly.

For children and adolescents, the type of physical activity may be particularly important for brain development. Activities that combine cardiovascular exercise with cognitive engagement and coordination challenges—such as team sports, dance, or martial arts—appear to yield the greatest benefits for executive function and frontal region gray matter volume (Khan and Hillman, 2014). These activities stimulate not only the motor cortex but also the regions responsible for planning, attention, and social cognition. The developing brain may respond differently to exercise than the adult brain, with potentially greater plasticity and specific sensitive periods during which different brain regions are most responsive to activity-dependent changes (Tomporowski et al., 2008).

2.3.2 Age group: 18–64 years old

Adults can engage in various physical activities, including sports, games, leisure activities, housework, occupational tasks, walking, cycling, and structured exercise within their daily routines and obligations.

Adults should perform at least 150 minutes of moderate-intensity aerobic physical activity weekly, or 75 minutes of vigorous-intensity aerobic physical activity, or an equivalent combination, in bouts of at least ten minutes.

For additional health benefits, adults should increase to 300 minutes of moderate-intensity aerobic activity weekly, 150 minutes of vigorous-intensity aerobic activity, or an equivalent combination.

Muscle-strengthening activities involving major muscle groups should be performed at least twice weekly.

For adults, exercise consistency appears to be as important as total volume. Regular, moderate exercise distributed throughout the week may be more effective for maintaining gray matter volume than less frequent, condensed high-intensity sessions (Roe et al., 2021). This pattern may optimize neuroplasticity molecular cascades, keeping neurotrophic factor levels more consistently elevated. For working-age adults, integrating physical activity into daily routines—such as active commuting or using standing desks—may provide significant benefits by reducing prolonged sedentary periods, which independently negatively affect brain health (Dempsey et al., 2018).

2.3.3 Age group: 65 years and above

Older adults should engage in at least 150 minutes of moderate-intensity aerobic physical activity weekly, 75 minutes of vigorous-intensity aerobic physical activity, or an equivalent combination.

Aerobic activity should be performed in segments of at least 10 minutes.

For additional health benefits, older adults should increase their activity to 300 minutes of moderate-intensity aerobic exercise, 150 minutes of vigorous-intensity aerobic exercise, or an equivalent combination.

Those with limited mobility should participate in balance-improving and fall-prevention activities at least three times weekly.

Muscle-strengthening activities involving major muscle groups should be performed at least twice weekly.

Those unable to meet the recommended levels due to health conditions should be as physically active as their abilities permit.

For older adults, multimodal exercise programs combining aerobic, resistance, and balance training show particular promise for maintaining gray matter volume and cognitive function (Lipardo and Tsang, 2018). The resistance component appears especially important for older individuals, helping maintain muscle mass and function while stimulating myokine and growth factor release, which support brain health. Balance and coordination exercises engage the cerebellum and vestibular systems, potentially preserving gray matter in these regions (Rogge et al., 2018). For those with mobility limitations, adapted exercises performed in seated or supported positions can still yield significant brain health benefits. Even light activities like tai chi and gentle yoga have been shown to preserve gray matter volume in aging populations, particularly when practiced consistently over long periods (Mortimer et al., 2012).

Across all age groups, the progressive overload principle applies to exercise for brain health—gradually increasing intensity or duration maintains neuroplastic stimulus and avoids benefit plateauing. However, excessive exercise without adequate recovery may be counterproductive, potentially increasing inflammation and oxidative stress (Mee-Inta et al., 2019).

3 Sedentary behavior and its consequences

3.1 The effects of sedentary lifestyle on gray matter volume

Numerous studies connect sedentary behavior to detrimental effects on gray matter volume. Longer periods of inactivity are linked to reduced hippocampal, gray matter, and total brain volumes, especially in middle-aged individuals (Arvanitakis et al., 2023). Among overweight or obese children, television watching has been associated with reduced gray matter volume in specific brain regions, potentially impacting intellectual development (Zavala-Crichton et al., 2020). Research on older individuals has shown interactions between time spent in various activities and frontal lobe volume, which altered cognitive outcomes. In addition, sedentary behavior substantially correlates with lower whole-brain gray and white matter sizes in individuals with multiple sclerosis (Maasakkers et al., 2021).

The relationship between sedentary behavior and gray matter volume appears dose-dependent, with greater sedentary time associated with more pronounced volumetric reductions (Arnardottir et al., 2016). Research using accelerometry to objectively measure sedentary time demonstrates that each additional daily sedentary hour is associated with approximately 0.2% lower hippocampal volume in middle-aged adults, independent of physical activity levels (Falck et al., 2017). This suggests that sedentary behavior may represent a distinct brain health risk factor, not merely the absence of physical activity.

Sedentary behavior patterns matter—prolonged, uninterrupted sitting appears more detrimental than the same total sedentary time distributed in shorter bouts interspersed with activity periods (Wheeler et al., 2020). This mirrors metabolic health research findings and suggests that regular breaks in sedentary time may help mitigate negative impacts on gray matter volume. Mechanistically, these breaks may help maintain cerebral blood flow, reduce inflammation, and prevent extended periods of reduced neuronal activity.

Specific sedentary behavior types may have differential effects on brain structure. Screen-based sedentary activities, such as television viewing, show stronger negative associations with gray matter volume compared to reading or socializing while seated (Bakrania et al., 2018). This distinction may be related to the level of cognitive engagement in different sedentary activities—mentally stimulating seated activities may potentially offset some of the negative effects of physical inactivity through cognitive demand.

Regional specificity is another important consideration. The hippocampus, prefrontal cortex, and certain subcortical structures appear particularly vulnerable to sedentary behavior-induced volume reductions, while other brain regions show more resilience (Siddarth et al., 2018). This pattern corresponds to regions most responsive to physical activity, suggesting common underlying mechanisms in both exercise-induced volume increases and sedentary behavior-related decreases.

3.2 Risk of neurodegenerative diseases associated with sedentary behavior

Sedentary behavior, involving prolonged sitting or lying with low energy output, has become an alarming neurodegenerative disease risk factor. Research proves that individuals participating in "sit-all-day syndrome," such as excessive TV watching and office work, face a higher risk of chronic diseases, including Alzheimer's disease, Parkinson's disease, and dementia (Busse and Ramdharry, 2020; He et al., 2022). This is critical since sedentary behavior is commonly fostered in modern society, with an average sitting time of 6–7 hours daily.

The pathophysiological mechanisms linking sedentary behavior to neurodegenerative diseases are increasingly understood. Prolonged sitting reduces cerebral blood flow and glucose metabolism, potentially contributing to amyloid accumulation and tau hyperphosphorylation—hallmarks of Alzheimer's disease pathology (Wennberg et al., 2017). Sedentary behavior also increases systemic inflammation and oxidative stress, potentially compromising blood-brain barrier integrity and accelerating neurodegeneration (Broskey et al., 2014). In Parkinson's disease, sedentary behavior may exacerbate dopaminergic neuron loss through reduced trophic factor support and impaired mitochondrial function (LaHue et al., 2016).

Epidemiological studies have quantified these relationships, showing that each additional daily hour of sedentary behavior is associated with approximately a 12% increased dementia risk, even after adjusting for physical activity levels and other risk factors (Livingston et al., 2020). Longitudinal neuroimaging studies demonstrate that individuals with high sedentary time show accelerated gray matter atrophy, particularly in regions affected early in Alzheimer's disease, such as the entorhinal cortex and hippocampus (Reas et al., 2017).

Clinical studies reinforce the idea that immobility exacerbates neurodegenerative diseases. Studies on older adults with MCI and early-stage Alzheimer's disease-related cognitive decline show that physical activity interventions, such as aerobic exercise and resistance training, increase functional ability and improve cognition (Morris et al., 2017a; Broadhouse et al., 2020). Physical inactivity is correlated with an increased risk of heart attacks, type 2 diabetes, and all-cause mortality (Mir, 2021; Park et al., 2020).

Sedentary behavior's cognitive consequences extend beyond diagnosed neurodegenerative conditions. Even in ostensibly healthy adults, greater sedentary time is associated with poorer performance on executive function, processing speed, and memory tests—cognitive domains that rely heavily on frontal and temporal lobe gray matter, which is most affected by sedentary behavior (Falck et al., 2017). This suggests that sedentary behavior may contribute to subclinical cognitive changes long before diagnosable neurodegenerative disease emerges.

Prolonged sitting, combined with generally low physical activity, negatively affects not only physical health but also cognitive ability. Researchers are developing interventions to decrease sedentary time and increase physical activity (Keadle et al., 2017). These programs focus on environmental cues and principles of replacing sitting time with low- to moderate-intensity activities. Efforts aim to enhance awareness of the threats posed by inactive behavior and inspire greater activity and exercise engagement.

Genetic factors may moderate the relationship between sedentary behavior and neurodegenerative risk. APOE ε4 allele carriers—the strongest genetic risk factor for Alzheimer's disease—appear particularly vulnerable to the negative effects of sedentary behavior on gray matter volume and cognitive function (Smith et al., 2016). This suggests that reducing sedentary behavior may be especially important for individuals with genetic predispositions to neurodegenerative conditions, potentially offering a modifiable pathway to reduce genetic risk.

3.3 Counteracting sedentary damage

Recovering from sedentarism requires programs that mitigate the negative effects of long-term sitting. Evidence reveals that sitting for extended periods harms various aspects of health, including chronic illnesses, insulin resistance, vascular dysfunction, and decreased muscle tissue size and strength (Park et al., 2020; Wullems et al., 2016).

Emerging research suggests several promising approaches to counteract the negative effects of sedentary behavior on gray matter volume. "Exercise snacking"—brief bursts of physical activity interspersed throughout the day—may be particularly effective for breaking up sedentary time and maintaining cerebral blood flow (Dempsey et al., 2016). Even short activity breaks of 2–3 minutes every 30 minutes of sitting improve brain health markers, including BDNF levels and cerebrovascular function (Carter et al., 2018). These micro-bouts may be more feasible to implement in workplace and home settings than structured exercise sessions.

Environmental modifications represent another promising approach. Height-adjustable desks, active workstations, and environmental prompts to encourage movement can significantly reduce sedentary time in office and home settings (Healy et al., 2016). Wearable technology providing real-time feedback on sedentary behavior has shown efficacy in reducing sitting time, with preliminary evidence suggesting positive effects on cognitive function and brain health markers (Hartman et al., 2021).

Exercise prevents or limits sedentary bouts. Light-intensity physical activity leads to statistically significant improvements in body weight, waist circumference, blood pressure, blood glucose levels, and vascular function in adults and the elderly (Mir, n.d.).

Cognitive engagement during sedentary periods may partially mitigate negative effects on gray matter volume. Activities requiring complex thought, problem-solving, or social interaction appear to offer some protection against the neurological consequences of physical inactivity (Vemuri et al., 2014). This aligns with the cognitive reserve hypothesis, which suggests that mentally stimulating activities build neural resources that can compensate for negative influences on brain structure. However, cognitive engagement alone cannot fully offset the benefits of physical activity or the harms of prolonged sitting.

Implementing routine exercises amplifies antioxidant capability and DNA repair mechanism levels, preventing oxidative stress and promoting whole-body health. Regular physical activity upregulates endogenous antioxidant defenses, including superoxide dismutase, catalase, and glutathione peroxidase, which neutralize reactive oxygen species that can damage neuronal DNA and cellular components (Radak et al., 2005). Exercise also enhances DNA repair capacity through increased expression of repair enzymes like OGG1 (8-oxoguanine DNA glycosylase), which removes oxidatively damaged DNA bases (Radak et al., 2011). These molecular adaptations may partially explain exercise's neuroprotective effects against age-related gray matter decline and neurodegenerative processes.

4 MRI in physical activity research

MRI has revolutionized our understanding of the relationship between physical activity and brain health by providing noninvasive, high-resolution imaging of brain structure and function. Various MRI modalities offer complementary information: structural MRI quantifies gray matter volume and morphology, functional MRI detects activity-related changes in blood flow, diffusion tensor imaging assesses white matter integrity, and magnetic resonance spectroscopy measures neurochemical concentrations (Ten Brinke et al., 2015). Together, these techniques provide a comprehensive assessment of how physical activity and sedentary behavior influence brain health.

MRI especially helps in physical activity performance studies by evaluating physical changes within different body parts (Kujawa et al., 2023). Advanced MRI sequences can detect subtle alterations in tissue composition, microstructure, and metabolic activity that may precede macroscopic volumetric changes. Magnetization transfer imaging can identify changes in myelin content following exercise interventions, while arterial spin labeling (ASL) measures exercise-induced improvements in cerebral perfusion (Alfini et al., 2016). These techniques allow researchers to track progressive adaptations occurring in response to physical activity and identify the earliest indicators of sedentary behavior-related decline.

MRI neuroimaging technique applications in educational studies have been invaluable for educators examining psychological processes, helping to understand cognitive function, including physical activities and brain development, and their connection with educational processes, learning outcomes, and brain mechanisms (Haeger et al., 2019). In this context, MRI scans can evaluate measurable changes in muscle structure, bone health, and overall body composition to comprehend the influence of physical activity on health (Towse et al., 2005). Anatomical changes can be identified by MRI scans, which look for morphological differences in the brain and help detect functional changes such as brain metabolism, blood flow, and hormone production. These changes may involve zone- or whole-brain alterations in volume, water relaxation and diffusion, magnetization transfer coefficient changes, and hyperechogenicity visualizations through transcranial ultrasonography. Secondly, structural and functional imaging techniques, such as single-photon emission computed tomography and proton magnetic resonance spectroscopy, provide insights into regional metabolism and chemical receptor binding changes. High-field MRI can accurately pinpoint modifications in brain structures and functions related to physical activities and inactive lifestyles. Moreover, MRI technology is crucial in understanding how physical exercises affect brain function and connections, providing a means to examine the consequences of exercise on cognitive and mental health.

Previous research using MRI has consistently demonstrated the positive effects of physical activity on brain health. Among elderly people with small-vessel disease, MRI scans revealed a positive correlation between physical activity and enhanced white matter microstructural integrity, suggesting that regular exercise may protect against cognitive decline by maintaining white matter health (Gons et al., 2015). A multimodal MRI study on healthy young males showed a selective increase in blood flow to the hippocampus following exercise, indicating enhanced vascular function in this region. The findings suggest that just one week of exercise is sufficient to induce significant microstructural and vascular changes in the brain, providing insights into the temporal dynamics of neuroplasticity and the potential therapeutic benefits of exercise (Steventon et al., 2021). Though some studies indicate modest benefits of exercise on gray matter volume, the overall evidence remains inconclusive, with many studies showing neutral effects. Despite promising findings, the variability in results, particularly regarding gray matter volume, indicates that the relationship between physical activity and brain health is multifaceted and requires further exploration to fully understand its implications.

While MRI is a powerful tool for studying the effects of physical activity on the brain, it has limitations. MRI can be expensive and time-consuming, which limits accessibility. Some individuals may be unable to undergo MRI scans due to medical conditions or claustrophobia. In addition, MRI studies often involve relatively small sample sizes, which can affect the generalizability of findings. Technical challenges exist in standardizing image acquisition and analysis protocols across studies, making direct comparisons difficult. Partial volume effects can confound volumetric measurements, particularly in regions with complex morphology. Longitudinal studies face additional challenges related to scanner drift and upgrades, which can introduce systematic biases in volume measurements over time (Morey et al., 2010).

Recent methodological advances have begun to address these limitations. Automated segmentation algorithms using machine learning approaches have improved gray matter volume quantification reliability and efficiency (Akkus et al., 2017). Multisite harmonization techniques can now account for scanner differences, enabling larger, more diverse study populations. Ultra-high-field MRI (7 T and above) offers unprecedented spatial resolution, allowing visualization of hippocampal subfields and cortical layers that may be differentially affected by physical activity (Düzel et al., 2011). These technological developments promise to enhance our understanding of exercise-induced changes in gray matter volume spatiotemporal dynamics.

5 Exercise, physical activity and sedentary behavior

Physical activity has been shown to have positive effects on brain health, particularly cognition. Studies suggest that physical activity can increase brain volume, specifically in regions like the hippocampus, which is crucial for memory and learning (Erickson et al., 2011; Killgore et al., 2013). The temporal dynamics of these volume changes reveal important insights about brain plasticity—detectable hippocampal volume increases can occur within 6–8 weeks of beginning regular aerobic exercise, with continued growth over 6–12 months of sustained activity (Kleemeyer et al., 2016). These structural changes correlate with improved memory test performance, particularly in tasks involving spatial navigation and pattern separation (Hötting and Röder, 2013).

Beyond the hippocampus, exercise-induced volumetric increases have been observed in the prefrontal cortex, anterior cingulate, and parietal regions—areas involved in executive function, attention, and multisensory integration (Colcombe et al., 2006). The distribution of these changes suggests that physical activity benefits neural networks supporting higher-order cognitive processes rather than simply affecting primary sensory or motor regions. This may explain why exercise interventions often show the most pronounced effects on executive functions like working memory, cognitive flexibility, and inhibitory control (Erickson et al., 2019).

Physical exercise influences brain health by releasing molecules from peripheral organs that impact neuronal functions. Aerobic exercise is considered a potential therapeutic approach for nervous system diseases and may reveal molecular targets for pharmacological interventions (Thomas et al., 2012). The "exercise–brain axis" represents a complex communication system whereby peripheral signals generated during physical activity—including myokines, metabolites, and hormones—cross the blood-brain barrier to influence neural function (Pedersen, 2019). This systemic perspective helps explain how exercise benefits brain regions not directly involved in motor control and highlights potential therapeutic targets for pharmacological intervention when exercise is not feasible.

Maintaining an active lifestyle, both mentally and physically, is crucial for preventing cognitive decline in the elderly. Brain physiology is regulated by intricate and interrelated cellular connections in both healthy and diseased states, in response to physical activity.

Conversely, sedentary behavior is linked to poorer cognitive functions and an increased risk of late-life cognitive decline. Longitudinal studies demonstrate that high levels of sedentary behavior predict faster rates of cognitive decline and earlier onset of dementia symptoms, independent of physical activity levels (Kivipelto et al., 2018). The cognitive domains most affected by sedentary behavior—executive function, processing speed, and episodic memory—mirror those most enhanced by physical activity, suggesting reciprocal but independent mechanisms (Falck et al., 2017c).

Interactive effects between sedentary behavior and genetic risk factors have been observed, with APOE ε4 carriers showing greater cognitive vulnerability to prolonged sedentary time (Langa and Levine, 2014). Similarly, individuals with existing cerebrovascular disease or cardiometabolic risk factors experience more pronounced cognitive effects from sedentary behavior, potentially due to compromised cerebral perfusion and reduced metabolic reserve.

The evidence confirms the important relationship between exercise, physical activity, brain health, and cognition, underlining the need to endorse an active lifestyle. The dose-response relationship between physical activity and cognitive outcomes appears nonlinear, with the greatest benefits observed when transitioning from sedentary to light-to-moderate activity levels (Erickson et al., 2019). This suggests that even modest increases in physical activity may yield substantial cognitive benefits, particularly for previously inactive individuals.

Sitting and reclining for extended periods are linked to several poor physiological outcomes, including insulin resistance and vascular dysfunction (Wheeler et al., 2017). These metabolic and vascular consequences may mediate many negative effects of sedentary behavior on gray matter volume. Insulin resistance reduces BDNF expression and impairs neuroplasticity, while vascular dysfunction compromises cerebral blood flow as well as oxygen and nutrient delivery to neural tissue. Thus, sedentary behavior creates a neurobiological environment that is both less supportive of new neural growth and more conducive to neurodegeneration. Table 6.1 reviews the impact of physical activity on brain volume based on t.

TABLE 6.1 Studies showing the impact of physical activity on brain volume.

Study	Participants	Intervention	Outcomes	Key findings
Voss et al. (2013)	Older adults	6-month exercise intervention	Cognitive function, hippocampal volume	Improved cognitive function and increased hippocampal volume
Erickson et al. (2011)	Older adults	12-week aerobic exercise program	Memory performance, hippocampal volume	Increased hippocampal volume and improved memory performance
Arnardottir et al. (2016)	Older adults	N/A (observational)	GMV, WMV, TPA	Gray matter and white matter at both baseline and follow-up are independently associated with more TPA
Morris et al. (2017)	Older adults with early-stage AD-related cognitive decline	6 months of aerobic exercise versus nonaerobic stretching and training control	Composite scores of memory and executive function, hippocampal volume, and total gray matter volume	AEx improved functional ability in early-stage AD

Study	Participants	Intervention	Outcomes	Key findings
Broadhouse et al. (2020)	Older adults with MCI	Physical exercise and cognitive training	Hippocampus and subregion volume	Resistance training protects AD-prone hippocampal subfields over the long term, in addition to gradually improving cognition
Ortega et al. (2022)	Children with overweight or obesity	90 min of exercise	Intelligence, executive function, academic performance, and hippocampal volume	Intelligence and cognitive flexibility improve after 20 weeks of high-intensity exercise
Aslanyan et al. (2023)	Older adults	N/A (observational)	Aβ burden, brain volumes, sleep duration, and physical activity	Longer sleep duration and more aerobic physical activity were associated with better cognitive performance among study participants through lower Aβ load and larger regional volumes, respectively

N/A- Not Applicable; GMV- Grey Matter Volume; WMV- White Matter Volume; TPA- Total Physical Activity; Aβ- Amyloid beta; AD- Alzheimer's Disease; MCI- Mild Cognitive Impairment; AEx- Aerobic Exercise.

6 Clinical relevance and case studies

The relationship between physical activity, sedentary behavior, and gray matter volume has significant clinical implications across various populations. In clinical settings, physical activity interventions are increasingly recognized as valuable adjunctive treatments for brain health conditions. This section explores case studies and clinical applications, illustrating the translational relevance of research findings.

7 Case studies

7.1 Case study 1: Early Alzheimer's disease

A 72-year-old female with MCI and early Alzheimer's disease biomarkers (elevated amyloid-β on PET imaging) participated in a 6-month aerobic exercise program consisting of supervised treadmill walking three times weekly. Preintervention and postintervention MRI scans revealed a 2.3% hippocampal volume increase compared to a 1.3% decrease in matched controls receiving standard care (Morris et al., 2017). This volumetric preservation corresponded with stabilized episodic memory test performance and a reduced dementia conversion rate at the 18-month follow-up. This case exemplifies how exercise can potentially modify disease progression even after pathological processes have begun.

7.2 Case study 2: Major depressive disorder

A 34-year-old male with treatment-resistant depression and significant anhedonia showed reduced gray matter volume in the prefrontal cortex and anterior cingulate during the initial assessment. Following an 8-week high-intensity interval training program combined with cognitive behavioral therapy, MRI revealed partial prefrontal volume restoration and increased connectivity between prefrontal and limbic regions (Gujral et al., 2017). These neuroanatomical changes paralleled clinical improvement, with a 40% reduction in depression severity scores and enhanced response to antidepressant medication treatment. This case highlights the potential of exercise to augment traditional psychiatric treatments through neuroplastic mechanisms.

7.3 Case study 3: Pediatric obesity and executive function

Children (ages 8–11) with obesity and executive function deficits participated in a 9-month after-school physical activity program emphasizing enjoyable, moderate-to-vigorous activity. Compared to waitlist controls, the intervention group showed increased gray matter volume in the dorsolateral prefrontal cortex and superior parietal regions, corresponding with improved cognitive control and working memory task performance (Ortega et al., 2022). Parents and teachers reported reduced impulsivity and improved classroom behavior, suggesting real-world functional benefits of the observed structural changes.

7.4 Clinical applications and implementation

These cases illustrate how physical activity interventions can be tailored to specific clinical populations, with measurable effects on brain structure and function. Translating research findings into clinical practice requires considering several factors:

1. Prescription parameters: Exercise recommendations must be specific to the condition and individual, considering factors such as baseline fitness, comorbidities, and medication effects. Intensity, duration, frequency, and activity type should be carefully calibrated to maximize benefits while ensuring safety and adherence (Northey et al., 2018).
2. Integration with conventional treatments: Physical activity interventions are most effective when integrated with standard medical care rather than being positioned as alternative treatments. Exercise can enhance the efficacy of pharmacological interventions, potentially allowing for dose reductions and minimizing side effects.
3. Monitoring and progression: Regular assessment of both brain structure (through MRI when feasible) and cognitive/functional outcomes allows for intervention refinement and provides motivational feedback to patients. Wearable technology can facilitate activity monitoring between clinical visits.
4. Barriers and facilitators: Clinicians must address practical implementation barriers, including motivation, physical limitations, time constraints, and facility access. Behavioral strategies, such as goal setting, activity scheduling, and social support, can improve adherence.

The growing evidence base supporting the effects of exercise on gray matter volume provides a compelling rationale for integrating physical activity prescriptions into standard care protocols for brain health conditions. As neuroimaging becomes more accessible in clinical settings, structural brain changes may eventually serve as biomarkers for monitoring treatment responses and tailoring interventions to individual patients.

8 Technological advances in MRI and brain imaging

Recent technological innovations in MRI technology have dramatically enhanced our ability to detect and quantify exercise-related changes in gray matter volume. These advances span hardware improvements, sequence development, and analytical approaches, collectively enabling more precise characterization of the brain's response to physical activity and sedentary behavior.

8.1 Hardware innovations

Ultra-high field MRI systems (7 T and above) now provide unprecedented spatial resolution, allowing visualization of fine anatomical details previously undetectable at lower field strengths. These systems can resolve hippocampal subfields (CA1, CA3, dentate gyrus) that may respond differentially to exercise interventions (Yassa and Stark, 2011). Evidence suggests that the dentate gyrus—a site of adult neurogenesis—may be particularly responsive to physical activity, with volumetric changes detectable within weeks of initiating exercise.

Advanced receiver coil arrays with higher channel counts improve the signal-to-noise ratio and enable accelerated acquisition protocols. These developments make it feasible to conduct comprehensive multimodal brain assessments within clinically practical timeframes, facilitating larger-scale studies and longitudinal monitoring.

Motion correction technologies, including prospective motion correction and real-time feedback systems, reduce the impact of participant movement on image quality. This is particularly valuable when studying populations that may struggle with traditional MRI protocols, such as children, older adults with dementia, or individuals with movement disorders.

8.2 Sequence development

Multiparametric mapping techniques simultaneously quantify multiple tissue properties (T1, T2, T2*, proton density) in a single scan session, providing more comprehensive gray matter microstructure characterization than conventional volumetric approaches. These methods can detect subtle exercise-induced changes in myelin content, iron concentration, and water compartmentalization that may precede or underlie volumetric alterations.

ASL sequences noninvasively measure cerebral blood flow, allowing assessment of vascular adaptations to exercise that may support or drive changes in gray matter volume. Recent studies using ASL have demonstrated that exercise-induced increases in hippocampal perfusion predict subsequent volumetric growth, suggesting a mechanistic relationship between vascular and structural adaptations (Pereira et al., 2007).

Diffusion MRI techniques, including neurite orientation dispersion and density imaging and diffusion kurtosis imaging, provide measures of microstructural integrity and complexity within gray matter regions. These methods can detect exercise-related changes in dendritic density and orientation that are not apparent in conventional volumetric analyses.

8.3 Analytical advances

Machine learning approaches to image analysis have revolutionized the detection and quantification of exercise effects on brain structure. Convolutional neural networks (CNNs) can now segment brain regions with accuracy approaching expert manual tracing, but with greater efficiency and reproducibility. These automated methods reduce measurement error and facilitate larger dataset analyses, improving statistical power to detect intervention effects.

Multivariate pattern analysis techniques can identify distributed structural change patterns associated with physical activity, moving beyond region-specific analyses to capture network-level effects. These approaches have revealed that exercise may induce coordinated structural changes across functionally connected brain regions, suggesting system-level reorganization rather than isolated regional effects.

These technological advances collectively enable more precise characterization of how physical activity and sedentary behavior affect brain structure across the lifespan. As these methods become more widely available, they promise to enhance our understanding of the dose-response relationships, mechanistic pathways, and individual variability in brain responses to exercise interventions.

9 The role of artificial intelligence and deep learning in brain volume analysis

Artificial intelligence (AI) and deep learning approaches are transforming how researchers analyze and interpret the relationships between physical activity, sedentary behavior, and gray matter volume. These computational methods offer numerous advantages over traditional analytical approaches, including improved accuracy, efficiency, and the ability to detect complex patterns not discernible through conventional analyses.

9.1 Enhanced segmentation and volumetric analysis

Deep learning algorithms, particularly CNNs, have dramatically improved brain structure segmentation accuracy and reliability—a critical first step in volumetric analysis. Traditional segmentation methods often struggle with partial volume effects, individual anatomical variability, and image artifacts. In contrast, well-trained neural networks can achieve segmentation accuracy comparable to expert manual tracing while processing images orders of magnitude faster (Akkus et al., 2017).

Recent developments include networks specifically optimized for detecting exercise-related changes in key regions like the hippocampus. These specialized algorithms can identify subtle volumetric differences that might be missed by general-purpose segmentation tools, increasing sensitivity to intervention effects. Some approaches now incorporate longitudinal consistency constraints, reducing measurement variability and improving the detection of true biological change over time.

9.2 Pattern recognition and multivariate analysis

Beyond simple volumetric measurements, AI methods excel at identifying distributed structural change patterns associated with physical activity levels. Machine learning classifiers can distinguish between physically active and sedentary individuals' brain scans with accuracy exceeding 80%, based on subtle gray matter distribution patterns. These multivariate approaches capture system-level effects that may be more functionally relevant than isolated regional changes.

Deep learning models can integrate information across brain regions to predict individual fitness levels or activity patterns from structural MRI scans. This "brain-based biomarker" approach offers potential applications for monitoring exercise adherence and intervention efficacy without relying solely on self-reported activity measures.

9.3 Predictive modeling and personalized interventions

Perhaps the most promising AI application in this field is predicting individual responses to physical activity interventions. By analyzing baseline brain structure alongside demographic, genetic, and lifestyle factors, machine learning algorithms can forecast who is likely to show the greatest volumetric benefits from specific exercise programs (Rosano et al., 2017). This predictive capability could enable personalized exercise prescriptions tailored to individual brain health needs and response patterns.

AI models trained on longitudinal datasets can simulate potential brain trajectories under different activity scenarios, allowing researchers to estimate the long-term impacts of physical activity habits on brain aging. These computational models suggest that maintaining consistent physical activity throughout adulthood could delay age-related gray matter atrophy by 4–7 years in specific regions.

9.4 Challenges and limitations

Despite their promise, AI approaches face several challenges in this application domain. Most deep learning models require large training datasets that may not be available for specialized applications like exercise neuroimaging. Transfer learning and data augmentation techniques can partially address this limitation but may not fully capture the unique characteristics of exercise-related brain changes.

The "black box" nature of complex neural networks can make it difficult to interpret how specific features influence predictions, thereby limiting mechanistic insights. Emerging explainable AI techniques aim to increase transparency by identifying the imaging features that most strongly drive predictions, potentially revealing new biological patterns.

As AI methods mature and become more accessible, they promise to enhance both research capabilities and clinical applications in promoting brain health through physical activity.

10 Future directions and recommendations

10.1 Research priorities

10.1.1 Mechanistic studies

Future research should prioritize identifying precise cellular and molecular mechanisms underlying exercise-induced changes in gray matter volume. While neurogenesis, angiogenesis, and glial proliferation have been implicated, their relative contributions remain unclear. Multimodal imaging combined with peripheral biomarkers could help elucidate these mechanisms in humans, while parallel animal studies using techniques like single-cell RNA sequencing may provide complementary insights.

10.1.2 Dose-response relationships

More precise characterization of dose-response relationships between physical activity parameters (intensity, duration, frequency, type) and gray matter outcomes would enable exercise prescription optimization. Adaptive intervention designs that systematically vary exercise parameters could help identify minimum effective doses and optimal regimens for different populations.

10.1.3 Individual variability

The considerable heterogeneity in brain responses to physical activity warrants further investigation. Genetic, epigenetic, demographic, and baseline fitness factors likely moderate exercise effects on gray matter volume. Large-scale studies incorporating these variables could help identify who benefits most from specific activity patterns and why some individuals show minimal structural response despite adherence to exercise regimens.

10.2 Methodological recommendations

10.2.1 Standardized protocols

Development and adoption of standardized image acquisition, processing, and analysis protocols would enhance comparability across studies and facilitate data pooling. Initiatives like the Human Connectome Project have demonstrated the value of harmonized neuroimaging approaches; similar standards should be established specifically for exercise neuroimaging research.

10.2.2 Longitudinal designs

Future studies should emphasize longitudinal designs with multiple imaging time points to capture exercise-induced changes, temporal dynamics, and their reversibility upon cessation. Such designs can distinguish acute from chronic effects and characterize the structural adaptation time course more precisely than cross-sectional comparisons.

10.3 Translation and implementation

10.3.1 Clinical integration

Translating research findings into clinical practice requires collaboration among neuroscientists, exercise physiologists, and healthcare providers. The development of standardized exercise prescriptions for specific neurological and psychiatric conditions, incorporating neuroimaging biomarkers to monitor efficacy, would enhance clinical applications.

10.3.2 Community-based interventions

Scalable, community-based physical activity programs designed specifically to target brain health represent an important application of this research. Such initiatives should consider accessibility, adherence factors, and cost-effectiveness while incorporating principles derived from controlled research studies.

10.4 Emerging technologies and approaches

10.4.1 Remote assessment

Development of remote neuroimaging assessment methods, including portable MRI technology and surrogate markers measurable outside traditional imaging facilities, could democratize access to brain health monitoring and enable larger-scale implementation studies in diverse settings.

10.4.2 Digital health integration

Integration of neuroimaging biomarkers with digital health technologies, including wearable activity monitors, smartphone-based cognitive assessments, and telehealth platforms, could create comprehensive systems for monitoring and promoting brain health through physical activity.

By pursuing these future directions, researchers and clinicians can build upon the substantial existing evidence linking physical activity, sedentary behavior, and gray matter volume to develop more effective, personalized approaches to promoting brain health across the lifespan.

11 Conclusion

The utilization of MRI for cerebral gray matter volume quantification has elucidated the connection between brain well-being, sedentary lifestyle, and physical activity. Regular physical exercise has been linked to gray matter volume improvements, especially in areas related to memory, cognitive function, and sensory processing, as indicated by consistent research results. These associations can be noninvasively evaluated using MRI methodologies, offering comprehensive insights into brain activity and structure, thereby facilitating an understanding of cerebral alterations.

The convergence of evidence from epidemiological, interventional, and mechanistic studies establishes a compelling case for physical activity as a powerful modulator of brain structure and function. The relationship between movement behaviors and gray matter volume reflects the brain's remarkable plasticity and its responsiveness to environmental influences throughout the lifespan. While sedentary behavior emerges as a distinct risk factor for volumetric decline and associated cognitive impairments, regular physical activity offers a countervailing force that can preserve and even enhance neural tissues, particularly in regions vulnerable to age-related deterioration and neurodegenerative processes.

The clinical implications of these findings extend across numerous populations, from healthy individuals seeking cognitive optimization to those managing or at risk for neurological and psychiatric conditions. Exercise interventions tailored to specific needs and capabilities offer a nonpharmacological approach to promoting brain health that complements conventional treatments while minimizing adverse effects. The dose-response relationships, though still being refined, suggest that even modest increases in physical activity can yield meaningful neuroanatomical and functional benefits, particularly when sustained over time.

The findings can be used to develop personalized exercise programs tailored to enhance brain health in different populations, including individuals with cognitive impairments or neurological disorders. As our understanding of the complex interplay between physical activity patterns and brain structure deepens, several key considerations emerge for both research and application. First, intervention timing may be critical, with evidence suggesting that different life stages present unique opportunities for optimizing brain structure through physical activity. Early-life physical activity may establish neural reserves, providing resilience against future insults, while midlife exercise may help maintain these reserves, and late-life activity may partially restore lost volume or function. Second, integrating physical activity with other lifestyle factors—including nutrition, sleep quality, cognitive engagement, and social connection—likely produces synergistic effects on brain health that exceed those of physical activity alone.

In conclusion, advanced neuroimaging integration with detailed assessments of physical activity and sedentary behavior has substantially advanced our understanding of how movement patterns shape brain structure and function across the lifespan. The evidence strongly supports the promotion of regular physical activity and the reduction of prolonged sedentary time as modifiable lifestyle factors that can enhance brain health and potentially mitigate cognitive decline and the risk of neurological disorders. As research methodologies continue to evolve, our ability to characterize these relationships with greater precision will inform increasingly targeted interventions to optimize brain health through movement behaviors.

References

Akkus, Z., Galimzianova, A., Hoogi, A., Rubin, D.L., Erickson, B.J., 2017. Deep learning for brain MRI segmentation: state of the art and future directions. J. Digit. Imaging. 1618727X. 30 (4), 449–459. doi:10.1007/s10278-017-9983-4.

Alfini, A.J., Weiss, L.R., Leitner, B.P., Smith, T.J., Hagberg, J.M., Smith, J.C., 2016. Hippocampal and cerebral blood flow after exercise cessation in master athletes. Front. Aging Neurosci. 16634365. 8. Available from: https://journal.frontiersin.org/article/10.3389/fnagi.2016.00184/full. doi:10.3389/fnagi.2016.00184.

Arnardottir, N.Y., Koster, A., Van Domelen, D.R., Brychta, R.J., Caserotti, P., Eiriksdottir, G., Sverrisdottir, J.E., Sigurdsson, S., Johannsson, E., Chen, K.Y., Gudnason, V., Harris, T.B., Launer, L.J., Sveinsson, T., 2016. Association of change in brain structure to objectively measured physical activity and sedentary behavior in older adults: age, gene/environment susceptibility-Reykjavik study. Behav. Brain Res. 18727549. 296, 118–124. Available from: www.elsevier.com/locate/bbr. doi:10.1016/j.bbr.2015.09.005.

Arvanitakis, Z., Morris, A.W.J., Golimstok,, Launer, L.J., 2023. Sedentary behavior, brain-derived neurotrophic factor and brain structure in midlife: a longitudinal brain MRI sub-study of the coronary artery risk development in young adults study. Front. Dement. Available from: https://www.cardia.dopm.uab.edu/images/.

Aslanyan, V., Ortega, N., Fenton, L., Harrison, T.M., Raman, R., Mack, W.J., Pa, J., 2023. Protective effects of sleep duration and physical activity on cognitive performance are influenced by β-amyloid and brain volume but not tau burden among cognitively unimpaired older adults. Neuroimage Clin. 22131582. 39. Available from: https://www.journals.elsevier.com/neuroimage-clinical/. doi:10.1016/j.nicl.2023.103460.

Baker, L.D., Frank, L.L., Foster-Schubert, K., Green, P.S., Wilkinson, C.W., McTiernan, A., Plymate, S.R., Fishel, M.A., Watson, G.S., Cholerton, B.A., Duncan, G.E., Mehta, P.D., Craft, S., 2010. Effects of aerobic exercise on mild cognitive impairment: a controlled trial. Arch. Neurol. 15383687. 67 (1), 71–79. Available from: https://archneur.ama-assn.org/cgi/reprint/67/1/71. doi:10.1001/archneurol.2009.307UnitedStates.

Bakrania, K., Edwardson, C.L., Khunti, K., Bandelow, S., Davies, M.J., Yates, T., 2018. Associations between sedentary behaviors and cognitive function: cross-sectional and prospective findings from the UK Biobank. Am. J. Epidemiol. 0002-9262. 187 (3), 441–454. doi:10.1093/aje/kwx273.

Bamidis, P.D., Vivas, A.B., Styliadis, C., Frantzidis, C., Klados, M., Schlee, W., Siountas, A., Papageorgiou, S.G., 2014. A review of physical and cognitive interventions in aging. Neurosci. Biobehav. Rev. 01497634. 44, 206–220. doi:10.1016/j.neubiorev.2014.03.019.

Barha, C.K., Davis, J.C., Falck, R.S., Nagamatsu, L.S., Liu-Ambrose, T., 2017. Sex differences in exercise efficacy to improve cognition: a systematic review and meta-analysis of randomized controlled trials in older humans. Front. Neuroendocrinol. 10956808. 46, 71–85. Available from: https://www.elsevier.com/inca/publications/store/6/2/2/8/3/4/index.htt. doi:10.1016/j.yfrne.2017.04.002.

Baxi, M., Di Biase, M.A., Lyall, A.E., Cetin-Karayumak, S., Seitz, J., Ning, L., Makris, N., Rosene, D., Kubicki, M., Rathi, Y., 2020. Quantifying genetic and environmental influence on gray matter microstructure using diffusion MRI. Cereb. Cortex. 14602199. 30 (12), 6191–6205. Available from: https://cercor.oxfordjournals.org/. doi:10.1093/cerscor/bhaa174.

Broadhouse, K.M., Singh, M.F., Suo, C., Gates, N., Wen, W., Brodaty, H., Jain, N., Wilson, G.C., Meiklejohn, J., Singh, N., Baune, B.T., Baker, M., Foroughi, N., Wang, Y., Kochan, N., Ashton, K., Brown, M., Li, Z., Mavros, Y., Sachdev, P.S., Valenzuela, M.J., 2020. Hippocampal plasticity underpins long-term cognitive gains from resistance exercise in MCI. Neuroimage Clin. 22131582. 25. Available from: https://www.journals.elsevier.com/neuroimage-clinical/. doi:10.1016/j.nicl.2020.102182.

Broskey, N.T., Greggio, C., Boss, A., Boutant, M., Dwyer, A., Schlueter, L., Hans, D., Gremion, G., Kreis, R., Boesch, C., Canto, C., Amati, F., 2014. Skeletal muscle mitochondria in the elderly: effects of physical fitness and exercise training. J. Clin. Endocrinol. Metab. 19457197. 99 (5), 1852–1861. Available from: https://press.endocrine.org/doi/pdf/10.1210/jc.2013-3983. doi:10.1210/jc.2013-3983.

Busse, M., Ramdharry, G., 2020. Targeting sedentary behaviour in neurological disease. Pract. Neurol. 14747766. 20 (3), 187–188. Available from: https://pn.bmj.com/content/by/year. doi:10.1136/practneurol-2019-002491.

Carter, S.E., Draijer, R., Holder, S.M., Brown, L., Thijssen, D.H.J., Hopkins, N.D., 2018. Regular walking breaks prevent the decline in cerebral blood flow associated with prolonged sitting. J. Appl. Physiol. 15221601. 125 (3), 790–798. Available from: https://www.physiology.org/doi/pdf/10.1152/japplphysiol.00310.2018. doi:10.1152/japplphysiol.00310.2018.

Cheng, S.T., 2016. Cognitive reserve and the prevention of dementia: the role of physical and cognitive activities. Curr. Psychiatry Rep. 15351645. 18 (9). Available from: https://www.springerlink.com/content/1523-3812/. doi:10.1007/s11920-016-0721-2.

Colcombe, S.J., Erickson, K.I., Scalf, P.E., Kim, J.S., Prakash, R., McAuley, E., Elavsky, S., Marquez, D.X., Hu, L., Kramer, A.F., 2006. Aerobic exercise training increases brain volume in aging humans. J. Gerontol. A Biol. Sci. Med. Sci. 10795006. 61 (11), 1166–1170. Available from: https://biomed-gerontology.oxfordjournals.org/. doi:10.1093/gerona/61.11.1166.

De Sousa Fernandes, M.S., Ordônio, T.F., Santos, G.C.J., Santos, L.E.R., Calazans, C.T., Gomes, D.A., Santos, T.M., 2020. Effects of physical exercise on neuroplasticity and brain function: a systematic review in human and animal studies. Neural Plast. 16875443. 2020. Available from: https://www.hindawi.com/journals/np/. doi:10.1155/2020/8856621.

Deistung, A., Schweser, F., Reichenbach, J.R., 2017. Overview of quantitative susceptibility mapping. NMR Biomed. 10991492. 30 (4). Available from: https://onlinelibrary.wiley.com/journal/10.1002/(ISSN)1099-1492. doi:10.1002/nbm.3569.

Demirakca, T., Brusniak, W., Tunc-Skarka, N., Wolf, I., Meier, S., Matthäus, F., Ende, G., Schulze, T.G., Diener, C., 2014. Does body shaping influence brain shape? Habitual physical activity is linked to brain morphology independent of age. World J. Biol. Psychiatry. 18141412. 15 (5), 387–396. Available from: https://informahealthcare.com/wbp. doi:10.3109/15622975.2013.803600.

Dempsey, P.C., Larsen, R.N., Dunstan, D.W., Owen, N., Kingwell, B.A., 2018. Sitting less and moving more implications for hypertension. Hypertension. 15244563. 72 (5), 1037–1076. Available from: https://hyper.ahajournals.org/. doi:10.1161/HYPERTENSIONAHA.118.11190.

Dempsey, P.C., Larsen, R.N., Sethi, P., Sacre, J.W., Straznicky, N.E., Cohen, N.D., Cerin, E., Lambert, G.W., Owen, N., Kingwell, B.A., Dunstan, D.W., 2016. Benefits for type 2 diabetes of interrupting prolonged sitting with brief bouts of light walking or simple resistance activities. Diabetes Care. 19355548. 39 (6), 964–972. Available from: https://care.diabetesjournals.org/content/diacare/39/6/964.full.pdf. doi:10.2337/dc15-2336.

Dunton, G.F., 2017. Ecological momentary assessment in physical activity research. Exerc. Sport Sci. Rev. 15383008. 45 (1), 48–54. Available from: https://journals.lww.com/acsm-essr/pages/default.aspx. doi:10.1249/JES.0000000000000092.

Düzel, E., Schütze, H., Yonelinas, A.P., Heinze, H.J., 2011. Functional phenotyping of successful aging in long-term memory: preserved performance in the absence of neural compensation. Hippocampus. 10981063. 21 (8), 803–814.

Erickson, K.I., Hillman, C., Stillman, C.M., Ballard, R.M., Bloodgood, B., Conroy, D.E., Macko, R., Marquez, D.X., Petruzzello, S.J., Powell, K.E., 2019. Physical activity, cognition, and brain outcomes: a review of the 2018 physical activity guidelines. Med. Sci. Sports Exerc. 15300315. 51 (6), 1242–1251. Available from: https://journals.lww.com/acsm-msse/Pages/issuelist.aspx. doi:10.1249/MSS.0000000000001936.

Erickson, K.I., Leckie, R.L., Weinstein, A.M., 2014. Physical activity, fitness, and gray matter volume. Neurobiol. Aging. 15581497. 35 (2), S20. Available from: www.elsevier.com/locate/neuaging. doi:10.1016/j.neurobiolaging.2014.03.034.

Erickson, K.I., Voss, M.W., Prakash, R.S., Basak, C., Szabo, A., Chaddock, L., Kim, J.S., Heo, S., Alves, H., White, S.M., Wojcicki, T.R., Mailey, E., Vieira, V.J., Martin, S.A., Pence, B.D., Woods, J.A., McAuley, E., Kramer, A.F., 2011. Exercise training increases size of hippocampus and improves memory. Proc. Natl. Acad. Sci. U. S. A. 10916490. 108 (7), 3017–3022. Available from: https://www.pnas.org/content/108/7/3017.full.pdf+html. doi:10.1073/pnas.1015950108.

Falck, R.S., Davis, J.C., Liu-Ambrose, T., 2017. What is the association between sedentary behaviour and cognitive function? A systematic review. Br. J. Sports Med. 14730480. 51 (10), 800–811. Available from: https://bjsm.bmjjournals.com/. doi:10.1136/bjsports-2015-095551.

Gomes-Osman, J., Cabral, D.F., Morris, T.P., McInerney, K., Cahalin, L.P., Rundek, T., Oliveira, A., Pascual-Leone, A., 2018. Exercise for cognitive brain health in aging: a systematic review for an evaluation of dose. Neurol. Clin. Pract. 21630933. 8 (3), 257–265. Available from: https://cp.neurology.org/. doi:10.1212/CPJ.0000000000000460.

Gons, R.A., Tuladhar, A.M., de Laat, K.F., van Norden, A.G., van Dijk, E.J., Norris, D.G., Zwiers, M.P., de Leeuw, F.E., 2015. Physical activity is related to the structural integrity of cerebral white matter in older adults: the AGES-Reykjavik study. Brain Imaging Behav. 9 (4), 734–743.

Gujral, S., Aizenstein, H., Reynolds, C.F., Butters, M.A., Erickson, K.I., 2017. Exercise effects on depression: possible neural mechanisms. Gen. Hosp. Psychiatry. 18737714. 49, 2–10. Available from: www.elsevier.com/locate/genhospsych. doi:10.1016/j.genhosppsych.2017.04.012.

Haeger, A., Costa, A.S., Schulz, J.B., Reetz, K., 2019. Cerebral changes improved by physical activity during cognitive decline: a systematic review on MRI studies. Neuroimage Clin. 22131582. 23. Available from: https://www.journals.elsevier.com/neuroimage-clinical/. doi:10.1016/j.nicl.2019.101933.

Hartman, Y.A.W., Tillmans, L.C.M., Benschop, D.L., Hermans, A.N.L., Nijssen, K.M.R., Eijsvogels, T.M.H., Willems, P.H.G.M., Tack, C.J., Hopman, M.T.E., Claassen, J.A.H.R., Thijssen, D.H.J., 2021. Long-term and acute benefits of reduced sitting on vascular flow and function. Med. Sci. Sports Exerc. 15300315. 53 (2), 341–350. Available from: https://journals.lww.com/acsm-msse/Pages/issuelist.aspx. doi:10.1249/MSS.0000000000002462.

He, Q., Bennett, A.N., Fan, B., Han, X., Liu, J., Wu, K.C.H., Huang, R., Chan, J.C.N., Chan, K.H.K., 2022. Assessment of bidirectional relationships between leisure sedentary behaviors and neuropsychiatric disorders: a two-sample mendelian randomization study. Genes. 20734425. 13 (6). Available from: https://www.mdpi.com/2073-4425/13/6/962/pdf?version=1653644975. doi:10.3390/genes13060962.

Healy, G.N., Eakin, E.G., Owen, N., LaMontagne, A.D., Moodie, M., Winkler, E.A.H., Fjeldsoe, B.S., Wiesner, G., Willenberg, L., Dunstan, D.W., 2016. A cluster randomized controlled trial to reduce office workers' sitting time: effect on activity outcomes. Med. Sci. Sports Exerc. 15300315. 48 (9), 1787–1797. Available from: https://www.lww.com/product/?0195-9131. doi:10.1249/MSS.0000000000000972.

Hoogman, M., Bralten, J., Hibar, D.P., Mennes, M., Zwiers, M.P., Schweren, L.S.J., van Hulzen, K.J.E., Medland, S.E., Shumskaya, E., Jahanshad, N., Zeeuw, P.D., Szekely, E., Sudre, G., Wolfers, T., Onnink, A.M.H., Dammers, J.T., Mostert, J.C., Vives-Gilabert, Y., Kohls, G., Oberwelland, E., Seitz, J., Schulte-Rüther, M., Ambrosino, S., Doyle, A.E., Høvik, M.F., Dramsdahl, M., Tamm, L., van Erp, T.G.M., Dale, A., Schork, A., Conzelmann, A., Zierhut, K., Baur, R., McCarthy, H., Yoncheva, Y.N., Cubillo, A., Chantiluke, K., Mehta, M.A., Paloyelis, Y., Hohmann, S., Baumeister, S., Bramati, I., Mattos, P., Tovar-Moll, F., Douglas, P., Banaschewski, T., Brandeis, D., Kuntsi, J., Asherson, P., Rubia, K., Kelly, C., Martino, A.D., Milham, M.P., Castellanos, F.X., Frodl, T., Zentis, M., Lesch, K.P., Reif, A., Pauli, P., Jernigan, T.L., Haavik, J., Plessen, K.J., Lundervold, A.J., Hugdahl, K., Seidman, L.J., Biederman, J., Rommelse, N., Heslenfeld, D.J., Hartman, C.A., Hoekstra, P.J., Oosterlaan, J., Polier, G.V., Konrad, K., Vilarroya, O., Ramos-Quiroga, J.A., Soliva, J.C., Durston, S., Buitelaar, J.K., Faraone, S.V., Shaw, P., Thompson, P.M., Franke, B., 2017. Subcortical brain volume differences in participants with attention deficit hyperactivity disorder in children and adults: a cross-sectional mega-analysis. Lancet Psychiatry. 22150374. 4 (4), 310–319. doi:10.1016/S2215-0366(17)30049-4.

Hörder, H., Johansson, L., Guo, X., Grimby, G., Kern, S., Östling, S., Skoog, I., 2018. Midlife cardiovascular fitness and dementia. Neurology. 1526632X. 90 (15), e1298. Available from: https://www.neurology.org. doi:10.1212/WNL.0000000000005290.

Hötting, K., Röder, B., 2013. Beneficial effects of physical exercise on neuroplasticity and cognition. Neurosci. Biobehav. Rev. 01497634. 37 (9), 2243–2257. doi:10.1016/j.neubiorev.2013.04.005.

Jack, C.R., Wiste, H.J., Weigand, S.D., Knopman, D.S., Vemuri, P., Mielke, M.M., Lowe, V., Senjem, M.L., Gunter, J.L., Machulda, M.M., Gregg, B.E., Pankratz, V.S., Rocca, W.A., Petersen, R.C., 2015. Age, sex, and APOE ε4 effects on memory, brain structure, and β-Amyloid across the adult life span. JAMA Neurol. 21686149. 72 (5), 511–519. Available from: https://archneur.jamanetwork.com/article.aspx?articleid=2193880. doi:10.1001/jamaneurol.2014.4821.

Keadle, S.K., Conroy, D.E., Buman, M.P., Dunstan, D.W., Matthews, C.E., 2017. Targeting reductions in sitting time to increase physical activity and improve health. Med. Sci. Sports Exerc. 15300315. 49 (8), 1572–1582. Available from: https://www.lww.com/product/?0195-9131. doi:10.1249/MSS.0000000000001257.

Khan, N.A., Hillman, C.H., 2014. The relation of childhood physical activity and aerobic fitness to brain function and cognition: a review. Pediatr. Exerc. Sci. 15432920. 26 (2), 138–146. Available from: https://journals.humankinetics.com/AcuCustom/Sitename/Documents/DocumentItem/04%20Khan%20PES_2013_0125_138-146-ej.pdf. doi:10.1123/pes.2013-0125.

Killgore, W.D.S., Olson, E.A., Weber, M., 2013. Physical exercise habits correlate with gray matter volume of the hippocampus in healthy adult humans. Sci. Rep. 20452322. 3. doi:10.1038/srep03457.

Kivipelto, M., Mangialasche, F., Ngandu, T., 2018. Lifestyle interventions to prevent cognitive impairment, dementia and Alzheimer disease. Nat. Rev. Neurol. 1759-4758. 14 (11), 653–666. doi:10.1038/s41582-018-0070-3.

Kleemeyer, M.M., Kühn, S., Prindle, J., Bodammer, N.C., Brechtel, L., Garthe, A., Kempermann, G., Schaefer, S., Lindenberger, U., 2016. Changes in fitness are associated with changes in hippocampal microstructure and hippocampal volume among older adults. NeuroImage. 10959572. 131, 155–161. Available from: https://www.elsevier.com/inca/publications/store/6/2/2/9/2/5/index.htt. doi:10.1016/j.neuroimage.2015.11.026.

Kujawa, M.J., Marcinkowska, A.B., Grzywińska, M., Waśkow, M., Romanowski, A., Szurowska, E., Winklewski, P.J., Szarmach, A., 2023. Physical activity and the brain myelin content in humans. Front. Cell. Neurosci. 16625102. 17. Available from: https://www.frontiersin.org/journals/cellular-neuroscience. doi:10.3389/fncel.2023.1198657.

LaHue, S.C., Comella, C.L., Tanner, C.M., 2016. The best medicine? The influence of physical activity and inactivity on Parkinson's disease. Mov. Disord. 15318257. 31 (10), 1444–1454. Available from: https://onlinelibrary.wiley.com/journal/10.1002/(ISSN)1531-8257. doi:10.1002/mds.26728.

Langa, K.M., Levine, D.A., 2014. The diagnosis and management of mild cognitive impairment: a clinical review. JAMA. 15383598. 312 (23), 2551–2561. Available from: https://jama.jamanetwork.com/article.aspx?articleid=2040164. doi:10.1001/jama.2014.13806.

Lentoor, A.G., Motsamai, T.B., Nxiweni, T., Mdletshe, B., Mdingi, S., 2023. Protocol for a systematic review of the effects of gardening physical activity on neuroplasticity and cognitive function. AIMS Neurosci. 23737972. 10 (2), 118–129. Available from: https://www.aimspress.com/journal/neuroscience. doi:10.3934/Neuroscience.2023009.

Li, C.S., Chen, Y., Ide, J.S., 2022. Gray matter volumetric correlates of attention deficit and hyperactivity traits in emerging adolescents. Sci. Rep. 20452322. 12 (1). Available from: www.nature.com/srep/index.html. doi:10.1038/s41598-022-15124-7.

Lipardo, D.S., Tsang, W.W.N., 2018. Effects of combined physical and cognitive training on fall prevention and risk reduction in older persons with mild cognitive impairment: a randomized controlled study. J. Alzheimers Dis. 63 (2), 651–660.

Liu, P.Z., Nusslock, R., 2018. Exercise-mediated neurogenesis in the hippocampus via BDNF. Front. Neurosci. 1662453X. 12. Available from: https://www.frontiersin.org/articles/10.3389/fnins.2018.00052/full. doi:10.3389/fnins.2018.00052.

Livingston, G., Huntley, J., Sommerlad, A., Ames, D., Ballard, C., Banerjee, S., Brayne, C., Burns, A., Cohen-Mansfield, J., Cooper, C., Costafreda, S.G., Dias, A., Fox, N., Gitlin, L.N., Howard, R., Kales, H.C., Kivimäki, M., Larson, E.B., Ogunniyi, A., Orgeta, V., Ritchie, K., Rockwood, K., Sampson, E.L., Samus, Q., Schneider, L.S., Selbæk, G., Teri, L., Mukadam, N., 2020. Dementia prevention, intervention, and care: 2020 report of the Lancet Commission. Lancet. 01406736. 396 (10248), 413–446. doi:10.1016/s0140-6736(20)30367-6.

Maasakkers, C.M., Thijssen, D.H.J., Knight, S.P., Newman, L., O'Connor, J.D., Scarlett, S., Carey, D., Buckley, A., McMorrow, J.P., Leidhin, C.N., Feeney, J., Melis, R.J.F., Kenny, R.A., Claassen, J.A.H.R., Looze, C.D., 2021. Hemodynamic and structural brain measures in high and low sedentary older adults. J. Cereb. Blood Flow Metab. 15597016. 41 (10), 2607–2616. Available from: https://journals.sagepub.com/loi/jcb. doi:10.1177/0271678X211009382.

Marusiak, J., Żeligowska, E., Mencel, J., Kisiel-Sajewicz, K., Majerczak, J., Zoladz, J., Jaskólska, A., Jaskólska, A., 2015. Interval training-induced alleviation of rigidity and hypertonia in patients with Parkinson's disease is accompanied by increased basal serum brain-derived neurotrophic factor. J. Rehabil. Med. 1650-1977. 47 (4), 372–375. doi:10.2340/16501977-1931.

McGee, S.L., Hargreaves, M., 2019. Epigenetics and exercise. Trends Endocrinol. Metab. 18793061. 30 (9), 636–645. Available from: www.elsevier.com/locate/tem. doi:10.1016/j.tem.2019.06.002.

Mee-Inta, O., Zhao, Z.W., Kuo, Y.M., 2019. Physical exercise inhibits inflammation and microglial activation. Cells. 20734409. 8 (7). Available from: https://www.mdpi.com/2073-4409/8/7/691/pdf. doi:10.3390/cells8070691.

Mir, I.A., 2021. Sedentary Behavior, Cardiovascular Risk and Importance of Physical Activity and Breaking-up Sedentary Behavior. Sage Journals. Available from: www.intechopen.com.

Morey, R.A., Selgrade, E.S., Wagner, H.R., Huettel, S.A., Wang, L., McCarthy, G., 2010. Scan-rescan reliability of subcortical brain volumes derived from automated segmentation. Hum. Brain Mapp. 10970193. 31 (11), 1751–1762. Available from: https://onlinelibrary.wiley.com/doi/10.1002/hbm.20973/pdf. doi:10.1002/hbm.20973.

Morris, J.K., Vidoni, E.D., Johnson, D.K., Van Sciver, A., Mahnken, J.D., Honea, R.A., Wilkins, H.M., Brooks, W.M., Billinger, S.A., Swerdlow, R.H., Burns, J.M., 2017. Aerobic exercise for Alzheimer's disease: a randomized controlled pilot trial. PLoS ONE. 19326203. 12 (2). Available from: https://journals.plos.org/plosone/article/file?id=10.1371/journal.pone.0170547&type=printable. doi:10.1371/journal.pone.0170547.

Mortimer, J.A., Ding, D., Borenstein, A.R., Decarli, C., Guo, Q., Wu, Y., Zhao, Q., Chu, S., 2012. Changes in brain volume and cognition in a randomized trial of exercise and social interaction in a community-based sample of non-demented Chinese elders. J. Alzheimers Dis. 18758908. 30 (4), 757–766. Available from: www.iospress.nl/journal/journal-of-alzheimers-disease/. doi:10.3233/JAD-2012-120079.

Northey, J.M., Cherbuin, N., Pumpa, K.L., Smee, D.J., Rattray, B., 2018. Exercise interventions for cognitive function in adults older than 50: a systematic review with meta-analysis. Br. J. Sports Med. 0306-3674. 52 (3), 154–160. doi:10.1136/bjsports-2016-096587.

Ortega, F.B., Mora-Gonzalez, J., Cadenas-Sanchez, C., Esteban-Cornejo, I., Migueles, J.H., Solis-Urra, P., Verdejo-Román, J., Rodriguez-Ayllon, M., Molina-Garcia, P., Ruiz, J.R., Martinez-Vizcaino, V., Hillman, C.H., Erickson, K.I., Kramer, A.F., Labayen, I., Catena, A., 2022. Effects of an exercise program on brain health outcomes for children with overweight or obesity: the ActiveBrains randomized clinical trial. JAMA Netw. Open. 25743805. 5 (8), e2227893. Available from: https://jamanetwork.com/journals/jamanetworkopen. doi:10.1001/jamanetworkopen.2022.27893.

Paillard, T., Rolland, Y., de Barreto, P.S., 2015. Protective effects of physical exercise in Alzheimer's disease and Parkinson's disease: a narrative review. J. Clin. Neurol. 20055013. 11 (3), 212–219. Available from: https://www.thejcn.com/Synapse/Data/PDFData/0145JCN/jcn-11-212.pdf. doi:10.3988/jcn.2015.11.3.212.

Park, J.H., Moon, J.H., Kim, H.J., Kong, M.H., Oh, Y.H., 2020. Sedentary lifestyle: overview of updated evidence of potential health risks. Korean J. Fam. Med. 20926715. 41 (6), 365–373. Available from: https://koreamed.org/JournalVolume.php?id=1&code=0001KJFM. doi:10.4082/KJFM.20.0165.

Pedersen, B.K., 2019. Physical activity and muscle–brain crosstalk. Nat. Rev. Endocrinol. 17595037. 15 (7), 383–392. Available from: https://www.nature.com/nrendo/index.html. doi:10.1038/s41574-019-0174-x.

Pereira, A.C., Huddleston, D.E., Brickman, A.M., Sosunov, A.A., Hen, R., McKhann, G.M., Sloan, R., Gage, F.H., Brown, T.R., Small, S.A., 2007. An in vivo correlate of exercise-induced neurogenesis in the adult dentate gyrus. Proc. Natl. Acad. Sci. U. S. A. 00278424. 104 (13), 5638–5643. doi:10.1073/pnas.0611721104.

Phillips, C., 2017. Lifestyle modulators of neuroplasticity: how physical activity, mental engagement, and diet promote cognitive health during aging. Neural Plast. 16875443. 2017. Available from: https://www.hindawi.com/journals/np/. doi:10.1155/2017/3589271.

Radak, Z., Bori, Z., Koltai, E., Fatouros, I.G., Jamurtas, A.Z., Douroudos, I.I., Terzis, G., Nikolaidis, M.G., Chatzinikolaou, A., Sovatzidis, A., Kumagai, S., Naito, H., Boldogh, I., 2011. Age-dependent changes in 8-oxoguanine-DNA glycosylase activity are modulated by adaptive responses to physical exercise in human skeletal muscle. Free Radic. Biol. Med. 18734596. 51 (2), 417–423. doi:10.1016/j.freeradbiomed.2011.04.018.

Radak, Z., Chung, H.Y., Goto, S., 2005. Exercise and hormesis: oxidative stress-related adaptation for successful aging. Biogerontology. 15736768. 6 (1), 71–75. doi:10.1007/s10522-004-7386-7.

Reas, E.T., Laughlin, G.A., Bergstrom, J., Kritz-Silverstein, D., Barrett-Connor, E., McEvoy, L.K., 2017. Effects of sex and education on cognitive change over a 27-year period in older adults: the Rancho Bernardo study. Am. J. Geriatr. Psychiatry. 15457214. 25 (8), 889–899. Available from: https://www.sciencedirect.com/science/journal/10647481. doi:10.1016/j.jagp.2017.03.008.

Roe, J.M., Vidal-Piñeiro, D., Sørensen, Ø., Brandmaier, A.M., Düzel, S., Gonzalez, H.A., Kievit, R.A., Knights, E., Kühn, S., Lindenberger, U., Mowinckel, A.M., Nyberg, L., Park, D.C., Pudas, S., Rundle, M.M., Walhovd, K.B., Fjell, A.M., Westerhausen, R., Masters, C.L., Bush, A.I., Fowler, C., Darby, D., Pertile, K., Restrepo, C., Roberts, B., Robertson, J., Rumble, R., Ryan, T., Collins, S., Thai, C., Trounson, B., Lennon, K., Li, Q.X., Ugarte, F.Y., Volitakis, I., Vovos, M., Williams, R., Baker, J., Russell, A., Peretti, M., Milicic, L., Lim, L., Rodrigues, M., Taddei, K., Taddei, T., Hone, E., Lim, F., Fernandez, S., Rainey-Smith, S., Pedrini, S., Martins, R., Doecke, J., Bourgeat, P., Fripp, J., Gibson, S., Leroux, H., Hanson, D., Dore, V., Zhang, P., Burnham, S., Rowe, C.C., Villemagne, V.L., Yates, P., Pejoska, S.B., Jones, G., Ames, D., Cyarto, E., Lautenschlager, N., Barnham, K., Cheng, L., Hill, A., Killeen, N., Maruff, P., Silbert, B., Brown, B., Sohrabi, H., Savage, G., Vacher, M., 2021. Asymmetric thinning of the cerebral cortex across the adult lifespan is accelerated in Alzheimer's disease. Nat. Commun. 20411723. 12 (1). Available from: https://www.nature.com/ncomms/index.html. doi:10.1038/s41467-021-21057-y.

Rogge, A.K., Röder, B., Zech, A., Hötting, K., 2018. Exercise-induced neuroplasticity: balance training increases cortical thickness in visual and vestibular cortical regions. NeuroImage. 10959572. 179, 471–479. Available from: https://www.elsevier.com/inca/publications/store/6/2/2/9/2/5/index.htt. doi:10.1016/j.neuroimage.2018.06.065.

Rosano, C., Guralnik, J., Pahor, M., Glynn, N.W., Newman, A.B., Ibrahim, T.S., Erickson, K., Cohen, R., Shaaban, C.E., MacCloud, R.L., Aizenstein, H.J., 2017. Hippocampal response to a 24-month physical activity intervention in sedentary older adults. Am. J. Geriatr. Psychiatry. 15457214. 25 (3), 209–217. Available from: https://www.sciencedirect.com/science/journal/10647481. doi:10.1016/j.jagp.2016.11.007.

Sanchis-Segura, C., Ibañez-Gual, M.V., Adrián-Ventura, J., Aguirre, N., Gómez-Cruz, Á.J., Avila, C., Forn, C., 2019. Sex differences in gray matter volume: how many and how large are they really? Biol. Sex Differ. 20426410. 10 (1). Available from: https://www.bsd-journal.com/. doi:10.1186/s13293-019-0245-7.

Siddarth, P., Burggren, A.C., Eyre, H.A., Small, G.W., Merrill, D.A., Ginsberg, S.D., 2018. Sedentary behavior associated with reduced medial temporal lobe thickness in middle-aged and older adults. PLOS ONE. 1932-6203. 13 (4), e0195549. doi:10.1371/journal.pone.0195549.

Smith, J.C., Lancaster, M.A., Nielson, K.A., Woodard, J.L., Seidenberg, M., Durgerian, S., Sakaie, K., Rao, S.M., 2016. Interactive effects of physical activity and APOE-ε4 on white matter tract diffusivity in healthy elders. NeuroImage. 10959572. 131, 102–112. Available from: https://www.elsevier.com/inca/publications/store/6/2/2/9/2/5/index.htt. doi:10.1016/j.neuroimage.2015.08.007.

Sølvsten, C.A.E., De Paoli, F., Christensen, J.H., Nielsen, A.L., 2018. Voluntary physical exercise induces expression and epigenetic remodeling of VegfA in the rat hippocampus. Mol. Neurobiol. 15591182. 55 (1), 567–582. Available from: https://www.springer.com/biomed/neuroscience/journal/12035. doi:10.1007/s12035-016-0344-y.

Steventon, J.J., Chandler, H.L., Foster, C., Dingsdale, H., Germuska, M., Massey, T., Parker, G., Wise, R.G., Murphy, K., 2021. Changes in white matter microstructure and MRI-derived cerebral blood flow after 1-week of exercise training. Sci. Rep. 20452322. 11 (1). Available from: www.nature.com/srep/index.html. doi:10.1038/s41598-021-01630-7.

Stillman, C.M., Cohen, J., Lehman, M.E., Erickson, K.I., 2016. Mediators of physical activity on neurocognitive function: a review at multiple levels of analysis. Front. Hum. Neurosci. 16625161. 10. Available from: https://journal.frontiersin.org/article/10.3389/fnhum.2016.00626/full. doi:10.3389/fnhum.2016.00626.

Tamnes, C.K., Herting, M.M., Goddings, A.L., Meuwese, R., Blakemore, S.J., Dahl, R.E., Güroğlu, B., Raznahan, A., Sowell, E.R., Crone, E.A., Mills, K.L., 2017. Development of the cerebral cortex across adolescence: a multisample study of inter-related longitudinal changes in cortical volume, surface area, and thickness. J. Neurosci. 15292401. 37 (12), 3402–3412. Available from: https://www.jneurosci.org/content/jneuro/37/12/3402.full.pdf. doi:10.1523/JNEUROSCI.3302-16.2017.

Teixeira, C.V.L., Rezende, T.J.R., Weiler, M., Nogueira, M.H., Campos, B.M., Pegoraro, L.F.L., Vicentini, J.E., Scriptore, G., Cendes, F., Balthazar, M.L.F., 2016. Relation between aerobic fitness and brain structures in amnestic mild cognitive impairment elderly. Age. 15744647. 38 (3). Available from: https://www.springerlink.com/openurl.asp?genre=journal&eissn=1574-4647. doi:10.1007/s11357-016-9912-3.

Ten Brinke, L.F., Bolandzadeh, N., Nagamatsu, L.S., Hsu, C.L., Davis, J.C., Miran-Khan, K., Liu-Ambrose, T., 2015. Aerobic exercise increases hippocampal volume in older women with probable mild cognitive impairment: a 6-month randomised controlled trial. Br. J. Sports Med. 14730480. 49 (4), 248–254. Available from: https://bjsm.bmj.com/content/49/4/248.full.pdf+html. doi:10.1136/bjsports-2013-093184.

Thomas, A.G., Dennis, A., Bandettini, P.A., Johansen-Berg, H., 2012. The effects of aerobic activity on brain structure. Front. Psychol. 16641078. 3. Available from: https://www.frontiersin.org/Journal/DownloadFile.ashx?pdf=1&FileId=2049&articleId=19477&Version=1&ContentTypeId=21&FileName=fpsyg-03-00086.pdf. doi:10.3389/fpsyg.2012.00086.

Tomporowski, P.D., Davis, C.L., Miller, P.H., Naglieri, J.A., 2008. Exercise and children's intelligence, cognition, and academic achievement. Educ. Psychol. Rev. 1573336X. 20 (2), 111–131. doi:10.1007/s10648-007-9057-0.

Towse, T.F., Slade, J.M., Meyer, R.A., 2005. Effect of physical activity on MRI-measured blood oxygen level-dependent transients in skeletal muscle after brief contractions. J. Appl. Physiol. 99 (2), 715–722. doi:10.1152/japplphysiol.00272.2005.

Van Praag, H., Shubert, T., Zhao, C., Gage, F.H., 2005. Exercise enhances learning and hippocampal neurogenesis in aged mice. J. Neurosci. 02706474. 25 (38), 8680–8685. doi:10.1523/JNEUROSCI.1731-05.2005.

van Praag, H., Wrann, C.D., 2022. Special issue: peripheral factors and neural plasticity. Brain Plast. 2213-6304. 8 (1), 1–3. doi:10.3233/bpl-229001.

Vauzour, D., Camprubi-Robles, M., Miquel-Kergoat, S., Andres-Lacueva, C., Bánáti, D., Barberger-Gateau, P., Bowman, G.L., Caberlotto, L., Clarke, R., Hogervorst, E., Kiliaan, A.J., Lucca, U., Manach, C., Minihane, A.-M., Mitchell, E.S., Perneczky, R., Perry, H., Roussel, A.-M., Schuermans, J., Sijben, J., Spencer, J.P.E., Thuret, S., van de Rest, O., Vandewoude, M., Wesnes, K., Williams, R.J., Williams, R.S.B., Ramirez, M., 2017. Nutrition for the ageing brain: towards evidence for an optimal diet. Ageing Res. Rev. 15681637. 35, 222–240. doi:10.1016/j.arr.2016.09.010.

Vaynman, S., Ying, Z., Gomez-Pinilla, F., 2004. Hippocampal BDNF mediates the efficacy of exercise on synaptic plasticity and cognition. Eur. J. Neurosci. 0953-816X. 20 (10), 2580–2590. doi:10.1111/j.1460-9568.2004.03720.x.

Vemuri, P., Lesnick, T.G., Przybelski, S.A., Machulda, M., Knopman, D.S., Mielke, M.M., Roberts, R.O., Geda, Y.E., Rocca, W.A., Petersen, R.C., Jack, C.R., 2014. Association of lifetime intellectual enrichment with cognitive decline in the older population. JAMA Neurol. 21686149. 71 (8), 1017–1024. Available from: https://archneur.jamanetwork.com/article.aspx?articleid=1883334. doi:10.1001/jamaneurol.2014.963.

Voss, M.W., Heo, S., Prakash, R.S., Erickson, K.I., Alves, H., Chaddock, L., Szabo, A.N., Mailey, E.L., Wójcicki, T.R., White, S.M., Gothe, N., Mcauley, E., Sutton, B.P., Kramer, A.F., 2013. The influence of aerobic fitness on cerebral white matter integrity and cognitive function in older adults: results of a one-year exercise intervention. Hum. Brain Mapp. 10970193. 34 (11), 2972–2985. doi:10.1002/hbm.22119.

Wang, Y., Liu, X., Hu, Y., Yu, Z., Wu, T., Wang, J., Liu, J., Liu, J., 2022. Impaired functional network properties contribute to white matter hyperintensity related cognitive decline in patients with cerebral small vessel disease. BMC Med. Imaging. 14712342. 22 (1). Available from: https://www.biomedcentral.com/bmcmedimaging/. doi:10.1186/s12880-022-00769-7.

Wennberg, A.M.V., Savica, R., Mielke, M.M., 2017. Association between various brain pathologies and gait disturbance. Dement. Geriatr. Cogn. Disord. 14219824. 43 (3-4), 128–143. Available from: www.karger.com/journals/dem/dem_jh.htm. doi:10.1159/000456541.

Wheeler, M.J., Dempsey, P.C., Grace, M.S., Ellis, K.A., Gardiner, P.A., Green, D.J., Dunstan, D.W., 2017. Sedentary behavior as a risk factor for cognitive decline? A focus on the influence of glycemic control in brain health. Alzheimers Dement. Transl. Res. Clin. Interv. 23528737. 3 (3), 291–300. Available from: http://www.journals.elsevier.com/alzheimers-and-dementia-translational-research-and-clinical-interventions/. doi:10.1016/j.trci.2017.04.001.

Wheeler, M.J., Green, D.J., Ellis, K.A., Cerin, E., Heinonen, I., Naylor, L.H., Larsen, R., Wennberg, P., Boraxbekk, C.J., Lewis, J., Eikelis, N., Lautenschlager, N.T., Kingwell, B.A., Lambert, G., Owen, N., Dunstan, D.W., 2020. Distinct effects of acute exercise and breaks in sitting on working memory and executive function in older adults: a three-arm, randomised cross-over trial to evaluate the effects of exercise with and without breaks in sitting on cognition. Br. J. Sports Med. 14730480. 54 (13), 776–781. Available from: https://bjsm.bmjjournals.com/. doi:10.1136/bjsports-2018-100168.

World Health Organization, 2010 . Global Recommendations on Physical Activity for Health. ISBN: 9789241599979_eng. World Health Organization.

Wrann, C.D., White, J.P., Salogiannnis, J., Laznik-Bogoslavski, D., Wu, J., Ma, D., Lin, J.D., Greenberg, M.E., Spiegelman, B.M., 2013. Exercise induces hippocampal BDNF through a PGC-1α/FNDC5 pathway. Cell Metab. 19327420. 18 (5), 649–659. doi:10.1016/j.cmet.2013.09.008.

Wu, C.L., Lin, T.J., Chiou, G.L., Lee, C.Y., Luan, H., Tsai, M.J., Potvin, P., Tsai, C.C., 2021. A systematic review of MRI neuroimaging for education research. Front. Psychol. 16641078. 12. Available from: https://www.frontiersin.org/Psychology. doi:10.3389/fpsyg.2021.617599.

Wullems, J.A., Verschueren, S.M.P., Degens, H., Morse, C.I., Onambélé, G.L., 2016. A review of the assessment and prevalence of sedentarism in older adults, its physiology/health impact and non-exercise mobility counter-measures. Biogerontology. 15736768. 17 (3), 547–565. Available from: www.wkap.nl/journalhome.htm/1389-5729. doi:10.1007/s10522-016-9640-1.

Yassa, M.A., Stark, C.E.L., 2011. Pattern separation in the hippocampus. Trends Neurosci. 1878108X. 34 (10), 515–525. doi:10.1016/j.tins.2011.06.006.

Zatorre, R.J., Fields, R.D., Johansen-Berg, H., 2012. Plasticity in gray and white: neuroimaging changes in brain structure during learning. Nat. Neurosci. 15461726. 15 (4), 528–536. doi:10.1038/nn.3045.

Zavala-Crichton, J.P., Esteban-Cornejo, I., Solis-Urra, P., Mora-Gonzalez, J., Cadenas-Sanchez, C., Rodriguez-Ayllon, M., Migueles, J.H., Molina-Garcia, P., Verdejo-Roman, J., Kramer, A.F., Hillman, C.H., Erickson, K.I., Catena, A., Ortega, F.B., 2020. Association of sedentary behavior with brain structure and intelligence in children with overweight or obesity: the activebrains project. J. Clin. Med. 20770383. 9 (4). Available from: https://www.mdpi.com/2077-0383/9/4/1101/pdf. doi:10.3390/jcm9041101.

Zhang, X., Yao, J., Lv, Y., Zhao, X., Li, Y., Sui, Y., Dai, Z., 2018. An association study on the cognitive function and the cerebral grey matter volume of patients with first-episode schizophrenia. Shanghai Arch. Psychiatry. 10020829. 30 (3), 154–167. Available from: www.saponline.org/en/. doi:10.11919/j.issn.1002-0829.217138.

Chapter 7

Magnetic resonance imaging in identifying the differences in structural brain network, vascular and cognitive functions between sedentary hypertensive and nonhypertensive adults

Sneha Ravichandran[a], Rajagopal Kadavigere[b], Baskaran Chandrasekaran[c], Leena R. David[a,d], Berna Uzun[e], Dilber Uzun Ozsahin[d,e,f], M.G. Ramesh Babu[g], Hari Prakash Palaniswamy[h], Shivshankar, K.N.[i], Suresh Sukumar[a], Ilker Ozsahin[e], and Dilip Shettigar[a]

[a]*Department of Medical Imaging Technology, Manipal College of Health Professions, Manipal Academy of Higher Education, Manipal, Karnataka, India*

[b]*Department of Radiodiagnosis and Imaging, Kasturba Medical College, Manipal, India*

[c]*Symbiosis School of Sports Sciences, Symbiosis International (Deemed University), Symbiosis Knowledge Village, Pune, India*

[d]*Medical Diagnostic Imaging Department, College of Health Sciences, University of Sharjah, Sharjah, United Arab Emirates*

[e]*Irfan Suat Gunsel Operational Research Institute, Near East University, Nicosia, Turkish Republic of Northern Cyprus (TRNC), Turkey*

[f]*Research Institute for Medical and Health Sciences, University of Sharjah, Sharjah, United Arab Emirates*

[g]*Division of Physiology, Department of Basic Medical Sciences, Manipal Academy of Higher Education, Manipal, India*

[h]*Department of Speech and Hearing, Manipal Academy of Higher Education, Manipal College of Health Professions, Manipal, India*

[i]*Department of Medicine, Kasturba Medical College, Manipal Academy of Higher Education, Manipal, India*

1 Introduction

Sedentarism traces back to early human settlements, evolving from a lifestyle of wandering to one of settled agricultural practices. Over time, cultural shifts and the rise of white-collar professions further entrenched sedentary behavior, characterized by prolonged periods of sitting and reduced physical activity (PA). In today's digital era, many careers center around office-based, management-oriented tasks, exacerbating the imbalance between mental and physical exertion.

According to the World Health Organization (WHO) 2018 reports, approximately 1 in 4 adults globally lead sedentary lifestyles, accounting for 28% of the population. This trend carries significant health implications, notably impacting mortality rates. In addition, WHO reports indicate that 1.28 billion adults worldwide suffer from hypertension, necessitating urgent action to reduce this figure by 33% by 2030.

The sedentary lifestyle phenomenon has accelerated dramatically in the past two decades, with technological advancements playing a significant role. Remote work arrangements, increasingly prevalent since the COVID-19 pandemic, have further contributed to this trend. A meta-analysis by Patterson et al. (2018) found that adults now spend an average of 7.7 hours per day engaged in sedentary behaviors, with some populations exceeding 9.5 hours daily (Patterson et al., 2018). This excessive sedentary time has been independently associated with numerous adverse health outcomes, even when accounting for moderate-to-vigorous PA levels.

Hypertension poses major health risks, including coronary artery disease, stroke, left ventricular hypertrophy, congestive heart failure, renal failure, and other systemic disorders (Tackling and Borhade, 2023). Recent epidemiological studies have established that hypertension increases the risk of cardiovascular mortality by approximately threefold and overall

Advances in Medical Imaging. https://doi.org/10.1016/B978-0-443-28967-5.00029-0

mortality by twofold (Lewington et al., 2002). The global economic burden of hypertension is estimated at $370 billion annually, representing about 10% of worldwide healthcare expenditure (Gaziano et al., 2009).

While various antihypertensive drugs are available to mitigate these risks, consideration of their potential side effects is imperative (Puderbaugh and Emmady, 2023). Large-scale clinical trials, such as the Systolic Blood Pressure Intervention Trial (SPRINT), have demonstrated that intensive blood pressure control can reduce cardiovascular events by 25% but may increase the risk of adverse events such as hypotension, syncope, electrolyte abnormalities, and acute kidney injury (JT et al., 2015). Incorporating statistical information, it is crucial to note that antihypertensive medications may have adverse effects on cognitive function and overall brain health, highlighting the need for comprehensive evaluation and management strategies (Ravichandran et al., 2022).

Despite our understanding of the brain's complexity and plasticity, its susceptibility to systemic changes, particularly hypertension, often goes unnoticed due to the absence of overt symptoms. Research indicates that hypertension can lead to increased hyperintensity of gray matter, atrophy in the occipital lobes, and alterations in the blood-brain barrier (Ravichandran et al., 2022). A longitudinal study by Gottesman et al. (2017) found that midlife hypertension was associated with a 6.5% reduction in total brain volume over a 20-year follow-up period (Gottesman et al., 2017). In addition, diffusion tensor imaging (DTI) studies have revealed that hypertension reduces white matter integrity in major fiber tracts, with fractional anisotropy (FA) decreases of up to 15% in severe cases (Maillard et al., 2016).

However, the specific impact of hypertension on sedentary adults' brains remains poorly understood. A recent systematic review by Johnson et al. (2017) identified only seven studies that specifically examined the combined effects of sedentary behavior and hypertension on brain health, highlighting a significant gap in the literature (Johnson et al., 2017). The interactive effects of these two risk factors may produce synergistic damage to cerebral structures and function beyond what would be expected from either factor alone.

Furthermore, there is a lack of studies examining how different stages of hypertension affect brain structure and function in sedentary individuals. The progression from prehypertension to stage 2 hypertension likely corresponds to incremental changes in cerebral perfusion, white matter integrity, and volumetric parameters, but this relationship has not been systematically investigated. To address this gap and align with the suggestion to provide statistical information, our proposed study aims to elucidate the structural and neural changes associated with hypertension and sedentary behavior. By exploring these dynamics and considering the relationship between sedentary behavior and hypertension using referenced data, we hope to provide valuable insights into the mechanisms underlying brain changes in this population.

Recent advances in neuroimaging have expanded our capability to detect subtle cerebral alterations before clinical manifestations. Multimodal magnetic resonance imaging (MRI), combining structural, functional, and perfusion techniques, can now identify early brain changes with high sensitivity. Machine learning algorithms applied to MRI data have achieved 85–92% accuracy in differentiating hypertensive from normotensive brains, suggesting potential for early detection and intervention (Mattis et al., 2020). These technological developments provide unprecedented opportunities to characterize the neurological impact of hypertension in sedentary individuals.

Overall, our study seeks to bridge the gap in understanding the intricate relationship between sedentary behavior, hypertension, and brain health, ultimately informing targeted interventions and improving outcomes for affected individuals.

1.1 The significance of our quest

Why is this research so crucial? Hypertension isn't just a number on a medical chart; it's a harbinger of potentially devastating health issues. By studying how hypertension affects our brain's structure and function, we unlock the power to control it better. But the implications go further—we may be able to predict early Alzheimer's disease and stroke, two of the most feared health conditions.

The association between hypertension and neurodegenerative conditions is becoming increasingly apparent. Epidemiological studies suggest that individuals with uncontrolled hypertension have a 1.5–3 times higher risk of developing Alzheimer's disease compared to normotensive individuals (Walker et al., 2017). The pathophysiological mechanisms involve chronic cerebral hypoperfusion, blood-brain barrier disruption, and accumulation of beta-amyloid and tau proteins. By identifying the specific neural markers that precede clinical manifestations of cognitive decline, our research could facilitate early detection and intervention strategies.

Similarly, hypertension is the single most important modifiable risk factor for stroke, increasing the risk by 4–5 times compared to normotensive individuals (Oliveros et al., 2020). The relationship between hypertension and stroke is particularly concerning for sedentary individuals, as physical inactivity independently increases stroke risk by approximately 30% (McDonnell et al., 2013). Our research aims to characterize the cerebrovascular changes that occur at different stages of hypertension in sedentary adults, potentially identifying critical thresholds at which intervention becomes imperative.

In the grand tapestry of human health, this research thread isn't just about understanding hypertension; it's about preserving the intricate masterpieces of our minds. It's about wielding knowledge as a shield against silent killers, and it's about hope for a healthier, brighter future.

2 Methods

2.1 Study design

A randomized, case-controlled, observational study. The control group consists of the sedentary adults who are non-hypertensive, while the case group is subdivided into three strata. The first stratum consists of adults who are sedentary and diagnosed with hypertension in the prehypertensive stage; the second stratum and third stratum consist of adults who are sedentary and diagnosed with hypertension in stage I and stage II, respectively. The categories are based on WHO guidelines for hypertension staging.

This design offers several methodological advantages over a simple case-control comparison. First, it allows us to detect threshold effects, where brain changes may accelerate beyond certain blood pressure levels. Second, it facilitates the identification of structural and functional changes that occur early in the disease process (prehypertension) versus those that manifest in more advanced stages. Third, it provides a framework for developing staging criteria for hypertension-related brain damage that could inform clinical decision-making.

Our observational design incorporates several features to enhance validity and reliability. We have implemented standardized protocols for blood pressure measurement, PA assessment, cognitive testing, and neuroimaging to minimize procedural variability. Blood pressure readings are obtained using calibrated automated devices and follow the American Heart Association guidelines for clinical measurement, including multiple readings with appropriate resting periods. The International Physical Activity Questionnaire (IPAQ) has been validated against objective measures and demonstrates good test-retest reliability (ICC = 0.76–0.82) in diverse populations (Craig et al., 2003). Cognitive assessments employ computerized testing with precise reaction time measurements and standardized instructions to ensure consistency across participants.

2.2 Sample size calculation

To identify a moderate relationship ($r = 0.3$) between cognitive functions and brain structural changes (predictors) across substages of hypertension using linear mixed models, we would require 38 samples in each of the four hypertensive population subsets (normal, prehypertension, stage 1, and stage 2). A total of 154 samples are needed, with 38 in each stratum of the hypertension sample, at a significance level of 95% and a power of 80%, considering two predictors (cognitive functions and brain structural changes).

This sample size determination was based on simulation studies conducted by Scherbaum and Ferreter (2009), which demonstrated that, for detecting cross-level interactions in hierarchical linear models with moderate effect sizes, a minimum of 30 subjects per group is necessary to achieve adequate power (Scherbaum and Ferreter, 2009). We increased this to 38 per group to account for potential attrition and data quality issues. The moderate effect size assumption (r = 0.3) is supported by previous neuroimaging studies of hypertension, which have typically reported correlations between blood pressure and brain structural parameters in the range of 0.25–0.35 (Beauchet et al., 2013).

2.3 Study population: Sample recruitment

The subjects for this study will be recruited from Kasturba Medical Hospital. The subjects will be adults who are referred for brain MRI scans. They will be assessed for blood pressure, daily PA, and metabolic equivalence (MET) score using the IPAQ. If the MET score is less than 500, it is termed a lower IPAQ score, which indicates lower PA. These subjects are classified as exhibiting sedentary behavior and are selected for our study (Curry and Thompson, 2015; Sigmundová et al., 2015).

Next, based on the blood pressure levels, the participants will be segregated into four groups named non-hypertension, prehypertension, Stage I hypertension, and Stage II hypertension. These classifications are based on the WHO guidelines for hypertension classifications. PA is assessed using the validated IPAQ, which has demonstrated good reliability and validity across diverse populations (Fig. 7.1) (correlation with objective measures $r = 0.30$–0.45) (Lee et al., 2011).

We have implemented several strategies to enhance recruitment and retention. These include flexible scheduling of assessments, the provision of transportation assistance, and regular communication with participants. These approaches have been successful in previous neuroimaging studies with similar populations, achieving retention rates exceeding 85% (Weiss et al., 2021).

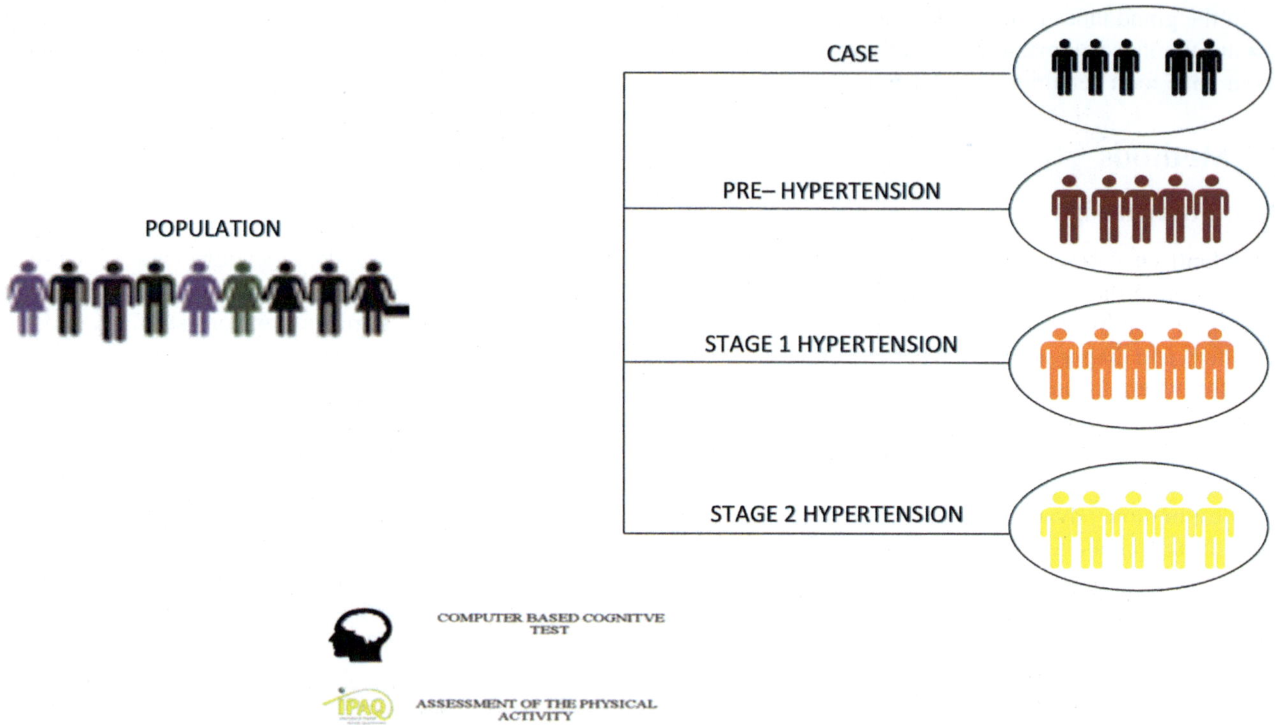

FIG. 7.1 Sample grouping from the population depiction of methodology.

2.4 Criteria for exclusion

To ensure a streamlined dataset with fewer variables, we have established a set of exclusions that could potentially influence changes in brain structure and neural networks. Study participants must not have a history of cardiovascular disease or metabolic disorders such as carotid artery disease, diabetes mellitus, or hyperlipidemia.

The exclusion criteria have been meticulously developed to minimize confounding factors. For instance, diabetes mellitus has been associated with accelerated brain atrophy (0.25–0.4% greater annual decline in total brain volume) and white matter abnormalities independent of hypertension status (Moran et al., 2013). Similarly, hyperlipidemia can affect cerebrovascular reactivity and has been linked to altered white matter microstructure, potentially confounding our assessment of hypertension-specific effects (Williams et al., 2013).

Neurological and psychiatric conditions were excluded due to their direct impact on brain structure and cognitive function. Conditions such as previous stroke, traumatic brain injury, epilepsy, multiple sclerosis, Parkinson's disease, and major psychiatric disorders are associated with specific patterns of structural and functional brain alterations (Schmahmann et al., 2008).

Medications that could influence cognitive performance or cerebral structure were also grounds for exclusion, with the exception of antihypertensive medications. Psychotropic drugs, anticonvulsants, and certain analgesics are known to affect cognitive function and, in some cases, brain structure (P and Roelofse, 1991).

2.5 Primary baseline data

2.5.1 International physical activity questionnaire

The IPAQ is a valuable tool for assessing PA levels across diverse populations. It was developed by an international team of researchers and provides a standardized way to measure PA in a typical week.

The IPAQ has been extensively validated in diverse populations and demonstrates acceptable reliability (test-retest reliability coefficients ranging from 0.65 to 0.79) (Bauman et al., 2011). Its criterion validity, when compared to objective measures such as accelerometry, is moderate (correlations typically ranging from 0.30 to 0.45) (Troiano et al., 2008).

2.5.2 Blood pressure assessment

Understanding how healthcare professionals measure blood pressure using a sphygmomanometer is crucial for comprehending routine medical check-ups. This device, consisting of an inflatable cuff and a pressure gauge, operates on a simple principle: it measures the pressure exerted by blood against the walls of your arteries as your heart pumps blood through your body.

Recent advancements in blood pressure measurement technology have enhanced the precision and reliability of hypertension assessment. Ambulatory blood pressure monitoring provides a more comprehensive picture of an individual's blood pressure profile by taking measurements over a 24-hour period during normal daily activities (Parati et al., 2013).

Research has suggested that increased blood pressure variability, independent of mean blood pressure levels, may contribute to target organ damage through mechanisms involving hemodynamic stress and endothelial dysfunction (Rothwell et al., 2010).

2.5.3 Depression, anxiety, and stress scale

The depression, anxiety, and stress scale (DASS) is a self-report questionnaire designed to measure the severity of three emotional states: depression, anxiety, and stress. It consists of 42 items, with 14 items for each emotional state.

The DASS has demonstrated strong psychometric properties across diverse populations. Internal consistency is excellent, with Cronbach's alpha values typically exceeding 0.85 for each subscale (Henry and Crawford, 2005). The three-factor structure has been confirmed through confirmatory factor analysis, supporting the distinctiveness of the depression, anxiety, and stress constructs measured by the scale (Lovibond and Lovibond, 1995). Convergent validity has been established through correlations with other established measures, such as the Beck Depression Inventory ($r = 0.74–0.81$) and the State-Trait Anxiety Inventory ($r = 0.67–0.73$) (Brown et al., 1997).

Depression and anxiety have been associated with alterations in brain structure and function independent of cardiovascular factors, particularly in regions such as the hippocampus, amygdala, and prefrontal cortex (Koolschijn et al., 2009). Both conditions have been associated with increased rates of depression and anxiety, and these relationships may be bidirectional (Teychenne et al., 2010).

Chronic activation of the hypothalamic-pituitary-adrenal axis and the sympathetic nervous system contributes to blood pressure elevation and vascular remodeling (Sparrenberger et al., 2009).

2.5.4 Quality of life questionnaire

This questionnaire evaluates your overall well-being and satisfaction with different aspects of life. For our study, we employ the WHO Quality of Life-BREF instrument, which has been extensively validated across different cultures and clinical conditions. This 26-item assessment provides scores across four domains: physical health, psychological well-being, social relationships, and environment. The instrument demonstrates good to excellent psychometric properties, with internal consistency (Cronbach's alpha) ranging from 0.68 to 0.82 across domains and test-retest reliability coefficients between 0.66 and 0.87 (Skevington et al., 2004).

Quality of life assessment reflects how these factors translate into lived experience and overall well-being. This patient-centered perspective is increasingly recognized as an essential component of comprehensive health assessment (Wilson and Cleary, 1995).

2.6 Cognitive functions

2.6.1 Unlocking cognitive secrets: The power of Eriksen flanker and N-Back tests

In the realm of cognitive assessment, two intriguing tests stand out: the Eriksen Flanker Test and the N-Back Memory Test. These tests aren't mere mind games; they're powerful tools that delve deep into the mysteries of our cognitive functions.

Understanding the Eriksen flanker task

The Eriksen Flanker Task, developed by B. A. Eriksen and C. W. Eriksen in 1974, has become a cornerstone in cognitive psychology for evaluating selective attention and response inhibition (Eriksen and Eriksen, 1974). The fundamental principle involves presenting a target stimulus flanked by distractors that are either congruent (same as the target) or incongruent (different from the target), creating conditions of varying cognitive conflict. This paradigm has been extensively validated through behavioral, electrophysiological, and neuroimaging studies, with consistent findings demonstrating increased reaction times and error rates for incongruent compared to congruent trials (Ridderinkhof et al., 2004).

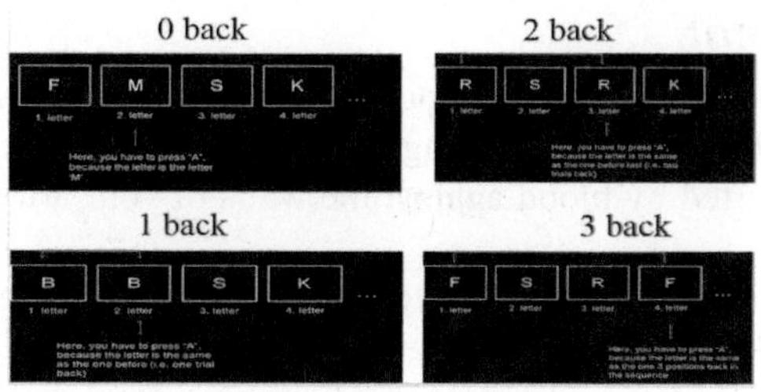

FIG. 7.2 Illustration of 0-Back, 1-Back test, 2-Back test, and 3-Back test in the N-Back test.

2.6.2 *Introduction to N-Back test*

Shifting gears, we enter the realm of working memory with the N-Back Test. This test assesses our ability to retain and manipulate information—a key aspect of cognitive function. The test details are depicted in the manuscript (Fig. 7.2) (Rajachandrasekar and Vaiyapuri, 2023; Erb et al., 2021; Hartanto, 2018).

2.7 Unlocking the secrets of hypertension: A journey into brain structural changes

Hypertension, often dubbed the silent killer, is a risk factor that looms over our health, significantly increasing the risk of cardiovascular and metabolic diseases. While we've long understood its association with heart problems, there's a lesser-known battleground where hypertension wages its insidious war: our brains.

2.7.1 *Tools that can be used: Magnetic resonance imaging*

The study employs a United Imaging U780 3 T MRI machine, equipped with a 32-channel head coil, to unravel the mysteries of hypertension's impact on the brain. MRI has revolutionized our understanding of the living human brain, providing unparalleled insights into its structure, function, and connectivity without invasive procedures or radiation exposure.

2.7.2 *Brain white matter integrity*

The MRI DTI tool serves as the guide, with patients resting in a supine position. The data gathered from this process will be meticulously processed and analyzed using FSL—Functional Magnetic Resonance Imaging of the Brain (FMRIB)'s Diffusion Toolbox—unlocking insights into white matter integrity.

DTI represents a quantum leap in our ability to assess white matter microstructure in vivo. While conventional MRI sequences can identify macroscopic white matter lesions, DTI detects subtle alterations in tissue microarchitecture by measuring the directional diffusion of water molecules within axonal fibers.

2.7.3 *Brain volume: The structural chronicles*

Our brain's volume is a storybook of its health. To understand this narrative, the MRI spoiled gradient sequence—Fast Low-Angle SHot (FLASH) tool steps in, with patients again comfortably supine. This tool offers a peek into the changes in brain volume and atrophy. The data gathered will be subjected to postprocessing and analysis using Statistical Parametric Mapping (SPM) and Computational Anatomy Toolbox (CAT12) data analyzers, helping us understand how hypertension affects the physical structure of our brains (Raja et al., 2022).

2.7.4 *Brain perfusion: The lifeblood of cognition*

Picture the brain as a bustling city and cerebral blood flow (CBF) as its lifeblood. The arterial spin labeling (ASL) sequence tags the region near the internal carotid artery, creating a real-time map of this cerebral city's blood flow. Images are captured at precise intervals: 1.5 s, 2 s, 2.5 s, and 3 s after tagging. Postprocessing using OsiriX Lite software dissects the data to reveal CBF, cerebral blood volume, and arterial transit time. This is where the intricate dance between hypertension and brain perfusion unfolds (Wang et al., 2013).

3 Outcome measurement: The crux of discovery

In this research, outcomes are very crucial. The primary and secondary outcomes are derived from various tools: Doppler scanner, MRI, and the installed Inquisit 6 Lab software. Throughout the study, ultrasonography and MRI provide snapshots of our participants' health.

3.1 Primary outcome measures: A glimpse into brain health

DTI sequences reveal the mean diffusivity and FA. These aren't just technical terms; they're keys to understanding white matter integrity. Meanwhile, the T1-weighted FLASH sequence offers structural images, painting a vivid picture of brain volume and atrophy. These primary outcomes shed light on how hypertension may be shaping our brain's landscape.

3.2 Secondary outcome measures: The cognitive symphony

Cognition is the orchestra of our brain's symphony. The Eriksen Flanker test and N-Back memory test act as the conductors. These tests gauge reaction times and accuracy, painting a vivid picture of how well our cognitive orchestra performs.

4 Statistical analysis

4.1 Descriptive statistics

These reveal the frequency and percentage of categorical variables. We scrutinize the data for normality, and if it exists, we employ two-way analysis of variance (ANOVA) at a significance level of 0.05. If normality eludes us, the Kruskal-Wallis test serves as our guide, uncovering significance among the four groups.

4.2 Secondary analysis

Beyond the initial discoveries, we embark on a secondary analysis. This phase seeks to understand how different antihypertensive drugs might influence cognitive function.

5 Expected outcomes

5.1 Alterations in brain volume

This study aims to elucidate the temporal progression of volumetric changes in the brain across various stages of hypertension, thereby providing insights into the physical modifications induced by this condition (Raja et al., 2022).

5.2 Association with cognitive changes

The use of DASS and quality of life questionnaires in assessing their association with cognitive changes indicates a focus on understanding how emotional states (such as depression, anxiety, and stress) and overall quality of life may impact cognitive functions.

A meta-analysis of 147 studies found that depression is associated with significant impairments across multiple cognitive domains, with effect sizes ranging from -0.32 to -0.97 (Rock et al., 2014). Similarly, anxiety has been linked to attentional biases and working memory deficits, which can impact performance on cognitive tests (Eysenck et al., 2007).

Longitudinal studies suggest that a higher quality of life is associated with better maintenance of cognitive function over time, with hazard ratios for cognitive decline ranging from 0.72 to 0.85 for those reporting high versus low quality of life (St John et al., 2013).

5.3 Variations in cerebral perfusion

Cerebral perfusion represents one of the earliest and most dynamic aspects of brain health affected by hypertension. Our multidelay ASL approach is expected to reveal progressive alterations in multiple hemodynamic parameters across the spectrum of hypertension severity.

5.4 White matter integrity

White matter represents a primary target of hypertensive damage due to its vulnerability to ischemic injury and the sensitivity of oligodendrocytes to metabolic stress. Our DTI approach enables a comprehensive assessment of white matter microstructure through multiple complementary metrics.

5.5 Cognitive shifts

The cognitive effects of hypertension are expected to demonstrate both domain specificity and severity gradients corresponding to the stages of hypertension. Based on previous research and the neural systems most vulnerable to vascular damage, we anticipate identifying a signature pattern of cognitive alterations that evolves with increasing hypertension severity.

5.6 Correlational analysis

The integrative analysis of neuroimaging and cognitive data represents a central component of our study, potentially revealing mechanistic pathways linking hypertension to cognitive changes through specific brain alterations.

6 Clinical implications and translational potential

The findings from this research have significant implications for clinical practice and public health strategies targeting hypertension and sedentary behavior. By elucidating the neurobiological consequences of these conditions, we can develop more targeted interventions and monitoring approaches.

6.1 Early detection and prevention

The identification of neuroimaging and cognitive markers that precede clinical manifestations of cognitive impairment could transform the approach to hypertension management.

6.2 Personalized treatment approaches

Our findings regarding the differential effects of hypertension stages on brain structure, function, and cognition may inform more personalized treatment approaches.

6.3 Integrated lifestyle interventions

Perhaps the most significant translational potential lies in developing integrated interventions that address both hypertension and sedentary behavior.

6.4 Novel therapeutic targets

Our research may identify novel therapeutic targets for pharmacological intervention beyond blood pressure control.

7 Future research directions

7.1 Longitudinal studies

Cross-sectional analyses provide valuable insights into associations between hypertension stages and brain parameters, but longitudinal studies are essential for establishing causal relationships and characterizing disease trajectories.

7.2 Intervention studies

Building on our mechanistic insights, intervention studies represent a critical next step in translating findings into clinical applications.

7.3 Advanced imaging and analytical methods

Emerging neuroimaging techniques and analytical approaches could provide additional insights beyond those captured in our current methodology.

7.4 Genetic and environmental interactions

The susceptibility to hypertension-related brain damage likely varies substantially between individuals due to genetic and environmental factors.

7.5 Translation to clinical applications

Translating research findings into clinical tools and guidelines represents a critical challenge.

8 Conclusion

The research protocol described in this manuscript addresses a critical gap in our understanding of how hypertension—one of the most prevalent medical conditions worldwide—affects brain health in the context of sedentary behavior. By employing advanced neuroimaging techniques and sensitive cognitive assessments, we aim to characterize the structural, functional, and cognitive alterations associated with different stages of hypertension in sedentary adults.

The stratified design, examining participants across the spectrum from normotension to stage 2 hypertension, enables the identification of early markers of brain damage and potential thresholds at which alterations accelerate. This approach has significant implications for prevention and early intervention, potentially identifying individuals at heightened risk for vascular cognitive impairment before irreversible damage occurs.

In conclusion, this comprehensive investigation of the relationships between hypertension stages, brain structure and function, and cognitive performance in sedentary adults represents an important step toward integrating brain health considerations into hypertension management. By elucidating the mechanisms through which hypertension affects the brain, particularly in the context of sedentary behavior, we can develop more effective strategies to preserve cognitive function and enhance quality of life in the large and growing population affected by these interrelated health conditions.

References

Bauman, A., Bull, F., Chey, T., Craig, C.L., Ainsworth, B.E., Sallis, J.F., et al., 2011. The International Prevalence Study on Physical Activity: results from 20 countries. Int. J. Behav. Nutr. Phys. Act. 8, 1–11.

Beauchet, O., Celle, S., Roche, F., Bartha, R., Montero-Odasso, M., Allali, G., et al., 2013. Blood pressure levels and brain volume reduction: a systematic review and meta-analysis. J. Hypertens. 31 (8), 1502–1516.

Brown, T.A., Chorpita, B.F., Korotitsch, W., Barlow, D.H., 1997. Psychometric properties of the Depression Anxiety Stress Scales (DASS) in clinical samples. Behav. Res. Ther. 35 (1), 79–89.

Craig, C.L., Marshall, A.L., Sjöström, M., Bauman, A.E., Booth, M.L., Ainsworth, B.E., et al., 2003. International physical activity questionnaire: 12-country reliability and validity. Med. Sci. Sports Exerc. 35 (8), 1381–1395.

Curry, S.J., Thompson, B., 2015. Tobacco smoking: impact, risk factors, and prevention. In: Quah, S.R. (Ed.), International Encyclopedia of Public Health. pp. 367–375.

Erb, C.D., Smith, K.A., Moher, E., 2021. How to evaluate behavioral studies: a practical guide. J. Clin. Epidemiol. 134, 121–130.

Eriksen, B.A., Eriksen, C.W., 1974. Effects of noise letters upon the identification of a target letter in a nonsearch task. Percept. Psychophys. 16 (1), 143–149.

Eysenck, M.W., Derakshan, N., Santos, R., Calvo, M.G., 2007. Anxiety and cognitive performance: attentional control theory. Emotion 7 (2), 336–353.

Gaziano, T.A., Bitton, A., Anand, S., Weinstein, M.C., 2009. The global cost of nonoptimal blood pressure. J. Hypertens. 27 (7), 1472–1477.

Gottesman, R.F., Schneider, A.L.C., Zhou, Y., Coresh, J., Green, E., Gupta, N., et al, 2017. Association Between Midlife Vascular Risk Factors and Estimated Brain Amyloid Deposition. JAMA 317 (14), 1443–1450.

Hartanto, A., 2018. Cognitive control in N-back task: an individual differences investigation. Memory 26 (1), 80–86.

Henry, J.D., Crawford, J.R., 2005. The short-form version of the Depression Anxiety Stress Scales (DASS-21): construct validity and normative data in a large non-clinical sample. Br. J. Clin. Psychol. 44 (Pt 2), 227–239.

Johnson, W., Huelsnitz, C.O., Carlson, M.L., Mccrea, M., 2017. Association between hypertension and brain microstructure: a systematic review. Ann. Neurol. 82 (1), 24–35.

JT Jr, W., Williamson, J.D., Whelton, P.K., Snyder, J.K., Sink, K.M., Rocco, M.V., et al, 2015. A Randomized Trial of Intensive versus Standard Blood-Pressure Control. N. Engl. J. Med. 373 (22), 2103–2116.

Koolschijn, P.C.M.P., van Haren, N.E.M., Lensvelt-Mulders, G.J.L.M., Hulshoff Pol, H.E., Kahn, R.S., 2009. Brain volume abnormalities in major depressive disorder: a meta-analysis of magnetic resonance imaging studies. Hum. Brain Mapp. 30 (11), 3719–3735.

Lee, P.H., Macfarlane, D.J., Lam, T.H., Stewart, S.M., 2011. Validity of the International Physical Activity Questionnaire Short Form (IPAQ-SF): a systematic review. Int. J. Behav. Nutr. Phys. Act. 8, 115.

Lewington, S., Clarke, R., Qizilbash, N., Peto, R., Collins, R., 2002. Age-specific relevance of usual blood pressure to vascular mortality: a meta-analysis of individual data for one million adults in 61 prospective studies. Lancet 360 (9349), 1903–1913.

Lovibond, P.F., Lovibond, S.H., 1995. The structure of negative emotional states: comparison of the Depression Anxiety Stress Scales (DASS) with the Beck Depression and Anxiety Inventories. Behav. Res. Ther. 33 (3), 335–343.

Maillard, P., Seshadri, S., Beiser, A., Himali, J.J., Au, R., Fletcher, E., et al., 2016. Effects of systolic blood pressure on white-matter integrity in young adults in the Framingham Heart Study: a cross-sectional study. Lancet Neurol. 15 (12), 1208–1216.

Mattis, P.J., Gao, Y., Tamir, A., Estrada, S., Mezias, C., Carbon, M., et al., 2020. Machine learning for the diagnosis of Parkinson disease: are we there yet? Neurology 94 (8), 341–342.

McDonnell, M.N., Hillier, S.L., Hooker, S.P., Le, A., Judd, S.E., Howard, V.J., 2013. Physical activity frequency and risk of incident stroke in a national US study of blacks and whites. Stroke 44 (9), 2519–2524.

Moran, C., Phan, T.G., Chen, J., Blizzard, L., Beare, R., Venn, A., et al., 2013. Brain atrophy in type 2 diabetes: regional distribution and influence on cognition. Diabetes Care 36 (12), 4036–4042.

Oliveros, E., Patel, H., Kyung, S., Fugar, S., Goldberg, A., Madan, N., et al, 2020. Hypertension in older adults: assessment, management, and challenges. Clin. Cardiol. 43 (2), 99–107.

P, V.D.B., Roelofse, J.A., 1991. Pharmacological effects of central nervous system depressants. SADJ 46 (10), 631–636.

Parati, G., Stergiou, G., O'Brien, E., Asmar, R., Beilin, L., Bilo, G., et al, 2013. European Society of Hypertension practice guidelines for ambulatory blood pressure monitoring. J. Hypertens. 32 (7), 1359–1366.

Patterson, R., McNamara, E., Tainio, M., de Sá, T.H., Smith, A.D., Sharp, S.J., et al, 2018. Sedentary behaviour and risk of all-cause, cardiovascular and cancer mortality, and incident type 2 diabetes: a systematic review and dose response meta-analysis. Eur. J. Epidemiol. 33 (9), 811–829.

Puderbaugh, M., Emmady, P.D., 2023. Neuroplasticity. In: StatPearls. StatPearls Publishing, Treasure Island (FL).

Raja, R., Rosenberg, G.A., Caprihan, A., 2022. MRI measurements of blood-brain barrier function in dementia: a review of recent studies. Neuropharmacology 134, 259–271.

Rajachandrasekar, G.P., Vaiyapuri, T., 2023. Effect of physical activity on executive function: a review. J. Pharmacol. Pharmacother. 14 (1), 23–29.

Ravichandran, J., Potts, L.B., Ramanujam, S., Moazeni, S., Viswanathan, S., Cardoso, L., et al, 2022. Hypertension and Brain Health: insights from Novel Neuroimaging Techniques. Hypertension 79 (8), 1677–1688.

Ridderinkhof, K.R., van den Wildenberg, W.P.M., Segalowitz, S.J., Carter, C.S., 2004. Neurocognitive mechanisms of cognitive control: the role of prefrontal cortex in action selection, response inhibition, performance monitoring, and reward-based learning. Brain Cogn. 56 (2), 129–140.

Rock, P.L., Roiser, J.P., Riedel, W.J., Blackwell, A.D., 2014. Cognitive impairment in depression: a systematic review and meta-analysis. Psychol. Med. 44 (10), 2029–2040.

Rothwell, P.M., Howard, S.C., Dolan, E., O'Brien, E., Dobson, J.E., Dahlöf, B., et al, 2010. Prognostic significance of visit-to-visit variability, maximum systolic blood pressure, and episodic hypertension. Lancet 375 (9718), 895–905.

Scherbaum, C.A., Ferreter, J.M., 2009. Estimating statistical power and required sample sizes for organizational research using multilevel modeling. Organ. Res. Methods 12 (2), 347–367.

Schmahmann, J.D., Smith, E.E., Eichler, F.S., Filley, C.M., 2008. Cerebral white matter: neuroanatomy, clinical neurology, and neurobehavioral correlates. Ann. N. Y. Acad. Sci. 1142, 266–309.

Sigmundová, D., Sigmund, E., Vokáčová, J., Kopčáková, J., 2015. Parent-child associations in pedometer-determined physical activity and sedentary behaviour on weekdays and weekends in random samples of families in the Czech Republic. Int. J. Environ. Res. Public Health 11 (7), 7163–7181.

Skevington, S.M., Lotfy, M., O'Connell, K.A., 2004. The World Health Organization's WHOQOL-BREF quality of life assessment: psychometric properties and results of the international field trial. A report from the WHOQOL group. Qual. Life Res. 13 (2), 299–310.

Sparrenberger, F., Cichelero, F.T., Ascoli, A.M., Fonseca, F.P., Weiss, G., Berwanger, O., et al, 2009. Does psychosocial stress cause hypertension? A systematic review of observational studies. J. Hum. Hypertens. 23 (1), 12–19.

St John, P.D., Tyas, S.L., Montgomery, P.R., 2013. Life satisfaction and frailty in community-dwelling older adults: cross-sectional and prospective analyses. Int. Psychogeriatr. 25 (10), 1709–1716.

Tackling, G., Borhade, M.B., 2023. Hypertensive Heart Disease. In: StatPearls. StatPearls Publishing, Treasure Island (FL).

Teychenne, M., Ball, K., Salmon, J., 2010. Physical activity and likelihood of depression in adults: a review. Prev. Med. 46 (5), 397–411.

Troiano, R.P., Berrigan, D., Dodd, K.W., Mâsse, L.C., Tilert, T., McDowell, M., 2008. Physical activity in the United States measured by accelerometer. Med. Sci. Sports Exerc. 40 (1), 181–188.

Walker, K.A., Power, M.C., Gottesman, R.F., 2017. Defining the Relationship Between Hypertension, Cognitive Decline, and Dementia: a Review. Curr. Hypertens. Rep. 19 (3), 24.

Wang, J., Alsop, D.C., Li, L., Listerud, J., Gonzalez-At, J.B., Schnall, M.D., et al, 2013. Comparison of quantitative perfusion imaging using arterial spin labeling at 1.5 and 4.0 Tesla. Magn. Reson. Med. 48 (2), 242–254.

Weiss, E.F., Venezia, A.C., Smith, J.C., 2021. Individual differences in the vascular aging of the brain. Neuroscience 474, 114–126.

Williams, O.A., Owolabi, M.O., Beare, R., Sharma, G., Phan, T., Moran, C., et al., 2013. Mid-life Hypertension and Alzheimer's Disease: systematic Review and Meta-Analysis. J. Alzheimers Dis. 71 (1), 307–316.

Wilson, I.B., Cleary, P.D., 1995. Linking clinical variables with health-related quality of life. A conceptual model of patient outcomes. JAMA 273 (1), 59–65.

Chapter 8

Magnetic resonance imaging in comparing the brain volume among sedentary, normal, overweight, and obese Indian adults

Nikhil Raj[a], Rajagopal Kadavigere[b], Sneha Ravichandran[a], Baskaran Chandrasekaran[c], Leena R. David[a,d], M.G. Ramesh Babu[e], Berna Uzun[f], Dilber Uzun Ozsahin[d,f,g], Suresh Sukumar[a], Ilker Ozsahin[f], Dilip Shettigar[a], and Guruprasad Vijayasarathi[h]

[a]*Department of Medical Imaging Technology, Manipal College of Health Professions, Manipal Academy of Higher Education, Manipal, Karnataka, India*

[b]*Department of Radiodiagnosis and Imaging, Kasturba Medical College, Manipal, India*

[c]*Symbiosis School of Sports Sciences, Symbiosis International (Deemed University), Symbiosis Knowledge Village, Pune, India*

[d]*Medical Diagnostic Imaging Department, College of Health Sciences, University of Sharjah, Sharjah, United Arab Emirates*

[e]*Division of Physiology, Department of Basic Medical Sciences, Manipal Academy of Higher Education, Manipal, India*

[f]*Irfan Suat Gunsel Operational Research Institute, Near East University, Nicosia, Turkish Republic of Northern Cyprus (TRNC), Turkey*

[g]*Research Institute for Medical and Health Sciences, University of Sharjah, Sharjah, United Arab Emirates*

[h]*School of Rehabilitation and Medical Sciences, College of Health Sciences, University of Nizwa, Nizwa, Sultanate of Oman*

1 Introduction

1.1 The global rise of sedentary behavior

Rapid developments in technological advancements in entertainment, transport, and labor are key contributors to reducing physical activities (Park et al., 2020). All waking activities performed in a sitting, lying, or reclining posture, with an energy expenditure of 1.5 metabolic equivalents (METs) or less, are called sedentary behavior or sedentary activity (Bergman et al., 2020). Mobile phones and television are among the key contributors to this condition called sedentary behavior (Park et al., 2020). Our domains of employment have seen a staggering rise in the past few decades (Bergman et al., 2020). Sedentary behavior has seen a significant rise in the past few decades and affects almost 31% of the human population (Arnardottir et al., 2016). On average, adults spend more than 90% of their time in sedentary activities (Wheeler et al., 2019).

The prevalence of sedentary behavior has reached alarming proportions globally and is recognized as a significant public health concern (Owen et al., 2010). Studies indicate that prolonged sitting time is associated with an increased risk of cardiovascular diseases, type 2 diabetes, and allcause mortality, independent of physical activity levels (Biswas et al., 2015). The sedentary lifestyle has become so pervasive that it has been termed by some researchers as "the sitting disease" (Hamilton et al., 2007).

1.2 Obesity: Prevalence and health implications

Body mass index (BMI) is defined as the ratio of weight measured in kilograms divided by height in meters squared (Ahirwar and Mondal, 2019). A person is considered to be overweight if he/she has a BMI between the range of 25–29.9, and BMI values >30 are considered obese (Lin and Li, 2021). A drastic increase in obesity from 29% to 38% has occurred

Advances in Medical Imaging. https://doi.org/10.1016/B978-0-443-28967-5.00010-1

in the last two decades (González et al., 2017). In men, obesity has increased from 29% to 36%, and in women, obesity has reached a whopping 38% from 29.8% (González et al., 2017).

Obesity is linked with increased mortality and a wide range of diseases, such as chronic renal disease, nonalcoholic fatty liver disease, osteoarthritis, a variety of malignancies, and even death. The cost of treating these diseases is increasing each decade, making obesity a socioeconomic problem (Lin and Li, 2021). For individuals with a BMI, each one-unit BMI increase was associated with an additional cost of $253 (approximately Rs 20,000). It takes roughly $5275 (approximately 4 Lakhs Indian rupee (INR)) for the treatment of a patient with obesity and a single comorbidity associated with it (Okunogbe et al., 2021).

1.3 Brain–obesity connection: An emerging field

Much about brain volume and obesity is not yet explored (Lin and Li, 2021). BMI is associated with brain atrophy in the elderly (Raji et al., 2009). Studies have shown that brain regions like the hippocampus, cingulate gyrus, and frontal lobes are at risk for obesity-induced brain atrophy (Fotuhi and Lubinski, 2013). The hippocampus regulates food consumption, such as experiencing hunger at a set time of day, and influences memory and the learning process. It also regulates appetite and the dietary habits of individuals (Hamer and Batty, 2019). Obesity can influence hippocampal functioning and volume (Dekkers et al., 2019).

2 Aim and objectives

2.1 Aim

This study was conducted to understand the effects obesity has on the brain volume of sedentary normal weight, overweight, and obese adults. A high-resolution 3 T magnetic resonance imaging (MRI) scanner from UNITED IMAGING was used to acquire images of the brain. The study was designed to assess and compare the brain structural changes caused by increased BMI in normal weight, overweight, and obese individuals.

2.2 Objective

Assessment of brain volume and structural changes in the brain among sedentary normal-weight, overweight, and obese sedentary adults.

3 Materials and methods

3.1 Study design and setting

We conducted a cross sectional comparative study at the Department of Radiodiagnosis and Imaging, Kasturba Medical College, MAHE, Manipal, Karnataka, India, spanning from December 2022 to March 2024. The study involved sedentary adults aged 18–60 years of all genders who met the inclusion criteria and lacked the comorbidities outlined in the exclusion criteria. We employed purposive sampling to investigate the impact of obesity on brain volume in this population, using a simple random sampling method to select participants from the overall population.

3.2 Ethics and registration

The Institutional Ethics Committee-2 of Kasturba Medical College and Kasturba Hospital granted the study formal approval on the eighth of August 2023. Furthermore, the trial has been duly registered with the Clinical Trials Registry of India (CTRI) and bears the reference number Trial/2023/08/072586, with the corresponding registration number CTRI/2023/09/057310.

3.3 Sample size calculation

We calculated the sample size using the one-way analysis of variance (ANOVA) methodology. Based on this formula, with a 0.42 mm difference considered clinically significant, the required sample size in each group was calculated to be 34 participants. With three groups (normal, overweight, and obese), the total sample size required was 102 participants.

3.4 Participant selection

3.4.1 Inclusion criteria

We included patients referred for MRI brain scans at KMC Hospital. Patients with mild head trauma with a Glasgow Coma Score >13 were also included. The key criteria for inclusion were the evaluation of physical activity using the International Physical Activity Questionnaire (IPAQ) in METs (metabolic equivalent units) and the BMI of the patient.

3.4.2 Exclusion criteria

We excluded individuals with major neurological, psychiatric, or psychological disorders such as schizophrenia and dementia; uncontrolled hypertension; substance or drug abuse; cardiovascular diseases; systemic brain diseases; those undergoing antiinflammatory drug therapy; and patients with hypoxic ischemic encephalopathy, as these factors have been documented to influence brain volume.

3.5 Data collection procedures

3.5.1 Informed consent process

We provided each participant with a participant information sheet and delivered a comprehensive explanation of the study procedure. Participation was contingent upon the patient's free will, affording the patient the freedom to opt out of the study at any point in time. We obtained informed consent from each patient, signifying their voluntary participation in the study.

3.5.2 Baseline data collection

We assessed previous physical activity using the IPAQ (Fig. 8.1). The questionnaire was administered to the participants after obtaining appropriate written consent. The questionnaire's validity is moderate to good (0.64–0.08) when compared with accelerometer data, and it took 5–7 minutes to complete.

FIG. 8.1 IPAQ webpage.

We recorded the patient's height, weight, and age, and calculated the patient's BMI. BMI was calculated by dividing the person's weight (in kg) by their height squared (in m^2), as BMI = weight (kg)/height^2(m^2).

We classified individuals based on their BMI as follows:

Severely underweight: BMI < 16.5 kg/m^2
Underweight: BMI 16.5–18.4 kg/m^2
Normal weight: BMI 18.5–24.9 kg/m^2
Overweight: BMI 25–29.9 kg/m^2
Obese: BMI ≥ 30 kg/m^2

3.6 MRI acquisition

3.6.1 *Instruments used*

We acquired structural brain MRI images for all participants using the UMR 780, 3 T MRI machine by United Imaging at KMC Manipal (Fig. 8.2). We used weighing machines to measure the patients' weight and height. A 32 channel brain coil was used to acquire the isometric 3D spoiled gradient recalled echo (SPGR) sequence with a slice thickness of 1 mm and an increment of 1.5 mm.

3.6.2 *MRI protocol*

We acquired two sequences for brain volume estimation: a 3-plane localizer and a 3D SPGR. The 3D SPGR had the following parameters:

TR: 7.8 ms
TI: 1010 ms
TE: 3.1 ms
Flip angle: 10 degrees

3.6.3 *Positioning and acquisition*

We positioned patients in the head first supine position in the MRI machine. The brain coil was placed on the patient's head, and the localizer was positioned at the glabella. We selected the thesis protocol, which included a 3-plane localizer and 3D SPGR sequences (spoiled gradient). The 3D sequence had an isometric voxel size, which was needed for the Voxel-Based Morphometry (VBM) software. The acquisition took approximately 20–30 minutes.

3.7 Image processing and analysis

3.7.1 *Brain volume measurement*

We used the MRI spoiled gradient (3D T1 FSE SPGR) sequence. Statistical parametric mapping (SPM) and computational anatomy toolbox version 12.8 (CAT12) tools were employed for postprocessing and data analysis of the VBM based data obtained from the participants.

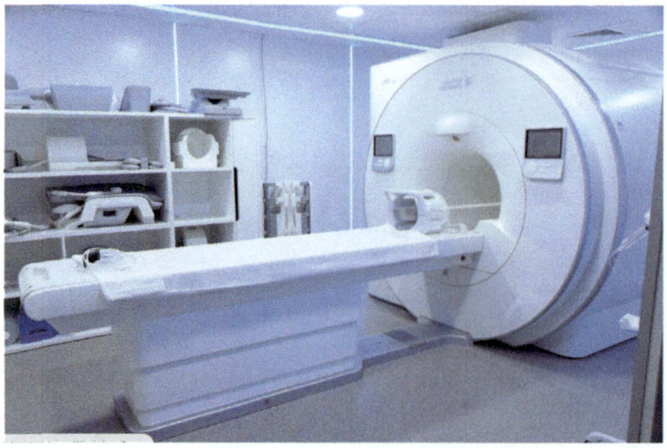

FIG. 8.2 MRI.

4 Outcome measures

We conducted MRI analysis for different brain structures prescribed in the Automated Anatomical Labeling atlas 3 (AAL3) Atlas, using the CONN toolbox software version 15 installed in MATLAB software version R2014a. The preprocessing steps were carried out with the help of CAT12 installed in the SPM12 (SPM version 12) software. We used MATLAB (version R2019a) as a common platform for this program.

5 Statistical analysis

We used summary statistics to characterize categorical variables, particularly weight categories. Based on the data's normality distribution, we employed either a one-way ANOVA or a Kruskal Wallis test to evaluate the data's significance for group comparisons. We used Statistical Package for the Social Sciences (SPSS) software (SPSS Inc., Chicago, IL, USA) for statistical data analysis, with continuous variables such as intracranial volumes summarized using arithmetic mean, standard deviation, median, and interquartile range.

6 Results

6.1 Study population

We recruited a total of 102 participants for this study, with equal distribution across three BMI categories: normal weight ($n = 34$), overweight ($n = 34$), and obese ($n = 34$). The study population included 54 males (52.9%) and 48 females (47.1%).

6.2 Demographic characteristics

6.2.1 Overall sample demographics

The mean age of all participants was 41.3 ± 11.7 years, ranging from 18 to 60 years. We observed no significant age differences between BMI groups ($p = 0.78$). Physical activity levels, as measured by the IPAQ, showed significant differences between BMI groups ($p < 0.001$), with lower activity levels observed in the overweight and obese groups compared to the normal weight group.

6.2.2 Group-specific demographics

Normal weight group (n = 34)

Mean age: 40.8 ± 12.1 years
Males: 18 (52.9%)
Females: 16 (47.1%)
Mean BMI: 22.1 ± 1.8 kg/m^2
Mean IPAQ score: 2856 ± 1245 MET-min/week

Overweight group (n = 34)

Mean age: 41.2 ± 11.5 years
Males: 18 (52.9%)
Females: 16 (47.1%)
Mean BMI: 27.3 ± 1.4 kg/m^2
Mean IPAQ score: 2134 ± 987 MET-min/week

Obese group (n = 34)

Mean age: 42.0 ± 11.6 years
Males: 18 (52.9%)
Females: 16 (47.1%)
Mean BMI: 33.7 ± 3.2 kg/m^2
Mean IPAQ score: 1789 ± 832 MET-min/week

6.3 Brain volumetric findings

6.3.1 *Male participants*

For male participants, we observed significant positive associations between BMI and various brain volume measures after adjusting for age. Specifically, a unit increase in BMI was associated with the following changes:

Total intracranial volume (TIV): increase of 6.911 units.

Cerebrospinal fluid: increase of 3.3 units.

Right middle temporal gyrus volume: increase of 0.10253 units.

Gray matter volume (GMV): increase of 2.473 units.

White matter volume: increase of 1.1475 units.

These results demonstrate a positive correlation between BMI and various brain volume measures in males, suggesting that a higher BMI is associated with increases in overall brain volume and specific brain regions.

Further region specific analyses revealed differential associations between BMI and brain volumes across various structures. In male participants, higher BMI was significantly associated with increased volumes in the thalamus ($\beta = 0.312$, $p = 0.007$), putamen ($\beta = 0.287$, $p = 0.013$), and cerebellum ($\beta = 0.254$, $p = 0.029$).

6.3.2 *Female participants*

In contrast, we observed no significant effects of BMI on brain volume in female participants. This lack of association remained consistent across all measured brain regions and tissue types (all $p > 0.05$), suggesting potential sex specific differences in how adiposity relates to brain structure.

6.3.3 *Age-related interactions*

Our analysis revealed significant interactions between age and BMI in relation to brain volumes. In both sexes, the relationship between BMI and brain volumes was modified by age, with stronger negative associations observed in older participants (age × BMI interaction: $p = 0.012$ for TIV, $p = 0.008$ for GMV).

6.3.4 *Physical activity as a moderator*

We found that physical activity levels moderated the relationship between BMI and brain volumes. Among participants with higher physical activity levels (above median IPAQ scores), the associations between BMI and brain volumes were attenuated compared to those with lower activity levels. This moderating effect was particularly pronounced for hippocampal volume (interaction $p = 0.031$).

7 Discussion

7.1 Sex-specific patterns in brain–obesity relationship

Our study revealed distinctive patterns in the relationship between BMI and brain volumes between male and female participants. In males, higher BMI was positively associated with various brain volumetric measures, while female participants showed no significant associations between BMI and brain volumes.

These sex specific patterns may reflect fundamental biological differences in adipose tissue distribution, metabolism, and neurobiological responses to obesity between males and females (Karastergiou et al., 2012). Men typically accumulate adipose tissue in visceral and abdominal regions, which has been associated with greater systemic inflammation and metabolic dysregulation compared to the subcutaneous and gluteofemoral fat distribution more common in women (Kuk et al., 2005). These distinct patterns of fat distribution may differentially impact brain structure through various mechanisms, including inflammation, insulin resistance, and hormonal pathways (Azad et al., 2007; Hara et al., 2015).

7.2 Age as a critical moderator

The significant interactions between age and BMI in our analyses highlight the importance of considering developmental stages when examining obesity's impact on brain structure. The stronger negative associations observed in older participants, particularly for hippocampal volume, suggest that aging brains may be more vulnerable to obesity related changes.

This finding aligns with the concept of inflammaging (Franceschi et al., 2018), where age related chronic low-grade inflammation combines with obesity induced inflammation to create a particularly detrimental neurobiological environment.

In addition, age related declines in neuroplasticity may reduce the brain's capacity to compensate for obesity-related insults (Barnes et al., 2000).

7.3 Physical activity as a protective factor

The moderating effect of physical activity on BMIbrain volume relationships observed in our study suggests potential neuroprotective benefits of exercise, even in the context of elevated BMI. This protective effect was particularly evident for hippocampal volume, a region known to be responsive to exercise through increased neurogenesis, angiogenesis, and neurotrophic factor production (Erickson et al., 2011).

These findings support the "fat but fit" paradigm (Barry et al., 2014; Ortega et al., 2018), which suggests that cardiorespiratory fitness may mitigate some of the adverse health effects associated with excess adiposity. Our results extend this concept to brain health, indicating that maintaining physical activity may help preserve brain structure despite elevated BMI. This has important implications for public health interventions, suggesting that promoting physical activity should be a priority even when weight loss is not achieved (Swift et al., 2013).

7.4 Implications for the Indian context

Our study provides novel insights into obesitybrain relationships in an Indian population, addressing an important gap in the literature, which has predominantly focused on Western populations. India's rapid urbanization and nutritional transition have created a context where traditional diets and lifestyle patterns are rapidly shifting, potentially exacerbating the impact of obesity on various health outcomes, including brain health (Ng and Popkin, 2012).

The increasing prevalence of sedentary behavior in India, particularly in urban areas, mirrors global trends but occurs within a unique cultural and socioeconomic context (Pratt et al., 2012). Workplace computerization has dramatically reduced physical activity requirements in many sectors, creating new challenges for maintaining brain health in the working-age population. Understanding these obesity–brain relationships in diverse populations is essential for developing culturally appropriate interventions.

8 Clinical implications

These findings suggest that:

- The relationship between obesity and brain structure is complex and sex dependent
- Age significantly modifies the obesity–brain relationship
- Physical activity may provide neuroprotective effects, regardless of BMI status
- Early intervention strategies should consider sex-specific approaches

9 Expected public health impact

In today's fast paced world, workplace computerization has significantly reduced the need for intensive labor, resulting in a substantial decrease in physical activity among employees. By conducting this study on the impact of obesity and sedentary behavior on brain health, we aim to advocate for the integration of physical activity into workplace routines. This initiative seeks to delay the adverse brain changes that can occur in employees and raise awareness about the importance of maintaining cognitive health.

This research holds national significance, particularly in the context of India's efforts to promote sustainable health practices to prevent conditions like cardiovascular diseases and diabetes. The outcomes of our study are poised to address the future need for corporate wellness programs in India, ultimately contributing to the overall well-being of the workforce (Goetzel et al., 2014).

10 Conclusion

We have demonstrated complex, sex specific relationships between BMI and brain volume in a sample of 102 sedentary Indian adults. Our findings reveal that, while male participants showed positive associations between BMI and various brain volumetric measures, female participants showed no significant associations. These results challenge simplistic narratives about obesity and brain health, suggesting more nuanced, sex-dependent relationships that vary across the lifespan.

The moderating effects of physical activity and age observed in our study highlight the importance of considering these factors in understanding obesity's impact on brain structure. The protective effect of physical activity, particularly on hippocampal volume, suggests that promoting exercise may benefit brain health even without significant weight loss.

Our study contributes to a growing body of evidence linking lifestyle factors, such as obesity and sedentary behavior, to brain structure (Siddarth et al., 2018; Falck et al., 2017). The complex patterns observed including sex differences, age interactions, and physical activity moderation highlight the need for nuanced, personalized approaches to promoting brain health in the context of modern lifestyles. By integrating physical activity into workplace routines and developing comprehensive wellness programs (Buckley et al., 2015), we may help preserve cognitive health in an increasingly sedentary world.

References

Ahirwar, R., Mondal, P.R., 2019. Prevalence of obesity in India: a systematic review. Diabetes Metab. Syndr. Clin. Res. Rev. 13 (1), 318–321.

Arnardottir, N.Y., Koster, A., Domelen, D.R.V., Brychta, R.J., Caserotti, P., Eiriksdottir, G., et al., 2016. Association of change in brain structure to objectively measured physical activity and sedentary behavior in older adults: age, gene/environment susceptibility-Reykjavik study. Behav. Brain Res. 296, 118–124.

Azad, N.A., Al Bugami, M., Loy-English, I., 2007. Gender differences in dementia risk factors. Gend. Med. 4 (2), 120–129.

Barnes, C.A., Rao, G., Houston, F.P., 2000. LTP induction threshold change in old rats at the perforant path–granule cell synapse. Neurobiol. Aging 21 (5), 613–620.

Barry, V.W., Baruth, M., Beets, M.W., Durstine, J.L., Liu, J., Blair, S.N., 2014. Fitness vs. fatness on all-cause mortality: a meta-analysis. Prog. Cardiovasc. Dis. 56 (4), 382–390.

Bergman, F., Matsson-Frost, T., Jonasson, L., Chorell, E., Sörlin, A., Wennberg, P., et al., 2020. Walking time is associated with hippocampal volume in overweight and obese office workers. Front. Hum. Neurosci. 14, 307.

Biswas, A., Oh, P.I., Faulkner, G.E., Bajaj, R.R., Silver, M.A., Mitchell, M.S., et al., 2015. Sedentary time and its association with risk for disease incidence, mortality, and hospitalization in adults: a systematic review and meta-analysis. Ann. Intern. Med. 162 (2), 123–132.

Buckley, J.P., Hedge, A., Yates, T., Copeland, R.J., Loosemore, M., Hamer, M., et al., 2015. The sedentary office: an expert statement on the growing case for change towards better health and productivity. Br. J. Sports Med. 49 (21), 1357–1362.

Dekkers, I.A., Jansen, P.R., Lamb, H.J., 2019. Obesity, brain volume, and white matter microstructure at MRI: a cross-sectional UK Biobank study. Radiology 291 (3), 763–771.

Erickson, K.I., Voss, M.W., Prakash, R.S., Basak, C., Szabo, A., Chaddock, L., et al., 2011. Exercise training increases size of hippocampus and improves memory. Proc. Natl. Acad. Sci. U. S. A. 108 (7), 3017–3022.

Falck, R.S., Davis, J.C., Liu-Ambrose, T., 2017. What is the association between sedentary behaviour and cognitive function? A systematic review. Br. J. Sports Med. 51 (10), 800–811.

Fotuhi, M., Lubinski, B., 2013. The Effects of Obesity on Brain Structure and Size [Internet]. Practical Neurology. Available from: https://practicalneurology.com/articles/2013-july-aug/the-effects-of-obesity-on-brain-structure-and-size/pdf.

Franceschi, C., Garagnani, P., Parini, P., Giuliani, C., Santoro, A., 2018. Inflammaging: a new immune-metabolic viewpoint for age-related diseases. Nat. Rev. Endocrinol. 14 (10), 576–590.

Goetzel, R.Z., Henke, R.M., Tabrizi, M., Pelletier, K.R., Loeppke, R., Ballard, D.W., et al., 2014. Do workplace health promotion (wellness) programs work? J. Occup. Environ. Med. 56 (9), 927–934.

González, K., Fuentes, J., Márquez, J.L., 2017. Physical inactivity, sedentary behavior and chronic diseases. Korean J Fam Med 38 (3), 111.

Hamer, M., Batty, G.D., 2019. Association of body mass index and waist-to-Hip ratio with brain structure. Neurology 92 (6), e594–e600.

Hamilton, M.T., Hamilton, D.G., Zderic, T.W., 2007. Role of low energy expenditure and sitting in obesity, metabolic syndrome, type 2 diabetes, and cardiovascular disease. Diabetes 56 (11), 2655–2667.

Hara, Y., Waters, E.M., McEwen, B.S., Morrison, J.H., 2015. Estrogen effects on cognitive and synaptic health over the lifecourse. Physiol. Rev. 95 (3), 785–807.

Karastergiou, K., Smith, S.R., Greenberg, A.S., Fried, S.K., 2012. Sex differences in human adipose tissues - the biology of pear shape. Biol. Sex. Differ. 3 (1), 13.

Kuk, J.L., Lee, S., Heymsfield, S.B., Ross, R., 2005. Waist circumference and abdominal adipose tissue distribution: influence of age and sex. Am. J. Clin. Nutr. 81 (6), 1330–1334.

Lin, X., Li, H., 2021. Obesity: epidemiology, pathophysiology, and therapeutics. Front. Endocrinol. (Lausanne) 12, 706978.

Ng, S.W., Popkin, B.M., 2012. Time use and physical activity: a shift away from movement across the globe. Obes. Rev. 13 (8), 659–680.

Okunogbe, A., Nugent, R., Spencer, G., Ralston, J., Wilding, J., 2021. Economic impacts of overweight and obesity: current and future estimates for eight countries. BMJ Glob. Health 6 (10), e006351.

Ortega, F.B., Ruiz, J.R., Labayen, I., Lavie, C.J., Blair, S.N., 2018. The Fat but Fit paradox: what we know and don't know about it. Br. J. Sports Med. 52 (3), 151–153.

Owen, N., Healy, G.N., Matthews, C.E., Dunstan, D.W., 2010. Too much sitting: the population-health science of sedentary behavior. Exerc. Sport Sci. Rev. 38 (3), 105–113.

Park, J.H., Moon, J.H., Kim, H.J., Kong, M.H., Oh, Y.H., 2020. Sedentary lifestyle: overview of updated evidence of potential health risks. Korean J Fam Med 41 (6), 365–373.

Pratt, M., Sarmiento, O.L., Montes, F., Ogilvie, D., Marcus, B.H., Perez, L.G., et al., 2012. The implications of megatrends in information and communication technology and transportation for changes in global physical activity. Lancet 380 (9838), 282–293.

Raji, C.A., Ho, A.J., Parikshak, N.N., Becker, J.T., Lopez, O.L., Kuller, L.H., et al., 2009. Brain structure and obesity. Hum. Brain Mapp. 31, 353–364.

Siddarth, P., Burggren, A.C., Eyre, H.A., Small, G.W., Merrill, D.A., 2018. Sedentary behavior associated with reduced medial temporal lobe thickness in middle-aged and older adults. PLoS One 13 (4), e0195549.

Swift, D.L., Lavie, C.J., Johannsen, N.M., Arena, R., Earnest, C.P., O'Keefe, J.H., et al, 2013. Physical activity, cardiorespiratory fitness, and exercise training in primary and secondary coronary prevention. Circ. J. 77 (2), 281–292.

Wheeler, M.J., Dunstan, D.W., Smith, B., Smith, K.J., Scheer, A., Lewis, J., et al., 2019. Morning exercise mitigates the impact of prolonged sitting on cerebral blood flow in older adults. J. Appl. Physiol. 126 (4), 1049–1055.

Part III

Vascular function and ultrasound applications

Chapter 9

Impact of sedentary behavior and hypertension on vascular and cognitive health in office workers: Imaging insights and interventions

Sathya Sabina Muthu[a], Priya Seagal V.K.[b], Sneha Ravichandran[a], Rajagopal Kadavigere[c], Baskaran Chandrasekaran[d], Leena R. David[a,e], Poovitha Shruthi Paramashiva[f], Dilber Uzun Ozsahin[e,g,h], Berna Uzun[h], Suresh Sukumar[a], Ilker Ozsahin[h], and Dilip Shettigar[a]

[a]*Department of Medical Imaging Technology, Manipal College of Health Professions, Manipal Academy of Higher Education, Manipal, Karnataka, India*

[b]*Adichunchanagiri University, Karnataka, India*

[c]*Department of Radiodiagnosis and Imaging, Kasturba Medical College, Kasturba Medical College, Manipal, India*

[d]*Symbiosis School Of Sports Sciences, Symbiosis International (Deemed University), Pune, Maharashtra, India*

[e]*Medical Diagnostic Imaging Department, College of Health Sciences, University of Sharjah, Sharjah, United Arab Emirates*

[f]*Manipal Academy of Higher Education, Division of Yoga, Centre for Integrative Medicine and Research, Manipal, India*

[g]*Research Institute for Medical and Health Sciences, University of Sharjah, Sharjah, United Arab Emirates*

[h]*Irfan Suat Gunsel Operational Research Institute, Near East University, Nicosia, Turkish Republic of Northern Cyprus (TRNC), Turkey*

1 Introduction

1.1 Modern workplaces and sedentarism

In today's technologically advanced world, the shift toward computerization and automation has drastically transformed workplace environments. While these advancements have significantly improved productivity and reduced the need for manual labor, they have also contributed to a growing public health concern—sedentarism. Sedentary behavior, defined as any waking behavior characterized by a seated or reclined posture with low energy expenditure (typically $\leq$1.5 metabolic equivalents (METs)), has become a common feature of modern work life (Tremblay et al., 2017). This shift is particularly evident among individuals engaged in white collar or desk-based occupations, where prolonged sitting is an integral part of daily routines. Numerous studies have linked sedentary behavior with an increased risk of chronic diseases. Mechanistically, sedentarism contributes to hyperglycemia (elevated blood glucose), hyperlipidemia (high levels of fat in the blood), and systemic inflammation (Dunstan et al., 2021). These metabolic disruptions are believed to stem from the body's decreased ability to regulate glucose and lipid metabolism due to prolonged physical inactivity. In addition, sitting for extended hours negatively impacts vascular health by altering shear stress and endothelial function, increasing the risk of cardiovascular diseases, including hypertension and atherosclerosis.

Modern office setups often involve employees sitting in the same position for hours, frequently with minimal movement. This static posture not only impacts musculoskeletal health but also restricts circulation and metabolism (Daneshmandi et al., 2017). The vascular system, particularly in the lower extremities, is especially vulnerable, as reduced muscle contractions during prolonged sitting decrease venous return and impair endothelial responsiveness. Over time, these changes may lead to arterial stiffness, reduced flow-mediated dilation (FMD), and increased cardiovascular risk (Stoner et al., 2021). Moreover, the effects of reduced physical activity extend beyond the cardiovascular and metabolic systems to brain health. Sedentary behavior has been associated with decreased cerebral blood flow and impaired autoregulation of cerebral vessels.

Advances in Medical Imaging. https://doi.org/10.1016/B978-0-443-28967-5.00034-4

These alterations may contribute to cognitive decline, particularly in aging adults. Reduced physical activity impairs the delivery of oxygen and nutrients to the brain, which are essential for maintaining cognitive functions such as memory, attention, and problem-solving abilities (Hillman et al., 2008).

While technological progress in the workplace has improved efficiency and convenience, it has inadvertently created a lifestyle conducive to sedentarism. To counter these adverse health effects, it is crucial to incorporate regular physical activity into daily routines, including short active breaks, standing desks, and encouraging movement throughout the workday. Promoting a more active work culture is essential for preserving overall health and preventing the long-term consequences of sedentary living.

2 Hypertension in office workers

With the rapid growth of digitization and an increasing shift toward sedentary desk-based occupations, office workers are becoming particularly vulnerable to lifestyle related diseases, including hypertension. The modern work environment often requires employees to sit for prolonged hours with minimal physical activity, which significantly contributes to poor cardiometabolic health (Reichel et al., 2022). Contemporary research strongly links occupational sedentary behavior with early markers of cardiometabolic disease, including elevated blood pressure and metabolic dysregulation (Tri Tjahjono and Arthamin, 2024). Globally, hypertension affects approximately 1.28 billion adults , making it one of the most prevalent metabolic disorders (Cheraghi et al., 2025). It is often termed the "silent killer" because its progression is generally asymptomatic until serious complications arise. Chronic high blood pressure imposes stress on the vascular system, damaging blood vessels and affecting multiple organs over time. Some of the common systemic manifestations of prolonged hypertension include hypertensive retinopathy, choroidopathy, optic neuropathy, and generalized vascular dysfunction (Mensah, 2016). One of the significant vascular changes observed is arterial stiffness, which further elevates cardiovascular risk.

The insidious nature of hypertension makes early diagnosis critical yet challenging. Research now focuses on subtle vascular changes, such as reduced blood flow velocity and increased vascular wall thickness, which may serve as early diagnostic markers. These changes can help identify individuals at risk before irreversible organ damage occurs. For example, vascular alterations in the kidneys can lead to hypertensive nephropathy and impaired renal function. Similarly, persistent hypertension adversely affects the brain, contributing to structural and functional changes. Cerebral consequences of hypertension are particularly alarming. Chronic high blood pressure is associated with reduced cerebral blood flow and impaired cerebral perfusion, which may result in cognitive decline, brain atrophy, and a reduction in brain volume over time (Li et al., 2023b). Neuroimaging studies have shown that hypertensive individuals often exhibit increased white matter hyperintensities (WMHs),areas of the brain that reflect microvascular damageand changes in gray matter integrity (Li et al., 2023a). These alterations are linked to cognitive deficits, such as reduced memory, slower processing speed, and executive dysfunction. Given the high prevalence and hidden nature of hypertension, especially among sedentary office workers, there is a pressing need to promote preventive strategies. These may include regular blood pressure screening, workplace wellness programs, active breaks during work hours, and ergonomic interventions such as standing desks. Addressing sedentary behavior and raising awareness about hypertension's early signs could play a crucial role in reducing the long-term burden of this widespread condition.

3 Vascular changes due to sedentarism

The growing physical inactivity caused by modern work systems, which require prolonged sitting for hours due to increased workloads and deadlines, has put the health of office workers at stake (Daneshmandi et al., 2017). Physical inactivity is a contributor to various diseases, including hypertension, atherosclerosis, and cardiovascular diseases, thus reducing life expectancy and leading to increased mortality across the world (Young et al., 2016). Recent studies have shown that uninterrupted prolonged sitting impairs endothelial function and induces arterial structural alterations, predominantly in the lower body vascular system (Thosar et al., 2015). Similar effects may occur in the cerebral vasculature, causing impairment of cerebral blood flow following prolonged sitting. Some of the vascular changes that occur due to sedentarism are mentioned below.

3.1 Endothelial damage

Endothelial dysfunction is a critical early marker of cardiovascular disease and is closely linked to sedentary behavior, particularly prolonged sitting. Research indicates that sitting continuously for long durations s can lead to a significant decline in endothelial function (Padilla and Fadel, 2017). This impairment is primarily assessed using a noninvasive technique called FMD. FMD measures the ability of blood vessels, particularly peripheral arteries like the femoral artery, to dilate in response to increased blood flow, a response largely mediated by nitric oxide (NO) produced by healthy endothelial cells (Thijssen et al., 2011). During prolonged sitting, reduced muscle activity leads to diminished blood flow and shear stress in the lower limbs, resulting in decreased NO production and compromised endothelial responsiveness. This decline in FMD values reflects endothelial dysfunction, which is an early indicator of atherosclerosis, the hardening and narrowing of arteries due to plaque buildup (Restaino et al., 2016). The reduction in endothelial function due to sedentary behavior is concerning because it suggests the onset of vascular changes even in healthy individuals. Therefore, frequent interruptions to sitting time, such as standing or engaging in light activity, are recommended to maintain vascular health. Continuous monitoring of FMD can be a valuable tool in assessing cardiovascular risk associated with prolonged sedentary behavior.

4 Blood flow hemodynamics

Prolonged sitting adversely affects key hemodynamic parameters such as blood flow and shear stress in both upper and lower limb vasculature. These parameters are essential for maintaining vascular tone and overall endothelial health. Shear stress, the frictional force exerted by blood flow on the vessel wall, plays a critical role in stimulating endothelial cells to release vasoactive substances, most notably nitric oxide (NO) (Peng et al., 2019). NO is a potent vasodilator that helps regulate vascular tone, inhibits platelet aggregation, and prevents inflammatory responses within the vascular wall. When an individual remains seated for extended periods, muscle inactivity leads to decreased venous return and reduced arterial blood flow, particularly in the lower limbs. This reduction in blood flow results in diminished shear stress, thereby impairing the endothelium's ability to release adequate levels of NO. Consequently, vascular tone regulation is disrupted, predisposing the blood vessels to dysfunction. Studies have consistently shown that prolonged sitting leads to a measurable decline in shear rate and blood flow in both upper and lower limbs, which may initiate or accelerate endothelial damage (Thosar et al., 2012). These changes contribute to a higher risk of cardiovascular diseases, highlighting the importance of frequent movement and posture changes to maintain vascular function and prevent long-term health consequences.

4.1 Arterial stiffness

A sedentary work life characterized by prolonged sitting is increasingly recognized as a risk factor for developing arterial stiffness, a condition where arteries lose their elasticity, impairing their ability to expand and contract with each heartbeat. One of the most reliable indicators of arterial stiffness is an increase in Pulse Wave Velocity (PWV), which measures the speed at which blood pressure waves move through the arteries. A higher PWV reflects stiffer arteries and is associated with elevated cardiovascular risk. Studies have shown that individuals with sedentary lifestyles, particularly those with extended periods of uninterrupted sitting, exhibit increased arterial stiffness (Moinuddin et al., 2025). This phenomenon is especially pronounced in obese young adults and hypertensive or obese middle-aged individuals. In these populations, the combination of excess body weight, elevated blood pressure, and physical inactivity creates a physiological environment that accelerates vascular aging and impairs arterial compliance. Prolonged sitting reduces muscle contractions and venous return, leading to lower shear stress on the arterial walls (Tamiya et al., 2024). This reduction in mechanical stimulation can cause endothelial dysfunction and promote structural changes in the vessel wall, such as increased collagen deposition and reduced elastin content, which contribute to stiffness. Incorporating regular movement breaks, standing, or light activity throughout the day can mitigate these adverse effects and help preserve arterial health.

4.2 Peripheral vasculature alterations

Reduced physical activity significantly affects arterial structure and function, particularly the lumen diameter and endothelial responsiveness. Physical inactivity leads to diminished mechanical stimuli, such as blood flow and shear stress, which are essential for maintaining vascular health. Studies have demonstrated that individuals who are physically inactive show a marked reduction in FMD responses, especially in peripheral arteries like the popliteal (in the lower limb) and brachial (in the upper limb) arteries (Boyle et al., 2013). This decline in FMD indicates compromised endothelial function, an early marker of vascular dysfunction and cardiovascular risk. The extent of vascular alterations due to inactivity is influenced by both the duration and intensity of inactivity. For instance, a sudden and sustained reduction in physical activity, such as in patients confined to bed rest, leads to significant vascular changes. These include an increase in diastolic blood pressure and a further decline in FMD responses, reflecting early signs of arterial stiffness and impaired endothelial regulation (Nosova et al., 2014). Moreover, endothelial cell injury from reduced physical activity may trigger the release of endothelial microparticles into the circulation (Horstman et al., 2004). These microparticles are considered biomarkers of endothelial dysfunction and vascular stress. Their elevated presence further supports the link between physical inactivity, endothelial damage, and increased cardiovascular disease risk.

4.3 Cerebral vasculature alterations

Desk-based occupations that involve prolonged sitting and reduced physical activity are increasingly associated with negative effects on cerebral circulation and brain health. Extended periods of sitting especially beyond 6 hours per day can lead to a measurable decline in cerebral blood flow (Jones et al., 2022). Doppler ultrasound imaging in desk-based workers with prolonged sitting for long hours has revealed reduced blood flow velocity and impaired cerebral autoregulation, the brain's mechanism for maintaining stable blood flow despite fluctuations in systemic blood pressure (Carter et al., 2018). This impairment suggests that sedentary behavior compromises the ability of cerebral vessels to adequately supply the brain with oxygen and nutrients during periods of need. Over time, prolonged sitting contributes to increased cerebrovascular resistance, which is particularly detrimental in older adults (Maasakkers et al., 2020). This resistance impairs nutrient delivery and waste removal, critical functions for maintaining cognitive health. Consequently, chronic sedentary behavior has been linked to declines in cognitive functions such as memory, attention, and processing speed (Gogniat, 2026). These effects are more pronounced in aging populations, who are already at increased risk for neurodegenerative conditions. Therefore, incorporating regular physical movement throughout the workday, even in the form of light walking or stretching, can help preserve cerebral autoregulation, enhance cognitive function, and protect long-term brain health in individuals with sedentary lifestyles.

5 Methods to assess vascular changes

Evidence has proven that sedentary behavior significantly damages the blood vessels; for example, it lowers FMD, increases vascular pressure, degrades shear rate, and raises PWV (Zheng et al., 2021). All of the research shows that uninterrupted prolonged sitting can impair endothelial vascular function in peripheral conduit arteries among children, young adults, and at-risk groups. Sitting particularly affects the arteries in the lower limbs, but vascular abnormalities in the upper limbs have not been identified. Since no studies have explicitly examined these groups e.g., at-risk populations (obesity), sex, and level of fitness within the same human trial, it remains unclear how severe sitting-induced endothelial dysfunction is compared to healthy young adults (Daniele et al., 2022).

Parameters such as blood flow and shear rate/stress are also affected by prolonged sitting in both lower and upper limb arteries, and it is currently unclear whether the differences are significant (Daniele et al., 2022). Studies have proven that a sedentary lifestyle can cause reactive hyperemia, which can be linked to reduced microvascular function and vasodilation (Restaino et al., 2015). Studies have also demonstrated a positive association between arterial stiffness and sedentary time (Huynh et al., 2014; Ahmadi-Abhari et al., 2017) which is associated with other comorbidities. Sedentarism can also affect another hemodynamic factor called the oscillatory shear index (OSI), which is used to assess disturbed blood flow patterns. A study reveals that three hours of sitting significantly reduces the antegrade shear rate in the superficial femoral artery and brachial artery (BA), which in turn affects FMD and an increase in the OSI is noted in the BA. This implies that sitting modifies blood hemodynamics and contributes to endothelial damage (Thosar et al., 2014). Another study compared the acute effects of cycling on carotid artery hemodynamics between basketball athletes and sedentary subjects. OSI values in basketball athletes were higher than those in sedentary controls (Liu et al., 2015). The relationship between OSI and sedentarism can be a subject for further investigation to understand its impact on vascular health. Certain investigations show no changes or increased blood pressure (systolic blood pressure and diastolic blood pressure) in subjects who sit for prolonged periods, which increases the risk of developing cardiovascular diseases in high-risk populations (Luo et al., 2020). Such vessel structures can be measured using ultrasonography (USG), Doppler ultrasound, computed tomography (CT) angiography (CTA), magnetic resonance angiography (MRA), and digital subtraction angiography. Modalities that offer blood vessel assessment are shown in Table 9.1.

6 Methods to assess cerebral alterations

Physical inactivity is strongly associated with an increased risk of hypertension. When focusing on hypertensive effects on the brain, the most commonly observed changes are decreased cerebral blood flow, reduced cerebral perfusion, and increased WMHs as well as gray matter changes. Cerebral blood flow can be assessed using transcranial Doppler ultrasound,

TABLE 9.1 Imaging modalities offering blood vessel assessment.

Modality	Advantages	Disadvantages
USG	• Noninvasive • Relatively inexpensive • Real time viewing • Easily accessible	• Operator dependent • Large body habitus can restrict vessel evaluation • Deep or small vessels are difficult to visualize
CTA	• Detailed high resolution images • Used to study deep or small blood vessels • Quick assessment	• Uses ionizing radiation • Expensive • Requires contrast agents and a higher risk of contrast media reactions • Technical limitations • Prepreparation is required
MRA	• Detailed high resolution images • Used to study deep or small blood vessels • No harmful radiation is used	• High cost • Not suitable for claustrophobic individuals or patients with metal implants • Time consuming
DSA	• Detailed high resolution images • Used to study deep or small blood vessels	• Invasive procedure • Uses ionizing radiation to study the vessels • Needs contrast injection • Prepreparation is required

TABLE 9.2 Imaging modalities to assess cerebral alterations.

Modality	Advantages	Disadvantages
CT	• Reduced scanning time • Less expensive • Quick assessment	• Uses ionizing radiation • Less precise compared to other modalities
MRI	• Detailed high resolution images • Better grey and matter differentiation	• High cost • Not suitable for claustrophobic individuals or patients with metal implants • No ionizing radiation is used
Synthetic MRI	• Reduced scan time compared to traditional MRI • Obtains multiple contrast images in a single acquisition—T1, T2, and proton density images	• Technical limitations • Requires additional software and applications

functional magnetic resonance imaging (MRI), positron emission tomography (PET), CT perfusion scans, xenon CT, single-photon emission CT (SPECT), and arterial spin labeling. Cerebral perfusion can be studied using MRI, CT, ultrasound, and nuclear imaging techniques such as SPECT and PET. WMHs, which indicate the extent of white matter damage, can be examined using MRI (T2 and fluid-attenuated inversion recovery, FLAIR sequences) and diffusion tensor imaging. Gray matter changes, which include the expansion of sulcal spaces and ventricles, decreased total and regional gray matter volume, and reduced total brain volume (Burgmans et al., 2010; Firbank et al., 2007; Goldstein et al., 2002; Longstreth et al., 1996; Raz et al., 2007), can be assessed using volumetric analysis by skull stripping and preprocessing data from CT and MRI images. The imaging modalities used to study cerebral alterations include CT, MRI, and synthetic MRI, with their respective advantages and limitations summarized in Table 9.2.

7 Cognitive decline and sedentary behavior

Cognition refers to a broad range of mental processes that include learning, memory, attention, reasoning, problem-solving, and decision-making (Harvey, 2019) . These functions are essential for individuals to effectively perform daily tasks and engage in complex work-related activities. In modern work environments, particularly among office workers, cognitive ability is a vital component of productivity, creativity, and job performance. However, recent studies have increasingly shown that a sedentary lifestyle marked by prolonged periods of physical inactivity and sitting may negatively impact cognitive functioning Cai et al., 2023). Sedentary behavior can lead to a decline in key cognitive skills, including memory, attention span, and executive functioning, all of which are critical in high-demand work settings. Office workers often sit for extended hours with limited physical movement, which can restrict blood flow to the brain and reduce oxygen and nutrient delivery. This can impair brain health over time, reducing neural efficiency and connectivity. In the long term, persistent physical inactivity may contribute to structural and functional brain changes, ultimately leading to cognitive decline (Gregory et al., 2012). Cognitive decline not only reduces work performance and productivity but may also signal the early stages of more serious neurological conditions such as dementia or Alzheimer's disease (Colita et al., 2024). These conditions are characterized by progressive deterioration of memory and thinking skills, often beginning subtly before manifesting more severely. Older office workers are particularly vulnerable, as age-related changes in the brain may be compounded by a sedentary lifestyle (Imboden, 2024). Hence, early detection of cognitive decline is crucial to initiate preventive strategies and protect long-term brain health.

Regular physical activity has been shown to counteract the negative effects of sedentarism. Exercise promotes increased blood circulation throughout the body and brain, enhancing oxygenation and nutrient supply to neural tissues. It also stimulates neurogenesis, the process of creating new neurons and supports synaptic plasticity, which is essential for learning and memory (van Praag, 2008). Moreover, physical activity helps reduce stress, improves mood, and enhances mental clarity. Incorporating physical movement into daily office routines—such as walking breaks, standing desks, or short bouts of stretching—can significantly benefit cognitive health. Encouraging an active workplace culture is essential not only for physical well-being but also for preserving mental sharpness, delaying age-related cognitive decline, and improving the overall quality of life in office workers.

8 Methods to measure cognitive functions

The phrase "cognitive function" is extensive and describes mental processes related to reasoning, information manipulation, and acquiring knowledge. Domains of cognitive functions are numerous, and they play an important role in day-to-day activities. Cognitive assessment includes the examination of attentiveness, intellectuality, judgment, learning ability, memory, orientation, perception, problem-solving, psychomotor skills, reaction time, and social intactness (Peddie et al., 2021). Several psychological measurement tools to evaluate cognitive function are available with different parameters; each set of evaluation tools or batteries to assess cognitive functions has its own merits and demerits.

8.1 Mini-mental state exam

The Mini-Mental State Examination (MMSE) is a widely recognized and validated tool used for the brief assessment of cognitive function. It is often considered one of the gold standards in clinical and research settings for detecting cognitive impairment and tracking changes over time (Cook et al., 2005). The MMSE is structured into two major components. The first component evaluates orientation, memory, and attention by asking questions related to time, place, immediate and short-term memory recall, and basic attention tasks such as serial subtraction. This section is essential for detecting early signs of disorientation or memory decline. The second component focuses on language and executive functions by assessing verbal and written abilities. Tasks in this section include naming common objects, following verbal and written commands, reading comprehension, writing a sentence, and copying a design. These items help assess more advanced cognitive functions related to communication and motor coordination. Despite its clinical utility and widespread acceptance, the MMSE is not frequently used in routine screenings due to practical limitations (Nouchi and Kawashima, 2014). The tool is copyrighted, which can lead to increased costs for its use in larger populations or routine assessments. These restrictions have limited its accessibility, especially in low-resource settings or general workplace screenings, despite its diagnostic value.

8.2 Mini Cog

A 3-minute evaluation tool appropriate for primary care centers consists of a clock-drawing and three-item recall test. The clock-drawing test assesses cognitive, language, executive function, and visuospatial skills, while the delayed three item recall analyzes memory (Gallegos et al., 2022). It has demonstrated high sensitivity and specificity for diagnosing cognitive impairment, although its accuracy varies depending on location and interpretation techniques (Lui et al., 2026; Trongsakul et al., 2015; Ricci et al., 2016; Dudas et al., 2005).

8.3 Montreal cognitive assessment

The Montreal Cognitive Assessment (MoCA) is a brief yet comprehensive cognitive screening tool designed to detect mild cognitive impairment and early signs of cognitive dysfunction. Developed as a more sensitive alternative to the MMSE, MoCA takes approximately 10 minutes to administer and evaluates multiple cognitive domains. These include attention and concentration, executive functions, memory, language, visuoconstructional skills, conceptual thinking, calculations, and orientation. Its wide range of test items makes it an effective tool for detecting subtle cognitive deficits that may not be captured by other brief assessments. MoCA has several advantages over the MMSE. One of the most notable differences lies in the memory component. MoCA includes more words in its memory tasks, uses minimal learning trials, and incorporates a longer delay period before the recall portion. These elements increase the test's sensitivity to memory impairments, particularly in individuals with early-stage dementia or Alzheimer's disease. In addition, MoCA includes more complex and diverse tasks, such as alternating trail making, clock drawing, and verbal fluency, which make it more effective in assessing higher-order cognitive functions. Another key advantage of MoCA is its accessibility. Unlike the MMSE, which is copyrighted and may incur usage costs, MoCA is freely available for use in clinical, educational, and research settings. It has been translated into over 56 languages, increasing its applicability across diverse populations and cultures (Nasreddine et al., 2005). This makes MoCA a preferred tool for both initial screenings and ongoing cognitive assessments, especially in older adults and at-risk groups.

8.4 Addenbrooke's cognitive examination

It is a concise, multilingual pencil-and-paper tool for assessing cognitive function, which can be used at the bedside (Mathuranath et al., 2007). This test is freely available and used for clinical and research purposes. The Addenbrooke's cognitive examination is intended to address the MMSE's neuropsychological omissions (Hodges and Larner, 2017). It includes tests for attention, memory, language, visual perception, visuospatial skills, and orientation.

8.5 Trial making test

It is a quick screening test for cognitive impairment. Divided into two parts, Part A assesses psychomotor speed and visual search, and Part B involves executive functions (Bhatia et al., 2007). The advantages of trial tests include their ease of use, short administration times, and sensitivity to executive dysfunction (Kim et al., 2014; Maj et al., 1993). However, they also have drawbacks, such as low specificity, high variability, measuring overall completion times without performing individual analysis, and a fixed spatial configuration for each condition.

8.6 A quick test of cognitive speed

A quick test of cognitive speed assesses processing speed and overall cognitive speed (Linari et al., 2022). The test stands out as a concise and user-friendly assessment tool, requiring minimal training for administration and interpretation. Comprising three conditions, the first two focus on single dimensions color or form naming, while the third integrates two dimensions color and form naming (Nielsen et al., 2004).

Certain well-accepted batteries used in India include the PGI-battery of brain dysfunction (PGI-BBD), the National Institute of Mental Health and Neuro Sciences (NIMHANS) Neuropsychological Battery (NNB), and the All India Institute of Medical Sciences (AIIMS) Neuropsychological Battery.

8.7 ThePostgraduate Institute battery of brain dysfunction (PGI-BBD)

PGI-BBD was developed at the Post Graduate Institute of Medical Education and Research, Chandigarh, India, by Dwarka Prasad and Santhosh K. Verma in 1990. It covers multiple areas of cognitive functions and includes five subtests: PGI Memory Scale, Bhatia's Short Scale, Verbal Adult Intelligence Scale, Nahor-Benson Test of Perceptual Acuity, and Bender Visual-Motor Gestalt Test (Classon et al., 2022). It involves variables that include verbal retention for common and uncommon pairs, visual retention, identification, delayed recall, immediate recall, mental balance, concentration, remote memory, and recent memory (Sahu et al., 2005).

8.8 NIMHANS neuropsychological battery

NNB is the earliest test battery developed in India. The neuropsychological battery is designed for children (5–15 years) with 28 subtests to assess brain-injured individuals, for adults (16–65 years) with 19 subtests to evaluate neurocognitive disorders, and for elderly individuals, which includes 28 subtests for children, 19 subtests for adults, and 7 domains for elderly subjects (Tripathi et al., 2013).

Other digital cognitive assessment testing includes Inquisit software, CANTAB, Cogstate, CogniFit, cCOG, and Internet-based Cognitive Assessment Tool (ICAT).

9 Methods to measure physical activity

Any bodily motion of the skeletal muscle that requires energy expenditure is termed physical activity, including movement during traveling, work, or domestic activities. The activity can be mild, moderate, or vigorous depending on the tasks performed and can affect the health of an individual. Employing a valid, effective, convenient, and reliable tool to measure physical activity is important, and the main methods to assess physical activity are mentioned below. The different methods for physical activity measurement are shown in Fig. 9.1 (Sylvia et al., 2014).

9.1 Self- report questionnaires

Selfreport questionnaires are one of the most widely used tools for assessing physical activity in both clinical and research settings. They are popular due to their low cost, ease of administration, and ability to collect data from large populations in a relatively short time. These questionnaires typically gather information about the frequency, duration, and intensity of various physical activities over a specified period, such as daily or weekly routines (Sallis, 1991; Jacobs et al., 1993). Their flexibility makes them useful for a broad range of studies and population groups. However, the accuracy and reliability of self-report questionnaires can be influenced by several factors. One primary limitation is their reliance on the participant's memory and ability to accurately recall past activities. The subjective nature of the responses can lead to overestimation or underestimation of activity levels. Furthermore, variations in how physical activity intensity is interpreted and reported can affect the consistency of the data (Shephard, 1999; Shephard, 2003 ; Baranowski et al., 1984). These tools are often less effective at capturing light or moderate activities and do not always provide accurate estimates of energy expenditure. They also depend on the respondent's literacy and comprehension skills, which can vary significantly across different populations. Other factors such as question complexity, the influence of social desirability (i.e., reporting what is perceived as acceptable), and the age of the respondent, can further affect the quality of the data collected (Klesges et al., 1990; Uitenbroek, 1993; Durante and Ainsworth, 1996; Vanhees et al., 2005). Despite these limitations, self-report questionnaires are more reliable when analyzing group-level trends rather than individual behaviors. Therefore, selecting the appropriate questionnaire should align with the research objectives, target population, and the type of physical activity information required.

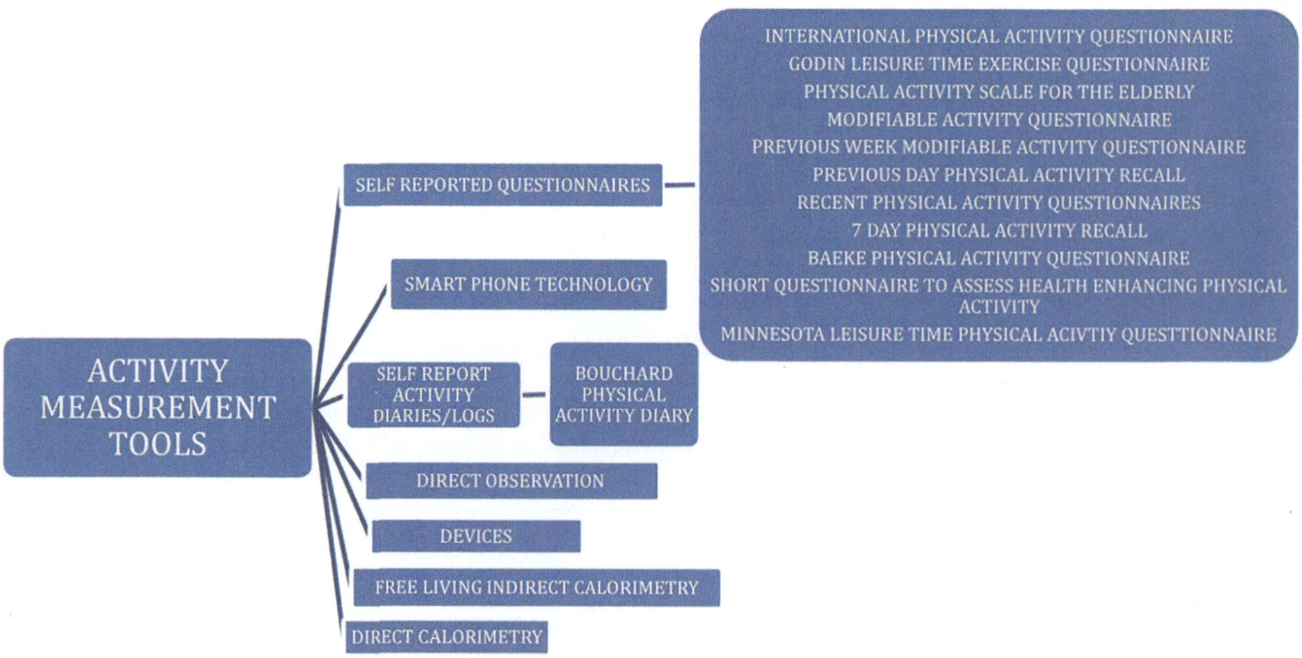

FIG. 9.1 Different methods for physical activity measurement.

9.2 Self-report activity diary/logs

Self-report activity diaries or logs are tools used to track physical activity by requiring individuals to document their movements and activities in real time or at regular intervals throughout the day (Vanhees et al., 2005; Van Der Ploeg et al., 2010). These logs offer a more detailed and accurate account of physical activity compared to retrospective questionnaires, as they minimize recall bias and allow for the recording of specific parameters such as the type, duration, intensity, and purpose of each activity. This real-time documentation enables researchers and healthcare providers to better assess patterns of physical activity and sedentary behavior. One of the key advantages of activity diaries is their ability to capture nuanced information, including light and moderate physical activities that are often missed in standard questionnaires. This makes them particularly valuable in studies that require detailed and individualized data. Moreover, they can be tailored to specific research needs and populations, providing flexibility in data collection. However, the effectiveness of activity diaries depends largely on participant compliance and cognitive capacity. For individuals with cognitive impairments, elderly populations, or those with low literacy, maintaining such records can be burdensome and prone to errors or omissions (Rachele et al., 2012). The constant need for monitoring and recording can also lead to participant fatigue over time. Despite these limitations, activity diaries remain a valuable tool in physical activity research when used with the appropriate population and support mechanisms.

9.3 Smart phone technology

Physical activity can be measured through mobile features provided in cellphones, applications specifically designed for activity measurement, or external devices linked to mobile phones. The regular usage of smartphones and their technology at present makes them very convenient and easily accessible, even in remote places. Mobiles are equipped with built-in sensors such as accelerometers, gyroscopes, and GPS technology, which help in monitoring the activity of participants. Smart phone applications such as Mobile Teen (Dunton et al., 2014) and the Moves App developed by ProtoGeo (Murphy et al., 2020) utilize the inbuilt motion sensors to measure physical activity.

9.4 Direct observation

Observers record detailed activities of the subject in real time; the assessment is more accurate, provides high quality details, is less expensive compared to activity monitoring devices, and allows specific activities to be monitored. Limitations include reliance on observer interpretation, the need for observer training, and challenges in managing and scoring the data. It can be utilized to evaluate the physical activity of groups within a controlled environment (Sallis, 2010).

9.5 Devices

In recent decades, numerous electronic monitoring devices, such as accelerometers, pedometers, heart rate monitors, and wearable fitness trackers, have become available. Some of the commonly used devices are listed below.

9.5.1 Accelerometer

They are motion sensors classified based on their working mechanism as piezoelectric or piezo-resistive sensors (Plasqui and Westerterp, 2007) and as uniaxial or triaxial accelerometers based on the direction of acceleration monitored. These tools are precise, retain a large amount of information, are simple to use (Westerterp, 2009) and can be placed at several sites on the body, including the hip, thigh, and waist. However, these sensors are expensive, require specific hardware and software, and cannot differentiate the intensity of work or body position (Shephard, 2003; Baranowski et al., 1984).

9.5.2 Pedometer

The device calculates the number of steps when the participant's hip moves vertically with a force beyond the threshold point (Sylvia et al., 2014). It measures the steps in real-time, is less expensive and convenient, but is unable to record horizontal motion, upper torso movements, or the timeframe, frequency, or intensity of physical activity (Trost, 2001; Freedson and Miller, 2000; Tudor-Locke et al., 2002; Kilanowski, Consalvi and Epstein, 1999).

9.5.3 Heart rate monitors

Monitors show a relationship between heart rate, physical activity, and energy consumption when worn on the wrist or chest (Welk and Corbin, 1995). They are more accurate compared to pedometers and accelerometers, easy to use (Hensen, 2017) and categorize the intensity of physical activity. However, related and extraneous factors such as age, gender, body mass, muscle mass, caffeine, and stress can affect the heart rate, which can lead to false readings (Trost, 2001; Livingstone, 1997).

9.5.4 Wearable fitness trackers

These are electronic trackers that allow self-monitoring, due to their increased use, becoming widely accepted tools in the current decade for monitoring steps taken, time frame, intensity, calorie intake and calories burned, screen time, active time, and sleep schedules.Their drawbacks include inaccurate readings, privacy-related concerns, and the cost of the trackers. Examples of fitness trackers include Fitbit, Apple Watch, Garmin, Google Pixel, Samsung Galaxy, Amazfit, and Nike.

9.5.5 Arm bands

The technology had been developed and uses movement and heat-related sensors, as well as galvanic skin response, to monitor the body and skin temperature and energy dissipation (Andre et al., 2006). Armbands, while they can be used regularly, cannot determine the type or duration of activity and are not suitable for higher-intensity activity (Drenowatz and Eisenmann, 2011). Examples include Sense wear Pro, Sense Wear Mini, Pro3, Model MF, HealthWear, and BodyBugg.

9.6 Indirect calorimetry

Doubly labeled water technique is a standard reference for energy expenditure and physical activity measurement. The method involves injecting water labeled with isotopes of oxygen and hydrogen, measuring an equilibrium state along with total body water, and then tracking the excretion of these isotopes from the body. The technique can be used for all age groups; however, the process (involving isotopes and a spectrometer) is expensive, time-consuming, and allows only a short period of data to be calculated. In addition, the type, duration, frequency, and intensity of physical activity cannot be assessed using this method (Ndahimana and Kim, 2017).

9.7 Direct calorimetry

It is a precise technique, but it is expensive, as it measures heat loss using a calorimeter (Kaiyala and Ramsay, 2011). Calorimeters are instruments used to measure the amount of heat and the volume produced over a specific period. The four main types of calorimeters include isothermal, heat sink, convection, and differential calorimeters. The four main types of calorimeters are shown in Fig. 9.2.

Heat conduction/isothermic calorimeters (Zhang, 2010): It maintains a constant-temperature fluid, such as water, either in a jacket or bath surrounding the chamber or through a system of copper tubing connected to the walls of the chamber to ensure a steady temperature.

Heat sink direct calorimeters (Webster et al., 1986): Constitutes a tank where heat is pumped out and measured using a liquid-cooled heat exchanger.

Direct convection calorimeters (Snellen et al., 1983; Levine, 2005): Are gas-filled containers that have a constant airflow rate. They determine the amount of heat lost by the subject by measuring both temperature and thermodynamic variations between the incoming and exiting air.

Direct differential calorimeters (Kaiyala and Ramsay, 2011): These involve two similar chambers, one with an electric heater and the other enclosing the individual. The subject's metabolic rate is obtained by comparing the differences in heat loss between the two chambers.

10 Methods to reduce sedentarism in office workers

Modern workspaces have evolved significantly with the integration of advanced technologies and digital tools that have revolutionized how work is performed. While these innovations have increased efficiency and productivity, they have also drastically reduced the need for physical movement during the workday. Most office jobs now involve prolonged

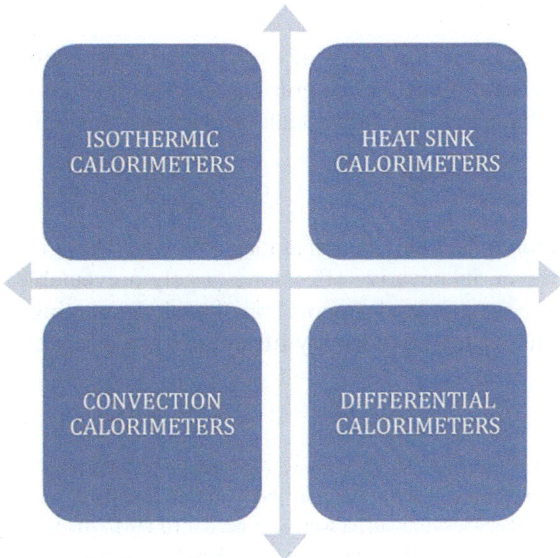

FIG. 9.2 Types of calorimeters.

periods of sitting at computer-based workstations, resulting in a predominantly sedentary lifestyle among office workers (Parry and Straker, 2013). The convenience of technology, such as email, video conferencing, and automated systems, has eliminated many of the physically engaging tasks that once required employees to move around the workplace. However, this shift toward sedentary work routines has serious long-term health implications. Research increasingly shows that prolonged sitting is associated with an elevated risk of chronic diseases such as hypertension, obesity, diabetes, cardiovascular issues, and cognitive decline. These conditions not only reduce the quality of life but also place a significant burden on healthcare systems. Moreover, sedentary behavior can negatively impact mental well-being, increasing stress and reducing overall job satisfaction. To mitigate these effects, it is essential to rethink and redesign office environments and work routines. Workspaces should incorporate ergonomically designed furniture, such as sit-stand desks, encourage regular breaks for stretching or short walks, and promote active commuting or on-site wellness programs. Employers can also support a healthier culture by organizing fitness challenges, providing access to standing meeting areas, or encouraging walking meetings. By making these small yet impactful changes, modern offices can foster healthier lifestyles, improve employee well-being, and enhance productivity in the long run (Shrestha et al., 2018). The following includes a list of methods that can be incorporated.

10.1 Awareness and training

The individuals involved in desk-based office work should be properly aware of the risks associated with sedentary behavior, which involves prolonged sitting, through workshops and webinars. Companies can promote physical activity by engaging employees in fitness camps or studios (Malik et al., 2014).

10.2 Sit-stand workstation

The conventional workstations should be replaced with adjustable sit-stand workstations that allow sitting and standing alternately. This type of workstation would reduce prolonged periods of continuous sitting (Chau et al., 2014).

10.3 Flexible work-breaks

The work environment should encourage breaks that allow employees to move around. The work should be more task-based than focused on time tracking. The work-breaks and movements involved would boost office workers' overall health, which would, in turn, increase the company's progress and productivity (Lyubykh et al., 2022).

10.4 Regular exercise

The office workers should exercise regularly. Any form of exercise or sports increases circulation and metabolism, thus reducing the stress on the vascular system. Older office-going adults who exercise regularly perform better and are less prone to a decline in brain cognitive functions than individuals who are physically inactive (Bliss et al., 2021).

11 Conclusion

Sedentarism is a significant contributor to declining health among office workers globally, linked to vascular issues, hypertension, and cognitive decline. However, its negative effects are largely preventable. Even small, consistent increases in physical activity can improve vascular function and support brain health. Promoting movement throughout the workday is essential for reversing the long-term impacts of a sedentary lifestyle and fostering overall well-being.

References

Ahmadi-Abhari, S., Sabia, S., Shipley, M.J., Kivimäki, M., Singh-Manoux, A., Tabak, A., McEniery, C., Wilkinson, I.B., Brunner, E.J., 2017. Physical activity, sedentary behavior, and long-term changes in aortic stiffness: the Whitehall II study. J. Am. Heart Assoc. 20479980. 6 (8). Available from: http://jaha.ahajournals.org/content/ahaoa/6/8/e005974full.pdf?download=true. doi:10.1161/JAHA.117.005974.

Andre, D., Pelletier, R., Farringdon, J., et al., 2006. The Development of the SenseWear® Armband, a Revolutionary Energy Assessment Device to Assess Physical Activity and Lifestyle. BodyMedia Inc. pp. 1–19.

Baranowski, T., Dworkin, R.J., Cieslik, C.J., Hooks, P., Clearman, D.R., Ray, L., Dunn, J.K., Nader, P.R., 1984. Reliability and validity of self report of aerobic activity. Fam. Health Proj. Res. Q. Exerc. Sport. 21683824. 55 (4), 309–317. doi:10.1080/02701367.1984.10608408.

Bhatia, T., Shriharsh, V., Adlakha, S., Bisht, V., Garg, K., Deshpande, S., 2007. The trail making test in India. Indian J. Psychiatry. 0019-5545. 49 (2), 113. doi:10.4103/0019-5545.33258.

Bliss, E.S., Wong, R.H.X., Howe, P.R.C., Mills, D.E., 2021. Benefits of exercise training on cerebrovascular and cognitive function in ageing. J. Cereb. Blood Flow Metab. 41 (3), 447–470. doi:10.1177/0271678X20957807.

Boyle, L.J., Credeur, D.P., Jenkins, N.T., Padilla, J., Leidy, H.J., Thyfault, J.P., Fadel, P.J., 2013. Impact of reduced daily physical activity on conduit artery flow-mediated dilation and circulating endothelial microparticles. J. Appl. Physiol. 115 (10), 1519–1525. doi:10.1152/japplphysiol.00837.2013.

Burgmans, S., van Boxtel, M.P.J., Gronenschild, E.H.B.M., Vuurman, E.F.P.M., Hofman, P., Uylings, H.B.M., Jolles, J., Raz, N., 2010. Multiple indicators of age-related differences in cerebral white matter and the modifying effects of hypertension. NeuroImage. 10538119. 49 (3), 2083–2093. doi:10.1016/j.neuroimage.2009.10.035.

Cai, X., Qian, G., Wang, F., Zhang, M., Da, Y., Liang, J., 2023. Association between sedentary behavior and risk of cognitive decline or mild cognitive impairment among the elderly: a systematic review and meta-analysis. Front. Neurosci. 17, 1221990. doi:10.3389/fnins.2023.1221990.

Carter, S.E., Draijer, R., Holder, S.M., Brown, L., Thijssen, D.H., Hopkins, N.D., 2018. Regular walking breaks prevent the decline in cerebral blood flow associated with prolonged sitting. J. Appl. Physiol. 125 (3), 790–798. doi:10.1152/japplphysiol.00310.2018.

Chau, J.Y., Daley, M., Dunn, S., Srinivasan, A., Do, A., Bauman, A.E., van der Ploeg, H.P., 2014. The effectiveness of sit-stand workstations for changing office workers' sitting time: results from the Stand@Work randomized controlled trial pilot. Int. J. Behav. Nutr. Phys. Act 11, 127. doi:10.1186/s12966-014-0127-7.

Cheraghi, Z., Azmi-Naei, B., Cheraghi, P., Doosti-Irani, A., 2025. The global prevalence of uncontrolled hypertension: a systematic review and meta-analysis. BMC Public Health 25 (1). doi:10.1186/s12889-025-25553-4.

Classon, E., van den Hurk, W., Wressle, E., Rehn, I., Johansson, M.M., 2022. A quick test of cognitive speed (AQT): regression-based norms for cognitively healthy 80 to 94-year olds. Aging, Neuropsychology, and Cognition. 17444128. 29 (5), 820–839. Available from: http://www.tandf.co.uk/journals/titles/13825585.asp. doi:10.1080/13825585.2021.1922585.

Colita, E., Mateescu, V.O., Olaru, D., Popa-Wagner, A., 2024. Cognitive decline in ageing and disease: risk factors, genetics and treatments. Curr. Health Sci. J. 50 (2), 170–180. doi:10.12865/CHSJ.50.02.02.

Cook, J.L., Kiss, Z.S., Ptasznik, R., Malliaras, P., 2005. Is vascularity more evident after exercise? Implications for tendon imaging. Am. J. Roentgenol. 0361803X. 185 (5), 1138–1140. doi:10.2214/AJR.04.1205.

Daneshmandi, H., Choobineh, A., Ghaem, H., Karimi, M., 2017. Adverse effects of prolonged sitting behavior on the general health of office workers. J. Lifestyle Med. 7 (2), 69–75. doi:10.15280/jlm.2017.7.2.69.

Daniele, A., Lucas, S.J.E., Rendeiro, C., 2022. Detrimental effects of physical inactivity on peripheral and brain vasculature in humans: insights into mechanisms, long-term health consequences and protective strategies. Front. Physiol. 1664042X. 13. Available from: http://www.frontiersin.org/Physiology/archive/. doi:10.3389/fphys.2022.998380.

Drenowatz, Eisenmann, J.C., 2011. Validation of the SenseWear Armband at high intensity exercise. Eur. J. Appl. Physiol. 111 (5), 883–887. doi:10.1007/s00421-010-1695-0.

Dudas, R.B., Clague, F., Thompson, S.A., Graham, K.S., Hodges, J.R., 2005. Episodic and semantic memory in mild cognitive impairment. Neuropsychologia. 00283932. 43 (9), 1266–1276. doi:10.1016/j.neuropsychologia.2004.12.005.

Dunstan, D.W., Dogra, S., Carter, S.E., Owen N., 2021. Sit less and move more for cardiovascular health: emerging insights and opportunities. Nat. Rev. Cardiol. 18 (9), 637–648. doi:10.1038/s41569-021-00547-y.

Dunton, G.F., Dzubur, E., Kawabata, K., Yanez, B., Bo, B., Intille, S., 2014. Development of a smartphone application to measure physical activity using sensor-assisted self-report. Front. Public Health. 22962565. 2. Available from: http://journal.frontiersin.org/article/10.3389/fpubh.2014.00012/pdf. doi:10.3389/fpubh.2014.00012.

Durante, R., Ainsworth, B.E., 1996. The recall of physical activity: using a cognitive model of the question-answering process. Med. Sci. Sports Exerc. 01959131. 28 (10), 1282–1291. doi:10.1097/00005768-199610000-00012.

Firbank, M.J., Wiseman, R.M., Burton, E.J., Saxby, B.K., O'Brien, J.T., Ford, G.A., 2007. Brain atrophy and white matter hyperintensity change in older adults and relationship to blood pressure: brain atrophy, WMH change and blood pressure. J. Neurol. 14321459. 254 (6), 713–721. doi:10.1007/s00415-006-0238-4.

Freedson, P.S., Miller, K., 2000. Objective monitoring of physical activity using motion sensors and heart rate. Res. Q. Exerc. Sport. 21683824. 71, 21–29. doi:10.1080/02701367.2000.11082782.

Gallegos, M., Morgan, M.L., Cervigni, M., Martino, P., Murray, J., Calandra, M., Razumovskiy, A., Caycho-Rodríguez, T., Gallegos, W.L.A., 2022. 45 Years of the mini-mental state examination (MMSE): a perspective from ibero-america. Dement. Neuropsychol. 19805764. 16 (4), 384–387. Available from: https://www.demneuropsy.com.br/wp-content/uploads/articles_xml/1980-5764-dn-2021-0097/1980-5764-dn-2021-0097.pdf. doi:10.1590/1980-5764-DN-2021-0097.

Gogniat, M.A., et al., 2026. The associations between sedentary behavior and cognition in a population cohort of older adults. J. Alzheimer's Dis. 109 (4), 1815–1823. doi:10.1177/13872877251410964.

Goldstein, I.B., Bartzokis, G., Guthrie, D., Shapiro, D., 2002. Ambulatory blood pressure and brain atrophy in the healthy elderly. Neurology. 00283878. 59 (5), 713–719. Available from: http://www.neurology.org. doi:10.1212/WNL.59.5.713.

Gregory, S., Parker, B., Thompson, P., 2012. Physical activity, cognitive function, and brain health: what is the role of exercise training in the prevention of dementia? Brain Sci. 2 (4), 684–708. doi:10.3390/brainsci2040684.

Harvey, P.D., 2019. Domains of cognition and their assessment. Dialogues Clin. Neurosci. 21 (3), 227–237. doi:10.31887/DCNS.2019.21.3/pharvey.

Hensen, S.J., 2017. Measuring physical activity with heart rate monitors. Am. J. Public Health. 15410048. 107 (12), e24. Available from: http://ajph.aphapublications.org/doi/pdf/10.2105/AJPH.2017.304121. doi:10.2105/AJPH.2017.304121.

Hillman, C.H., Erickson, K.I., Kramer, A.F., 2008. Be smart, exercise your heart: exercise effects on brain and cognition. Nat. Rev. Neurosci. 9 (1), 58–65. doi:10.1038/nrn2298.

Hodges, J.R., Larner, A.J., 2017. Addenbrooke's Cognitive Examinations: ACE, ACE-R, ACE-III, ACEapp, and M-ACE. Springer Science and Business Media LLC, pp. 109–137. doi:10.1007/978-3-319-44775-9_6.

Horstman, L.L., Jy, W., Jimenez, J.J., Ahn, Y.S., 2004. Endothelial microparticles as markers of endothelial dysfunction. Front. Biosci. 9, 1118–1135. doi:10.2741/1270.

Huynh, Q.L., Blizzard, C.L., Sharman, J.E., Magnussen, C.G., Dwyer, T., Venn, A.J., 2014. The cross-sectional association of sitting time with carotid artery stiffness in young adults. BMJ Open. 20446055. 4 (3). Available from: http://bmjopen.bmj.com/content/4/3/e004384full.pdf+html. doi:10.1136/bmjopen-2013-004384.

Imboden, M., 2024. Maintaining brain health: an imperative for successful aging and business performance. Am. J. Health Promot. 38 (4), 576–589. doi:10.1177/08901171241232042.

Jacobs, D.R., Ainsworth, B.E., Hartman, T.J., Arthur, S.L., 1993. A simultaneous evaluation of 10 commonly used physical activity questionnaires. Med. Sci. Sports Exerc. 15300315. 25 (1), 81–91. doi:10.1249/00005768-199301000-00012.

Jones, R., McArthur, D., McCoy, S.M., Stoner, L., Fryer, S., Credeur, D.P., 2022. Impact of acute uninterrupted sitting on cerebrovascular hemodynamics. Int. J. Exerc. Sci. 15 (2), 1156–1167. doi:10.70252/BZQJ6904.

Kaiyala, K.J., Ramsay, D.S., 2011. Direct animal calorimetry, the underused gold standard for quantifying the fire of life. Comp. Biochem. Physiol. A Mol. Integr. Physiol. 15314332. 158 (3), 252–264. Available from: https://www.journals.elsevier.com/comparative-biochemistry-and-physiology-part-a-molecular-and-integrative-physiology. doi:10.1016/j.cbpa.2010.04.013.

Kilanowski, C.K., Consalvi, A.R., Epstein, L.H., 1999. Validation of an electronic pedometer for measurement of physical activity in children. Pediatr. Exerc. Sci. 08998493. 11 (1), 63–68. Available from: http://journals.humankinetics.com/PES. doi:10.1123/pes.11.1.63.

Kim, H.J., Baek, M.J., Kim, S., Zhan, W., 2014. Alternative type of the trail making test in nonnative English-speakers: the trail making test-Black & White. PLoS ONE. 1932-6203. 9 (2), e89078. doi:10.1371/journal.pone.0089078.

Klesges, R.C., Eck, L.H., Mellon, M.W., Fulliton, W., Somes, G.W., Hanson, C.L., 1990. The accuracy of self-reports of physical activity. Med. Sci. Sports Exerc. 01959131. 22 (5), 690–697. doi:10.1249/00005768-199010000-00022.

Levine, J.A., 2005. Measurement of energy expenditure. Public Health Nutr. 8 (7 A), 1123–1132. doi:10.1079/PHN2005800.

Li, X., Hui, Y., Shi, H., Zhao, X., Li, R., Chen, Q., Lv, H., Li, J., Chen, S., Zhao, P., Wu, Y., Wu, S., Wang, Z., 2023b. Association of blood pressure with brain perfusion and structure: a population-based prospective study. Eur. J. Radiol. 165, 110889. doi:10.1016/j.ejrad.2023.110889.

Li, Z., Wang, W., Sang, F., Zhang, Z., Li, X., 2023a. White matter changes underlie hypertension-related cognitive decline in older adults. NeuroImage Clin. 38, 103389. doi:10.1016/j.nicl.2023.103389.

Linari, I., Juantorena, G.E., Ibáñez, A., Petroni, A., Kamienkowski, J.E., 2022. Unveiling trail making test: visual and manual trajectories indexing multiple executive processes. Sci. Rep. 20452322. 12 (1). Available from: https://www.nature.com/srep/index.html. doi:10.1038/s41598-022-16431-9.

Liu, H.B., Yuan, W.X., Qin, K.R., Hou, J., 2015. Acute effect of cycling intervention on carotid arterial hemodynamics: basketball athletes versus sedentary controls. Biomed. Eng. Online 14, S17. doi:10.1186/1475-925X-14-S1-S17.

Livingstone, M.B.E., 1997. Heart-rate monitoring: the answer for assessing energy expenditure and physical activity in population studies? Br. J. Nutr. 00071145. 78 (6), 869–871. doi:10.1079/BJN19970205.

Longstreth, W.T., Manolio, T.A., Arnold, A., Burke, G.L., Bryan, N., Jungreis, C.A., Enright, P.L., O'Leary, D., Fried, L., 1996. Clinical correlates of white matter findings on cranial magnetic resonance imaging of 3301 elderly people: the cardiovascular health study. Stroke. 00392499. 27 (8), 1274–1282. Available from: http://stroke.ahajournals.org/. doi:10.1161/01.STR.27.8.1274.

Lui, F., Gonzalez, K.I., Launico, M.V., 2026. Cognitive ssessment. In: StatPearls [Internet]. StatPearls Publishing, Treasure Island (FL). Available from: https://www.ncbi.nlm.nih.gov/books/NBK556049/.

Luo, D., Cheng, Y., Zhang, H., Ba, M., Chen, P., Li, H., Chen, K., Sha, W., Zhang, C., Chen, H., 2020. Association between high blood pressure and long term cardiovascular events in young adults: systematic review and meta-analysis. BMJ. 17561833. 370. Available from: http://www.bmj.com/. doi:10.1136/bmj.m3222.

Lyubykh, Z., Gulseren, D., Premji, Z., Wingate, T.G., Deng, C., Bélanger, L.J., Turner, N., 2022. Role of work breaks in well-being and performance: a systematic review and future research agenda. J. Occup. Health Psychol. 27 (5), 470–487. doi:10.1037/ocp0000337.

Maasakkers, C.M., Melis, R.J., Kessels, R.P., Gardiner, P.A., Olde Rikkert, M.G., Thijssen, D.H., Claassen, J.A., 2020. The short-term effects of sedentary behaviour on cerebral hemodynamics and cognitive performance in older adults: a cross-over design on the potential impact of mental and/or physical activity. Alzheimers Res. Ther. 12 (1), 76. doi:10.1186/s13195-020-00644-z.

Maj, M., D'Elia, L., Satz, P., Janssen, R., Zaudig, M., Uchiyama, C., Starace, F., Galderisi, S., Chervinsky, A., 1993. Evaluation of two new neuropsychological tests designed to minimize cultural bias in the assessment of HIV-1 seropositive persons: a WHO study. Arch. Clin. Neuropsychol. 08876177. 8 (2), 123–135. doi:10.1016/0887-6177(93)90030-5.

Malik, S.H., Blake, H., Suggs, L.S., 2014. A systematic review of workplace health promotion interventions for increasing physical activity. Br. J. Health Psychol. 19, 149–180. doi:10.1111/bjhp.12052.

Mathuranath, P., Cherian, J., Mathew, R., George, A., Alexander, A., Sarma, S., Mini, 2007. Mental state examination and the addenbrooke's cognitive examination: effect of education and norms for a multicultural population. Neurol. India 55 (2), 106–110. doi:10.4103/0028-3886.32779.

Mensah, G.A., 2016. Hypertension and target organ damage: don't believe everything you think! Ethn. Dis. 26 (3), 275–278. doi:10.18865/ed.26.3.275.

Moinuddin, A., Paterson, C., Turner, L., Stone, K., Lucas, S.J.E., Griffiths, T., Fryer, S., 2025. The effects of acute exposure to prolonged sitting, with and without interruption, on central and peripheral pulse wave velocity: a systematic review and meta-analysis. Vasc. Med. 30 (4), 403–413. doi:10.1177/1358863X251333672.

Murphy, J., Uttamlal, T., Schmidtke, K.A., Vlaev, I., Taylor, D., Ahmad, M., Alsters, S., Purkayastha, P., Scholtz, S., Ramezani, R., Ahmed, A.R., Chahal, H., Darzi, A., Blakemore, A.I.F., 2020. Tracking physical activity using smart phone apps: assessing the ability of a current app and systematically collecting patient recommendations for future development. BMC Med. Inform. Decis. Mak. 1472-6947. 20 (1). doi:10.1186/s12911-020-1025-3.

Nasreddine, Z.S., Phillips, N.A., Bédirian, V., Charbonneau, S., Whitehead, V., Collin, I., Cummings, J.L., Chertkow, H., 2005. The Montreal Cognitive Assessment, MoCA: a brief screening tool for mild cognitive impairment. J. Am. Geriatr. Soc. 53 (4), 695–699. doi:10.1111/j.1532-5415.2005.53221.x.

Ndahimana, D., Kim, E.-K., 2017. Measurement methods for physical activity and energy expenditure: a review. Clin. Nutr. Res. 2287-3732. 6 (2), 68. doi:10.7762/cnr.2017.6.2.68.

Nielsen, N.P., Wiig, E.H., Warkentin, S., Minthon, L., 2004. Clinical utility of color-form naming in Alzheimer's disease: preliminary evidence. Percept. Mot. Skills. 00315125. 99 (3), 1201–1204. Available from: http://pms.sagepub.com/content/by/year. doi:10.2466/pms.99.3f.1201-1204.

Nosova, E.V., Yen, P., Chong, K.C., Alley, H.F., Stock, E.O., Quinn, A., Hellmann, J., Conte, M.S., Owens, C.D., Spite, M., Grenon, S.M., 2014. Short-term physical inactivity impairs vascular function. J. Surg. Res. 190 (2), 672–682. doi:10.1016/j.jss.2014.02.001.

Nouchi, R., Kawashima, R., 2014. Improving cognitive function from children to old age: a systematic review of recent smart ageing intervention studies. Adv. Neurosci. 2356-6787. 2014, 1–15. doi:10.1155/2014/235479.

Padilla, J., Fadel, P.J., 2017. Prolonged sitting leg vasculopathy: contributing factors and clinical implications. Am. J. Physiol. Heart Circ. Physiol. 313 (4), H722–H728. doi:10.1152/ajpheart.00326.2017.

Parry, S., Straker, L., 2013. The contribution of office work to sedentary behaviour associated risk. BMC Public Health 13, 296. doi:10.1186/1471-2458-13-296.

Peddie, M.C., Kessell, C., Bergen, T., Gibbons, T.D., Campbell, H.A., Cotter, J.D., Rehrer, N.J., Thomas, K.N., 2021. The effects of prolonged sitting, prolonged standing, and activity breaks on vascular function, and postprandial glucose and insulin responses: a randomised crossover trial. PLoS ONE. 19326203. 16 (1). Available from: https://journals.plos.org/plosone/article/file?id=10.1371/journal.pone.0244841&type=printable. doi:10.1371/journal.pone.0244841.

Peng, Z., Shu, B., Zhang, Y., Wang, M., 2019. Endothelial response to pathophysiological stress. Arterioscler. Thromb. Vasc. Biol. 39 (11), e233–e243. doi:10.1161/ATVBAHA.119.312580.

Plasqui, G., Westerterp, K.R., 2007. Physical activity assessment with accelerometers: an evaluation against doubly labeled water. Obesity. 1930739X. 15 (10), 2371–2379. doi:10.1038/oby.2007.281.

Rachele, J.N., Mcphail, S.M., Washington, T.L., Cuddihy, T.F., 2012. Practical physical activity measurement in youth: a review of contemporary approaches. World J. Pediatr. 18670687. 8 (3), 207–216. doi:10.1007/s12519-012-0359-z.

Raz, N., Rodrigue, K.M., Haacke, E.M., 2007. Brain aging and its modifiers: insights from in vivo neuromorphometry and susceptibility weighted imaging. Ann. N.Y. Acad. Sci. 17496632, 84–93.1097. http://onlinelibrary.wiley.com/journal/10.1111/(ISSN)1749-6632.10.1196/annals.1379.018.

Reichel, K., Prigge, M., Latza, U., Kurth, T., Backé, E., 2022. Association of occupational sitting with cardiovascular outcomes and cardiometabolic risk factors: a systematic review with a sex-sensitive/gender-sensitive perspective. BMJ Open 12 (2), e048017. doi:10.1136/bmjopen-2020-048017.

Restaino, R.M., Holwerda, S.W., Credeur, D.P., Fadel, P.J., Padilla, J., 2015. Impact of prolonged sitting on lower and upper limb micro- and macrovascular dilator function. Exp. Physiol. 1469445X. 100 (7), 829–838. Available from: http://onlinelibrary.wiley.com/journal/10.1111/(ISSN)1469-445X. doi:10.1113/EP085238.

Restaino, R.M., Walsh, L.K., Morishima, T., Vranish, J.R., Martinez-Lemus, L.A., Fadel, P.J., Padilla, J., 2016. Endothelial dysfunction following prolonged sitting is mediated by a reduction in shear stress. Am. J. Physiol. Heart Circ. Physiol. 310 (5), H648–H653. doi:10.1152/ajpheart.00943.2015.

Ricci, M., Pigliautile, M., D'Ambrosio, V., Ercolani, S., Bianchini, C., Ruggiero, C., Vanacore, N., Mecocci, P., 2016. The clock drawing test as a screening tool in mild cognitive impairment and very mild dementia: a new brief method of scoring and normative data in the elderly. Neurol. Sci. 1590-1874. 37 (6), 867–873. doi:10.1007/s10072-016-2480-6.

Sahu, R.N., Naik, G.P., Dusad, A., Agrawal, V., 2005. Neurocognitive function in women affected by the Bhopal gas disaster. Indian J. Psychiatry. 0019-5545. 47 (1), 51. doi:10.4103/0019-5545.46076.

Sallis, J.F., 1991. Self-report measures of children's physical activity. J. Sch. Health. 61 (5), 215–219. doi:10.1111/j.1746-1561.1991.tb06017.x.

Sallis, J.F., 2010. Measuring physical activity: practical approaches for program evaluation in native American communities. J. Public Health Manag. Pract. 10784659. 16 (5), 404–410. Available from: http://journals.lww.com/jphmp. doi:10.1097/PHH.0b013e3181d52804.

Shephard, R.J., 1999. How much physical activity is needed for good health? Int. J. Sports Med. 01724622. 20 (1), 23–27. doi:10.1055/s-2007-971086.

Shephard, R.J., 2003. Limits to the measurement of habitual physical activity by questionnaires. Br. J. Sports Med. 03063674. 37 (3), 197–206. Available from: http://bjsm.bmjjournals.com/. doi:10.1136/bjsm.37.3.197.

Shephard, R.J., 2003. Limits to the measurement of habitual physical activity by questionnaires. Br. J. Sports Med. 37 (3), 197–206. doi:10.1136/bjsm.37.3.197.

Shrestha, N., Kukkonen-Harjula, K.T., Verbeek, J.H., Ijaz, S., Hermans, V., Pedisic, Z., 2018. Workplace interventions for reducing sitting at work. Cochr. Database Syst. Rev. 6 (6), CD010912. doi:10.1002/14651858.CD010912.pub5.

Snellen, J.W., Chang, K.S., Smith, W., 1983. Technical description and performance characteristics of a human whole-body calorimeter. Med. Biol. Eng. Comput. 17410444. 21 (1), 9–20. doi:10.1007/BF02446401.

Stoner, L., Barone, G.B., Meyer, M.L., Fryer, S., Credeur, D., Paterson, C., Stone, K., Hanson, E.D., Kowalsky, R.J., Horiuchi, M., Mack, C.P., Dave, G., 2021. A primer on repeated sitting exposure and the cardiovascular system: considerations for study design, analysis, interpretation, and translation. Front. Cardiovasc. Med. 8, 716938. doi:10.3389/fcvm.2021.716938.

Sylvia, L.G., Bernstein, E.E., Hubbard, J.L., Keating, L., Anderson, E.J., 2014. Practical guide to measuring physical activity. J. Acad. Nutr. Diet. 22122672. 114 (2), 199–208. doi:10.1016/j.jand.2013.09.018.

Tamiya, H., Hoshiai, M., Abe, T., Watanabe, H., Fujii, Y., Tsubaki, A., 2024. Prolonged sitting induces elevated blood pressure in healthy young men: a randomized crossover trial. Cureus 16 (2), e55224. doi:10.7759/cureus.55224.

Thijssen, D.H., Black, M.A., Pyke, K.E., Padilla, J., Atkinson, G., Harris, R.A., Parker, B., Widlansky, M.E., Tschakovsky, M.E., Green, D.J., 2011. Assessment of flow-mediated dilation in humans: a methodological and physiological guideline. Am. J. Physiol. Heart Circ. Physiol. 300 (1), H2–H12. doi:10.1152/ajpheart.00471.2010.

Thosar, S.S., Bielko, S.L., Mather, K.J., Johnston, J.D., Wallace, J.P., 2015. Effect of prolonged sitting and breaks in sitting time on endothelial function. Med. Sci. Sports Exerc. 47 (4), 843–849. doi:10.1249/MSS.0000000000000479.

Thosar, S.S., Bielko, S.L., Wiggins, C.C., Wallace, J.P., 2014. Differences in brachial and femoral artery responses to prolonged sitting. Cardiovasc. Ultrasound 12 (50). Available from: https://www.cardiovascularultrasound.com/home/. doi:10.1186/1476-7120-12-50.

Thosar, S.S., Johnson, B.D., Johnston, J.D., Wallace, J.P., 2012. Sitting and endothelial dysfunction: the role of shear stress. Med. Sci. Monit. 18 (12), RA173–RA180. doi:10.12659/MSM.883589.

Tremblay, M.S., Aubert, S., Barnes, J.D., Saunders, T.J., Carson, V., Latimer-Cheung, A.E., Chastin, S.F.M., Altenburg, T.M., Chinapaw, M.J.M., 2017. SBRN terminology consensus project participants. Sedentary Behavior Research Network (SBRN) - terminology consensus project process and outcome. Int. J. Behav. Nutr. Phys. Act 14 (1), 75. doi:10.1186/s12966-017-0525-8.

Tri Tjahjono, C., Arthamin, M., 2024. Sedentary lifestyle as a leading cause of some diseases and disability. CRJIM 5 (1), 60–85. doi:10.21776/ub.crjim.2024.005.01.09.

Tripathi, R., Kumar, J.K., Bharath, S., Marimuthu, P., Varghese, M., 2013. Clinical validity of NIMHANS neuropsychological battery for elderly: a preliminary report. Indian J. Psychiatry. 19983794. 55 (3), 279–282. doi:10.4103/0019-5545.117149.

Trongsakul, S., Lambert, R., Clark, A., Wongpakaran, N., Cross, J., 2015. Development of the T hai version of M ini- C og, a brief cognitive screening test. Geriatr. Gerontol. Int. 1444-1586. 15 (5), 594–600. doi:10.1111/ggi.12318.

Trost, S.G., 2001. Objective measurement of physical activity in youth: current issues, future directions. Exerc. Sport Sci. Rev. 00916331. 29 (1), 32–36. Available from: http://journals.lww.com/acsm-essr/pages/default.aspx. doi:10.1097/00003677-200101000-00007.

Tudor-Locke, C., Ainsworth, B.E., Thompson, R.W., Matthews, C.E., 2002. Comparison of pedometer and accelerometer measures of free-living physical activity. Med. Sci. Sports Exerc. 01959131. 34 (12), 2045–2051. Available from: http://www.lww.com/product/?0195-9131. doi:10.1097/00005768-200212000-00027.

Uitenbroek, D.G., 1993. Seasonal variation in leisure time physical activity. Med. Sci. Sports Exerc. 15300315. 25 (6), 755–760. doi:10.1249/00005768-199306000-00017.

Van Der Ploeg, H.P., Merom, D., Chau, J.Y., Bittman, M., Trost, S.G., Bauman, A.E., 2010. Advances in population surveillance for physical activity and sedentary behavior: reliability and validity of time use surveys. Am. J. Epidemiol. 14766256. 172 (10), 1199–1206. doi:10.1093/aje/kwq265.

van Praag, H., 2008. Neurogenesis and exercise: past and future directions. Neuromol. Med. 10 (2), 128–140. doi:10.1007/s12017-008-8028-z.

Vanhees, L., Lefevre, J., Philippaerts, R., Martens, M., Huygens, W., Troosters, T., Beunen, G., 2005. How to assess physical activity? How to assess physical fitness? Eur. J. Cardiovasc. Prev. Rehabil. 1741-8267. 12 (2), 102–114. doi:10.1097/00149831-200504000-00004.

Webster, J.D., Welsh, G., Pacy, P., Garrow, J.S., 1986. Description of a human direct calorimeter, with a note on the energy cost of clerical work. Br. J. Nutr. 14752662. 55 (1), 1–6. doi:10.1079/BJN19860003.

Welk, G.J., Corbin, C.B., 1995. The validity of the tritrac-R3D activity monitor for the assessment of physical activity in children. Res. Q. Exerc. Sport. 21683824. 66 (3), 202–209. doi:10.1080/02701367.1995.10608834.

Westerterp, K.R., 2009. Assessment of physical activity: a critical appraisal. Eur. J. Appl. Physiol. 14396319. 105 (6), 823–828. doi:10.1007/s00421-009-1000-2.

Young, D.R., et al., 2016. Sedentary behavior and cardiovascular morbidity and mortality: a science advisory from the American Heart Association. Circulation 134 (13), e262–e279. doi:10.1161/CIR.0000000000000440.

Zhang, W.S., 2010. Construction, calibration and testing of a decimeter-size heat-flow calorimeter. Thermochim. Acta. 00406031. 499 (1-2), 128–132. doi:10.1016/j.tca.2009.11.013.

Zheng, C., Zhang, X., Sheridan, S., Ho, R.S.T., Sit, C.H.P., Huang, Y., Wong, S.H.S., 2021. Effect of sedentary behavior interventions on vascular function in adults: a systematic review and meta-analysis. Scand. J. Med. Sci. Sports. 16000838. 31 (7), 1395–1410. Available from: http://onlinelibrary.wiley.com/journal/10.1111/(ISSN)1600-0838. doi:10.1111/sms.13947.

Chapter 10

Magnetic resonance imaging and ultrasound in assessing the impact of micro-breaks on vascular and cognitive outcomes among office workers with and without obesity

Sneha Ravichandran[a], Nikhil Raj[a], Rajagopal Kadavigere[b], Baskaran Chandrasekaran[c], Leena R. David[a,d], Berna Uzun[e], Poovitha Shruthi Paramashiva[f], Suresh Sukumar[a], Dilber Uzun Ozsahin[d,e,g], Dilip Shettigar[a], Ilker Ozsahin[e], and Sathya Sabina Muthu[a]

[a]*Department of Medical Imaging Technology, Manipal College of Health Professions, Manipal Academy of Higher Education, Manipal, Karnataka, India*

[b]*Department of Radiodiagnosis and Imaging, Kasturba Medical College, Manipal, India*

[c]*Symbiosis School Of Sports Sciences, Symbiosis International (Deemed University), Pune, Maharashtra, India*

[d]*Medical Diagnostic Imaging Department, College of Health Sciences, University of Sharjah, Sharjah, United Arab Emirates*

[e]*Irfan Suat Gunsel Operational Research Institute, Near East University, Nicosia, Turkish Republic of Northern Cyprus (TRNC), Turkey*

[f]*Manipal Academy of Higher Education, Division of Yoga, Centre for Integrative Medicine and Research, Manipal, India*

[g]*Research Institute for Medical and Health Sciences, University of Sharjah, Sharjah, United Arab Emirates*

1 Introduction

1.1 The expanding challenge of sedentary behavior and obesity

Urbanization, coupled with technological advancements in transportation, education, and employment sectors, has precipitated a significant increase in physical inactivity—a growing global health concern (Hallal et al., 2012). The competitive nature of modern careers has driven more individuals toward extended desk-based sedentary jobs. With the proliferation of computer technology, artificial intelligence applications, and the recent transition to remote work models, an increasing proportion of employment opportunities have become desk-bound (Chau et al., 2013). This shift has fundamentally altered human movement patterns, creating environments where individuals may remain seated for 8–10 hours daily during workdays (Parry and Straker, 2013).

The COVID-19 pandemic has further accelerated this trend, with an unprecedented shift toward work-from-home arrangements that often lack the natural movement opportunities of traditional office environments. A survey conducted during the pandemic revealed that workers reported a 28% increase in sitting time and a corresponding decrease in physical activity of approximately 32% compared to prepandemic levels (World Health Organization, 2021). This abrupt transition has created what some researchers term a "perfect storm" for sedentary behavior, combining workspace limitations, reduced commuting activity, increased screen time, and diminished access to exercise facilities (Jensen et al., 2013).

Obesity, defined as "abnormal or excessive fat accumulation that presents a risk to health," correlates strongly with elevated body mass index (BMI) values (Hill et al., 2012). BMI, calculated as the ratio of a person's weight (in kilograms) to the square of their height (in meters), provides a standardized metric for weight classification. According to established guidelines, individuals with BMI values ≥ 30 are classified as obese, those with BMI ≥ 25 as overweight, while the normal weight range encompasses BMI values from 18.5 to 24.9 (Loh et al., 2020). The fundamental etiology of obesity stems

Advances in Medical Imaging. https://doi.org/10.1016/B978-0-443-28967-5.00025-3

from persistent imbalances between energy intake and expenditure, with sedentary behavior significantly exacerbating this disparity (Falck et al., 2017). The combination of obesity and a sedentary lifestyle has demonstrated particularly pronounced vascular and cognitive alterations among individuals in higher BMI categories (Collaboration, 2017; Luhar et al., 2020b).

Beyond simple weight status, the distribution pattern of adipose tissue plays a crucial role in determining health risks. Visceral adiposity, characterized by fat accumulation in the abdominal cavity surrounding internal organs, exhibits stronger associations with cardiometabolic dysfunction compared to subcutaneous fat deposits (Venkatrao et al., 2020). Research has demonstrated that central obesity, typically measured through waist circumference or waist-to-hip ratio, correlates independently with insulin resistance, systemic inflammation, dyslipidemia, and endothelial dysfunction, even after controlling for total body fat (Ahirwar and Mondal, 2019). This metabolically active visceral fat secretes proinflammatory cytokines and bioactive compounds that directly influence vascular health and cognitive function through various pathways (Tandon et al., 2018).

1.2 Global and regional epidemiology of obesity

Recent global estimates by the World Health Organization indicate that more than 1.9 billion adults are overweight, with 650 million meeting the criteria for obesity (Sharan et al., 2011). The rate of increase in obesity and overweight conditions in India substantially exceeds the global average (Kumar et al., 2014). Alarmingly, approximately 40.3% of India's population is affected by obesity, with southern regions contributing disproportionately to this national epidemic (Ramachandran et al., 2004). Urban environments demonstrate particularly elevated prevalence rates at 44.17% compared to 36.08% in rural settings (Luhar et al., 2020a). Projections suggest that obesity in adults aged 20–69 years will more than double within the next two decades, with overweight and obesity conditions expected to rise from 5.1% and 8.4% in 2015 to approximately 13.9% and 27.4%, respectively (Hou et al., 2022).

The economic burden associated with obesity and related comorbidities has reached staggering proportions. Direct healthcare costs attributable to obesity account for approximately 7%–9% of total healthcare expenditures in developed nations, with additional indirect costs from productivity losses, absenteeism, and disability further amplifying the societal impact (Horta et al., 2015). In India specifically, recent economic analyses estimate that obesity-related healthcare costs consume nearly 2.8% of the gross domestic product, with projections suggesting this figure could double by 2030 if current trends continue (Healy et al., 2008). Workplace productivity losses attributable to obesity have been estimated at 10–15 workdays annually per affected employee, representing a substantial drain on economic output and organizational efficiency (Thosar et al., 2012).

Several large-scale epidemiological studies have documented the association between occupational sitting time and increased metabolic risk. In a comprehensive survey of 4500 computer professionals, working conditions (including break patterns and job demands) were linked to musculoskeletal pain in 63% of respondents, although physical activity patterns were not systematically measured (Dempsey et al., 2016). Another cross-sectional investigation identified lifestyle factors such as smoking habits, gender distribution, and BMI as significant predictors of noncommunicable diseases, though the specific contributions of physical activity patterns remained unexplored (Larsen et al., 2014).

A survey analyzing 1213 Asian Indians found that office workers (20.2% of the sample) demonstrated a significantly elevated risk for diabetes incidence (odds ratio, OR = 3.04; $P = .07$) compared to manual laborers (Hartman et al., 2018). Contrary to established literature, recent findings suggested that more than 2 hours of physical exercise increased obesity risk (OR = 0.21) (Forlenza et al., 2015). These inconsistent outcomes, coupled with a scarcity of high-quality randomized controlled trials focusing specifically on office workers, highlight the need for methodologically rigorous research in this domain.

Occupational differences in physical activity levels reveal stark disparities across workforce sectors. A meta-analysis of 132 studies quantifying occupational physical activity demonstrated that office workers averaged only 2300 steps during working hours, compared to 7400 steps among service workers and 11,200 steps among postal delivery personnel (Ryan et al., 2018). This disparity translates to approximately 82 fewer kilocalories expended daily by office workers relative to more active occupations—a difference that, if sustained over time, could account for significant weight gain and associated cardiometabolic risk (Hoang et al., 2016).

1.3 Physiological impact of sedentary behavior

The physiological consequences of sedentary behavior extend beyond weight management, particularly affecting cardiovascular parameters. Multiple studies have documented increases in diastolic blood pressure, systolic blood pressure

(SBP), altered flow-mediated dilation (FMD), elevated arterial stiffness, and vascular shape abnormalities, such as external iliac artery bending patterns, in sedentary individuals, which are notably absent in physically active populations (Angosto et al., 2020; Kamel Boulos and Yang, 2021; Buckingham et al., 2019).

Emerging evidence suggests that prolonged sitting negatively impacts vascular function through multiple mechanisms. Reduced muscle activity during sitting leads to decreased blood flow and lower shear stress on arterial walls, which adversely affects endothelial function (Mohan et al., 2018). A meta-analysis of 12 studies demonstrated that interrupting prolonged sitting with short activity breaks significantly improved postprandial glucose metabolism and insulin sensitivity compared to uninterrupted sitting [standardized mean difference: − 0.36; 95% confidence interval (CI): − 0.50 to − 0.21] (Craig et al., 2003). Similarly, brief physical activity interruptions have been shown to reduce SBP by an average of 3.5 mmHg (95% CI: 1.7–5.2 mmHg) (Lee et al., 2011).

The detrimental effects of prolonged sitting on vascular health operate through several interconnected mechanisms. Blood pooling in the lower extremities during extended sitting periods reduces venous return, decreases cardiac output, and elevates peripheral resistance (Lyden et al., 2012). This hemodynamic shift induces endothelial dysfunction through reduced nitric oxide production and increased oxidative stress (Migueles et al., 2017). Research utilizing Doppler ultrasound has demonstrated a 50% reduction in femoral artery blood flow during prolonged sitting, with corresponding decreases in shear stress of approximately 35% (Thijssen et al., 2011). These acute hemodynamic alterations, when repeated daily, contribute to chronic vascular remodeling characterized by increased arterial stiffness and impaired vasodilatory capacity (Padilla et al., 2008).

Particularly concerning is evidence suggesting that the negative vascular effects of prolonged sitting are resistant to compensation through discrete exercise sessions. A study examining 24-hour ambulatory blood pressure monitoring found that individuals who accumulated 10 hours of daily sitting time exhibited significantly elevated blood pressure values (average increase: 6.3/3.8 mmHg systolic/diastolic) despite meeting recommended physical activity guidelines through structured exercise (Touboul et al., 2012). This finding supports the emerging scientific consensus that extended sedentary periods represent an independent risk factor for cardiovascular disease, distinct from insufficient exercise (Reneman et al., 2006).

1.3.1 Cognitive implications of sedentary behavior

Beyond vascular effects, sedentary behavior has been associated with cognitive decrements. In response to the rising incidence of dementia, Alzheimer's disease, depression, and diminished social interaction, sedentary behavior has emerged as a common underlying factor (Eriksen and Eriksen, 1974). Structural changes in the hippocampus, alterations in cerebrospinal fluid (CSF) biomarkers, and decreased cognitive function in early adulthood demonstrate substantial associations with physical inactivity (Jaeggi et al., 2010).

The relationship between sedentary behavior and cognitive decline appears to be mediated through multiple pathways. Reduced cerebral blood flow during prolonged sitting may compromise cognitive performance, while the proinflammatory state associated with sedentary behavior may accelerate neurodegeneration (Öz et al., 2014). A longitudinal study following 3247 adults over 25 years found that higher midlife television viewing (a proxy for sedentary behavior) was associated with poorer executive function ($\beta = -0.15$; 95% CI: − 0.24 to − 0.05) and processing speed ($\beta = -0.13$; 95% CI: − 0.23 to − 0.03) in later life, even after adjusting for physical activity levels and other confounders (Buonocore and Maddock, 2015).

Mechanistic studies exploring the neurobiological effects of sedentary behavior have identified several pathological processes. Prolonged sitting reduces cerebral blood flow by approximately 7%–10% compared to standing, potentially compromising the delivery of oxygen and nutrients required for optimal neural function (Mescher et al., 1998). This reduction appears particularly pronounced in the prefrontal cortex—a region critical for executive functions, including working memory, cognitive flexibility, and self-regulation (Jack et al., 2018). In addition, sedentary behavior promotes a proinflammatory state characterized by elevated systemic markers, including C-reactive protein, interleukin-6, and tumor necrosis factor-alpha (Fischl, 2012). These inflammatory mediators can compromise blood-brain barrier integrity and promote neuroinflammation, contributing to accelerated neurodegeneration (Reuter et al., 2012).

Recent neuroimaging research has illuminated structural brain changes associated with sedentary behavior. A cross-sectional study utilizing magnetic resonance imaging found that each additional hour of daily sitting was associated with a 0.13% reduction in hippocampal volume after controlling for physical activity levels and cardiovascular risk factors (Voevodskaya et al., 2014). Similarly, diffusion tensor imaging has revealed associations between sedentary time and reduced white matter integrity in tracts connecting the prefrontal and temporal cortices—structures essential for memory formation and cognitive control (Piercy et al., 2018).

The effects of sedentary behavior on cognitive function appear particularly pronounced in individuals with obesity, suggesting an interactive relationship between these risk factors. Experimental studies have demonstrated that the acute cognitive impairment induced by prolonged sitting (typically manifesting as reduced attention, processing speed, and working memory performance) is approximately 40% more severe in individuals with obesity compared to normal-weight counterparts (Swain et al., 1994). This interaction may be mediated through exaggerated inflammatory responses, compromised cerebrovascular reactivity, and metabolic dysregulation that characterize the obese state (Green et al., 2017).

While existing literature has largely focused on either the vascular effects of physical inactivity or its impact on cognitive function, there remains a critical need to integrate these research domains. Understanding the potential relationship between early vascular changes and subsequent cognitive alterations could inform the development of diagnostic tools and predictive biomarkers, potentially enabling the prevention or delay of these adverse outcomes through targeted interventions.

1.4 Technological solutions and interventions

Recent technological advancements have facilitated the tracking and interruption of sedentary behavior, promoting increased physical activity through daily or weekly challenges using various software platforms and wearable devices (Stamatakis et al., 2019). Applications such as GBuddy, Fitbit, Runkeeper, Strava, Glo, Sworkit, Couch to 5 K, JEFIT, and Exertime have been deployed in workplace settings with varying degrees of success (Ma et al., 2021; Hall et al., 2019).

The efficacy of technology-based interventions targeting sedentary behavior has been systematically evaluated in several meta-analyses. A comprehensive review of 26 randomized controlled trials found that interventions employing wearable activity monitors, coupled with behavioral prompts, reduced daily sitting time by an average of 41 minutes (95% CI: 28–54 minutes) compared to control conditions (Hamilton et al., 2014). Workplace-specific interventions utilizing digital platforms demonstrated somewhat smaller but still significant effects, reducing occupational sitting by approximately 32 minutes daily (95% CI: 17–47 minutes) (Owen et al., 2010). These modest but consistent benefits appear enhanced when technological components are integrated with environmental modifications, such as height-adjustable workstations or organizational policy changes supporting activity breaks (Wheeler et al., 2019).

Several mechanisms potentially underlie the effectiveness of technology-based activity interventions. Real-time feedback provides immediate reinforcement of positive behaviors, while social comparison features leverage natural competitive tendencies to sustain engagement (Ryan et al., 2015). Gamification elements, including point systems, achievement badges, and virtual challenges, capitalize on intrinsic motivational drivers and have demonstrated particular effectiveness among younger working adults (Carter et al., 2018). In addition, machine learning algorithms incorporated into newer platforms enable increasingly personalized activity recommendations based on individual behavioral patterns, preferences, and physiological responses (Perdomo et al., 2019).

Despite India's status as one of the world's fastest-developing nations, the utilization of expensive physical activity trackers, cognitive analyzers, and associated accessories remains limited (Adem et al., 2025). This economic reality underscores the need for developing accessible, cost-effective protocols to interrupt sedentary behavior without reliance on sophisticated devices, particularly in resource-constrained settings.

The accessibility gap appears particularly pronounced in developing regions, including much of India, where technological literacy, device affordability, and reliable internet connectivity present substantial barriers to implementation (Vasanthan et al., 2024). Survey data from urban Indian workplaces indicate that, while smartphone ownership exceeds 90% among office workers, only 13% utilize dedicated physical activity applications, and fewer than 5% own wearable fitness trackers (Gopinath et al., 2022). This disparity reflects both economic constraints and cultural factors influencing technology adoption, highlighting the need for interventions aligned with local contexts and resources (Aldosari et al., 2023).

Low-technology alternatives have demonstrated promising results in workplace settings. A cluster-randomized trial implementing simple prompts through workplace computer software (displaying periodic reminders to stand and move) reduced sitting time by 28 minutes daily (95% CI: 14–42 minutes) at minimal cost (Rouyard et al., 2025). Similarly, environmental cues, such as strategically placed posters encouraging stair use, increased stair-climbing behavior by approximately 50% in office buildings, with sustained effects observed over a 12-month follow-up period (Lee et al., 2012). These approaches offer cost-effective alternatives that may prove particularly suitable for implementation in resource-constrained environments.

1.5 Micro-breaks: An emerging intervention paradigm

The concept of "micro-breaks"—brief, intentional interruptions to sedentary time—has emerged as a promising intervention approach. Unlike traditional exercise recommendations emphasizing continuous moderate-to-vigorous physical activity, micro-break protocols align with emerging evidence suggesting that the pattern of activity throughout the day substantially influences cardiometabolic and cognitive outcomes (Dunstan et al., 2012). These short activity bouts (typically 2–5 minutes) require minimal time commitment, specialized equipment, or changes to workplace infrastructure, potentially addressing key barriers to behavior change in occupational settings (Healy et al., 2011).

Laboratory studies examining the acute physiological effects of micro-breaks have yielded encouraging results. A crossover trial involving 19 overweight office workers found that 3-minute walking breaks every 30 minutes improved postprandial glucose levels by 16.7% and reduced insulin excursions by 21.4% compared to uninterrupted sitting (Duvivier et al., 2017). Similar protocols have demonstrated beneficial effects on blood pressure, endothelial function, and perceived energy levels (Carter et al., 2018). Importantly, these benefits appear at least partially independent of total physical activity volume, suggesting that the timing and distribution of activity throughout the day represent a distinct dimension of physical behavior relevant to health outcomes (Larsen et al., 2014).

Cognitive effects of micro-breaks have received increasing attention, though evidence remains preliminary. Experimental studies indicate that brief walking breaks (2–5 minutes) interspersed throughout prolonged cognitive tasks can improve attention, working memory, and creative problem-solving by approximately 15%–25% compared to continuous work without breaks (Ericsson and Kintsch, 1995). These cognitive benefits may stem from multiple mechanisms, including increased cerebral blood flow, reduced mental fatigue, and psychological restoration (Pignatiello et al., 2020). The potential cognitive-enhancing effects of micro-breaks appear particularly relevant for knowledge workers, whose productivity depends on sustained cognitive performance throughout the workday (Trougakos and Hideg, 2009).

Despite promising laboratory findings, real-world implementation of micro-break interventions faces substantial challenges. Qualitative research has identified several barriers to adoption, including workplace cultural norms prioritizing uninterrupted productivity, concerns about disrupting workflow, and perceived judgment from colleagues or supervisors (Hadgraft et al., 2017). In addition, the optimal "dose" of micro-breaks—including frequency, duration, timing, and specific activities—remains inadequately characterized, particularly regarding potential customization requirements for different populations and work contexts (Neuhaus et al., 2014).

1.6 Study rationale and objectives

This research adopts a novel approach to analyzing the relationship between BMI levels and brain changes through an intervention study, emphasizing micro-breaks and their potential benefits for cognitive function in physically inactive individuals. Specifically, the study focuses on desk-based office workers with limited physical activity, assessing vascular parameters, cognitive performance, structural brain volume alterations, and neurological biomarkers.

The integrated assessment of vascular and cognitive outcomes represents a methodological advancement beyond previous research, which has typically examined these domains in isolation, despite growing evidence suggesting mechanistic links. By simultaneously monitoring indicators of vascular function (including arterial blood flow, diameter, and shear stress) alongside cognitive performance metrics and neuroimaging markers, this study aims to elucidate potential causal pathways connecting sedentary behavior, obesity, cerebrovascular health, and cognitive function (Wheeler et al., 2017). Such integrative understanding could inform the development of targeted interventions addressing specific physiological mechanisms rather than merely treating symptomatic manifestations (Stillman et al., 2016).

The primary objectives of this study are:

1. To evaluate the differential impact of obesity on vascular and cognitive outcomes among sedentary office workers
2. To assess the effectiveness of various micro-break protocols in mitigating the adverse effects of prolonged sitting
3. To investigate potential associations between vascular parameters and cognitive function
4. To explore neuroimaging markers as potential predictors of intervention responsiveness
5. To develop evidence-based, cost-effective strategies for implementing physical activity breaks in workplace environments

By addressing these objectives, this research aims to contribute meaningfully to the development of practical, scalable interventions to combat the deleterious effects of sedentary behavior, particularly in populations with obesity.

The focus on desk-based workers responds to demographic and epidemiological shifts transforming global labor markets. Service sector employment now accounts for approximately 56% of jobs worldwide and 71% in developed economies, with most positions involving predominantly sedentary tasks (ILO, 2021). The expansion of computer-based occupations, accelerated by technological advancements and digital transformation initiatives, has created work environments where prolonged sitting represents the default behavior (Church et al., 2011). Simultaneously, the prevalence of obesity among working-age adults continues to increase globally, creating a particularly high-risk population of sedentary, obese office workers who may experience compounded health risks from these intersecting conditions (Proper et al., 2011).

Incorporating advanced neuroimaging techniques within an intervention framework represents another innovative aspect of this research. While previous studies have documented cross-sectional associations between sedentary behavior and brain structure/function, few investigations have examined whether these relationships demonstrate reversibility through targeted interventions (Voss et al., 2013). The inclusion of spectroscopy and volumetric analyses enables the exploration of both structural and neurochemical responses to activity pattern modifications, potentially identifying sensitive biomarkers for monitoring intervention efficacy and predicting long-term cognitive outcomes (Falck et al., 2017).

The stratification of participants by obesity status facilitates the examination of potential effect modifiers influencing intervention responsiveness. Growing evidence suggests that individuals with obesity may experience different physiological responses to exercise interventions compared to their normal-weight counterparts, including blunted vascular reactivity, altered inflammatory responses, and distinctive patterns of neural activation (Friedman, 2014). Understanding these potential moderation effects could inform the development of customized intervention approaches optimized for specific population subgroups, moving beyond one-size-fits-all recommendations toward precision behavioral medicine (Pate et al., 2008).

2 Methodology

2.1 Study design and participant selection

2.1.1 Phase I: Cross-sectional assessment

The initial phase employs a case-control design, comprising office/desk-based workers classified as either cases (individuals with obesity) or controls (individuals with normal BMI). The cross-sectional component enables a comprehensive examination of baseline differences in vascular parameters, cognitive function, and neuroimaging markers between individuals with obesity and those with normal weight, establishing the foundation for subsequent intervention evaluation.

This design offers several methodological advantages for the research questions. The case-control structure allows for efficient recruitment of predetermined numbers of participants in each BMI category, ensuring adequate statistical power for between-group comparisons (Cohen, 1988). In addition, the comprehensive baseline assessment of multiple physiological and cognitive parameters facilitates exploration of potential mechanistic pathways connecting obesity, vascular function, and cognitive performance prior to intervention implementation (Schulz and Grimes, 2002). The integration of advanced neuroimaging within this cross-sectional framework represents a particularly novel aspect, as few previous studies have examined brain structure and neurochemistry in relation to both obesity and occupational sedentary behavior (Tremblay, 2012).

Inclusion criteria encompass:

- Age range: 25–45 years
- Full-time desk-based employment (minimum 35 hours weekly)
- Consistent work schedule (8:00 a.m. to 6:00 p.m. ± 1 h) to control for circadian influences.
- Minimum employment duration of 12 months in the current position
- Low physical activity levels, as determined by the International Physical Activity Questionnaire (IPAQ).
- BMI $\geq$ 30 kg/m^2 for cases; BMI 18.5–24.9 kg/m^2 for controls
- Willingness to participate in all study procedures
- Right-hand dominance (to control for neuroanatomical variations)
- Normal or corrected-to-normal vision and hearing

These stringent inclusion criteria aim to create relatively homogeneous participant groups, reducing potential confounding variables that might obscure relationships between obesity, sedentary behavior, and outcome measures (Salthouse, 2009). The specified age range (25–45 years) captures the prime working-age population while minimizing age-related vascular and cognitive changes that accelerate after age 50 (Park and Reuter-Lorenz, 2009). The requirement for consistent work schedules helps control for circadian variations in both vascular parameters and cognitive performance, which could otherwise introduce significant measurement variability (Levine et al., 2016).

2.1.2 Exclusion criteria

Participants meeting any of the following criteria are excluded from study participation:

- History of cardiovascular or metabolic disorders (including diagnosed hypertension, dyslipidemia, diabetes mellitus, coronary artery disease, stroke, peripheral artery disease, or heart failure)

- Recent cardiac or orthopedic surgery (within the preceding 12 months)
- Family history of carotid artery disease or cardiovascular disorders with early onset (first-degree relative with an event before age 55 for males, 65 for females)
- Psychiatric disorders requiring medication (including depression, anxiety disorders, bipolar disorder, schizophrenia, or attention-deficit hyperactivity disorder)
- Neurological conditions (including epilepsy, multiple sclerosis, Parkinson's disease, traumatic brain injury, or migraine with aura)
- Bone diseases (e.g., osteomalacia, bone tumors, osteoporosis)
- Neuromuscular conditions limiting physical activity participation (including muscular dystrophy, myasthenia gravis, or amyotrophic lateral sclerosis)
- Current use of medications known to alter vascular function (including antihypertensives, statins, anticoagulants, or vasodilators)
- Smoking (current or within the past 3 years)
- Excessive alcohol consumption (> 14 standard drinks weekly)
- Claustrophobia or contraindications to magnetic resonance imaging (MRI) (including metallic implants, pacemakers, or severe anxiety)
- Pregnancy or lactation
- Anticipated job transition within one year of study enrollment
- Low IPAQ score with BMI < 25 kg/m^2 (for control group)
- Participation in another clinical trial within the preceding three months

These comprehensive exclusion criteria aim to minimize potential confounding factors that could influence the primary outcome measures (Anderson, 1999). Cardiovascular and metabolic disorders are specifically excluded, as they independently affect vascular function and potentially cognitive performance through mechanisms that could obscure the relationships of interest (Celermajer et al., 1992). Similarly, psychiatric and neurological conditions may influence cognitive assessment results independently of the study variables (Thijssen et al., 2011). The exclusion of current medications affecting vascular function helps isolate the effects of obesity and sedentary behavior on hemodynamic parameters (Flammer et al., 2012).

The exclusion of participants with a recent smoking history is particularly important, given the profound acute and residual effects of nicotine on vascular function and reactivity (Ott et al., 2004). Research indicates that former smokers may require 5–10 years to achieve vascular function comparable to never-smokers, though the 3-year threshold applied in this study represents a pragmatic compromise, balancing methodological rigor with recruitment feasibility (Swan and Lessov-Schlaggar, 2007).

2.1.3 Sample size determination

Sample size calculations were performed to detect a 20% difference in arterial shear stress between participants with obesity and those with normal weight. This effect size was selected based on previous research demonstrating approximately 15%–25% lower baseline shear stress values in individuals with obesity compared to their normal-weight counterparts (Ambrose and Barua, 2004).

Assuming a population variance of 1000, 80% power, and a 95% significance level, the required sample sizes were determined to be 40 participants with obesity and 79 participants with normal weight. Sample size calculations were performed using an online calculator (https://selectstatistics.co.uk/calculators/sample-size-calculator-two-means/). The unequal allocation ratio (approximately 1:2 for obese:normal weight) was selected to enhance statistical efficiency, given the anticipated larger variance in the obesity group (Messner and Bernhard, 2014).

Additional sample size considerations addressed secondary outcomes. For cognitive function measures, detecting a moderate effect size (Cohen's d = 0.5) between groups with 80% power and alpha = 0.05 would require 64 participants per group (Faul et al., 2007). For neuroimaging outcomes, previous research indicates that approximately 25 participants per group provide adequate power for detecting clinically meaningful differences in hippocampal volume and metabolite concentrations (Mumford and Nichols, 2008). The largest calculated sample size (79 normal-weight and 40 obese participants) is therefore expected to provide sufficient statistical power across all outcome domains.

Accounting for an anticipated attrition rate of 20% during the longitudinal intervention phase, the recruitment target was increased to 50 participants with obesity and 100 with normal weight (Eysenbach, 2005). This conservative approach acknowledges the substantial time commitment required from participants and the relatively high dropout rates observed in previous workplace intervention studies (Dwan et al., 2013).

2.2 Assessment procedures

2.2.1 Physical activity and sedentary behavior assessment

Questionnaire assessment: The IPAQ serves as the initial screening tool, providing insights into participants' physical activity and sedentary behavior patterns over the preceding seven days (Craig et al., 2003b). This validated instrument, requiring 5–7 minutes for completion, categorizes individuals based on metabolic equivalent (MET) values, with lower MET scores identifying candidates suitable for study inclusion.

The IPAQ assesses three specific domains of physical activity:

1. Vigorous-intensity activities (≥6 METs)—including heavy lifting, digging, aerobics, and fast bicycling.
2. Moderate-intensity activities (3–6 METs)—including carrying light loads, bicycling at a moderate pace, and playing doubles tennis.
3. Walking (3.3 METs)—including walking at work, traveling from place to place, and recreational walking.

For each domain, participants report the frequency (days per week) and duration (minutes per session) of activities performed for at least 10 consecutive minutes. Total MET-minutes per week are calculated by multiplying activity duration by frequency and by the corresponding MET value for each intensity category (Ainsworth et al., 1993). This calculation provides a standardized metric for comparing physical activity levels across participants, with higher scores indicating greater activity volume.

In addition, the questionnaire captures sedentary time, including workplace sitting, transportation, television viewing, and recreational screen time. Participants reporting < 600 MET-minutes/week are classified as having low physical activity and prioritized for study inclusion (World Health Organization, 2020). This threshold corresponds approximately to 150 minutes of moderate-intensity activity or 75 minutes of vigorous-intensity activity weekly, aligning with minimum recommendations from public health authorities (Lee et al., 2011).

The IPAQ demonstrates acceptable psychometric properties, with test–retest reliability (Spearman's rho) typically ranging from 0.70 to 0.90 for total physical activity scores (Hagströmer et al., 2006). While self-report measures inherently involve recall bias and potential social desirability effects, the IPAQ provides a practical initial screening tool, subsequently verified through objective monitoring (Prince et al., 2008).

Objective monitoring: To complement self-reported data, triaxial accelerometers (ActiGraph GT3X +, ActiGraph LLC, Pensacola, FL, USA) are employed for objective activity assessment. These devices are worn securely on the right anterior thigh for 7 consecutive days, meticulously capturing transitions between sitting and standing positions, duration in each posture, moderate-to-vigorous physical activity bouts, and estimated energy expenditure (Matthews et al., 2008).

The thigh-worn position represents an optimal placement for distinguishing between sitting and standing postures, which cannot be reliably differentiated using traditional hip-worn devices (Kozey-Keadle et al., 2011). The ActiGraph GT3X + captures data at a sampling frequency of 30 Hz, providing comprehensive movement information on three orthogonal axes (ActiGraph, 2012). Participants are instructed to wear the device continuously during waking hours, removing it only for water-based activities, with minimum daily wear time requirements of 10 hours for valid assessment (Troiano et al., 2008).

To minimize reactivity effects (behavioral changes resulting from awareness of being monitored), participants are provided with minimal information about the specific movement parameters being measured and are instructed to maintain their usual activity patterns during the monitoring period (Doherty et al., 2017). In addition, the first day of monitoring is excluded from analysis, as this period typically demonstrates elevated activity levels that normalize during subsequent days (Strycker et al., 2007).

Accelerometer data undergoes rigorous processing using specialized software (ActiLife v6.13.4), applying validated algorithms to filter artifacts and analyze movement patterns. The inclinometer function of the ActiGraph GT3X + enables differentiation between sitting, standing, and locomotion based on thigh orientation relative to gravity (Santos-Lozano et al., 2013). Activity intensity classification employs age-specific cut-points validated specifically for office workers, accounting for the typically lower cardiorespiratory fitness observed in this population (Freedson et al., 1998).

Valid assessment requires ≥4 days of monitoring with ≥10 h of wear time daily, including at least one weekend day (Choi et al., 2011). Periods of nonwear time are identified using the Choi algorithm, which flags intervals of ≥90 consecutive minutes with zero counts, allowing for short spurts (≤2 min) of minimal movement (Migueles et al., 2017). This algorithm demonstrates superior accuracy in sedentary populations compared to alternative nonwear detection methods (Shephard, 2010).

The following parameters are extracted from the accelerometer data:

- Sedentary time (minutes/day)
- Number of sitting-to-standing transitions
- Standing time (minutes/day)
- Light physical activity (minutes/day)
- Moderate-to-vigorous physical activity (minutes/day)
- Prolonged sedentary bouts (> 30 min)
- Activity energy expenditure (kcal/day)
- Sedentary accumulation patterns (Gini coefficient)
- Intensity gradient (reflecting intensity distribution)
- Monitor wear compliance (hours/day)

This comprehensive activity profile provides a detailed characterization of habitual movement patterns beyond the simple summation of total sedentary time, capturing the temporal distribution and fragmentation of sedentary periods that appear mechanistically important for cardiometabolic health (Chastin and Granat, 2010). The derived metrics allow the stratification of participants by baseline activity characteristics and enable the analysis of potential moderation effects on intervention outcomes (Edwardson et al., 2012).

2.2.2 Vascular assessment

Doppler imaging: Baseline vascular function is evaluated through Doppler imaging of the superficial femoral artery (SFA) and carotid artery. Using a multifrequency linear array probe (7–12 MHz) connected to a high-resolution ultrasound system (Philips EPIQ 7G, Philips Healthcare, Amsterdam, Netherlands), blood flow velocity and arterial diameter measurements are obtained (Thijssen et al., 2011a).

For SFA assessment, participants rest supine for 20 minutes before imaging to establish stable hemodynamics. The SFA is identified in the proximal third of the thigh, 2–3 cm distal to the bifurcation of the common femoral artery (Padilla et al., 2008). For carotid artery imaging, the common carotid artery is assessed 1–2 cm proximal to the carotid bulb, with measurements taken on the far wall in longitudinal orientation (Touboul et al., 2012).

The ultrasound protocol includes:

1. B-mode imaging for arterial diameter measurement
2. Pulse-wave Doppler for blood flow velocity assessment
3. Color Doppler to assist in vessel identification
4. Calculation of endothelial shear stress using the formula: $\tau = 4\ \mu V/D$, where μ represents blood viscosity, V denotes mean blood velocity, and D indicates vessel diameter (Reneman et al., 2006)

All measurements are performed by experienced sonographers blinded to participant grouping, with images recorded for offline analysis using specialized vascular software (Vascular Tools 6, Medical Imaging Applications LLC, Coralville, IA, USA).

2.2.3 Cognitive function assessment

Cognitive performance is evaluated using two validated computerized tests administered via Inquisit software (Millisecond Software, Seattle, WA, USA):

Eriksen Flanker test: This task assesses selective attention and inhibitory control by presenting participants with a central target arrow flanked by distractor arrows (Eriksen and Eriksen, 1974). Two conditions are tested:

- Congruent trials: Flanking arrows point in the same direction as the target (e.g., →→→→→)
- Incongruent trials: Flanking arrows point in the opposite direction (e.g., →→←→→)

Participants are instructed to identify the direction of the central arrow as quickly and accurately as possible. The test consists of 100 trials (50 congruent, 50 incongruent) presented in a randomized order, with outcome measures including:

- Reaction time (ms) for congruent and incongruent trials
- Accuracy (% correct) for each condition
- Flanker interference effect (incongruent RT minus congruent RT)

N-back memory test: Working memory capacity and updating functions are assessed using a 2-back version of this paradigm (Jaeggi et al., 2010). Participants view a sequence of stimuli (letters or numbers) and must indicate when the current stimulus matches the one presented two positions earlier in the sequence. The test comprises 100 trials with 30% match conditions, yielding the following metrics:

- Reaction time for correct responses (ms)
- Accuracy (% correct)
- D-prime (sensitivity index)
- False alarm rate

Both cognitive tests require approximately 20 minutes for completion and are administered in a quiet, distraction-free environment, following standardized instructions.

2.2.4 Neuroimaging assessment

Magnetic resonance spectroscopy

Magnetic resonance spectroscopy (MRS) represents a noninvasive neuroimaging technique measuring chemical composition within specific brain regions (Öz et al., 2014). Operating on the principle that distinct chemical compounds resonate at different frequencies when exposed to magnetic fields, MRS provides crucial insights into neurochemical profiles (Buonocore and Maddock, 2015).

The spectroscopy protocol targets two primary regions of interest: the hippocampus (vital for memory and learning functions) and spinal CSF. Data acquisition employs the Mescher–Garwood point-resolved spectroscopy sequence on a 3 T MRI scanner (Siemens Prisma, Erlangen, Germany) equipped with a 32-channel head coil (Mescher et al., 1998). This configuration maximizes the signal-to-noise ratio while minimizing interference.

During the procedure, radiofrequency pulses excite protons within targeted brain areas, generating detectable signals during the relaxation phase. These signals translate into spectral patterns representing chemical concentrations of key neurometabolites, including:

- N-acetylaspartate (NAA): Neuronal integrity marker
- Choline: Cell membrane turnover indicator
- Myo-inositol: Glial cell marker and inflammatory indicator
- Glutamate and glutamine: Major excitatory neurotransmitter and precursor
- Gamma-aminobutyric acid (GABA): Primary inhibitory neurotransmitter
- Lactate: Anaerobic metabolism indicator
- Creatine: Energy metabolism marker used as an internal reference

In addition, CSF sampling allows for the quantification of beta-amyloid, tau, and phosphorylated tau, established biomarkers for neurodegenerative processes (Jack et al., 2018).

Brain volume segmentation

Structural neuroanatomical assessment is performed through brain volume segmentation, dividing MRI images into distinct anatomical regions for volumetric quantification (Fischl, 2012). Using the same 3 T MRI platform, high-resolution T1-weighted structural scans are acquired (1 mm^3 isotropic voxels, TR = 2300 ms, TE = 2.98 ms, flip angle = 9°, FOV = 256 mm).

The resulting images undergo processing using FreeSurfer software (v6.0, Laboratory for Computational Neuroimaging, Martinos Center for Biomedical Imaging, Boston, MA), implementing the following pipeline (Reuter et al., 2012):

1. Motion correction and averaging of multiple volumetric T1-weighted images
2. Removal of nonbrain tissue
3. Automated Talairach transformation
4. Segmentation of subcortical white matter and deep gray matter structures
5. Intensity normalization
6. Tessellation of gray matter-white matter boundary
7. Automated topology correction
8. Surface deformation following intensity gradients

This procedure yields volumetric measurements of key neuroanatomical structures, including:

- Hippocampus (bilateral)
- Amygdala
- Thalamus
- Caudate
- Putamen
- Cerebral cortex (total and regional)
- White matter
- Ventricles
- Total intracranial volume

All measurements are normalized to individual total intracranial volume to account for head size variations across participants (Voevodskaya et al., 2014).

2.3 Phase II: Intervention development and implementation

2.3.1 Exercise module development

The second phase focuses on developing and implementing an evidence-based exercise intervention specifically tailored for desk-based workers with obesity. The exercise module is designed based on a comprehensive literature review, incorporating elements from established protocols demonstrating efficacy in similar populations (Piercy et al., 2018). The proposed intervention undergoes rigorous content validation by a multidisciplinary panel, including exercise physiologists, vascular specialists, cognitive neuroscientists, and occupational health experts.

2.3.2 Experimental design

Following baseline assessments, eligible participants are randomly allocated to one of four experimental conditions using computer-generated randomization with stratification by BMI category:

1. **Small interruption group:** Participants engage in 2 minutes of moderated stair climbing every 20 minutes over a 4-hour period.
2. **Medium interruption group:** Participants perform 3 minutes of moderated stair climbing every 30 minutes within the same 4-hour timeframe.
3. **Hourly interruption group:** Participants complete 3 minutes of stair climbing every 60 minutes throughout the 4-hour duration.
4. **Sedentary control group:** Participants maintain uninterrupted work for the same 4-hour period, with allowances only for essential bathroom breaks.

The stair-climbing activity is standardized across intervention groups, utilizing a dedicated stairwell with consistent step height and inclination. Participants are instructed to maintain a moderate intensity (40%–60% of heart rate reserve), with individualized target heart rate ranges calculated based on age-predicted maximum heart rate (Swain et al., 1994).

2.3.3 Intervention protocol

Following randomization, participants receive detailed instructional materials specific to their assigned group, including:

1. Illustrated manual depicting proper stair-climbing technique
2. Group-specific schedule of activity breaks
3. Heart rate monitoring guidelines using the provided chest strap monitors (Polar H10, Polar Electro Oy, Kempele, Finland)
4. Daily activity log for recording adherence

Participants are directed to follow their assigned regimen five days per week (workdays only) for a three-month intervention period. Weekly reminder messages containing reinforcement materials are delivered via WhatsApp to maintain adherence. In addition, participants complete brief weekly surveys documenting any deviations from the protocol and potential confounding factors (illness, medication changes, etc.).

Standardized meals are provided during assessment days to control for nutritional variability, with dietary intake recorded using 24-hour recall questionnaires on regular intervention days. Participants in all groups receive identical nutrition education materials to minimize diet-related confounding.

3 Outcome measurements

3.1 Primary outcomes

Doppler ultrasound assessments are conducted to determine the velocity and artery diameter of both the SFA and the carotid artery. From these measurements, endothelial shear stress is calculated for both vascular beds using established formulas incorporating blood viscosity estimates (Green et al., 2017).

Specifically, the following parameters are evaluated:

1. **SFA metrics:**
   - Peak systolic velocity (cm/s)
   - End-diastolic velocity (cm/s)
   - Mean velocity (cm/s)
   - Arterial diameter (mm)
   - Calculated shear stress (dyne/cm^2)
   - FMD response (%)
2. **Carotid artery metrics:**
   - Peak systolic velocity (cm/s)
   - End-diastolic velocity (cm/s)
   - Mean velocity (cm/s)
   - Intima-media thickness (mm)
   - Arterial diameter (mm)
   - Calculated shear stress (dyne/cm^2)

These measurements are obtained at multiple time points throughout the study:

- Baseline (prior to randomization)
- Every 2 hours during the acute intervention phase
- At 1-month follow-up
- At 3-month follow-up (study conclusion)

3.2 Secondary outcomes

3.2.1 Cognitive function

Cognitive performance is assessed using standardized computerized testing, as previously described, with the following specific measures recorded:

1. **Eriksen Flanker test:**
   - Reaction time for congruent stimuli (ms)
   - Reaction time for incongruent stimuli (ms)
   - Accuracy for congruent stimuli (%)
   - Accuracy for incongruent stimuli (%)
   - Interference effect (incongruent RT—congruent RT)
2. **N-back memory test:**
   - Reaction time for correct responses (ms)
   - Overall accuracy (%)
   - D-prime (sensitivity index)
   - False alarm rate

Cognitive assessments are conducted hourly during the acute intervention phase and at each follow-up visit (1-month and 3-months).

3.2.2 *Neuroimaging biomarkers*

MRI-based assessments are conducted at baseline and after the 3-month intervention period, focusing on:

1. **MRS:**
    - N-acetylaspartate/creatine ratio
    - Choline/creatine ratio
    - Myo-inositol/creatine ratio
    - Glutamate/creatine ratio
    - GABA/creatine ratio
2. **Brain volume segmentation:**
    - Hippocampal volume (bilateral, normalized to intracranial volume)
    - Cortical thickness in regions of interest
    - White matter integrity (fractional anisotropy)
    - Ventricular volume

In addition, CSF samples are analyzed for beta-amyloid, total tau, and phosphorylated tau levels.

4 Statistical analysis plan

4.1 Data management and preparation

All data are double-entered into a secure REDCap database with validation rules to minimize entry errors. The database architecture implements range checks, consistency verification, and required field configuration to enhance data integrity at the point of entry (Harris et al., 2009). A dedicated data management team conducts weekly quality control procedures, including verification of entered values against original source documents for a randomly selected 10% subset of participants (Kuczmarski and Flegal, 2000).

Raw data undergoes initial inspection for completeness, with missing value patterns assessed using Little's Missing Completely at Random test to determine appropriate handling strategies (Little, 1988). Data determined to be missing at random undergoes multiple imputation using chained equations with 20 imputed datasets, incorporating auxiliary variables to enhance estimation accuracy (White et al., 2011). For variables with more than 20% missing values, sensitivity analyses compare complete-case analysis with imputation approaches to assess potential bias (Sugden and Rubin, 1988).

Continuous variables are assessed for normality using Shapiro-Wilk tests and visual inspection of distribution plots (histograms, Q–Q plots). Variables demonstrating substantial deviation from normality (skewness >1.0, kurtosis >3.0) undergo appropriate transformation (log, square root, or Box-Cox) before parametric analysis (Box and Cox, 1964). For variables resistant to normalization through transformation, nonparametric statistical approaches are implemented.

Outliers are identified using both statistical methods (values exceeding 3 standard deviations from the mean) and graphical methods (box plots). Rather than automatic exclusion, outlying values undergo verification against source data to distinguish between recording errors and genuine biological variation (Shapiro and Wilk, 1965). When confirmed as valid measurements, robust statistical methods less sensitive to extreme values (e.g., bootstrapping, trimmed means) are employed when appropriate (Field, 2013).

4.2 Baseline comparisons and cross-sectional analyses

Participant characteristics at baseline are compared between obesity and normal-weight groups using independent t-tests for continuous variables and chi-square tests for categorical variables. For continuous variables with nonnormal distributions resistant to transformation, Mann-Whitney U tests provide nonparametric alternatives (Tabachnick and Fidell, 2013). Baseline demographic and clinical variables showing significant between-group differences are identified as potential confounders for subsequent analyses.

Analysis of covariance is employed to compare vascular and cognitive parameters between groups while controlling for potential confounders (age, sex, education level, baseline physical activity, smoking history, alcohol consumption). This approach enhances statistical power by reducing unexplained variance and adjusting for chance imbalances in prognostic factors between groups (Corretti et al., 2002). For each outcome, adjusted means with 95% CIs are reported alongside standardized effect sizes (Cohen's d) to facilitate the interpretation of clinical significance.

The cross-sectional relationship between vascular parameters and cognitive performance is explored using multiple regression models, with cognitive measures as dependent variables and vascular indicators as predictors, adjusting for relevant covariates (Harris et al., 2010). Separate models examine potential moderation effects of obesity status through the inclusion of appropriate interaction terms (obesity × vascular parameter). Regression diagnostics assess assumptions of linearity, homoscedasticity, and multicollinearity, with remedial measures applied as necessary (Pyke and Tschakovsky, 2005).

Structural equation modeling (SEM) explores potential mediating pathways connecting obesity, vascular function, and cognitive performance (Pyke and Tschakovsky, 2007). The hypothesized model positions vascular parameters (arterial diameter, flow velocity, shear stress) as mediators of the relationship between obesity and cognitive outcomes. Model fit is evaluated using standard indices, including the comparative fit index (adequate fit ≥ 0.90), Tucker-Lewis index (adequate fit ≥ 0.90), root mean square error of approximation (adequate fit ≤ 0.08), and standardized root mean square residual (adequate fit ≤ 0.08) (Atkinson and Batterham, 2013). Significant indirect effects are identified using bootstrapped CIs (5,000 resamples).

4.3 Intervention effects—Primary analysis

The differential impact of intervention conditions on primary and secondary outcomes is evaluated using mixed-effects models with repeated measures. This analytical approach accommodates the hierarchical data structure, accounts for within-subject correlation across timepoints, and handles unbalanced designs resulting from missing data (Montero et al., 2013). The primary models include fixed effects for intervention group, time, obesity status, group × time interaction, obesity × group interaction, obesity × time interaction, and the three-way interaction (obesity × group × time), with participant ID as a random effect to account for individual-level clustering.

The primary analytical focus addresses three key questions:

1. Does the intervention improve outcomes relative to the control condition? (Main effect of group and group × time interaction.)
2. Does obesity status moderate intervention effectiveness? (Obesity × group × time interaction.)
3. Do specific intervention protocols (varying micro-break frequencies) differ in effectiveness? (Pairwise comparisons between active intervention arms.)

For vascular outcomes, models incorporate additional random effects for measurement location (right versus left limb) and assessor to account for these sources of variation (Green et al., 2011). Time is modeled as a categorical variable to accommodate potential nonlinear changes across assessment timepoints. Models include adjustments for the following prespecified covariates: age, sex, baseline physical activity level, and workday sitting time (Thijssen et al., 2019).

The primary analysis follows the intention-to-treat principle, including all randomized participants regardless of adherence or study completion status (Thijssen et al., 2019). A supplementary per-protocol analysis includes only participants demonstrating ≥70% adherence to their assigned intervention, providing insight into efficacy under optimal implementation conditions (Widlansky et al., 2003).

Pairwise comparisons between intervention groups are conducted with appropriate adjustments for multiple comparisons using the Benjamini-Hochberg procedure to control the false discovery rate (Ras et al., 2013). Effect sizes are reported as Cohen's d with 95% CIs for between-group comparisons and as standardized mean differences from baseline for within-group changes.

4.4 Intervention effects: Secondary and exploratory analyses

For neuroimaging outcomes, region-of-interest analyses are complemented by exploratory whole-brain approaches using voxel-based morphometry and tract-based spatial statistics, with appropriate correction for multiple comparisons using the false discovery rate method ($q < 0.05$) (Inaba et al., 2010). Longitudinal changes in brain structure (regional volumes, cortical thickness) and neurochemistry (metabolite ratios) are analyzed using linear mixed models with the same fixed and random effects structure specified for primary outcomes (Laurent et al., 2006).

Dose-response relationships between intervention adherence (percentage of completed micro-breaks) and outcome changes are explored using polynomial regression, allowing for potential nonlinear associations (Vlachopoulos et al., 2010). These analyses help determine whether threshold effects exist and inform minimum effective "doses" for future implementation efforts.

Individual-level response heterogeneity is examined through the calculation of reliable change indices for each outcome, classifying participants as positive responders, nonresponders, or negative responders based on changes exceeding measurement error (Mitchell et al., 2010). Logistic regression models then identify baseline characteristics associated with the probability of a positive response, potentially informing future approaches to personalized intervention targeting (Townsend et al., 2015).

Ecological momentary assessment data collected via smartphone throughout the intervention period enable time-varying effect modeling to examine acute impacts of micro-breaks on cognitive performance (Mattace-Raso et al., 2006). These analyses reveal whether cognitive benefits manifest immediately following activity breaks and determine the temporal decay of such effects, informing the optimal timing of breaks for cognitive enhancement.

4.5 Mediator and moderator analyses

SEM is employed to explore potential mediating effects of vascular parameters on the relationship between intervention condition and cognitive outcomes (Mattace-Raso et al., 2010). The hypothesized mediation model positions changes in vascular function (improved blood flow, increased arterial diameter, enhanced endothelial function) as mechanisms through which physical activity breaks influence cognitive performance (Van Bortel et al., 2012). Bootstrap resampling (5,000 iterations) generates CIs for indirect effects, with significant mediation indicated by intervals excluding zero (Cruickshank et al., 2002).

Moderator analyses investigate whether baseline characteristics modify intervention effectiveness using appropriate interaction terms in mixed-effects models (Sutton-Tyrrell et al., 2005). Candidate moderators include:

1. **Demographic factors:** Age, sex, educational attainment, socioeconomic status
2. **Anthropometric variables:** BMI, waist circumference, waist-to-hip ratio, body fat percentage
3. **Baseline activity patterns:** Sedentary time, prolonged sedentary bouts, physical activity level
4. **Vascular parameters:** Resting blood pressure, arterial stiffness, baseline shear stress
5. **Cognitive status:** Baseline performance on executive function and memory tasks
6. **Workplace factors:** Job control, supervisor support, workload, physical environment

For continuous moderators, Johnson-Neyman techniques identify regions of significance—specific values of the moderator at which intervention effects transition between statistical significance and nonsignificance (Hansen et al., 2006). This approach offers greater precision than arbitrary categorization and provides clinically meaningful thresholds for intervention targeting (Ben-Shlomo et al., 2014).

In addition, latent class analysis identifies potential response phenotypes—subgroups of participants demonstrating similar patterns of change across multiple outcomes (Shirwany and Zou, 2010). This person-centered analytical approach may reveal distinct response typologies not apparent through traditional variable-centered methods, potentially informing the future development of tailored intervention strategies (Booth et al., 2012).

4.6 Implementation and process evaluation

Implementation fidelity is assessed using a structured framework examining five core dimensions: adherence, quality of delivery, participant responsiveness, program differentiation, and adaptation (Joyner and Green, 2009). Mixed methods analysis integrates quantitative adherence metrics (percentage of completed micro-breaks) with qualitative data from participant interviews and workplace observations to comprehensively evaluate implementation processes (Hamburg et al., 2007).

Economic evaluation employs cost-effectiveness and cost-utility analyses from both employer and societal perspectives (Thosar et al., 2012). Direct intervention costs (personnel time, materials, training) and indirect costs (productivity impacts, healthcare utilization) are systematically captured, with quality-adjusted life years serving as the primary effectiveness metric for cost-utility calculations (Restaino et al., 2016). Sensitivity analyses explore the impact of varying assumptions regarding discount rates, time horizons, and implementation scalability (Thosar et al., 2015).

All analyses are conducted using appropriate statistical software (R version 4.1.0, SPSS version 28, and Mplus version 8.6), with the significance threshold set at $\alpha = 0.05$ (two-tailed), unless otherwise specified for multiple comparison adjustments. Analysis code will be publicly archived to ensure analytical transparency and reproducibility (Morishima et al., 2016).

5 Discussion

5.1 Significance and innovation

This study addresses several critical gaps in the current understanding of the relationship between sedentary behavior, obesity, and vascular and cognitive health. By implementing and evaluating structured micro-break interventions in a real-world office environment, the research offers practical insights for workplace health promotion strategies.

The integration of advanced vascular imaging with cognitive assessment and neuroimaging represents a methodological advancement, enabling the exploration of potential mechanisms linking physical activity patterns with brain health. This multimodal approach may identify early biomarkers of neurovascular dysfunction that are amenable to intervention.

5.2 Potential implications

5.2.1 Clinical implications

Findings from this study could inform clinical recommendations for patients with obesity, particularly regarding the integration of brief physical activity bouts throughout the workday. The delineation of optimal micro-break frequency and duration may facilitate the development of personalized activity prescriptions based on individual cardiometabolic profiles.

5.2.2 Workplace policy implications

Results may support the implementation of organizational policies promoting regular activity breaks, potentially through restructured work schedules, modified physical environments, or technology-based prompting systems. Economic analyses incorporating productivity metrics and healthcare utilization could strengthen the business case for workplace wellness initiatives.

5.2.3 Public health implications

On a broader scale, this research contributes to evolving public health messaging regarding the importance of breaking up sedentary time, independent of structured exercise participation. The development of accessible, low-cost intervention strategies holds particular relevance for resource-limited settings, where sophisticated activity monitoring technologies may be unavailable.

5.3 Limitations and mitigation strategies

Several limitations warrant consideration. The relatively short intervention duration (3 months) may underestimate the long-term benefits of micro-break implementation. Future studies should incorporate extended follow-up periods to assess sustainability and cumulative effects.

The focus on office workers limits generalizability to other occupational settings with different constraints on movement opportunities. However, this population represents a substantial and growing segment of the workforce, experiencing high levels of sedentary behavior.

Potential Hawthorne effects may influence participant behavior, particularly in the control group, who may compensate by increasing nonwork physical activity. Activity monitoring during nonwork hours will help identify such compensatory behaviors.

5.4 Future directions

Building on this foundation, subsequent research might explore:

1. Technology-enhanced interventions using wearable devices and smartphone applications to deliver personalized prompts and feedback
2. Dose-response relationships across more varied intervention frequencies and durations
3. Comparative effectiveness against other workplace interventions (e.g., standing desks, treadmill workstations)
4. Long-term cardiometabolic and cognitive outcomes extending over years rather than months
5. Implementation science approaches to identify barriers and facilitators to sustainable program adoption in diverse workplace settings

6 Conclusion

The escalating prevalence of obesity, coupled with increasingly sedentary occupational demands, presents a significant public health challenge with implications for both vascular and cognitive health. This study protocol outlines a comprehensive approach to evaluating the effectiveness of strategically implemented micro-breaks in mitigating the adverse effects of prolonged sitting among office workers, with and without obesity.

By integrating multiple assessment modalities spanning vascular function, cognitive performance, and neuroimaging biomarkers, this research aims to illuminate potential pathways linking physical activity patterns with brain health. The development of accessible, evidence-based intervention strategies holds promise for enhancing workplace wellness initiatives and informing clinical recommendations for individuals at elevated cardiometabolic risk.

The findings from this study will contribute to the expanding literature on "exercise snacking" as a practical approach to health promotion in contemporary work environments, where continuous movement may be constrained by occupational demands. Ultimately, these insights may inform broader public health strategies addressing the interconnected challenges of sedentary behavior and obesity.

References

ActiGraph, 2012. GT3X + and wGT3X + Device Manual. ActiGraph L. The ActiGraph. 2012 May 30. GT3X+ and WGT3X+ Device Manual URL: https://Dl.Theactigraph.Com/GT3Xp_wGT3Xp_Device_Manual.Pdf (Accessed 06-08-2017).

Adem, J.B., Alhur, A.A., Walle, A.D., Mamo, D.N., Kebede, S.D., Degefa, S.M., 2025. Systematic review of barriers and facilitators to digital health technology interventions for chronic disease management in Ethiopia: insights for implementing digital health in developing countries. Health Inform. J. 31 (3). doi:10.1177/14604582251381262.

Ahirwar, R., Mondal, P.R., 2019. Prevalence of obesity in India: a systematic review. Diabetes Metab. Syndr. 18780334. 13 (1), 318–321. Available from: http://www.journals.elsevier.com/diabetes-and-metabolic-syndrome-clinical-research-and-reviews/. doi:10.1016/j.dsx.2018.08.032.

Ainsworth, B.E., Haskell, W.L., Leon, A.S., Jacobs, D.R., Montoye, H.J., Sallis, J.F., Paffenbarger, R.S., 1993. Compendium of physical activities: classification of energy costs of human physical activities. Med. Sci. Sports Exerc. 25 (1), 71–74. doi:10.1249/00005768-199301000-00011.

Aldosari, N., Ahmed, S., McDermott, J., Stanmore, E., 2023. The use of digital health by south asian communities: scoping review. J. Med. Internet Res. 25. doi:10.2196/40425

Ambrose, J.A., Barua, R.S., 2004. The pathophysiology of cigarette smoking and cardiovascular disease: an update. J. Am. Coll. Cardiol. 43 (10). doi:10.1016/j.jacc.2003.12.047.

Anderson, T.J., 1999. Assessment and treatment of endothelial dysfunction in humans. J. Am. Coll. Cardiol. 34 (3). doi:10.1016/S0735-1097(99)00259-4.

Angosto, S., García-Fernández, J., Valantine, I., Grimaldi-Puyana, M., 2020. The intention to use fitness and physical activity apps: a systematic review. Sustainability. 2071-1050. 12 (16), 6641. doi:10.3390/su12166641.

Atkinson, G., Batterham, A.M., 2013. The percentage flow-mediated dilation index: a large-sample investigation of its appropriateness, potential for bias and causal nexus in vascular medicine. Vasc. Med. 18 (6). doi:10.1177/1358863X13508446.

Ben-Shlomo, Y., Spears, M., Boustred, C., May, M., Anderson, S.G., Benjamin, E.J., Boutouyrie, P., Cameron, J., Chen, C.H., Cruickshank, J.K., Hwang, S.J., Lakatta, E.G., Laurent, S., Maldonado, J., Mitchell, G.F., Najjar, S.S., Newman, A.B., Ohishi, M., Pannier, B., Wilkinson, I.B., 2014. Aortic pulse wave velocity improves cardiovascular event prediction: an individual participant meta-analysis of prospective observational data from 17,635 subjects. J. Am. Coll. Cardiol. 63 (7). doi:10.1016/j.jacc.2013.09.063.

Booth, F.W., Roberts, C.K., Laye, M.J., 2012. Lack of exercise is a major cause of chronic diseases. Compr. Physiol. 2 (2). doi:10.1002/cphy.c110025.

Box, G.E.P., Cox, D.R., 1964. An analysis of transformations. J. R. Stat. Soc. Ser. B Stat. Methodol. 26 (2). doi:10.1111/j.2517-6161.1964.tb00553.x.

Buckingham, S.A., Williams, A.J., Morrissey, K., Price, L., Harrison, J., 2019. Mobile health interventions to promote physical activity and reduce sedentary behaviour in the workplace: a systematic review. Digit. Health. 20552076. 5. Available from: http://dhj.sagepub.com/. doi:10.1177/2055207619839883.

Buonocore, M.H., Maddock, R.J., 2015. Magnetic resonance spectroscopy of the brain: a review of physical principles and technical methods. Rev. Neurosci. 03341763. 26 (6), 609–632. Available from: http://www.reference-global.com/loi/revneuro. doi:10.1515/revneuro-2015-0010.

Carter, S.E., Draijer, R., Holder, S.M., Brown, L., Thijssen, D.H.J., Hopkins, N.D., 2018. Regular walking breaks prevent the decline in cerebral blood flow associated with prolonged sitting. J. Appl. Physiol. 15221601. 125 (3), 790–798. Available from: https://www.physiology.org/doi/pdf/10.1152/japplphysiol.00310.2018. doi:10.1152/japplphysiol.00310.2018.

Celermajer, D.S., Sorensen, K.E., Gooch, V.M., Miller, O.I., Sullivan, I.D., Lloyd, J.K., Deanfield, J.E., Spiegelhalter, D.J., 1992. Non-invasive detection of endothelial dysfunction in children and adults at risk of atherosclerosis. Lancet 340 (8828). doi:10.1016/0140-6736(92)93147-F.

Chastin, S.F.M., Granat, M.H., 2010. Methods for objective measure, quantification and analysis of sedentary behaviour and inactivity. Gait Posture 31 (1). doi:10.1016/j.gaitpost.2009.09.002.

Chau, J.Y., Grunseit, A.C., Chey, T., Stamatakis, E., Brown, W.J., Matthews, C.E., Bauman, A.E., Van Der Ploeg, H.P., 2013. Daily sitting time and all-cause mortality: a meta-analysis. PLoS ONE. 19326203. 8 (11). Available from: http://www.plosone.org/article/fetchObject.action?uri=info%3Adoi%2F10.1371%2Fjournal.pone.0080000&representation=PDF. doi:10.1371/journal.pone.0080000Australia.

Choi, L., Liu, Z., Matthews, C.E., Buchowski, M.S., 2011. Validation of accelerometer wear and nonwear time classification algorithm. Med. Sci. Sports Exerc. 43 (2). doi:10.1249/MSS.0b013e3181ed61a3.

Church, T.S., Thomas, D.M., Tudor-Locke, C., Katzmarzyk, P.T., Earnest, C.P., Rodarte, R.Q., Martin, C.K., Blair, S.N., Bouchard, C., 2011. Trends over 5 decades in U.S. occupation-related physical activity and their associations with obesity. PLoS ONE 6 (5). doi:10.1371/journal.pone.0019657.

Cohen, J. (1988). Statistical Power Analysis for the Behavioural Sciences, Hillside, NJ: Lawrence Earlbaum Associates.

Collaboration, 2017. Worldwide trends in body-mass index, underweight, overweight, and obesity from 1975 to 2016: a pooled analysis of 2416 population-based measurement studies in 128·9 million children, adolescents, and adults. Lancet 390, 2627–2642.

Corretti, M.C., Anderson, T.J., Benjamin, E.J., Celermajer, D., Charbonneau, F., Creager, M.A., Deanfield, J., Drexler, H., Gerhard-Herman, M., Herrington, D., Vallance, P., Vita, J., Vogel, R., 2002. Guidelines for the ultrasound assessment of endothelial-dependent flow-mediated vasodilation of the brachial artery: a report of the international brachial artery reactivity task force. J. Am. Coll. Cardiol. 39 (2). doi:10.1016/S0735-1097(01)01746-6.

Craig, C.L., Marshall, A.L., Sjöström, M., Bauman, A.E., Booth, M.L., Ainsworth, B.E., Pratt, M., Ekelund, U., Yngve, A., Sallis, J.F., Sallis, J.F., Oja, P., 2003. International physical activity questionnaire: 12-country reliability and validity. Med. Sci. Sports Exerc. 35 (8), 1381–1395. doi:10.1249/01.MSS.0000078924.61453.FB.

Cruickshank, K., Riste, L., Anderson, S.G., Wright, J.S., Dunn, G., Gosling, R.G., 2002. Aortic pulse-wave velocity and its relationship to mortality in diabetes and glucose intolerance: an integrated index of vascular function? Circulation 106 (16). doi:10.1161/01.CIR.0000033824.02722.F7.

Dempsey, P.C., Larsen, R.N., Sethi, P., Sacre, J.W., Straznicky, N.E., Cohen, N.D., Cerin, E., Lambert, G.W., Owen, N., Kingwell, B.A., Dunstan, D.W., 2016. Benefits for type 2 diabetes of interrupting prolonged sitting with brief bouts of light walking or simple resistance activities. Diabetes Care. 19355548. 39 (6), 964–972. Available from: http://care.diabetesjournals.org/content/diacare/39/6/964.full.pdf. doi:10.2337/dc15-2336.

Doherty, A., Jackson, D., Hammerla, N., Plötz, T., Olivier, P., Granat, M.H., White, T., Van Hees, V.T., Trenell, M.I., Owen, C.G., Preece, S.J., Gillions, R., Sheard, S., Peakman, T., Brage, S., Wareham, N.J., 2017. Large scale population assessment of physical activity using wrist worn accelerometers: the UK biobank study. PLoS ONE 12 (2). doi:10.1371/journal.pone.0169649.

Dunstan, D.W., Kingwell, B.A., Larsen, R., Healy, G.N., Cerin, E., Hamilton, M.T., Shaw, J.E., Bertovic, D.A., Zimmet, P.Z., Salmon, J., Owen, N., 2012. Breaking up prolonged sitting reduces postprandial glucose and insulin responses. Diabetes Care 35 (5). doi:10.2337/dc11-1931.

Duvivier, B.M.F.M., Schaper, N.C., Hesselink, M.K.C., van Kan, L., Stienen, N., Winkens, B., Koster, A., Savelberg, H.H.C.M., 2017. Breaking sitting with light activities vs structured exercise: a randomised crossover study demonstrating benefits for glycaemic control and insulin sensitivity in type 2 diabetes. Diabetologia 60 (3), 490–498. doi:10.1007/s00125-016-4161-7.

Dwan, K., Gamble, C., Williamson, P.R., Kirkham, J.J., 2013. Systematic review of the empirical evidence of study publication bias and outcome reporting bias - an updated review. PLoS ONE 8 (7). doi:10.1371/journal.pone.0066844.

Edwardson, C.L., Gorely, T., Davies, M.J., Gray, L.J., Khunti, K., Wilmot, E.G., Yates, T., Biddle, S.J.H., 2012. Association of sedentary behaviour with metabolic syndrome: a meta-analysis. PLoS ONE 7 (4). doi:10.1371/journal.pone.0034916.

Ericsson, K.A., Kintsch, W., 1995. Long-term working memory. Psychol. Rev. 102 (2). doi:10.1037/0033-295x.102.2.211.

Eriksen, B.A., Eriksen, C.W., 1974. Effects of noise letters upon the identification of a target letter in a nonsearch task. Percept. Psychophys. 15325962. 16 (1), 143–149. doi:10.3758/BF03203267.

Eysenbach, G., 2005. The law of attrition. J. Med. Internet Res. 7 (1). doi:10.2196/jmir.7.1.e11.

Falck, R.S., Davis, J.C., Liu-Ambrose, T., 2017. What is the association between sedentary behaviour and cognitive function? A systematic review. Br. J. Sports Med. 14730480. 51 (10), 800–811. Available from: http://bjsm.bmjjournals.com/. doi:10.1136/bjsports-2015-095551.

Faul, F., Erdfelder, E., Lang, A.G., Buchner, A., 2007. G*Power 3: a flexible statistical power analysis program for the social, behavioral, and biomedical sciences. Behav. Res. Methods 39 (2). doi:10.3758/BF03193146.

Field, A., 2013. Discovering Statistics Using IBM SPSS Statistics, 4th ed. Sage. Grosso, P. J.A. In Sustainability (Switzerland) (Vol. 11, Number 1).

Fischl, B., 2012. FreeSurfer. Neuroimage 62 (2), 774–781.

Flammer, A.J., Anderson, T., Celermajer, D.S., Creager, M.A., Deanfield, J., Ganz, P., Hamburg, N.M., Lüscher, T.F., Shechter, M., Taddei, S., Vita, J.A., Lerman, A., 2012. The assessment of endothelial function: from research into clinical practice. Circulation 126 (6). doi:10.1161/CIRCULATIONAHA.112.093245.

Forlenza, O.V., Radanovic, M., Talib, L.L., Aprahamian, I., Diniz, B.S., Zetterberg, H., Gattaz, W.F., 2015. Cerebrospinal fluid biomarkers in Alzheimer's disease: diagnostic accuracy and prediction of dementia. Alzheimer's and Dementia: Diagnosis, Assessment and Disease Monitoring. 23528729. 1 (4), 455–463. Available from: http://www.journals.elsevier.com/alzheimers-and-dementia-diagnosis-assessment-and-disease-monitoring. doi:10.1016/j.dadm.2015.09.003.

Freedson, P.S., Melanson, E., Sirard, J., 1998. Calibration of the Computer Science and Applications, Inc. accelerometer. Med. Sci. Sports Exerc. 30 (5). doi:10.1097/00005768-199805000-00021.

Friedman, J., 2014. 20 YEARS OF LEPTIN: leptin at 20: an overview. J. Endocrinol. 223 (1). doi:10.1530/joe-14-0405.

Gopinath, K., Selvam, G., Narayanamurthy, G., 2022. Determinants of the Adoption of Wearable Devices for Health and Fitness: a Meta-analytical Study. Commun. Assoc. Inf. Syst. 50 (1). doi:10.17705/1CAIS.05019.

Green, D.J., Hopman, M.T.E., Padilla, J., Laughlin, M.H., Thijssen, D.H.J., 2017. Vascular adaptation to exercise in humans: role of hemodynamic stimuli. Physiol. Rev. 15221210. 97 (2), 495–528. Available from: http://physrev.physiology.org/content/97/2/495.full.pdf. doi:10.1152/physrev.00014.2016.

Green, D.J., Jones, H., Thijssen, D., Cable, N.T., Atkinson, G., 2011. Flow-mediated dilation and cardiovascular event prediction: does nitric oxide matter? Hypertension 57 (3). doi:10.1161/HYPERTENSIONAHA.110.167015.

Hadgraft, N.T., Winkler, E.A.H., Healy, G.N., Lynch, B.M., Neuhaus, M., Eakin, E.G., Dunstan, D.W., Owen, N., Fjeldsoe, B.S., 2017. Intervening to reduce workplace sitting: mediating role of social-cognitive constructs during a cluster randomised controlled trial. Int. J. Behav. Nutr. Phys. Act. 14 (1). doi:10.1186/s12966-017-0483-1.

Hagströmer, M., Oja, P., Sjöström, M., 2006. The International Physical Activity Questionnaire (IPAQ): a study of concurrent and construct validity. Public Health Nutr. 9 (6). doi:10.1079/phn2005898.

Hall, J., Kay, T., McConnell, A.K., Mansfield, L., 2019. Implementation of sit-stand desks as a workplace health initiative: stakeholder views. Int. J. Workplace Health Manag. 1753-8351. 12 (5), 369–386. doi:10.1108/ijwhm-02-2019-0026.

Hallal, P.C., Andersen, L.B., Bull, F.C., Guthold, R., Haskell, W., Ekelund, U., Alkandari, J.R., Bauman, A.E., Blair, S.N., Brownson, R.C., Craig, C.L., Goenka, S., Heath, G.W., Inoue, S., Kahlmeier, S., Katzmarzyk, P.T., Kohl, H.W., Lambert, E.V., Lee, I.M., Leetongin, G., Lobelo, F., Loos, R.J.F., Marcus, B., Martin, B.W., Owen, N., Parra, D.C., Pratt, M., Puska, P., Ogilvie, D., Reis, R.S., Sallis, J.F., Sarmiento, O.L., Wells, J.C., 2012. Global physical activity levels: surveillance progress, pitfalls, and prospects. Lancet. 1474547X. 380 (9838), 247–257. Available from: http://www.journals.elsevier.com/the-lancet/. doi:10.1016/S0140-6736(12)60646-1.

Hamburg, N.M., McMackin, C.J., Huang, A.L., Shenouda, S.M., Widlansky, M.E., Schulz, E., Gokce, N., Ruderman, N.B., Keaney, J.F., Vita, J.A., 2007. Physical inactivity rapidly induces insulin resistance and microvascular dysfunction in healthy volunteers. Arteriosclerosis, Thrombosis, and Vascular Biology 27 (12). doi:10.1161/ATVBAHA.107.153288.

Hamilton, M.T., Hamilton, D.G., Zderic, T.W., 2014. Sedentary behavior as a mediator of type 2 diabetes. Med. Sport Sci. 16622812. 60, 11–26. Available from: https://www.karger.com/MSPSC. doi:10.1159/000357332.

Hansen, T.W., Staessen, J.A., Torp-Pedersen, C., Rasmussen, S., Thijs, L., Ibsen, H., Jeppesen, J., 2006. Prognostic value of aortic pulse wave velocity as index of arterial stiffness in the general population. Circulation 113 (5). doi:10.1161/CIRCULATIONAHA.105.579342.

Harris, P.A., Taylor, R., Thielke, R., Payne, J., Gonzalez, N., Conde, J.G., 2009. Research electronic data capture (REDCap)-A metadata-driven methodology and workflow process for providing translational research informatics support. J. Biomed. Inform. 42 (2). doi:10.1016/j.jbi.2008.08.010.

Harris, R.A., Nishiyama, S.K., Wray, D.W., Richardson, R.S., 2010. Ultrasound assessment of flow-mediated dilation. Hypertension 55 (5). doi:10.1161/HYPERTENSIONAHA.110.150821.

Hartman, Y.A.W., Karssemeijer, E.G.A., Van Diepen, L.A.M., Olde Rikkert, M.G.M., Thijssen, D.H.J., 2018. Dementia patients are more sedentary and less physically active than age-and sex-matched cognitively healthy older adults. Dement. Geriatr. Cogn. Disord. 14219824. 46 (1-2), 81–89. Available from: https://www.karger.com/journals/dem/dem_jh.htm. doi:10.1159/000491995.

Healy, G.N., Matthews, C.E., Dunstan, D.W., Winkler, E.A.H., Owen, N., 2011. Sedentary time and cardio-metabolic biomarkers in US adults: NHANES 200306. Eur. Heart J. 32 (5), 590–597. doi:10.1093/eurheartj/ehq451.

Healy, Dunstan, Salmon, Cerin, Shaw, Zimmet, et al, 2008. Breaks in sedentary time: beneficial associations with metabolic risk. Diabetes Care 31 (4), 661–666.

Hill, J.O., Wyatt, H.R., Peters, J.C., 2012. Energy balance and obesity. Circulation. 15244539. 126 (1), 126–132. doi:10.1161/CIRCULATIONAHA.111.087213.

Hoang, T.D., Reis, J., Zhu, N., Jacobs, D.R., Launer, L.J., Whitmer, R.A., Sidney, S., Yaffe, K., 2016. Effect of early adult patterns of physical activity and television viewing on midlife cognitive function. JAMA Psychiatry. 2168622X. 73 (1), 73–79. Available from: http://archpsyc.jamanetwork.com/data/Journals/PSYCH/934825/yoi150075.pdf. doi:10.1001/jamapsychiatry.2015.2468.

Horta, B.L., Schaan, B.D., Bielemann, R.M., Vianna, C.Á., Gigante, D.P., Barros, F.C., Ekelund, U., Hallal, P.C., 2015. Objectively measured physical activity and sedentary-time are associated with arterial stiffness in Brazilian young adults. Atherosclerosis. 18791484. 243 (1), 148–154. Available from: https://www.elsevier.com/locate/atherosclerosis. doi:10.1016/j.atherosclerosis.2015.09.005.

Hou, J., Li, X., Li, Z., Yin, L., Chen, X., Liang, F., 2022. An in vivo data-based computational study on sitting-induced hemodynamic changes in the external iliac artery. J. Biomech. Eng. 0148-0731. 144 (2). doi:10.1115/1.4052292.

ILO, 2021. World Employment and Social Outlook Trends 2021. In International Labour Organization.

Inaba, Y., Chen, J.A., Bergmann, S.R., 2010. Prediction of future cardiovascular outcomes by flow-mediated vasodilatation of brachial artery: a meta-analysis. Int. J. Cardiovasc. Imaging 26 (6). doi:10.1007/s10554-010-9616-1.

Jack, C.R., Bennett, D.A., Blennow, K., Carrillo, M.C., Dunn, B., Haeberlein, S.B., Holtzman, D.M., Jagust, W., Jessen, F., Karlawish, J., Liu, E., Molinuevo, J.L., Montine, T., Phelps, C., Rankin, K.P., Rowe, C.C., Scheltens, P., Siemers, E., Snyder, H.M., Sperling, R., Elliott, C., Masliah, E., Ryan, L., Silverberg, N., 2018. NIA-AA research framework: toward a biological definition of Alzheimer's disease. Alzheimers Dement. 15525279. 14 (4), 535–562. Available from: http://www.elsevier.com/wps/find/journaldescription.cws_home/704180/description#description. doi:10.1016/j.jalz.2018.02.018.

Jaeggi, S.M., Buschkuehl, M., Perrig, W.J., Meier, B., 2010. The concurrent validity of the N-back task as a working memory measure. Memory. 14640686. 18 (4), 394–412. doi:10.1080/09658211003702171.

Jensen, M.D., Ryan, D.J., Apovian, C.M., Comuzzie, A.G., Donato, K.A., et al., 2013. AHA/ACC/TOS guideline for the management of overweight and obesity in adults: a report of the American College of Cardiology/American Heart Association Task Force on Practice Guidelines and The Obesity Society. Circulation 129, S102–S138.

Joyner, M.J., Green, D.J., 2009. Exercise protects the cardiovascular system: effects beyond traditional risk factors. J. Physiol. 587 (23). doi:10.1113/jphysiol.2009.179432.

Kamel Boulos, M.N., Yang, S.P., 2021. Mobile physical activity planning and tracking: a brief overview of current options and desiderata for future solutions. mHealth. 23069740. 7. Available from: https://mhealth.amegroups.com/article/view/38687/pdf. doi:10.21037/mhealth.2020.01.01.

Kozey-Keadle, S., Libertine, A., Lyden, K., Staudenmayer, J., Freedson, P.S., 2011. Validation of wearable monitors for assessing sedentary behavior. Med. Sci. Sports Exerc. 43 (8). doi:10.1249/MSS.0b013e31820ce174.

Kuczmarski, R.J., Flegal, K.M., 2000. Criteria for definition of overweight in transition: background and recommendations for the United States. Am. J. Clin. Nutr. 72 (5). doi:10.1093/ajcn/72.5.1074.

Kumar, G., Sundaram, D., 2014. Prevalence and risk factors of hypertension among bank employees in urban Puducherry, India. Int. J. Occup. Environ. Med. 5 (2), 94–100.

Larsen, R.N., Kingwell, B.A., Sethi, P., Cerin, E., Owen, N., Dunstan, D.W., 2014. Breaking up prolonged sitting reduces resting blood pressure in overweight/obese adults. Nutr. Metab. Cardiovasc. Dis. 09394753. 24 (9), 976–982. doi:10.1016/j.numecd.2014.04.011.

Laurent, S., Cockcroft, J., Van Bortel, L., Boutouyrie, P., Giannattasio, C., Hayoz, D., Pannier, B., Vlachopoulos, C., Wilkinson, I., Struijker-Boudier, H., 2006. Expert consensus document on arterial stiffness: methodological issues and clinical applications. Eur. Heart J. 27 (21). doi:10.1093/eurheartj/ehl254.

Lee, K.K., Perry, A.S., Wolf, S.A., Agarwal, R., Rosenblum, R., Fischer, S., Grimshaw, V.E., Wener, R.E., Silver, L.D., 2012. Promoting routine stair use: evaluating the impact of a stair prompt across buildings. Am. J. Prev. Med. 42 (2). doi:10.1016/j.amepre.2011.10.005.

Lee, P.H., Macfarlane, D.J., Lam, T.H., Stewart, S.M., 2011. Validity of the international physical activity questionnaire short form (IPAQ-SF): a systematic review. Int. J. Behav. Nutr. Phys. Act. 1479-5868. 8 (1). doi:10.1186/1479-5868-8-115.

Levine, M.E., Lu, A.T., Chen, B.H., Hernandez, D.G., Singleton, A.B., Ferrucci, L., Bandinelli, S., Salfati, E., Manson, J.A.E., Quach, A., Kusters, C.D.J., Kuhi, D., Wong, A., Teschendorff, A.E., Widschwendter, M., Ritz, B.R., Absher, D., Assimes, T.L., Horvath, S., 2016. Menopause accelerates biological aging. Proc. Natl. Acad. Sci. U. S. A. 113 (33). doi:10.1073/pnas.1604558113.

Little, R.J.A., 1988. A test of missing completely at random for multivariate data with missing values. J. Am. Stat. Assoc. 83 (404). doi:10.1080/01621459.1988.10478722.

Loh, R., Stamatakis, E., Folkerts, D., Allgrove, J.E., Moir, H.J., 2020. Effects of Interrupting prolonged sitting with physical activity breaks on blood glucose, insulin and triacylglycerol measures: a systematic review and meta-analysis. Sports Med. 11792035. 50 (2), 295–330. Available from: http://rd.springer.com/journal/40279. doi:10.1007/s40279-019-01183-w.

Luhar, S., Timæus, I.M., Jones, R., Cunningham, S., Patel, S.A., Kinra, S., Clarke, L., Houben, R., 2020a. Forecasting the prevalence of overweight and obesity in India to 2040. PLoS ONE. 19326203. 15 (2). Available from: https://journals.plos.org/plosone/article/file?id=10.1371/journal.pone.0229438&type=printable. doi:10.1371/journal.pone.0229438.

Luhar, S., Timæus, I.M., Jones, R., Cunningham, S., Patel, S.A., Kinra, S., Clarke, L., Houben, R., Joe, W., 2020b. Forecasting the prevalence of overweight and obesity in India to 2040. PLOS ONE. 1932-6203. 15 (2), e0229438. doi:10.1371/journal.pone.0229438.

Lyden, K., Kozey Keadle, S.L., Staudenmayer, J.W., Freedson, P.S., 2012. Validity of two wearable monitors to estimate breaks from sedentary time. Med. Sci. Sports Exerc. 15300315. 44 (11), 2243–2252. doi:10.1249/MSS.0b013e318260c477.

Ma, J., Ma, D., Li, Z., Kim, H., 2021. Effects of a workplace sit-stand desk intervention on health and productivity. Int. J. Environ. Res. Public Health 18 (21), 11604.

Mattace-Raso, F.U.S., Hofman, A., Verwoert, G.C., Wittemana, J.C.M., Wilkinson, I., Cockcroft, J., McEniery, C., Yasmina, Laurent, S., Boutouyrie, P., Bozec, E., Hansen, T.W., Torp-Pedersen, C., Ibsen, H., Jeppesen, J., Vermeersch, S.J., Rietzschel, E., de Buyzere, M., Gillebert, T.C., Dolejsova, M., 2010. Determinants of pulse wave velocity in healthy people and in the presence of cardiovascular risk factors: 'Establishing normal and reference values'. Eur. Heart J. 31 (19). doi:10.1093/eurheartj/ehq165.

Mattace-Raso, F.U.S., Van Der Cammen, T.J.M., Hofman, A., Van Popele, N.M., Bos, M.L., Schalekamp, M.A.D.H., Asmar, R., Reneman, R.S., Hoeks, A.P.G., Breteler, M.M.B., Witteman, J.C.M., 2006. Arterial stiffness and risk of coronary heart disease and stroke: the Rotterdam Study. Circulation 113 (5). doi:10.1161/CIRCULATIONAHA.105.555235.

Matthews, C.E., Chen, K.Y., Freedson, P.S., Buchowski, M.S., Beech, B.M., Pate, R.R., Troiano, R.P., 2008. Amount of time spent in sedentary behaviors in the United States, 2003-2004. Am. J. Epidemiol. 167 (7). doi:10.1093/aje/kwm390.

Mescher, M., Merkle, H., Kirsch, J., Garwood, M., Gruetter, R., 1998. Simultaneous in vivo spectral editing and water suppression. NMR Biomed. 09523480. 11 (6), 266–272. doi:10.1002/(SICI)1099-1492(199810)11:6<266::AID-NBM530>3.3.CO;2-A.

Messner, B., Bernhard, D., 2014. Smoking and cardiovascular disease: mechanisms of endothelial dysfunction and early atherogenesis. Arteriosclerosis, Thrombosis, and Vascular Biology 34 (3). doi:10.1161/ATVBAHA.113.300156.

Migueles, J.H., Cadenas-Sanchez, C., Ekelund, U., Delisle Nyström, C., Mora-Gonzalez, J., Löf, M., Labayen, I., Ruiz, J.R., Ortega, F.B., 2017. Accelerometer data collection and processing criteria to assess physical activity and other outcomes: a systematic review and practical considerations. Sports Med. 11792035. 47 (9), 1821–1845. Available from: http://rd.springer.com/journal/40279. doi:10.1007/s40279-017-0716-0.

Mitchell, G.F., Hwang, S.J., Vasan, R.S., Larson, M.G., Pencina, M.J., Hamburg, N.M., Vita, J.A., Levy, D., Benjamin, E.J., 2010. Arterial stiffness and cardiovascular events: the framingham heart study. Circulation 121 (4). doi:10.1161/CIRCULATIONAHA.109.886655.

Mohan, S., Jarhyan, P., Ghosh, S., Venkateshmurthy, N.S., Gupta, R., Rana, R., Malhotra, C., Rao, M.B., Kalra, S., Tandon, N., Reddy, K.S., Prabhakaran, D., 2018. UDAY: a comprehensive diabetes and hypertension prevention and management program in India. BMJ Open. 2044-6055. 8 (6), e015919. doi:10.1136/bmjopen-2017-015919.

Montero, D., Walther, G., Benamo, E., Perez-Martin, A., Vinet, A., 2013. Effects of exercise training on arterial function in type 2 diabetes mellitus: a systematic review and meta-analysis. Sports Med. 43 (11). doi:10.1007/s40279-013-0085-2.

Morishima, T., Restaino, R.M., Walsh, L.K., Kanaley, J.A., Fadel, P.J., Padilla, J., 2016. Prolonged sitting-induced leg endothelial dysfunction is prevented by fidgeting. Am. J. Physiol. Heart Circ. Physiol. 311 (1). doi:10.1152/ajpheart.00297.2016.

Mumford, J.A., Nichols, T.E., 2008. Power calculation for group fMRI studies accounting for arbitrary design and temporal autocorrelation. NeuroImage 39 (1). doi:10.1016/j.neuroimage.2007.07.061.

Neuhaus, M., Eakin, E.G., Straker, L., Owen, N., Dunstan, D.W., Reid, N., Healy, G.N., 2014. Reducing occupational sedentary time: a systematic review and meta-analysis of evidence on activity-permissive workstations. Obes. Rev. 15 (10). doi:10.1111/obr.12201.

Ott, A., Andersen, K., Dewey, M.E., Letenneur, L., Brayne, C., Copeland, J.R.M., Dartigues, J.F., Kragh-Sorensen, P., Lobo, A., Martinez-Lage, J.M., Stijnen, T., Hofman, A., Launer, L.J., 2004. Effect of smoking on global cognitive function in nondemented elderly. Neurology 62 (6). doi:10.1212/01.WNL.0000115110.35610.80.

Owen, N., Sparling, P.B., Healy, G.N., Dunstan, D.W., Matthews, C.E., 2010. Sedentary behavior: emerging evidence for a new health risk. Mayo Clin. Proc. 00256196. 85 (12), 1138–1141. Available from: http://www.mayoclinicproceedings.com/content/85/12/1138.full.pdf+html. doi:10.4065/mcp.2010.0444.

Öz, G., Alger, J.R., Barker, P.B., Bartha, R., Bizzi, A., Boesch, C., Bolan, P.J., Brindle, K.M., Cudalbu, C., Dinçer, A., Dydak, U., Emir, U.E., Frahm, J., Gilberto González, R., Gruber, S., Gruetter, R., Gupta, R.K., Heerschap, A., Henning, A., Hetherington, H.P., Howe, F.A., Hüppi, P.S., Hurd, R.E., Kantarci, K., Klomp, D.W.J., Kreis, R., Kruiskamp, M.J., Leach, M.O., Lin, A.P., Luijten, P.R., Marjańska, M., Maudsley, A.A., Meyerhoff, D.J., Mountford, C.E., Nelson, S.J., Pamir, N., Pan, J.W., Peet, A.C., Poptani, H., Poptani, H., Posse, S., Pouwels, P.J.W., Ratai, E.M., Ross, B.D., Scheenen, T.W.J., Schuster, C., Smith, I.C.P., Soher, B.J., Tkáč, I., Vigneron, D.B., Kauppinen, R.A., 2014. Clinical proton MR spectroscopy in central nervous system disorders. Radiology. 15271315. 270 (3), 658–679. Available from: http://pubs.rsna.org/doi/pdf/10.1148/radiol.13130531. doi:10.1148/radiol.13130531.

Padilla, J., Johnson, B.D., Newcomer, S.C., Wilhite, D.P., Mickleborough, T.D., Fly, A.D., Mather, K.J., Wallace, J.P., 2008. Normalization of flow-mediated dilation to shear stress area under the curve eliminates the impact of variable hyperemic stimulus. Cardiovasc. Ultrasound. 14767120. 6. doi:10.1186/1476-7120-6-44.

Park, D.C., Reuter-Lorenz, P., 2009. The adaptive brain: aging and neurocognitive scaffolding. Annu. Rev. Psychol. 60. doi:10.1146/annurev.psych.59.103006.093656

Parry, S., Straker, L., 2013. The contribution of office work to sedentary behaviour associated risk. BMC Public Health. 1471-2458. 13 (1). doi:10.1186/1471-2458-13-296.

Pate, R.R., O'Neill, J.R., Lobelo, F., 2008. The evolving definition of "sedentary. Exerc. Sport Sci. Rev. 36 (4). doi:10.1097/JES.0b013e3181877d1a.

Perdomo, S.J., Gibbs, B.B., Kowalsky, R.J., Taormina, J.M., Balzer, J.R., 2019. Effects of alternating standing and sitting compared to prolonged sitting on cerebrovascular hemodynamics. Sport Sci. Health. 18247490. 15 (2), 375–383. Available from: http://www.springerlink.com/openurl.asp?genre=journal&issn=1824-7490. doi:10.1007/s11332-019-00526-4.

Piercy, K.L., Troiano, R.P., Ballard, R.M., Carlson, S.A., Fulton, J.E., Galuska, D.A., George, S.M., Olson, R.D., 2018. The physical activity guidelines for Americans. JAMA. 15383598. 320 (19), 2020–2028. Available from: http://jama.jamanetwork.com/journal.aspx. doi:10.1001/jama.2018.14854.

Pignatiello, G.A., Martin, R.J., Hickman, R.L., 2020. Decision fatigue: a conceptual analysis. J. Health Psychol. 25 (1). doi:10.1177/1359105318763510.

Prince, S.A., Adamo, K.B., Hamel, M.E., Hardt, J., Connor Gorber, S., Tremblay, M., 2008. A comparison of direct versus self-report measures for assessing physical activity in adults: a systematic review. Int. J. Behav. Nutr. Phys. Act. 5. doi:10.1186/1479-5868-5-56

Proper, K.I., Singh, A.S., Van Mechelen, W., Chinapaw, M.J.M., 2011. Sedentary behaviors and health outcomes among adults: a systematic review of prospective studies. Am. J. Prev. Med. 40 (2). doi:10.1016/j.amepre.2010.10.015.

Pyke, K.E., Tschakovsky, M.E., 2005. The relationship between shear stress and flow-mediated dilatation: implications for the assessment of endothelial function. J. Physiol. 568 (2). doi:10.1113/jphysiol.2005.089755.

Pyke, K.E., Tschakovsky, M.E., 2007. Peak vs. total reactive hyperemia: which determines the magnitude of flow-mediated dilation? J. Appl. Physiol. 102 (4). doi:10.1152/japplphysiol.01024.2006.

Ramachandran, A., Snehalatha, C., Baskar, A.D.S., Mary, S., Sathish Kumar, C.K., Selvam, S., Catherine, S., Vijay, V., 2004. Temporal changes in prevalence of diabetes and impaired glucose tolerance associated with lifestyle transition occurring in the rural population in India. Diabetologia 47 (5), 860–865. doi:10.1007/s00125-004-1387-6.

Ras, R.T., Streppel, M.T., Draijer, R., Zock, P.L., 2013. Flow-mediated dilation and cardiovascular risk prediction: a systematic review with meta-analysis. Int. J. Cardiol. 168 (1). doi:10.1016/j.ijcard.2012.09.047.

Reneman, Arts, Hoeks, 2006. Wall shear stress–an important determinant of endothelial cell function and structure–in the arterial system in vivo. Discrepancies with theory. J. Vasc. Res. 43 (3), 251–269.

Restaino, R.M., Walsh, L.K., Morishima, T., Vranish, J.R., Martinez-Lemus, L.A., Fadel, P.J., Padilla, J., 2016. Endothelial dysfunction following prolonged sitting is mediated by a reduction in shear stress. Am. J. Physiol. Heart Circ. Physiol. 310 (5). doi:10.1152/ajpheart.00943.2015.

Reuter, M., Schmansky, N.J., Rosas, H.D., Fischl, B., 2012. Within-subject template estimation for unbiased longitudinal image analysis. NeuroImage. 10959572. 61 (4), 1402–1418. doi:10.1016/j.neuroimage.2012.02.084.

Rouyard, T., Yoda, E., Akksilp, K., Dieterich, A.V., Kc, S., Dabak, S.V., Müller, A.M., 2025. Effects of workplace interventions on sedentary behaviour and physical activity: an umbrella review with meta-analyses and narrative synthesis. Lancet Public Health 10 (4). doi:10.1016/S2468-2667(25)00038-6.

Ryan, D.J., Stebbings, G.K., Onambele, G.L., 2015. The emergence of sedentary behaviour physiology and its effects on the cardiometabolic profile in young and older adults. Age. 15744647. 37 (5). Available from: http://www.springerlink.com/openurl.asp?genre=journal&eissn=1574-4647. doi:10.1007/s11357-015-9832-7.

Ryan, D.J., Wullems, J.A., Stebbings, G.K., Morse, C.I., Stewart, C.E., Onambele-Pearson, G.L., 2018. Segregating the distinct effects of sedentary behavior and physical activity on older adults' cardiovascular structure and function: part 1-linear regression analysis approach. J. Sports Sci. 36 (24), 537–542.

Salthouse, T.A., 2009. When does age-related cognitive decline begin? Neurobiol. Aging 30 (4). doi:10.1016/j.neurobiolaging.2008.09.023.

Santos-Lozano, A., Santín-Medeiros, F., Cardon, G., Torres-Luque, G., Bailón, R., Bergmeir, C., Ruiz, J.R., Lucia, A., Garatachea, N., 2013. Actigraph GT3X: validation and determination of physical activity intensity cut points. Int. J. Sports Med. 34 (11). doi:10.1055/s-0033-1337945.

Schulz, K.F., Grimes, D.A., 2002. Sample size slippages in randomised trials: exclusions and the lost and wayward. Lancet 359 (9308). doi:10.1016/S0140-6736(02)07882-0.

Shapiro, S.S., Wilk, M.B., 1965. An analysis of variance test for normality (complete samples). Biometrika 52 (3/4). doi:10.2307/2333709.

Sharan, D., Parijat, P., Sasidharan, A.P., Ranganathan, R., Mohandoss, M., Jose, J., 2011. Workstyle risk factors for work related musculoskeletal symptoms among computer professionals in India. J. Occup. Rehabil. 10530487. 21 (4), 520–525. doi:10.1007/s10926-011-9294-4.

Shephard, R.J., 2010. Breaks in Sedentary Time: beneficial associations with metabolic risk. In: Yearbook of Sports Medicine, Vol. 2010. doi:10.1016/s0162-0908(09)79483-8.

Shirwany, N.A., Zou, M.H., 2010. Arterial stiffness: a brief review. Acta Pharmacol. Sin. 31 (10). doi:10.1038/aps.2010.123.

Stamatakis, E., Johnson, N.A., Powell, L., Hamer, M., Rangul, V., Holtermann, A., 2019. Short and sporadic bouts in the 2018 US physical activity guidelines: is high-intensity incidental physical activity the new HIIT? Br. J. Sports Med. 0306-3674. 53 (18), 1137–1139. doi:10.1136/bjsports-2018-100397.

Stillman, C.M., Cohen, J., Lehman, M.E., Erickson, K.I., 2016. Mediators of physical activity on neurocognitive function: a review at multiple levels of analysis. Front. Hum. Neurosci. 10 (Dec 2016). doi:10.3389/fnhum.2016.00626.

Strycker, L.A., Duncan, S.C., Chaumeton, N.R., Duncan, T.E., Toobert, D.J., 2007. Reliability of pedometer data in samples of youth and older women. Int. J. Behav. Nutr. Phys. Act. 4. doi:10.1186/1479-5868-4-4

Sugden, R.A., Rubin, D.B., 1988. Multiple imputation for nonresponse in surveys. J. R. Stat. Soc. Ser. A Stat. Soc. 151 (3). doi:10.2307/2983027.

Sutton-Tyrrell, K., Najjar, S.S., Boudreau, R.M., Venkitachalam, L., Kupelian, V., Simonsick, E.M., Havlik, R., Lakatta, E.G., Spurgeon, H., Kritchevsky, S., Pahor, M., Bauer, D., Newman, A., 2005. Elevated aortic pulse wave velocity, a marker of arterial stiffness, predicts cardiovascular events in well-functioning older adults. Circulation 111 (25). doi:10.1161/CIRCULATIONAHA.104.483628.

Swain, D.P., Abernathy, K.S., Smith, C.S., Lee, S.J., Bunn, S.A., 1994. Target heart rates for the development of cardiorespiratory fitness. Med. Sci. Sports Exerc. 15300315. 26 (1), 112–116. doi:10.1249/00005768-199401000-00019.

Swan, G.E., Lessov-Schlaggar, C.N., 2007. The effects of tobacco smoke and nicotine on cognition and the brain. Neuropsychol. Rev. 17 (3). doi:10.1007/s11065-007-9035-9.

Tabachnick, B.G., Fidell, L.S., 2013. Using Multivariate Statistics, sixth ed. Pearson. In Contemporary Psychology: A Journal of Reviews (Vol. 28, Number 8).

Tandon, N., Anjana, R.M., Mohan, V., Kaur, T., Afshin, A., Ong, K., Mukhopadhyay, S., Thomas, N., Bhatia, E., Krishnan, A., Mathur, P., Dhaliwal, R.S., Shukla, D.K., Bhansali, A., Prabhakaran, D., Rao, P.V., Yajnik, C.S., Kumar, G.A., Varghese, C.M., Furtado, M., Agarwal, S.K., Arora, M., Bhardwaj, D., Chakma, J.K., Cornaby, L., Dutta, E., Glenn, S., Gopalakrishnan, N., Gupta, R., Jeemon, P., Johnson, S.C., Khanna, T., Kinra, S., Kutz, M., Muraleedharan, P., Naik, N., Odell, C.M., Oommen, A.M., Pandian, J.D., Parameswaran, S., Pati, S., Prasad, N., Raju, D.S., Roy, A., Sharma, M., Shekhar, C., Shukla, S.R., Singh, N.P., Thakur, J.S., Unnikrishnan, R., Varughese, S., Xavier, D., Zachariah, G., Lim, S.S., Naghavi, M., Dandona, R., Vos, T., Murray, C.J.L., Reddy, K.S., Swaminathan, S., Dandona, L., 2018. The increasing burden of diabetes and variations among the states of India: the Global Burden of Disease Study 1990–2016. Lancet Glob. Health. 2214109X. 6 (12), e1352. Available from: http://www.elsevier.com/journals/the-lancet-global-health/2214-109x. doi:10.1016/S2214-109X(18)30387-5.

Thijssen, D.H.J., Black, M.A., Pyke, K.E., Padilla, J., Atkinson, G., Harris, R.A., Parker, B., Widlansky, M.E., Tschakovsky, M.E., Green, D.J., 2011. Assessment of flow-mediated dilation in humans: a methodological and physiological guideline. Am. J. Physiol. Heart Circ. Physiol. 15221539. 300 (1), H2. Available from: http://ajpheart.physiology.org/content/300/1/H2.full.pdf+html. doi:10.1152/ajpheart.00471.2010.

Thijssen, D.H.J., Bruno, R.M., Van Mil, A.C.C.M., Holder, S.M., Faita, F., Greyling, A., Zock, P.L., Taddei, S., Deanfield, J.E., Luscher, T., Green, D.J., Ghiadoni, L., 2019. Expert consensus and evidence-based recommendations for the assessment of flow-mediated dilation in humans. Eur. Heart J. 40 (30), 2534–2547. doi:10.1093/eurheartj/ehz350.

Thosar, S.S., Bielko, S.L., Mather, K.J., Johnston, J.D., Wallace, J.P., 2015. Effect of prolonged sitting and breaks in sitting time on endothelial function. Med. Sci. Sports Exerc. 47 (4), 843–849. doi:10.1249/MSS.0000000000000479.

Thosar, S.S., Johnson, B.D., Johnston, J.D., Wallace, J.P., 2012. Sitting and endothelial dysfunction: the role of shear stress. Med. Sci. Monit. 1234-1010. 18 (12) RA173. doi:10.12659/msm.883589.

Touboul, P.J., Hennerici, M.G., Meairs, S., Adams, H., Amarenco, P., Bornstein, N., Csiba, L., Desvarieux, M., Ebrahim, S., Hernandez, R.H., Jaff, M., Kownator, S., Naqvi, T., Prati, P., Rundek, T., Sitzer, M., Schminke, U., Tardif, J.C., Taylor, A., Vicaut, E., Woo, K.S., 2012. Mannheim carotid intima-media thickness and plaque consensus (2004-2006-2011). Cerebrovasc. Dis. 14219786. 34 (4), 290–296. doi:10.1159/000343145.

Townsend, R.R., Wilkinson, I.B., Schiffrin, E.L., Avolio, A.P., Chirinos, J.A., Cockcroft, J.R., Heffernan, K.S., Lakatta, E.G., McEniery, C.M., Mitchell, G.F., Najjar, S.S., Nichols, W.W., Urbina, E.M., Weber, T., 2015. Recommendations for improving and standardizing vascular research on arterial stiffness: a scientific statement from the American Heart Association. Hypertension 66 (3). doi:10.1161/HYP.0000000000000033.

Tremblay, M., 2012. Letter to the editor: standardized use of the terms "sedentary" and "sedentary behaviours. Applied Physiology, Nutrition and Metabolism 37 (3). doi:10.1139/H2012-024.

Troiano, R.P., Berrigan, D., Dodd, K.W., Mâsse, L.C., Tilert, T., Mcdowell, M., 2008. Physical activity in the United States measured by accelerometer. Med. Sci. Sports Exerc. 40 (1). doi:10.1249/mss.0b013e31815a51b3.

Trougakos, J.P., Hideg, I., 2009. Momentary work recovery: the role of within-day work breaks. Res. Occup. Stress Well Being 7. doi:10.1108/S1479-3555(2009)0000007005.

Van Bortel, L.M., Laurent, S., Boutouyrie, P., Chowienczyk, P., Cruickshank, J.K., De Backer, T., Filipovsky, J., Huybrechts, S., Mattace-Raso, F.U.S., Protogerou, A.D., Schillaci, G., Segers, P., Vermeersch, S., Weber, T., 2012. Expert consensus document on the measurement of aortic stiffness in daily practice using carotid-femoral pulse wave velocity. J. Hypertens. 30 (3). doi:10.1097/HJH.0b013e32834fa8b0.

Vasanthan, L., Natarajan, S.K., Babu, A., Kamath, M.S., Kamalakannan, S., 2024. Digital health interventions for improving access to primary care in India: a scoping review. PLOS Glob. Public Health 4 (5 May). doi:10.1371/journal.pgph.0002645.

Venkatrao, M., Nagarathna, R., Majumdar, V., Patil, S.S., Rathi, S., Nagendra, H., 2020. Prevalence of obesity in India and its neurological implications: a multifactor analysis of a nationwide cross-sectional study. Ann. Neurosci. 0972-7531. 27 (3-4), 153–161. doi:10.1177/0972753120987465.

Vlachopoulos, C., Aznaouridis, K., Stefanadis, C., 2010. Prediction of cardiovascular events and all-cause mortality with arterial stiffness. a systematic review and meta-analysis. J. Am. Coll. Cardiol. 55 (13), 1318–1327. doi:10.1016/j.jacc.2009.10.061.

Voevodskaya, O., Simmons, A., Nordenskjöld, R., Kullberg, J., Ahlström, H., Lind, L., Wahlund, L.-O., Larsson, E.-M., Westman, E., 2014. The effects of intracranial volume adjustment approaches on multiple regional MRI volumes in healthy aging and Alzheimer's disease. Front. Aging Neurosci. 1663-4365. 6. doi:10.3389/fnagi.2014.00264.

Voss, M.W., Vivar, C., Kramer, A.F., van Praag, H., 2013. Bridging animal and human models of exercise-induced brain plasticity. Trends Cogn. Sci. 17 (10), 525–544. doi:10.1016/j.tics.2013.08.001.

Wheeler, M.J., Dempsey, P.C., Grace, M.S., Ellis, K.A., Gardiner, P.A., Green, D.J., Dunstan, D.W., 2017. Sedentary behavior as a risk factor for cognitive decline? A focus on the influence of glycemic control in brain health. Alzheimers Dement. Transl. Res. Clin. Interv. 3 (3), 291–300. doi:10.1016/j.trci.2017.04.001.

Wheeler, M.J., Dunstan, D.W., Ellis, K.A., Cerin, E., Phillips, S., Lambert, G., Naylor, L.H., Dempsey, P.C., Kingwell, B.A., Green, D.J., 2019. Effect of morning exercise with or without breaks in prolonged sitting on blood pressure in older overweight/obese adults. Hypertension. 0194-911X. 73 (4), 859–867. doi:10.1161/hypertensionaha.118.12373.

White, I.R., Royston, P., Wood, A.M., 2011. Multiple imputation using chained equations: issues and guidance for practice. Stat. Med. 30 (4). doi:10.1002/sim.4067.

Widlansky, M.E., Gokce, N., Keaney, J.F., Vita, J.A., 2003. The clinical implications of endothelial dysfunction. J. Am. Coll. Cardiol. 42 (7). doi:10.1016/S0735-1097(03)00994-X.

World Health Organization, 2020. *Who Guidelines On Physical Activity And Sedentary Behaviour*. https://www.who.int/publications/i/item/9789240015128

World Health Organization, 2021. Obesity and overweight [Internet]. [cited. World Health Organization, (2022)].

Chapter 11

Ultrasound in assessing the association of self-reported physical activity and sitting time with the diaphragm and lower limb muscle thickness

Aishwarya Shetty[a], Baskaran Chandrasekaran[b], Leena R. David[a,c], Rajagopal Kadavigere[d], Sneha Ravichandran[a], Berna Uzun[e], Dilber Uzun Ozsahin[c,e,f], Dilip Shettigar[a], Ilker Ozsahin[e], Sathya Sabina Muthu[a], and Suresh Sukumar[a]

[a]*Department of Medical Imaging Technology, Manipal College of Health Professions, Manipal Academy of Higher Education, Manipal, Karnataka, India*
[b]*Symbiosis School of Sports Sciences, Symbiosis International (Deemed University), Pune, Maharashtra, India*
[c]*Medical Diagnostic Imaging Department, College of Health Sciences, University of Sharjah, Sharjah, United Arab Emirates*
[d]*Kasturba Medical College, Department of Radiodiagnosis and Imaging, Kasturba Medical College, Manipal, India*
[e]*Irfan Suat Gunsel Operational Research Institute, Near East University, Nicosia, Turkish Republic of Northern Cyprus (TRNC), Turkey*
[f]*Research Institute for Medical and Health Sciences, University of Sharjah, Sharjah, United Arab Emirates*

1 Introduction

Physical activity (PA) has long been recognized for its numerous health benefits, including the prevention of chronic diseases such as obesity, type 2 diabetes, and cardiovascular disease (1). Regular exercise improves overall wellness, boosting not only cardiovascular health but also enhancing muscle strength, function, and growth. Conversely, individuals who engage in infrequent PA may experience detrimental effects, such as a decline in muscle mass, strength, and function (Frontera et al., 1988). The relationship between exercise and muscle growth is cumulative, with regular and sustained PA leading to gradual increases in muscle thickness and strength. However, the specific impact of different levels of PA on muscle thickness remains an important question: how much does muscle thickness differ across individuals with low, moderate, and high levels of PA? Understanding this relationship could provide valuable insights into how exercise intensity and frequency contribute to changes in muscle morphology and overall muscular health.

There is a well-established association between increased intensity of PA and overall muscular strength in young, healthy individuals. Conversely, age-related declines in muscle thickness and strength are often linked to reduced PA levels. Strength tests, such as those performed using a hand-held dynamometer, measure the capacity of a muscle group to generate significant force during a specific coordinated effort. These tests provide insight into muscle function and can signal functional impairments. However, because these tests assess the collective output of a group of muscles that are neuromuscularly activated, they do not capture changes in a specific muscle's morphology or function (Goodpaster et al., 2006).

Variations in muscle structure and function can be attributed to several factors, including aging or reduced PA. Such changes, including muscle degeneration and fatty infiltration, can be detected using advanced visualization techniques that assess morphological alterations in the muscle (Semciw et al., 2014). These techniques are essential for understanding how functional activity affects muscle adaptation. For example, during the stance phase of walking, the anterior fibers of the gluteus minimus are stimulated after the other gluteal muscles, playing a key role in stabilizing the anterior hip joint. These variations in muscle activation patterns are critical for understanding how muscles adapt and function differently under various conditions.

As far as current research suggests, while moderate to vigorous PA (MVPA) may not be directly associated with significant muscle development or strength gains, it is nonetheless strongly linked to a range of important physiological benefits. These include improvements in cardiorespiratory fitness, enhanced metabolic function, and increased overall functional

Advances in Medical Imaging. https://doi.org/10.1016/B978-0-443-28967-5.00011-3

capacity. Higher levels of MVPA contribute to more efficient oxygen uptake and delivery, better endurance, and a reduced risk of chronic diseases, all of which support a person's ability to perform daily tasks and maintain long-term health (Drenowatz et al., 2016). It is only possible to compare the effects of PA on muscle growth and power within the same group.

PA can be assessed using both subjective and objective methods, with quantification often based on key parameters such as frequency, intensity, duration, and type. Objective measurements offer a more reliable and accurate representation of an individual's activity levels compared to self-reported data, which may be prone to recall bias or social desirability bias. Devices such as accelerometers and pedometers are commonly used objective tools. Accelerometers measure movement across multiple planes and can capture data related to the intensity and duration of activity, while pedometers primarily count steps. These tools allow for the monitoring of PA over varying time frames, ranging from a few hours to several days or even weeks, providing valuable insights into an individual's physical fitness patterns and overall activity levels.

Ultrasound imaging has emerged as a useful tool for assessing muscle morphology; however, the existing evidence exploring the relationship between PA levels and the thickness of lower limb muscles—such as the soleus and gastrocnemius—as well as the diaphragm, remains inconclusive. In this study, we hypothesized that: (1) individuals who engage in regular PA would demonstrate greater muscle thickness and strength compared to those with a sedentary lifestyle, and (2) muscle thickness would exhibit age-related changes. Accordingly, we aimed to investigate the associations between various dimensions of self-reported PA and sedentary behavior (sitting time) with diaphragm and lower limb muscle thickness. By examining these relationships, the study seeks to contribute to a deeper understanding of how activity patterns may influence muscular health, particularly in muscles critical for posture, locomotion, and respiration.

2 Methods

2.1 Study design

This research was designed as a prospective, single-center, randomized crossover trial aimed at evaluating the relationship between PA levels and muscle thickness in a specific age group. The study was conducted at Kasturba Medical Hospital, a tertiary care interdisciplinary teaching hospital in Manipal, India, over an 11-month period from January 2022 to November 2022. The prospective nature of the study allowed for real-time data collection and monitoring, enhancing the quality and relevance of the findings.

As part of ethical compliance, the study received formal approval from the Institutional Ethics Committee of Kasturba Hospital (IEC2: 125/2022) and was registered with the Clinical Trial Registry of India (CTRI/2022/10/046187) in accordance with national and international research standards (19). The registration ensured transparency in study objectives, methodology, and outcome reporting.

The inclusion process for participants followed a structured and standardized protocol to ensure consistency and reliability in subject selection. Patients were carefully screened based on predefined eligibility criteria, with suitable candidates enrolled after obtaining informed consent. The step-by-step inclusion procedure is visually represented in Fig. 11.1, which outlines the recruitment, screening, and allocation process followed throughout the trial.

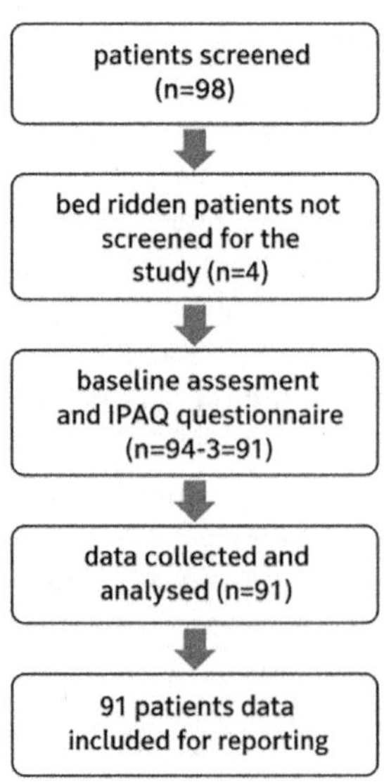

FIG. 11.1 STROBE flow diagram illustrating the inclusion of participants.

2.2 Participants

Potential participants for this study were recruited from patients awaiting radiological screening at the radiology department of an interdisciplinary teaching hospital. Prior to enrollment, all individuals were provided with written informed consent forms, ensuring full ethical compliance and voluntary participation in the study.

To maintain a homogeneous study population and reduce confounding variables, participants were carefully screened against specific exclusion criteria. These criteria included recent trauma, orthopedic surgeries, bedridden conditions, disabilities, osteoarthritis, and chronic heart or lung disorders, as these factors could potentially interfere with diaphragm thickness measurements and overall muscle assessment.

After applying these exclusion criteria, we selected both male and female patients aged 18 to 35 years to participate in the study. This age range was chosen to focus on a population with relatively stable musculoskeletal and respiratory function, ensuring the reliability and validity of the data collected. The characteristics of the selected study population are outlined in Table 11.1.

2.3 Physical activity

For young and middle-aged participants, the Short International PA Questionnaire (IPAQ) was used to assess self-reported PA over the previous seven days. The questionnaire evaluated the time spent engaging in various activity levels, including vigorous and moderate-intensity activities, walking, and sedentary behaviors such as sitting. The intensity of each activity was quantified using metabolic equivalents (METs). Specifically, vigorous activity was assigned 8 METs, moderate activity was assigned 4 METs, and walking was assigned 3.3 METs. This method allowed for the calculation of each participant's overall PA levels and provided an estimate of their energy expenditure based on the frequency and duration of each activity type.

2.4 Sample size calculation

A total sample size of 91 participants was calculated to be sufficient to detect a moderate correlation ($r > 0.4$) between PA levels and muscle thickness. This estimation was based on a significance level of 95% ($\alpha = 0.05$) and a statistical power of 80% ($\beta = 0.20$), ensuring that the study was adequately powered to identify statistically meaningful associations. The sample size calculation was guided by established statistical conventions, aiming to minimize the risks of both Type I and Type II errors while maintaining the reliability and validity of the study's findings.

To enhance the precision and reliability of the statistical analyses, an advanced algorithm was employed to compute the cumulative distribution of the correlation coefficients across the entire dataset. This methodological approach allowed for a comprehensive assessment of the strength and direction of associations between key study variables, particularly between PA levels and muscle thickness measurements. By leveraging this analytical technique, the study was better positioned to identify subtle patterns and relationships that may not be evident through conventional correlation analysis alone, thereby increasing the robustness of the findings.

By incorporating this approach, the study established a rigorous and statistically robust framework for data interpretation. This enhanced analytical precision enabled the identification of meaningful trends and relationships, thereby supporting

TABLE 11.1 Demographic and various factors of the patients.

Variables		N, Mean ± SD	(%)
Lifestyle	Smoking	Chronic N = 31 [28.01 ± 4.509] Occasional N = 15 [27.61 ± 4.338] Nonsmoker N = 45 [27.9 ± 4.885]	34.06 16.48 49.45
	Alcohol	Chronic N = 7 [27.85 ± 3.109] Occasional N = 39 [27.90 ± 4.29] Nonalcoholic N = 45 [27.97 ± 4.67]	7.6942.8549.45
Occupation	Employed	N = 68 [27.9 ± 3.951]	74.7
	Unemployed	N = 23 [27.97 ± 3.514]	25
Physical activity levels	Vigorous	N = 39 [27.9 ± 4.375]	42.85
	Moderate	N = 46 [27.83 ± 4.749]	50.54
	Walking	N = 6 [27.77 ± 2.516]	6.59

more confident conclusions regarding the significance and consistency of the observed outcomes. Ultimately, this methodological rigor contributed to the overall validity and reliability of the study's findings (Bujang and Baharum, 2016).

2.5 Muscle thickness

In this study, the assessment of lower limb musculature focused on key muscle groups essential for mobility and postural control. Specifically, the soleus muscle and two components of the quadriceps femoris—the rectus femoris (RF) and the vastus intermedius (VI)—were evaluated bilaterally (i.e., in both the right and left lower limbs). These muscles were selected due to their fundamental roles in locomotion, balance maintenance, and overall lower limb function.

To assess respiratory muscle function, particular attention was given to the diaphragm, a primary muscle of respiration. Participants were positioned in a standardized supine posture to ensure uniformity and reproducibility of measurement conditions. Diaphragmatic evaluation was performed using M-mode ultrasonography, a technique that enables high-resolution, real-time visualization of muscle motion. Diaphragm thickness was measured during both phases of the respiratory cycle—inhalation and exhalation—to accurately reflect diaphragmatic excursion and contractile activity. This methodological approach provided a dynamic and precise assessment of diaphragm function. The specific measurement protocol and anatomical reference points are illustrated in Fig. 11.2.

3 Procedure

For this study, a high-resolution ultrasound system equipped with both linear and convex (curvilinear) transducers was utilized to ensure precise and consistent imaging of muscle architecture. The linear transducer, known for its high spatial resolution, was employed to assess superficial musculature, including the anterior quadriceps—specifically the RF and

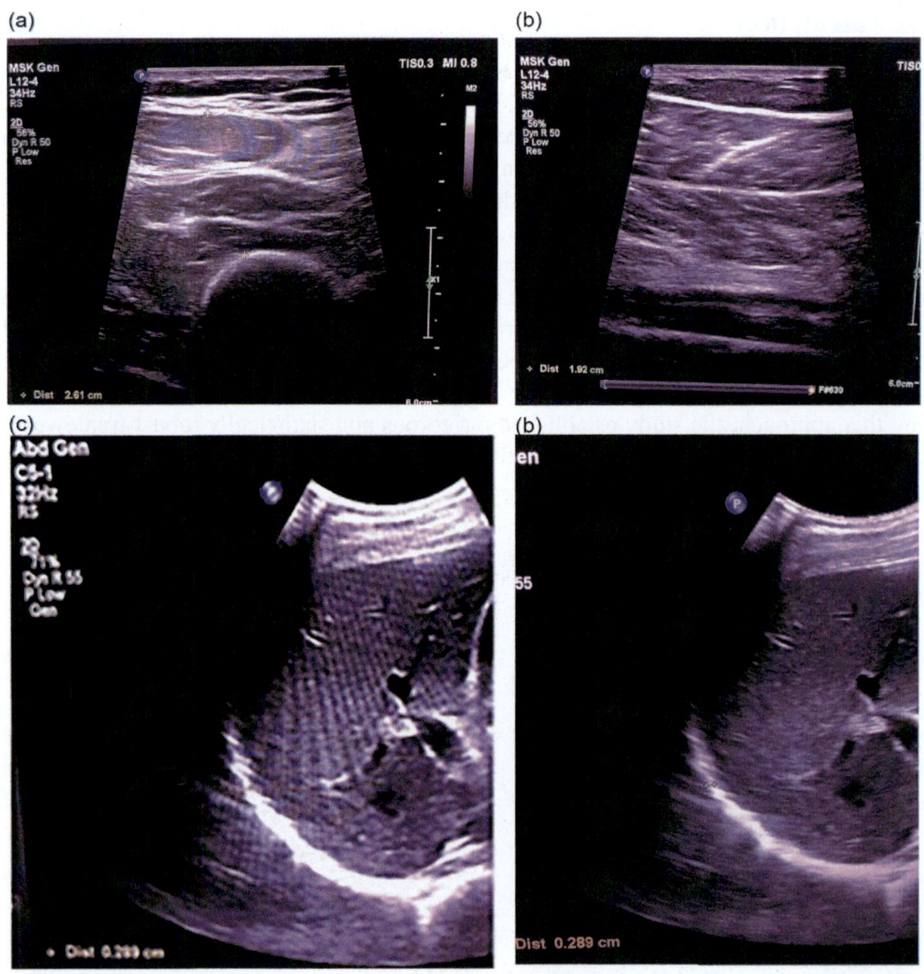

FIG. 11.2 The thickness measurements of all the muscles. (A) Soleus, (B) Quadriceps (rectus femoris and vastus intermedialis), (C) Diaphragm during inhalation, (D) Diaphragm during exhalation.

VI—as well as the soleus muscle. These muscles are located relatively close to the skin surface, making them well-suited for evaluation with a linear probe. In contrast, the convex transducer, which provides greater depth penetration at the expense of some resolution, was selected for the assessment of the diaphragm, a deeper-lying respiratory muscle. This dual-transducer approach allowed for optimized visualization and accurate measurement of both superficial and deep muscle structures, enhancing the overall quality and reliability of the ultrasound assessments.

All participants underwent standardized ultrasound assessments targeting three key muscle groups: the anterior quadriceps, the soleus, and the diaphragm. This multimodal imaging approach enabled comprehensive and uniform data collection, facilitating accurate comparisons among participants with differing levels of PA.

To enhance the methodological rigor of the study, consistent imaging protocols were rigorously applied. These included standardized transducer positioning, precise identification of anatomical landmarks, and uniform measurement techniques across all assessments. Such standardization was implemented to minimize interoperator variability and to improve the reproducibility and reliability of the muscle thickness measurements. Collectively, these procedures contributed to the internal validity of the study and ensured that the observed associations could be interpreted with greater confidence (Bujang and Baharum, 2016).

3.1 To measure the quadriceps

A 13 MHz linear array transducer was employed to assess the anterior thigh musculature in all participants. B-mode ultrasonography was used to visualize the anterior quadriceps muscle group, enabling high-resolution imaging of the RF and VI muscles. To ensure consistency and reproducibility of measurements, participants were positioned in a standardized supine posture, with knees fully extended and feet maintained in a neutral anatomical position. This controlled positioning minimized variability related to muscle length or tension, thereby enhancing the reliability of the acquired muscle thickness measurements across the study population.

The thickness of the anterior thigh musculature, referred to as total muscle thickness (TMT), was measured by determining the perpendicular distance from the anterior fascia of the RF muscle to the posterior fascia of the VI muscle. This precise anatomical landmarking technique allowed for clear differentiation between muscle layers, thereby enhancing the accuracy of the muscle thickness measurements. For each participant, axial cross-sectional images of the anterior quadriceps were acquired bilaterally and systematically recorded. These images facilitated detailed morphological analysis, enabling a robust comparative evaluation of muscle structure between the right and left limbs (Takahashi et al., 2021).

3.2 To measure the soleus

To assess the soleus muscle, ultrasonographic imaging was performed using a 13 MHz linear array transducer, with the probe positioned at the distal third of the calf, a region where the soleus muscle is optimally visualized due to its anatomical location and accessibility. B-mode ultrasonography was employed to accurately identify the soleus muscle, providing high-resolution images that clearly delineated its anatomical structure and muscle boundaries.

Participants were positioned in a prone posture, with a pillow placed beneath the ankles or shins to ensure comfort and proper alignment throughout the procedure. Two standardized variations of this positioning were used to evaluate the soleus muscle under different biomechanical conditions. In the first position, the knees were fully extended, with the ankle maintained at 0° dorsiflexion (neutral position). In the second position, the knees were flexed to approximately 30°, while the ankle remained in the neutral dorsiflexion position (0°). These variations were intended to investigate potential changes in muscle morphology and contractile behavior as a function of knee flexion and its impact on the soleus muscle's functional capacity.

In addition to structural analysis, dynamic muscle function was assessed by switching the ultrasound device to M-mode. This allowed for continuous monitoring and real-time recording of soleus muscle motion during both voluntary contractions and passive movements. M-mode ultrasonography provided valuable data on muscle activity patterns, enabling a detailed understanding of the soleus muscle's functional behavior in response to different loading conditions. This combined approach ensured a thorough and precise assessment of both the structural and functional characteristics of the soleus muscle, offering insights into its role in mobility and posture (Fujiwara et al., 2010).

3.3 To measure the diaphragm

To assess diaphragmatic thickness and movement accurately, a high-frequency 13 MHz linear array transducer was utilized. The transducer was positioned perpendicularly to the chest wall at the zone of apposition—typically between the eighth and 10th intercostal spaces along the mid-axillary line—where the diaphragm lies adjacent to the inner aspect of the lower rib cage. This anatomical location offers optimal acoustic access for capturing the diaphragm in a relaxed and contractile state during respiration.

M-mode (motion mode) ultrasonography was employed for the evaluation, as it provides high-resolution, real-time imaging of tissue movement along a single scan line. This technique enabled continuous recording of diaphragmatic activity, capturing both end-expiratory and peak-inspiratory phases across multiple respiratory cycles.

Diaphragmatic thickness was measured as the vertical distance between the echogenic pleural and peritoneal membranes. End-expiratory thickness (Tdi, ee) and peak-inspiratory thickness (Tdi, pi) were recorded consistently, allowing for an accurate assessment of diaphragmatic contractility and excursion. By visualizing these changes dynamically, M-mode ultrasonography offers a reliable and noninvasive means to assess diaphragmatic function and quantify its correlation with varying PA levels.

Measurements of diaphragmatic thickness at end-expiration (Tdi, ee) and at peak inspiration (Tdi, pi) were obtained across multiple consecutive respiratory cycles to enhance the reliability and consistency of the data. These measurements were captured within a single M-mode ultrasound image frame, allowing for precise visualization of diaphragmatic motion throughout the respiratory cycle. This method enabled accurate calculation of the diaphragm thickening ratio, which reflects the muscle's contractile activity and functional capacity during respiration.

By evaluating diaphragm thickness both at rest and during peak contraction, the study provided detailed insights into the dynamic function of the diaphragm, including its adaptability and responsiveness to varying levels of PA. This approach is particularly valuable in understanding respiratory muscle performance in healthy individuals and in populations at risk for diaphragmatic dysfunction.

Importantly, M-mode ultrasonography served as a noninvasive, real-time imaging modality that is both reproducible and clinically practical. The standardized protocol for transducer placement and measurement ensured minimal operator-dependent variability, thereby enhancing the overall accuracy and clinical applicability of the findings (Boussuges et al., 2021; Goligher et al., 2015).

4 Results

A total of 91 participants, aged between 18 and 35 years, were enrolled in the study. Of these, 78 were male, with a mean age of 27 years (standard deviation [SD] = 4.67), and 13 were female, with a mean age of 28 years (SD = 4.67). This gender distribution ensured a sufficiently diverse and representative sample, allowing for comparative analysis and interpretation of the study's findings across both male and female participants. The inclusion of both genders facilitated a comprehensive exploration of the research questions, enhancing the generalizability of the results.

4.1 Association between muscle thickness and PA

A statistically significant relationship was identified between PA levels and lower limb muscle thickness, with particular emphasis on the quadriceps muscle group. The analysis revealed that as participants' PA levels increased, from low to moderate and high, there was a corresponding increase in the thickness of the RF and VI muscles in both limbs. This trend suggests that higher levels of PA are positively associated with muscle hypertrophy in these key muscle groups involved in lower limb mobility and stability.

The relationship was found to be statistically significant, with a p-value of < 0.01, indicating a strong and reliable correlation between PA and muscle development in the lower extremities. The robustness of this association supports the hypothesis that regular PA, particularly activities that engage the quadriceps, plays a crucial role in stimulating muscle growth and improving muscle architecture. Furthermore, the significant increase in muscle thickness observed as PA levels rose suggests that even moderate increases in PA could yield beneficial effects on muscle morphology, potentially leading to enhanced functional capacity and overall lower-limb strength.

These findings have important implications for understanding how varying levels of PA can influence muscle health, particularly in terms of lower extremity function. Given the quadriceps' essential role in locomotion, balance, and posture, the observed correlation between higher PA levels and increased muscle thickness emphasizes the importance of regular PA for maintaining or improving muscle strength and function across different age groups and fitness levels.

In contrast, the relationship between PA and diaphragm thickness, although positively correlated, did not reach statistical significance. The p-value for diaphragm thickness during inhalation was 0.35, while that during exhalation was 0.17. These results suggest that while PA may contribute to improved diaphragmatic function or hypertrophy to some extent, the effect may not be strong enough to be detected as statistically significant within the current sample size or study design. Factors such as variability in breathing patterns, measurement sensitivity, or differential training effects on respiratory versus skeletal muscles may account for these findings, as seen in Table 11.2. Fig. 11.3 presents the Pearson correlation

TABLE 11.2 Thickness values in comparison with PA.

Muscle	Low PA	Moderate PA	High PA	Volume (Met/Min/ Week)	Pearson's correlation coefficient	p-value
Left quadriceps (rectus femoris and vastus intermedialis) (cm)	1.3	1.79	2.8	0.651	0.653	<0.001
Right quadriceps (rectus femoris and vastus intermedialis) (cm)	1.3	1.78	2.8	0.647	0.709	<0.001
Leftsoleus (cm)	1.0	1.56	2.2	0.706	0.68	<0.001
right soleus (cm)	1.0	1.55	2.2	0.665	0.646	<0.001
Inspiration diaphragm (mm)	0.19	0.25	0.29	0.057	0.097	0.358
Expiration diaphragm (mm)	0.18	0.23	0.27	−0.106	−0.143	0.178

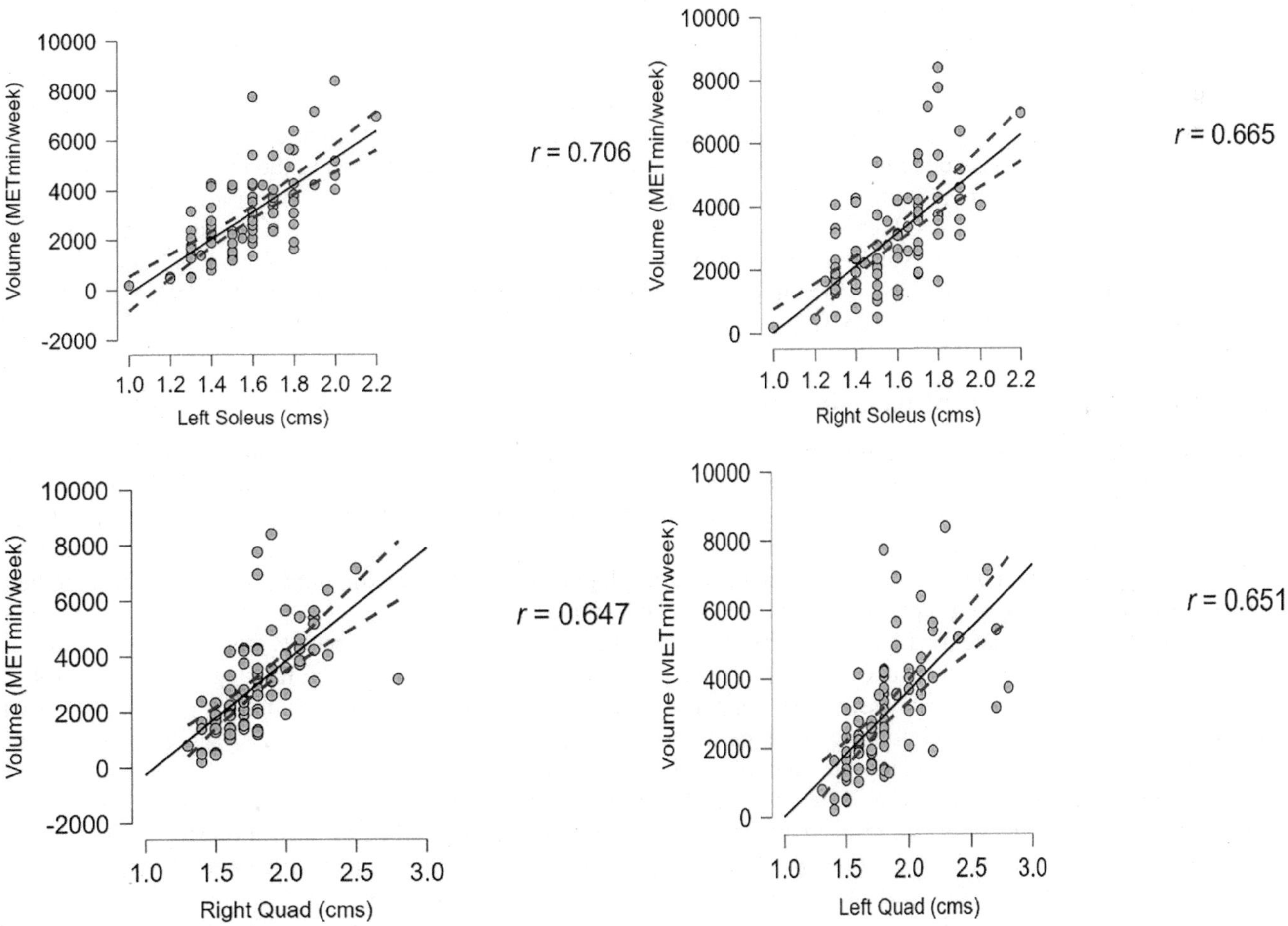

FIG. 11.3 The correlation between lower limb muscle thickness and the PA.

results between PA levels and lower limb muscle thickness. Each scatter plot demonstrates a consistent positive linear trend, highlighting that individuals with higher PA scores—based on IPAQ classification tended to exhibit greater muscle thickness. These findings underscore the role of PA in promoting skeletal muscle development, particularly in weight-bearing muscle groups such as the quadriceps.

Collectively, these results reinforce the physiological impact of regular PA on muscle morphology in the lower limbs, while suggesting a more complex or less direct influence on respiratory muscle adaptation, such as that of the diaphragm.

Further investigation with larger sample sizes or targeted respiratory muscle assessments may be necessary to clarify the relationship between PA and diaphragm morphology.

5 Discussion

The purpose of our research was to investigate the association between muscle thickness and different levels of PA. Our study focused on several aspects, including age, appropriate muscles for examination, and potential implications based on our research concept (Van Domelen et al., 2011).

5.1 PA levels in the participants

In our study, a total of 91 participants were included, categorized into three groups based on their PA levels: 6 individuals in the low PA (LPA) group, 46 in the moderate PA (MPA) group, and 39 in the very high PA (VPA) group. This distribution indicates that approximately 7% of participants engaged in LPA, 50% participated in MPA, and 43% were involved in VPA. These proportions offer valuable insights into the activity patterns of our sample population.

When compared to findings from previous studies conducted in other populations, the distribution of PA levels in our study shows notable differences. For example, research conducted in India among a broader population revealed that approximately 54% of participants were classified as physically inactive, with only 14% reporting high levels of PA. These contrasts may reflect differences in cultural, environmental, or socio-economic factors that influence PA participation across different regions. Additionally, the higher proportion of individuals engaging in VPA in our study may be indicative of a population with greater access to structured PA programs or different health behavior patterns.

These differences emphasize the need for context-specific approaches when interpreting the relationship between PA levels and health outcomes, highlighting potential areas for targeted interventions to promote higher levels of PA in populations with lower participation rates (Ikezoe et al., 2011). On a global scale, data from East and Southeast Asia suggest that about 15.8% of individuals are physically inactive (Anjana et al., 2014). These comparative figures highlight the varying levels of PA across regions and underscore the importance of understanding regional differences when interpreting PA and its effects on health and muscle morphology.

5.2 Muscle thickness in the participants

In our study, significant differences in muscle thickness were observed between participants in the LPA and VPA groups. Specifically, the soleus muscle thickness was measured at 1.0 cm in the LPA group and 2.2 cm in the VPA group ($P = .001$), indicating a substantial increase in muscle thickness with higher levels of PA. Similarly, quadriceps muscle thickness, measured in both the RF and VI muscles, was 1.3 cm in the LPA group and 2.8 cm in the VPA group. These results suggest a notable positive association between PA levels and muscle development in the lower extremities, supporting the hypothesis that more frequent and intense PA promotes hypertrophy in key muscle groups.

These findings are consistent with existing research highlighting the benefits of increased PA for muscle development. For example, Schoenfeld's study on resistance exercise demonstrated that high-resistance training can lead to a 9.5% increase in quadriceps muscle thickness, further supporting the notion that higher PA levels contribute to greater muscle mass and strength. Such studies reinforce the idea that engaging in regular, high-intensity PA is an effective means of stimulating muscle growth and enhancing muscle function (Guthold et al., 2018).

However, no statistically significant difference was found in diaphragm thickness between the LPA and VPA groups, with values of 0.19 cm and 0.29 cm, respectively ($P = .358$). This suggests that while PA may influence skeletal muscle development, its effect on respiratory muscle thickness, such as the diaphragm, might be less pronounced or influenced by other factors, such as respiratory muscle training or inherent physiological differences.

Further comparison with the broader population reveals interesting regional variations. According to Silva et al.'s (2010) study, Asians tend to have lower skeletal muscle mass compared to African Americans, Whites, and Hispanics (Schoenfeld et al., 2015). In our study, without accounting for PA levels, the average muscle thickness of the quadriceps (RF and VI) was 1.78 cm on the right and 1.79 cm on the left. The soleus muscle thickness was slightly thinner, with measurements of 1.55 cm on the right and 1.56 cm on the left. These baseline measurements provide important context for understanding how PA influences muscle thickness within this specific cohort and how these figures compare to other populations globally.

The quadriceps, soleus, and diaphragm were selected for this study due to their critical roles in both skeletal muscle function and respiratory mechanics. Additionally, previous research has demonstrated that muscle thickness can change with age and vary by gender, making these muscles particularly relevant for examining the impact of PA and aging on muscle morphology. For example, Fujiwara et al. (2010) observed that individuals over the age of 60 exhibited significantly smaller gastrocnemius muscles compared to those in their 20 s. However, in contrast to the gastrocnemius, no significant age-related differences in soleus thickness were observed for either sex.

The gastrocnemius muscle, which plays a key role in locomotion, exhibited more pronounced atrophy from the ages of 40–79, while the soleus muscle remained relatively stable in thickness across the same age range. These findings support the hypothesis that the gastrocnemius muscle deteriorates and atrophies more rapidly with age compared to the soleus. This differential aging pattern may have important implications for maintaining lower limb function, particularly in aging populations. Given that the soleus is primarily involved in postural control and endurance activities, its relative preservation of thickness could suggest that it is less susceptible to age-related atrophy than muscles like the gastrocnemius.

Aging, in general, is a well-established factor contributing to muscle strength loss, with muscle mass and strength typically decreasing as individuals age. Understanding the differences in atrophy rates between key muscles like the gastrocnemius and soleus can inform strategies aimed at preserving lower limb function and improving the quality of life for older adults (Drenowatz et al., 2016). Notably, previous studies have indicated that skeletal muscle degeneration in men may begin around the age of 27 (Fujiwara et al., 2010). With these considerations, our study specifically focused on a younger age group, limiting participants to those aged 18–35 years to reduce the confounding effects of aging on muscle thickness.

In contrast to the lower limb muscles, the diaphragm muscle thickness exhibited less variation in response to PA, both during inhalation and exhalation. Specifically, the analysis revealed a negative correlation between diaphragm thickness in these two phases, suggesting that diaphragmatic muscle thickness may not fluctuate as significantly with PA compared to the more pronounced changes observed in lower limb muscles. This finding implies that, unlike the quadriceps or soleus, diaphragm thickness may be less sensitive to variations in PA levels.

While the diaphragm did not show significant changes in thickness in this study, previous research has indicated that certain types of PA, such as weight training, can lead to an increase in diaphragmatic dimensions in healthy adults. Studies have suggested that regular resistance training, which involves substantial breathing effort, can enhance respiratory muscle function and potentially lead to hypertrophy of the diaphragm. This contrast highlights the need for further investigation into the specific types of PA that may stimulate diaphragmatic muscle growth and how these adaptations contribute to overall respiratory efficiency and endurance (Enright et al., 2006; Silva et al., 2010). Previous studies have observed increases in diaphragm thickness with inspiratory muscle training (IMT) in healthy individuals, suggesting that such training can enhance pulmonary mechanics and improve the efficiency of inspiratory muscles. This aligns with the broader understanding that regular respiratory muscle training can lead to hypertrophy of the diaphragm, which may, in turn, improve overall respiratory function, including increased tidal volume and enhanced endurance during sustained PA. However, in our study, the effects of PA on diaphragm thickness were less pronounced, indicating that, unlike other muscles such as the quadriceps or soleus, the diaphragm may not respond as significantly to general PA in the same way.

This discrepancy suggests that specialized training, such as IMT, might be more effective in inducing measurable changes in diaphragm muscle thickness and function compared to more generalized forms of exercise. While PA levels did not show a strong correlation with diaphragm thickness in this study, the potential for respiratory muscle training to enhance inspiratory muscle efficiency remains an important area for future research, especially in individuals with respiratory impairments or those aiming to improve pulmonary function.

In this study, we focused on the lower limb muscles and diaphragm to better understand their association with PA. The lower body is typically the most engaged during PA, particularly in exercises like walking and running, which are among the most common forms of exercise. These activities primarily involve the lower limbs and place significant demand on muscles such as the quadriceps and soleus, leading to observable changes in muscle thickness. Additionally, such physical exertion influences diaphragm function, especially in terms of how breathing patterns adapt to the increased demands of PA.

Recent research has demonstrated that increased PA, including weight training, can positively influence diaphragm thickness and function. For instance, weight training and other resistance exercises have been shown to enhance the dimensions of the diaphragm, improving respiratory efficiency and contributing to better overall pulmonary mechanics. While our study did not observe significant changes in diaphragm thickness with general PA, the potential role of specialized respiratory muscle training or more intense exercise regimens in promoting diaphragmatic hypertrophy remains an important consideration. This underscores the intricate relationship between respiratory muscle adaptations and PA, particularly in relation to exercise types that engage both the lower limbs and respiratory muscles in tandem (Enright et al., 2006). Moreover, in a stationary standing posture, the quadriceps, soleus, and gastrocnemius muscles play a critical role in maintaining stability and supporting the body's weight. These muscles are constantly engaged to counteract gravity and ensure postural control. During walking, the activation of these lower limb muscles continues, particularly during the stance phase, where they generate the necessary forces to propel the body forward. The quadriceps, soleus, and gastrocnemius muscles work together to support the body's weight and initiate movement. Meanwhile, the diaphragm adjusts to the increasing demands of respiration, coordinating with the movement of the lower limbs to ensure efficient breathing and oxygen delivery during physical exertion.

This interplay between lower limb muscle activity and diaphragmatic function underscores the dynamic nature of physical exertion, where both muscle groups are involved in maintaining stability and facilitating movement. Understanding how these muscles interact during common activities like walking can provide deeper insights into the physiological adaptations that occur with regular PA.

Considering both constant and variable factors, our research demonstrated a significant increase in the thickness of the quadriceps (RF and VI), soleus muscle, and diaphragm in response to varying levels of PA. Specifically, the mean muscle thickness for the quadriceps was 1.3 cm in the LPA group, 1.78 cm in the MPA group, and 2.8 cm in the VPA group. This clear positive correlation between PA levels and muscle development underscores the role of higher PA in promoting muscle hypertrophy. Similarly, the soleus muscle thickness followed a similar pattern of increasing thickness with higher PA, with mean values of 1.0 cm in the LPA group, 1.56 cm in the MPA group, and 2.2 cm in the VPA group.

These findings were statistically significant, with the observed differences across groups indicating that PA—particularly at higher levels—contributes to measurable changes in the thickness of both lower limb muscles. The results provide strong evidence supporting the hypothesis that increased PA leads to greater hypertrophy of muscles like the quadriceps and soleus. This highlights the importance of regular and sustained PA in promoting muscle health, particularly in key muscle groups that contribute to mobility and stability.

6 Conclusions

Previous studies have consistently demonstrated a correlation between PA levels and peripheral muscle thickness. Our findings align with this body of research, revealing significant increases in muscle thickness in response to higher levels of PA. These results further reinforce the notion that regular PA, particularly at higher intensities, plays a key role in promoting muscle hypertrophy. However, to gain a more nuanced and accurate understanding of this relationship, future studies should incorporate objectively measured PA levels, such as those obtained through accelerometers or direct observation. This would help reduce the potential biases associated with self-reported data, which can often be influenced by recall bias or social desirability.

Moreover, future investigations could explore the underlying mechanisms that govern how different types of PA—such as endurance training versus resistance training—impact muscle morphology across various muscle groups. By examining these mechanisms, researchers could better understand the specific adaptations that occur in response to different exercise modalities and identify which types of PA are most effective for promoting muscle growth in particular muscle groups. This could ultimately inform exercise prescription guidelines for enhancing muscle health and function.

7 Limitations and recommendations

Since the PA measures employed in this study were self-reported through the IPAQ, there is an inherent risk of memory and response biases. These biases may have influenced participants' reporting accuracy, leading to potential discrepancies between actual and perceived PA levels. Future studies could address this limitation by utilizing objectively measured PA, such as through accelerometers or direct observation, to obtain more reliable and accurate data on PA levels.

In the current investigation, PA parameters were assessed subjectively using the IPAQ, a common tool in observational studies that often relies on self-perceived PA. However, as self-reported data tend to be prone to inaccuracies and bias, the true relationship between PA and muscle thickness remains somewhat unclear. To gain a more precise understanding, future research should focus on objectively measuring PA to mitigate these limitations and enhance the validity of the findings.

Additionally, the cross-sectional nature of this study limits our ability to infer causal relationships between PA and muscle thickness changes. While cross-sectional studies are valuable for identifying associations, they do not provide insight into the temporal dynamics of these relationships. Longitudinal studies would allow for a more comprehensive understanding of how sustained PA influences muscle strength and thickness over time. Future research should explore the impact of various lifestyle factors—such as diet, PA habits, and overall health status—on muscle development. Understanding how PA and lifestyle choices influence muscle strength and morphology requires studies that assess not only changes in muscle thickness but also the long-term effects of consistent PA on muscle adaptation.

References

Anjana, R.M., Pradeepa, R., Das, A.K., Deepa, M., Bhansali, A., Joshi, S.R., Joshi, P.P., Dhandhania, V.K., Rao, P.V., Sudha, V., Subashini, R., Unnikrishnan, R., Madhu, S.V., Kaur, T., Mohan, V., Shukla, D.K., 2014. Physical activity and inactivity patterns in India – results from the ICMR-INDIAB study (Phase-1) [ICMR-INDIAB-5]. Int. J. Behav. Nutr. Phys. Act. 1479-5868. 11 (1). doi:10.1186/1479-5868-11-26.

Boussuges, A., Rives, S., Finance, J., Chaumet, G., Vallée, N., Risso, J.J., Brégeon, F., 2021. Ultrasound assessment of diaphragm thickness and thickening: reference values and limits of normality when in a seated position. Front. Med. 2296858X. 8. Available from: journal.frontiersin.org/journal/medicine. doi:10.3389/fmed.2021.742703.

Bujang, M.A., Baharum, N., 2016. Sample Size Guideline for Correlation Analysis. World J. Soc. Sci. Res. 2332-5534. 3 (1), 37. doi:10.22158/wjssr.v3n1p37.

Drenowatz, C., Prasad, V.K., Hand, G.A., Shook, R.P., Blair, S.N., 2016. Effects of moderate and vigorous physical activity on fitness and body composition. J. Behav. Med. 15733521. 39 (4), 624–632. Available from: https://www.wkap.nl/journalhome.htm/0160-7715. doi:10.1007/s10865-016-9740-z.

Enright, S.J., Unnitham, V.B., Heward, C., Withnall, L., Davies, D.H., 2006. Effect of high-intensity inspiratory muscle training on lung volumes, diaphragm thickness, and exercise capacity in subjects who are healthy. Phys. Ther. 00319023. 86 (3), 345–354. Available from: https://academic.oup.com/ptj/issue. doi:10.1093/ptj/86.3.345.

Frontera, W.R., Meredith, C.N., O'Reilly, K.P., Knuttgen, H.G., Evans, W.J., 1988. Strength conditioning in older men: skeletal muscle hypertrophy and improved function. J. Appl. Physiol. 64 (3), 1038–1044. doi:10.1152/jappl.1988.64.3.1038.

Fujiwara, K., Asai, H., Toyama, H., Kunita, K., Yaguchi, C., Kiyota, N., Tomita, H., Jacobs, J.V., 2010. Changes in muscle thickness of gastrocnemius and soleus associated with age and sex. Aging Clin. Exp. Res. 1594-0667. 22 (1), 24–30. doi:10.1007/bf03324811.

Goligher, E.C., Laghi, F., Detsky, M.E., Farias, P., Murray, A., Brace, D., Brochard, L.J., Sebastien-Bolz, S., Rubenfeld, G.D., Kavanagh, B.P., Ferguson, N.D., 2015. Measuring diaphragm thickness with ultrasound in mechanically ventilated patients: feasibility, reproducibility and validity. Intensive Care Med. 14321238. 41 (4), 642–649. Available from: https://link.springer.de/link/service/journals/00134/index.htm. doi:10.1007/s00134-015-3687-3.

Goodpaster, B.H., Park, S.W., Harris, T.B., Kritchevsky, S.B., Nevitt, M., Schwartz, A.V., Simonsick, E.M., Tylavsky, F.A., Visser, M., Newman, A.B., 2006. The loss of skeletal muscle strength, mass, and quality in older adults: the health, aging and body composition study. J. Gerontol. A Biol. Sci. Med. Sci. 10795006. 61 (10), 1059–1064. Available from: http://biomedgerontology.oxfordjournals.org/. doi:10.1093/gerona/61.10.1059.

Guthold, R., Stevens, G.A., Riley, L.M., Bull, F.C., 2018. Worldwide trends in insufficient physical activity from 2001 to 2016: a pooled analysis of 358 population-based surveys with 1·9 million participants. Lancet Glob. Health. 2214109X. 6 (10), e1077. Available from: http://www.elsevier.com/journals/the-lancet-global-health/2214-109x. doi:10.1016/S2214-109X(18)30357-7.

Ikezoe, T., Mori, N., Nakamura, M., Ichihashi, N., 2011. Age-related muscle atrophy in the lower extremities and daily physical activity in elderly women. Arch. Gerontol. Geriatr. 18726976. 53 (2), e153. doi:10.1016/j.archger.2010.08.003.

Schoenfeld, B.J., Peterson, M.D., Ogborn, D., Contreras, B., Sonmez, G.T., 2015. Effects of low- vs. High-load resistance training on muscle strength and hypertrophy in well-trained men. J. Strength Cond. Res. 15334295. 29 (10), 2954–2963. Available from: http://journals.lww.com/nsca-jscr. doi:10.1519/JSC.0000000000000958.

Semciw, A.I., Green, R.A., Murley, G.S., Pizzari, T., 2014. Gluteus minimus: an intramuscular EMG investigation of anterior and posterior segments during gait. Gait Posture. 18792219. 39 (2), 822–826. doi:10.1016/j.gaitpost.2013.11.008.

Silva, A.M., Shen, W., Heo, M., Gallagher, D., Wang, Z., Sardinha, L.B., Heymsfield, S.B., 2010. Ethnicity-related skeletal muscle differences across the lifespan. Am. J. Hum. Biol. 1042-0533. 22 (1), 76–82. doi:10.1002/ajhb.20956.

Takahashi, Y., Fujino, Y., Miura, K., Toida, A., Matsuda, T., Makita, S., 2021. Intra- and inter-rater reliability of rectus femoris muscle thickness measured using ultrasonography in healthy individuals. Ultrasound J. 25248987. 13 (1). Available from: https://link.springer.com/journal/13089. doi:10.1186/s13089-021-00224-8.

Van Domelen, D.R., Koster, A., Caserotti, P., Brychta, R.J., Chen, K.Y., McClain, J.J., Troiano, R.P., Berrigan, D., Harris, T.B., 2011. Employment and physical activity in the U.S. Am. J. Prev. Med. 18732607. 41 (2), 136–145. doi:10.1016/j.amepre.2011.03.019.

Chapter 12

Ultrasound-based assessment of vascular endothelial function among overweight people

K.B. Shishira[a], K. Vaishali[a], Leena R. David[b,c], Suresh Sukumar[c], Berna Uzun[d], Dilber Uzun Ozsahin[b,d,e], Ilker Ozsahin[d], and Rajagopal Kadavigere[f]

[a]*Department of Medical Imaging Technology, Manipal College of Health Professions, Manipal Academy of Higher Education, Manipal, Karnataka, India*

[b]*Medical Diagnostic Imaging Department, College of Health Sciences, University of Sharjah, Sharjah, United Arab Emirates*

[c]*Department of Medical Imaging Technology, Manipal College of Health Professions, Manipal Academy of Higher Education, Manipal, India*

[d]*Irfan Suat Gunsel Operational Research Institute, Near East University, Nicosia, Turkish Republic of Northern Cyprus (TRNC), Turkey*

[e]*Research Institute for Medical and Health Sciences, University of Sharjah, Sharjah, United Arab Emirates*

[f]*Department of Radiodiagnosis and Imaging, Kasturba Medical College, Manipal Academy of Higher Education, Manipal, India*

1 Introduction

Cardiovascular health has emerged as a critical area of medical research and public health concern in the contemporary world. A sedentary lifestyle, characterized by insufficient physical activity, has been widely acknowledged as a primary risk factor for the development and progression of cardiovascular diseases (CVDs) (Sesso et al., 2000). The circulatory system, which encompasses the heart and blood vessels, plays a fundamental role in maintaining the overall well-being and functional integrity of the human body (Rippe, 2019). Of particular concern is the cardiovascular health of overweight individuals, who face a significantly increased risk of developing cardiovascular complications (Ingelsson et al., 2007).

In the realm of cardiovascular health research, technological advancements have revolutionized our ability to assess vascular function noninvasively. Among these innovations, ultrasound-based techniques have emerged as pivotal tools for understanding cardiovascular risk, particularly in populations with increased vulnerability, such as overweight and obese individuals (Kapuku et al., 2006; Garcia et al., 2012). Flow-mediated dilation (FMD) of the brachial artery represents a groundbreaking diagnostic approach that provides unprecedented insights into vascular endothelial function. This sophisticated ultrasound technique measures arterial diameter changes in response to increased blood flow, serving as a critical indicator of cardiovascular health (Kaspar et al., 2019; Ramalli et al., 2019). By offering comprehensive pathophysiological, diagnostic, and prognostic information, FMD has become an invaluable tool in modern cardiovascular assessment (Garcia et al., 2012).

The global landscape of overweight and obesity presents a compelling narrative of public health challenges. In 2016, approximately 39% of adults aged 18 and older were classified as overweight, with projections suggesting a dramatic increase to 57.8% by 2030 (Kelly et al., 2008). Notably, regions like India are expected to experience a twofold increase in overweight populations between 2010 and 2040, further escalating cardiovascular risk (Kelly et al., 2008). Being overweight can precipitate numerous detrimental health conditions, including hypertension, atherosclerotic plaque development, diabetes mellitus, and altered lipid profiles (Rippe, 2019).

The versatility of FMD lies in its adaptability across various clinical settings. Researchers and clinicians can perform these assessments using both standard and portable ultrasound devices, making it exceptionally feasible for ambulatory and research environments (Alsheikh-Ali et al., 2006). However, the technique is not without its complexities. The accuracy of FMD measurements depends critically on precise technical considerations, including meticulous measurement of wall shear rate and arterial diameter changes (Ramalli et al., 2019). A significant challenge in FMD assessments is the inherent

Advances in Medical Imaging. https://doi.org/10.1016/B978-0-443-28967-5.00018-6

operator dependency. The repeatability and reliability of measurements can vary substantially based on the skill and experience of the healthcare professional conducting the examination (Jakubowski et al., 2020).

CVDs remain the leading cause of mortality worldwide, profoundly impacting individual quality of life and life expectancy (Csige et al., 2018; Ingelsson et al., 2007). The relationship between excess weight and vascular health presents a complex and evolving physiological narrative. Research consistently demonstrates that overweight and obese individuals frequently exhibit impaired endothelial function, characteristically manifested through reduced FMD compared to their lean counterparts (Kose et al., 2024; Olson et al., 2006). Promising research indicates that weight loss interventions—whether medical or surgical—can significantly improve endothelial function. These improvements are particularly noteworthy, with studies revealing strong correlations between enhanced vascular function and improved glycemic control (Dengel et al., 2006; Gokce et al., 2005).

The cardiovascular implications of excess weight extend beyond adult populations. Pediatric research has revealed alarming associations between childhood obesity and early vascular modifications. Increased carotid intima-media thickness (CIMT) and reduced arterial elasticity have been documented in children with excess weight and elevated blood pressure, highlighting the critical importance of early intervention (Bonafini et al., 2019; Ciccone et al., 2016). While FMD provides invaluable insights into vascular health, researchers continue to investigate its incremental predictive value beyond traditional cardiovascular risk models. The technique offers a nuanced approach to identifying early vascular changes, potentially guiding targeted interventions to mitigate cardiovascular risk in vulnerable populations (Kaspar et al., 2019; Kapuku et al., 2006; Kose et al., 2024).

Physical activity emerges as a critical intervention strategy in managing cardiovascular health. Research has consistently demonstrated that interventions targeting adiposity and excess weight can significantly mitigate the development of cardiovascular risk factors (Wilson et al., 2002). Regular exercise has been shown to effectively reduce the severity of multiple cardiovascular risk factors, including obesity, abdominal fat accumulation, elevated blood pressure, and systemic inflammatory conditions (Bull et al., 2020). The World Health Organization recommends a structured approach to physical activity for optimal health benefits: adults should engage in 150–300 minutes of moderate-intensity exercise or 75–150 minutes of vigorous-intensity exercise weekly (Bull et al., 2020).

The broader context of cardiovascular risk remains intrinsically linked to lifestyle factors, with sedentary behaviors and obesity continuing to pose significant global health challenges. Understanding cardiovascular health requires a holistic perspective that integrates advanced diagnostic techniques, lifestyle interventions, and personalized medical approaches. The emergence of sophisticated assessment tools like FMD represents a significant stride in our ability to detect, understand, and potentially mitigate cardiovascular risk. As research continues to evolve, the intersection of technological innovation, medical understanding, and lifestyle interventions promises increasingly sophisticated strategies for maintaining cardiovascular health across diverse populations.

2 Literature review

The impact of several parameters, including dyslipidemia, hypertension, glucose intolerance, inflammatory indicators, and others, has a considerable negative effect on people with increased body mass index (BMI), especially regarding their cardiovascular function (Poirier et al., 2006). A substantial accumulation of body fat has the potential to negatively affect the structure as well as the function of the heart, even in the absence of concomitant medical problems (Poirier et al., 2004).

The unique eating behaviors, habits, situations, and health problems of each person have a significant impact on their cardiovascular health. These are classified into two distinct groups: modifiable risk factors and nonmodifiable risk factors. Nonmodifiable risk factors refer to those that cannot be altered, such as a person's age, ethnicity, or family history of illness. Modifiable risk factors, on the other hand, refer to those that can be minimized or managed through behavioral changes. These may include excess body weight, poor nutrition, physical inactivity, smoking, alcohol use, stress, and inadequate sleep (Gregg et al., 2005).

The study examined vascular endothelial function in 42 individuals with sepsis, most of whom had respiratory or abdominal origins, utilizing ultrasound-based assessment. When compared to their contemporaries in good health, it revealed a significant reduction in flow-mediated vasodilation (FMD). Crucially, FMD degradation was more marked in the nonsurvivors over time, and it was especially noticeable 72 hours after the onset of sepsis. These results suggest that in individuals who are septic and have unstable hemodynamics, altered brachial FMD may function as an early signal for a poor prognosis. The study also emphasizes the value of noninvasive methods for detecting vascular abnormalities, particularly in groups with aberrant glucose metabolism or advanced age, where endothelial dysfunction may indicate an increased risk of CVD (Becker et al., 2012; Bruno et al., 2017; Königstein et al., 2021; Farouque and Meredith, 2001).

Vascular endothelial function is a strong indicator of cardiovascular health, which is important for maintaining the homeostasis of vascular tone. Endothelial cells (ECs), vascular smooth muscle cells (VSMCs), adventitial tissues with inflammatory cells, the autonomic nervous system, and the vasa vasorum are the cellular components involved in the control of vascular function. The interaction of these cellular components is crucial for vascular health and disease. A number of growth factors, cytokines, and extracellular matrix components are secreted by these cellular components, contributing to vascular homeostasis (Shimokawa and Satoh, 2014). lumens of blood vessels are lined by ECs, vascular smooth muscles, and adventitial connective tissue. ECs release mediators like nitric oxide (NO), prostacyclin, acetylcholine, angiotensin, and other hyperpolarizing factors that relax or contract vascular smooth muscles. NO is an important gas released by the endothelium that causes dilation and increased blood flow. Reduced bioavailability of nitric oxide is a significant factor causing endothelial dysfunction, which is associated with decreased flow velocity (FV), changes in wall shear stress, and arterial diameter. These are key early events of atherogenesis and have been considered signs of coronary artery diseases. In addition, oxidative stress is a crucial stimulus that affects VSMC activity and proliferation (Maturana et al., 2007).

3 Pathophysiology of vascular dysfunction among overweight individuals

Excess adipose tissue, particularly visceral fat, represents far more than a simple energy storage system. In overweight individuals, adipose tissue functions as an active endocrine organ that produces a complex network of proinfammatory chemicals, primarily adipokines, which fundamentally alter vascular physiology. These inflammatory mediators initiate a chronic inflammatory state and oxidative stress that profoundly impact vascular health, setting the stage for significant cardiovascular complications (Zhou et al., 2021).

The inflammatory cascade triggered by excess adipose tissue directly impairs EC function through multiple interconnected pathways. One of the most critical mechanisms involves the substantial reduction of nitric oxide (NO) synthesis. Nitric oxide normally regulates vascular tone and maintains vessel elasticity. When its production is compromised, blood vessels become less responsive, leading to reduced vessel elasticity, decreased blood flow, increased peripheral resistance, and elevated blood pressure.

Proinflammatory adipokines generate excessive reactive oxygen species (ROS) that cause significant damage to the vascular system. These reactive molecules interfere with EC membranes, disrupt nitric oxide bioavailability, promote EC dysfunction, and accelerate atherosclerotic processes. The oxidative stress creates a destructive cycle that progressively undermines vascular health and function.

Overweight conditions are frequently accompanied by significant metabolic alterations, including insulin resistance, dyslipidemia, and altered glucose metabolism. These metabolic changes further compound endothelial dysfunction by promoting inflammatory responses, increasing oxidative stress, reducing vascular responsiveness, and predisposing individuals to atherosclerosis. The intricate interplay between adipose tissue, inflammatory markers, and metabolic processes creates a complex environment that continuously challenges vascular health.

Chronic inflammatory states associated with excess adiposity lead to progressive vascular structural modifications. These changes include increased arterial wall thickness, reduced vessel elasticity, impaired vasodilation mechanisms, and enhanced susceptibility to plaque formation. The vascular system undergoes remodeling that fundamentally alters its ability to respond effectively to physiological demands.

Adipose tissue secretes multiple bioactive molecules that disrupt vascular homeostasis. These include proinflammatoryinflammatory cytokines like TNF-α and IL-6, adipokines such as leptin and adiponectin, ROS, and hormonal modulators that affect vascular tone. Each of these molecules contributes to a complex inflammatory landscape that continuously challenges the integrity of the vascular system.

The pathophysiological progression of vascular dysfunction in overweight individuals follows a multifaceted journey. It begins with excess adipose tissue accumulation, progresses through increased inflammatory mediator production, leads to EC dysfunction, reduces nitric oxide bioavailability, impairs vascular responsiveness, and ultimately elevates cardiovascular risk. This process sets the stage for numerous clinical complications, including hypertension, atherosclerosis, and the potential development of metabolic syndrome.

Understanding these intricate mechanisms provides crucial insights into potential intervention strategies targeting cardiovascular health. The complex interactions between adipose tissue, inflammatory processes, and vascular function highlight the critical importance of comprehensive approaches to managing weight and supporting vascular health. By addressing the underlying pathophysiological mechanisms, healthcare professionals can develop more targeted interventions to mitigate cardiovascular risks associated with excess weight.

4 Measurement methods

4.1 Vascular function assessment techniques

Vascular endothelial function could be assessed using various types of procedures, both invasively and noninvasively. Invasive approaches for assessing endothelial function include:

- Quantitative coronary angiography: One of the most precise and reliable techniques to evaluate the endothelium in coronary arteries is quantitative coronary angiography. This involves the infusion of vasoactive agents.
- Intravascular ultrasonography (IVUS): It is a Doppler technique that uses a Doppler-guided wire to measure vessels, and acetylcholine (ACh) or other vasoactive infusions to assess coronary blood flow.
- Venous occlusion plethysmography: This is the most traditional method for measuring blood flow. This technique is employed to evaluate vasodilator responsiveness, endothelial function, and blood flow regulation.

The few noninvasive procedures that are used to determine vascular function are:

- Doppler ultrasound: It involves the ultrasonographic measurement of the structure and function of blood vessels. It aids in the identification and classification of vascular diseases, such as deep vein thrombosis and atherosclerotic plaques.
- FMD: This technique enables the assessment of reactive hyperemia-induced endothelial dysfunction in the brachial artery.
- Arterial tonometry: This combines FMD with disposable pneumatic finger-tip probes to measure arterial pulse wave amplitude and obtain the reactive hyperemia index that reflects microvascular endothelial function.
- CIMT: It is an indirect sonographic measure of the extent of atheromatous plaque deposition. The common carotid artery's intima and media thickness, which vary depending on several factors, can be evaluated.

The best method for assessing endothelium function requires it to be noninvasive, secure, repeatable, reproducible, affordable, and standardized (Sena et al., 2022). Compared to other procedures that evaluate global vascular changes, ultrasound techniques, being noninvasive, are beneficial in detecting early changes in vascular alterations and evaluating local vascular changes (Yang and Ha, 2015).

4.2 Ultrasound assessment of vascular endothelial function

Various ultrasound techniques that are used in assessing vascular function are presented below.

1. IVUS: It is a Doppler technique that evaluates coronary blood FV using acetylcholine (Ach) or other vasoactive infusions and a Doppler-directed wire to collect vessel parameters (Voros et al., 2014). This is ideal for identifying an atherosclerotic plaque. The downside is that it is invasive, less available, requires the use of an iodine contrast agent to place the catheter, and does not assess the adventitia (Libby et al., 2019).
2. Carotid intima-media thickness evaluation: The most reliable imaging biomarker that complies with international standards for determining atherosclerosis risk is ultrasonography of the common carotid artery (CCA) obtained with an echocardiograph utilizing a 12 L linear probe (Stein et al., 2008; Simova, 2015). The main benefits are that it is easy to understand, precise, and repeatable. The drawbacks include operator dependence and the absence of measurement for adventitia thickness, a key factor in the development of atherosclerosis (Bersi et al., 2016).
3. FMD: A noninvasive procedure called FMD makes it possible to identify vascular dysfunction in the brachial artery in response to reactive hyperemia. Brachial artery diameter is traditionally measured using high-resolution ultrasonography equipment with a linear artery transducer (7.5–12 MHz), monitoring baseline measures at rest and during reactive hyperemia (Maruhashi et al., 2018). Even in asymptomatic individuals, impaired FMD is an essential instrument for identifying endothelial dysfunction and can foresee and detect risk factors for CVDs (Shimbo et al., 2007). Furthermore, FMD has long-term prognostic significance in cardiovascular patients, and FMD allows the stratification of individuals for impending cardiometabolic events. The gold-standard approach, FMD, has been thoroughly tested and shows a high correlation with coronary artery function (Fathi et al., 2004; Gokce et al., 2003). The limitations of FMD include a lack of uniformity, variances in cuff positioning on the wrists and arms, and variations in vessel sizes that make it challenging to evaluate results between laboratories. Limiting factors may also include changes in blood vessel structure and poor dilatation. The main drawbacks of this technology include its lack of precision relative to the size of the artery, intermeasurement variations, low consistency, and operator reliance (Sorensen et al., 1995).
4. Doppler ultrasonography: It is a noninvasive technology that measures arterial diameter and blood flow rate with high-resolution ultrasound equipment using a 7–15 MHz multifrequency linear array probe. It is reproducible, accurate,

and simple to use. Limitations on measurement depth are among the drawbacks, along with the need for highly skilled personnel, precise placement over large blood vessels at an ideal angle, and interference artifacts at high blood flow velocities (Sena et al., 2022).

It is essential to evaluate human vascular endothelial function to identify early indications of CVD. Although there are several approaches, none are yet reliable enough for use in individual clinical decision-making. In clinical settings, developing a noninvasive, safe, and accurate technique is essential for early diagnosis and treatment. FMD, which is measured by brachial artery ultrasonography, is a good indicator of endothelial function, despite limitations such as observer variability and manual readings. New automated analysis technologies, on the other hand, promise to improve productivity and uniformity, making it easier to conduct extensive screening and reduce variation across labs. To further improve the evaluation of endothelial function, the Doppler ultrasound methodology has the advantage of directly visualizing arterial diameter and quantifying blood flow. These ultrasonography methods have great potential to enhance clinical practice in terms of patient follow-up, early diagnosis, and stratification (Bruno et al., 2017; Farouque and Meredith, 2001; Sena et al., 2022; Preik et al., 2003).

5 Role of exercise in cardiovascular health among overweight

Exercise training plays a pivotal role in improving body weight and overall health outcomes in overweight and obese individuals. Regular physical activity is not only essential for weight management but also serves as a preventive measure against a range of chronic illnesses, including diabetes, hypertension, obesity, cancer, coronary heart disease, CVD, and depression (Atakan et al., 2021; Booth et al., 2012; Warburton, 2006). The positive association between higher levels of physical activity and improved cardiovascular health is well-documented, emphasizing the role of exercise in mitigating risk factors associated with metabolic and cardiovascular disorders in overweight and obese adults.

Incorporating both greater physical activity and dietary calorie restriction has been shown to yield a multitude of cardiovascular benefits. These include improved insulin sensitivity, lowered blood pressure, enhanced lipid profiles, reduced fat mass, and decreased waist circumference, all of which contribute to overall cardiovascular health (Tuomilehto et al., 2001; Stevens et al., 2001; Zhang et al., 2020; Baillot et al., 2015). Several longitudinal studies further highlight that increased physical activity is associated with a reduced risk of developing cardiovascular and respiratory diseases. In addition, regular exercise is known to significantly enhance longevity and reduce all-cause mortality, reinforcing the importance of sustained physical activity over time (Paffenbarger et al., 1986). Moreover, the elevated rates of CVD-related deaths and increased all-cause mortality linked to a high BMI in both men and women can be mitigated through consistent exercise engagement (Hu et al., 2004; Stevens, 2002). This suggests that lifestyle modifications incorporating structured physical activity could serve as a key intervention for reducing obesity-related health risks.

Research has increasingly focused on the cardiovascular effects of different exercise modalities, particularly high-intensity interval training (HIIT) and moderate-intensity continuous training (MICT), in overweight and obese populations. Both exercise types have demonstrated varying degrees of efficacy in improving vascular function, as assessed by FMD and vascular flow metrics (BÆkkerud et al., 2016; Hovsepian et al., 2021; Shenouda et al., 2017). Notably, Hovsepian et al. (2021) reported a significant improvement in FMD following a 10-week HIIT intervention, indicating enhanced endothelial function. Shenouda et al. (2017) similarly found that MICT led to a 2.2% increase in brachial artery FMD, suggesting that sustained moderate-intensity exercise can induce beneficial vascular adaptations. However, Bækkerud et al. (2016) observed no significant changes in FMD following either HIIT or MICT, though MICT was associated with reductions in both venous outflow and arterial inflow. These findings suggest that while exercise interventions can improve cardiovascular health, their specific effects may vary based on exercise intensity, duration, and individual physiological responses.

Furthermore, arterial stiffness, a key marker of cardiovascular health, has also been examined in relation to different exercise modalities. Chin et al. (2020) reported no significant changes in arterial stiffness following HIIT or MICT, raising questions about the extent to which these training forms influence arterial compliance. However, other studies indicate that the type and characteristics of physical activity may play a critical role in determining its impact on various CVD risk factors. For instance, Vanhees et al. (2012) emphasized that the structure and intensity of an exercise program can significantly modulate its effects on vascular function, lipid metabolism, and overall cardiovascular health. This highlights the need for tailored exercise prescriptions that align with individual health statuses and fitness levels to optimize cardiovascular benefits.

The mechanisms by which exercise promotes cardiovascular health are multifaceted. Exercise enhances endothelial function by increasing nitric oxide bioavailability, which helps maintain vascular tone and reduces arterial stiffness. Physical activity also modulates inflammation by decreasing proinflammatoryinflammatory cytokines and increasing antiinflammatory markers, which collectively reduce atherosclerotic risk. In addition, exercise improves autonomic balance by enhancing parasympathetic activity and reducing sympathetic overactivity, leading to better heart rate variability and reduced cardiovascular stress. These physiological adaptations contribute to an overall reduction in CVD risk and improved cardiac efficiency, particularly in overweight and obese individuals who are at an elevated risk of cardiovascular complications.

Another important consideration is the impact of exercise on metabolic health, which is closely linked to cardiovascular outcomes. Overweight and obese individuals often present with metabolic syndrome, characterized by insulin resistance, dyslipidemia, hypertension, and central adiposity. Regular physical activity has been shown to enhance glucose uptake by skeletal muscles, improve insulin sensitivity, and regulate blood sugar levels, thereby reducing the likelihood of developing type 2 diabetes, a significant risk factor for CVD. Exercise also positively influences lipid metabolism by increasing high-density lipoprotein (HDL) cholesterol levels while lowering low-density lipoprotein (LDL) cholesterol and triglycerides, which collectively reduce atherosclerosis progression.

In addition to traditional exercise modalities such as HIIT and MICT, other forms of physical activity, including resistance training and flexibility exercises, have been explored for their cardiovascular benefits. Resistance training, in particular, has been found to improve muscular strength, enhance metabolic rate, and contribute to weight management, all of which play a role in cardiovascular health. Studies indicate that combining aerobic and resistance training yields superior outcomes in reducing CVD risk factors compared to aerobic exercise alone. Flexibility exercises, such as yoga, have also gained attention due to their potential to improve vascular function through stress reduction and enhanced autonomic regulation.

Beyond physiological benefits, exercise also has profound psychological effects that indirectly contribute to cardiovascular health. Overweight and obese individuals often experience mental health challenges, including anxiety, depression, and low self-esteem, which can influence lifestyle choices and adherence to health-promoting behaviors. Regular physical activity has been shown to enhance mood, reduce stress, and promote psychological well-being, thereby encouraging long-term commitment to an active lifestyle. The bidirectional relationship between mental health and physical health underscores the importance of addressing both aspects to achieve optimal cardiovascular outcomes.

Despite the well-established benefits of exercise, adherence to physical activity guidelines remains a challenge among overweight and obese individuals. Barriers such as lack of motivation, physical discomfort, and time constraints often hinder long-term engagement in exercise programs. Addressing these barriers through personalized exercise prescriptions, behavioral interventions, and community-based support systems is essential for promoting sustained physical activity. In addition, incorporating technology-based interventions, such as wearable fitness trackers and mobile health applications, can enhance motivation and adherence by providing real-time feedback and personalized goal-setting.

In conclusion, exercise is an indispensable component of cardiovascular health management in overweight and obese individuals. While both HIIT and MICT demonstrate potential benefits for vascular function, the results remain somewhat inconclusive, necessitating further investigation. The variability in study findings suggests that individualized exercise regimens, considering factors such as duration, intensity, and personal health conditions, may be required to maximize cardiovascular benefits. Moreover, the integration of resistance training and flexibility exercises into exercise programs could further enhance cardiovascular and metabolic health outcomes. Despite some discrepancies in findings, the overarching consensus remains that regular physical activity significantly contributes to reducing CVD risk, improving overall health outcomes, and enhancing longevity in overweight populations. Future research should focus on optimizing exercise interventions tailored to individual needs and addressing barriers to adherence, ensuring that the cardiovascular benefits of physical activity are accessible to a broader population.

6 Effects of exercise on vascular function

Exercise represents a powerful modality for improving vascular health, with profound physiological impacts that extend far beyond simple cardiovascular conditioning. Long-term exercise training has been consistently demonstrated to significantly slow the progression of atherosclerosis by improving vascular function and reducing arterial wall thickness. Research has provided compelling evidence of the transformative effects of various exercise modalities on vascular health across different populations.

High-intensity exercise interventions have shown remarkable potential in cardiovascular rehabilitation. A notable study among cardiac allograft recipients revealed that one year of high-intensity exercise led to significant improvements in IVUS measurements (Nytrøen et al., 2013). This finding underscores the potential of targeted exercise interventions to positively modify vascular structures, even in individuals with complex medical histories.

The impact of combined exercise approaches has been particularly promising. A twelve-week program combining aerobic and resistance training demonstrated significant improvements in CIMT among middle-aged women with abdominal obesity. The study observed a notable reduction in CIMT from 0.61 ± 0.13 mm to 0.58 ± 0.12 mm, indicating a meaningful improvement in vascular health. This reduction is clinically significant, as decreased CIMT is associated with reduced cardiovascular risk (Kim et al., 2021).

One of the most critical mechanisms by which exercise improves vascular function is through enhanced nitric oxide (NO) generation. Increased NO production directly improves vasodilation, a crucial aspect of vascular health. Research has established that a 1.0% increase in FMD correlates with an 8%–13% decreased risk of CVD, highlighting the profound protective potential of regular exercise (Ras et al., 2013).

Resistance training has demonstrated particularly impressive effects on vascular parameters. An eight-week study involving healthy volunteers revealed significant improvements in arterial characteristics. Participants experienced substantial increases in resting artery diameters, with the superficial femoral artery showing a 27% increase and the carotid artery a 13% increase. Moreover, mean blood flow elevated dramatically by 53% and 55%, respectively, demonstrating exercise's remarkable capacity to enhance blood circulation and vascular function (Stebbings et al., 2013).

Different exercise modalities have shown varied but generally positive impacts on vascular health. HIIT and MICT have both demonstrated potential in improving vascular function. While some studies show significant improvements in FMD, others report more nuanced results. For instance, a 10-week HIIT intervention significantly improved FMD, while some studies found mixed or minimal changes, underscoring the complexity of exercise's vascular effects.

The physiological mechanisms underlying exercise's vascular benefits are multifaceted. Regular physical activity promotes:

- Improved endothelial function (Roitman and LaFontaine, 2005)
- Enhanced nitric oxide production (Pellegrin et al., 2009)
- Reduced vascular resistance (Francescomarino et al., 2009)
- Increased blood FV (Lima et al., 2014)
- Improved arterial elasticity (Okamoto et al., 2007)
- Reduced inflammatory markers (Gostic and Blatt, 2013)
- Enhanced metabolic efficiency (Oral, 2021)

Importantly, these vascular improvements are not limited to any specific demographic. Research has demonstrated beneficial effects across various populations, including overweight individuals, cardiac patients, and healthy adults. The consistent finding is that sustained physical activity can significantly modify vascular health, offering a powerful nonpharmacological intervention for CVD prevention.

The potential of exercise extends beyond immediate physiological changes. Long-term studies have consistently linked increased physical activity with reduced cardiovascular and respiratory disease risks. Moreover, exercise has been associated with increased longevity and reduced all-cause mortality, particularly among individuals with an elevated BMI.

In conclusion, exercise emerges as a potent therapeutic intervention for vascular health. Its ability to modify vascular structure, improve endothelial function, and reduce cardiovascular risk factors positions it as a critical strategy in preventive and rehabilitative cardiovascular medicine. The diverse mechanisms through which exercise influences vascular function underscore the importance of maintaining an active lifestyle as a fundamental approach to cardiovascular wellness.

7 Conclusion

In summary, sustained physical activity has a profound impact on cardiovascular health, particularly in overweight and obese individuals. By considerably delaying the progression of atherosclerosis, exercise serves as a powerful tool in reducing CVD risk. One of the primary mechanisms through which exercise exerts its protective effects is the improvement of endothelial function, as evidenced by increased vasodilation and nitric oxide production. Enhanced nitric oxide bioavailability plays a crucial role in maintaining vascular tone, reducing arterial stiffness, and improving overall circulatory efficiency. Studies utilizing ultrasonography have further demonstrated that regular physical activity, including HIIT, MICT, and resistance training, improves several vascular function markers. These include increased blood FV, enhanced endothelial function, and reduced vascular resistance, all of which contribute to a lower risk of cardiovascular complications.

Exercise also positively influences metabolic health, a key factor in cardiovascular risk reduction. Overweight and obese individuals often present with metabolic syndrome, characterized by insulin resistance, hypertension, dyslipidemia, and central adiposity. Regular physical activity has been shown to enhance glucose uptake by skeletal muscles, increase insulin sensitivity, and regulate blood sugar levels, thereby reducing the risk of type 2 diabetes—a significant contributor to CVD. In addition, exercise has been found to modulate lipid metabolism by increasing HDL cholesterol while reducing LDL cholesterol and triglyceride levels. These improvements in lipid profiles directly contribute to the prevention of atherosclerosis and other cardiovascular complications.

Beyond metabolic benefits, physical activity influences autonomic function, which plays a vital role in cardiovascular health. Regular exercise enhances parasympathetic activity while reducing sympathetic overactivity, leading to improved heart rate variability and reduced cardiovascular stress. These autonomic adaptations are associated with a lower risk of arrhythmias, sudden cardiac events, and overall cardiovascular mortality. Furthermore, the antiinflammatory effects of exercise contribute significantly to cardiovascular protection. Chronic inflammation is a well-known factor in the progression of CVD, and physical activity has been shown to decrease proinflammatory inflammatory cytokines while increasing antiinflammatory markers, thereby reducing atherosclerotic plaque formation and vascular dysfunction.

Resistance training, in addition to aerobic exercise, has gained increasing attention for its role in cardiovascular health. While traditionally associated with improvements in muscular strength and functional fitness, resistance training also contributes to cardiovascular benefits by enhancing vascular function, reducing arterial stiffness, and improving metabolic health. Combining aerobic and resistance training has been shown to yield superior outcomes in managing cardiovascular risk factors compared to either modality alone. In addition, flexibility and mind-body exercises, such as yoga, have been found to promote cardiovascular health by reducing stress, enhancing autonomic regulation, and improving overall vascular function.

Mental health is another crucial aspect of CVD prevention that is influenced by exercise. Overweight and obese individuals often experience psychological challenges, including anxiety, depression, and low self-esteem, which can negatively impact lifestyle choices and adherence to health-promoting behaviors. Regular physical activity has been shown to enhance mood, reduce stress, and promote psychological well-being, thereby encouraging long-term commitment to a physically active lifestyle. The interplay between mental and physical health underscores the importance of holistic approaches in CVD prevention and management.

Despite the overwhelming evidence supporting the benefits of exercise, adherence to physical activity guidelines remains a challenge for many overweight and obese individuals. Various barriers, including lack of motivation, physical discomfort, and time constraints, hinder long-term engagement in exercise programs. Addressing these barriers through personalized exercise prescriptions, behavioral interventions, and community-based support systems is crucial for promoting sustained physical activity. In addition, the integration of technology-based interventions, such as wearable fitness trackers and mobile health applications, has emerged as a promising strategy to enhance motivation, track progress, and provide personalized feedback to individuals aiming to improve their cardiovascular health.

It is important to recognize that the effects of exercise on cardiovascular health are not uniform across individuals. Factors such as age, genetic predisposition, baseline fitness levels, and comorbid conditions influence the degree of benefit derived from physical activity. Therefore, personalized exercise programs tailored to individual needs and health conditions are essential to maximize cardiovascular benefits. Future research should focus on optimizing exercise prescriptions, exploring novel training modalities, and investigating the long-term sustainability of various exercise interventions in diverse populations.

Public health strategies should also prioritize increasing awareness of the cardiovascular benefits of exercise and promoting active lifestyles across all age groups. Workplace wellness programs, community-based exercise initiatives, and policy-driven interventions that encourage physical activity in schools and urban planning can significantly contribute to reducing the burden of CVD. In addition, healthcare professionals play a critical role in advocating for exercise as a preventive and therapeutic intervention by incorporating physical activity counseling into routine medical practice.

In conclusion, regular physical activity is a cornerstone of CVD prevention and management. Whether through aerobic exercise, resistance training, or flexibility exercises, the positive impact of sustained physical activity on vascular health, metabolic function, autonomic regulation, and psychological well-being is well established. The cumulative effects of exercise—including improvements in lipid profiles, blood pressure regulation, insulin sensitivity, and endothelial function—underscore its indispensable role in reducing cardiovascular risk and enhancing overall health outcomes. While challenges in adherence remain, ongoing efforts to integrate exercise into daily life, personalize training programs, and leverage technological advancements can help maximize the cardiovascular benefits of physical activity. Future research should continue to refine exercise recommendations, explore novel interventions, and ensure that individuals of all backgrounds have access to the resources needed to maintain an active lifestyle. By prioritizing exercise as a fundamental component of public health initiatives, we can work toward a healthier, longer-living population with a reduced CVD burden.

References

Alsheikh-Ali, A.A., Ouda, H.Z., Pandian, N., Karas, R.H., Kuvin, J.T., 2006. Evaluation of peripheral vascular endothelial function with a portable ultrasound device. Echocardiography. 15408175. 23 (8), 623–626. doi:10.1111/j.1540-8175.2006.00280.x.

Atakan, M.M., Koşar, Ş.N., Güzel, Y., Tin, H.T., Yan, X., 2021. The role of exercise, diet, and cytokines in preventing obesity and improving adipose tissue. Nutrients. 20726643. 13 (5). Available from: https://www.mdpi.com/2072-6643/13/5/1459/pdf. doi:10.3390/nu13051459.

BÆkkerud, F.H., Solberg, F., Leinan, I.M., WislØff, U., Karlsen, T., Rognmo, Ø., 2016. Comparison of three popular exercise modalities on VO2max in overweight and obese. Med. Sci. Sports Exerc. 15300315. 48 (3), 491–498. Available from: http://www.lww.com/product/?0195-9131. doi:10.1249/MSS.0000000000000777.

Baillot, A., Romain, A.J., Boisvert-Vigneault, K., Audet, M., Baillargeon, J.P., Dionne, I.J., Valiquette, L., Chakra, C.N.A., Avignon, A., Langlois, M.-F., Kaser, S., 2015. Effects of lifestyle interventions that include a physical activity component in class II and III obese individuals: a systematic review and meta-analysis. PLOS ONE. 1932-6203. 10 (4), e0119017. doi:10.1371/journal.pone.0119017.

Becker, L., Prado, K., Foppa, M., Martinelli, N., Aguiar, C., Furian, T., Clausell, N., Rohde, L.E., 2012. Endothelial dysfunction assessed by brachial artery ultrasound in severe sepsis and septic shock. J. Crit. Care. 08839441. 27 (3), 316.e9. doi:10.1016/j.jcrc.2011.08.002.

Bersi, M.R., Bellini, C., Wu, J., Montaniel, K.R.C., Harrison, D.G., Humphrey, J.D., 2016. Excessive adventitial remodeling leads to early aortic maladaptation in angiotensin-induced hypertension. Hypertension. 15244563. 67 (5), 890–896. Available from: http://hyper.ahajournals.org/. doi:10.1161/HYPERTENSIONAHA.115.06262.

Bonafini, S., Giontella, A., Tagetti, A., Marcon, D., Montagnana, M., Benati, M., Danese, E., Minuz, P., Maffeis, C., Antoniazzi, F., Fava, C., 2019. Author correction: markers of subclinical vascular damages associate with indices of adiposity and blood pressure in obese children. Hypertens. Res. 0916-9636. 42 (7), 1097. doi:10.1038/s41440-019-0262-2.

Booth, F.W., Roberts, C.K., Laye, M.J., 2012. Lack of exercise is a major cause of chronic diseases. Compr. Physiol. 20404603. 2 (2), 1143–1211. doi:10.1002/cphy.c110025.

Bruno, R.M., Reesink, K.D., Ghiadoni, L., 2017. Advances in the non-invasive assessment of vascular dysfunction in metabolic syndrome and diabetes: focus on endothelium, carotid mechanics and renal vessels. Nutr. Metab. Cardiovasc. Dis. 15903729. 27 (2), 121–128. Available from: http://www.elsevier.com/wps/find/journaldescription.cws_home/704955/description#description. doi:10.1016/j.numecd.2016.09.004.

Bull, F.C., Al-Ansari, S.S., Biddle, S., Borodulin, K., Buman, M.P., Cardon, G., Carty, C., Chaput, J.P., Chastin, S., Chou, R., Dempsey, P.C., Dipietro, L., Ekelund, U., Firth, J., Friedenreich, C.M., Garcia, L., Gichu, M., Jago, R., Katzmarzyk, P.T., Lambert, E., Leitzmann, M., Milton, K., Ortega, F.B., Ranasinghe, C., Stamatakis, E., Tiedemann, A., Troiano, R.P., Ploeg, H.P.V.D., Wari, V., Willumsen, J.F., 2020. World Health Organization 2020 guidelines on physical activity and sedentary behaviour. Br. J. Sports Med. 14730480. 54 (24), 1451–1462. Available from: http://bjsm.bmjjournals.com/. doi:10.1136/bjsports-2020-102955.

Chin, E.C., Yu, A.P., Lai, C.W., Fong, D.Y., Chan, D.K., Wong, S.H., Sun, F., Ngai, H.H., Yung, P.S.H., Siu, P.M., 2020. Low-frequency HIIT improves body composition and aerobic capacity in overweight men. Med. Sci. Sports Exerc. 15300315. 52 (1), 56–66. Available from: https://journals.lww.com/acsm-msse/Pages/issuelist.aspx. doi:10.1249/MSS.0000000000002097.

Ciccone, M.M., Faienza, M.F., Altomare, M., Nacci, C., Montagnani, M., Federica, V., Cortese, F., Gesualdo, M., Zito, A., Mancarella, R., Leogrande, D., Viola, D., Scicchitano, P., Giordano, P., 2016. Endothelial and metabolic function interactions in overweight/obese children: the role of high-molecular weight adiponectin. J. Atheroscler. Thromb. 18803873. 23. Available from: http://www.jstage.jst.go.jp/browse/.

Csige, I., Ujvárosy, D., Szabó, Z., Lőrincz, I., Paragh, G., Harangi, M., Somodi, S., 2018. The impact of obesity on the cardiovascular system. J. Diabetes Res. 2314-6745. 2018, 1–12. doi:10.1155/2018/3407306.

Dengel, D.R., Kelly, A.S., Olson, T.P., Kaiser, D.R., Dengel, J.L., Bank, A.J., 2006. Effects of weight loss on insulin sensitivity and arterial stiffness in overweight adults. Metabolism. 00260495. 55 (7), 907–911. doi:10.1016/j.metabol.2006.02.018.

Farouque, H.M.O., Meredith, I.T., 2001. The assessment of endothelial function in humans. Coron. Artery Dis. 09546928. 12 (6), 445–454. doi:10.1097/00019501-200109000-00003.

Fathi, R., Haluska, B., Isbel, N., Short, L., Marwick, T.H., 2004. The relative importance of vascular structure and function in predicting cardiovascular events. J. Am. Coll. Cardiol. 07351097. 43 (4), 616–623. Available from: www.elsevier.com/locate/jacc. doi:10.1016/j.jacc.2003.09.042.

Francescomarino, S.D., Sciartilli, A., Valerio, V.D., Baldassarre, A.D., Gallina, S., 2009. The effect of physical exercise on endothelial function. Sports Med. 11792035. 39 (10), 797–812. Available from: https://www.scopus.com/inward/record.uri?eid=2-s2.0-70349194901&doi=10.2165%2f11317750-000000000-00000&partnerID=40&md5=74c14aa7cfd7e68fd3156391d7de4c60. doi:10.2165/11317750-000000000-00000.

Garcia, M.M.O., Lima, P.R.P., Correia, L.C.L., 2012. Valor prognóstico da função endotelial em portadores de aterosclerose: revisão sistemática. Arquivos Brasileiros de Cardiologia. 16784170. 99 (3), 857–864. Available from: http://www.scielo.br/pdf/abc/v99n3/en_aop07712.pdf. doi:10.1590/S0066-782X2012005000078Brazil.

Gokce, N., Keaney, J.F., Hunter, L.M., Watkins, M.T., Nedeljkovic, Z.S., Menzoian, J.O., Vita, J.A., 2003. Predictive value of noninvasivelydetermined endothelial dysfunction for long-term cardiovascular events inpatients with peripheral vascular disease. J. Am. Coll. Cardiol. 07351097. 41 (10), 1769–1775. doi:10.1016/s0735-1097(03)00333-4.

Gokce, N., Vita, J.A., McDonnell, M., Forse, A.R., Istfan, N., Stoeckl, M., Lipinska, I., Keaney, J.F., Apovian, C.M., 2005. Effect of medical and surgical weight loss on endothelial vasomotor function in obese patients. Am. J. Cardiol. 00029149. 95 (2), 266–268. doi:10.1016/j.amjcard.2004.09.016.

Gostic, C.L., Blatt, D., 2013. The Role of Physical Activity in Healthy Living: Its Anti-Inflammatory Effects Inflammation, Advancing Age and Nutrition: Research and Clinical Interventions. Elsevier Inc., pp. 279–286. Available from: https://www.scopus.com/inward/record.uri?eid=2-s2.0-84902388805&doi=10.1016%2fB978-0-12-397803-5.00023-X&partnerID=40&md5=d03913a69506366dbcdd91e22092c5df. doi:10.1016/B978-0-12-397803-5.00023-X.

Gregg, E.W., Cheng, Y.J., Cadwell, B.L., Imperatore, G., Williams, D.E., Flegal, K.M., Narayan, K.M.V., Williamson, D.F., 2005. Secular trends in cardiovascular desease risk factors according to body mass index in US adults. JAMA. 00987484. 293 (15), 1868–1874. Available from: http://jama.jamanetwork.com/journal.aspx. doi:10.1001/jama.293.15.1868.

Hovsepian, V., Marandi, S., Esfarjani, F., Zavar, R., Sadeghi, M., 2021. The effect of all extremity high intensity interval training on athero-protective factors and endothelial function in overweight and obese women. Int. J. Prev. Med. 20088213. 12 (1). Available from: http://www.ijpvmjournal.net/browse.asp. doi:10.4103/ijpvm.IJPVM_248_19.

Hu, F.B., Willett, W.C., Li, T., Stampfer, M.J., Colditz, G.A., Manson, J.A.E., 2004. Adiposity as compared with physical activity in predicting mortality among women. N. Engl. J. Med. 00284793. 351 (26), 2694–2703. doi:10.1056/NEJMoa042135.

Ingelsson, E., Sullivan, L.M., Fox, C.S., Murabito, J.M., Benjamin, E.J., Polak, J.F., Meigs, J.B., Keyes, M.J., O'Donnell, C.J., Wang, T.J., D'Agostino, R.B., Wolf, P.A., Vasan, R.S., 2007. Burden and prognostic importance of subclinical cardiovascular disease in overweight and obese individuals. Circulation. 00097322. 116 (4), 375–384. doi:10.1161/CIRCULATIONAHA.107.688788.

Jakubowski, M., Turek-Jakubowska, A., Szahidewicz-Krupska, E., Gawrys, K., Gawrys, J., Doroszko, A., 2020. Profiling the endothelial function using both peripheral artery tonometry (EndoPAT) and Laser Doppler Flowmetry (LD)—complementary studies or waste of time? Microvasc. Res. 00262862. 130, 104008. doi:10.1016/j.mvr.2020.104008.

Kapuku, G.K., Harshfield, G.A., Davis, H.C., Treiber, F.A., 2006. Early markers of cardiovascular disease. Vasc. Pharmacol. 15371891. 45 (5), 277–280. doi:10.1016/j.vph.2006.08.009.

Kaspar, M., Baumgartner, I., Staub, D., Drexel, H., Thalhammer, C., 2019. Non-invasive ultrasound-based imaging of atherosclerosis. Vasa. 16642872. 48 (2), 126–133. Available from: https://econtent.hogrefe.com/loi/vas. doi:10.1024/0301-1526/a000747.

Kelly, T., Yang, W., Chen, C.S., Reynolds, K., He, J., 2008. Global burden of obesity in 2005 and projections to 2030. Int. J. Obes. 14765497. 32 (9), 1431–1437. doi:10.1038/ijo.2008.102.

Kim, D.K., Jee, J.H., Park, W.H., 2021. Effects of aerobic and resistance exercise training on carotid intima-media thickness in abdominal obese women. Metab. Syndr. Relat. Disord. 15578518. 19 (4), 200–204. Available from: www.liebertonline.com/met/. doi:10.1089/met.2020.0118.

Königstein, K., Wagner, J., Frei, M., Knaier, R., Klenk, C., Carrard, J., Schwarz, A., Hinrichs, T., Schmidt-Trucksäss, A., 2021. Endothelial function of healthy adults from 20 to 91 years of age: prediction of cardiovascular risk by vasoactive range. J. Hypertens. 14735598. 39 (7), 1361–1369. Available from: http://journals.lww.com/jhypertension. doi:10.1097/HJH.0000000000002798.

Kose, M.P., Piskinpasa, M.E., Hacioglu, Y., Karabag, T., 2024. The relationship of visceral adiposity with endothelial functions and subclinical atherosclerosis in obese individuals. Romanian J. Internal Med. = Revue Roumaine de Medecine Interne. 2501062X. 62 (4), 404–413. doi:10.2478/rjim-2024-0024.

Libby, P., Buring, J.E., Badimon, L., et al., 2019. Atherosclerosis. Nat. Rev. Dis. Primers 5, 56. doi:10.1038/s41572-019-0106-z.

Lima, A., Ritti-Dias, R., Forjaz, C.L.M., Correia, M., Miranda, A., Brasileiro-Santos, M., Santos, A., Filho, D.S., Silva, A., 2014. A session of resistance exercise increases vasodilation in intermittent claudication patients. Appl. Physiol. Nutr. Metab. 17155312. 40 (1), 59–64. Available from: https://www.scopus.com/inward/record.uri?eid=2-s2.0-84920667789&doi=10.1139%2fapnm-2014-0342&partnerID=40&md5=d02056b5fb1e7fe132ad3b5863fc97fd. doi:10.1139/apnm-2014-0342.

Maruhashi, T., Iwamoto, Y., Kajikawa, M., Oda, N., Kishimoto, S., Matsui, S., Hashimoto, H., Aibara, Y., Yusoff, F.M., Hidaka, T., Kihara, Y., Chayama, K., Noma, K., Nakashima, A., Goto, C., Hida, E., Higashi, Y., 2018. Interrelationships among flow-mediated vasodilation, nitroglycerine- induced vasodilation, baseline brachial artery diameter, hyperemic shear stress, and cardiovascular risk factors. J. Am. Heart Assoc. 20479980. 7 (1). Available from: http://jaha.ahajournals.org/content/ahaoa/7/1/e006797full.pdf?download=true. doi:10.1161/JAHA.117.006797.

Maturana, M.A., Irigoyen, M.C., Spritzer, P.M., 2007. Menopause, estrogens, and endothelial dysfunction: current concepts. Clinics. 18075932. 62 (1), 77–86. Available from: http://www.scielo.br/pdf/clin/v62n1/a12v62n1.pdf. doi:10.1590/S1807-59322007000100012undefined.

Nytrøen, K., Rustad, L.A., Erikstad, I., Aukrust, P., Ueland, T., Lekva, T., Gude, E., Wilhelmsen, N., Hervold, A., Aakhus, S., Gullestad, L., Arora, S., 2013. Effect of high-intensity interval training on progression of cardiac allograft vasculopathy. J. Heart Lung Transplant. 15573117. 32 (11), 1073–1080. doi:10.1016/j.healun.2013.06.023.

Okamoto, T., Masuhara, M., Ikuta, K., 2007. Combined aerobic and resistance training and vascular function: effect of aerobic exercise before and after resistance training. J. Appl. Physiol. 15221601. 103 (5), 1655–1661. Available from: https://www.scopus.com/inward/record.uri?eid=2-s2.0-35649022156&doi=10.1152%2fjapplphysiol.00327.2007&partnerID=40&md5=b6dd3a7e3e4719e195ec8df0458e1e76. doi:10.1152/japplphysiol.00327.2007.

Olson, T.P., Schmitz, K.H., Leon, A.S., Dengel, D.R., 2006. Vascular structure and function in women. relationship with body mass index. Am. J. Prev. Med. 07493797. 30 (6), 487–492. doi:10.1016/j.amepre.2006.02.006.

Oral, O., 2021. Nitric oxide and its role in exercise physiology. J. Sports Med. Phys. Fitness. 00224707. 61 (9), 1208–1211. Available from: https://www.scopus.com/inward/record.uri?eid=2-s2.0-85117920489&doi=10.23736%2fS0022-4707.21.11640-8&partnerID=40&md5=056de2904bda084313a8a2cf2f9bc5a9. doi:10.23736/S0022-4707.21.11640-8.

Paffenbarger, R.S., Hyde, R., Wing, A.L., Hsieh, C.-C., 1986. All-cause mortality, and longevity of college alumni. N. Engl. J. Med. 314 (10), 605–613.

Pellegrin, M., Mazzolai, L., Berthelot, A., Laurant, P., 2009. Dysfonction endothéliale et risque cardiovasculaire. L'exercice protège la fonction endothéliale et prévient la maladie cardiovasculaire. Sci. Sports. 0765-1597. 24 (2), 63–73. Available from: https://www.sciencedirect.com/science/article/pii/S0765159708001779. doi:10.1016/j.scispo.2008.10.003.

Poirier, P., Giles, T.D., Bray, G.A., Hong, Y., Stern, J.S., Pi-Sunyer, F.X., Eckel, R.H., 2006. Obesity and cardiovascular disease: pathophysiology, evaluation, and effect of weight loss. Circulation. 0009-7322. 113 (6), 898–918. doi:10.1161/circulationaha.106.171016.

Poirier, P., Martin, J., Marceau, P., Biron, S., Marceau, S., 2004. Impact of bariatric surgery on cardiac structure, function and clinical manifestations in morbid obesity. Expert Rev. Cardiovasc. Ther. 2 (2), 193–201. doi:10.1586/14779072.2.2.193.

Preik, M., Lauer, T., Heib, C., Tabery, S., Strauer, B.E., Kelm M., 2003. Automated ultrasonic measurement of human arteries for the determination of endothelial function. Ultraschall Med. 21 (5), 195–198.

Ramalli, A., Aizawa, K., Shore, A.C., Morizzo, C., Palombo, C., Lenge, M., Tortoli, P., 2019. Continuous simultaneous recording of brachial artery distension and wall shear rate: a new boost for flow-mediated vasodilation. IEEE Trans. Ultrason. Ferroelectr. Frequency Control. 15258955. 66 (3), 463–471. doi:10.1109/TUFFC.2018.2889111.

Ras, R.T., Streppel, M.T., Draijer, R., Zock, P.L., 2013. Flow-mediated dilation and cardiovascular risk prediction: a systematic review with meta-analysis. Int. J. Cardiol. 18741754. 168 (1), 344–351. Available from: www.elsevier.com/locate/ijcard. doi:10.1016/j.ijcard.2012.09.047.

Rippe, J.M., 2019. Lifestyle strategies for risk factor reduction, prevention, and treatment of cardiovascular disease. Am. J. Lifestyle Med. 15598284. 13 (2), 204–212. Available from: http://www.sagepub.com/journalsProdDesc.nav?prodId=Journal201781. doi:10.1177/1559827618812395.

Roitman, J.L., LaFontaine, T., 2005. Exercise, atherosclerosis, and the endothelium: where the action is! Part 1. Strength Cond. J. 15241602. 27 (5), 31–33. Available from: http://journals.lww.com/nsca-scj/pages/default.aspx. doi:10.1519/00126548-200510000-00004.

Sena, C.M., Gonçalves, L., Seiça, R., 2022. Methods to evaluate vascular function: a crucial approach towards predictive, preventive, and personalised medicine. EPMA J. 18785085. 13 (2), 209–235. Available from: http://www.springer.com/journal/13167/about. doi:10.1007/s13167-022-00280-7.

Sesso, H.D., Paffenbarger, R.S., Lee, I.-M., 2000. Physical activity and coronary heart disease in men. Circulation. 0009-7322. 102 (9), 975–980. doi:10.1161/01.cir.102.9.975.

Shenouda, N., Gillen, J.B., Gibala, M.J., MacDonald, M.J., 2017. Changes in brachial artery endothelial function and resting diameter with moderate-intensity continuous but not sprint interval training in sedentary men. J. Appl. Physiol. 15221601. 123 (4), 773–780. Available from: http://jap.physiology.org/content/123/4/773.full.pdf. doi:10.1152/japplphysiol.00058.2017.

Shimbo, D., Grahame-Clarke, C., Miyake, Y., Rodriguez, C., Sciacca, R., Di Tullio, M., Boden-Albala, B., Sacco, R., Homma, S., 2007. The association between endothelial dysfunction and cardiovascular outcomes in a population-based multi-ethnic cohort. Atherosclerosis. 00219150. 192 (1), 197–203. doi:10.1016/j.atherosclerosis.2006.05.005.

Shimokawa, H., Satoh, K., 2014. Vascular function. Arteriosc. Thromb. Vasc. Biol. 15244636. 34 (11), 2359–2362. Available from: http://atvb.ahajournals.org/. doi:10.1161/ATVBAHA.114.304119.

Simova, I., 2015. Intima-media thickness: appropriate evaluation and proper measurement. e-J. Cardiol. Pract. 13 (21).

Sorensen, K.E., Celermajer, D.S., Spiegelhalter, D.J., Georgakopoulos, D., Robinson, J., Thomas, O., Deanfield, J.E., 1995. Non-invasive measurement of human endothelium dependent arterial responses: accuracy and reproducibility. Heart. 13556037. 74 (3), 247–253. doi:10.1136/hrt.74.3.247.

Stebbings, G.K., Morse, C.I., McMahon, G.E., Onambele, G.L., 2013. Resting arterial diameter and blood flow changes with resistance training and detraining in healthy young individuals. J. Athl. Train. 10626050. 48 (2), 209–219. Available from: http://natajournals.org/doi/pdf/10.4085/1062-6050-48.1.17. doi:10.4085/1062-6050-48.1.17.

Stein, J.H., Korcarz, C.E., Hurst, R.T., Lonn, E., Kendall, C.B., Mohler, E.R., Najjar, S.S., Rembold, C.M., Post, W.S., 2008. Use of carotid ultrasound to identify subclinical vascular disease and evaluate cardiovascular disease risk: a consensus statement from the American Society of echocardiography carotid intima-media thickness task force endorsed by the society for vascular medicine. J. Am. Soc. Echocardiogr. 08947317. 21 (2), 93–111. Available from: http://www.elsevier.com/inca/publications/store/6/2/3/3/1/5/index.htt. doi:10.1016/j.echo.2007.11.011.

Stevens, J., 2002. Fitness and fatness as predictors of mortality from all causes and from cardiovascular disease in men and women in the lipid research clinics study. Am. J. Epidemiol. 156 (9), 832–841.

Stevens, V.J., Obarzanek, E., Cook, N.R, Lee, I.M., Appel, L.J., Smith West, D., Milas, N.C., Mattfeldt-Beman, M., Belden, L., Bragg, C., Millstone, M., 2001. Long-term weight loss and changes in blood pressure: results of the Trials of Hypertension Prevention, phase II. Ann. Intern. Med. 134 (1), 1–11.

Tuomilehto, J., Lindström, J., Eriksson, J.G., Valle, T.T., Hämäläinen, H., Ilanne-Parikka, P., Keinänen-Kiukaanniemi, S., Laakso, M., Louheranta, A., Rastas, M., Salminen, V., Aunola, S., Cepaitis, Z., Moltchanov, V., Hakumäki, M., Mannelin, M., Martikkala, V., Sundvall, J., Uusitupa, M., 2001. Prevention of type 2 diabetes mellitus by changes in lifestyle among subjects with impaired glucose tolerance. N. Engl. J. Med. 0028-4793. 344 (18), 1343–1350. doi:10.1056/NEJM200105033441801.

Vanhees, L., Geladas, N., Hansen, D., Kouidi, E., Niebauer, J., Reiner, Ž., Cornelissen, V., Adamopoulos, S., Prescott, E., Börjesson, M., 2012. Importance of characteristics and modalities of physical activity and exercise in the management of cardiovascular health in individuals with cardiovascular risk factors: recommendations from the EACPR (Part II). Eur. J. Prev. Cardiol. 2047-4873. 19 (5), 1005–1033. doi:10.1177/1741826711430926.

Voros, S., Rinehart, S., Vazquez-Figueroa, J.G., Kalynych, A., Karmpaliotis, D., Qian, Z., Joshi, P.H., Anderson, H., Murrieta, L., Wilmer, C., Carlson, H., Ballard, W., Brown, C., 2014. Prospective, head-to-head comparison of quantitative coronary angiography, quantitative computed tomography angiography, and intravascular ultrasound for the prediction of hemodynamic significance in intermediate and severe lesions, using fractional flow reserve as reference standard (from the Atlanta i and II Study). Am. J. Cardiol. 18791913. 113 (1), 23–29. doi:10.1016/j.amjcard.2013.09.010.

Warburton, D.E.R., 2006. Health benefits of physical activity: the evidence. Can. Med. Assoc. J. 174 (6), 801–809.

Wilson, P.W.F., D'Agostino, R.B., Sullivan, L., Parise, H., Kannel, W.B., 2002. Overweight and obesity as determinants of cardiovascular risk. Arch. Intern. Med. 0003-9926. 162 (16), 1867. doi:10.1001/archinte.162.16.1867.

Yang, W.-I., Ha, J.-W., 2015. Non-invasive assessment of vascular alteration using ultrasound. Clin. Hypertens. 2056-5909. 21 (1). doi:10.1186/s40885-015-0035-4.

Zhang, X., Cash, R.E., Bower, J.K., Focht, B.C., Paskett, E.D., 2020. Physical activity and risk of cardiovascular disease by weight status among U.S adults. PLoS ONE. 19326203. 15 (5). Available from: https://journals.plos.org/plosone/article/file?id=10.1371/journal.pone.0232893&type=printable. doi:10.1371/journal.pone.0232893.

Zhou, Y., Li, H., Xia, N., 2021. The interplay between adipose tissue and vasculature: role of oxidative stress in obesity. Front. Cardiovasc. Med. 2297055X. 8. Available from: www.frontiersin.org/journals/cardiovascular-medicine. doi:10.3389/fcvm.2021.650214.

Part IV

Frailty, musculoskeletal health, and imaging

Chapter 13

Influence of physical activity and exercise on frail older adults using ultrasound

Tina Agnes Dsouza[a], Rajesh Navada[a], Suresh Sukumar[b], Leena R. David[b,c], Rajagopal Kadavigere[d], Sneha Ravichandran[b], Berna Uzun[e], Dilber Uzun Ozsahin[c,e,f], and Ilker Ozsahin[e]

[a]*Department of Physiotherapy, Manipal Academy of Higher Education, Manipal College of Health Professions, Manipal, India*

[b]*Department of Medical Imaging Technology, Manipal College of Health Professions, Manipal Academy of Higher Education, Manipal, Karnataka, India*

[c]*Medical Diagnostic Imaging Department, College of Health Sciences, University of Sharjah, Sharjah, United Arab Emirates*

[d]*Department of Radiodiagnosis and Imaging, Kasturba Medical College, Manipal, India*

[e]*Irfan Suat Gunsel Operational Research Institute, Near East University, Nicosia, Turkish Republic of Northern Cyprus (TRNC), Turkey*

[f]*Research Institute for Medical and Health Sciences, University of Sharjah, Sharjah, United Arab Emirates*

1 Introduction

Aging is a natural and inevitable process that brings about various physiological, psychological, and social changes. The concept of healthy aging has gained increasing attention due to the rising global aging population. According to *Healthy aging: A review of what works* (2010), healthy aging is defined as the capacity to lead a healthy, socially inclusive existence, mostly free from sickness or impairment. This definition underscores the importance of maintaining functional independence and quality of life as people grow older. Healthy aging is more likely in individuals who actively participate in activities that prevent falls, reduce the risk of disability and hospitalization, and promote overall health and well-being.

Among older adults, particularly those who are frail, maintaining independence is a key challenge. Frailty is a state of increased vulnerability to stressors due to declines in physiological reserves and function across multiple organ systems (Cesari et al., 2018). It is characterized by weakness, fatigue, slowed mobility, unintentional weight loss, and decreased physical activity levels (Apóstolo et al., 2018). These factors contribute to an increased risk of falls, hospitalization, disability, and mortality. Frailty is often considered an irreversible condition that is difficult to manage due to its multisystem involvement and wide array of physical, psychological, and social problems. As a result, frailty leads to increased dependency and a reduced quality of life for affected individuals.

Despite its complexities, frailty can be mitigated or managed through targeted interventions, with physical activity and exercise playing a crucial role (Cadore et al., 2013; de Labra et al., 2015). Regular physical activity is essential for preserving muscle mass, bone density, cardiovascular health, and overall functional capacity in older adults. Exercise interventions can help improve strength, balance, coordination, and flexibility, which are essential for reducing the risk of falls and maintaining mobility (Theou et al., 2011). Moreover, engaging in structured exercise programs can contribute to improved psychological well-being, enhanced cognitive function, and greater social interaction, all of which are vital for healthy aging.

One of the primary concerns in frail older adults is musculoskeletal deterioration, which includes sarcopenia (age-related loss of muscle mass and function) and osteoporosis (loss of bone density and strength). These conditions can significantly impact an individual's ability to perform activities of daily living independently. Regular resistance training and weight-bearing exercises have been shown to slow down or even reverse some of these degenerative changes by stimulating muscle hypertrophy and enhancing bone mineral density (Cruz-Jentoft et al., 2019). Furthermore, physical activity improves joint health, reducing stiffness and the risk of osteoarthritis related pain and immobility.

In addition to musculoskeletal health, cardiovascular function is a critical determinant of an individual's ability to remain active and independent. Frail older adults often experience diminished cardiovascular efficiency, characterized by reduced cardiac output, arterial stiffness, and impaired endothelial function. Engaging in aerobic exercises such as walking, cycling, or swimming has been shown to improve cardiovascular endurance, enhance circulation, and promote better

Advances in Medical Imaging. https://doi.org/10.1016/B978-0-443-28967-5.00003-4

oxygen delivery to tissues (Cesari et al., 2009). These adaptations contribute to a lower risk of cardiovascular disease and improved overall well-being in older individuals.

Another important aspect of physical activity in frail older adults is its impact on mental health and cognitive function. Aging is often associated with an increased risk of cognitive decline and neurodegenerative disorders such as dementia and Alzheimer's disease. Research suggests that regular exercise can have neuroprotective effects, promoting neurogenesis, enhancing synaptic plasticity, and improving cerebral blood flow (Northey et al., 2018). Physical activity has also been linked to lower levels of depression and anxiety, which are prevalent among older adults and can significantly affect their quality of life. Social engagement through group exercise programs further contributes to emotional well-being and reduces feelings of isolation and loneliness.

Given the significance of physical activity in improving the health of frail older adults, it is essential to have reliable methods for assessing and monitoring physiological changes associated with these interventions. Ultrasound imaging has emerged as a valuable, noninvasive tool for evaluating various aspects of musculoskeletal and cardiovascular health in aging individuals (Perkisas et al., 2018; Ticinesi et al., 2017). Ultrasound offers real-time imaging and allows clinicians and researchers to assess muscle architecture, bone quality, arterial stiffness, and overall vascular function. This modality provides an effective means of tracking the effects of exercise interventions over time and identifying early markers of decline in frail populations.

One of the key applications of ultrasound in aging research is the assessment of muscle quality and quantity. Sarcopenia is a major contributor to frailty, and early detection is crucial for implementing timely interventions (Cruz-Jentoft et al., 2019). Ultrasound imaging can measure muscle thickness, cross-sectional area, and echogenicity, which are indicators of muscle health (Perkisas et al., 2021). Increased echogenicity, for example, is associated with higher fat infiltration and fibrosis in muscles, signifying reduced muscle function. By utilizing ultrasound, healthcare providers can tailor exercise programs to target specific muscle groups and monitor improvements in muscle mass and function over time.

In addition to muscle assessment, ultrasound can be used to evaluate bone health. Traditional methods, such as dual-energy X-ray absorptiometry (DEXA), are commonly used for bone density assessment, but ultrasound offers a radiation-free alternative that can be particularly beneficial for frail older adults. Ultrasound can assess bone stiffness and quality, providing insights into osteoporosis risk and fracture susceptibility. Exercise interventions, particularly resistance training and weight-bearing activities, have been shown to improve bone health, and ultrasound imaging allows for noninvasive monitoring of these changes.

Cardiovascular health is another area where ultrasound plays a crucial role. Age-related changes in vascular function, such as arterial stiffness and endothelial dysfunction, contribute to an increased risk of hypertension, stroke, and cardiovascular disease. Ultrasound techniques, including Doppler imaging and flow-mediated dilation, enable the assessment of arterial health and blood flow dynamics. These measures help identify individuals at risk of cardiovascular complications and evaluate the effectiveness of aerobic exercise interventions in improving vascular function.

Beyond its diagnostic capabilities, ultrasound is also a valuable tool for guiding rehabilitation and exercise prescription in frail older adults. By visualizing musculoskeletal structures in real-time, clinicians can use ultrasound to provide biofeedback during exercise sessions, ensuring proper muscle activation and movement patterns. This approach enhances the safety and effectiveness of exercise interventions, particularly for individuals with mobility limitations or musculoskeletal conditions.

Overall, the integration of physical activity, exercise interventions, and ultrasound imaging represents a comprehensive approach to promoting healthy aging in frail older adults. As the global population continues to age, the need for effective strategies to maintain independence, prevent disability, and enhance quality of life becomes increasingly important. Regular exercise, coupled with advanced imaging techniques like ultrasound, offers a promising pathway toward achieving these goals.

Healthy aging is a multidimensional process that involves maintaining physical, cognitive, and social well-being. Frailty presents significant challenges, but evidence suggests that targeted exercise interventions can mitigate many of its adverse effects (Apóstolo et al., 2018; Theou et al., 2011). Physical activity plays a fundamental role in preserving musculoskeletal and cardiovascular health, reducing the risk of falls, improving mental well-being, and enhancing overall functionality. Ultrasound imaging serves as a valuable tool for assessing and monitoring these changes, enabling personalized exercise interventions that address the specific needs of frail older adults (Perkisas et al., 2021). By embracing a holistic approach that combines exercise and advanced assessment techniques, we can foster healthier, more independent aging populations and improve the quality of life for older individuals worldwide. Physical activity and exercise can significantly benefit frail older adults, and ultrasound can be a valuable tool to assess and monitor the effects of these interventions on various aspects of their musculoskeletal and cardiovascular health.

1.1 Muscle mass and strength

- **Increase in muscle mass**: To better understand the relationship between muscle characteristics and frailty among older individuals, it is important to analyze the clinical characteristics, muscle mass, and physical performance in this population. Previous studies have primarily focused on frailty-related factors in combined populations of males and females, with few studies specifically exploring older males. However, it has been observed that the age-related decrease in skeletal muscle mass and strength is faster in older males compared to women (Meng et al., 2019). By examining the possible relationship between frailty and muscle mass, as well as physical function-related factors, we can potentially identify early signs of prefrailty/frailty during health examinations of older individuals. Moreover, it is well-documented that resistance training can have positive effects on muscle mass and strength, even in frail older adults (Makizako et al., 2015).
- Contrary to popular belief, resistance training is a safe and effective way to increase muscular strength and tone. In fact, research has shown that regular resistance exercise can increase muscle size by up to 9% in just eight weeks of training, even in frail elderly individuals (Peterson et al., 2010). Resistance training works by stimulating the muscles to contract against an external resistance, which in turn leads to an increase in strength, tone, mass, and endurance. This is achieved through the production and release of certain substances (Hughes et al., 2021). Tissue characteristics provide valuable insights into the structure and composition of muscles. Ultrasound imaging techniques employ sound waves to generate detailed pictures of underlying tissues. By analyzing ultrasound images, researchers are able to assess muscle thickness, cross-sectional area, and architecture, all of which play crucial roles in understanding muscle function and performance. One of the primary advantages of ultrasound is its ability to provide real-time imaging. This allows for the dynamic assessment of muscle behavior during various movements, such as muscle contraction or elongation.
- In addition, ultrasound can be used to visualize muscle fascicles, the individual muscle fibers that make up the muscle tissue. Regular resistance training exercises can lead to an increase in muscle mass, even in frail older adults. Ultrasound can be used to measure muscle thickness and cross-sectional area, providing quantitative data on muscle growth (Ashir et al., 2023; Sarto et al., 2021).
- **Improvement in muscle strength**: Resistance training is widely acknowledged as a highly effective strategy for maximizing muscle strength. By integrating resistance exercises into their workout routines, individuals can enhance and augment their muscular capacities to a significant degree, rather than opting for other activities such as climbing, brisk walking, cycling, or dancing. Ultrasound possesses the ability to comprehensively analyze alterations in muscle architecture and effectively identify advancements in muscle quality (Krzysztofik et al., 2019).
- **Bone health**: The following are key factors to maintain and improve bone health, which is indeed essential for overall well-being and quality of life.

1.2 Bone density

Participating in weight-bearing exercises, such as walking, running, or engaging in resistance training, has been extensively documented as an effective strategy for preserving and enhancing bone density. These forms of physical activity subject the bones to mechanical stress, which triggers an adaptive physiological response known as bone remodeling. This process involves the activation of osteoblasts, the bone-forming cells responsible for increasing bone mass and strengthening the skeletal structure over time. Engaging in consistent weight-bearing exercises can help counteract the natural decline in bone mineral density that occurs with aging, thereby reducing the risk of osteoporosis and fractures.

Resistance training, which includes activities such as weightlifting, bodyweight exercises, and resistance band workouts, is particularly beneficial for maintaining bone health. The application of controlled loads through muscle contractions exerts forces on the skeletal system, stimulating bone formation and enhancing bone architecture. Studies have shown that individuals who regularly engage in strength training experience improvements in bone mineral density, particularly in weight-bearing areas such as the spine, hips, and lower extremities. Furthermore, resistance exercises contribute to overall musculoskeletal health by improving muscle strength and coordination, which further reduces fall risk and associated fractures.

Weight-bearing aerobic exercises, such as dancing, stair climbing, and hiking, also contribute to bone health by increasing impact forces on the skeletal system. These activities promote bone remodeling by stimulating osteogenic responses, leading to greater bone mass and improved structural integrity. In contrast, nonweight-bearing activities, such as swimming and cycling, while beneficial for cardiovascular fitness and joint health, do not provide the same osteogenic stimulus as weight-bearing exercises.

Although DEXA scans remain the gold standard for assessing bone mineral density, ultrasound has emerged as a complementary tool for evaluating bone quality. Quantitative ultrasound (QUS) is a noninvasive imaging technique that measures parameters such as bone stiffness, elasticity, and microarchitecture. Unlike DEXA, which primarily assesses bone density, ultrasound can provide insights into bone strength and fracture susceptibility by evaluating trabecular structure and cortical thickness. In addition, QUS has the advantage of being radiation-free, portable, and cost-effective, making it a suitable option for preliminary bone health assessments, particularly in resource-limited settings.

Ultrasound technology also holds potential in detecting stress fractures and other bone irregularities (Benedetti et al., 2018; Hong and Kim, 2018). Stress fractures, which result from repetitive mechanical loading and inadequate bone recovery, are a common concern in both active older adults and individuals with reduced bone density. High-resolution ultrasound imaging can help identify early-stage stress fractures by detecting disruptions in bone continuity and localized cortical thickening. This early detection capability enables timely intervention, including activity modification, nutritional support, and targeted exercise regimens to prevent further bone damage and complications.

Integrating weight-bearing exercises into a regular fitness regimen, alongside the use of ultrasound for monitoring bone quality, can be highly advantageous in promoting long-term skeletal health. By engaging in appropriate physical activities and leveraging modern imaging technologies, individuals can take proactive steps to maintain bone density, reduce fracture risk, and support overall mobility and independence as they age.

1.3 Body composition

- **Fat mass reduction:** Physical activity plays a crucial role in reducing fat mass and maintaining optimal body composition. Exercise contributes to these outcomes through several mechanisms.
- **Enhanced energy expenditure:** Engaging in physical activities leads to the burning of calories for energy by our bodies. This elevated energy expenditure can create what is known as a calorie deficit, which is essential for fat loss. By consistently burning more calories than you consume, your body gradually taps into stored fat reserves for energy, leading to a reduction in overall fat mass (Hill and Millan, 2014).
- **Promotion of muscle development:** Exercise, particularly strength training and resistance exercises, stimulates the growth and preservation of lean muscle mass within our bodies. Since muscle tissue has higher metabolic demands compared to fatty tissue, having increased muscle mass can elevate your resting metabolic rate—even when at rest or during periods without physical exertion (Krzysztofik et al., 2019). This heightened metabolism translates into more efficient caloric burn throughout the day, making it easier for individuals to maintain control over their body composition. A common misconception is that diet alone is sufficient for reducing overall body weight. While diet does play a significant role, research has shown that exercise combined with a balanced, healthy diet is more effective for reducing visceral body fat (Benton and Young, 2017). Moderate to vigorous regular exercise, when coupled with a reasonable reduction in caloric intake, can lead to 5%–10% weight loss. This can be achieved through aerobic exercise, resistance training, or a combination of both. It is worth noting that the benefits of exercise extend beyond weight loss and fat reduction. Exercise helps in the maintenance of muscle mass, which is particularly important as we age. One way to measure subcutaneous fat is by using ultrasound, which involves sending high-frequency sound waves into the body. These waves bounce back when they encounter different tissues and can be analyzed to create an image that shows the thickness of the subcutaneous fat layer at specific locations on the body. This method is commonly used to assess how fat is distributed in areas such as the abdomen, thighs, or arms. Visceral adipose tissue thickness is a measure of the fat that accumulates around organs within the abdominal cavity. Ultrasound imaging techniques can be employed to estimate this thickness by visually examining the abdomen and calculating the distance between its anterior wall and the spine. This computed value serves as an indirect indicator for assessing visceral fat deposition. Due to its association with multiple health risks, monitoring visceral fat levels holds significance in evaluating overall health status. Ultrasound is frequently employed in clinical and research settings to estimate the thickness of body fat due to its noninvasive nature and relatively low risk. Nonetheless, it is crucial to acknowledge that while ultrasound provides useful insights into fat distribution, more advanced techniques such as dual-energy X-ray absorptiometry or magnetic resonance imaging may be more accurate for precise measurement of fat mass. The selection of a method should consider specific objectives and available resources for the assessment at hand (Wagner, 2013 ; Nadeem et al., 2018).

1.4 Mobility as joint health

Exercise and physical therapy have the potential to significantly enhance joint mobility and alleviate pain associated with conditions such as osteoarthritis. It is important to note several key factors: Firstly, incorporating low-impact aerobic exercises into one's routine can contribute positively by fostering cardiovascular fitness and promoting better joint mobility while minimizing strain on affected joints. In addition, these exercises aid in maintaining a healthy weight, an essential aspect of mitigating joint discomfort (Kerschan-Schindl and MacHold, 2011; Skou et al., 2018).

Incorporating exercises that aim to strengthen the muscles surrounding the impacted joint can offer enhanced support, leading to decreased pain. Strengthened muscles contribute to stabilizing the joint and reducing strain on its structures.

Engaging in targeted exercises designed to improve joint flexibility and range of motion brings notable benefits. These specific exercises prevent joints from becoming rigid, preserving their functionality. Physical therapy entails the development of a personalized exercise regimen catered to your unique requirements and the affected joints. In addition, manual therapy techniques can be employed by physical therapists to enhance joint mobility and alleviate pain (Suzuki et al., 2019).

Musculoskeletal ultrasound, a form of imaging modality, can be utilized for visualizing joint structures and evaluating diverse joint-related conditions, such as joint effusion (accumulation of fluid within the joint) and inflammation. Musculoskeletal ultrasound offers healthcare professionals real-time, dynamic images that permit an assessment of the integrity and health status of soft tissues, ligaments, tendons, and cartilage surrounding joints (Lento and Primack, 2008; Patil and Dasgupta, 2012).

Musculoskeletal ultrasound is an effective tool for identifying the presence of excessive fluid accumulation within a joint, a condition known as joint effusion. This finding frequently indicates inflammation or other related joint ailments. Through this imaging technique, healthcare professionals can visually analyze the quantity and location of the fluid, facilitating accurate diagnosis and treatment choices (Basra and Humphries, 2017; Burke et al., 2019).

With its ability to visualize changes occurring in the synovium—the lining of joints—ultrasound plays a crucial role in assessing joint inflammation. Conditions characterized by inflammatory processes, like rheumatoid arthritis, often involve synovial inflammation; hence, using ultrasound examination enables observation of key signs, including synovial thickening (synovial hypertrophy) and increased blood flow (hyperemia) (Patil and Dasgupta, 2012).

1.5 Utilization of ultrasound for injection guidance

Ultrasound has become an indispensable tool in medical practice, particularly for guiding injections and aspirations in various musculoskeletal and joint-related conditions. Medical professionals commonly employ ultrasound to enhance precision during joint injections or aspirations, significantly improving the accuracy of needle placement. By utilizing real-time visualization capabilities, healthcare providers can ensure that therapeutic agents, such as corticosteroids, hyaluronic acid, or platelet-rich plasma (PRP), are delivered precisely to the intended location within the joint space. This targeted approach maximizes the effectiveness of the treatment while minimizing potential complications, such as inadvertent injection into surrounding soft tissues or neurovascular structures. Furthermore, ultrasound-guided aspirations allow for the efficient removal of excess joint fluid in cases of effusion, infection, or inflammation, thereby providing both diagnostic and therapeutic benefits.

The advantages of ultrasound-guided injections extend beyond accuracy. Studies have demonstrated that using ultrasound guidance improves patient outcomes by reducing pain during the procedure, decreasing the number of needle passes required, and enhancing overall patient satisfaction. Unlike blind or palpation-guided techniques, which rely on anatomical landmarks and clinical experience alone, ultrasound ensures direct visualization of anatomical structures, making it particularly beneficial for deep or complex joints such as the hip and shoulder. In addition, ultrasound guidance is highly valuable when administering injections in patients with anatomical variations, significant joint deformities, or postsurgical changes that may obscure traditional landmark-based approaches. An example of an ultrasound-guided injection for shoulder pain is illustrated in Fig. 13.1.

1.5.1 *Disease progression monitoring*

For individuals with chronic joint conditions, such as osteoarthritis, rheumatoid arthritis, or gout, ultrasound serves as an effective imaging modality for monitoring disease progression over time. One of its key advantages is its ability to detect early pathological changes that may not yet be visible on conventional radiographs. Ultrasound imaging allows healthcare practitioners to assess alterations in joint structure, synovial hypertrophy, effusion, cartilage degeneration, and the presence of inflammatory markers such as synovitis and tenosynovitis. This level of detail is particularly important in inflammatory arthritis, where early intervention can help prevent long-term joint damage and disability.

In rheumatoid arthritis, for example, ultrasound can be used to track disease activity by evaluating synovial inflammation, joint erosions, and increased vascularization using power Doppler imaging. These findings help rheumatologists adjust treatment plans, monitor the response to disease-modifying antirheumatic drugs, and make informed decisions regarding

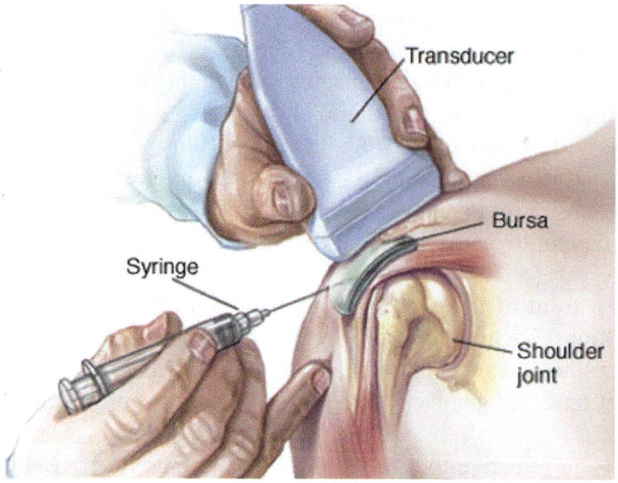

FIG. 13.1 Ultrasound-guided injection for shoulder pain. *(From Ultrasound or Fluoroscopy Guided Steroid injections, n.d. Retrieved August 7, 2025, from https://www.completepaincare.com/patient-education/services-provided/ultrasound-fluoroscopy-guided-steroid-injections/.)*

therapy escalation or tapering. Similarly, in osteoarthritis, ultrasound can reveal osteophytes, joint space narrowing, and subchondral bone changes, which contribute to pain and functional limitations. By tracking these changes longitudinally, clinicians can personalize treatment strategies, including physical therapy, pharmacological management, and lifestyle modifications, to slow disease progression and improve patient outcomes. Fig. 13.2 and 13.3 (Kiel et al., n.d.) show the ultrasound imaging of the knee joint.

1.5.2 Noninvasive and safe imaging

One of the major advantages of ultrasound as a diagnostic and monitoring tool is its noninvasive nature. Unlike imaging modalities such as X-rays or computed tomography (CT) scans, which expose patients to ionizing radiation, ultrasound employs high-frequency sound waves to generate real-time images of internal structures. This makes it a safe and repeatable imaging technique that can be used frequently without concerns about cumulative radiation exposure. As a result, ultrasound is particularly beneficial for long-term monitoring of chronic conditions, as it allows for regular follow-up assessments without additional risk to the patient.

Furthermore, ultrasound is a portable and cost-effective imaging modality, making it accessible for use in various healthcare settings, including outpatient clinics, bedside evaluations, and remote or underserved areas. Its ability to provide immediate, real-time feedback enables rapid clinical decision-making, facilitating prompt interventions when necessary. In addition, ultrasound-guided procedures typically require fewer resources and shorter procedure times compared to fluoroscopy- or magnetic resonance imaging (MRI)-guided interventions, making them a more efficient and patient-friendly option.

Overall, the utilization of ultrasound for injection guidance, disease progression monitoring, and safe imaging has revolutionized musculoskeletal and rheumatologic care. Its precision, accessibility, and noninvasive nature make it an invaluable tool for clinicians seeking to enhance diagnostic accuracy, optimize treatment strategies, and improve patient outcomes in the management of joint-related conditions.

1.6 Assessment of tissue damage

Frail older adults are at a significantly higher risk of sustaining injuries when engaging in physical activities, particularly if they overexert themselves or perform exercises with improper techniques. Frailty is often characterized by a decline in muscle mass (sarcopenia), reduced strength, and decreased bone density, which collectively contribute to a diminished ability to tolerate physical stress. These physiological changes result in an increased susceptibility to musculoskeletal injuries, such as muscle strains, ligament sprains, fractures, and falls. In addition, frail individuals may experience slower healing times due to impaired circulation, reduced collagen synthesis, and chronic conditions such as osteoporosis or arthritis, which further compromise tissue integrity.

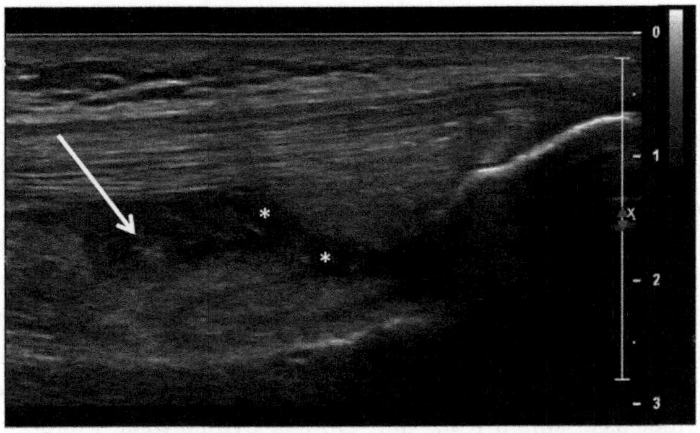

FIG. 13.2 Ultrasound image (B-mode), longitudinal suprapatellar view, showing joint effusion (*asterisks*) and synovial inflammatory thickening (*arrow*) within the subquadricipital recess. *(From D'Agostino, V., Sorriento, A., Cafarelli, A., Donati, D., Papalexis, N., Russo, A., Lisignoli, G., Ricotti, L., Spinnato, P., 2024. Ultrasound imaging in knee osteoarthritis: current role, recent advancements, and future perspectives. J. Clin. Med. 13 (16). https://doi.org/10.3390/jcm13164930.)*

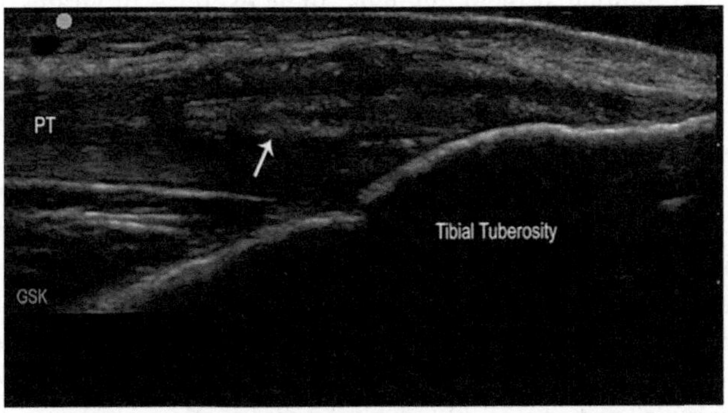

FIG. 13.3 Abnormal patellar tendon in long axis with tophaceous deposits (*arrow*) consistent with gout. *(From Kiel, J., Denq, W., Mueller, L. (Eds.), n.d., Gout (By J.K. Do, G. Kaeley).)*

Given these risks, it is crucial for frail older adults to participate in structured and supervised exercise programs designed to enhance their strength, balance, flexibility, and cardiovascular fitness while minimizing the likelihood of injury (Thakkar et al., 2022). Healthcare professionals, such as physiotherapists, geriatric specialists, and exercise physiologists, play a key role in developing personalized exercise regimens tailored to an individual's specific health status, mobility limitations, and overall fitness goals. These professionals ensure that exercises are performed safely and effectively, helping to reduce the chances of overuse injuries, falls, or exacerbation of preexisting conditions.

1.6.1 Utilization of ultrasound in tissue damage assessment

In cases where injuries do occur, diagnostic imaging technologies such as ultrasound provide invaluable assistance in assessing the extent of tissue damage. Ultrasound imaging allows healthcare professionals to visualize soft tissues in real time, enabling precise identification of muscle strains, ligament tears, joint effusions, and other musculoskeletal abnormalities. Unlike X-rays, which primarily detect fractures and bony structures, ultrasound offers superior imaging capabilities for soft tissues, making it an effective tool for diagnosing muscle and tendon injuries. This real-time visualization is particularly beneficial in distinguishing between mild, moderate, and severe injuries, which informs treatment decisions and rehabilitation planning.

Beyond its role in initial diagnosis, ultrasound also aids in guiding clinical interventions. For example, if a muscle tear is detected, ultrasound can help determine whether conservative management, such as physical therapy and rest, is sufficient or if more invasive treatments, like PRP injections or surgical intervention, may be required. In addition, in cases of chronic pain or joint inflammation, ultrasound-guided injections of corticosteroids or hyaluronic acid can provide targeted relief, minimizing systemic side effects and improving functional outcomes.

1.6.2 Monitoring injury recovery and rehabilitation

Throughout the rehabilitation process, ultrasound serves as a valuable tool for tracking healing progress and making informed adjustments to therapy protocols. By periodically assessing the affected tissues, healthcare providers can monitor signs of recovery, such as reduced inflammation, improved tissue integrity, and muscle regeneration. Ultrasound imaging also helps in detecting complications, such as excessive scar tissue formation, persistent fluid accumulation, or incomplete healing, which could hinder recovery if left unaddressed.

For instance, if an individual recovering from a muscle tear exhibits prolonged swelling or echogenic changes indicative of fibrosis, their rehabilitation plan may be modified to incorporate techniques such as manual therapy, myofascial release, or progressive resistance training. In addition, ultrasound-guided biofeedback can be employed to ensure proper muscle activation during rehabilitative exercises, helping patients regain strength and function more effectively.

It is important to emphasize that, while ultrasound provides critical insights into tissue health and recovery, it is typically used in conjunction with other assessment tools and clinical evaluations. Functional assessments, range-of-motion measurements, and strength testing remain essential components of a comprehensive rehabilitation plan. Healthcare providers integrate these multiple evaluation methods to develop well-rounded, individualized care strategies that prioritize both safety and effectiveness (Shah and Bhatnagar, 2019).

1.6.3 Importance of tailored exercise programs for frail older adults

Given the heightened risk of injury among frail individuals, their exercise programs must be carefully structured and continuously adapted based on ongoing assessments. Programs should incorporate low-impact strength training, balance exercises, and flexibility routines to build resilience while minimizing strain on vulnerable tissues. For example, resistance band exercises, chair-based workouts, and aquatic therapy offer effective ways to enhance muscular strength and joint stability without imposing excessive mechanical stress.

In addition, gradual progression is crucial to prevent overuse injuries. Older adults should begin with low-intensity exercises and slowly increase their activity levels under the supervision of a qualified professional. Regular monitoring through tools such as ultrasound, alongside functional assessments, ensures that their exercise regimen remains safe and beneficial.

In conclusion, frail older adults require a comprehensive approach to physical activity that balances the benefits of exercise with the potential risks of musculoskeletal injuries. Ultrasound imaging plays a pivotal role in assessing tissue damage, guiding interventions, and monitoring recovery, helping to ensure that exercise programs are both safe and effective. By integrating evidence-based exercise protocols with advanced diagnostic techniques, healthcare providers can promote healthier aging and improved quality of life for frail individuals (Shah and Bhatnagar, 2019; Thakkar et al., 2022).

2 Limitations

Though ultrasound has its merits in monitoring physical activity, it is not without its drawbacks. Researchers and practitioners must consider these limitations when deciding on monitoring technologies and planning their studies or systems. Depending on the requirements of the monitoring task at hand, alternative technologies such as accelerometers, gyroscopes, or computer vision systems may offer more appropriate solutions. Ultrasound's monitoring capabilities are restricted by its limited range, making it unsuitable for tracking physical activity that takes place beyond this range. Activities such as long-distance running or engagements in large open spaces may not be effectively monitored using ultrasound technology. In addition, the presence of obstacles like walls, buildings, or the user's own body can obstruct the line of sight required for accurate measurement and result in inaccurate readings. Establishing a monitoring system based on ultrasound can be intricate, necessitating meticulous calibration and precise positioning of transmitters and receivers. Ultrasound systems generally necessitate higher power consumption compared to alternative monitoring technologies, thereby posing a potential constraint for the portability and wearable application of such devices. The procurement and upkeep costs associated with ultrasound equipment can be substantial, potentially impeding its wide adoption in physical activity monitoring (Do Nascimento et al., 2020; Paul et al., 2023).

3 Future recommendations

Tailored Exercise Programs should emphasize creating personalized, low-impact physical activity programs tailored to the specific needs of frail older adults. A collaborative approach between various healthcare providers should plan the programs, considering bone-strengthening activities, balance and coordination exercises, nutritional support, individual health status, mobility limitations, and risk of injury to optimize bone health benefits.

Ultrasound should be incorporated into routine investigations to assess bone health over time in frail older adults, to keep track of improvements or defects, and to modify the physical activity plan accordingly. Extensive research may be focused on validating the use of ultrasound in clinical practice.

4 Conclusion

Engaging in regular physical activity plays a crucial role in promoting overall musculoskeletal health in frail older adults. Weight-bearing exercises, such as walking, stair climbing, and resistance training, have been shown to improve bone density, enhance muscle strength, and support balance, thereby reducing the risk of osteoporosis, fractures, and falls. By incorporating structured exercise programs tailored to the needs of frail individuals, healthcare providers can help mitigate age-related declines in bone and muscle health, ultimately enhancing mobility, independence, and quality of life.

Bone health is a critical concern in aging populations, particularly for those classified as frail. Frailty often results in decreased bone mineral density, making bones more susceptible to fractures even with minimal trauma. The decline in muscle mass and strength further exacerbates the risk of falls, which can lead to severe consequences such as hip fractures, loss of independence, and increased mortality rates. However, consistent participation in appropriately designed physical activity programs can stimulate bone remodeling and reinforce skeletal integrity. Resistance exercises, including the use of light weights or resistance bands, contribute to increased bone formation and improved overall musculoskeletal resilience. In addition, exercises focusing on flexibility and balance, such as yoga or tai chi, help enhance postural stability, reducing fall risk in frail older adults.

Beyond the benefits of physical activity, advancements in medical imaging have made it possible to assess and monitor bone health in a noninvasive and efficient manner. Ultrasound imaging has emerged as a valuable tool in evaluating bone quality and detecting early signs of deterioration. Unlike traditional methods such as DEXA, ultrasound does not expose patients to radiation, making it a safe and repeatable option for ongoing bone health assessments. It provides insights into bone microarchitecture, strength, and elasticity, helping healthcare professionals identify individuals at risk for osteoporosis or fractures before significant damage occurs.

In addition to its diagnostic capabilities, ultrasound plays an essential role in guiding interventions aimed at preserving bone health. By tracking changes in bone density and quality over time, healthcare providers can adjust exercise regimens, nutritional strategies, and medical treatments to better support skeletal health. For instance, individuals demonstrating early signs of bone weakening may benefit from targeted weight-bearing exercises, vitamin D and calcium supplementation, or pharmacological interventions to slow bone loss. Furthermore, ultrasound can assist in the early detection of stress fractures and other musculoskeletal abnormalities, allowing for timely intervention and prevention of further complications.

In conclusion, the integration of physical activity and ultrasound-based monitoring presents a comprehensive approach to enhancing bone health and overall well-being in frail older adults. Exercise remains a cornerstone of osteoporosis prevention and fall risk reduction, while ultrasound serves as a valuable tool for assessing bone changes and optimizing intervention strategies. By combining these strategies, healthcare professionals can empower older adults to maintain independence, improve mobility, and prevent further frailty-related complications, ultimately supporting healthier aging and a higher quality of life.

References

Healthy aging: a review of what works, 2010.

Apóstolo, J., Cooke, R., Bobrowicz-Campos, E., Santana, S., Marcucci, M., Cano, A., Holland, C., 2018. Predicting risk and outcomes for frail older adults: an umbrella review of frailty screening tools. JBI Database Syst. Rev. Implement. Rep. 16 (1), 1–77.

Ashir, A., Jerban, S., Barrère, V., Wu, Y., Shah, S.B., Andre, M.P., Chang, E.Y., 2023. Skeletal muscle assessment using quantitative ultrasound: a narrative review. Sensors. 14248220. 23 (10). Available from: http://www.mdpi.com/journal/sensors. doi:10.3390/s23104763.

Basra, H.A.S., Humphries, P.D., 2017. Juvenile idiopathic arthritis: what is the utility of ultrasound? Br. J. Radiol. 1748880X. 90 (1073). Available from: http://www.birpublications.org/doi/pdf/10.1259/bjr.20160920. doi:10.1259/bjr.20160920.

Benedetti, M.G., Furlini, G., Zati, A., Mauro, G.L., 2018. The effectiveness of physical exercise on bone density in osteoporotic patients. Biomed Res. Int. 23146141. 2018. Available from: http://www.hindawi.com/journals/biomed/. doi:10.1155/2018/4840531.

Benton, D., Young, H.A., 2017. Reducing calorie intake may not help you lose body weight. Perspect. Psychol. Sci. 17456924. 12 (5), 703–714. Available from: http://pps.sagepub.com/content/by/year. doi:10.1177/1745691617690878.

Burke, C.J., Alizai, H., Beltran, L.S., Regatte, R.R., 2019. MRI of synovitis and joint fluid. J. Magn. Reson. Imaging. 15222586. 49 (6), 1512–1527. Available from: http://onlinelibrary.wiley.com/journal/10.1002/(ISSN)1522-2586. doi:10.1002/jmri.26618.

Cadore, E.L., Rodríguez-Mañas, L., Sinclair, A., Izquierdo, M., 2013. Effects of different exercise interventions on risk of falls, gait ability, and balance in physically frail older adults: a systematic review. Rejuvenation Res. 15578577. 16 (2), 105–114. doi:10.1089/rej.2012.1397.

Cesari, M., de Carvalho, I.A., Thiyagarajan, J.A., Cooper, C., Martin, F.C., Reginster, J.-Y., Vellas, B., Beard, J.R., 2018. Evidence for the domains supporting the construct of intrinsic capacity. J. Gerontol. A Biol. Sci. Med. Sci. 1079-5006. 73 (12), 1653–1660. doi:10.1093/gerona/gly011.

Cesari, M., Kritchevsky, S.B., Newman, A.B., Simonsick, E.M., Harris, T.B., Penninx, B.W., Brach, J.S., Tylavsky, F.A., Satterfield, S., Bauer, D.C., Rubin, S.M., Visser, M., Pahor, M., 2009. Added value of physical performance measures in predicting adverse health-related events: results from the health, aging and body composition study. J. Am. Geriatr. Soc. 15325415. 57 (2), 251–259. doi:10.1111/j.1532-5415.2008.02126.x.

Cruz-Jentoft, A.J., Bahat, G., Bauer, J., Boirie, Y., Bruyère, O., Cederholm, T., Cooper, C., Landi, F., Rolland, Y., Sayer, A.A., Schneider, S.M., Sieber, C.C., Topinkova, E., Vandewoude, M., Visser, M., Zamboni, M., Bautmans, I., Baeyens, J.P., Cesari, M., Cherubini, A., Kanis, J., Maggio, M., Martin, F., Michel, J.P., Pitkala, K., Reginster, J.Y., Rizzoli, R., Sánchez-Rodríguez, D., Schols, J., 2019. Sarcopenia: revised European consensus on definition and diagnosis. Age Ageing. 14682834. 48 (1), 16–31. Available from: http://ageing.oxfordjournals.org/. doi:10.1093/ageing/afy169.

de Labra, C., Guimaraes-Pinheiro, C., Maseda, A., Lorenzo, T., Millán-Calenti, J.C., 2015. Effects of physical exercise interventions in frail older adults: a systematic review of randomized controlled trials. BMC Geriatr. 1471-2318. 15 (1). doi:10.1186/s12877-015-0155-4.

Do Nascimento, L.M.S., Bonfati, L.V., Freitas, M.L.B., Junior, J.J.A.M., Siqueira, H.V., Stevan, S.L., 2020. Sensors and systems for physical rehabilitation and health monitoring—a review. Sensors (Switzerland). 14248220. 20 (15), 1–28. Available from: https://www.mdpi.com/1424-8220/20/15/4063/pdf. doi:10.3390/s20154063.

Hill, J.C., Millan, I.S., 2014. Validation of musculoskeletal ultrasound to assess and quantify muscle glycogen content. A novel approach. Physician Sportsmed. 42 (3), 45–52. doi:10.3810/psm.2014.09.2075.

Hong, A.R., Kim, S.W., 2018. Effects of resistance exercise on bone health. Endocrinol. Metab. 20935978. 33 (4), 435–444. Available from: https://e-enm.org/. doi:10.3803/EnM.2018.33.4.435.

Hughes, M., Bruni, C., Cuomo, G., Delle Sedie, A., Gargani, L., Gutierrez, M., et al., 2021. The role of ultrasound in systemic sclerosis: on the cutting edge to foster clinical and research advancement. J. Scleroderma Relat. Disord. 6 (2), 123–132.

Kerschan-Schindl, K., MacHold, K., 2011. Rehabilitation von Patienten mit rheumatoider Arthritis. Physikalische Medizin Rehabilitationsmedizin Kurortmedizin. 1439085X. 21 (6), 297–310. doi:10.4061/2011/681640.

Kiel, J., Denq, W., Mueller, L., n.d. Internet Book of Musculoskeletal Ultrasound Gout. In: Kiel, J., Denq, W., Mueller, L.

Krzysztofik, M., Wilk, M., Wojdała, G., Gołaś, A., 2019. Maximizing muscle hypertrophy: a systematic review of advanced resistance training techniques and methods. Int. J. Environ. Res. Public Health. 1660-4601. 16 (24), 4897. doi:10.3390/ijerph16244897.

Lento, P.H., Primack, S., 2008. Advances and utility of diagnostic ultrasound in musculoskeletal medicine. Curr. Rev. Musculoskelet. Med. 19359748. 1 (1), 24–31. Available from: http://www.springer.com/humana+press/orthopedics/journal/12178. doi:10.1007/s12178-007-9002-3.

Makizako, H., Shimada, H., Doi, T., Tsutsumimoto, K., Suzuki, T., 2015. Impact of physical frailty on disability in community-dwelling older adults: a prospective cohort study. BMJ Open. 20446055. 5 (9). Available from: http://bmjopen.bmj.com/content/5/9/e008462full.pdf+html. doi:10.1136/bmjopen-2015-008462.

Meng, L., Shi, H., Shi, J., Yu, P.L., Xi, H., 2019. Differences in clinical characteristics, muscle mass, and physical performance among different frailty levels in Chinese older men. Chin. Med. J. 25425641. 132 (3), 352–355. Available from: https://journals.lww.com/cmj/pages/default.aspx. doi:10.1097/CM9.0000000000000035.

Nadeem, B., Bacha, R., Gilani, S.A., 2018. Correlation of subcutaneous fat measured on ultrasound with body mass index. J. Med. Ultrasound 26 (4), 205–209.

Northey, J.M., Cherbuin, N., Pumpa, K.L., Smee, D.J., Rattray, B., 2018. Exercise interventions for cognitive function in adults older than 50: a systematic review with meta-Analysis. Br. J. Sports Med. 14730480. 52 (3), 154–160. Available from: http://bjsm.bmjjournals.com/. doi:10.1136/bjsports-2016-096587.

Patil, P., Dasgupta, B., 2012. Role of diagnostic ultrasound in the assessment of musculoskeletal diseases. Ther. Adv. Musculoskelet. Dis. 1759-720X. 4 (5), 341–355. doi:10.1177/1759720x12442112.

Paul, M., Maglaras, L., Ferrag, M.A., Almomani, I., 2023. Digitization of healthcare sector: a study on privacy and security concerns. ICT Express. 24059595. 9 (4), 571–588. Available from: https://www.journals.elsevier.com/ict-express/. doi:10.1016/j.icte.2023.02.007.

Perkisas, S., Bastijns, S., Baudry, S., Bauer, J., Beaudart, C., Beckwée, D., Cruz-Jentoft, A., Gasowski, J., Hobbelen, H., Jager-Wittenaar, H., Kasiukiewicz, A., Landi, F., Małek, M., Marco, E., Martone, A.M., de Miguel, A.M., Piotrowicz, K., Sanchez, E., Sanchez-Rodriguez, D., Scafoglieri, A., Vandewoude, M., Verhoeven, V., Wojszel, Z.B., De Cock, A.-M., 2021. Application of ultrasound for muscle assessment in sarcopenia: 2020 SARCUS update. Eur. Geriatr. Med. 1878-7657. 12 (1), 45–59. doi:10.1007/s41999-020-00433-9.

Perkisas, S., Vandewoude, M., Dewulf, P., Van Hoye, K., 2018. Role of muscle ultrasound for sarcopenia evaluation in clinical practice. Rev. Endocr. Metab. Disord. 19 (4), 293–304.

Peterson, M.D., Rhea, M.R., Sen, A., Gordon, P.M., 2010. Resistance exercise for muscular strength in older adults: a meta-analysis. Ageing Res. Rev. 15681637. 9 (3), 226–237. doi:10.1016/j.arr.2010.03.004.

Sarto, F., Spörri, J., Fitze, D.P., Quinlan, J.I., Narici, M.V., Franchi, M.V., 2021. Implementing ultrasound imaging for the assessment of muscle and tendon properties in elite sports: practical aspects, methodological considerations and future directions. Sports Med. 11792035. 51 (6), 1151–1170. Available from: http://rd.springer.com/journal/40279. doi:10.1007/s40279-021-01436-7.

Shah, A.B., Bhatnagar, N., 2019. Ultrasound imaging in musculoskeletal injuries-what the Orthopaedic surgeon needs to know. J. Clin. Orthop. Trauma. 22133445. 10 (4), 659–665. Available from: http://www.elsevier.com/wps/find/journaldescription.cws_home/724754/description#description. doi:10.1016/j.jcot.2019.05.010.

Skou, S.T., Pedersen, B.K., Abbott, J.H., Patterson, B., Barton, C., 2018. Physical activity and exercise therapy benefit more than just symptoms and impairments in people with Hip and knee osteoarthritis. J. Orthop. Sports Phys. Ther. 01906011. 48 (6), 439–447. Available from: https://www.jospt.org/doi/pdf/10.2519/jospt.2018.7877. doi:10.2519/jospt.2018.7877.

Suzuki, Y., Iijima, H., Tashiro, Y., Kajiwara, Y., Zeidan, H., Shimoura, K., Nishida, Y., Bito, T., Nakai, K., Tatsumi, M., Yoshimi, S., Tsuboyama, T., Aoyama, T., 2019. Home exercise therapy to improve muscle strength and joint flexibility effectively treats pre-radiographic knee OA in community-dwelling elderly: a randomized controlled trial. Clin. Rheumatol. 0770-3198. 38 (1), 133–141. doi:10.1007/s10067-018-4263-3.

Thakkar, S., Muhammad, T., Srivastava, S., Greenhalgh, D.G., 2022. Cross-sectional associations of physical frailty with fall, multiple falls and fall-injury among older Indian adults: findings from LASI, 2018. PLOS ONE. 1932-6203. 17 (8), e0272669. doi:10.1371/journal.pone.0272669.

Theou, O., Stathokostas, L., Roland, K.P., Jakobi, J.M., Patterson, C., Vandervoort, A.A., Jones, G.R., 2011. The effectiveness of exercise interventions for the management of frailty: a systematic review. J. Aging Res. 20902212. 2011. doi:10.4061/2011/569194.

Ticinesi, A., Meschi, T., Narici, M.V., Lauretani, F., Maggio, M., 2017. Muscle ultrasound and sarcopenia in older individuals: a clinical perspective. J. Am. Med. Dir. Assoc. 15389375. 18 (4), 290–300. doi:10.1016/j.jamda.2016.11.013.

Wagner, D.R., 2013. Ultrasound as a tool to assess body fat. J. Obes. 20900716. 2013. doi:10.1155/2013/280713.

Chapter 14

Radiological assessment as a surrogate marker of frailty

Revati Amin[a], K. Vaishali[b], Suresh Sukumar[c], Leena R. David[c,d], Berna Uzun[e], Dilber Uzun Ozsahin[d,e,f], Ilker Ozsahin[e], and Rajagopal Kadavigere[g]

[a]*Department of Physiotherapy, Kasturba Medical College Mangalore, Manipal Academy of Higher Education, Manipal, India*
[b]*Department of Physiotherapy, Manipal College of Health Professions, Manipal Academy of Higher Education, Manipal, India*
[c]*Department of Medical Imaging Technology, Manipal College of Health Professions, Manipal Academy of Higher Education, Manipal, Karnataka, India*
[d]*Medical Diagnostic Imaging Department, College of Health Sciences, University of Sharjah, Sharjah, United Arab Emirates*
[e]*Irfan Suat Gunsel Operational Research Institute, Near East University, Nicosia, Turkish Republic of Northern Cyprus (TRNC), Turkey*
[f]*Research Institute for Medical and Health Sciences, University of Sharjah, Sharjah, United Arab Emirates*
[g]*Department of Radiodiagnosis and Imaging, Kasturba Medical College, Manipal Academy of Higher Education, Manipal, India*

1 Introduction

1.1 Definition of frailty

Aging is associated with a gradual decline in the capacity of multiple organ systems, raising the probability of poor health outcomes following sickness or injury (Liguori et al., 2018). Frailty prevalence was determined to be 30% among people aged 45 and above in India. After meticulously adjusting for a wide range of sociodemographic, economic, and lifestyle characteristics, it became apparent that women aged 60 and older were twice as likely as their male counterparts in the same age group to have frailty. Notably, the gender disparity was particularly prominent among individuals aged 45–59 (Ghosh et al., 2023). Unplanned hospitalizations and admissions to critical care units are more common among elderly people, which eventually increases the risk of mortality as a whole (Frost et al., 2009). Frailty is a biological phenomenon caused by the progressive decline of several physiological systems and increased susceptibility to adverse events (Fulop et al., 2010). Although comorbidity and disability share some characteristics with frailty, they are not the same condition. As an illustration, older people with the same underlying medical issues may display substantially varied degrees of disability and frailty (O'Donovan et al., 2019).

Nearly 20 years ago, professionals developed a way to categorize frailty by considering factors like unintentional weight loss, excessive fatigue, weak muscles, inadequate walking, and low levels of activity. Most medical tests and evaluations examine a person's ability to move, overall health, mental and emotional well-being, and how well their brain functions. Frailty is a multifaceted condition marked by low physiological reserve and poor tolerance to stress, according to general consensus among interdisciplinary experts (Wleklik et al., 2020). Remember, there is no universally acknowledged unified definition of frailty. Frailty is a complex and multifaceted concept that holds profound significance in the fields of healthcare, gerontology, and human well-being (Wleklik et al., 2020). It represents a state of vulnerability and decreased resilience in individuals, often associated with advancing age but not limited to it (MacLeod et al., 2016). This condition transcends the natural process of aging and instead embodies a distinct clinical syndrome characterized by an increased susceptibility to adverse health outcomes (Brown et al., 2016). Understanding frailty is crucial, as it has far-reaching implications for individuals' health, quality of life, and the healthcare systems that support them.

Frailty encompasses various dimensions, including physical, psychological, and social elements (Gobbens et al., 2010). Physically, frail individuals may experience muscle weakness, decreased endurance, and increased susceptibility to illness and injury (Strandberg et al., 2011). Psychologically, they might exhibit cognitive decline, depression, and anxiety (Ní Mhaoláin et al., 2012; Brown et al., 2014). Socially, frailty can lead to social isolation and diminished social engagement

Advances in Medical Imaging. https://doi.org/10.1016/B978-0-443-28967-5.00007-1

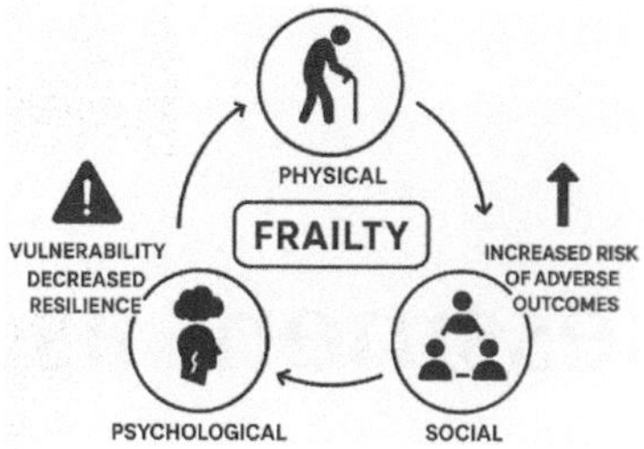

FIG. 14.1 A conceptual diagram showing the *multidimensional nature of frailty* and key risk factors.

(Mehrabi and Béland, 2020). These interconnected dimensions contribute to an overall state of vulnerability that affects one's ability to maintain independence and cope with stressors (Aldwin, 2018). Identifying frailty is of paramount importance, as it can serve as a crucial indicator for healthcare providers and caregivers (Van Damme et al., 2020). Recognizing frailty allows for targeted interventions and personalized care plans that can enhance the individual's quality of life and reduce the risk of adverse events such as falls, hospitalizations, and disability (Lee, 2015). Moreover, understanding frailty aids in resource allocation within healthcare systems, helping to optimize the allocation of services and support for those who need it most (Bentov et al., 2019; Cecchini et al., 2021). Fig. 14.1 presents a conceptual diagram illustrating the *multidimensional nature of frailty* and key risk factors.

1.2 Importance of assessing frailty

Over 50 different frailty assessment tools that are customized for diverse healthcare settings have been developed as a result of the growing recognition of frailty assessment as a useful healthcare tool. However, there is often limited agreement between these assessments, primarily because they tend to focus on different aspects of frailty. These assessments can be broadly categorized into three methodological approaches: phenotypic, multidimensional, and deficit accumulation. Phenotypic approaches concentrate on physical criteria like unintentional weight loss, low physical activity, and reduced muscle strength to identify frailty. Critics argue that these approaches may be too narrow and recommend considering a broader range of criteria, including cognitive and psychological factors. Multidimensional approaches have been developed to capture frailty beyond physical characteristics (Wleklik et al., 2020).

The deficit accumulation strategy identifies frailty by analyzing a variety of factors, including signs and symptoms, mental or physical limitations, or the existence of ailments that raise the risk of unfavorable health outcomes. Incorporating frailty assessment into clinical practice is considered essential for identifying individuals at risk of health decline, especially when exposed to health stressors. This is particularly significant in an aging population, where there is a push to incorporate it into clinical practice to more precisely predict unfavorable health outcomes and improve patient care (Rahman, 2018; Dent et al., 2016). Despite obstacles such as uncertainty in selecting the right assessment tool, there is ongoing support for integrating frailty assessments across healthcare settings (Dent et al., 2019).

Physical inactivity and social isolation associated with aging accelerate the deterioration of muscle mass and strength and exacerbate chronic conditions like diabetes, depression, dementia, cardiovascular, and cerebrovascular diseases. There are currently no known medications or established care standards to slow down the aging process. Even in the elderly, physical activity, particularly exercise training, has been shown to have an effect on critical aging markers such as chronic inflammation, mitochondrial dysfunction, myokine release, autophagy, oxidative damage, and insulin-like growth factor (IDF-1) signaling. Exercise improves physical function and quality of life, reduces the prevalence of noncommunicable diseases, and lowers overall premature mortality. Aging is associated with increased levels of reactive oxygen species and inflammation, which impair muscle function. Aging is also linked to reduced insulin growth factor (IGF-1) levels, which reduce protein synthesis and muscle growth via the phosphatidylinositol-3-kinase (PI3K)/AKT/mammalian target of rapamycin pathway. Exercise has strong anti-inflammatory and anti-oxidative stress effects, stimulating nuclear erythroid-2-like factor-2 (Nrf-2) and consequently enhancing muscular function. Furthermore, exercise increases protein synthesis by stimulating the IGF-1 pathway and promotes myokines, which reduce protein breakdown. Exercising boosts transcription factor peroxisome proliferator-activated receptor γ coactivator-1α (PGC-1α) signaling, which improves mitochondrial function and reduces inflammation caused by nuclear factor κB. Guidelines on physical activity and sedentary behavior highly recommend moderate or higher-intensity multicomponent physical activity on at least three days each week. Functional balance and strength training are emphasized in order to improve functional ability and reduce falls in older adults and those with chronic illnesses (Angulo et al., 2020; Izquierdo et al., 2021; Kalyani et al., 2015; O'Donovan et al., 2020).

1.3 Role of radiological assessment

Radiological imaging, such as X-rays, computed tomography (CT) scans, and magnetic resonance imaging (MRIs), plays a crucial role in identifying musculoskeletal issues like osteoporosis, osteoarthritis, and muscle loss, often observed in frail elderly individuals, leading to diminished physical abilities and functional decline. Dual-energy X-ray absorptiometry (DXA) scans are commonly used to gauge fracture risk, especially in frail individuals with weakened bones. Detecting vascular problems, such as atherosclerosis or vascular calcifications, through radiological imaging is vital, as these issues can contribute to cardiovascular and cerebrovascular diseases that worsen frailty (La Tegola et al., 2018; Cruz-Jentoft et al., 2019; LeBoff et al., 2022).

Sarcopenia, characterized by age-related muscle mass and function loss, is a significant aspect of frailty, and radiological techniques can track muscle changes over time. Radiological evaluations visualize various organs and systems, aiding in the identification of conditions like lung, liver, or gastrointestinal problems that can impact or complicate frailty management.

Moreover, radiological assessments are essential for diagnosing fall-related fractures, which are common in frail individuals and heighten frailty risk. Brain imaging, such as MRI, reveals structural and functional central nervous system changes linked to cognitive decline and neurological conditions often associated with frailty. Lastly, radiological evaluations enable the long-term monitoring of specific conditions, which is vital for assessing intervention effectiveness and adjusting treatment plans (Taylor et al., 2023; Bentov et al., 2019). However, it is crucial to understand that frailty is a multifaceted state influenced by a range of factors, such as social, psychological, and functional variables. Therefore, a comprehensive frailty assessment should encompass clinical, functional, and cognitive evaluations, and consider social determinants of health. Integrating radiological assessments with these evaluations is crucial for a holistic understanding of an older adult's health status and frailty risk (Cecchini et al., 2021; Maltais et al., 2020).

2 Radiological imaging modalities

2.1 Dual-energy X-ray absorptiometry

DXA is a medical procedure that involves the use of two X-ray beams with varying energy levels. One of these beams is predominantly absorbed by lean mass, which consists mainly of muscles (making it a valuable tool for assessing sarcopenia), while the other beam is absorbed by bones, enabling the measurement of bone mineral density (BMD). DXA utilizes a very low dose of ionizing radiation, and a single full-body scan exposes an individual to roughly the same amount of background radiation they might naturally encounter in a day, typically amounting to <10 mSv. DXA is advised for osteoporosis screening in women 65 years of age and older, as well as in older males at increased risk of having weak bones. Using DXA, sarcopenia can be detected if specific criteria are met, such as <5.45 kg/m^2 for women and <7.26 kg/m^2 for males. DXA is not frequently used as a preoperative technique or as a typical method to evaluate muscle mass, nevertheless. Although DXA evaluates total lean mass, including several tissues, it lacks the specificity required to evaluate particular muscles or muscle groups. It is often used for bone health and body composition assessments. Furthermore, it fails to distinguish between muscle quality and quantity, offering no information on aspects like muscle strength or composition, which are critical for preoperative evaluations. DXA is not the recommended approach for measuring muscle mass in preoperative settings due to factors such as a lack of functional information, economic considerations, and radiation exposure (Tavoian et al., 2019; Scafoglieri and Clarys, 2018; Uchai et al., 2007; Deshpande et al., 2023). Fig. 14.2 shows the DXA scan of the right hip joint of a 70-year-old female patient (Knipe, 2023).

2.2 Computed tomography scans

CT is a medical imaging technique that involves taking X-ray images from various angles as they revolve around a single axis, ultimately creating a three-dimensional computer-generated representation of the body. These X-rays undergo changes in intensity as they pass through different tissues, and these variations are relative to how air and water attenuate them. The degree of intensity change is quantified using Hounsfield units (HU), where air is assigned a value of roughly −1000 HU, and water is considered zero. Assessing sarcopenia from CT images can be accomplished using specialized software packages like SliceOmatic®, MATLAB, or ImageJ, although these tools are not obligatory.

The easiest technique to quantify sarcopenia is to visually define specific components like muscles and fat and calculate their cross-sectional area (CSA) using the free Digital Imaging and Communications in Medicine program. The calculation of muscle mass using a certain Hounsfield Unit (HU) range, typically between −29 and 150 HU, within the defined area is a more complicated procedure. This method evaluates the condition of the muscle indirectly if the indicated area

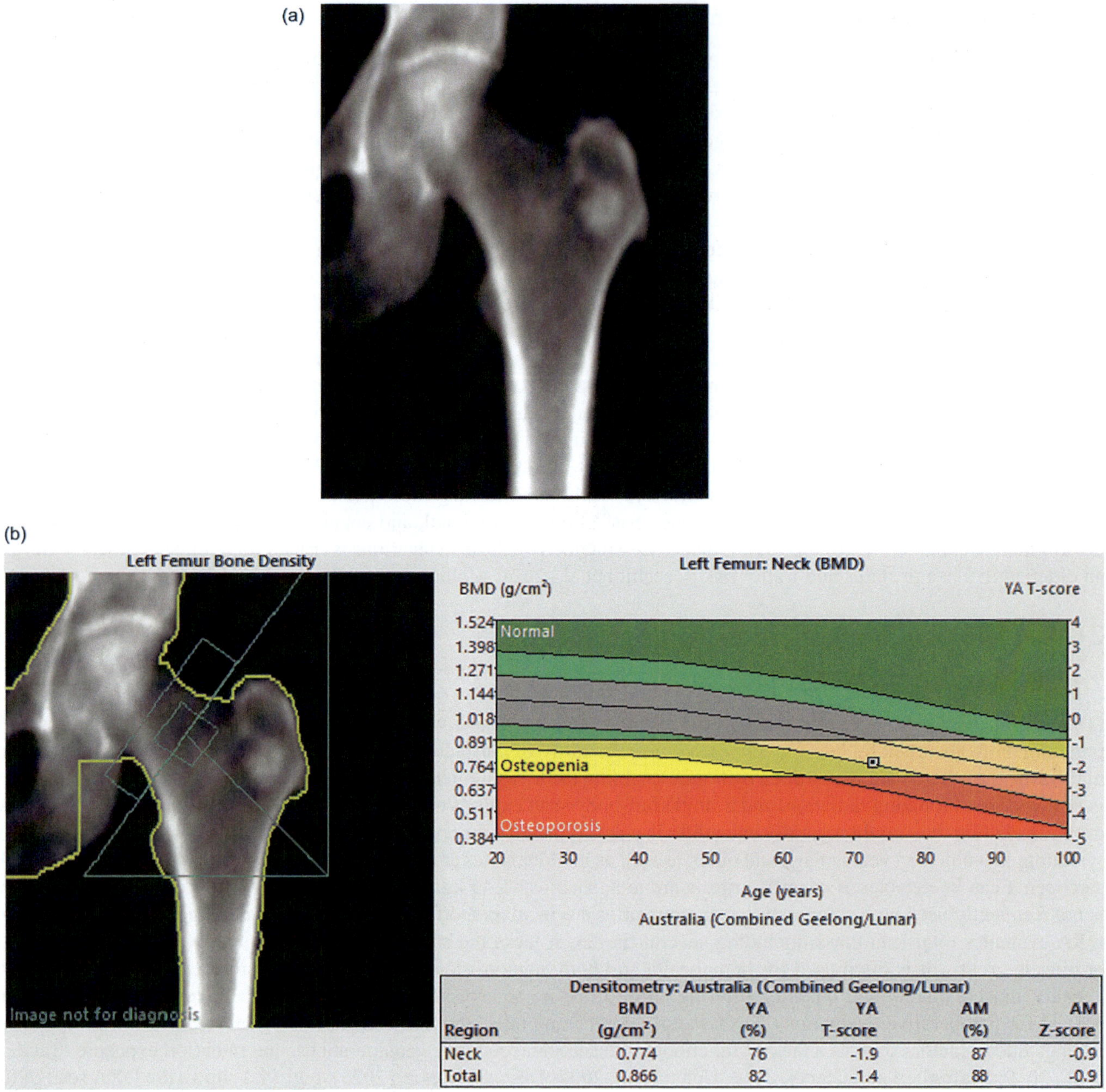

FIG. 14.2 DEXA scan showing bone mineral density within the osteopenic range. Focal ovoid density in the right greater trochanter. *(From Knipe, H., 2023. Enchondroma (DEXA). In: Radiopaedia.org. https://doi.org/10.53347/rID-160909.)*

has numerous components, such as fatty infiltration of the muscles (also known as myosteatosis). By establishing HU criteria for bone density, which impact the assessment's accuracy and sensitivity, a comparable technique can be used to evaluate disorders like osteopenia or osteoporosis. Automated systems that can determine body composition, including muscle, bone, and fat, from chest or abdominal CT images are currently being developed by researchers. Phase contrast, a newly developed CT technique, delivers better soft tissue resolution. In addition, although they are not frequently used for diagnosing sarcopenia, other CT scan modalities, such as positron emission tomography and single-photon emission CT employing gamma rays, can be utilized to evaluate muscle metabolism.

Sarcopenia identified through CT scans is linked to adverse outcomes after surgery and in the intensive care setting. Abdomino-pelvic CT scans are routinely employed to investigate potential visceral injuries, acute abdominal pain, or complex abdominal conditions that may necessitate surgery. Sarcopenia or osteopenia can be detected by a number of anatomical features, which can then help identify people who are at risk of experiencing unfavorable outcomes.

The skeletal muscle index is calculated by first measuring the total CSA of muscles at the level of the third lumbar vertebra (L3) in square centimeters (cm^2). Then, this measurement is divided by the patient's height in square meters (m^2). This division helps standardize the CSA relative to the patient's height. To assist in this measurement process, image processing software can be employed. At the L3 level, various muscles are identified, including the psoas, erector spinae, quadratus lumborum, transversus abdominis, external and internal obliques, and rectus abdominis. To determine the skeletal muscle index, all muscles within a single slice are identified using specific thresholds based on HU, which distinguish muscle tissue (typically falling within the HU range of −29 to +150) while excluding areas with fatty infiltration.

Although consensus guidelines do not provide specific diagnostic thresholds for sarcopenia using CT measurements, there is a clear relationship between the amount of muscle in a single lumbar CT image and the overall fat-free mass in the body, as well as the skeletal muscle in the limbs measured by DXA. In patients who have had both DXA and CT scans, the corresponding criteria for identifying sarcopenia based on the skeletal muscle index (estimated from DXA) are $<55.4\ cm^2/m^2$ for men and $<38.9\ cm^2/m^2$ for women. Another study, which aimed to predict mortality in patients with solid tumors in the respiratory or gastrointestinal tract, found very similar values: $<52.4\ cm^2/m^2$ for men and $<38.5\ cm^2/m^2$ for women. For patients undergoing colorectal cancer surgery, having a low skeletal muscle index is linked to a higher risk of postoperative infection and delayed recovery. In patients undergoing transcatheter aortic valve replacement, for each increase of $14\ cm^2/m^2$ in the skeletal muscle index, there was a one-day shorter hospital stay. Additionally, in mechanically ventilated intensive care unit (ICU) patients, having a low skeletal muscle index is associated with higher mortality rates at the time of hospital discharge, at 30 days, and at 6 months. It is also connected to one-year mortality in older trauma patients in the ICU.

The psoas muscle plays a crucial role in hip movement and provides support to various parts of the lower back and hip joints. Typically, when we measure the CSA of the psoas muscle (by outlining its shape in a cross-sectional view), it reflects the combined area of both psoas muscles, not just one side. Studies have shown that having low psoas muscle density or volume is associated with unfavorable outcomes after surgery. For example, in surgeries to remove liver metastases, psoas muscle sarcopenia is linked to a higher risk of complications. However, in older patients undergoing routine lumbar spine surgery, psoas muscle sarcopenia may not be a reliable predictor because of its close relationship with spine function. In older trauma patients, having psoas muscle sarcopenia is associated with higher mortality rates both six months and two years after they leave the hospital.

Thoracic CT scans are commonly conducted before chest surgeries for tasks like preoperative cancer staging, evaluating major trauma, or diagnosing pulmonary embolisms. As a result, many patients in the ICU or those preparing for surgery have thoracic imaging data available, which anesthesiologists can easily access. Similar techniques and tools are employed to assess sarcopenia and osteopenia in thoracic CT scans as those used in abdomino-pelvic CT scans. The process of quantifying CT scans in the thoracic region closely resembles the method employed to calculate the abdominal skeletal mass index.

In particular, the CSA of thoracic muscles is determined using a single axial CT slice at the level of the carina, and this measurement demonstrates a strong correlation with thoracic muscle volume. A more robust correlation is achieved by averaging the measurements from three slices, which include the carina level and two additional slices situated slightly above and below it (usually spaced at intervals of 1–1.25 mm). These slices encompass various muscles, such as the pectoralis major and minor, intercostals, serratus anterior, paraspinal muscles, and latissimus dorsi.

Thoracic CT scans can be used to identify sarcopenia in specific muscles like the pectoralis and erector spinae, providing valuable prognostic information, especially in patients with chronic obstructive pulmonary disease.

Cranial imaging is commonly performed in older individuals, even after minor head injuries, and it provides insights into sarcopenia. It also indirectly indicates cognitive decline associated with brain shrinkage. To assess sarcopenia, the CSA of muscles like the temporalis and masseter can be measured. In older adults who have sustained injuries, having a smaller masseter muscle CSA, as determined through cranial CT scans, has been associated with higher mortality rates. Brain shrinkage can be evaluated by calculating the brain atrophy index using cranial CT scans. This index involves measuring the ratio of the width of the two lateral ventricles at the head of the caudate nucleus to the width of the skull at a specific cranial-caudal level. An increase in this index indicates the shrinkage of deep frontal subcortical white matter. Researchers have explored the use of cranial CT scans to predict outcomes in injured older adults and have found that those with masseter muscle sarcopenia and a high brain atrophy index tend to have higher long-term mortality rates (Bunch et al., 2024; Hussain et al., 2022; Laur et al., 2022; Lortie et al., 2022). Figure 14.3 compares the CT images of typical sarcopenic and nonsarcopenic patients (Chianca et al., 2022).

2.3 Magnetic resonance imaging

MRI operates based on the concept that certain atomic nuclei emit radio-frequency energy when they encounter an external magnetic field. Through the use of varying timing sequences of radio-frequency pulses, referred to as T1-weighted and

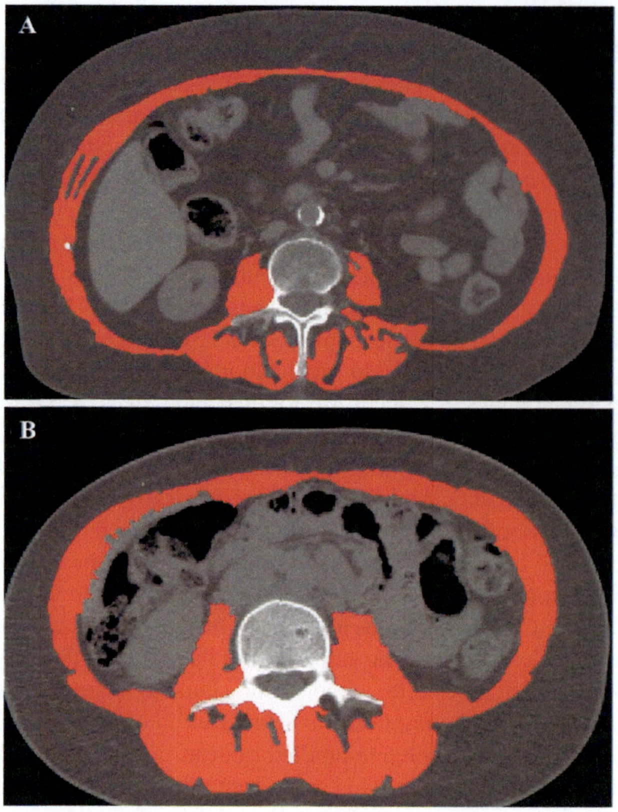

FIG. 14.3 Computed tomography images of a typical sarcopenic (image A) and nonsarcopenic (image B) patient. Skeletal muscle area (shown in *red*) is reduced in patient A, and this is especially visible in the paraspinal muscles, which are larger and less infiltrated by intramuscular fat. A skeletal muscle index of 38 cm^2/m^2 and 58 cm^2/m^2 was calculated using the CSA and height (CSA/height^2) of patients A and B, respectively. *(From Chianca, V., Albano, D., Messina, C., Gitto, S., Ruffo, G., Guarino, S., Grande, F.D., Sconfienza, L.M., 2022. Sarcopenia: imaging assessment and clinical application. Abdom. Radiol. 2366-0058. 47(9), 3205–3216. https://doi.org/10.1007/s00261-021-03294-3.)*

T2-weighted, it stimulates the energy transition in these atomic nuclei, while magnetic field gradients help precisely locate the resulting signals. This technology enables the generation of highly detailed images capable of distinguishing various types of tissues, such as muscle, bone, water, or fat, with exceptional accuracy. It is worth noting that MRI does not involve ionizing radiation, in contrast to X-rays or CT scans. However, it can be expensive and may not be suitable for patients with certain metallic implants, such as cochlear implants or cardiac pacemakers.

In contrast to CT scans, MRI is less commonly utilized as a preoperative procedure and typically concentrates on specific regions of the body. For example, when assessing lumbar spinal stenosis, which is the most frequent reason for lumbar spinal surgery in individuals over 65, MRI primarily visualizes the dural sac and central canal, with less focus on the evaluation of the paraspinal muscles. Despite its limitations, researchers are investigating the use of MRI as a means to evaluate sarcopenia as a predictive indicator in specific patient populations. This includes measuring fat-free muscle area to gauge muscle quality in individuals dealing with decompensated cirrhosis (Maltais et al., 2019; Huber et al., 2020). Figure 14.4 shows an example of sarcopenia assessment using MRI (Chianca et al., 2022).

2.4 Ultrasound

Ultrasound is a safe, affordable, easily transportable, and radiation-free imaging method. The low-intensity ultrasound employed for diagnosis generally does not create heat or pressure. Numerous studies have investigated the use of ultrasound to detect sarcopenia, and the findings have been promising. To evaluate muscle thickness using ultrasound, measurements are made with on-screen calipers, which determine the distance between the upper edge of the underlying bone and the lower boundary of the ventral fascia of the muscle group under examination. A comprehensive assessment is usually conducted with the patient in a standing position, involving an examination of both anterior and posterior muscle groups. This comprehensive approach is highly correlated with total body muscle mass. Some protocols involve measuring muscle mass at multiple sites, including the forearm, biceps, triceps, abdomen, subscapularis, quadriceps, hamstrings,

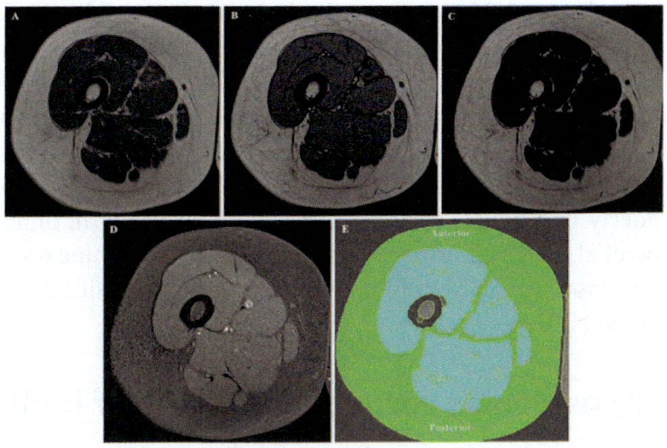

FIG. 14.4 Dixon magnetic resonance sequence acquired on the right thigh of a 62-year-old woman. Axial T1W in-phase (image A), out-of-phase (image B), 100% fat images (image C), and 100% water images (image D). After postprocessing, a semiautomatic segmentation can be obtained (image E) showing fat (*green*) and muscle tissue (turquoise) separately in the same slice. *(From Chianca, V., Albano, D., Messina, C., Gitto, S., Ruffo, G., Guarino, S., Grande, F.D., Sconfienza, L.M., 2022. Sarcopenia: imaging assessment and clinical application. Abdom. Radiol. 2366-0058. 47 (9), 3205–3216. https://doi.org/10.1007/s00261-021-03294-3.)*

gastrocnemius, and tibialis anterior, known as a nine-site protocol. However, this extensive protocol may not be practical in acute care settings. Therefore, a four-site protocol has been developed for bedside assessments, with the patient lying supine. In this method, a transverse approach is used to measure the bilateral anterior muscle thickness of the rectus femoris and vastus intermedius muscles at the midpoint of the femur and two-thirds of the distance between the anterior superior iliac spine and the upper pole of the patella. Alternatively, a single measurement point has been proposed, focusing on the mid-anterior thigh, situated halfway between the lateral condyle of the femur and the greater trochanter.

Numerous studies have examined the use of ultrasound to detect sarcopenia, with promising results. While there is no standardized clinical assessment, ultrasound has been applied to identify fat, bone, and muscle. Sarcopenia identified through ultrasound has been strongly associated with clinical frailty in older outpatient populations. In surgical ICU patients, sarcopenia diagnosed through ultrasound measurements of rectus femoris CSA has been linked to outcomes such as discharge to nursing facilities or in-hospital mortality (Simó-Servat et al., 2023; Anderson et al., 2023).

3 Blood indicators associated with the underlying biological factors of frailty

Frailty is a condition characterized by physical weakness and vulnerability that often affects older adults. It is associated with various biological processes in the body, including inflammation, oxidative stress, mitochondrial dysfunction, cellular aging (senescence), and genetic instability. Researchers have been exploring specific markers within the body that are connected to these biological processes and play a role in the development of frailty. For instance, one significant factor contributing to frailty is disrupted inflammation, often referred to as "inflamm-aging." In older individuals, certain pro-inflammatory substances, such as interleukin-6 (IL-6) and tumor necrosis factor-alpha, become elevated, and these elevations are linked to frailty and worse health outcomes. Another important factor is oxidative stress, which leads to damage to lipids and proteins within the body. Oxidative stress hampers the body's ability to respond effectively to crucial hormones and growth factors like IGF-1. IGF-1 plays a critical role in maintaining muscle mass, promoting recovery from injuries, and serving as a sensitive indicator of overall metabolism and nutrition. Lower levels of IGF-1 are associated with frailty. Furthermore, frailty is also connected to a reduction in the activity of proteins known as sirtuins, which play a role in regulating cellular metabolism. Activating sirtuins has been found to replicate the effects of caloric restriction, a dietary practice known to improve lifespan and overall health in animals (Fabrício et al., 2020).

4 Serum markers of frailty in routine clinical practice

In routine medical practice, several blood tests have been explored to detect frailty, but none of them are precise enough for regular use. Frailty appears to be closely related to nutritional and inflammatory factors, as frail individuals commonly exhibit reduced levels of serum albumin and elevated levels of C-reactive protein in their blood. Additionally, frailty is associated with higher counts of neutrophils and monocytes in older disabled women, regardless of their IL-6 levels. Conversely, nursing home residents with higher hemoglobin levels tend to experience lower mortality rates, whereas frail older individuals typically have lower hemoglobin levels, and this cannot be explained by issues like bone marrow suppression or iron deficiency. In older women, frailty is linked to vitamin B12 deficiency, although it remains uncertain whether vitamin supplements can effectively slow down or reverse the progression of frailty. Frail individuals also tend to exhibit elevated levels of transferrin and fibrinogen, two glycoproteins assessed in tests for anemia and blood clotting disorders.

In the case of men, frailty is associated with insufficient vitamin D levels, which result in increased parathyroid hormone levels. However, this relationship does not seem to hold true for women. Low cholesterol levels are connected to a higher risk of mortality among elderly individuals and also correlate with frailty in older patients admitted to the hospital (Onder et al., 2003; Félix-Redondo et al., 2013). Furthermore, lower levels of alanine transaminase, a marker for liver function, are linked to frailty and greater mortality in men living in the community, with this association being more consistent among older individuals (Ballew et al., 2023).

5 Changes in body composition as surrogate markers of frailty

As people age, their bodies go through various changes in terms of how they are composed. This includes a decrease in muscle and lean body mass, a reduction in bone density (which can range from mild osteopenia to more severe osteoporosis), and an increase in fat. While changes in mass do not always directly reflect changes in how the body functions, they usually move in the same direction. There are multiple ways to evaluate changes in body composition. One approach, known as anthropometry, involves taking measurements during a clinical examination. These measurements may include parameters such as Body Mass Index, skinfold thickness, and circumferences around the waist and limbs. However, anthropometry lacks precision, especially when applied to females. Another method is called bioelectrical impedance analysis, a noninvasive technique that estimates muscle mass by passing a small electrical current through the body. This method relies on the fact that muscles contain a high proportion of water, and electrical currents move differently through water-rich tissues. Nevertheless, factors like hydration levels, swelling (edema), and nutritional status can impact the accuracy of these measurements (Laur et al., 2022; Ferriolli and Fernandes, 2024).

6 Sarcopenia

Age-related declines in muscle mass, strength, and function that result in decreased muscle volume, decreased force production, and deteriorating physical performance are the hallmarks of sarcopenia. The root cause encompasses various elements such as advanced age, fluctuations in hormone levels, inadequate dietary intake, inactive lifestyle, and the existence of long-term medical conditions. Frailty, on the other hand, refers to a larger concept that includes a variety of physical and cognitive factors that contribute to increased vulnerability and decreased physiological reserves. Physical signs of frailty include weakness, fatigue, and decreased physical activity, but cognitive frailty can cause cognitive deterioration. Nutritional frailty is connected with poor dietary habits, and generally, frail individuals demonstrate lower physiological resilience, making them more vulnerable to adverse health outcomes. Frailty is generally caused by a complex interaction of factors such as age, chronic diseases, social isolation, insufficient nutrition, and a deteriorating ability to cope with physiological stressors. Both frailty and sarcopenia have common underlying factors linked to adverse health events, and there is potential for them to be reversible (Keevil and Romero-Ortuno, 2015; Cooper et al., 2012).

Initially, sarcopenia was described as the natural decline in muscle tissue that occurs as people age. However, this definition has evolved to include not just a reduction in muscle mass but also a decline in muscle function. The decrease in physical abilities, often assessed through tests like gait speed and muscle strength, closely relates to sarcopenia and forms a crucial aspect of physical frailty. Currently, the consensus for diagnosing age-related sarcopenia requires the presence of both decreased muscle mass and impaired muscle function. Specific criteria have been established for common assessments, such as slow gait speed (a speed <0.8 m per second is considered slow), weak grip strength (<26 kg for men and <16 kg for women), and reduced lean body mass measured by whole-body DXA. Another approach to estimate sarcopenia involves a biochemical measure, the serum creatinine-to-serum-cystatin-C ratio, which provides an estimate of muscle mass and is associated with outcomes in intensive care units (Daly et al., 2022; Bernabeu-Wittel et al., 2020).

7 Osteopenia/osteoporosis

The aging-related loss of muscle mass, strength, and functionality in sarcopenia causes a reduction in muscle mass, a decrease in force generation, and a compromise in physical performance. Aging, hormonal fluctuations, poor nutrition, a sedentary lifestyle, and chronic illnesses are the main factors. Osteopenia, on the other hand, is characterized by lower-than-normal BMD, which indicates a decrease in calcium and mineral content in the bones. Individuals with osteopenia have a higher risk of fractures than people with normal bone density, but it is lower than the risk associated with osteoporosis. Osteopenia is a common precursor to osteoporosis and can be caused by aging, hormonal imbalances, insufficient calcium and vitamin D consumption, a lack of physical exercise, or certain drugs (Edwards et al., 2015; Nielsen et al., 2018). In older women, frailty is correlated with both osteopenia and osteoporosis, and these associations may arise from common underlying factors. Furthermore, sarcopenia, characterized by muscle loss, is associated with bones becoming narrower and having thinner outer layers. In older men, decreased grip strength and slower walking speed are individually linked to reduced bone density in the femoral neck. When it comes to diagnosing and assessing osteopenia and osteoporosis, it is widely recognized that employing imaging methods is the preferred approach (Rolland et al., 2021; Inoue et al., 2022).

8 Discussion

In recent years, there has been a growing acknowledgment of the importance of frailty on the health and well-being of people aged 45 and older. This comprehensive research sheds light on the prevalence of frailty, with a focus on women aged 60 and older, who are twice as likely as males in the same age range to encounter the condition. Comprehending the subtleties of frailty within this population is essential for formulating focused therapies and enhancing their overall well-being. Frailty is a multidimensional condition that includes physical, psychological, and social components, resulting in susceptibility and poor resilience. Frailty assessment is critical in healthcare, with numerous techniques available, divided into three categories: phenotypic, multidimensional, and deficit accumulation (Ghosh et al., 2023; Rockwood and Howlett, 2018).

Physical activity, particularly exercise training, is highlighted as a way to reduce the consequences of frailty, improve physical function, and lower the risk of adverse outcomes. Radiological evaluations, such as X-rays, CT scans, and MRIs, are critical in detecting musculoskeletal conditions linked with frailty, such as osteoporosis and muscle loss. DXA and CT scans are used to determine muscle mass and diagnose sarcopenia. MRI and ultrasound are additional imaging modalities being investigated for measuring body composition and detecting sarcopenia. Frailty is linked to several blood markers, including inflammation, oxidative stress, and genetic instability. However, current blood tests are not sufficiently precise for routine clinical usage (de Labra et al., 2015; Oliveira et al., 2020; Boutin et al., 2015; Beaudart et al., 2016; Oliveira et al., 2020; Xie et al., 2020).

Frailty and sarcopenia, which are both marked by age-related muscle loss, have similar underlying causes. Osteopenia and osteoporosis, both of which are characterized by low bone density, have also been related to frailty, particularly in older women. The review emphasizes the significance of taking a comprehensive approach to frailty assessment, including clinical, functional, cognitive, and social variables. Overall, understanding frailty has far-reaching implications for enhancing quality of life and the provision of healthcare delivery (Polito et al., 2022; Laskou et al., 2022; Huang et al., 2023; Gielen et al., 2023; Greco et al., 2019).

9 Conclusion

Personalized therapies customized to individual frailty profiles address specific physical, psychological, and social components, improving efficacy in frailty management for individuals aged 45 and above. The use of advanced frailty assessment approaches, such as wearable devices and artificial intelligence, enables real-time monitoring and early interventions, resulting in improved healthcare precision. An integrative approach to frailty management, which combines interventions such as physical activity and cognitive training, necessitates collaboration among geriatric physicians, physiotherapists, psychologists, and social workers to promote a comprehensive understanding. Exploring nutritional interventions and understanding the complex relationship between nutrition and frailty is critical for sustaining muscle mass and overall health in older adults. Imaging technology, clinical precision, blood markers, and longitudinal research will help advance our understanding of frailty, enabling targeted interventions and health policies that promote healthy aging.

References

Aldwin, C.M., 2018. The Elders Life Stress Inventory: Egocentric and Nonegocentric Stress. Informa UK Limited, pp. 49–69. doi:10.4324/9781315803074-4.

Anderson, B.M., Wilson, D.V., Qasim, M., Correa, G., Evison, F., Gallier, S., Ferro, C.J., Jackson, T.A., Sharif, A., 2023. Ultrasound quadriceps muscle thickness is variably associated with frailty in haemodialysis recipients. BMC Nephrol. 14712369. 24 (1). Available from: https://bmcnephrol.biomedcentral.com/. doi:10.1186/s12882-022-03043-8.

Angulo, J., El Assar, M., Álvarez-Bustos, A., Rodríguez-Mañas, L., 2020. Physical activity and exercise: strategies to manage frailty. Redox Biol. 22132317. 35, 101513. doi:10.1016/j.redox.2020.101513.

Ballew, S.H., Zhou, L., Surapaneni, A., Grams, M.E., Windham, B.G., Selvin, E., Coresh, J., Miao, S., Inker, L.A., Levey, A.S., 2023. A novel creatinine muscle index based on creatinine filtration: associations with frailty and mortality. J. Am. Soc. Nephrol. 15333450. 34 (3), 495–504. Available from: https://journals.lww.com/jasn/pages/default.aspx. doi:10.1681/ASN.0000000000000037.

Beaudart, C., McCloskey, E., Bruyère, O., Cesari, M., Rolland, Y., Rizzoli, R., de Carvalho, I.A., Thiyagarajan, J.A., Bautmans, I., Bertière, M.C., Brandi, M.L., Al-Daghri, N.M., Burlet, N., Cavalier, E., Cerreta, F., Cherubini, A., Fielding, R., Gielen, E., Landi, F., Petermans, J., Reginster, J.Y., Visser, M., Kanis, J., Cooper, C., 2016. Sarcopenia in daily practice: assessment and management. BMC Geriatr. 14712318. 16 (1), 1–10. Available from: http://www.biomedcentral.com/bmcgeriatr/. doi:10.1186/s12877-016-0349-4.

Bentov, I., Kaplan, S.J., Pham, T.N., Reed, M.J., 2019. Frailty assessment: from clinical to radiological tools. Br. J. Anaesth. 14716771. 123 (1), 37–50. Available from: https://www.journals.elsevier.com/british-journal-of-anaesthesia. doi:10.1016/j.bja.2019.03.034.

Bernabeu-Wittel, M., Gómez-Díaz, R., González-Molina, Á., Vidal-Serrano, S., Díez-Manglano, J., Salgado, F., Soto-Martín, M., Ollero-Baturone, M., 2020. Oxidative stress, telomere shortening, and apoptosis associated to sarcopenia and frailty in patients with multimorbidity. J. Clin. Med. 20770383. 9 (8), 1–12. Available from: https://www.mdpi.com/2077-0383/9/8/2669/pdf. doi:10.3390/jcm9082669.

Boutin, R.D., Yao, L., Canter, R.J., Lenchik, L., 2015. Sarcopenia: current concepts and imaging implications. Am. J. Roentgenol. 15463141. 205 (3), W255. Available from: http://www.ajronline.org/doi/pdf/10.2214/AJR.15.14635. doi:10.2214/AJR.15.14635.

Brown, P.J., Roose, S.P., Fieo, R., Liu, X., Rantanen, T., Sneed, J.R., Rutherford, B.R., Devanand, D.P., Avlund, K., 2014. Frailty and depression in older adults: a high-risk clinical population. Am. J. Geriatr. Psychiatry. 15457214. 22 (11), 1083–1095. Available from: http://www.sciencedirect.com/science/journal/10647481. doi:10.1016/j.jagp.2013.04.010.

Brown, P.J., Rutherford, B.R., Yaffe, K., Tandler, J.M., Ray, J.L., Pott, E., Chung, S., Roose, S.P., 2016. The depressed frail phenotype: the clinical manifestation of increased biological aging. Am. J. Geriatr. Psychiatry. 15457214. 24 (11), 1084–1094. Available from: http://www.sciencedirect.com/science/journal/10647481. doi:10.1016/j.jagp.2016.06.005.

Bunch, P.M., Rigdon, J., Niazi, M.K.K., Barnard, R.T., Boutin, R.D., Houston, D.K., Lenchik, L., 2024. Association of CT-derived skeletal muscle and adipose tissue metrics with frailty in older adults. Acad. Radiol. 18784046. 31 (2), 596–604. Available from: https://www.sciencedirect.com/science/journal/10766332. doi:10.1016/j.acra.2023.06.003.

Cecchini, S., Di Rosa, M., Soraci, L., Fumagalli, A., Misuraca, C., Colombo, D., Piomboni, I., Carnevali, F., Paci, E., Galeazzi, R., Giordano, P., Fedecostante, M., Cherubini, A., Lattanzio, F., 2021. Chest X-ray score and frailty as predictors of in-hospital mortality in older adults with COVID-19. J. Clin. Med. 2077-0383. 10 (13), 2965. doi:10.3390/jcm10132965.

Chianca, V., Albano, D., Messina, C., Gitto, S., Ruffo, G., Guarino, S., Grande, F.D., Sconfienza, L.M., 2022. Sarcopenia: imaging assessment and clinical application. Abdom. Radiol. 2366-0058. 47 (9), 3205–3216. doi:10.1007/s00261-021-03294-3.

Cooper, C., Dere, W., Evans, W., Kanis, J.A., Rizzoli, R., Sayer, A.A., Sieber, C.C., Kaufman, J.-M., van Kan, G.A., Boonen, S., Adachi, J., Mitlak, B., Tsouderos, Y., Rolland, Y., Reginster, J.-Y.L., 2012. Frailty and sarcopenia: definitions and outcome parameters. Osteoporos. Int. 0937-941X. 23 (7), 1839–1848. doi:10.1007/s00198-012-1913-1.

Cruz-Jentoft, A.J., Bahat, G., Bauer, J., Boirie, Y., Bruyère, O., Cederholm, T., Cooper, C., Landi, F., Rolland, Y., Sayer, A.A., Schneider, S.M., Sieber, C.C., Topinkova, E., Vandewoude, M., Visser, M., Zamboni, M., Bautmans, I., Baeyens, J.P., Cesari, M., Cherubini, A., Kanis, J., Maggio, M., Martin, F., Michel, J.P., Pitkala, K., Reginster, J.Y., Rizzoli, R., Sánchez-Rodríguez, D., Schols, J., 2019. Sarcopenia: revised European consensus on definition and diagnosis. Age Ageing. 14682834. 48 (1), 16–31. Available from: http://ageing.oxfordjournals.org/. doi:10.1093/ageing/afy169.

Daly, R.M., Iuliano, S., Fyfe, J.J., Scott, D., Kirk, B., Thompson, M.Q., Dent, E., Fetterplace, K., Wright, O.R.L., Lynch, G.S., Zanker, J., Yu, S., Kurrle, S., Visvanathan, R., Maier, A.B., 2022. Screening, diagnosis and management of sarcopenia and frailty in hospitalized older adults: recommendations from the Australian and New Zealand Society for sarcopenia and frailty research (ANZSSFR) Expert Working Group. J. Nutr. Health Aging. 17604788. 26 (6), 637–651. Available from: http://www.springerlink.com/content/121281/. doi:10.1007/s12603-022-1801-0.

de Labra, C., Guimaraes-Pinheiro, C., Maseda, A., Lorenzo, T., Millán-Calenti, J.C., 2015. Effects of physical exercise interventions in frail older adults: a systematic review of randomized controlled trials. BMC Geriatr. 1471-2318. 15 (1). doi:10.1186/s12877-015-0155-4.

Dent, E., Kowal, P., Hoogendijk, E.O., 2016. Frailty measurement in research and clinical practice: a review. Eur. J. Intern. Med. 31, 3–10. doi:10.1016/j.ejim.2016.03.007.

Dent, E., Martin, F.C., Bergman, H., Woo, J., Romero-Ortuno, R., Walston, J.D., 2019. Management of frailty: opportunities, challenges, and future directions. Lancet. 1474547X. 394 (10206), 1376–1386. Available from: http://www.journals.elsevier.com/the-lancet/. doi:10.1016/S0140-6736(19)31785-4.

Deshpande, N., Hadi, M.S., Lillard, J.C., Passias, P.G., Linzey, J.R., Saadeh, Y.S., LaBagnara, M., Park, P., 2023. Alternatives to DEXA for the assessment of bone density: a systematic review of the literature and future recommendations. J. Neurosurg. Spine. 15475646. 38 (4), 436–445. Available from: https://thejns.org/spine/view/journals/j-neurosurg-spine/38/4/article-p436.xml. doi:10.3171/2022.11.SPINE22875.

Edwards, M.H., Dennison, E.M., Sayer, A.A., Fielding, R., Cooper, C., 2015. Osteoporosis and sarcopenia in older age. Bone. 87563282. 80, 126–130. doi:10.1016/j.bone.2015.04.016.

Fabrício, D.D., Chagas, M.H., Diniz, B.S., 2020. Frailty and cognitive decline. Transl. Res. 221, 58–64. doi:10.1016/j.trsl.2020.01.002.

Félix-Redondo, F.J., Grau, M., Fernández-Bergés, D., 2013. Cholesterol and cardiovascular disease in the elderly. Facts and gaps. Aging Dis. 4 (3), 154–169.

Ferriolli, E., Fernandes, P.M., 2024. Frailty syndrome and healthcare for older adults. Sao Paulo Med. J. 142 (4). doi:10.1590/1516-3180.2024.1424.21052024.

Frost, S.A., Alexandrou, E., Bogdanovski, T., Salamonson, Y., Parr, M.J., Hillman, K.M., 2009. Unplanned admission to intensive care after emergency hospitalisation: risk factors and development of a nomogram for individualising risk. Resuscitation. 03009572. 80 (2), 224–230. doi:10.1016/j.resuscitation.2008.10.030.

Fulop, T., Larbi, A., Witkowski, J.M., McElhaney, J., Loeb, M., Mitnitski, A., Pawelec, G., 2010. Aging, frailty and age-related diseases. Biogerontology. 1389-5729. 11 (5), 547–563. doi:10.1007/s10522-010-9287-2.

Ghosh, A., Kundu, M., Devasenapathy, N., Woodward, M., Jha, V., 2023. Frailty among middle-aged and older women and men in India: findings from wave 1 of the longitudinal Ageing study in India. BMJ Open. 2044-6055. 13 (7), e071842. doi:10.1136/bmjopen-2023-071842.

Gielen, E., Dupont, J., Dejaeger, M., Laurent, M.R., 2023. Sarcopenia, osteoporosis and frailty. Metabolism.

Gobbens, R.J.J., van Assen, M.A.L.M., Luijkx, K.G., Wijnen-Sponselee, M.T., Schols, J.M.G.A., 2010. Determinants of frailty. J. Am. Med. Dir. Assoc. 15389375. 11 (5), 356–364. doi:10.1016/j.jamda.2009.11.008.

Greco, E.A., Pietschmann, P., Migliaccio, S., 2019. Osteoporosis and sarcopenia increase frailty syndrome in the elderly. Front. Endocrinol. 10. doi:10.3389/fendo.2019.00255.

Huang, T., Li, C., Chen, F., Xie, D., Yang, C., Chen, Y., 2023. Prevalence and risk factors of osteosarcopenia: a systematic review and meta-analysis. BMC Geriatr. 23 (1), 1–15.

Huber, F.A., Grande, F.D., Rizzo, S., Guglielmi, G., Guggenberger, R., 2020. MRI in the assessment of adipose tissues and muscle composition: how to use it. Quant. Imaging Med. Surg. 22234306. 10 (8), 1636–1649. Available from: http://qims.amegroups.com/article/view/38300/html. doi:10.21037/QIMS.2020.02.06.

Hussain, S., Mubeen, I., Ullah, N., Shah, S.S.U.D., Khan, B.A., Zahoor, M., Ullah, R., Khan, F.A., Sultan, M.A., 2022. Modern diagnostic imaging technique applications and risk factors in the medical field: a review. Biomed. Res. Int. 23146141. 2022. Available from: http://www.hindawi.com/journals/biomed/. doi:10.1155/2022/5164970.

Inoue, T., Maeda, K., Satake, S., Matsui, Y., Arai, H., 2022. Osteosarcopenia, the co-existence of osteoporosis and sarcopenia, is associated with social frailty in older adults. Aging Clin. Exp. Res. 1720-8319. 34 (3), 535–543. doi:10.1007/s40520-021-01968-y.

Izquierdo, M., Merchant, R.A., Morley, J.E., Anker, S.D., Aprahamian, I., Arai, H., Aubertin-Leheudre, M., Bernabei, R., Cadore, E.L., Cesari, M., Chen, L.-K., de Souto Barreto, P., Duque, G., Ferrucci, L., Fielding, R.A., García-Hermoso, A., Gutiérrez-Robledo, L.M., Harridge, S.D.R., Kirk, B., Kritchevsky, S., Landi, F., Lazarus, N., Martin, F.C., Marzetti, E., Pahor, M., Ramírez-Vélez, R., Rodriguez-Mañas, L., Rolland, Y., Ruiz, J.G., Theou, O., Villareal, D.T., Waters, D.L., Won, C.W., Woo, J., Vellas, B., Singh, M.F., 2021. International exercise recommendations in older adults (ICFSR): expert consensus guidelines. J. Nutr. Health Aging. 12797707. 25 (7), 824–853. doi:10.1007/s12603-021-1665-8.

Kalyani, R.R., Metter, E.J., Egan, J., Golden, S.H., Ferrucci, L., 2015. Hyperglycemia predicts persistently lower muscle strength with aging. Diabetes Care 38 (1), 82–90. doi:10.2337/dc14-1166.

Keevil, V.L., Romero-Ortuno, R., 2015. Ageing well: a review of sarcopenia and frailty. Proc. Nutr. Soc. 14752719. 74 (4), 337–347. http://journals.cambridge.org/PNS. doi:10.1017/S0029665115002037.

Knipe, H., 2023. Enchondroma (DEXA). In: Radiopaedia.org. doi: 10.53347/rID-160909.

La Tegola, L., Mattera, M., Cornacchia, S., Cheng, X., Guglielmi, G., 2018. Diagnostic imaging of two related chronic diseases: sarcopenia and osteoporosis. J. Frailty Sarcopenia Falls. 24594148. 03 (3), 139–147. doi:10.22540/jfsf-03-138.

Laskou, F., Fuggle, N.R., Patel, H.P., Jameson, K., Cooper, C., Dennison, E., 2022. Associations of osteoporosis and sarcopenia with frailty and multimorbidity among participants of the Hertfordshire Cohort Study. J. Cachexia Sarcopenia Muscle. 21906009. 13 (1), 220–229. Available from: http://onlinelibrary.wiley.com/journal/10.1007/13539.2190-6009. doi:10.1002/jcsm.12870.

Laur, O., Weaver, M.J., Bridge, C., Chow, E., Rosenthal, M., Bay, C., Javedan, H., Harris, M.B., Khurana, B., 2022. Computed tomography-based body composition profile as a screening tool for geriatric frailty detection. Skeletal Radiol. 14322161. 51 (7), 1371–1380. Available from: https://www.springer.com/journal/256. doi:10.1007/s00256-021-03951-0.

LeBoff, M.S., Greenspan, S.L., Insogna, K.L., Lewiecki, E.M., Saag, K.G., Singer, A.J., Siris, E.S., 2022. The clinician's guide to prevention and treatment of osteoporosis. Osteoporos. Int. 0937-941X. 33 (10), 2049–2102. doi:10.1007/s00198-021-05900-y.

Lee, Y., 2015. Evidence-based prevention of frailty in older adults. J. Korean Geriatr. Soc. 19 (3), 121–129. doi:10.4235/jkgs.2015.19.3.121.

Liguori, I., Russo, G., Curcio, F., Bulli, G., Aran, L., Della-Morte, D., Gargiulo, G., Testa, G., Cacciatore, F., Bonaduce, D., Abete, P., 2018. Oxidative stress, aging, and diseases. Clin. Interv. Aging. 1178-1998. 13, 757–772. doi:10.2147/cia.s158513.

Lortie, J., Gage, G., Rush, B., Heymsfield, S.B., Szczykutowicz, T.P., Kuchnia, A.J., 2022. The effect of computed tomography parameters on sarcopenia and myosteatosis assessment: a scoping review. J. Cachexia, Sarcopenia Muscle. 21906009. 13 (6), 2807–2819. Available from: http://onlinelibrary.wiley.com/journal/10.1007/13539.2190-6009. doi:10.1002/jcsm.13068.

MacLeod, S., Musich, S., Hawkins, K., Alsgaard, K., Wicker, E.R., 2016. The impact of resilience among older adults. Geriatr. Nurs. 15283984. 37 (4), 266–272. Available from: http://www.elsevier.com/inca/publications/store/6/2/3/0/8/8/index.htt. doi:10.1016/j.gerinurse.2016.02.014.

Maltais, M., de Souto Barreto, P., Moon, S.Y., Rolland, Y., Vellas, B., 2019. Prospective association of white matter hyperintensity volume and frailty in older adults. Exp. Gerontol. 18736815. 118, 51–54. Available from: www.elsevier.com/locate/expgero. doi:10.1016/j.exger.2019.01.007.

Maltais, M., de Souto Barreto, P., Perus, L., Mangin, J.F., Grigis, A., Chupin, M., Bouyahia, A., Gabelle, A., Delrieux, J., Rolland, Y., Vellas, B., Guyonnet, S., Carrié, I., Brigitte, L., Faisant, C., Lala, F., Delrieu, J., Villars, H., Combrouze, E., Badufle, C., Zueras, A., Andrieu, S., Cantet, C., Morin, C., Van Kan, G.A., Dupuy, C., Rolland, Y., Caillaud, C., Ousset, P.J., Lala, F., Dartigues, J.F., Marcet, I., Delva, F., Foubert, A., Cerda, S., Marie-Noëlle-Cuffi, C.C., Rouaud, O., Manckoundia, P., Quipourt, V., Marilier, S., Franon, E., Bories, L., Pader, M.L., Basset, M.F., Lapoujade, B., Faure, V., LiYungTong, M., Malick-Loiseau, C., Cazaban-Campistron, E., Desclaux, F., Blatge, C., Dantoine, T., Laubarie-Mouret, C., Saulnier, I., Clément, J.P., Picat, M.A., Bernard-Bourzeix, L., Willebois, S., Désormais, I., Cardinaud, N., Bonnefoy, M., Livet, P., Rebaudet, P., Gédéon, C., Burdet, C., Terracol, F., Pesce, A., Roth, S., Chaillou, S., Louchart, S., Sudres, K., Lebrun, N., Barro-Belaygues, N., Touchon, J., Bennys, K., Gabelle, A., Romano, A., Touati, L., Marelli, C., Pays, C., Robert, P., Duff, F.L., Gervais, C., Gonfrier, S., Gasnier, Y., Bordes, S., Begorre, D., Carpuat, C., Khales, K., Lefebvre, J.F., Idrissi, S.M.E., Skolil, P., Salles, J.P., Dufouil, C., Lehéricy, S., Chupin, M., Mangin, J.F., Bouhayia, A., Allard, M., Ricolfi, F., Dubois, D., Martel, M.P.B., Cotton, F., Bonafé, A., Chanalet, S., Hugon, F., Bonneville, F., Cognard, C., Chollet, F., Payoux, P., Voisin, T., Delrieu, J., Peiffer, S., Hitzel, A., Allard, M., Zanca, M., Monteil, J., Darcourt, J., Molinier, L., Derumeaux, H., Costa, N., Perret, B., Vinel, C., Caspar-Bauguil, S., 2020. Prospective associations between diffusion tensor imaging parameters and frailty in older adults. J. Am. Geriatr. Soc. 15325415. 68 (5), 1050–1055. Available from: http://onlinelibrary.wiley.com/journal/10.1111/(ISSN)1532-5415. doi:10.1111/jgs.16343.

Mehrabi, F., Béland, F., 2020. Effects of social isolation, loneliness and frailty on health outcomes and their possible mediators and moderators in community-dwelling older adults: a scoping review. Arch. Gerontol. Geriatr. 90, 104119. doi:10.1016/j.archger.2020.104119.

Ní Mhaoláin, A.M., Fan, C.W., Romero-Ortuno, R., Cogan, L., Cunningham, C., Kenny, R., Lawlor, B., 2012. Frailty, depression, and anxiety in later life. Int. Psychogeriatr. 24 (8), 1265–1274. doi:10.1017/S1041610211002110.

Nielsen, B.R., Abdulla, J., Andersen, H.E., Schwarz, P., Suetta, C., 2018. Sarcopenia and osteoporosis in older people: a systematic review and meta-analysis. Eur. Geriatr. Med. 18787657. 9 (4), 419–434. Available from: http://www.springer.com/journal/41999. doi:10.1007/s41999-018-0079-6.

O'Donovan, M.R., Sezgin, D., Liew, A., O'Caoimh, R., 2019. Burden of disease, disability-adjusted life years and frailty prevalence. QJM. 1460-2725. 112 (4), 261–267. doi:10.1093/qjmed/hcy291.

O'Donovan, M., Sezgin, D., Kabir, Z., Liew, A., O'Caoimh, R., 2020. Assessing global frailty scores: development of a Global Burden of Disease-Frailty Index (GBD-FI). IJERPH 17 (16), 5695. 10.3390/ijerph17165695

Oliveira, J.S., Pinheiro, M.B., Fairhall, N., Walsh, S., Franks, T.C., Kwok, W., Bauman, A., Sherrington, C., 2020. Evidence on physical activity and the prevention of frailty and sarcopenia among older people: a systematic review to inform the World Health Organization physical activity guidelines. J. Phys. Act. Health. 15435474. 17 (12), 1247–1258. Available from: https://journals.humankinetics.com/view/journals/jpah/17/12/article-p1247.xml. doi:10.1123/jpah.2020-0323.

Onder, G., Landi, F., Volpato, S., Fellin, R., Carbonin, P., Gambassi, G., Bernabei, R., 2003. Serum cholesterol levels and in-hospital mortality in the elderly. Am. J. Med. 00029343. 115 (4), 265–271. doi:10.1016/s0002-9343(03)00354-1.

Polito, A., Barnaba, L., Ciarapica, D., Azzini, E., 2022. Osteosarcopenia: a Narrative Review on Clinical Studies. Int. J. Mol. Sci. 1422-0067. 23 (10), 5591. doi:10.3390/ijms23105591.

Rahman, S., 2018. Living Well with Frailty: From Assets and Deficits to Resilience Living Well with Frailty: From Assets and Deficits to Resilience. Taylor and Francis Inc, pp. 1–176. Available from: https://www.routledge.com/products/9781351402118. doi:10.4324/9780203732694.

Rockwood, K., Howlett, S.E., 2018. Fifteen years of progress in understanding frailty and health in aging. BMC Med. 17417015. 16 (1). Available from: http://www.biomedcentral.com/bmcmed/. doi:10.1186/s12916-018-1223-3.

Rolland, Y., Cesari, M., Fielding, R.A., Reginster, J.Y., Vellas, B., Cruz-Jentoft, A.J., 2021. Osteoporosis in frail older adults: recommendations for research from the ICFSR Task Force 2020. J. Frailty Aging. 22734309. 10 (2), 1–8. doi:10.14283/jfa.2021.4.

Scafoglieri, A., Clarys, J.P., 2018. Dual energy X-ray absorptiometry: gold standard for muscle mass? J. Cachexia Sarcopenia Muscle. 21906009. 9 (4), 786–787. Available from: http://onlinelibrary.wiley.com/journal/10.1007/13539.2190-6009. doi:10.1002/jcsm.12308.

Simó-Servat, A., Guevara, E., Perea, V., Alonso, N., Quirós, C., Puig-Jové, C., Barahona, M.J., 2023. Role of muscle ultrasound for the study of frailty in elderly patients with diabetes: a pilot study. Biology. 20797737. 12 (6). Available from: www.mdpi.com/journal/biology. doi:10.3390/biology12060884.

Strandberg, T., Pitkälä, K., Tilvis, R., 2011. Frailty in older people. Eur. Geriatr. Med. 2 (6), 344–355. doi:10.1016/j.eurger.2011.08.003.

Tavoian, D., Ampomah, K., Amano, S., Law, T.D., Clark, B.C., 2019. Changes in DXA-derived lean mass and MRI-derived cross-sectional area of the thigh are modestly associated. Sci. Rep. 20452322. 9 (1). Available from: www.nature.com/srep/index.html. doi:10.1038/s41598-019-46428-w.

Taylor, J.A., Greenhaff, P.L., Bartlett, D.B., Jackson, T.A., Duggal, N.A., Lord, J.M., 2023. Multisystemphysiological perspective of human frailty and itsmodulation by physical activity. Physiol. Rev. 15221210. 103 (2), 1137–1191. Available from: https://physrev.physiology.org. doi:10.1152/physrev.00037.2021.

Uchai, S., Andersen, L.F., Johansson, J., Hopstock, L.A., Hjartåker, A., 2007. Dual-energy X-ray absorptiometry derived adiposity measures and pre-frailty/frailty among norwegian adults: the Tromsø Study. J. Nutr., 1–8.

Van Damme, J., Neiterman, E., Oremus, M., Lemmon, K., Stolee, P., 2020. Perspectives of older adults, caregivers, and healthcare providers on frailty screening: a qualitative study. BMC Geriatr. 1471-2318. 20 (1). doi:10.1186/s12877-020-1459-6.

Wleklik, M., Uchmanowicz, I., Jankowska, E.A., Vitale, C., Lisiak, M., Drozd, M., Pobrotyn, P., Tkaczyszyn, M., Lee, C., 2020. Multidimensional approach to frailty. Front. Psychol. 11. doi:10.3389/fpsyg.2020.00564.

Xie, W.Q., Xiao, G.L., Hu, P.W., He, Y.Q., Lv, S., Xiao, W.F., 2020. Possible sarcopenia: early screening and intervention-narrative review. Ann. Palliat. Med. 22245839. 9 (6), 4283–4293. Available from: http://apm.amegroups.com/article/viewFile/55264/pdf. doi:10.21037/apm-20-967.

Chapter 15

Evaluation of the effects of exercise on osteoporosis using ultrasound

Prateek Srivastav[a], Leena R. David[b,c], Rajagopal Kadavigere[d], Suresh Sukumar[c], Sneha Ravichandran[c], Berna Uzun[e], Dilber Uzun Ozsahin[b,e,f], and Ilker Ozsahin[e]

[a]*Department of Physiotherapy, Manipal Academy of Higher Education, Manipal College of Health Professions, Manipal, India*
[b]*Medical Diagnostic Imaging Department, College of Health Sciences, University of Sharjah, Sharjah, United Arab Emirates*
[c]*Department of Medical Imaging Technology, Manipal College of Health Professions, Manipal Academy of Higher Education, Manipal, Karnataka, India*
[d]*Department of Radiodiagnosis and Imaging, Kasturba Medical College, Manipal, India*
[e]*Irfan Suat Gunsel Operational Research Institute, Near East University, Nicosia, Turkish Republic of Northern Cyprus (TRNC), Turkey*
[f]*Research Institute for Medical and Health Sciences, University of Sharjah, Sharjah, United Arab Emirates*

1 Introduction

Osteoporosis (OP) is a silent skeletal epidemic caused by low bone mass and microstructural deterioration of bone, resulting in increased fragility and fracture susceptibility (Sözen et al., 2019). This progressive condition is characterized by the loss of bone density and quality, leading to a heightened risk of fractures, particularly in the hip, spine, and wrist. OP primarily affects older adults, especially postmenopausal women, due to hormonal changes that accelerate bone loss. However, it is not limited to this demographic, as men and younger individuals with certain medical conditions or lifestyle factors can also develop OP. The disease results from periodic bone loss and reduced bone strength, mainly in the cancellous bone, which is more porous and metabolically active than the cortical bone (Sözen et al., 2019).

The etiology of OP is multifactorial, involving genetic predisposition, hormonal changes, nutritional deficiencies, and lifestyle factors. Reduced levels of estrogen in postmenopausal women are a significant contributor to bone loss, as estrogen plays a critical role in maintaining bone density. In addition, inadequate intake of calcium and vitamin D, excessive alcohol consumption, smoking, sedentary lifestyle, and chronic diseases such as rheumatoid arthritis or diabetes can further exacerbate bone loss. Long-term use of certain medications, including corticosteroids, proton pump inhibitors, and anticonvulsants, can also contribute to OP by affecting bone metabolism.

Currently, the clinical management of OP primarily revolves around pharmacological interventions. Medications such as bisphosphonates, selective estrogen receptor modulators, denosumab, and teriparatide are commonly prescribed to slow bone loss, enhance bone formation, and reduce fracture risk. While these treatments are effective, they come with potential side effects, such as gastrointestinal issues, atypical fractures, and osteonecrosis of the jaw, which have raised concerns about long-term use. Therefore, nonpharmacological approaches, particularly exercise, have gained attention as complementary strategies for OP management.

Exercise is considered the second line of treatment, while medicine remains the mainstream treatment for OP. Research suggests that exercise can have effects comparable to those of pharmacological interventions in osteoporotic patients (Hoffmann et al., 2023). Studies have demonstrated that engaging in regular physical activity can enhance bone mineral density (BMD) through new bone formation and remineralization. This is particularly important as increased BMD can help mitigate the risk of fractures and improve overall skeletal health.

Weight-bearing and resistance exercises are particularly beneficial for individuals with OP. Weight-bearing exercises—such as walking, jogging, dancing, and stair climbing—stimulate bone formation by subjecting the skeletal system to mechanical loading. This mechanical stress activates osteoblasts, the bone-forming cells, promoting bone deposition and strengthening the bone matrix. Similarly, resistance exercises—including weightlifting and resistance band

Advances in Medical Imaging. https://doi.org/10.1016/B978-0-443-28967-5.00016-2

training—enhance muscle strength, which in turn exerts positive forces on bones, leading to improved bone density and structural integrity.

Beyond its direct impact on bone health, exercise offers several additional benefits for individuals with OP. It can increase muscle strength and flexibility, which helps offload the osteoporotic bone and thereby reduces mechanical wear and tear in joints (Hoffmann et al., 2023). Moreover, improved muscle function can enhance postural stability, reducing the likelihood of falls—a major concern for individuals with OP. Falls are a leading cause of osteoporotic fractures, particularly hip fractures, which can have devastating consequences, including loss of mobility, reduced independence, and increased mortality risk.

Exercise has also been shown to improve balance, coordination, and proprioception (Bergland et al., 2011). These improvements further reduce the risk of falls and fractures, thereby enhancing overall mobility and quality of life. In addition, engaging in regular physical activity can alleviate pain associated with OP. Many osteoporotic patients experience chronic pain due to vertebral fractures or musculoskeletal strain. Exercise-induced endorphin release, combined with improved muscular support around fragile bones, can help reduce pain perception and enhance overall well-being.

Given these benefits, it is crucial to explore the impact of exercise on OP using qualitative assessment tools. Researchers and healthcare providers employ various qualitative methods, such as patient interviews, focus groups, and observational studies, to gain insights into patients' experiences, adherence to exercise regimens, and perceived benefits. Understanding patients' perspectives can inform the development of tailored exercise programs that are more effective and sustainable in real-world settings.

At present, the clinical diagnosis of OP relies mainly on BMD measurements. BMD can be precisely and accurately assessed using various imaging techniques, including single gamma photon ultrasound (US), single photon absorptiometry, dual-energy X-ray absorptiometry (DXA), quantitative X-ray computed tomography (QCT), and dual gamma photon absorptiometry (McClung et al., 2021). DXA is considered the gold standard for diagnosing OP due to its high precision and reliability. It measures BMD at critical sites such as the lumbar spine and hip, providing valuable information for assessing fracture risk and monitoring disease progression.

Despite their accuracy, these imaging techniques have certain limitations. They are often expensive, require specialized equipment, and expose patients to varying levels of radiation. In addition, accessibility can be a concern, particularly in resource-limited settings where advanced imaging modalities may not be readily available. As a result, alternative diagnostic approaches, such as biochemical markers of bone turnover and clinical risk assessment tools like the FRAX score, are sometimes used to supplement BMD measurements and improve OP risk stratification.

Magnetic resonance imaging (MRI) and conventional X-ray imaging are also utilized in the clinical evaluation of OP. While MRI is primarily used for assessing bone marrow composition and detecting osteoporotic fractures, X-ray imaging can help identify vertebral compression fractures, which are common in individuals with OP (McClung et al., 2021). These imaging modalities are more accessible and financially viable compared to some of the more advanced techniques, making them valuable tools in OP management.

In addition to medical diagnostics and treatments, lifestyle modifications play a crucial role in preventing and managing OP. A balanced diet rich in calcium and vitamin D is essential for maintaining bone health. Dairy products, leafy greens, fortified foods, and exposure to sunlight can help ensure adequate calcium and vitamin D intake. In addition, limiting alcohol consumption, quitting smoking, and engaging in regular physical activity can further support bone health and reduce the risk of OP-related fractures.

OP is a prevalent and potentially debilitating condition characterized by low bone mass and structural deterioration of bone tissue. While pharmacological treatments remain the cornerstone of OP management, exercise has emerged as a vital complementary approach. Regular physical activity, particularly weight-bearing and resistance exercises, can enhance BMD, improve muscle strength and balance, reduce pain, and lower fracture risk. Given the limitations of traditional diagnostic methods, alternative assessment tools, including MRI and X-ray, offer accessible options for evaluating OP. Lifestyle modifications, including a nutrient-rich diet and healthy lifestyle choices, further contribute to bone health. Understanding the impact of exercise on OP through qualitative assessment tools can provide valuable insights into patient experiences and help tailor effective intervention strategies. By integrating exercise, appropriate medical treatments, and lifestyle modifications, individuals with OP can achieve better health outcomes and an improved quality of life.

2 Ultrasound in osteoporosis

Quantitative US (QUS) is an imaging technique that does not require invasive procedures and is used to evaluate bone health, particularly in diagnosing and monitoring OP (Glüer, 1997). It assesses bone density and structural integrity by measuring the transmission and reflection of high-frequency sound waves through the tissue (Cloutier et al., 2021). QUS is a radiation-free and cost-effective alternative to DXA, which exposes individuals to ionizing radiation, making it widely utilized (Cloutier et al., 2021). Key points regarding QUS's application in assessing OP include:

- **Measurement parameters:** QUS offers various parameters such as echo density, texture, shape and contour, echogenicity, boundary definition, and dynamic assessment for bone health evaluation.
- **Site of measurement:** QUS examinations can be performed at different sites, including the calcaneus (heel) or tibia (shin). Calcaneus-based assessments have emerged as more prevalent and have proven efficacy in predicting fracture risk (Barkmann, no date).
- **Advantages of QUS:** Compared to other imaging modalities, QUS is portable, cost-effective, and does not expose patients to radiation. This makes it particularly valuable in large-scale screenings and for use in regions with limited access to advanced diagnostic tools.
- **Limitations:** While QUS is effective in assessing bone quality, it does not provide the same level of detail as DXA or QCT. Therefore, it is often used as a preliminary screening tool rather than a definitive diagnostic method.

In addition to medical diagnostics and treatments, lifestyle modifications play a crucial role in preventing and managing OP. A balanced diet rich in calcium and vitamin D is essential for maintaining bone health. Dairy products, leafy greens, fortified foods, and exposure to sunlight can help ensure adequate calcium and vitamin D intake. In addition, limiting alcohol consumption, quitting smoking, and engaging in regular physical activity can further support bone health and reduce the risk of OP-related fractures.

Understanding the impact of exercise on OP through qualitative assessment tools can provide valuable insights into patient experiences and help tailor effective intervention strategies. By integrating exercise, appropriate medical treatments, and lifestyle modifications, individuals with OP can achieve better health outcomes and an improved quality of life.

3 Advantages of using quantitative ultrasound for OP

QUS offers numerous significant advantages for the assessment of OP, establishing it as an increasingly valuable tool in clinical practice. Most notably, QUS is distinguished by its noninvasive nature, as it does not involve ionizing radiation (Hans et al., 1996). This radiation-free characteristic makes QUS particularly suitable for vulnerable populations such as pregnant women, children, and patients requiring frequent monitoring. The absence of radiation exposure allows for repeated measurements without cumulative health risks, enabling clinicians to safely track bone changes over time and evaluate treatment efficacy at regular intervals.

The portability of QUS devices represents another considerable advantage in the clinical management of OP. Unlike large, stationary imaging equipment such as DXA or CT scanners, QUS machines are compact and lightweight, facilitating their deployment in various healthcare settings including primary care facilities, community health centers, and rural clinics. This portability dramatically improves accessibility to bone health assessment in underserved areas and resource-limited regions where traditional imaging modalities might be unavailable or impractical. QUS devices can also be effectively used in home-based healthcare settings, particularly beneficial for elderly or immobile patients who face challenges in traveling to specialized imaging centers.

From an economic perspective, QUS offers significant cost advantages compared to other bone assessment technologies. The equipment is substantially less expensive to purchase, operate, and maintain than DXA or CT scanners. QUS examinations are typically more affordable for patients and healthcare systems, reducing financial barriers to OP screening and monitoring. This cost-effectiveness makes QUS an attractive option for large-scale population screening programs, especially in health systems with limited financial resources.

Beyond these practical advantages, QUS provides unique insights into bone quality that complement the information obtained from traditional BMD measurements. While DXA primarily assesses bone quantity (mineral density), QUS parameters such as Broadband US Attenuation (BUA) and Speed of Sound (SOS) offer information about bone microarchitecture, elasticity, and strength. These parameters reflect the structural properties of bone that contribute to its mechanical competence and fracture resistance. Studies have demonstrated that QUS measurements can predict fracture risk independently of BMD, suggesting that QUS captures aspects of bone fragility not identified by DXA alone (Njeh et al., 1997).

QUS measurements at peripheral skeletal sites, particularly the calcaneus (heel bone), have shown strong correlations with fracture risk at major osteoporotic fracture sites such as the hip and spine. The calcaneal QUS measurement is particularly advantageous because the heel bone contains a high proportion of trabecular bone, which is metabolically active and responds more rapidly to changes in bone metabolism than cortical bone. This makes calcaneal QUS sensitive to early changes in bone status and potentially valuable for early detection of OP.

The simplicity and speed of QUS examinations enhance patient comfort and compliance. Procedures typically take only a few minutes to complete and do not require special preparation or recovery time. This efficiency improves clinical workflow and allows for higher patient throughput in busy clinical settings. In addition, the nonintimidating nature of US technology, compared to more imposing imaging equipment, may reduce anxiety in patients undergoing bone assessment, potentially increasing screening participation rates.

In summary, QUS technology offers a radiation-free, portable, cost-effective, and patient-friendly approach to OP assessment that provides valuable complementary information about bone quality and fracture risk. These advantages position QUS as an increasingly important tool in comprehensive bone health evaluation, particularly in settings where access to traditional imaging modalities is limited.

4 Limitations of using QUS for OP

While QUS offers numerous advantages in OP assessment, it also presents several significant limitations that must be carefully considered when interpreting results and making clinical decisions. The primary limitation of QUS lies in its standardization challenges across different devices and manufacturers (Guglielmi and de Terlizzi, 2009). Unlike DXA, which has well-established standardization protocols, QUS devices vary considerably in their technology, measurement parameters, and reference databases. This heterogeneity makes direct comparison of results between different QUS devices problematic, potentially leading to inconsistent diagnoses and treatment recommendations. The lack of universal calibration standards complicates the establishment of broadly applicable diagnostic thresholds and treatment initiation criteria, limiting the interchangeability of QUS measurements in clinical practice.

Technical limitations further constrain the utility of QUS in certain populations. In obese individuals or those with significant edema or soft tissue covering the measurement sites, US waves may encounter increased attenuation, potentially compromising measurement accuracy (Guglielmi and de Terlizzi, 2009). The variability in soft tissue composition between individuals introduces another layer of complexity in standardizing QUS measurements. In addition, conditions affecting the peripheral skeleton, such as localized inflammation, fractures, or deformities at the measurement site, can significantly distort QUS readings and lead to erroneous interpretations of bone status.

From a diagnostic perspective, QUS faces considerable limitations in its ability to provide comprehensive skeletal assessment. Unlike DXA or QCT, which can evaluate clinically relevant central skeletal sites like the lumbar spine and hip, QUS typically measures peripheral sites such as the calcaneus, radius, or phalanges. While these peripheral measurements correlate with overall fracture risk to some degree, they may not accurately reflect bone quality at the axial skeleton where osteoporotic fractures most commonly occur and have the greatest clinical impact. This site-specific limitation reduces the diagnostic precision of QUS for identifying individuals at highest risk for clinically significant fractures.

Another critical limitation is QUS's inability to provide detailed information about bone microarchitecture (Guglielmi and de Terlizzi, 2009). While QUS parameters indirectly reflect some aspects of bone structure, they cannot visualize or quantify specific microarchitectural features such as trabecular thickness, connectivity, and cortical porosity. These microarchitectural parameters are increasingly recognized as crucial determinants of bone strength and fracture resistance independent of BMD. The inability to assess these features represents a significant constraint on QUS's capacity to comprehensively evaluate fracture risk.

QUS also demonstrates limited sensitivity in monitoring treatment-induced changes in bone quality. The relatively modest changes in bone properties that occur with OP treatments may fall below the detection threshold of current QUS technology. This reduced sensitivity, combined with measurement variability, makes QUS less reliable than DXA for monitoring treatment response and guiding therapeutic decisions in clinical practice.

Interpretation challenges further complicate the clinical application of QUS. The World Health Organization diagnostic criteria for OP, based on T-scores derived from DXA measurements, cannot be directly applied to QUS parameters due to different measurement technologies and bone properties being assessed. While device-specific diagnostic thresholds have been proposed, their clinical validation remains incomplete, leading to uncertainty in diagnostic classification and treatment decisions based solely on QUS measurements.

Given these limitations, QUS is generally considered a screening tool rather than a definitive diagnostic method for OP. DXA remains the gold standard for diagnosis and is often required to confirm findings from QUS assessments before initiating treatment (Guglielmi and de Terlizzi, 2009). This two-step approach adds complexity to the diagnostic pathway and potentially increases healthcare costs and patient burden.

5 Clinical uses of using QUs for OP

QUS has established several important roles in clinical practice for OP management, offering unique capabilities that complement traditional assessment methods. One of the most significant clinical applications of QUS is in fracture risk assessment, where it provides valuable predictive information independent of BMD measurements (Guglielmi and de Terlizzi, 2009). Multiple large-scale prospective studies have demonstrated that QUS parameters, particularly calcaneal measurements, can effectively predict the risk of major osteoporotic fractures including hip, vertebral, and wrist fractures. The Stiffness Index (SI), BUA, and SOS measurements have shown strong associations with fracture incidence, with predictive capabilities comparable to DXA in some populations. This predictive value makes QUS particularly useful in initial risk stratification to identify individuals who may benefit from more comprehensive assessment or preventive interventions.

In primary healthcare settings that have limited access to specialized imaging technologies such as DXA, QUS serves as an invaluable first-line screening tool for OP (Lin, 2000). The accessibility, affordability, and ease of use of QUS devices enable widespread implementation of bone health assessment in community clinics, rural health centers, and other primary care facilities. This expanded access is crucial for addressing the significant underdiagnosis of OP, particularly in medically underserved areas. Screening protocols typically involve using device-specific cutoff values to categorize patients into risk groups, with those exceeding threshold values being referred for further assessment with DXA or other modalities. This tiered approach optimizes resource utilization while ensuring that high-risk individuals receive appropriate diagnostic evaluation.

QUS also plays an important role in monitoring skeletal changes in response to therapeutic interventions or lifestyle modifications. While not as sensitive as DXA for detecting small changes in bone properties, QUS can track significant alterations in bone quality over time, providing valuable feedback on treatment efficacy. This monitoring capability is particularly useful in settings where repeated DXA measurements are impractical due to limited availability, cost constraints, or concerns about cumulative radiation exposure. Serial QUS measurements can help clinicians assess whether interventions are producing the desired effects on bone health and guide decisions regarding treatment continuation or modification.

The interpretation of QUS results in clinical practice typically involves comparing patient measurements against population-based reference data (Lin, 2000). Most modern QUS devices include integrated software that automatically calculates T-scores (comparison to young adult mean values) and Z-scores (comparison to age-matched mean values), similar to the reporting framework used in DXA. These standardized scores facilitate risk classification and clinical decision-making. However, it is important to note that QUS-derived T-scores cannot be directly equated with DXA-derived T-scores for diagnostic purposes, as they reflect different bone properties measured at different skeletal sites. Device-specific thresholds have been established to guide interpretation, though these vary between manufacturers and technologies.

QUS has demonstrated particular utility in specific patient populations where conventional assessment methods present challenges. In pediatric patients, QUS offers radiation-free bone assessment during critical developmental periods when radiation exposure is especially concerning. In pregnant women, QUS provides a safe method for evaluating bone status without fetal radiation exposure. For very elderly or immobile patients who may have difficulty positioning for DXA scans, the simpler positioning requirements of QUS enhance accessibility. In addition, QUS has proven valuable in longitudinal epidemiological studies where its portability and lack of radiation exposure facilitate repeated measurements in large cohorts.

Beyond OP, QUS has expanded into assessing bone quality in related conditions such as osteopenia (lower than normal bone density that has not reached OP thresholds), secondary OP due to medications or medical conditions, and monitoring bone development in adolescents. The integration of QUS into comprehensive fracture risk assessment algorithms that incorporate clinical risk factors has further enhanced its clinical utility, providing a more holistic approach to bone health evaluation that considers multiple determinants of fracture risk.

QUS provides valuable insights into bone health when other methods, such as DXA, might not be feasible or repeated measurements are required (Hans et al., 1996). However, seeking guidance from healthcare professionals for personalized assessment and management of OP is essential (Henderson et al., 1998).

6 Exercise-induced changes in OP

OP is a condition characterized by reduced bone mass and structural deterioration, leading to an increased risk of fractures. Exercise plays a crucial role in both the prevention and management of OP by enhancing bone density, strength, and overall musculoskeletal health (Kesaniemi et al., 2001). The beneficial effects of exercise on OP are extensive and include the following:

1. **Enhanced bone density:** Engaging in weight-bearing exercises such as walking, running, and resistance training promotes bone remodeling, ultimately leading to increased bone density (Hong and Kim, 2018). This effect is particularly

significant for individuals with OP, as higher bone density strengthens bones and reduces fracture risk (Henderson et al., 1998). Studies indicate that consistent participation in high-impact activities enhances BMD more effectively than nonweight-bearing exercises.

2. **Improved bone strength:** Bone strength is crucial for reducing susceptibility to fractures. Resistance training, including weightlifting and resistance bands, stimulates mechanical stress responses in bones, encouraging bone tissue development and improving structural integrity (Kannus et al., 1997). Over time, these adaptations lead to stronger bones, reducing the likelihood of fractures and skeletal injuries.
3. **Enhanced muscular strength:** Developing strong muscles provides better support for bones, lowering the risk of falls and fractures (Benichou and Lord, 2016). Resistance exercises such as squats, lunges, and push-ups increase muscle mass and improve overall strength, helping individuals with OP maintain mobility and independence.
4. **Improved balance and coordination:** Exercise programs incorporating balance and coordination exercises, such as tai chi and yoga, enhance proprioception and reduce fall risk among individuals with OP (Thomas et al., 2019). Since falls often result in fractures, particularly in older adults, improving balance through targeted activities can be a critical component of OP management.
5. **Decreased fracture risk:** Regular physical activity positively influences bone density, muscular strength, and balance, all of which contribute to a lower risk of fractures (Rosal et al., 2010). Studies have shown that individuals who engage in consistent weight-bearing and strength-training exercises experience a significantly lower incidence of osteoporotic fractures.
6. **Pain management:** Exercise can serve as a natural pain management strategy for individuals with OP. Strengthening muscles and improving joint flexibility through controlled movements can alleviate discomfort associated with OP-related bone loss (Greenwell, 2010). Low-impact exercises such as swimming and cycling can also relieve stiffness and reduce pain intensity.
7. **Improved joint health:** Weight-bearing and low-impact exercises help maintain joint health by increasing flexibility and reducing stiffness, which is especially beneficial for individuals with OP (Clarkson et al., 1992). Stretching routines, yoga, and pilates can aid in maintaining joint mobility, reducing inflammation, and preventing further degeneration.
8. **Enhanced bone remodeling:** Bone remodeling is an ongoing process where old bone tissue is replaced with new, healthier bone. Exercise accelerates this process, helping to remove weakened bone and encourage the formation of stronger bone structures (Drinkwater, 1993). Resistance training and weight-bearing activities stimulate osteoblast activity, leading to improved skeletal health.
9. **Hormonal advantages:** Physical activity has a positive effect on hormonal balance, particularly in postmenopausal women, who are at a higher risk for OP due to declining estrogen levels (Nguyen, 2017). Regular exercise can help maintain estrogen levels, which are essential for bone metabolism and overall bone health. Weight-bearing activities, in particular, have been shown to have protective effects against hormone-related bone loss.
10. **Psychological well-being:** Engaging in regular exercise not only improves physical health but also enhances psychological well-being. Physical activity has been linked to elevated mood, reduced stress levels, and lower anxiety, all of which are important for individuals managing chronic conditions such as OP (NA, 1995). Exercise-induced endorphin release promotes a sense of well-being, improving overall quality of life.

Exercise plays a fundamental role in both preventing and managing OP. By improving bone density, strength, muscular support, and coordination, exercise reduces fracture risk and enhances overall mobility. In addition, it contributes to pain relief, joint health, hormonal balance, and psychological well-being. Therefore, incorporating a structured exercise program—including weight-bearing, resistance, balance, and flexibility exercises—into daily routines is essential for individuals at risk of or living with OP. Continued research and tailored exercise interventions will further optimize bone health strategies for those affected by this condition.

7 Case study

A case study demonstrating the utility of QUS in tracking bone changes with exercise.

Patient profile

- Name: Sarah
- Age: 55 years
- Gender: Female
- Medical history: No history of fractures or chronic medical conditions. Family history of OP.

Baseline assessment

- At baseline, Sarah underwent a QUS assessment using a calcaneal US device. The following parameters were measured:
- SOS: 1550 m/s
- BUA: 70 dB/MHz
- SI: 75

Exercise intervention

- Sarah participated in a 12-month exercise program under the supervision of a certified fitness trainer.
- Exercise included weight-bearing activities (e.g., brisk walking, stair climbing) and resistance training (e.g., weight lifting).
- She exercised three times a week for 60 min per session.

Follow-up assessments

QUS assessments were conducted at 3-month intervals during the exercise program.

Results

After 3 months, the SOS was measured at 1570 m/s, the BUA was 72 dB/MHz, and the SI was 80. After 6 months, the SOS increased to 1585 m/s, the BUA to 74 dB/MHz, and the SI to 85. After 9 months, the SOS further increased to 1595 m/s, the BUA to 76 dB/MHz, and the SI to 90. Finally, after 12 months, the SOS reached 1600 m/s, the BUA was 78 dB/MHz, and the SI was 95.

Interpretation

The QUS measurements show a progressive increase in bone quality parameters (SOS, BUA, and SI) over the 12-month exercise program; in a 55-year-old female, the normal values for SOS, BUA, and SI may still vary depending upon ethnicity, body composition, and overall health. This improvement suggests an increase in bone density and structural integrity. The changes observed in QUS parameters correlate with the expected positive effects of exercise on bone health. The patient's risk of osteoporotic fractures has likely decreased due to these improvements.

This case study illustrates the utility of QUS in tracking bone changes in response to exercise. Sarah's QUS measurements consistently improved over the 12-month exercise program, suggesting enhanced bone density and quality. QUS can serve as a valuable tool for monitoring the effectiveness of exercise interventions in improving bone health and reducing the risk of OP-related fractures.

8 Limitations

Despite the valuable insights gained from this review, several limitations must be acknowledged. One of the primary concerns is the sample size of the included studies. Many studies analyzing the effects of traumatic brain injury (TBI) have relatively small sample populations, which may limit the generalizability of the findings. Small sample sizes increase the risk of statistical errors and reduce the ability to detect subtle but clinically significant differences in outcomes. Future studies should focus on larger, more diverse cohorts to improve the robustness of the results.

Another key limitation is the heterogeneity in study design and methodology. The included studies vary in terms of population characteristics, injury severity classifications, treatment protocols, and outcome measures. This variability makes it challenging to compare results across studies and may introduce inconsistencies in the overall conclusions. Standardized assessment tools and uniform criteria for evaluating TBI-related outcomes are essential for improving the comparability and reliability of future research.

Potential biases also pose a challenge to the validity of findings. Selection bias, reporting bias, and publication bias may have influenced the overall conclusions of this review. Studies with positive findings are more likely to be published than those with negative or inconclusive results, which can skew the overall interpretation of TBI outcomes and interventions. In addition, retrospective studies relying on self-reported data may introduce recall bias, further impacting the accuracy of results. Addressing these biases requires greater transparency in study design, rigorous peer review processes, and the inclusion of unpublished or negative results to provide a more balanced perspective.

A significant limitation of existing research is the lack of long-term follow-up data. While many studies focus on the immediate and short-term effects of TBI, fewer have examined long-term recovery patterns and secondary complications that may arise years after the initial injury. Without sufficient longitudinal data, it is difficult to fully understand the prolonged impact of TBI on cognitive function, psychological well-being, and overall quality of life. Future research should prioritize extended follow-up periods to better capture the long-term consequences of TBI and identify factors that contribute to sustained recovery or decline.

Lastly, external factors such as differences in healthcare access and treatment protocols across regions may affect study outcomes. Variations in medical care, rehabilitation services, and socioeconomic conditions can influence recovery trajectories, making it challenging to generalize findings globally. Efforts should be made to ensure that future research considers these external influences and includes diverse populations to enhance the applicability of results across different healthcare settings.

9 Future directions

US's role in evaluating OP is an area of research that continues to advance, aiming to enhance its precision, broaden its applications, and address existing limitations. Several future directions and areas of investigation can be explored for US assessment of OP. Researchers can focus on refining US technology to provide more detailed insights into bone quality by examining trabecular microarchitecture and cortical bone properties. Increasing the portability and user-friendliness of US devices can facilitate their use in various healthcare settings, including primary care. Ongoing efforts are required to establish standardized procedures and measurements for QUS assessments. This will ensure consistency and reliability across different devices as well as operators. Further research is needed to validate US measurements' accuracy and clinical usefulness in assessing OP. Validation studies comparing US with established methods such as DXA and high-resolution imaging techniques would provide valuable insights into its effectiveness. It is important to develop robust predictive models incorporating US measurements and clinical risk factors to improve fracture risk prediction. This would enable a more accurate estimation of fracture risk. Long-term studies are necessary to determine the ability of US to predict future fractures over time and its role in guiding treatment decisions for OP patients. In addition, investigating the use of US in assessing bone health and monitoring changes in pediatric and adolescent populations, especially those with conditions affecting bone development, could provide valuable information about its efficacy within specific population groups. Research can be conducted on the use of US as a potential tool for monitoring bone changes in postmenopausal women, who face an elevated risk for OP. In addition, studies explore how US can be utilized to track the effects of various treatments for OP, such as medications and exercise interventions. Assessing bone quality and fracture risk reduction is another area where combining US with other imaging modalities like MRI or high-resolution peripheral quantitative computed tomography(HR-pQCT)could lead to a more comprehensive evaluation of bone health, including microarchitecture. Moreover, need to leverage artificial intelligence (AI) and machine learning algorithms to analyze US data effectively and extract valuable information regarding bone quality and fracture risk. Integration of AI technology may enhance accuracy and efficiency during US assessments. Patient-centered outcomes, including quality of life and functional outcomes in relation to US measurements, can be useful. In addition, the cost-effectiveness evaluation of incorporating US into OP screening and monitoring programs can be helpful compared to other imaging methods. Furthermore, there is a need to improve global accessibility by expanding the use of US for OP assessment in resource-limited settings where DXA and advanced imaging techniques may be less accessible.

10 Conclusion

TBI remains a critical public health issue with widespread implications for individuals, families, and healthcare systems. The findings from this review highlight the significant short- and long-term consequences of TBI, including cognitive impairments, emotional disturbances, and functional disabilities. These outcomes emphasize the urgent need for enhanced preventive strategies, improved acute care management, and comprehensive rehabilitation programs to support recovery and improve the quality of life for affected individuals.

Despite advancements in medical research, many challenges remain in understanding the full spectrum of TBI effects, particularly concerning long-term recovery trajectories and secondary complications. The variability in study methodologies, patient populations, and outcome measures underscores the necessity for standardized research approaches to ensure more reliable and comparable data. In addition, emerging therapeutic interventions, such as neuroprotective treatments and cognitive rehabilitation strategies, require further investigation through large-scale, well-controlled clinical trials.

Future research should prioritize longitudinal studies to assess the long-term effects of TBI and identify factors that contribute to optimal recovery. There is also a need for multidisciplinary collaboration among neurologists, psychologists, rehabilitation specialists, and public health professionals to develop more effective treatment frameworks. Furthermore, increasing public awareness about TBI prevention, particularly in high-risk groups such as athletes and military personnel, is crucial in reducing the overall incidence of these injuries.

In conclusion, while significant progress has been made in understanding TBI, there is still much to be explored to improve patient outcomes. Continued research, along with advancements in medical technology and rehabilitation strategies, will be essential in addressing the complexities of TBI and ensuring better long-term care for those affected.

References

Benichou, O., Lord, S.R., 2016. Rationale for strengthening muscle to prevent falls and fractures: a review of the evidence. Calcif. Tissue Int. 98 (6), 531–545. doi:10.1007/s00223-016-0107-9.

Bergland, A., Thorsen, H., Kåresen, R., 2011. Effect of exercise on mobility, balance, and health-related quality of life in osteoporotic women with a history of vertebral fracture: a randomized, controlled trial. Osteoporos. Int. 0937-941X. 22 (6), 1863–1871. doi:10.1007/s00198-010-1435-7.

Clarkson, P.M., Nosaka, K., Braun, B., 1992. Muscle function after exercise-induced muscle damage and rapid adaptation. Med. Sci. Sports Exerc. 0195-9131. 24 (5), 512–520. doi:10.1249/00005768-199205000-00004.

Cloutier, G., Destrempes, F., François, Y., 2021. An Tang, Quantitative ultrasound imaging of soft biological tissues: a primer for radiologists and medical physicists. Insights Imaging. 1869-4101. 12 (1). doi:10.1186/s13244-021-01071-w.

Drinkwater, B.L., 1993. Exercise in the prevention of osteoporosis. Osteoporos. Int. 0937-941X. 3 (S1), 169–171. doi:10.1007/bf01621898.

Glüer, C.C., 1997. Quantitative ultrasound techniques for the assessment of osteoporosis: expert agreement on current status. J. Bone Miner. Res. 08840431. 12 (8), 1280–1288. doi:10.1359/jbmr.1997.12.8.1280.

Greenwell, G.R., 2010. Optimum Exercise Intensity for Maximum Benefits. Med. Sci. Sports Exerc. 0195-9131. 42 (10), 73. doi:10.1249/01.mss.0000389395.64041.07.

Guglielmi, G., de Terlizzi, F., 2009. Quantitative Ultrasond in the assessment of Osteoporosis. Eur. J. Radiol. 71 (3), 425–431. doi:10.1016/j.ejrad.2008.04.060.

Hans, D., Dargent-Molina, P., Schott, A.M., Sebert, J.L., Cormier, C., Kotzki, P.O., Delmas, P.D., Pouilles, J.M., Breart, G., Meunier, P.J., 1996. Ultrasonographic heel measurements to predict Hip fracture in elderly women: the EPIDOS prospective study. Lancet. 01406736. 348 (9026), 511–514. doi:10.1016/s0140-6736(95)11456-4.

Henderson, N.K., White, C.P., Eisman, J.A., 1998. The roles of exercise and fall risk reduction in the prevention of osteoporosis. W.B. Saunders, Australia Endocrinology and Metabolism Clinics of North America. 08898529. 27 (2), 369–387. Available from: http://www.endo.theclinics.com/. doi:10.1016/S0889-8529(05)70010-4.

Hoffmann, I., Kohl, M., von Stengel, S., Jakob, F., Kerschan-Schindl, K., Lange, U., Peters, S., Schoene, D., Sieber, C., Thomasius, F., Bischoff-Ferrari, H.A., Uder, M., Kemmler, W., 2023. Exercise and the prevention of major osteoporotic fractures in adults: a systematic review and meta-analysis with special emphasis on intensity progression and study duration. Osteoporos. Int. 0937-941X. 34 (1), 15–28. doi:10.1007/s00198-022-06592-8.

Hong, A.R., Kim, S.W., 2018. Effects of resistance exercise on bone health. Korean Endocrine Society, South Korea Endocrinology and Metabolism. 20935978. 33 (4), 435–444. Available from: https://e-enm.org/. doi:10.3803/EnM.2018.33.4.435.

Kannus, P., Józsa, L., Natri, A., Järvinen, M., 1997. Effects of training, immobilization and remobilization on tendons. Scand. J. Med. Sci. Sports. 0905-7188. 7 (2), 67–71. doi:10.1111/j.1600-0838.1997.tb00121.x.

Kesaniemi, Y.A., Danforth, E.J., Jensen, M.D., Kopelman, P.G., Lefebvre, P., Reeder, B.A., 2001. Dose-response issues concerning physical activity and health: an evidence-based symposium. Med. Sci. Sports Exerc. 01959131. 6, S351-S358S351. https://journals.lww.com/acsm-msse/Pages/issuelist.aspx. doi:10.1097/00005768-200106001-00003.

Lin, E., 2000. Quantitative ultrasound: assessment of osteoporosis and bone status. Med. Sci. Sports Exerc. 0195-9131. 32 (4), 879. doi:10.1097/00005768-200004000-00026.

McClung, M.R., Pinkerton, J.V., Blake, J., et al., 2021. Management of osteoporosis in postmenopausal women: the 2021 position statement of The North American Menopause Society. Menopause. 1530-0374. 28 (9), 973–997. Available from: https://journals.lww.com/menopausejournal. doi:10.1097/GME.0000000000001831.

NA, 1995. ACSM Position Stand on Osteoporosis and Exercise. Med. Sci. Sports Exerc. 0195-9131. 27 (4), i. doi:10.1249/00005768-199504000-00025.

Nguyen, V.H., 2017. Osteoporosis prevention and osteoporosis exercise in community-based public health programs. Osteoporos. Sarcopenia. 24055255. 3 (1), 18–31. doi:10.1016/j.afos.2016.11.004.

Njeh, C.F., Boivin, C.M., Langton, C.M., 1997. The role of ultrasound in the assessment of osteoporosis: a review. Osteoporos. Int. 0937-941X. 7 (1), 7–22. doi:10.1007/bf01623454.

Rosal, M.C., White, M.J., Borg, A., Scavron, J., Candib, L., Ockene, I., Magner, R., 2010. Translational research at community health centers: challenges and successes in recruiting and retaining low-income Latino patients with type 2 diabetes into a randomized clinical trial. Diabetes Educ. 0145-7217. 36 (5), 733–749. doi:10.1177/0145721710380146.

Sözen, T., Özışık, L., Çalık Başaran, N., 2019. An overview and management of osteoporosis. Eur. J. Rheumatol. 2148-4279. 4 (1), 46–56. doi:10.5152/eurjrheum.2016.048.

Thomas, E., Battaglia, G., Patti, A., Brusa, J., Leonardi, V., Palma, A., Bellafiore, M., 2019. Physical activity programs for balance and fall prevention in elderly. Medicine. 0025-7974. 98 (27), e16218. doi:10.1097/md.0000000000016218.

Chapter 16

Role of physical activity and imaging in musculoskeletal stress

Nitika C. Panakkal[a], T. Visakh[a], Nayak Ullas U.[b], Leena R. David[a,c], Rajagopal Kadavigere[d], Berna Uzun[e], Dilber Uzun Ozsahin[c,e,f], Suresh Sukumar[a], Ilker Ozsahin[e], and Sidhiprada Mohapatra[g]

[a]*Department of Medical Imaging Technology, Manipal College of Health Professions, Manipal Academy of Higher Education, Manipal, Karnataka, India*

[b]*Division of Anatomy, Manipal Academy of Higher Education, Manipal, India*

[c]*Medical Diagnostic Imaging Department, College of Health Sciences, University of Sharjah, Sharjah, United Arab Emirates*

[d]*Department of Radiodiagnosis and Imaging, Kasturba Medical College, Manipal, India*

[e]*Irfan Suat Gunsel Operational Research Institute, Near East University, Nicosia, Turkish Republic of Northern Cyprus (TRNC), Turkey*

[f]*Research Institute for Medical and Health Sciences, University of Sharjah, Sharjah, United Arab Emirates*

[g]*Department of Physiotherapy, Manipal College of Health Professions, Manipal Academy of Higher Education, Manipal, India*

1 Introduction

Chronic musculoskeletal conditions, such as Work-Related Musculoskeletal Disorders (WRMSDs), encompass a spectrum of debilitating and often progressive disorders affecting muscles, bones, tendons, ligaments, nerves, blood vessels, and joints. At their core, these conditions manifest as a result of sustained or repetitive mechanical stressors, primarily attributed to occupational activities. WRMSDs are globally challenging chronic disorders, causing long-term disabilities, economic loss, and increasing healthcare expenditures. The global number of musculoskeletal disorders reached 1.3 billion in 2017, leading to 121.3 thousand deaths and 138.7 million disability-adjusted life years (Safiri et al., 2021).

WRMSDs have been defined by the United States Department of Labor "as disorders of the muscles, nerves, tendons, ligaments, joints, cartilage, blood vessels, or spinal discs, in areas such as the neck, shoulder, elbow, forearm, wrist, hand, abdomen (hernia only), back, knee, ankle, and foot associated with exposure to risk factors" (Barr et al., 2004b, as cited in Barr et al., 2004a). These disorders may include muscle strains and tears, ligament sprains, joint and tendon inflammation, pinched nerves, spinal disc degeneration, and medical conditions such as low back pain, tension neck syndrome, carpal tunnel syndrome, rotator cuff syndrome, De Quervain syndrome, trigger finger, tarsal tunnel syndrome, sciatica, epicondylitis, tendinitis, Raynaud's phenomenon, hand-arm vibration syndrome, bursitis, and herniated spinal disc, among others (U.S. Bureau of Labor Statistics, 2020; Garg and Kapellusch, 2011).

In the intricate world of musculoskeletal health, the saga of WRMSDs unfolds at the macroscopic level. Pain and dysfunction manifest within the hidden confines of a person's cells. WRMSDs are prevalent and often debilitating group of conditions. They take root in the very essence of our musculoskeletal tissues, leading to a cascade of cellular-level changes. These microscopic changes form the basis of our understanding of the pathophysiology of MSDs. Thereby, shedding light on how seemingly harmless occupational activities can result in significant harm (Greggi et al., 2024).

This chapter guides readers through an evidence-based and holistic understanding of WRMSDs. It presents a comprehensive view of WRMSDs including evidence-based findings on cellular and systemic changes, as well as the impact of WRMSDs on function, disability, and participation. In addition, the chapter presents the importance of early diagnosis and appropriate imaging techniques. Finally, the chapter ends by summarizing the role of physical activity in maintaining the health and well-being of the workforce, along with the development of targeted interventions to address WRMSDs.

Advances in Medical Imaging. https://doi.org/10.1016/B978-0-443-28967-5.00023-X

2 Physiological changes in work-related musculoskeletal disorders

2.1 Cellular-level changes in WRMSDs

The continuous mechanical stressors encountered in the workplace lead to a series of responses at the cellular level, resulting in cellular adaptations and, at times, damage. These adaptations often begin with alterations in gene expression patterns in musculoskeletal tissues (Aguilar-Agon et al., 2019). Myocytes, the contractile cells of muscles, may undergo either hypertrophy or atrophy, depending on the work demands (Sartori et al., 2021). In cases requiring sustained muscle contractions, hypertrophy is a protective mechanism, enhancing the ability to produce force. Conversely, repetitive micro-trauma can trigger muscle atrophy, a pathological response characterized by reduced muscle fiber size and strength, thereby increasing the risk of injury (Sartori et al., 2021).

Moreover, persistent mechanical stressors can induce inflammation, triggering the release of pro-inflammatory cytokines and chemokines (Barbe and Barr, 2006). These inflammatory mediators can lead to increased vascular permeability, tissue edema, and the infiltration of immune cells, causing a chronic inflammatory state within musculoskeletal tissues (Royall et al., 1989). Inflammation exacerbates pain and contributes to tissue remodeling and fibrosis, impairing standard tissue functionality and sustaining the pathological cycle. Over time, these chronic inflammatory responses can lead to the activation of fibroblasts and the deposition of excessive collagen, resulting in the development of fibrotic tissue, which is less compliant and less functional than healthy musculoskeletal tissue (Barbe and Barr, 2006; Barr and Barbe, 2002).

A study on the effects of high-force tasks on rats suggests that high-force tasks are more physiologically demanding than low-force tasks. The researchers observed increased pathological multijoint movement patterns and inflammatory cytokines in the forearm muscles of the rats. These findings suggest that peripheral tissue inflammation can influence the motor cortex (Barbe and Barr, 2006). The study also explores the role of Tumor Necrosis Factor-alpha (TNF-α) protein levels in muscle mechanical hyperalgesia (Barbe and Barr, 2006). The researchers detected a reduction in grip strength in young rats performing repetitive tasks, possibly due to myofiber degeneration, tendon degenerative changes, and pathological bone formation. The correlation between increased muscle inflammation and reduced grip strength was also discussed in this study (Abdelmagid et al., 2012). Moreover, the effects of the high repetition negligible force task on muscle inflammation and grip strength were also examined. The researchers observed a lowgrade, cyclical inflammatory response in musculotendinous tissues with the performance of the high repetition low force task (Barr and Barbe, 2002).

At the cellular level, the consequences of prolonged or repetitive work related stress on the musculoskeletal system are profound. These changes are influenced by a complex interplay of mechanical stress, inflammation, and the body's adaptive mechanisms. Understanding these cellular level changes is crucial for diagnosing and managing MSDs. It helps to develop effective preventive strategies for overall health and well-being of individuals at risk of WRMSDs (Barr et al., 2004a). Cellular level changes, such as inflammation and sensitization of nociceptors, can trigger systemic responses and contribute to central sensitization (Kuner and Kuner, 2021). The subsequent section will discuss systemic changes (Barbe and Barr, 2006).

2.2 Systemic changes associated with chronic inflammation due to WRMSDs

The pathological alterations linked to WRMSDs encompass changes at the structural level, impacting the integrity of joints, tendons, and ligaments. In particular, the repetitive loading of joints has the potential to induce changes in cartilage composition and structure. This is a classical feature in conditions like osteoarthritis (OA). Cartilage acts as a protective cushion between bones and undergoes degeneration, which leads to a decline in proteoglycan and collagen. Consequently, this results in joint pain, stiffness, and limitations in the range of motion (Loeser et al., 2012). Tendons and ligaments, which connect muscles to bones and stabilize joints, are also susceptible to pathological changes. Tendinopathy, a common feature of WRMSDs, includes microtears, disorganization of collagen, and increased inflammatory markers within tendons. These pathological changes weaken the tendon structure, making it less capable of withstanding mechanical stress and increasing the risk of tendon ruptures. Conversely, ligaments may undergo laxity or even partial tears due to chronic mechanical stress, compromising joint stability and leading to an increased susceptibility to traumatic injuries (Millar et al., 2021).

The activation of inflammatory mediators in peripheral tissues can trigger a systemic inflammatory reaction. These systemic responses can impact the entire body, resulting in physiological changes such as elevated heart rate, altered immune system function (Sanabria-Mazo et al., 2022), cortical changes (Kuner and Kuner, 2021), and psychological responses such as anxiety and depression (Karshikoff et al., 2023). The shift from cellular to systemic changes highlights chronic pain's intricate and interconnected characteristics. Thus, underscoring the importance of a holistic approach to pain management that addresses both local and systemic components.

3 Symptoms in WRMSDs

3.1 Pain in WRMSDs

3.1.1 Definition of pain

Pain is defined as "an unpleasant sensory and emotional experience associated with, or resembling that associated with, actual or potential tissue damage" by the International Association for the Study of Pain (Raja et al., 2020).

3.1.2 Categories of pain

Based on its underlying causes and characteristics, pain can be categorized into several broad categories:

Nociceptive pain arises from the activation of pain receptors, known as nociceptors, in response to tissue damage or injury. It is the most common type of pain. It can be further classified into two categories: somatic nociceptive pain and visceral nociceptive pain. Somatic pain originates from the skin, muscles, or bones, while visceral pain originates from the internal organs (Larocca, 1992; Nwonu, 2022; Abd-Elsayed et al., 2023).

Neuropathic pain: This pain arises from either nerve damage or nerve dysfunction. Pain may manifest as shooting, burning, or electric shock-like sensations. Disorders such as diabetic neuropathy, nerve impingements, and postherpetic neuralgia fall within the purview of this classification (Mitsikostas et al., 2022).

Nociplastic pain is the most recent classification, encompassing pain without discernible structural or nerve-related causes. This category involves alterations in the pain processing of the central nervous system, leading to an amplification of pain and a modification of pain perception. Work-related low back pain and fibromyalgia are conditions that fall under this umbrella (Nijs et al., 2023).

Psychogenic Pain: It is characterized by a significant emotional or psychological component and manifests when emotional distress or psychological factors contribute to the perception of pain, even in the absence of apparent physical causes (Lumley et al., 2011).

Acute and chronic pain: Based on duration, acute pain, typically of brief duration, serves as a significant warning sign of injury or illness. Its presence is vital for survival and tends to subside once the underlying cause has been addressed. In contrast, chronic pain endures for an extended period, often surpassing the anticipated time required for tissue healing. It profoundly impacts an individual's quality of life and necessitates ongoing management and treatment (Treede et al., 2019; Raja et al., 2020).

3.1.3 Peripheral mechanisms of chronic pain

Peripheral pain mechanisms involve processes of the peripheral nervous system and associated tissues. These mechanisms play a crucial role in the initiation and maintenance of chronic pain. Chronic pain often involves the sensitization of peripheral nerves. Inflammatory mediators, released as part of the inflammatory response to tissue injury, can activate nociceptors (sensory nerve fibers in pain affected areas) increasing their sensitivity and response to noxious stimuli (Zheng et al., 2022). Chronic pain can be initiated or exacerbated by ongoing tissue damage and inflammation. Injured or inflamed tissues release prostaglandins and cytokines that sensitize nociceptors and lower their activation threshold (Kidd and Urban, 2001). In some chronic pain conditions, nerve damage or injury (neuropathic pain) leads to abnormal signaling, which can lead to ectopic discharges or spontaneous pain signals originating from damaged nerves (Zheng et al., 2022). Prolonged exposure to noxious stimuli can result in peripheral sensitization, where nociceptors become hyperexcitable.In this state, nociceptors are more likely to fire pain signals even in response to normally nonpainful stimuli. This phenomenon is known as allodynia (Kidd and Urban, 2001). The immune system can also contribute to chronic pain by releasing inflammatory substances and activating immune cells at the site of injury or inflammation. This immune response can perpetuate pain and lead to further tissue damage (Fang et al., 2023). Chronic pain often involves structural and functional changes in peripheral tissues. This can include muscle contracture, fibrosis, joint abnormalities, and changes in skin sensitivity (Zheng et al., 2022). Neurotransmitter release and receptor expression changes at the peripheral nerve endings can modulate pain signals. Substances like substance P and glutamate play a role in amplifying pain signals (Dehghan et al., 2024). Chronic pain conditions, particularly those associated with inflammation, can lead to alterations in blood flow and microcirculation in affected tissues, contributing to pain and tissue damage (Zheng et al., 2022).

There has been a paradigm shift in understanding pain, and its management in recent years. The understanding of chronic pain has paved the way for central sensitization. This concept is of great importance in managing pain associated with WRMSDs. It highlights the need for comprehensive, multimodal approaches that address not only peripheral nociception but also the central mechanisms perpetuating pain. In the subsequent sections, we will explore emerging domains linked to chronic pain, such as nociplastic pain, neuroplasticity, and consequent central sensitization.

3.1.4 *Nociplastic pain: an emerging understanding of neuroplasticity and central sensitization in WRMSDs*

Nociplastic pain is a relatively recent addition to the vocabulary of pain medicine. It has contributed to a significant shift in our understanding and approach to pain. In contrast to nociceptive pain (resulting from actual tissue damage) or neuropathic pain (stemming from nerve injury or dysfunction), nociplastic pain deviates from traditional models. It signifies an evolving understanding that pain can persist and intensify without clear structural or nerve-related etiologies. Nociplastic pain is characterized by alterations in the processing of pain signals by the nervous system. It involves a heightened sensitivity of the central nervous system, known as central sensitization (Nijs et al., 2023).

Central sensitization is a fundamental concept in this domain, highlighting the complex mechanisms contributing to pain's persistence and intensification beyond the initial injury. It is an enhanced sensitivity within the central nervous system, triggered by noxious stimuli.The pain-processing pathways become overly sensitive, amplifying pain signals and prolonging their duration (Woolf and Salter, 2000; Ji et al., 2003). Central sensitization is not solely a consequence of ongoing pain but often plays a significant role in the shift from acute to chronic pain. Factors, such as persistent inflammation, recurrent injuries, and neurological changes contribute to central sensitization of pain. These neurological changes in prolonged exposure can lead to neuroplasticity in the brain (Woolf and Salter, 2000).

Neuroplasticity is the brain's capacity to reorganize itself in response to experiences presenting a dual challenge in chronic pain. In chronic pain, persistent nociceptive stimuli and ongoing inflammation can instigate maladaptive neuroplastic changes within the central nervous system. Thus, increasing sensitivity to pain signals and causing alterations in pain perception (Kuner and Flor, 2017). Central sensitization, a manifestation of neuroplasticity, include amplifying pain signals within the central nervous system. The neurons become hypersensitive to pain stimuli leading to increased pain intensity, even in response to nonpainful stimuli which is referred to as allodynia (Kidd and Urban, 2001; Ji et al., 2003). Moreover, chronic pain can reshape neural pathways linked to emotional regulation. Thus, leading to mood disturbances such as anxiety and depression. These emotional factors, may intensify the perception of pain, thereby establishing a cycle of distress (Kuner and Flor, 2017; Ji et al., 2003; Woolf and Salter, 2000).

This brings us to the endocannabinoid system. This system facilitates intercellular communication that is dispersed throughout the human body. It assists in regulating diverse physiological processes such as mood regulation, perception of pain, appetite control, immune responses, and quality of sleep. The endocannabinoid system is a complex network of receptors, endogenous cannabinoids, and enzymes that regulate the perception of pain, inflammation, and neuronal excitability (Guindon and Hohmann, 2009). However, chronic pain conditions can disturb this finely tuned system. When discomfort persists for an extended period, the endocannabinoid system components may undergo dysregulation, leading to imbalances in endocannabinoid signaling. This disruption exacerbate central sensitization, amplifying pain signals. Thus, contributing to the perpetuation of chronic pain states (Guindon and Hohmann, 2009; Zogopoulos et al., 2013).

Understanding the dynamic interplay between neuroplasticity and chronic pain underscores the significance of comprehensive pain management strategies. Neuroplasticity is harnessed by cognitive-behavioral therapy (CBT) and mindfulness-based interventions. These interventions promote more adaptive pain processing patterns, helping people retrain their brains to cope with chronic pain (Kuner and Flor, 2017; Woolf and Salter, 2000).

Chronic pain, characterized by nociplastic mechanisms and neuroplastic changes, poses a challenge to the traditional reliance on diagnostic tools like X-rays or magnetic resonance imaging (MRI) to diagnose pain conditions. Instead, it emphasizes the importance of the patient's subjective experience and highlights the need for holistic, patient-centered approaches to pain management. Addressing pain often entails a person-centered blend of interventions, including CBT, medications, and lifestyle modifications (Nijs et al., 2023). These will be discussed in the subsequent sections of the chapter.

3.2 Motor behavior associated with chronic inflammation due to WRMSDs

Chronic pain transforms an individual, affecting their movement and interaction with their environment. These changes often arise from the chronic inflammation associated with WRMSDs and are further exacerbated by pain. One notable consequence is the adoption of modified movement patterns. Individuals instinctively alter their postures and motions to minimize discomfort unknowingly alter their biomechanics (Hodges and Tucker, 2011). When faced with persistent pain, the body often undergoes a series of changes in motor behavior as it attempts to alleviate discomfort. These changes can manifest in various forms. Individuals may instinctively guard the affected area to prevent worsening pain, leading to protective postures and limited mobility (Hodges and Tucker, 2011).

Moreover, chronic pain frequently triggers changes in movement patterns to minimize discomfort. The modifications include favoring one side of the body, adopting awkward postures, or relying on compensatory movements impacting the biomechanics of posture and movement (Hodges and Tucker, 2011). Although these adaptations offer short-term relief, they maycause musculoskeletal imbalances and exacerbate long-term pain (Kantak et al., 2022). Prolonged pain can cause muscle

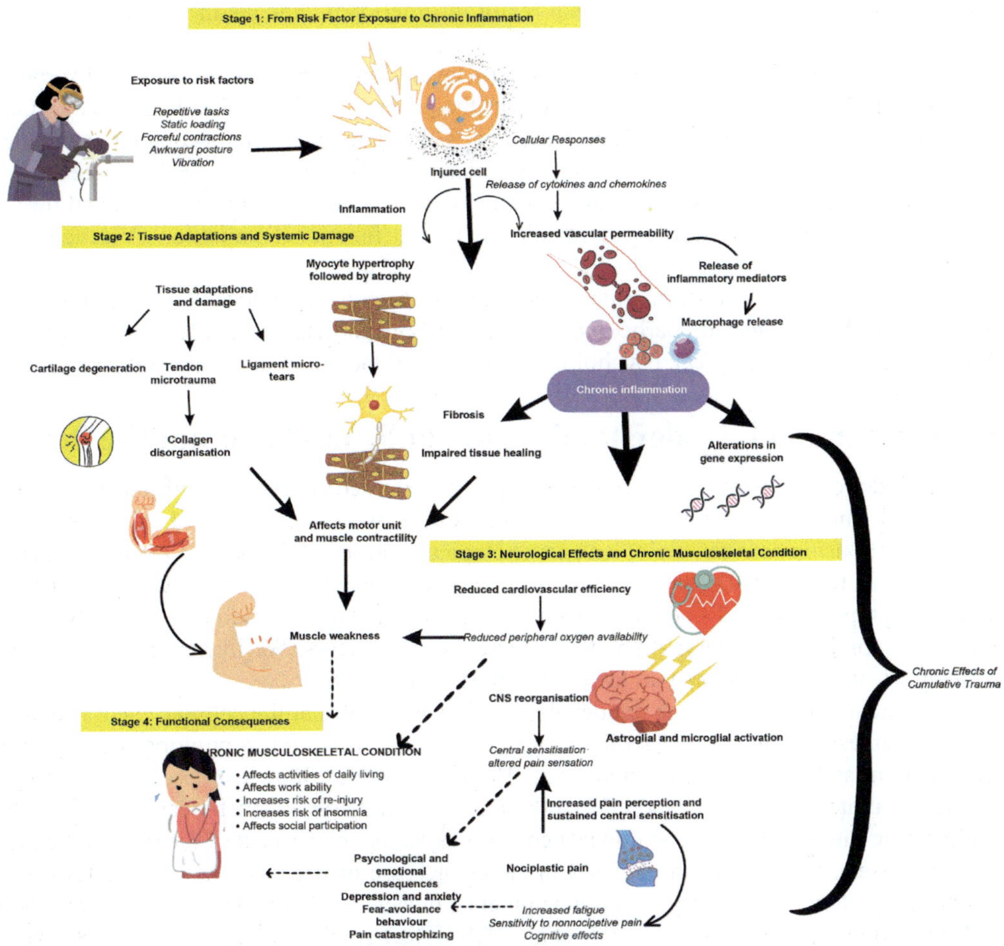

FIG. 16.1 Pathophysiology of work-related musculoskeletal disorders. *(Adapted from: Putsa, B., Jalayondeja, W., Mekhora, K., Bhuanantanondh, P., Jalayondeja, C., 2022. Factors associated with reduced risk of musculoskeletal disorders among office workers: a cross-sectional study 2017 to 2020. BMC Public Health 22 (1), 1503. doi:10.1186/s12889-022-13940-0; Kosek, E., Clauw, D., Nijs, J., Baron, R., Gilron, I., Harris, R.E., Mico, J.A., Rice, A.S.C., Sterling, M., 2021. Chronic nociplastic pain affecting the musculoskeletal system: clinical criteria and grading system. Pain 162 (11), 2629–2634. doi:10.1097/j.pain.0000000000002324; Nijs, J., Lahousse, A., Kapreli, E., Bilika, P., Saraçoğlu, İ., Malfliet, A., Coppieters, I., De Baets, L., Leysen, L., Roose, E., Clark, J., Voogt, L., Huysmans, E., 2021. Nociplastic pain criteria or recognition of central sensitization? pain phenotyping in the past, present and future. J. Clin. Med. 10 (15), 3203. doi:10.3390/jcm10153203; Davydov, D.M., Perlo, S., 2015. Cardiovascular activity and chronic pain severity. Physiol Behav. 152, 203–216. doi:10.1016/j.physbeh.2015.09.029; Galvez-Sánchez, C.M., Montoro, C.I., Duschek, S., Del Paso, G.A.R., 2020. Pain catastrophizing mediates the negative influence of pain and trait-anxiety on health-related quality of life in fibromyalgia. Qual. Life Res. 29 (7), 1871–1881. doi:10.1007/s11136-020-02457-x; Perlo, S., Davydov, D.M., 2016. "Chronic pain and the brain" impairment: introducing a translational neuroscience-based metric. Pain Med. 17 (5), 799–802. doi:10.1093/pm/pnw024.)*

disuse and atrophy in the affected area, reducing stability and support (Sartori et al., 2021). This increases risk (re)injury and impaire motor function. Pain and injury induced restrictions in joint mobility can reduce the range of motion and cause stiffness (Barbe and Barr, 2006). Chronic pain can cause psychological and physical changes that induce motor behavior. These psychological changes include movementrelated anxiety and fear, which can result in avoidance behaviors and decreased activity levels (Vlaeyen and Linton, 2000). In addition to causing bodily changes, chronic pain can impair motor control and coordination, cause avoidance behaviors to prevent discomfort from worsening, and contribute to fatigue (Kantak et al., 2022).

Furthermore, the fatigue commonly associated with chronic pain can diminish motor performance and endurance (Andrews et al., 2012). An individual may depend on aids like lumbar belts or cane and crutches to manage their pain. These devices though necessary can also change a person's motor behavior (Shahvarpour et al., 2018).

Understanding these mechanisms through which chronic pain can trigger alterations in motor behaviors is imperative for healthcare professionals. This highlights theneed for holistic pain management strategies encompassing lifestyle modifications, physical therapy, rehabilitation, ergonomic consultations, and psychological assistance (Stanos, 2012). The goal is to restore functional capacities and improve their overall well-being. Fig. 16.1 provides an overview of the cellular and systemic changes that result in functional changes in workers with chronic musculoskeletal disorders.

Apart from these changes at the cellular, tissue, and structural levels, long-term musculoskeletal disorders like WRMSDs can cause a wide range of clinical symptoms. Pain emerges as the primary symptom (Barbe and Barr, 2006). Consequent, decline in functional capability, adversely impacts an individual's ability to engage in occupational and daily tasks (Barbe and Barr, 2006; Barr and Barbe, 2002). These impairments and activity limitations can further increase the risk of reinjuries and compromise their overall well-being. Chronic pain and disability can also induce emotional turmoil, encompassing feelings of anxiety and depression, further aggravating pain and functional limitation (Vlaeyen and Linton, 2000; Karshikoff et al., 2023).

To summarize, the pathological transformations observed in chronic musculoskeletal conditions characteristic of WRMSDs are profound and multifaceted (Barbe and Barr, 2006; Barr and Barbe, 2002). In the subsequent section, we guide readers through interplay of biological, psychological, and social factors in the pain experience.

4 Biopsychosocial model: Understanding the individual and not the impairment

The biopsychosocial model of health is a comprehensive approach that considers the multifaceted interactions between biological, psychological, and social factors impacting health. Originating from the work of Engel G. L. in 1977, this framework highlights relationship between biological, psychological, and social factors in shaping health outcomes (Engel, 1977, 1980). When applied to WRMSDs, this model offers important insights into the underlying factors affecting pain and dysfunction. By adopting this model's perspective on chronic pain and dysfunction ue to WRMSDs, professionals can develop targeted and comprehensive interventions to mitigate the risk factors. We illustrate the complex interactions of biological, psychological, and social factors that lead to WRMSDs in Fig. 16.2.

Biological factors fundamentally influence the development of WRMSDs at the cellular level. Prolonged or repetitive tasks in the workplace can result in mechanical strain on musculoskeletal structures, prompting cellular reactions like tissue remodeling, inflammation, and even cellular impairment (Barbe and Barr, 2006). Comprehending these biological processes is essential for accurately diagnosing and efficiently managing WRMSDs. The biopsychosocial paradigm serves as a reminder that addressing these biological components constitutes merely one aspect of the overall picture. Psychological variables, like stress, anxiety, and coping strategies, can either increase or decrease the cellular-level changes linked to WRMSDs. For instance, prolonged stress can result in heightened muscle tension and contribute to muscle imbalancespredisposing to WRMSDs (Bongers et al., 1993; Gatchel et al., 2007). Recognizing and managing these psychological factors are imperative for a comprehensive approach to MSD management (Gatchel et al., 2007).

Equally crucial is the social context surrounding WRMSDs. Factors including occupational requirements, workplace ergonomics, support from coworkers and supervisors, and accessibility to healthcare facilities collectively

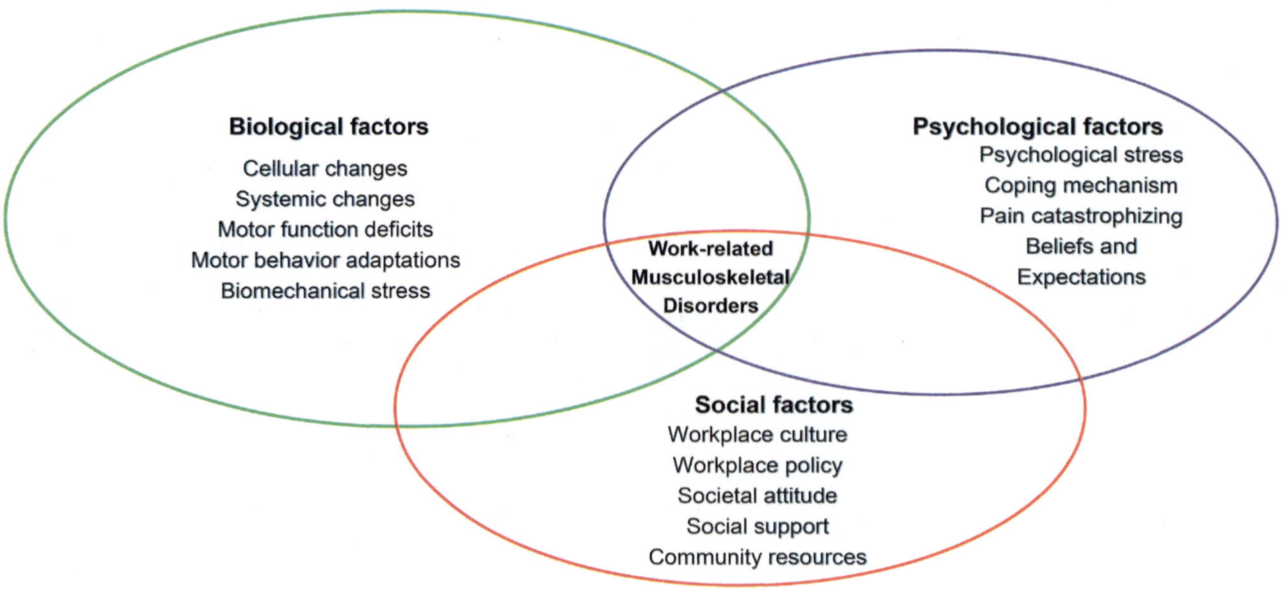

FIG. 16.2 The biophysiological model of work-related musculoskeletal disorders.

influence the onset and advancement of MSDs (Hartvigsen et al., 2018; Bezzina et al., 2023; Wakida et al., 2018). In certain instances, social factors may contribute to procrastination in seeking medical intervention, consequently affecting the cellular-level alterations within musculoskeletal tissues (Gatchel et al., 2007). The biopsychosocial model urges us to contemplate these social determinants of health when evaluating and managing MSDs (Gatchel et al., 2007).

WRMSDs do not arise solely from cellular-level changes in isolation; instead, they are a consequence of multifactorial interplay of biological, psychological, and social factors (). Recognizing and managing these factors is essential for prevention, diagnosis, and treatment of WRMSDs.

4.1 International classification of function, health and disability: Exploring WRMSDs beyond impairments and disabilities

The World Health Organization (WHO) developed the International Classification of Functioning, Disability, and Health (ICF) for a comprehensive and standardized approach to understanding and classifying health and health-related domains (World Health Organization, 2001). The WHO officially declared the inception of the ICF in 2001. The ICF was devised to transcend from a purely medical framework of health to incorporate physical, psychological, and social domains influencing health. ICF promotes collaboration among diverse healthcare practitioners and other stakeholders impacting health and illness such as employer involvement in risk factor reduction at work (World Health Organization, 2001; Martins, 2015; Kreuger et al., 2026).

ICF also promotes person centered intervention for WRMSDs (World Health Organization, 2001; Hutting et al., 2022). It shapes public health strategies by understanding the broader social determinants of WRMSDs (World Health Organization, 2001; Hartvigsen et al., 2018). The ICF comprises two primary elements: the functioning and disability segment and contextual factors. The functioning and disability components encompass bodily functions, body structures, activities, and participation. Contextual factors encompass environmental and personal factors. Environmental factors involve physical, social, and mental environments that have the potential to either facilitate or impede an individual's ability to function. Personal factors incorporate individual attributes, comprising beliefs, attitudes, coping mechanisms, and background (World Health Organization, 2001).

Three of the commonly reported conditions classified under WRMSDs; low back pain; neck pain; and carpal tunnel syndrome; have an ICF core set (Cieza et al., 2004). An ICF core set represents a subset of ICF categories deemed crucial for assessing a specific condition or population. These core sets are established through a consensus process involving experts from various fields, such as clinicians, researchers, and individuals with lived experience of the condition (Cieza et al., 2004). They offer a standardized method to evaluate the functioning and disability of individuals with WRMSDs, thereby informing clinical practice, research, and policy. The ICF core set comprises second-level categories encompassing body functions like pain, sensation, movement, and muscle strength; body structures such as the spine, muscles, and nerves; and activities and participation like walking, standing, sitting, lifting, dressing, working, and engaging in leisure activities (Cieza et al., 2004). All three core sets emphasize the contextual factors associated with work. For example, the core set for low back pain includes aspects like "acquiring, keeping and terminating a job",underscoring the vital interaction between work and WRMSDs (Cieza et al., 2004). As illustrated in Fig. 16.3, the ICF framework explains the case of a 32-year-old woman with low back pain.

The contextual factors highlighted in the ICF framework underscore that social determinants of health can significantly influence pain and dysfunction in individuals experiencing WRMSDs. Factors such as occupational conditions, workplace support, work adjustments, insurance coverage, health benefits, healthcare services accessibility, cultural and societal attitudes, and safety policies at work substantially impact the onset and exacerbation of WRMSDs (Hartvigsen et al., 2018; Buchbinder et al., 2018).

In essence, the ICF framework serves as a valuable instrument for enhancing the care and treatment of individuals with pain and dysfunction due to WRMSDs. It offers a standardized method for assessing functioning and disability, which can guide clinical practice, research, and policy-making efforts. Furthermore, the framework sheds light on social determinants and their repercussions on the disability dynamics linked to WRMSDs (Magalhães et al., 2019; World Health Organization, 2001).

To summarise, WRMSDs now encompasses microscopic cellular-level biological and structural changes, functional and behavioral changes, as well as factors involving contextual elements. These influence an employee's health, welfare, and efficiency. In the subsequent section, we delve into the significance of investigations and physical activity in diagnosing and managing WRMSDs.

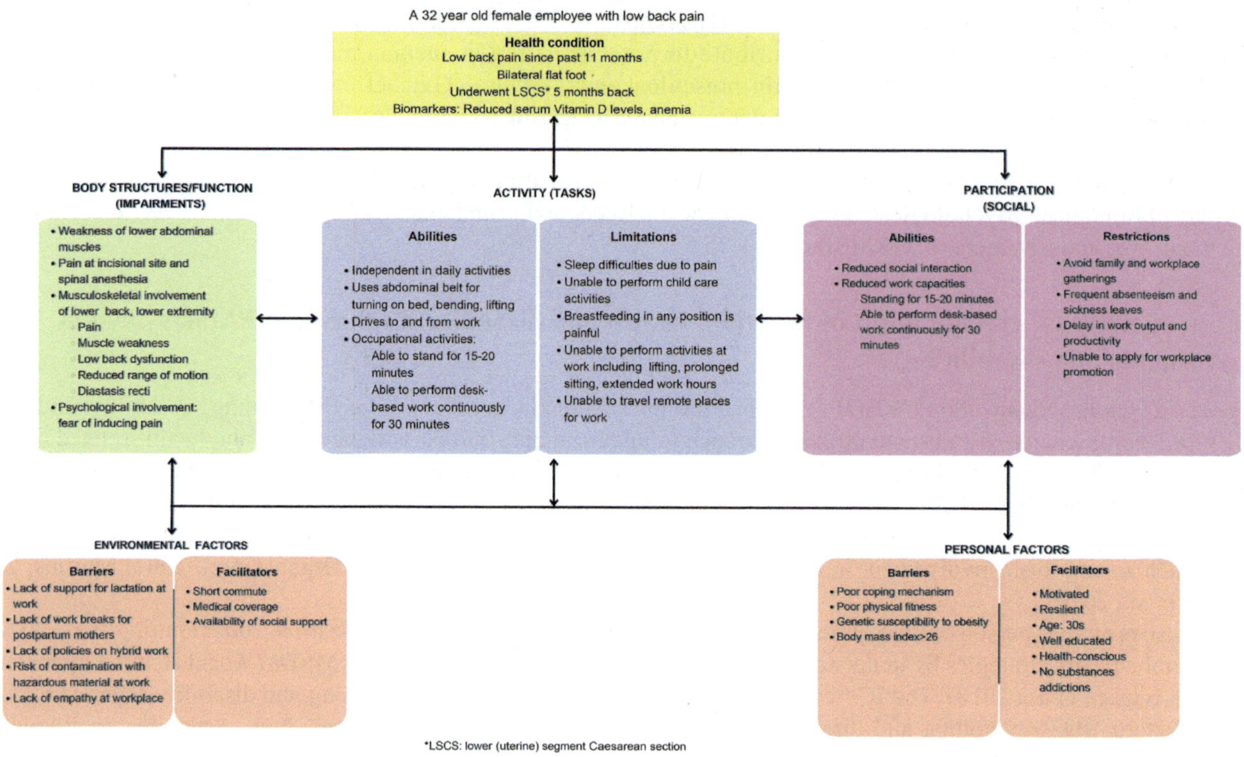

FIG. 16.3 The ICF framework for the case of a 32-year-old woman with low back pain.

5 Role of imaging in workplace caused musculoskeletal disorders

Most common causes of WRMSDs are repetitive movements, forceful exertions, staying in the same position for long periods, and working in extreme or awkward postures (International Labour Organization, 2010). According to the Global Burden of Disease in 2016, lower back pain was the leading cause of the number of years of disability since 1990 (Vos et al., 2017). As per a review conducted in 2021, the most prevalent WRMSDs are nonspecific lower back ache, lumbosacral radicular syndrome, subacromial pain, carpal tunnel syndrome, elbow tendinopathies, and knee and hip arthritis (van der Molen et al., 2021). In the United States, around $87.6 billion was spent on treating neck and lower back aches between 1996 and 2013 (Dieleman et al., 2016). Early diagnosis is essential in preventing unnecessary health costs and improving the effectiveness of treatment (Botchu and Gupta, 2021). Although subjective evaluation, physical examination, and biochemical tests are available to screen the musculoskeletal conditions, imaging can be essential in diagnosing and influencing the treatment choice. Imaging tests can be costly and involve radiation; however, the benefits can outweigh this. Benefits include improving life expectancy, better quality of life, and avoiding further unnecessary tests (Botchu and Gupta, 2021; Choudhury et al., 2025). The various imaging modalities include traditional radiography, computed tomography (CT), ultrasound, and magnetic ressonance imaging (MRI).

5.1 Radiography

X-ray radiography is the most common and traditionally used imaging modality to diagnose loss of disc space or degenerative changes. It is one of the feasible options for screening musculoskeletal (MSK) disorders. A typical effective dose for a lumbar spine radiograph is approximately 1.4-1.6 mSv, comparable to 6 months of exposure to natural background radiation (Mettler et al., 2008; American College of Radiology & Radiological Society of North America, 2025). Anteroposterior (AP) and lateral radiographs can typically evaluate a low back ache to study the disc space or the presence of osteophytes. Lateral radiographs are ideal for demonstrating anterior/posterior slipping of the vertebra. Weight-bearing radiographs allow physicians to determine malalignment or hypermobility (Simmons et al., 2003). Furthermore, subsequent studies report that weight-bearing radiographs are more sensitive and accurate in identifying joint space narrowing (Leach et al., 1970). Carpel tunnel syndrome is associated with compression of the median nerve caused by repetitive motions and awkward postures, resulting in swelling of tissues around the nerve. These regions can be visualized well by tangential radiographic projections of the carpal canal. Identifying these characteristic radiographic changes caused by MSD can inform a specific exercise plan or posture modifications. This may help improve the prognosis of the client. Shoulder radiographs can also help identify calcium deposits due to chronic tendonitis caused by prolonged exposure to microtrauma (Chianca et al., 2018). Conventional radiography can help identify these deposits (Albano et al., 2021). For osteoarthritis (OA), radiographs can demonstrate cartilage loss, bone damage, or spurs (Katz et al., 2021). Radiographs are reliable in evaluating arthritic changes in joints (Jacobson et al., 2008).

5.2 Computed tomography

CT scans are commonly done to evaluate musculoskeletal conditions because they can obtain a three-dimensional (3D) view of anatomical structures. However, it involves a higher amount of radiation and can contribute up to 40% of the cumulative radiation dose in diagnostic imaging (Jessen et al., 1999), but the values change based on setting and population studied (Brenner and Hall, 2007). Nevertheless, CT can be beneficial in depicting disorders of bony structures, detecting chronic morphological changes, and diagnosing spinal canal stenosis, disc herniation, and facet joint abnormalities (Hall, 1980; van Rijn et al., 2012). A CT scan is considered a 'should-do' imaging modality when physical examinations and history are consistent with disc herniation. It is also recommended when there is low back ache for 4–6 weeks, when surgery is being considered, or when any progressive neurological signs or symptoms are present (Khorami et al., 2021; Patel et al., 2016).

5.3 Ultrasound

As imaging equipment that uses nonionizing radiation and has reasonable costs, ultrasonography is the imaging modality used to evaluate tendon and ligament abnormalities. Tendons consist of numerous fibers, primarily type I collagen, connecting the muscles to the bones. Repetitive stress movements can damage the tendons, resulting in tendinopathy (Hodgson et al., 2012). In tendinopathy, there is reduced echogenicity with increased spacing of hyperechoic fibrillar lines in ultrasound due to the thickening of tendons. Power Doppler may show neovascularisation, usually associated with injury (Hodgson et al., 2012). Though ligaments are less echogenic than tendons, ultrasound is a useful tool in the diagnosis of ligament tears (Lee and Healy, 2005; Morvan et al., 2001). Dynamic imaging during passive movement can help demonstrate laxity. Ligament tears often appear in ultrasound as thickened ligaments with hypoechoic areas and surrounding fluid (Morvan et al., 2001). However, one disadvantage of ultrasound is its inability to image bone, making it challenging to study deep-lying tendons or any intraarticular damage (Hodgson et al., 2012). Also, ultrasound is operator-dependent, and slight changes in transducer angulation can mimic artifacts similar to tendon pathology. However, a systematic review conducted by Westacott et al. reported that ultrasound has a higher sensitivity (0.79–1.0) and can be used as the initial imaging procedure to evaluate gluteal tendon tears (Westacott et al., 2011). Another technique, known as ultrasound elastography, can be beneficial in assessing the mechanical properties of tissue. It produces a visual strain map and a conventional B-mode to visually represent tissue stiffness (De Zordo et al., 2009).

5.4 Magnetic resonance imaging

MRI is known for its excellent soft tissue detail and 3D viewing capabilities with excellent reproducibility. Although there is no ionizing radiation, MRI is costly and may not be suitable for claustrophobic patients. Nevertheless, higher magnetic fields provide a higher signal-to-noise ratio, resulting in better resolution of ligaments and tendons (Magee and Williams, 2006). T2-weighted images or short tau inversion recovery (STIR) sequences may demonstrate hyperintense fluid signals in tendon or ligament tears and changes in the surrounding tissue (Calleja and Connell, 2010; Hodgson et al., 2012). In gradient echo images, tendon abnormalities can appear as increased signal intensity. MRI is considered a 'should do' imaging modality when history and physical examinations are consistent with disc herniation with radiculopathy post 6 weeks. It is also recommended when surgery is considered for severe neurological signs/symptoms or when an epidural glucocorticosteroid injection is contemplated (Patel et al., 2016). For conditions like OA, an MRI can help demonstrate damage to joints, articular cartilage, or subchondral bone tissues. MRI can also facilitate the diagnosis, prognosis, and intervention for chronic MSK disease. The effectiveness of therapy can be analyzed by assessing inflammatory lesions and bone edema (Maksymowych, 2009).

5.5 Challenges to the use of imaging in the investigation and diagnosis of WRMSDs

Imaging investigations, such as MRI and CT scans, are valuable tools in assessing WRMSDs, but they come with limitations that must be considered in clinical practice.

One of the primary drawbacks of imaging investigations in WRMSDs is the potential for overdiagnosis and overtreatment. (Punnett and Wegman, 2004) underscore the significance of differentiating between incidental findings detected on imaging and clinically relevant pathology. Identifying minor abnormalities that may lack symptomatic manifestations can result in unnecessary interventions, thus escalating healthcare expenditures and patient distress (Punnett and Wegman, 2004).

The accuracy and reliability of imaging modalities exhibit variability, as illustrated by Terzis et al. (2021), who explored the interrater reliability of MRI compared to CT and arthroscopy for wrist pathologies. Variations in interpretation among radiologists and the potential occurrence of false positives or negatives may influence the level of diagnostic certainty and subsequent decisions regarding treatment for WRMSDs (Terzis et al., 2021).

Furthermore, imaging investigations like CT scans entail radiation exposure, as Alawneh et al. highlighted. The prolonged or repeated exposure raises concerns about possible long-term health effects of ionizing radiation from CT scans, especially in young or vulnerable groups. To reduce the risks associated with radiation exposure, other imaging modalities like MRI may be chosen in some instances (Choudhury et al., 2026).

Furthermore, the cost-effectiveness of imaging tests in WRMSDs is an essential consideration. Imaging techniques like MRI may be financially burdensome for patients and healthcare systems due to the high costs involved. It is imperative to strike a balance between the clinical advantages offered by imaging and the financial implications to ensure the appropriate allocation of resources for WRMSDs management (Deyo et al., 2009; Sharpe et al., 2012). In 2010, CT utilization in nonemergency outpatient departments declined by 1.7%, with more significant drops observed in in-patients, hospital outpatient departments, and private offices (Levin et al., 2014).

To summarize, although imaging examinations play a pivotal role in assessing and diagnosing WRMSDs, it is crucial to acknowledge and tackle the limitations of these imaging modalities and restrict their usage. Awareness of the potential for overdiagnosis, variations in interpretation, exposure to radiation, and financial implications can assist healthcare providers in making well-informed choices regarding the incorporation of imaging assessments in evaluating musculoskeletal conditions.

6 Role of physical activity in WRMSDs

6.1 Sedentary behavior as a risk factor for WRMSDs

Sedentary behavior is operationally defined as any form of waking behavior characterized by an energy expenditure equal to or ≤ 1.5 METs while assuming a seated or reclined position (Tremblay et al., 2017; Mansoubi et al., 2015). This encompasses tasks like working at a desk, driving, or watching television and has been correlated with a variety of unfavorable health consequences, including an elevated risk of WRMSDs (Dunlop et al., 2015; Dzakpasu et al., 2021; Loef et al., 2022). According to recent studies, there is an association between prolonged periods of sedentary behavior and a higher incidence of muscular discomfort, which might lead to the development of WRMSDs (Dzakpasu et al., 2021). Furthermore, sedentary behavior can cause physiological changes like altered muscle activity and strength, increased strain on the musculoskeletal system, and increased pain sensitivity; this explains the association between working from home and musculoskeletal pain (Loef et al., 2022).

Moreover, research conducted among individuals employed in sedentary desk occupations, report a high prevalence of WRMSDs (Demissie et al., 2024; Dabholkar et al., 2017). This draws attention to the fact that sedentary behavior increases the chance of developing musculoskeletal diseases at work. In addition, sedentary behavior has been correlated with restrictions in daily activities, regardless of levels of physical activity, highlighting its detrimental effects on musculoskeletal well-being (Dunlop et al., 2015).

Physical activity is pivotal in counteracting the adverse effects of sedentary behavior commonly witnessed in work environments (Dzakpasu et al., 2021). Subsequent sections will delve into physical activity and its contribution to reducing pain and motor impairment in WRMSDs.

6.2 Physical activity and its role in WRMSDs

Physical activity constitutes any voluntary movement of the skeletal muscles to generate energy expenditure. This encompasses activities pursued during leisure time, commuting, and as part of occupational tasks (Caspersen et al., 1985). The significance of physical activity lies in its vital role in preserving health. Higher levels of physical activity across ccupational, leisure and household domains, is inversely associated with all-cause mortality. An additional hour of moderate-intensity physical activity per week an reduce the relative risk of mortality by approximately 4% (Samitz et al., 2011).

Moreover, physical activity has been linked to favorable impacts on physical and mental well-being (Samitz et al., 2011; World Health Organization, 2020). Furthermore, the WHO recommends that adults engage in 150–300 minutes of moderate-intensity physical activity per week to achieve health benefits, including reductions in cardiovascular disease, type 2 diabetes, and some cancers (World Health Organization, 2020).

Several studies have highlighted the benefits of physical activity in preventing WRMSDs. Mansfield et al. (2018) systematically reviewed the association between physical activity participation and work-related upper quadrant disorders, further underscoring the relevance of physical activity in WRMSD prevention (Mansfield et al., 2018). Similarly, Morken et al. (2007) demonstrated a connection between physical activity and a decreased prevalence of musculoskeletal ailments, highlighting physical activity as a protective strategy for prevention of WRMSDs (Morken et al., 2007). Similarly, Campo et al. (2008) suggested that physical therapists believed WRMSDs could be averted by focusing on mechanical aspects and body mechanics, stressing the relevance of physical activity in upholding proper body mechanics for WRMSD prevention (Campo et al., 2008).

Moreover, Cromie et al. (2000) delved into the frequency and seriousness of WRMSDs among physical therapists, pinpointing contributing risk elements and underscoring the necessity of addressing physical activity as a preventive strategy (Cromie et al., 2000). In addition, Vieira et al. (2016) systematically reviewed work-related musculoskeletal disorders among physical therapists, shedding light on the potential influence of physical activity in sustaining work capacity and preventing WRMSDs (Vieira et al., 2016). Furthermore, physical activity when combined with multimodal program like ergonomic interventions help in preventing pain and disability due to WRMSDs (Moreira-Silva et al., 2016; Shakerian et al., 2016; Sharifi et al., 2022; Bullo et al., 2024).

7 Conclusion

To conclude, this chapter has offered an in-depth understanding of WRMSDs, from the physiological basis to assessment frameworks, the role of diagnostic imaging, and treatment strategies. WRMSDs encompass various conditions characterized by pain, motor dysfunction, and functional limitations, often originating from repetitive tasks, inadequate ergonomics, and sedentary behaviors in occupational settings.

Understanding the complex interaction of factors contributing to WRMSDs is crucial for successful prevention, diagnosis, and treatment. Physiological alterations at the cellular level and systemic responses to persistent pain, with or without tissue damage emphasize the need for holistic interventions. Diagnostic imaging methods are pivotal in recognizing musculoskeletal disorders and guiding therapeutic choices. However, over-reliance on imaging must be avoided. To effectively analyze and diagnose these disorders, healthcare practitioners must manage WRMSDs from a biopsychosocial approach. Using the ICF and grounding WRMSDs in a multidimensional approach involving physical activity, workplace based support, and ergonomic interventions can help promote the overall health and quality of life of workers.

References

Abd-Elsayed, A., Cutler, T., Zaky, S., 2023. Anatomy of the nervous system. In: Advanced Anesthesia Review. Oxford University Press (OUP), p. 145. doi:10.1093/med/9780197584521.003.0055.

Abdelmagid, S.M., Barr, A.E., Rico, M., Amin, M., Litvin, J., Popoff, S.N., Safadi, F.F., Barbe, M.F., 2012. Performance of repetitive tasks induces decreased grip strength and increased fibrogenic proteins in skeletal muscle: role of force and inflammation. PloS One 7 (5), e38359. doi:10.1371/journal.pone.0038359.

Aguilar-Agon, K.W., Capel, A.J., Martin, N.R.W., Player, D.J., Lewis, M.P., 2019. Mechanical loading stimulates hypertrophy in tissue-engineered skeletal muscle: molecular and phenotypic responses. J. Cell. Physiol. 234 (12), 23547–23558. doi:10.1002/jcp.28923.

Akodu, A.K., Ashalejo, Z.O., 2019. Work-related musculoskeletal disorders and work ability among hospital nurses. J. Taibah Univ. Med. Sci. 1658-03612. 14 (3), 252–261. Available from: http://www.elsevier.com/journals/journal-of-taibah-university-medical-sciences/1658-3612. doi:10.1016/j.jtumed.2019.02.009.

Alawneh, K.Z., Raffee, L.A., Oqlat, A.A., Oglat, A.A., Qawasmeh, M.A., Ali, M.K., Okour, A.M., Al-Mistarehi, A.H., 2021. The utility of brain CT scan modality in the management of dizziness at the emergency department: a retrospective single-center study. Ann. Med. Surg. 2049-0801. 64. Available from: http://www.elsevier.com/journals/annals-of-medicine-and-surgery/2049-0801. doi:10.1016/j.amsu.2021.102220.

Albano, D., Coppola, A., Gitto, S., Rapisarda, S., Messina, C., Sconfienza, L.M., 2021. Imaging of calcific tendinopathy around the shoulder: usual and unusual presentations and common pitfalls. Radiol. Med. 126 (4), 608–619. doi:10.1007/s11547-020-01300-0.

Allampallam, K., Chakraborty, J., Bose, K.K., Robinson, J., 1996. Explant culture, immunofluorescence and electron-microscopic study of flexor retinaculum in carpal tunnel syndrome. J. Occup. Environ. Med. 1076-2752. 38 (3), 264–271. doi:10.1097/00043764-199603000-00011.

Allen, M.S., Walter, E.E., McDermott, M.S., 2017. Personality and sedentary behavior: a systematic review and meta-analysis. Health Psychol. 1930-7810. 36 (3), 255–263. Available from: www.apa.org/journals/hea.html. doi:10.1037/hea0000429.

American College of Radiology & Radiological Society of North America, 2025. Radiation Dose to Adults From Common Imaging Examinations. p. 18. Available from: https://www.radiologyinfo.org/-/media/files/radinfo/new-dose-reference-card.pdf.

Andrews, N.E., Strong, J., Meredith, P.J., 2012. Activity pacing, avoidance, endurance, and associations with patient functioning in chronic pain: a systematic review and meta-analysis. Archives of physical medicine and rehabilitation, 93 (11), 2109–2121.e7. doi:10.1016/j.apmr.2012.05.029.

Arnaudo, E., 2021. Psychogenic pain as imaginary pain. Riv. Internazionale Filos. Psicol. 2239-2629. 12 (2), 190–199. Available from: https://www.rifp.it/ojs/index.php/rifp/article/download/rifp.2021.0017/1157. doi:10.4453/rifp.2021.0017.

Barbe, M.F., Barr, A.E., 2006. Inflammation and the pathophysiology of work-related musculoskeletal disorders. Brain, Behav., Immun. 0889-1591. 20 (5), 423–429. doi:10.1016/j.bbi.2006.03.001.

Barr, A.E., Barbe, M.F., 2002. Pathophysiological tissue changes associated with repetitive movement: a review of the evidence. Phys. Ther. 0031-9023. 82 (2), 173–187. Available from: https://academic.oup.com/ptj/issue. doi:10.1093/ptj/82.2.173.

Barr, A.E., Barbe, M.F., Clark, B.D., 2004a. Work-related musculoskeletal disorders of the hand and wrist: epidemiology, pathophysiology, and sensorimotor changes. J. Orthop. Sports Phys. Ther. 0190-6011. 34 (10), 610–627. doi:10.2519/jospt.2004.34.10.610.

Barr, et al., 2004b. Lost-Worktime Injuries and Illnesses: Characteristics and Resulting Days Away From Work, 2001.United States Department of Labor, Bureau of Labor Statistics. Available from: http://www.bls.gov/iif/home.htm.

Bezzina, A., Austin, E., Nguyen, H., James, C., 2023. Workplace psychosocial factors and their association with musculoskeletal disorders: a systematic review of longitudinal studies. Workplace Health Saf. 71 (12), 578–588. doi:10.1177/21650799231193578.

Bongers, P.M., de Winter, C.R., Kompier, M.A., Hildebrandt, V.H., 1993. Psychosocial factors at work and musculoskeletal disease. Scand. J. Work Environ. Health 19 (5), 297–312. doi:10.5271/sjweh.1470.

Botchu, R., Gupta, H., 2021. Updates of the imaging of musculoskeletal problems. J. Clin. Orthop. Trauma 22, 101612. doi:10.1016/j.jcot.2021.101612.

Brenner, D.J., Hall, E.J., 2007. Computed tomography—an increasing source of radiation exposure. N. Engl. J. Med. 357 (22), 2277–2284. doi:10.1056/NEJMra072149.

Buchbinder, R., van Tulder, M., Öberg, B., Costa, L.M., Woolf, A., Schoene, M., Croft, P., 2018. Low back pain: a call for action. Lancet 391 (10137), 2384–2388. doi:10.1016/S0140-6736(18)30488-4.

Bullo, V., Favro, F., Pavan, D., Bortoletto, A., Gobbo, S., De Palma, G., Mattioli, S., Sala, E., Cugusi, L., Di Blasio, A., Cruz-Diaz, D.C., Sales Bocalini, D., Bergamin, M., 2024. The role of physical exercise in the prevention of musculoskeletal disorders in manual workers: a systematic review and meta-analysis. Med. Lav. 115 (1), e2024008. doi:10.23749/mdl.v115i1.15404.

Burns, J.W., Jensen, M.P., Gerhart, J., Thorn, B.E., Lillis, T.A., Carmody, J., Keefe, F., 2023. Cognitive therapy, mindfulness-based stress reduction, and behavior therapy for people with Chronic low back pain: a comparative mechanisms study. J. Consult. Clin. Psychol. 1939-2117. 91 (3), 171–187. Available from: http://www.apa.org/pubs/journals/ccp/. doi:10.1037/ccp0000801.

Calleja, M., Connell, D., 2010. The achilles tendon. Semin. Musculoskelet. Radiol. 1089-7860. 14 (03), 307–322. doi:10.1055/s-0030-1254520.

Campo, M., Weiser, S., Koenig, K.L., Nordin, M., 2008. Work-related musculoskeletal disorders in physical therapists: a prospective cohort study with 1-year follow-up. Phys. Ther. 0031-9023. 88 (5), 608–619. doi:10.2522/ptj.20070127.

Carpenter, J.E., Flanagan, C.L., Thomopoulos, S., Yian, E.H., Soslowsky, L.J., 1998. The effects of overuse combined with intrinsic or extrinsic alterations in an animal model of rotator cuff tendinosis. Am. J. Sports Med. 0363-5465. 26 (6), 801–807. Available from: http://ajs.sagepub.com/archive/2008.dtl. doi:10.1177/03635465980260061101.

Caspersen, C.J., Powell, K.E., Christenson, G.M., 1985. Physical activity, exercise, and physical fitness: definitions and distinctions for health-related research. Public Health Rep. 100 (2), 126–131. Available from: https://www.ncbi.nlm.nih.gov/pmc/articles/PMC1424733/.

Chianca, V., Albano, D., Messina, C., Midiri, F., Mauri, G., Aliprandi, A., Catapano, M., Pescatori, L.C., Monaco, C.G., Gitto, S., Pisani Mainini, A., Corazza, A., Rapisarda, S., Pozzi, G., Barile, A., Masciocchi, C., Sconfienza, L.M., 2018. Rotator cuff calcific tendinopathy: from diagnosis to treatment. Acta. Biomed. 89 (1-S), 186–196. doi:10.23750/abm.v89i1-S.7022.

Choudhury, R., Choudhury, S.R., Kenaria, T., Jagalpathy, J., Vemuri, O., Muruhaiah, V., Papineni, V.R.K., et al., 2025, Radiation safety in musculoskeletal radiology: addressing challenges and implementing solutions. Apollo Med. 23 (1), 35–45. doi:10.1177/09760016251372680.

Choudhury, S.R., Kenaria, T., Jagalpathy, J., Vemuri, O., Muruhaiah, V., Papineni, V.R.K., 2026. Radiation safety in musculoskeletal radiology: addressing challenges and implementing solutions. Apollo Med. 23 (1), 35–45. doi:10.1177/09760016251372680.

Cieza, A., Stucki, G., Weigl, M., Disler, P., Jäckel, W., van der Linden, S., Kostanjsek, N., de Bie, R., 2004. ICF core sets for low back pain. J. Rehabil. Med. (44), 69–74. doi:10.1080/16501960410016037.

Cromie, J.E., Robertson, V.J., Best, M.O., 2000. Work-related musculoskeletal disorders in physical therapists: prevalence, severity, risks, and responses. Phys. Ther. 0031-9023. 80 (4), 336–351. Available from: http://ptjournal.apta.org/content/by/year. doi:10.1093/ptj/80.4.336.

Dabholkar, T., Yardi, S., Dabholkar, Y.G., Velankar, H.K., Ghuge, G., 2017. A survey of work-related musculoskeletal disorders among otolaryngologists. Indian J. Otolaryngol. Head Neck Surg. 0973-7707. 69 (2), 230–238. Available from: http://www.springerlink.com/content/0019-5421. doi:10.1007/s12070-017-1106-5.

De Blasiis, P., de Sena, G., Signoriello, E., Sirico, F., Imamura, M., Lus, G., 2022. Nociplastic pain in multiple sclerosis spasticity: dermatomal evaluation, treatment with intradermal saline injection and outcomes assessed by 3D gait analysis: review and a case report. Int. J. Environ. Res. Public Health. 1660-4601. 19 (13), 7872. doi:10.3390/ijerph19137872.

de Brito Magalhães, F., de Lima, M.A.G., da Fonseca Neves, R., Black, K.C., de Araújo, T.M., Porto, L.A., 2019. Disability and functioning assessment of women with RSI/WRMSDs: the use of the ICF checklist. Rev. Bras. Med. Trab. 1679-4435. 17 (4), 545–556. doi:10.5327/z1679443520190443.

De Zordo, T., Fink, C., Feuchtner, G.M., Smekal, V., Reindl, M., Klauser, A.S., 2009. Real-time sonoelastography findings in healthy achilles tendons. AJR Am. J. Roentgenol 193 (2), W134–W138. doi:10.2214/AJR.08.1843.

Dehghan, B., Abolhasanzadeh, N., Shademan, B., Nourazarian, A., 2024. Deciphering pain: molecular mechanisms and neurochemical pathways-challenges and future opportunities. Front. Mol. Biosci. 11, 1382555. doi:10.3389/fmolb.2024.1382555.

Demissie, B., Bayih, E.T., Demmelash, A.A., 2024. A systematic review of work-related musculoskeletal disorders and risk factors among computer users. Heliyon 10 (3), e25075. doi:10.1016/j.heliyon.2024.e25075.

Deyo, R.A., Mirza, S.K., Turner, J.A., Martin, B.I., 2009. Overtreating chronic back pain: time to back off? J. Am. Board. Fam. Med. 22 (1), 62–68. doi:10.3122/jabfm.2009.01.080102.

Dieleman, J.L., Baral, R., Birger, M., Bui, A.L., Bulchis, A., Chapin, A., Hamavid, H., Horst, C., Johnson, E.K., Joseph, J., Lavado, R., Lomsadze, L., Reynolds, A., Squires, E., Campbell, M., DeCenso, B., Dicker, D., Flaxman, A.D., Gabert, R., Highfill, T., Naghavi, M., Nightingale, N., Templin, T., Tobias, M.I., Vos, T., Murray, C.J.L., 2016. US spending on personal health care and public health, 1996-2013. JAMA. 1538-3598. 316 (24), 2627–2646. Available from: http://jamanetwork.com/journals/jama/data/Journals/JAMA/935941/joi160128.pdf. doi:10.1001/jama.2016.16885.

Dunlop, D.D., Song, J., Arnston, E.K., Semanik, P.A., Lee, J., Chang, R.W., Hootman, J.M., 2015. Sedentary time in US older adults associated with disability in activities of daily living independent of physical activity. J. Phys. Act. Health 12 (1), 93–101. doi:10.1123/jpah.2013-0311.

Dunlop, D.D., Song, J., Arntson, E.K., Semanik, P.A., Lee, J., Chang, R.W., Hootman, J.M., 2015. Sedentary time in US older adults associated with disability in activities of daily living independent of physical activity. J. Phys. Act. Health. 1543-5474. 12 (1), 93–101. Available from: http://journals.humankinetics.com/AcuCustom/Sitename/Documents/DocumentItem/14_Dunlop_JPAH_2013-0311_93_101-ej.pdf. doi:10.1123/jpah.2013-0311.

Dzakpasu, F.Q.S., Carver, A., Brakenridge, C.J., Cicuttini, F., Urquhart, D.M., Owen, N., Dunstan, D.W., 2021. Musculoskeletal pain and sedentary behaviour in occupational and non-occupational settings: a systematic review with meta-analysis. Int. J. Behav. Nutr. Phys. Act. 18 (1), 159. doi :10.1186/s12966-021-01191-y.

Engel, G.L., 1977. The need for a new medical model: a challenge for biomedicine. Science. 0036-8075. 196 (4286), 129–136. doi:10.1126/science.847460.

Engel, G.L., 1980. The clinical application of the biopsychosocial model. Am. J. Psychiatry 137 (5), 535–544. doi:10.1176/ajp.137.5.535.

Fang, X.X., Zhai, M.N., Zhu, M., He, C., Wang, H., Wang, J., Zhang, Z.J., 2023. Inflammation in pathogenesis of chronic pain: foe and friend. Mol. Pain 19.doi:10.1177/17448069231178176.

Fernandes, K., Dos Reis, C.S., Serranheira, F., 2023. Radiographers' musculoskeletal health in Western Switzerland: WRMSDs symptoms prevalence and risk factors. Work. 1051-9815. 74 (4), 1527–1537. doi:10.3233/wor-211379.

Flor, H., Kuner, R., 2022. Brain-based interventions for chronic pain. Neuroforum. 0947-0875. 28 (2), 95–104. doi:10.1515/nf-2021-0037.

Garg, A., Kapellusch, J.M., 2011. Job analysis techniques for distal upper extremity disorders. Rev. Hum. Factors Ergon. 1557-234X. 7 (1), 149–196. do i:10.1177/1557234X11410386.

Gatchel, R.J., Neblett, R., 2018. Central sensitization: a brief overview. J. Appl. Biobehav. Res. 1751-9861. 23 (2). Available from: http://onlinelibrary.wiley.com/journal/10.1111/(ISSN)1751-9861. doi:10.1111/jabr.12138.

Gatchel, R.J., Peng, Y.B., Peters, M.L., Fuchs, P.N., Turk, D.C., 2007. The biopsychosocial approach to chronic pain: scientific advances and future directions. Psychol. Bull. 0033-2909. 133 (4), 581–624. doi:10.1037/0033-2909.133.4.581.

Greggi, C., Visconti, V.V., Albanese, M., Gasperini, B., Chiavoghilefu, A., Prezioso, C., Persechino, B., Iavicoli, S., Gasbarra, E., Iundusi, R., Tarantino, U., 2024. Work-related musculoskeletal disorders: a systematic review and meta-analysis. J. Clin. Med. 13 (13), 3964. doi: 10.3390/jcm13133964.

Guindon, J., Hohmann, A., 2009. The endocannabinoid system and pain. CNS Neurol. Disord. Drug Targets. 1871-5273. 8 (6), 403–421. doi:10.2174/187152709789824660.

Hall, F.M., 1980. Back pain and the radiologist. Radiology. 0033-8419. 137 (3), 861–863. doi:10.1148/radiology.137.3.6449718.

Hart, D.A., Archambault, J.M., Kydd, A., Reno, C., Frank, C.B., Herzog, W., 1998. Gender and neurogenic variables in tendon biology and repetitive motion disorders. Clin. Orthop. Relat. Res. 1528-1132. 351, 44–56. https://journals.lww.com/clinorthop/pages/default.aspx. doi:10.1097/00003086-199806000-00007.

Hartvigsen, J., Hancock, M.J., Kongsted, A., Louw, Q., Ferreira, M.L., Genevay, S., Hoy, D., Karppinen, J., Pransky, G., Sieper, J., Smeets, R.J., Underwood, M., Low, L., 2018. Back Pain Series Working Group. What low back pain is and why we need to pay attention. Lancet 391 (10137), 2356–2367. doi:10.1016/S0140-6736(18)30480-X.

Heinonen, A., Sipilä, S., 2007. Physical activity and health. Adv. Physiother. 1651-1948. 9 (2), 49. doi:10.1080/14038190701433407.

Hodges, P.W., Tucker, K., 2011. Moving differently in pain: a new theory to explain the adaptation to pain. Pain 152 (3 Suppl), S90–S98. doi:10.1016/j.pain.2010.10.020.

Hodgson, R.J., O'Connor, P.J., Grainger, A.J., 2012. Tendon and ligament imaging. Br. J. Radiol. 0007-1285. 85 (1016), 1157–1172. doi:10.1259/bjr/34786470.

Hooper, J.M., Leoni, E., 1996. A physical activity continuum and the surgeon general's report. J. Phys. Educ., Recreat. Dance. 0730-3084. 67 (9), 62–63. doi:10.1080/07303084.1996.10604859.

Hutting, N., Caneiro, J.P., Ong'wen, O.M., Miciak, M., Roberts, L., 2022. Person-centered care for musculoskeletal pain: putting principles into practice. Musculoskelet. Sci. Pract. 62, doi:10.1016/j.msksp.2022.102663.

International Labour Organization, 2010. List of occupational diseases recommendation. Available from: https://www.ilo.org/publications/list-occupational-diseases-revised-2010-identification-and-recognition.

Jacobson, J.A., Girish, G., Jiang, Y., Sabb, B.J., 2008. Radiographic evaluation of arthritis: degenerative joint disease and variations. Radiology 248 (3), 737–747. doi:10.1148/radiol.2483062112.

Jeng, O.J., Radwin, R.G., Rodriquez, A.A., 1994. Functional psychomotor deficits associated with carpal tunnel syndrome. Ergonomics. 1366-5847. 37 (6), 1055–1069. doi:10.1080/00140139408963718.

Jessen, K.A., Shrimpton, P.C., Geleijns, J., Panzer, W., Tosi, G., 1999. Dosimetry for optimisation of patient protection in computed tomography. Appl. Radiat. Isot. 0969-8043. 50 (1), 165–172. doi:10.1016/s0969-8043(98)00024-4.

Ji, R.R., Kohno, T., Moore, K.A., Woolf, C.J., 2003. Central sensitization and LTP: do pain and memory share similar mechanisms? Trends Neurosci. 0166-2236. 26 (12), 696–705. Available from: www.elsevier.com/locate/tins. doi:10.1016/j.tins.2003.09.017.

Kantak, S.S., Johnson, T., Zarzycki, R., 2022. Linking pain and motor control: conceptualization of movement deficits in patients with painful conditions. Physical therapy 102 (4), pzab289. doi:10.1093/ptj/pzab289.

Karshikoff, B., Wåhlén, K., Åström, J., Lekander, M., Holmström, L., Wicksell, R.K., 2023. Inflammatory blood signature related to common psychological comorbidity in chronic Pain. Biomedicines 11 (3), 713. doi:10.3390/biomedicines11030713.

Katz, J.N., Arant, K.R., Loeser, R.F., 2021. Diagnosis and treatment of hip and knee osteoarthritis: a review. JAMA 325 (6), 568–578. doi:10.1001/jama.2020.22171.

Khorami, A.K., Oliveira, C.B., Maher, C.G., Bindels, P.J.E., Machado, G.C., Pinto, R.Z., Koes, B.W., Chiarotto, A., 2021. Recommendations for diagnosis and treatment of lumbosacral radicular pain: a systematic review of clinical practice guidelines. J. Clin. Med. 2077-0383. 10 (11). Available from: https://www.mdpi.com/2077-0383/10/11/2482/pdf. doi:10.3390/jcm10112482.

Kidd, B.L., Urban, L.A., 2001. Mechanisms of inflammatory pain. Br. J. Anaesth. 87 (1), 3–11. doi:10.1093/bja/87.1.3.

Kreuger, D., Carlier, B., Donker-Cools, B.H., Anema, J.R., Schaafsma, F.G., Oomens, S., 2026. Exploring the implementation of an ICF-based instrument and guideline for interprofessional collaboration in the return-to-work process. Work 84 (1), 250–263. doi:10.1177/10519815251409721.

Kuner, R., Flor, H., 2017. Structural plasticity and reorganisation in chronic pain. Nature reviews. Neuroscience 18 (2), 113. doi:10.1038/nrn.2017.5.

Kuner, R., Kuner, T., 2021. Cellular circuits in the brain and their modulation in acute and chronic pain. Physiol. Rev. 101 (1), 213–258. doi:10.1152/physrev.00040.2019.

Larocca H., 1992. A taxonomy of chronic pain syndromes. Spine 17 (10 Suppl), S344–S355. doi:10.1097/00007632-199210001-00002.

Leach, R.E., Gregg, T., Siber, F.J., 1970. Weight-bearing radiography in osteoarthritis of the knee. Radiology. 0033-8419. 97 (2), 265–268. doi:10.1148/97.2.265.

Lee, J.C., Healy, J.C., 2005. Normal sonographic anatomy of the wrist and hand. Radiographics. 1527-1323. 25 (6), 1577–1590. Available from: http://radiographics.rsnajnls.org/cgi/reprint/25/6/1577. doi:10.1148/rg.256055028.

Levin, D.C., Rao, V.M., Parker, L. Frangos A.J., 2014. Continued growth in emergency department imaging is bucking the overall trends. J. Am. Coll. Radiol. 11 (11), 1044–1047 . doi:10.1016/j.jacr.2014.07.008.

Lewis, R., Gómez Álvarez, C.B., Rayman, M., Lanham-New, S., Woolf, A., Mobasheri, A., 2019. Strategies for optimising musculoskeletal health in the 21 st century. BMC Musculoskelet. Disord. 1471-2474. 20 (1). Available from: http://www.biomedcentral.com/bmcmusculoskeletdisord/. doi:10.1186/s12891-019-2510-7.

Loef, B., van Oostrom, S.H., Bosma, E., Proper, K.I., 2022. The mediating role of physical activity and sedentary behavior in the association between working from home and musculoskeletal pain during the COVID-19 pandemic. Front. Public Health. 2296-2565. 10. Available from: http://journal.frontiersin.org/journal/public-health/section/public-health-education-and-promotion#archive. doi:10.3389/fpubh.2022.1072030.

Loeser, R.F., Goldring, S.R., Scanzello, C.R., Goldring, M.B., 2012. Osteoarthritis: a disease of the joint as an organ. Arthritis Rheum. 64 (6), 1697–1707. doi:10.1002/art.34453.

Lu, M.L., Lowe, B.D., Howard, N.L., Meyers, A.R., Fox, R.R., Dong, R.G., Baker, B.A., 2022. Work-related musculoskeletal disorders. In: Modern Occupational Diseases Diagnosis, Epidemiology, Management and Prevention. Bentham Science Publishers, United States, pp. 287–353. Available from: https://benthambooks.com/book/9789815049138/.

Lumley, M.A., Cohen, J.L., Borszcz, G.S., Cano, A., Radcliffe, A.M., Porter, L.S., Schubiner, H., Keefe, F.J., 2011. Pain and emotion: a biopsychosocial review of recent research. J. Clin. Psychol. 67 (9), 942–968. doi:10.1002/jclp.20816.

Magee, T., Williams, D., 2006. 3.0-T MRI of the supraspinatus tendon. Am. J. Roentgenol. 0361-803X. 187 (4), 881–886. doi:10.2214/ajr.05.1047.

Maksymowych, W.P., 2009. MRI in ankylosing spondylitis. Curr. Opin. Rheumatol. 1040-8711. 21 (4), 313–317. doi:10.1097/BOR.0b013e32832af481.

Mansfield, M., Thacker, M., Smith, T., 2018. Physical activity participation and the association with work-related upper quadrant disorders (WRUQDs): a systematic review. Musculoskelet. Care 16 (1), 178–187. doi:10.1002/msc.1204.

Mansoubi, M., Pearson, N., Clemes, S.A., Biddle, S.J.H., Bodicoat, D.H., Tolfrey, K., Edwardson, C.L., Yates, T., 2015. Energy expenditure during common sitting and standing tasks: examining the 1.5 MET definition of sedentary behaviour. BMC Public Health. 1471-2458. 15 (1). Available from: http://www.biomedcentral.com/bmcpublichealth. doi:10.1186/s12889-015-1851-x.

Martins A.C., 2015. Using the International Classification of Functioning, disability and health (ICF) to address facilitators and barriers to participation at work. Work 50 (4), 585–593. doi:10.3233/WOR-141965.

Meeus, M., Nijs, J., 2007. Central sensitization: a biopsychosocial explanation for chronic widespread pain in patients with fibromyalgia and chronic fatigue syndrome. Clin. Rheumatol. 0770-3198. 26 (4), 465–473. doi:10.1007/s10067-006-0433-9.

Mettler, F.A., Jr, Huda, W., Yoshizumi, T.T., Mahesh, M., 2008. Effective doses in radiology and diagnostic nuclear medicine: a catalog. Radiology 248 (1), 254–263. doi:10.1148/radiol.2481071451.

Mikaelsson, K., Eliasson, K., Lysholm, J., Nyberg, L., Michaelson, P., 2011. Physical capacity in physically active and non-active adolescents. J. Public Health. 0943-1853. 19 (2), 131–138. doi:10.1007/s10389-010-0371-5.

Millar, N.L., Silbernagel, K.G., Thorborg, K., Kirwan, P.D., Galatz, L.M., Abrams, G.D., Murrell, G.A.C., McInnes, I.B., Rodeo, S.A., 2021. Tendinopathy. Nat. Rev. Dis. Primers 7 (1), 1. doi:10.1038/s41572-020-00234-1.

Mitsikostas, D.-D., Moka, E., Orrillo, E., Aurilio, C., Vadalouca, A., Paladini, A., Varrassi, G., 2022. Neuropathic pain in neurologic disorders: a narrative review. Cureus. 2168-8184. 14 (2). doi:10.7759/cureus.22419.

Moreira-Silva, I., Teixeira, P.M., Santos, R., Abreu, S., Moreira, C., Mota, J., 2016. The effects of workplace physical activity programs on musculoskeletal pain: a systematic review and meta-analysis. Workplace Health Saf. 5 (64), 210–222. doi:10.1177/2165079916629688.

Morken, T., Magerøy, N., Moen, B.E., 2007. Physical activity is associated with a low prevalence of musculoskeletal disorders in the Royal Norwegian Navy: a cross sectional study. BMC Musculoskelet. Disord. 1471-2474. 8. doi:10.1186/1471-2474-8-56.

Morvan, G., Busson, J., Wybier, M., Mathieu, P., 2001. Ultrasound of the ankle. Eur. J. Ultrasound. 0929-8266. 14 (1), 73–82. doi:10.1016/s0929-8266(01)00147-1.

Mouilly, M., Ahami, A.O.T., Bouhali, B.E., 2022. Effects of COVID-19 home confinement on musculoskeletal health and physical performance in a cross-sectional study: results of the COVID-19 era online survey. Pan Afr. Med. J. 1937-8688. 43. doi:10.11604/pamj.2022.43.82.29068.

Nijs, J., Malfliet, A., Nishigami, T., 2023. Nociplastic pain and central sensitization in patients with chronic pain conditions: a terminology update for clinicians. Braz. J. Phys. Ther. 27 (3), 100518. doi:10.1016/j.bjpt.2023.100518.

Nwonu, C.N.S., 2022. Neuronal cell mechanisms of pain. West Afr. J. Med. 39 (10), 1075–1983. Available from: https://pubmed.ncbi.nlm.nih.gov/36260866/.

Patel, N.D., Broderick, D.F., Burns, J., Deshmukh, T.K., Fries, I.B., Harvey, H.B., Holly, L., Hunt, C.H., Jagadeesan, B.D., Kennedy, T.A., O'Toole, J.E., Perlmutter, J.S., Policeni, B., Rosenow, J.M., Schroeder, J.W., Whitehead, M.T., Cornelius, R.S., Corey, A.S., 2016. ACR appropriateness criteria low back pain. J. Am. Coll. Radiol. 13 (9), 1069–1078. doi:10.1016/j.jacr.2016.06.008.

Pérez-Fernández, J.I., Salaberria, K., de Ocenda, Á.R., 2022. Mindfulness-based pain management (MBPM) for chronic pain: a randomized clinical trial. Mindfulness. 1868-8535. 13 (12), 3153–3165. Available from: https://www.springer.com/journal/12671. doi:10.1007/s12671-022-02023-1.

Pol, R., Hristovski, R., Medina, D., Balague, N., 2019. From microscopic to macroscopic sports injuries. Applying the complex dynamic systems approach to sports medicine: a narrative review. Br. J. Sports Med. 0306-3674. 53 (19), 1214–1220. doi:10.1136/bjsports-2016-097395.

Punnett, L., Wegman, D.H., 2004. Work-related musculoskeletal disorders: the epidemiologic evidence and the debate. J. Electromyogr. Kinesiol. 1050-6411. 14 (1), 13–23. Available from: www.elsevier.com/locate/jelekin. doi:10.1016/j.jelekin.2003.09.015.

Raj, V.V., 2006. Occupational stress and radiography. Radiol. Technol. 0033-8397. 78 (2), 113–122.

Raja, S.N., Carr, D.B., Cohen, M., Finnerup, N.B., Flor, H., Gibson, S., Keefe, F.J., Mogil, J.S., Ringkamp, M., Sluka, K.A., Song, X.J., Stevens, B., Sullivan, M.D., Tutelman, P.R., Ushida, T., Vader, K., 2020. The revised international association for the study of pain definition of pain: concepts, challenges, and compromises. Pain. 1872-6623. 161 (9), 1976–1982. Available from: https://journals.lww.com/pain/Pages/currenttoc.aspx. doi:10.1097/j.pain.0000000000001939.

Royall, J.A., Berkow, R.L., Beckman, J.S., Cunningham, M.K., Matalon, S., Freeman, B.A., 1989. Tumor necrosis factor and interleukin 1 alpha increase vascular endothelial permeability. Am. J. Physiol. 257 (6), L399–L410. doi:10.1152/ajplung.1989.257.6.L399.

Safiri, S., Kolahi, A., Cross, M., Hill, C., Smith, E., Carson-Chahhoud, K., Mansournia, M.A., Almasi-Hashiani, A., Ashrafi-Asgarabad, A., Kaufman, J., Sepidarkish, M., Shakouri, S.K., Hoy, D., Woolf, A.D., March, L., Collins, G., Buchbinder, R., 2021. Prevalence, deaths, and disability-adjusted life years due to musculoskeletal disorders for 195 countries and territories 1990–2017. Arthritis Rheumatol. 2326-5191. 73 (4), 702–714. doi:10.1002/art.41571.

Samitz, G., Egger, M., Zwahlen, M., 2011. Domains of physical activity and all-cause mortality: systematic review and dose–response meta-analysis of cohort studies. Int. J. Epidemiol. 1464-3685. 40 (5), 1382–1400. doi:10.1093/ije/dyr112.

Sanabria-Mazo, J.P., Colomer-Carbonell, A., Carmona-Cervelló, M., Feliu-Soler, A., Borràs, X., Grasa, M., Esteve, M., Maes, M., Edo, S., Sanz, A., Luciano, J.V., 2022. Immune-inflammatory and hypothalamic-pituitary-adrenal axis biomarkers are altered in patients with non-specific low back pain: a systematic review. Front. Immunol. 13, 945513. doi:10.3389/fimmu.2022.945513.

Sartori, R., Romanello, V., Sandri, M., 2021. Mechanisms of muscle atrophy and hypertrophy: implications in health and disease. Nat. Commun. 12 (1), 330. doi:10.1038/s41467-020-20123-1.

Schuind, F., Ventura, M., Pasteels, J.L., 1990. Idiopathic carpal tunnel syndrome: histologic study of flexor tendon synovium. J. Hand Surg. 0363-5023. 15 (3), 497–503. doi:10.1016/0363-5023(90)90070-8.

Shahvarpour, A., Preuss, R., Sullivan, M.J.L., Negrini, A., Larivière, C., 2018. The effect of wearing a lumbar belt on biomechanical and psychological outcomes related to maximal flexion-extension motion and manual material handling. Appl. Ergon. 69, 17–24. doi:10.1016/j.apergo.2018.01.001.

Shakerian, M., Rismanchian, M., Khalili, P., Torki, A., 2016. Effect of physical activity on musculoskeletal discomforts among handicraft workers. J. Educ. Health Promot. 2277-9531. 5 (1), 8. doi:10.4103/2277-9531.184546.

Shakerian, M., Rismanchian, M., Khalili, P., Torki, A., 2016. Effect of physical activity on musculoskeletal discomforts among handicraft workers. J. Educ. Health Promot. 5, 8, doi:10.4103/2277-9531.184546.

Sharifi, A.S., Danesh, M.K., Gholamnia, R., 2022. Improvements in musculoskeletal symptoms, mental workload and mental fatigue: effects of a multi-component ergonomic intervention among call center workers. Work 72 (2), 765–774. doi:10.3233/WOR-213641.

Sharpe, R.E., Nazarian, L.N., Parker, L., Rao, V.M., Levin, D.C., 2012. Dramatically increased musculoskeletal ultrasound utilization from 2000 to 2009, especially by podiatrists in private offices. J. Am. Coll. Radiol. 9 (2), 141–146. doi:10.1016/j.jacr.2011.09.008.

Simmons, E.D., Guyer, R.D., Graham-Smith, A., Herzog, R., 2003. Radiograph assessment for patients with low back pain. Spine 3 (3), 3S–5S. doi:10.1016/s1529-9430(02)00454-0.

Stanos, S., 2012. Focused review of interdisciplinary pain rehabilitation programs for chronic pain management. Curr. Pain Headache Rep. 1531-3433. 16 (2), 147–152. doi:10.1007/s11916-012-0252-4.

Talmage, J.B., 2023. Nociplastic pain—A new name for an old and underrecognized concept. AMA Guides Newsl. 2642-0880. 28 (1), 16–30. doi:10.1001/amaguidesnewsletters.2023.janfeb02.

Tang, K.H.D., 2022. The prevalence, causes and prevention of occupational musculoskeletal disorders. Glob. Acad. J. Med. Sci. 2706-9036. 4 (2), 56–68. doi:10.36348/gajms.2022.v04i02.004.

Terzis, A., Klinger, A., Seegmüller, J., Sauerbier, M., 2021. Inter-rater reliability of magnetic resonance imaging in comparison to computed tomography and wrist arthroscopy in SLAC and SNAC wrist. J. Clin. Med. 2077-0383. 10 (16), 3592. doi:10.3390/jcm10163592.

Treede, R.D., Rief, W., Barke, A., Aziz, Q., Bennett, M.I., Benoliel, R., Cohen, M., Evers, S., Finnerup, N.B., First, M.B., Giamberardino, M.A., Kaasa, S., Korwisi, B., Kosek, E., Lavand'homme, P., Nicholas, M., Perrot, S., Scholz, J., Schug, S., Smith, B.H., Wang, S.J., 2019. Chronic pain as a symptom or a disease: the IASP classification of chronic pain for the International Classification of Diseases (ICD-11). Pain 160 (1), 19–27. doi:10.1097/j.pain.0000000000001384.

Tremblay, M.S., Aubert, S., Barnes, J.D., Saunders, T.J., Carson, V., Latimer-Cheung, A.E., Chastin, S.F.M., Altenburg, T.M., Chinapaw, M.J.M., 2017. Sedentary behavior research network (SBRN) - terminology consensus project process and outcome. Int. J. Behav. Nutr. Phys. Act. 14 (1), 75. doi:10.1186/s12966-017-0525-8.

U.S. Bureau of Labor Statistics, 2020. Occupational injuries and illnesses resulting in musculoskeletal disorders (MSDs). U.S. Department of Labor. Available from: https://www.bls.gov/iif/factsheets/msds.htm.

van der Molen, H.F., Visser, S., Alfonso, J.H., Curti, S., Mattioli, S., Rempel, D., Roquelaure, Y., Kuijer, P.P.F.M., Tamminga, S.J., 2021. Diagnostic criteria for musculoskeletal disorders for use in occupational healthcare or research: a scoping review of consensus- and synthesised-based case definitions. BMC Musculoskelet. Disord. 1471-2474. 22 (1). Available from: http://www.biomedcentral.com/bmcmusculoskeletdisord/. doi:10.1186/s12891-021-04031-z.

van Rijn, R.M., Wassenaar, M., Verhagen, A.P., Ostelo, R.W., Ginai, A.Z., de Boer, M.R., van Tulder, M.W., Koes, B.W., 2012. Computed tomography for the diagnosis of lumbar spinal pathology in adult patients with low back pain or sciatica: a diagnostic systematic review. Eur. Spine J. 21 (2), 228–239. doi:10.1007/s00586-011-2012-2.

Vieira, E.R., Schneider, P., Guidera, C., Gadotti, I.C., Brunt, D., 2016. Work-related musculoskeletal disorders among physical therapists: a systematic review. J. Back Musculoskelet. Rehabil. 29 (3), 417–428. doi:10.3233/BMR-150649.

Vlaeyen, J.W.S., Linton, S.J., 2000. Fear-avoidance and its consequences in chronic musculoskeletal pain: a state of the art. Pain 85 (3), 317–332. doi:10.1016/S0304-3959(99)00242-0.

Vos, T., Abajobir, A.A., Abbafati, C., Abbas, K.M., Abate, K.H., Abd-Allah, F., Abdulle, A.M., Abebo, T.A., Abera, S.F., Aboyans, V., Abu-Raddad, L.J., Ackerman, I.N., Adamu, A.A., Adetokunboh, O., Afarideh, M., Afshin, A., Agarwal, S.K., Aggarwal, R., Agrawal, A., Agrawal, S., Kiadaliri, A.A., Ahmadieh, H., Ahmed, M.B., Aichour, A.N., Aichour, I., Aichour, M.T.E., Aiyar, S., Akinyemi, R.O., Akseer, N., Lami, F.H.A., Alahdab, F., Al-Aly, Z., Alam, K., Alam, N., Alam, T., Alasfoor, D., Alene, K.A., Ali, R., Alizadeh-Navaei, R., Alkerwi, A., Alla, F., Allebeck, P., Allen, C., Al-Maskari, F., Al-Raddadi, R., Alsharif, U., Alsowaidi, S., Altirkawi, K.A., Amare, A.T., Amini, E., Ammar, W., Amoako, Y.A., Andersen, H.H., Antonio, C.A.T., Anwari, P., Ärnlöv, J., Artaman, A., Aryal, K.K., Asayesh, H., Asgedom, S.W., Assadi, R., Atey, T.M., Atnafu, N.T., Atre, S.R., Avila-Burgos, L., Avokpaho, E.F.G.A., Awasthi, A., Quintanilla, B.P.A., Saleem, H.O.B., Bacha, U., Badawi, A., Balakrishnan, K., Banerjee, A., Bannick, M.S., Barac, A., Barber, R.M., Barker-Collo, S.L., Bärnighausen, T., Barquera, S., Barregard, L., Barrero, L.H., Basu, S., Battista, B., Battle, K.E., Baune, B.T., Bazargan-Hejazi, S., Beardsley, J., Bedi, N., Beghi, E., Béjot, Y., Bekele, B.B., Bell, M.L., Bennett, D.A., Bensenor, I.M., Benson, J., Berhane, A., Berhe, D.F., Bernabé, E., Betsu, B.D., Beuran, M., Beyene, A.S., Bhala, N., Bhansali, A., Bhatt, S., Bhutta, Z.A., Biadgilign, S., Bienhoff, K., Bikbov, B., Birungi, C., Biryukov, S., Bisanzio, D., Bizuayehu, H.M., Boneya, D.J., Boufous, S., Bourne, R.R.A., Brazinova, A., Brugha, T.S., Buchbinder, R., Bulto, L.N.B., Bumgarner, B.R., Butt, Z.A., Cahuana-Hurtado, L., Cameron, E., Car, M., Carabin, H., Carapetis, J.R., Cárdenas, R., Carpenter, D.O., Carrero, J.J., Carter, A., Carvalho, F., Casey, D.C., Caso, V., Castañeda-Orjuela, C.A., Castle, C.D., Catalá-López, F., Chang, H.Y., Chang, J.C., Charlson, F.J., Chen, H., Chibalabala, M., Chibueze, C.E., Chisumpa, V.H., Chitheer, A.A., Christopher, D.J., Ciobanu, L.G., Cirillo, M., Colombara, D., Cooper, C., Cortesi, P.A., Criqui, M.H., Crump, J.A., Dadi, A.F., Dalal, K., Dandona, L., Dandona, R., Neves, J.D., Davitoiu, D.V., De Courten, B., De Leo, D., Degenhardt, L., Deiparine, S., Dellavalle, R.P., Deribe, K., Jarlais, D.C.D., Dey, S., Dharmaratne, S.D., Dhillon, P.K., Dicker, D., Ding, E.L., Djalalinia, S., Do, H.P., Dorsey, E.R., Santos, K.P.B.D., Douwes-Schultz, D., Doyle, K.E., Driscoll, T.R., Dubey, M., Duncan, B.B., El-Khatib, Z.Z., Ellerstrand, J., Enayati, A., Endries, A.Y., Ermakov, S.P., Erskine, H.E., Eshrati, B., Eskandarieh, S., Esteghamati, A., Estep, K., Fanuel, F.B.B., Farinha, C.S.E.S., Faro, A., Farzadfar, F., Fazeli, M.S., Feigin, V.L., Fereshtehnejad, S.M., Fernandes, J.C., Ferrari, A.J., Feyissa, T.R., Filip, I., Fischer, F., Fitzmaurice, C., Flaxman, A.D., Flor, L.S., Foigt, N., Foreman, K.J., Franklin, R.C., Fullman, N., Fürst, T., Furtado, J.M., Futran, N.D., Gakidou, E., Ganji, M., Garcia-Basteiro, A.L., Gebre, T., Gebrehiwot, T.T., Geleto, A., Gemechu, B.L., Gesesew, H.A., Gething, P.W., Ghajar, A., Gibney, K.B., Gill, P.S., Gillum, R.F., Ginawi, I.A.M., Giref, A.Z., Gishu, M.D., Giussani, G., Godwin, W.W., Gold, A.L., Goldberg, E.M., Gona, P.N., Goodridge, A., Gopalani, S.V., Goto, A., Goulart, A.C., Griswold, M., Gugnani, H.C., Gupta, R., Gupta, R., Gupta, T., Gupta, V., Hafezi-Nejad, N., Hailu, A.D., Hailu, G.B., Hamadeh, R.R., Hamidi, S., Handal, A.J., Hankey, G.J., Hao, Y., Harb, H.L., Hareri, H.A., Haro, J.M., Harvey, J., Hassanvand, M.S., Havmoeller, R., Hawley, C., Hay, R.J., Hay, S.I., Henry, N.J., Heredia-Pi, I.B., Heydarpour, P., Hoek, H.W., Hoffman, H.J., Horita, N., Hosgood, H.D., Hostiuc, S., Hotez, P.J., Hoy, D.G., Htet, A.S., Hu, G., Huang, H., Huynh, C., Iburg, K.M., Igumbor, E.U., Ikeda, C., Irvine, C.M.S., Jacobsen, K.H., Jahanmehr, N., Jakovljevic, M.B., Jassal, S.K., Javanbakht, M., Jayaraman, S.P., Jeemon, P., Jensen, P.N., Jha, V., Jiang, G., John, D., Johnson, C.O., Johnson, S.C., Jonas, J.B., Jürisson, M., Kabir, Z., Kadel, R., Kahsay, A., Kamal, R., Kan, H., Karam, N.E., Karch, A., Karema, C.K., Kasaeian, A., Kassa, G.M., Kassaw, N.A., Kassebaum, N.J., Kastor, A., Katikireddi, S.V., Kaul, A., Kawakami, N., Keiyoro, P.N., Kengne, A.P., Keren, A., Khader, Y.S., Khalil, I.A., Khan, E.A., Khang, Y.H., Khosravi, A., Khubchandani, J., Kieling, C., Kim, D., Kim, P., Kim, Y.J., Kimokoti, R.W., Kinfu, Y., Kisa, A., Kissimova-Skarbek, K.A., Kivimaki, M., Knudsen, A.K., Kokubo, Y., Kolte, D., Kopec, J.A., Kosen, S., Koul, P.A., Koyanagi, A., Kravchenko, M., Krishnaswami, S., Krohn, K.J., Defo, B.K., Bicer, B.K., Kumar, G.A., Kumar, P., Kumar, S., Kyu, H.H., Lal, D.K., Lalloo, R., Lambert, N., Lan, Q., Larsson, A., Lavados, P.M., Leasher, J.L., Lee, J.T., Lee, P.H., Leigh, J., Leshargie, C.T., Leung, J., Leung, R., Levi, M., Li, Y., Li, Y., Kappe, D.L., Liang, X., Liben, M.L., Lim, S.S., Linn, S., Liu, A., Liu, P.Y., Liu, S., Liu, Y., Lodha, R., Logroscino, G., London, S.J., Looker, K.J., Lopez, A.D., Lorkowski, S., Lotufo, P.A., Low, N., Lozano, R., Lucas, T.C.D., Macarayan, E.R.K., Razek, H.M.A.E., Razek, M.M.A.E., Mahdavi, M., Majdan, M., Majdzadeh, R., Majeed, A., Malekzadeh, R., Malhotra, R., Malta, D.C., Mamun, A.A., Manguerra, H., Manhertz, T., Mantilla, A., Mantovani, L.G., Mapoma, C.C., Marczak, L.B., Martinez-Raga, J., Martins-Melo, F.R., Martopullo, I., März, W., Mathur, M.R., Mazidi, M., McAlinden, C., McGaughey, M., McGrath, J.J., McKee, M., McNellan, C., Mehata, S., Mehndiratta, M.M., Mekonnen, T.C., Memiah, P., Memish, Z.A., Mendoza, W., Mengistie, M.A., Mengistu, D.T., Mensah, G.A., Meretoja, A., Meretoja, T.J., Mezgebe, H.B., Micha, R.,

Millear, A., Miller, T.R., Mills, E.J., Mirarefin, M., Mirrakhimov, E.M., Misganaw, A., Mishra, S.R., Mitchell, P.B., Mohammad, K.A., Mohammadi, A., Mohammed, K.E., Mohammed, S., Mohanty, S.K., Mokdad, A.H., Mollenkopf, S.K., Monasta, L., Hernandez, J.M., Montico, M., Moradi-Lakeh, M., Moraga, P., Mori, R., Morozoff, C., Morrison, S.D., Moses, M., Mountjoy-Venning, C., Mruts, K.B., Mueller, U.O., Muller, K., Murdoch, M.E., Murthy, G.V.S., Musa, K.I., Nachega, J.B., Nagel, G., Naghavi, M., Naheed, A., Naidoo, K.S., Naldi, L., Nangia, V., Natarajan, G., Negasa, D.E., Negoi, I., Negoi, R.I., Newton, C.R., Ngunjiri, J.W., Nguyen, C.T., Nguyen, G., Nguyen, M., Nguyen, Q.L., Nguyen, T.H., Nichols, E., Ningrum, D.N.A., Nolte, S., Nong, V.M., Norrving, B., Noubiap, J.J.N., O'Donnell, M.J., Ogbo, F.A., Oh, I.H., Okoro, A., Oladimeji, O., Olagunju, A.T., Olagunju, T.O., Olsen, H.E., Olusanya, B.O., Olusanya, J.O., Ong, K., Opio, J.N., Oren, E., Ortiz, A., Osgood-Zimmerman, A., Osman, M., Owolabi, M.O., Pa, M., Pacella, R.E., Pana, A., Panda, B.K., Papachristou, C., Park, E.K., Parry, C.D., Parsaeian, M., Patten, S.B., Patton, G.C., Paulson, K., Pearce, N., Pereira, D.M., Perico, N., Pesudovs, K., Peterson, C.B., Petzold, M., Phillips, M.R., Pigott, D.M., Pillay, J.D., Pinho, C., Plass, D., Pletcher, M.A., Popova, S., Poulton, R.G., Pourmalek, F., Prabhakaran, D., Prasad, N., Prasad, N.M., Purcell, C., Qorbani, M., Quansah, R., Rabiee, R.H.S., Radfar, A., Rafay, A., Rahimi, K., Rahimi-Movaghar, A., Rahimi-Movaghar, V., Rahman, M., Rahman, M.H.U., Rai, R.K., Rajsic, S., Ram, U., Ranabhat, C.L., Rankin, Z., Rao, P.V., Rao, P.C., Rawaf, S., Ray, S.E., Reiner, R.C., Reinig, N., Reitsma, M.B., Remuzzi, G., Renzaho, A.M.N., Resnikoff, S., Rezaei, S., Ribeiro, A.L., Ronfani, L., Roshandel, G., Roth, G.A., Roy, A., Rubagotti, E., Ruhago, G.M., Saadat, S., Sadat, N., Safdarian, M., Safi, S., Safiri, S., Sagar, R., Sahathevan, R., Salama, J., Salomon, J.A., Salvi, S.S., Samy, A.M., Sanabria, J.R., Santomauro, D., Santos, I.S., Santos, J.V., Milicevic, M.M.S., Sartorius, B., Satpathy, M., Sawhney, M., Saxena, S., Schmidt, M.I., Schneider, I.J.C., Schöttker, B., Schwebel, D.C., Schwendicke, F., Seedat, S., Sepanlou, S.G., Servan-Mori, E.E., Setegn, T., Shackelford, K.A., Shaheen, A., Shaikh, M.A., Shamsipour, M., Islam, S.M.S., Sharma, J., Sharma, R., She, J., Shi, P., Shields, C., Shigematsu, M., Shinohara, Y., Shiri, R., Shirkoohi, R., Shirude, S., Shishani, K., Shrime, M.G., Sibai, A.M., Sigfusdottir, I.D., Silva, D.A.S., Silva, J.P., Silveira, D.G.A., Singh, J.A., Singh, N.P., Sinha, D.N., Skiadaresi, E., Skirbekk, V., Slepak, E.L., Sligar, A., Smith, D.L., Smith, M., Sobaih, B.H.A., Sobngwi, E., Sorensen, R.J.D., Sousa, T.C.M., Sposato, L.A., Sreeramareddy, C.T., Srinivasan, V., Stanaway, J.D., Stathopoulou, V., Steel, N., Stein, D.J., Stein, M.B., Steiner, C., Steiner, T.J., Steinke, S., Stokes, M.A., Stovner, L.J., Strub, B., Subart, M., Sufiyan, M.B., Abdulkader, R.S., Sunguya, B.F., Sur, P.J., Swaminathan, S., Sykes, B.L., Sylte, D.O., Tabarés-Seisdedos, R., Taffere, G.R., Takala, J.S., Tandon, N., Tavakkoli, M., Taveira, N., Taylor, H.R., Tehrani-Banihashemi, A., Tekelab, T., Shifa, G.T., Terkawi, A.S., Tesfaye, D.J., Tesssema, B., Thamsuwan, O., Thomas, K.E., Thrift, A.G., Tiruye, T.Y., Tobe-Gai, R., Tollanes, M.C., Tonelli, M., Topor-Madry, R., Tortajada, M., Touvier, M., Tran, B.X., Tripathi, S., Troeger, C., Truelsen, T., Tsoi, D., Tuem, K.B., Tuzcu, E.M., Tyrovolas, S., Ukwaja, K.N., Undurraga, E.A., Uneke, C.J., Updike, R., Uthman, O.A., Uzochukwu, B.S.C., Van Boven, J.F.M., Varughese, S., Vasankari, T., Venkatesh, S., Venketasubramanian, N., Vidavalur, R., Violante, F.S., Vladimirov, S.K., Vlassov, V.V., Vollset, S.E., Wadilo, F., Wakayo, T., Wang, Y.P., Weaver, M., Weichenthal, S., Weiderpass, E., Weintraub, R.G., Werdecker, A., Westerman, R., Whiteford, H.A., Wijeratne, T., Wiysonge, C.S., Wolfe, C.D.A., Woodbrook, R., Woolf, A.D., Workicho, A., Hanson, S.W., Xavier, D., Xu, G., Yadgir, S., Yaghoubi, M., Yakob, B., Yan, L.L., Yano, Y., Ye, P., Yimam, H.H., Yip, P., Yonemoto, N., Yoon, S.J., Yotebieng, M., Younis, M.Z., Zaidi, Z., Zaki, M.E.S., Zegeye, E.A., Zenebe, Z.M., Zhang, X., Zhou, M., Zipkin, B., Zodpey, S., Zuhlke, L.J., Murray, C.J.L., 2017. Global, regional, and national incidence, prevalence, and years lived with disability for 328 diseases and injuries for 195 countries, 1990-2016: a systematic analysis for the global burden of disease study 2016. Lancet. 1474-547X. 390 (10100), 1211–1259. Available from: http://www.journals.elsevier.com/the-lancet/. doi:10.1016/S0140-6736(17)32154-2.

Wakida, Y., Mafune, K., Noguchi, Y., Ido, H., Hino, A., Hiro, H., 2018. Relationship between organisational resources and occupational stress among japanese nurses. Occup. Health Nurs. A355 (2), 864. doi:10.1136/oemed-2018-ICOHabstracts.1014.

Weinstein, S.M., Herring, S.A., 1992. Nerve problems and compartment syndromes in the hand, wrist, and forearm. Clin. Sports Med. 11 (1), 161–188.

Westacott, D.J., Minns, J.I., Foguet, P., 2011. The diagnostic accuracy of magnetic resonance imaging and ultrasonography in gluteal tendon tears—a systematic review. HIP Int. 1120-7000. 21 (6), 637–645. Available from: http://www.hip-int.com/public/HIP/issue/Article.action?cmd=navigate&urlkey=Public_Details&uid=213EDAD3-DCDA-4660-A839-F32E6729765F&t=hip. doi:10.5301/HIP.2011.8759.

Woolf, C.J., Salter, M.W., 2000. Neuronal plasticity: increasing the gain in pain. Science 288 (5472), 1765–1768. doi:10.1126/science.288.5472.1765.

World Health Organization, 2001. International Classification of Functioning, Disability and Health (ICF). Available from: https://www.who.int/standards/classifications/international-classification-of-functioning-disability-and-health.

World Health Organization, 2020. WHO guidelines on physical activity and sedentary behaviour. Available from: https://www.who.int/publications/i/item/9789240015128.

Zheng, Q., Dong, X., Green, D.P., Dong, X., 2022. Peripheral mechanisms of chronic pain. Med. Rev. 2749-9642. 2 (3), 251–270. Available from: https://www.degruyter.com/journal/key/mr/html?lang=en. doi:10.1515/mr-2022-0013.

Zogopoulos, P., Vasileiou, I., Patsouris, E., Theocharis, S.E., 2013. The role of endocannabinoids in pain modulation. Fundam. Clin. Pharmacol. 1472-8206. 27 (1), 64–80. doi:10.1111/fcp.12008.

Chapter 17

The role of ultrasound in assessing repetitive motion injuries in musculofascial system

Kalyan Chakravarthi Kosuri[a], Venu Madhav Nelluri[a], Rajagopal Kadavigere[b], Suresh Sukumar[c], Sneha Ravichandran[c], Leena R. David Berna Uzun[c,d], Dilber Uzun Ozsahin[d,e,f], and Ilker Ozsahin[e]

[a]*Department of Anatomy, Manipal Univeristy College Malaysia, Melaka, Malaysia*

[b]*Department of Radiodiagnosis and Imaging, Kasturba Medical College, Kasturba Medical College, Manipal, India*

[c]*Department of Medical Imaging Technology, Manipal College of Health Professions, Manipal Academy of Higher Education, Manipal, Karnataka, India*

[d]*Medical Diagnostic Imaging Department, College of Health Sciences, University of Sharjah, Sharjah, United Arab Emirates*

[e]*Irfan Suat Gunsel Operational Research Institute, Near East University, Nicosia, Turkish Republic of Northern Cyprus (TRNC), Turkey*

[f]*Research Institute for Medical and Health Sciences, University of Sharjah, Sharjah, United Arab Emirates*

1 Musculofascial ultrasound

1.1 Introduction

Ultrasound plays a crucial role in the assessment and management of musculofascial injuries in the field of medicine, offering real-time imaging, dynamic evaluation, and the ability to guide interventions accurately. It plays a significant role in diagnosing injuries, monitoring healing progress, and optimizing treatment strategies to facilitate recovery and restore function. Human anatomy serves as the foundation of ultrasound imaging, guiding image acquisition, interpretation, and clinical decision-making. A comprehensive knowledge of human anatomy is essential for maximizing the accurate diagnostic utility of ultrasound and providing high-quality patient care.

Musculofascial ultrasound is a diagnostic imaging technique that utilizes high-frequency sound waves to produce real-time images of the musculofascial system. Unlike other imaging modalities such as X-rays, computed tomography (CT) scans, or magnetic resonance imaging (MRI), musculoskeletal ultrasound does not involve ionizing radiation, making it safe for repeated use and ideal for use in pediatrics, pregnant patients, and certain patient populations. Musculofascial ultrasound can be performed at the bedside or in the clinic, providing immediate evaluation and timely intervention.

1.2 Principles

Musculofascial ultrasound does not require incisions or injections; it reduces the risk of complications and patient discomfort. Ultrasound waves are emitted by a transducer, which is placed directly on the skin over the area of interest. The sound waves penetrate the tissues and are reflected back to the transducer, producing echoes. The echoes are then processed by the ultrasound machine to generate detailed images of the internal structures, including muscles, tendons, ligaments, joints, and nerves. Musculoskeletal ultrasound offers the advantage of real-time imaging, allowing for dynamic assessment of structures during movement or stress.

1.3 Applications

Musculofascial ultrasound is used to diagnose a wide range of conditions affecting the muscles, tendons, ligaments, joints, and soft tissues. Common indications include:

Advances in Medical Imaging. https://doi.org/10.1016/B978-0-443-28967-5.00014-9

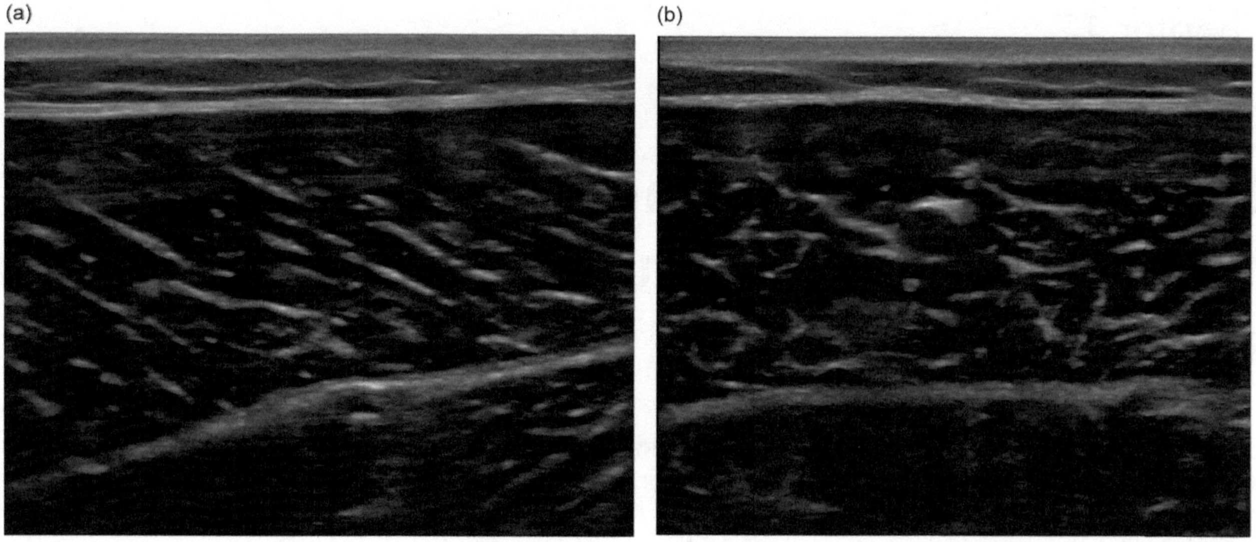

FIG. 17.1 Ultrasound anatomy of skeletal muscle. *(From Paoletta, M., Moretti, A., Liguori, S., Snichelotto, F., Menditto, I., Toro, G., Gimigliano, F., Iolascon, G., 2021. Ultrasound imaging in sport-related muscle injuries: Pitfalls and opportunities. Medicina (Lithuania), 57(10). https://doi.org/10.3390/medicina57101040)*

1. Evaluation of soft tissue injuries, such as muscle strains, tendon tears, ligament injuries, and bursitis.
2. Diagnosis and monitoring of joint disorders, such as arthritis, synovitis, effusions, and cartilage abnormalities.
3. Assessment of tendon pathology, including tendinopathy, tendinitis, and tendon tears.
4. Visualization of peripheral nerves for entrapment, compression, or neuropathy.
5. Musculoskeletal ultrasound is also used to guide therapeutic interventions, such as injections, aspirations, biopsies, and minimally invasive procedures.

Anatomical knowledge plays a pivotal role in the ultrasound assessment of repetitive motion injuries in the musculofascial system by providing a foundational understanding of normal anatomical structures and their dynamic interactions. This knowledge is essential for accurately identifying and interpreting ultrasound images, distinguishing between healthy and injured tissues, and localizing specific abnormalities within the intricate musculofascial network. Recognizing normal variations, understanding the biomechanics of muscles and tendons during repetitive movements, and differentiating between muscle groups contribute to precise diagnosis and effective treatment planning (Fig. 17.1).

2 Repetitive motion injuries in the musculofascial system and ultrasound imaging

2.1 Introduction

Repetitive motion injuries, also known as motion injuries or repetitive strain injuries, occur when repeated movements strain muscles, tendons, ligaments, nerves, or other soft tissues. These injuries often result from overuse or misuse of a particular body part during activities such as typing, using a computer mouse, playing musical instruments, or engaging in sports. Understanding the underlying anatomy of these motion injuries helps healthcare professionals accurately diagnose the condition and develop appropriate treatment plans, which may include rest, physical therapy, antiinflammatory medications, corticosteroid injections, or, in severe cases, surgical intervention.

2.2 Skeletal muscles

2.2.1 Introduction

Muscles are the contractile tissues in the human body responsible for generating force and producing movement. Comprised of muscle fibers arranged in bundles, skeletal muscle, also known as striated muscle or voluntary muscle, is one of the three major types of muscle tissue found in the human body. These muscles are attached to bones by tendons and play a fundamental role in movement, posture, and stability. Ultrasound imaging is a valuable tool for diagnosing injuries of skeletal muscles, providing real-time visualization of muscle structure and detecting abnormalities such as tears, strains, and contusions.

2.2.2 Anatomical structure

Skeletal muscle is composed of long, cylindrical cells called muscle fibers. These fibers are multinucleated and contain specialized structures called myofibrils, which are responsible for muscle contraction. Skeletal muscle fibers are organized into bundles called fascicles, surrounded by perimysium, forming the muscle belly. These muscles connect to bones via tendons or aponeuroses, transmitting force from muscle contractions to bone movement. Each muscle has an origin, typically the less movable attachment closer to the body's midline, and an insertion, which moves during contraction. Highly vascularized, skeletal muscles receive oxygen and nutrients from blood vessels, which are crucial for energy production and waste removal. Understanding this anatomical arrangement is fundamental for grasping muscle function in movement and the diagnosis and treatment of musculoskeletal injuries and conditions.

2.2.3 Muscle tears

Muscle tears, also known as muscle strains or muscle ruptures, occur when muscle fibers are overstretched or torn due to sudden or excessive force. Ultrasound can reveal discontinuity or disruption of muscle fibers, hematoma formation, and surrounding tissue inflammation. The severity of the tear can vary from mild (Grade I) to severe (Grade III) based on the extent of muscle fiber disruption.

Grade I muscle tear: In a mild muscle tear, there may be minimal disruption of muscle fibers, leading to subtle changes in muscle architecture. On ultrasound, a Grade I tear may appear as focal hypoechoic areas within the muscle belly, indicating localized muscle fiber damage. There may be mild surrounding edema or inflammation, but the overall muscle structure remains relatively intact.

Grade II muscle tear: A moderate muscle tear involves a more significant disruption of muscle fibers, leading to a partial tear of the muscle. Ultrasound imaging may reveal a larger hypoechoic area within the muscle, with evidence of muscle fiber discontinuity or disruption. The tear may appear as a well-defined defect within the muscle belly, surrounded by echogenic borders representing torn muscle fibers.

Grade III muscle tear: A severe muscle tear involves a complete rupture of the muscle fibers, resulting in a loss of continuity within the muscle. On ultrasound, a Grade III tear may appear as a complete gap or discontinuity in the muscle tissue, with retraction of the torn muscle ends. There may be a hematoma present within the torn muscle, indicated by a hypoechoic or anechoic fluid collection surrounded by echogenic debris or clot.

2.2.4 Muscle-tendon junction tear

Some muscle tears occur at the junction between the muscle belly and its tendon, known as the muscle–tendon junction (MTJ). Ultrasound imaging can visualize discontinuity or separation between the muscle and tendon, indicating a tear at the MTJ. The tear may be characterized by a hypoechoic defect at the junction site, with surrounding edema or hematoma extending into the muscle belly or tendon.

Surrounding the site of a muscle tear, there may be reactive changes such as edema, hemorrhage, or inflammation. Ultrasound may show increased echogenicity or swelling of the adjacent tissue, indicative of reactive changes secondary to the injury. Muscles commonly affected by tears and examined using ultrasound include the quadriceps, hamstrings, calf muscles, and rotator cuff muscles. Ultrasound imaging provides real-time visualization and assessment of muscle tears, allowing for accurate diagnosis, grading, and monitoring of the injury.

2.2.5 Muscle contusions

A muscle contusion, commonly known as a bruise, results from blunt trauma to the muscle tissue, leading to localized bleeding and inflammation. Muscle contusions are common injuries in sports and recreational activities but can also occur in everyday situations. Prompt recognition and appropriate management of muscle contusions are essential to facilitate optimal healing and minimize complications.

In ultrasound imaging, muscle contusions typically appear as hypoechoic (darker) regions within the muscle tissue due to the presence of blood, fluid, and disrupted muscle fibers. Initially, a fresh contusion may appear as a focal area of increased echogenicity (brightness) surrounded by hypoechoic fluid, representing acute bleeding and edema. As the contusion evolves over time, it may present as a heterogeneous area with mixed echogenicity, reflecting the organization of hematoma, inflammatory response, and tissue repair. Depending on the severity of the contusion, ultrasound may also detect associated findings such as muscle fiber disruption, hematoma formation, and surrounding tissue edema. In the case of a quadriceps contusion sustained during a football game, ultrasound can help visualize the depth and severity of the injury, facilitating appropriate management strategies such as rest, ice, compression, and physical therapy. Similarly, in a hamstring contusion resulting from a soccer collision, ultrasound can assist in monitoring the healing process, identifying any complications such as muscle tears or scar tissue formation, and guiding rehabilitation protocols to optimize recovery. Overall, ultrasound provides valuable insights into muscle contusions, allowing healthcare professionals to tailor treatment plans and monitor progress effectively.

2.2.6 Muscle hematomas

Hematomas are collections of blood within the muscle tissue, often resulting from trauma or injury. Ultrasound can visualize the presence of fluid collections with variable echogenicity, depending on the stage of hematoma formation. Acute hematomas may appear hypoechoic or anechoic, while chronic hematomas may exhibit internal septations or echogenic debris.

For instance, in the case of a quadriceps muscle hematoma, ultrasound may reveal a hypoechoic area within the muscle tissue, indicating the presence of blood accumulation. This area might exhibit irregular borders and possibly evidence of active bleeding, visualized as blood flow within the hematoma. Similarly, hamstring muscle hematomas may appear as hypoechoic masses within the hamstring muscles, potentially disrupting the normal muscle architecture. Biceps muscle hematomas could manifest as heterogeneous masses within the biceps muscle, while calf muscle hematomas might present as hypoechoic regions surrounded by echogenic muscle fibers. In the abdominal region, rectus abdominis muscle hematomas could be observed as hypoechoic or heterogeneous masses within the muscle tissue of the abdominal wall, providing crucial diagnostic information for appropriate management.

2.2.7 Muscle edema

Muscle edema, characterized by increased fluid accumulation within the muscle tissue, can occur in response to various injuries, including strains, contusions, or inflammatory conditions. Ultrasound may reveal diffuse or focal areas of increased echogenicity within the muscle, indicative of edema, as well as surrounding tissue inflammation. For instance, in the context of muscle trauma or injury, such as a strain or contusion, ultrasound may reveal diffuse or focal areas of increased echogenicity within the affected muscle, indicative of edema. This increased echogenicity results from the presence of fluid within the muscle fibers, disrupting their normal appearance. In addition, muscle edema may be accompanied by other signs, such as muscle thickening or loss of the typical fibrillar pattern. Examples of muscle edema visualized on ultrasound include findings such as increased echogenicity within the quadriceps or hamstring muscles following a strain injury, or within the calf muscles after trauma or overexertion . In cases of inflammatory conditions like myositis, ultrasound may show diffuse muscle edema affecting multiple muscle groups. Utilizing ultrasound imaging allows for a noninvasive and real-time assessment of muscle edema, aiding in diagnosis, treatment planning, and monitoring of the condition over time.

2.2.8 Muscle atrophy

Muscle atrophy refers to the decrease in muscle mass and size, often accompanied by a loss of muscle strength and function. It can occur because of various factors, including disuse, aging, injury, neurological conditions, or systemic diseases. Anatomical knowledge of muscles plays a crucial role in identifying muscle atrophy in ultrasound imaging. Understanding the normal appearance and architecture of muscles allows clinicians to recognize deviations indicative of atrophy.

Muscle size and thickness

Muscle size and thickness are key parameters used to assess muscle morphology, function, and health. Accurate measurement of muscle dimensions provides valuable information for understanding muscle status, diagnosing muscle-related conditions, and guiding interventions aimed at optimizing muscle health and performance. Example—Quadriceps muscle atrophy The quadriceps muscle is a large group of muscles located in the anterior compartment of the thigh. It consists of four individual muscles: vastus medialis, vastus lateralis, vastus intermedius, and rectus femoris, that work together to extend the knee joint, allowing for actions such as standing up from a seated position, walking, running, and jumping. In addition, the rectus femoris assists in flexing the hip joint. In cases of quadriceps muscle atrophy, ultrasound may reveal a reduction in the size and thickness of the affected muscle group compared to the contralateral side. Anatomical landmarks, such as the patella and femur, can be used as reference points to assess muscle size and identify atrophy.

Muscle architecture

Muscle architecture refers to the arrangement of muscle fibers within a muscle and their orientation relative to the muscle's line of action. It plays a crucial role in determining the muscle's functional properties, such as force generation, range of motion, and muscle contraction dynamics. Understanding the structural organization of muscles facilitates the prediction of muscle performance, guides biomechanical modeling, informs training and rehabilitation strategies, and enhances our overall comprehension of human movement and motor control.

Biceps brachii muscle atrophy refers to the wasting or loss of muscle mass in the biceps brachii. The biceps brachii is a two-headed muscle (long head and short head) located in the anterior compartment of the arm (Moore et al., 2017). It is one of the most well-known and easily recognizable muscles in the human body due to its distinctive appearance when flexed. The primary function of the biceps brachii is to flex the elbow joint, bringing the forearm closer to the upper arm, and to supinate the forearm (rotating the palm from a palm-down (pronated) to a palm-up (supinated) position). The long head of the biceps helps stabilize the shoulder joint, particularly during activities involving overhead movements or shoulder abduction (Terry and Chopp, 2000). In cases of biceps brachii muscle atrophy, muscle architecture may show alterations, such as decreased fascicle length or disruption in the normal arrangement of muscle fibers. In severe cases of atrophy, there may be disorganization of muscle fibers, leading to an irregular or heterogeneous appearance Anatomical landmarks such as the humerus and brachialis muscle can be used to assess muscle architecture and identify abnormalities.

Echogenicity and texture

Echogenicity refers to the brightness or darkness of a structure, and texture refers to the internal structural pattern or appearance on ultrasound imaging. In skeletal muscle, echogenicity is primarily influenced by the composition of the tissue and the density of its components. Healthy skeletal muscle typically appears hypoechoic (dark) on ultrasound due to its high-water content and relatively low reflectivity (Walker, 2004). In skeletal muscle, texture is determined by the arrangement of muscle fibers, connective tissue, and other components within the muscle. Healthy skeletal muscle typically exhibits a homogeneous or uniform texture, with evenly distributed muscle fibers and minimal internal echoes.

The gastrocnemius muscle is a two-headed muscle located in the posterior calf. The gastrocnemius muscle has two heads: the medial head originates from the medial condyle of the femur, just above the knee joint, and the lateral head originates from the lateral condyle of the femur, also just above the knee joint. Both heads of the gastrocnemius merge to form the Achilles tendon, which inserts into the calcaneus (heel bone) of the foot. The gastrocnemius muscle is primarily innervated by the tibial nerve, a branch of the sciatic nerve (specifically the posterior division of the sacral plexus). The gastrocnemius muscle is a powerful muscle with multiple actions. Its primary function is plantar flexion of the ankle joint, which is important for activities such as walking, running, jumping, and pushing off during propulsion. It also contributes to flexion of the knee joint and stabilization of the ankle, particularly during tasks that involve maintaining balance and posture. Its actions are essential for various activities of daily living and athletic performance, highlighting the importance of this muscle in lower limb function. Assessing the echogenicity and texture of the gastrocnemius muscle using ultrasound imaging can help identify structural abnormalities, evaluate tissue composition, and detect pathological conditions affecting the muscle Atrophy of the gastrocnemius muscle may result in increased echogenicity and altered muscle texture, reflecting fatty infiltration or fibrosis. Anatomical landmarks such as the calcaneus and tibia can be used to evaluate muscle echogenicity and texture.

Anatomical knowledge of echogenicity and texture of muscles helps clinicians in diagnosing muscle pathologies, monitoring disease progression, and guiding treatment strategies, such as rehabilitation exercises or medical interventions.

Comparison with contralateral side

Comparison with the contralateral or unaffected side is important for assessing asymmetry and identifying atrophy. Example—Deltoid muscle atrophy: The deltoid muscle is a triangular-shaped muscle located in the shoulder region, consisting of anterior, middle, and posterior heads. Comparison of the deltoid muscle on the affected side with the contralateral side can reveal asymmetry and differences in muscle size and appearance, indicating atrophy. Anatomical landmarks such as the acromion and humerus can be used for side-to-side comparison.

By integrating anatomical knowledge with ultrasound imaging, clinicians can accurately identify and characterize muscle atrophy, facilitating early detection, monitoring, and management of musculoskeletal conditions.

2.3 Entrapment neuropathy

In some cases, injuries or conditions affecting skeletal muscles may involve compression or entrapment of nearby nerves, leading to symptoms such as pain, tingling, weakness, or other neurological symptoms. Ultrasound can identify structural abnormalities, such as muscle hypertrophy or mass lesions, that may be contributing to nerve compression.

2.3.1 Carpal tunnel syndrome

Carpal tunnel syndrome (CTS) is one of the most common nerve entrapment syndromes, commonly caused by repetitive wrist movements or conditions like rheumatoid arthritis and anatomical variations of the flexor tendons of the forearm or lumbricals. It is characterized by compression of the median nerve as it passes through the carpal tunnel in the wrist. Ultrasound imaging is a valuable tool in diagnosing and assessing CTS.

Evaluation of CTS: Ultrasound imaging is an effective and noninvasive tool for diagnosing CTS, providing detailed anatomical information about the median nerve and aiding in treatment planning and monitoring. Ultrasound allows for direct visualization of the median nerve as it passes through the carpal tunnel. The ultrasound probe is placed transversely at the wrist crease, and the median nerve is identified as a hyperechoic structure surrounded by the flexor tendons. In CTS, the median nerve often becomes compressed, leading to swelling and increased cross-sectional area. Ultrasound can measure the CSA of the median nerve at the level of the carpal tunnel, providing quantitative data for diagnosis and severity assessment. Ultrasound can assess the morphology of the median nerve, including its shape, echogenicity, and the presence of focal abnormalities such as hypoechoic areas or flattening. Changes in nerve morphology can indicate nerve compression and help differentiate between acute and chronic CTS.

Dynamic ultrasound involves moving the wrist or fingers while imaging the median nerve, allowing for the assessment of nerve mobility and the presence of dynamic compression. Dynamic ultrasound can reveal changes in nerve morphology and cross-sectional area with wrist flexion and extension, aiding in the diagnosis of CTS and the assessment of its severity . Ultrasound imaging of the unaffected contralateral wrist can serve as a reference for comparison, helping to identify asymmetry in nerve morphology and cross-sectional area, which may indicate CTS. Ultrasound can guide the placement of corticosteroid injections into the carpal tunnel for therapeutic purposes). Real-time visualization provided by ultrasound ensures accurate needle placement and minimizes the risk of complications.

2.3.2 Cubital tunnel syndrome

Cubital tunnel syndrome is a condition that affects the ulnar nerve as it passes through the cubital tunnel, which is a narrow passageway on the inside of the elbow. The ulnar nerve is one of the three main nerves in the arm and is responsible for providing sensation to the little finger and half of the ring finger, as well as controlling some of the muscles in the hand. Compression of the ulnar nerve at the elbow is often caused by prolonged elbow flexion or direct trauma.

Evaluation of cubital tunnel syndrome by ultrasound involves assessing the ulnar nerve's morphology, size, and position as it traverses the cubital tunnel, examining for any signs of compression or irritation, such as swelling, hypoechoic areas indicative of edema, or abnormalities in nerve shape or mobility during elbow movement. In addition, ultrasound allows for real-time dynamic assessment, enabling observation of nerve behavior with elbow flexion and extension to determine if there's any impingement or subluxation. Comparative analysis with the contralateral side aids in identifying asymmetry or pathology. Ultrasound can also evaluate surrounding structures, like muscles, tendons, and ligaments, for contributing factors to nerve compression (Chakravarthi et al., 2024; Volpe et al., 2009). This imaging modality offers a noninvasive, cost-effective means to diagnose and monitor cubital tunnel syndrome, guiding treatment decisions and facilitating targeted interventions.

2.3.3 Radial tunnel syndrome

Radial tunnel syndrome is a condition characterized by compression or irritation of the radial nerve as it passes through the radial tunnel, which is a narrow passageway located in the forearm. The radial nerve is one of the three main nerves in the arm, responsible for providing sensation to the back of the hand, thumb, and first three fingers, as well as controlling certain muscles that enable movement of the wrist, hand, and fingers. Compression of the radial nerve in the radial tunnel can occur due to various factors, including repetitive use of the forearm muscles, direct trauma to the area, or anatomical abnormalities. Symptoms of radial tunnel syndrome may include pain or tenderness in the forearm, especially with gripping or twisting motions, weakness in the wrist or hand, and occasionally numbness or tingling in the fingers.

Ultrasound allows for visualization of the radial nerve and assessment of its size, shape, and integrity . Any changes in nerve morphology, such as enlargement, compression, or focal abnormalities, may indicate radial tunnel syndrome . Ultrasound can identify areas of compression or entrapment along the course of the radial nerve in the radial tunnel . This includes assessing for any compressive structures, such as fibrous bands, muscle bellies, or bony prominences, that may be impinging on the nerve . Real-time ultrasound imaging enables dynamic assessment of the radial nerve during wrist and forearm movement (Bodner et al., 2001). This can help identify areas of increased pressure or friction on the nerve with specific activities or positions . Ultrasound allows for evaluation of the surrounding soft tissues, including muscles, tendons, and ligaments. This can help identify any contributing factors to nerve compression, such as muscle hypertrophy or inflammation. In some cases, ultrasound may also assess the vascular supply to the radial nerve to rule out any vascular abnormalities contributing to symptoms.

2.3.4 *Meralgia paresthetica*

Meralgia paresthetica is a nerve condition characterized by tingling, numbness, or burning pain in the outer thigh area, typically resulting from compression or irritation of the lateral femoral cutaneous nerve (LFCN) as it passes under the inguinal ligament. Common causes include tight clothing, obesity, pregnancy, or injury to the hip or thigh region. Symptoms may worsen with activities that put pressure on the affected area.

Ultrasound evaluation for meralgia paresthetica involves a thorough assessment of the LFCN and surrounding structures in the inguinal region and thigh . The ultrasound examination aims to identify any abnormalities or compression of the LFCN, typically where it passes under the inguinal ligament . High-resolution ultrasound imaging allows for visualization of the nerve's course, size, and shape, as well as any signs of entrapment or compression by adjacent structures, such as the inguinal ligament or nearby blood vessels. Dynamic assessment during movement of the hip and thigh can provide valuable insights into nerve mobility and any changes in symptoms In addition, ultrasound can assess other potential causes of nerve compression, such as masses or hernias in the inguinal region (Seror and Seror, 2006). Overall, ultrasound serves as a valuable diagnostic tool in evaluating meralgia paresthetica, providing real-time imaging that can help guide treatment decisions and interventions.

2.3.5 *Tarsal tunnel syndrome*

Tarsal tunnel syndrome is a condition characterized by the compression or irritation of the tibial nerve as it passes through the tarsal tunnel, a narrow passageway located on the inside of the ankle. The tibial nerve is one of the major nerves in the leg, responsible for providing sensation to the bottom of the foot and controlling certain muscles that enable the movement of the foot and toes. Compression of the tibial nerve within the tarsal tunnel can occur due to various factors, including injury, inflammation, or anatomical abnormalities. Symptoms of tarsal tunnel syndrome may include pain, burning, tingling, or numbness in the ankle, heel, or sole of the foot, which may worsen with activity or prolonged standing.

Ultrasound examination for tarsal tunnel syndrome involves assessing the tibial nerve and adjacent structures within the tarsal tunnel. Key findings typically include enlargement or swelling of the tibial nerve, often indicating compression or irritation . Ultrasound can detect flattening or distortion of the nerve at the site of compression, providing valuable information about the severity and location of nerve entrapment . In addition, ultrasound can identify fluid accumulation (edema) around the tibial nerve, suggesting inflammation or irritation. Anatomical variations, such as ganglia, cysts, or bony prominences within the tarsal tunnel, may also be visualized, contributing to nerve compression . Real-time ultrasound imaging allows for dynamic assessment of the tibial nerve and surrounding structures during ankle movement, providing insights into nerve mobility and any changes in symptoms with specific movements.

2.3.6 *Piriformis syndrome*

Piriformis syndrome is a condition characterized by compression or irritation of the sciatic nerve as it passes through or under the piriformis muscle in the buttock region The piriformis muscle is a small muscle located deep in the buttocks, and the sciatic nerve is the largest nerve in the body, extending from the lower back down through the buttocks and into the legs When the piriformis muscle spasms, tightens, or swells, it can compress or irritate the sciatic nerve, leading to symptoms such as pain, tingling, or numbness that radiate down the back of the leg, mimicking sciatica The exact cause of piriformis syndrome is not always clear, but contributing factors may include overuse or repetitive strain, muscle imbalances, trauma, or anatomical variations.

Ultrasound evaluation for piriformis syndrome involves a thorough examination of the piriformis muscle, sciatic nerve, and surrounding structures in the buttock region . Key aspects of the evaluation include assessing the size, shape, and integrity of the piriformis muscle, as well as identifying any signs of inflammation, muscle hypertrophy, or abnormalities (Miller et al., 2012). Ultrasound can visualize the course of the sciatic nerve as it passes beneath or through the piriformis muscle, allowing for dynamic assessment of nerve mobility and any signs of compression or irritation In addition, ultrasound can assess other potential causes of sciatic nerve compression, such as adjacent structures or anatomical variations . Real-time imaging provides valuable insights into muscle and nerve behavior during movement, aiding in the diagnosis and management of piriformis syndrome.

In each of these nerve entrapment syndromes, ultrasound imaging provides valuable information about nerve morphology, compression sites, and dynamic changes during movement, aiding in diagnosis and guiding treatment decisions.

2.4 Muscular variations or anomalies

Muscular variations or anomalies encompass a spectrum of deviations from the typical anatomy, presenting diverse morphological alterations in muscle structure, location, or development . These variations can include the presence of accessory muscles, where additional muscle structures are found beyond the usual anatomy (Chakravarthi et al., 2023). Conversely, some individuals may exhibit absent muscles or experience abnormalities such as muscle fusion, where adjacent muscles merge, or muscle splitting, resulting in the division of a single muscle into multiple segments (Chakravarthi and Shetty, 2025). Abnormal muscle insertions, hypertrophy, atrophy, and variations in muscle size can also occur. In addition, rare anomalies may involve polymelia or polydactyly, presenting with extra limbs or digits accompanied by associated muscles. While many of these anatomical variations are typically benign and asymptomatic, their recognition is essential for healthcare professionals in fields such as anatomy, surgery, and radiology to ensure accurate assessments and informed clinical decision-making. Ultrasound plays a pivotal role in identifying anatomical muscular variations or anomalies due to its real-time imaging capabilities and noninvasive nature. This imaging modality allows for detailed visualization of muscle structures, facilitating the detection of variations such as accessory muscles, absent muscles, muscle fusion, or splitting. In addition, abnormalities in muscle insertions and the presence of polymelia or polydactyly can be effectively visualized through ultrasound imaging. This diagnostic tool provides valuable insights for healthcare professionals in various fields, aiding in accurate anatomical assessments, preoperative planning, and ongoing monitoring of muscular variations, ensuring comprehensive and informed patient care.

3 Tendons

3.1 Structure and function

Tendons are dense, fibrous connective tissues that connect muscles to bones. They are composed primarily of high-collagen fibers, mostly Type-1, arranged in a cross-linked triple-helix structure, providing strength and resilience. Tendons also contain other extracellular matrix components such as elastin, proteoglycans, and water molecules that bridge the helix. The complex hierarchical organization of collagen macromolecules is grouped into fibrils and fibers, which, in turn, are bundled into fibers and fascicles surrounded by a thin layer of vascularized connective tissue called endotendon. Fascicles are bundled together to form the tendon proper, which is the functional unit of the tendon. The hierarchical arrangement of collagen molecules, fibrils, fibers, fascicles, and tendon proper units provides them with the necessary strength and resilience to withstand tensile forces and transmit muscle forces to bones.

The Achilles tendon, the largest tendon in the human body, primarily consists of longitudinally oriented collagen fibers. These fibers run parallel to the longitudinal axis of the tendon, providing strength and stability during activities such as walking, running, and jumping (Komi et al., 1992). The tendons of the rotator cuff, including the supraspinatus, infraspinatus, teres minor, and subscapularis tendons, have a more complex fiber orientation. While they also contain longitudinal fibers, they additionally have a significant component of transverse and obliquely oriented fibers. This arrangement provides multidirectional stability to the shoulder joint, crucial for activities involving overhead motion, like throwing or lifting. The patellar ligament, connecting the patella to the tibia, primarily consists of longitudinally oriented collagen fibers. These fibers run parallel to the long axis of the tendon, facilitating the extension of the knee joint during activities such as jumping or climbing stairs (Visnes and Bahr, 2007). Flexor tendons of the hand, such as the flexor digitorum profundus and flexor digitorum superficialis tendons, have a combination of longitudinal and oblique fiber orientations. These tendons enable flexion of the fingers and thumb and are crucial for fine motor tasks, such as grasping objects or typing. Extensor tendons, responsible for extending the fingers and thumb, also exhibit a combination of longitudinal and oblique fiber orientations (Verdan, 1975). These tendons run from the forearm muscles to the fingers and thumb, allowing actions such as straightening the fingers or thumb to grasp objects.

Anatomical knowledge of the hierarchical arrangement of tendons is clinically significant in ultrasound imaging for diagnosing tendon pathology, monitoring treatment response, guiding interventions (tendon injections, percutaneous treatments, or minimally invasive procedures), and facilitating rehabilitation. It provides valuable insights into the structural integrity, function, and healing processes of tendons, aiding in the management of musculoskeletal conditions affecting these vital structures.

Tendons are covered by specialized connective tissue layers that provide protection, support, and facilitate movement. The primary coverings of tendons are the epitenon, paratenon, endotenon, and peritenon. The epitenon is the outermost layer of loose connective tissue surrounding the entire tendon and protects the tendon from external forces while facilitating gliding between the tendon and surrounding tissues. In the Achilles tendon, the epitenon covers the entire length of the tendon, allowing it to slide smoothly within the retrocalcaneal bursa and the paratenon. Injury or inflammation of the epitenon can lead to conditions like tenosynovitis, where the tendon sheath becomes inflamed, causing pain and limited movement. The paratenon is a thin sheath of connective tissue that surrounds certain tendons, particularly those with a more superficial location or tendons that experience significant movement. It provides lubrication and reduces friction between the tendon and its surrounding tissues. The tendons of the wrist extensors in the forearm are enveloped by the paratenon, allowing smooth movement of the tendons within the tendon sheaths during wrist and finger movements. Injuries to the paratenon can result in conditions such as paratenonitis, characterized by inflammation and pain around the tendon. The endotenon is a layer of connective tissue that surrounds individual fascicles within the tendon. It provides support and helps to maintain the structural integrity of the tendon. In the rotator cuff tendons, such as the supraspinatus tendon, the endotenon surrounds each fascicle within the tendon, contributing to the overall strength and function of the tendon. Injuries affecting the endotenon may lead to disruptions in tendon structure and function. The peritenon is a specialized connective tissue layer that surrounds tendons where they pass through synovial sheaths. It provides protection and facilitates movement within the synovial sheath. The flexor tendons of the hand, such as the flexor digitorum profundus and flexor digitorum superficialis tendons, are covered by the peritenon as they pass through the synovial sheaths of the fingers, allowing smooth movement during flexion and extension of the fingers.

Injuries or inflammation of the peritenon can result in conditions such as tenosynovitis, where the synovial sheath becomes inflamed, causing pain and restricted movement. The coverings of tendons help to maintain the structural integrity of tendons, reduce friction during movement, and protect them from external forces, contributing to their overall function in the musculoskeletal system. Thus, understanding the anatomy and function of tendon coverings is essential for interpreting ultrasound images of tendons accurately. It enables clinicians to assess tendon health, diagnose pathology, guide interventions, and monitor treatment responses effectively.

Ultrasound imaging serves a crucial role in assessing tendon injuries, tears, damage, and inflammation. It allows for real-time visualization of tendons, enabling healthcare professionals to identify abnormalities such as tears, ruptures, thickening, irregularity, or hypoechoic areas indicative of inflammation or damage. Ultrasound can accurately detect tendon tears and ruptures by visualizing disruptions in the tendon fibers and assess the extent of the injury, guiding treatment decisions. In addition, ultrasound is sensitive to signs of tendon inflammation, including swelling, increased vascularity, and hypervascularity, aiding in the diagnosis of conditions like tendinitis or tenosynovitis. Furthermore, ultrasound-guided procedures, such as corticosteroid injections, can be performed to provide targeted therapy for tendon injuries or inflammation. Overall, ultrasound imaging offers a noninvasive, cost-effective, and accessible method for evaluating tendon conditions, facilitating timely diagnosis and appropriate management strategies.

4 Ligaments

4.1 Structure and functions

Ligaments are tough bands of fibrous connective tissue that differ from tendons in function and share similarities in their collagen-based structure. Ligaments connect bones to bones and provide stability, support, and limit excessive movement of the joints (Chakravarthi, 2024; Chakravarthi et al., 2017; Woo et al., 1991). Ligaments are primarily composed of collagen fibers, specifically type I collagen, which provides strength and flexibility. These fibers are arranged in parallel bundles, giving ligaments their characteristic tensile strength. Example: The anterior cruciate ligament (ACL) of the knee is predominantly composed of collagen fibers, providing stability to the knee joint during activities such as walking, running, and jumping. Ligaments also contain other extracellular matrix components, such as elastin and proteoglycans, which contribute to their strength and flexibility . Ligaments are relatively avascular, receiving nutrients and oxygen through diffusion from surrounding tissues. They may contain blood vessels and nerve endings at their periphery, providing sensory feedback and proprioception.

4.2 Ultrasound imaging

The anatomy and structure of ligaments are critical considerations in ultrasound imaging of the musculoskeletal system. Understanding these aspects helps clinicians accurately identify, assess, and diagnose various ligamentous conditions. The ACL is one of the most injured ligaments in the knee, often resulting from sports-related trauma or sudden twisting motions. Ultrasound imaging can detect ACL tears by visualizing discontinuity or hypoechoic areas within the ligament. Assessing the integrity of the ACL is crucial for guiding treatment decisions, such as surgical reconstruction or conservative management. Medial collateral ligament (MCL) injuries are common, especially in activities involving valgus stress on the knee joint, such as contact sports or sudden changes in direction. Ultrasound can assess MCL injuries by evaluating ligament thickness, echogenicity, and continuity. Detection of MCL injuries on ultrasound guides appropriate management strategies, including conservative treatment or surgical repair. The rotator cuff is a group of four tendons and associated muscles that stabilize the shoulder joint and facilitate arm movement. Ultrasound imaging can identify rotator cuff tears, including partial- or full-thickness tears, by visualizing disruptions or irregularities in tendon structure. Assessing the integrity of the rotator cuff on ultrasound guides treatment decisions, such as physical therapy, corticosteroid injections, or surgical repair. Posterior cruciate ligament (PCL) injuries are less common than ACL injuries but can lead to significant knee instability and functional impairment. Ultrasound imaging can assess PCL integrity by visualizing the ligament and evaluating for signs of discontinuity or abnormal echogenicity. Detecting PCL injuries on ultrasound helps clinicians determine the extent of ligament damage and plan appropriate treatment strategies. Ankle ligament sprains are among the most prevalent musculoskeletal injuries, often resulting from inversion or twisting of the ankle. Ultrasound can detect ankle ligament sprains by assessing ligament thickness, echogenicity, and integrity. Identifying the extent of ligamentous injury on ultrasound guides treatment decisions, such as rest, immobilization, physical therapy, or surgical repair. In summary, the anatomy and structure of ligaments are essential considerations in ultrasound imaging for diagnosing and managing various musculoskeletal conditions. Examples such as ACL tears, MCL injuries, rotator cuff tears, PCL integrity, and ankle ligament sprains illustrate the clinical significance of ligament anatomy and structure in ultrasound assessment.

4.3 Calcification

Calcification is the process by which calcium salts, primarily hydroxyapatite, are deposited in tissues, leading to the formation of calcified structures (Kosuri et al., 2014). This phenomenon can occur in various tissues throughout the body, including bones, cartilage, blood vessels, and soft tissues such as tendons and ligaments (Chakravarthi et al., 2025). Calcification is a normal part of bone formation and remodeling, but abnormal or excessive calcification can be associated with various pathological conditions (Chakravarthi and Gaddam, 2026; Vattikuti and Towler, 2004). This process can occur due to various factors, including aging, trauma, inflammation, or metabolic disorders. In tendons, calcification can lead to conditions like calcific tendinitis, where calcium deposits accumulate, causing pain and reduced joint mobility. Ligaments may also experience calcification, affecting their elasticity and potentially contributing to joint stiffness. In joints, calcification can be associated with conditions such as osteoarthritis, where the articular cartilage undergoes degeneration, and calcium deposits may form in response to joint inflammation or wear and tear. Ultrasound plays a crucial role in identifying calcification of ligaments, tendons, and joints due to its ability to provide real-time, high-resolution imaging of soft tissues. In cases of calcific tendinitis, ultrasound can visualize the presence of calcium deposits within tendons, such as the rotator cuff in the shoulder or the Achilles tendon in the ankle. These deposits appear as hyperechoic foci within the tendon substance, often accompanied by posterior acoustic shadowing. Ultrasound can also assess the size, location, and distribution of calcifications, aiding in the diagnosis and evaluation of the severity of the condition. In addition, ultrasound can detect calcification within ligaments and the joint capsule, helping to identify conditions such as ligament calcification or calcific deposits associated with osteoarthritis. By providing detailed anatomical information in real-time, ultrasound facilitates accurate diagnosis and guides treatment decisions, including the selection of appropriate interventions such as targeted injections, physical therapy, or surgical management (Hurt and Baker, 2003).

5 Fascia

5.1 Structure and function

In the human body, fascia is a complex network of connective tissue that surrounds and supports muscles, organs, blood vessels, and nerves. It plays a crucial role in providing structural integrity, transmitting forces, and facilitating movement. Fascia can be classified into several types based on its location, function, and composition.

5.2 Superficial fascia

Superficial fascia, also known as subcutaneous tissue or hypodermis, is a layer of connective tissue located beneath the skin and above the deeper layers of muscle and fascia. It serves several functions, including insulation, protection, and energy storage. The structure of superficial fascia varies depending on its location in the body and can contain a combination of adipose tissue, collagen fibers, blood vessels, and nerves. Adipose tissue, or fat, is a major component of superficial fascia and serves as a site for energy storage and thermal insulation . It consists of adipocytes (fat cells) embedded within a network of collagen and elastin fibers. Example: In areas with thicker layers of adipose tissue, such as the buttocks or abdomen, the superficial fascia contains a higher proportion of adipocytes. Superficial fascia typically appears hypoechoic (darker) on ultrasound compared to surrounding tissues. This characteristic appearance is due to the relatively low density of collagen fibers and adipose tissue within the superficial fascia. Collagen and elastin fibers provide structural support and elasticity to the superficial fascia. Collagen fibers contribute to the strength and resilience of the tissue, while elastin fibers allow for flexibility and recoil. In areas where the skin undergoes frequent movement or stretching, such as the face or neck, the superficial fascia contains more elastin fibers to accommodate these movements. Superficial fascia contains a network of blood vessels, including arteries, veins, and capillaries, that supply oxygen and nutrients to the overlying skin and underlying tissues. These blood vessels also play a role in thermoregulation by regulating blood flow to dissipate or retain heat. Superficial fascia in the limbs contains a rich network of blood vessels that contribute to skin coloration and temperature regulation. Superficial fascia contains sensory nerves responsible for detecting touch, pressure, temperature, and pain. These nerves innervate the overlying skin and transmit sensory information to the central nervous system. Superficial fascia in areas with high tactile sensitivity, such as the fingertips or lips, contains a dense network of sensory nerves. The extracellular matrix of superficial fascia consists of a gel-like substance composed of glycosaminoglycans, proteoglycans, and other glycoproteins. This matrix provides a scaffold for cells and fibers within the tissue and facilitates cell-cell communication and tissue repair. In areas with a high degree of mechanical stress, such as the palms or soles of the feet, the connective tissue matrix of superficial fascia is denser and more fibrous (Stark et al., 1977). Overall, ultrasound imaging provides valuable information about the structure and characteristics of superficial fascia, including its thickness, composition, and integrity (Pillen et al., 2006). Understanding the appearance of superficial fascia on ultrasound helps clinicians evaluate superficial structures, diagnose pathology, and guide treatment decisions in various clinical settings (Alfen et al., 2007). Knowledge of the anatomical structure of superficial fascia in ultrasound imaging is essential for accurate interpretation and diagnosis of soft tissue conditions, including trauma, inflammation, and pathology affecting the subcutaneous layers of the body (Zaidman et al., 2017).

5.3 Deep fascia

5.3.1 Structure and functions

Deep fascia is a dense, fibrous connective tissue layer that surrounds muscles, bones, and joints in the body. It provides structural support, compartmentalization, and organization to the musculoskeletal system. Deep fascia can be classified into different types based on its location and function (Stecco et al., 2007). Investing fascia surrounds entire muscles or muscle groups, investing them like a sheath, and provides structural support while separating muscles from surrounding tissues. The investing fascia of the thigh (fascia lata) surrounds the entire thigh muscles, including the quadriceps and hamstrings (Standring, 2016). Deep fascia forms intermuscular fascia, which lies between adjacent muscles or muscle compartments, dividing them into separate muscle compartments, providing support, and facilitating muscle movement. Example—the intermuscular septa of the forearm separate the flexor and extensor muscle compartments, allowing independent movement of the wrist and fingers. Epimysium surrounds individual muscles, enclosing them in a continuous sheath. It provides structural support, facilitates muscle contraction, and transmits force to tendons. Perimysium lies within muscles, dividing them into smaller bundles or fascicles. It provides structural support, houses blood vessels and nerves, and facilitates nutrient and waste exchange. Endomysium surrounds individual muscle fibers within fascicles. It provides structural support to individual muscle fibers, houses capillaries, and facilitates nutrient exchange. Deep fascia collectively contributes to the structural integrity, organization, and function of the musculoskeletal system. Understanding the anatomy and function of deep fascia is essential for clinicians in various medical specialties, including surgery, sports medicine, and rehabilitation.

5.3.2 Deep fascia modifications

An aponeurosis is a flat, sheet-like tendon that connects muscles to each other or to bones, providing support, distributing forces, and facilitating muscle contraction. Examples of aponeuroses in the human body include the linea alba, a tough fibrous structure that runs vertically along the midline of the abdomen and serves as the attachment site for the abdominal muscles; the palmar aponeurosis, a thick fibrous layer in the palm of the hand that supports the soft tissues (Rayan and Tomasek, 1994); the galea aponeurotica, a tough sheet of connective tissue beneath the scalp that attaches various scalp muscles (Romo et al., 2005); the plantar aponeurosis, also known as the plantar fascia, which supports the arch of the foot (Wearing et al., 2006); and the scapular aponeurosis, a thin fibrous sheet on the dorsal aspect of the scapula that attaches several shoulder muscles.

5.3.3 Deep fascia abnormalities

The deep fascia plays a crucial role in maintaining the structural integrity and function of the musculoskeletal system. Its clinical importance in ultrasound imaging lies in its involvement in various pathological conditions and its role as a reference point for evaluating deeper structures. Deep fascia can be involved in musculoskeletal injuries such as strains, sprains, and tears. Ultrasound imaging can visualize disruptions or abnormalities in the deep fascia, aiding in the diagnosis and management of these injuries. Deep fascia abnormalities encompass a spectrum of structural and functional deviations from normal, including fibrosis, thickening, adhesions, inflammation, edema, neoplastic growths, compartment syndrome, vascular changes, atrophy, and congenital anomalies. These abnormalities can arise from various causes, such as injury, inflammation, overuse, systemic conditions, or congenital factors, and they can lead to symptoms such as pain, swelling, restricted movement, and functional impairment (Chakravarthi, 2022).

Abnormal thickening of the deep fascia may result from chronic inflammation, trauma, or repetitive stress. This thickening can restrict movement and cause pain. Plantar fasciitis is a common condition characterized by pain and inflammation of the plantar fascia, a thick band of tissue that runs along the bottom of the foot, connecting the heel bone to the toes It is one of the most prevalent causes of heel pain, particularly among runners, individuals who are overweight, and those who wear improper footwear (Irving et al., 2007). Plantar fasciitis typically presents with sharp or stabbing pain in the heel or along the bottom of the foot, particularly with the first steps in the morning or after prolonged periods of rest (Buchbinder, 2004). The pain may decrease with activity but often worsens with prolonged standing or walking (Thomas et al., 2010). Plantar fasciitis can result from various factors, including overuse, high-impact activities, tight calf muscles, poor foot biomechanics, or inadequate arch support. Diagnosis is usually based on clinical examination and may be confirmed with imaging studies such as X-rays or ultrasound. Iliotibial Band Syndrome (ITBS) is a common overuse injury characterized by pain on the outer side of the knee, typically occurring in runners and cyclists (Van Der Worp et al., 2012). It involves inflammation and irritation of the iliotibial band (IT band), a thick band of fascia that runs along the outside of the thigh from the hip to the knee (Fairclough et al., 2006). ITBS often develops due to repetitive friction between the IT band and the lateral femoral condyle as the knee flexes and extends during activities such as running or cycling (Fredericson et al., 2000). Factors contributing to ITBS include biomechanical imbalances, muscle weakness or tightness, improper training techniques, and changes in terrain or footwear. Symptoms of ITBS typically include pain and tenderness on the outer knee, especially during activities like running or descending stairs (Orchard et al., 1996).

Fibrosis of the deep fascia refers to the excessive deposition of fibrous tissue within the layers of connective tissue that surround muscles, tendons, and other structures. This condition can occur due to various factors, including injury, inflammation, overuse, or chronic conditions. Fibrosis leads to the thickening and stiffening of the deep fascia, which can restrict movement, impair function, and cause pain or discomfort (Fede et al., 2018). Common causes of fibrosis of the deep fascia include repetitive stress or strain on the affected area, such as in overuse injuries or repetitive motions. Trauma, surgery, or prolonged immobilization can also contribute to the development of fibrosis (Stecco et al., 2019). In addition, chronic inflammation due to conditions like fasciitis or autoimmune disorders may lead to fibrosis of the deep fascia over time (Bordoni et al., 2023). Symptoms of fibrosis of the deep fascia may include stiffness, tightness, or a feeling of "heaviness" in the affected area (Stecco et al., 2009a). There may also be pain or discomfort, particularly with movement or pressure on the affected tissue (Stecco et al., 2010). In some cases, fibrosis can lead to restricted range of motion and functional limitations (Krause et al., 2016). Treatment for fibrosis of the deep fascia typically focuses on relieving symptoms and improving mobility. This may involve techniques such as manual therapy, stretching exercises, and therapeutic modalities to reduce inflammation and promote tissue healing (Myers, 2014). In more severe cases, interventions such as ultrasound-guided fascial release or surgical procedures may be considered to address the fibrotic tissue and restore function. Early intervention and appropriate management of injuries or conditions affecting the deep fascia can help prevent the development of fibrosis and promote optimal musculoskeletal health (Stecco et al., 2016).

Adhesions in the deep fascia are abnormal connections or bands of scar tissue that form between layers of fascia or between fascia and adjacent structures (Liakakos et al., 2001). These adhesions can develop following surgery, trauma, inflammation, or prolonged immobilization, leading to restricted movement, pain, and functional impairment (Diamond and Freeman, 2001). Examples of adhesions include post-surgical adhesions in the thigh, leg, or foot; frozen shoulder resulting from adhesions around the shoulder joint; abdominal adhesions tethering organs to the abdominal wall; pelvic adhesions causing pelvic pain and infertility; spinal adhesions restricting spinal movement and exacerbating pain; and hand adhesions impairing finger dexterity and function (van Goor, 2007). Management of adhesions typically involves physical therapy, manual techniques to break up scar tissue, and, in severe cases, surgical intervention to release the adhesions and restore mobility and function (Holmdahl et al., 1997).

Ultrasound imaging is commonly used to detect and evaluate adhesions. In ultrasound imaging, adhesions may appear as hyperechoic (bright) bands or strands connecting adjacent structures, disrupting the normal architecture and movement of tissues (Levy and Hobbs, 2004). Doppler ultrasound can also assess blood flow within and around adhesions, aiding in the diagnosis and differentiation of adhesions from other pathologies. Ultrasound-guided procedures, such as adhesiolysis (breaking down adhesions) or drainage of fluid collections associated with adhesions, can be performed to alleviate symptoms and improve organ function Overall, ultrasound is a valuable tool for assessing adhesions due to its real-time imaging capabilities and lack of ionizing radiation, allowing for accurate diagnosis and guided interventions (Albanese and Rowe, 1995).

Edema, often referred to as swelling, occurs when excess fluid accumulates in the body's tissues (Trayes et al., 2013). This buildup can occur throughout the body but is commonly seen in the hands, arms, feet, ankles, and legs (Cho and Atwood, 2002). Edema can result from various factors, including injury, inflammation, medical conditions such as heart failure or kidney disease, pregnancy, or medications (Ely et al., 2006). Symptoms may include swelling, puffiness, and stretched or shiny skin (Blann et al., 2005). In some cases, edema can be a sign of a more serious underlying condition, so it's important to consult a healthcare professional for proper diagnosis and treatment (O'Brien et al., 2005). Treatment may involve addressing the underlying cause, such as managing heart or kidney conditions, reducing salt intake, elevating the affected area, wearing compression garments, and sometimes using medications to help remove excess fluid from the body (Eaton et al., 1999). Detecting edema in the deep fascia using ultrasound involves visualizing changes in the thickness and echogenicity (brightness) of the fascial layer (Pillen et al., 2009). In ultrasound imaging, edema appears as a hypoechoic (darker) or anechoic (black) area within the deep fascia, indicating the presence of fluid accumulation (Maurits et al., 2003). This can be observed as a widening or thickening of the fascial layer compared to normal (Walker et al., 2004). In addition, ultrasound may reveal other signs of inflammation, such as increased vascularity (hyperemia) or loss of normal fascial architecture (Koppel et al., 1988). By assessing these ultrasound findings along with clinical symptoms and history, healthcare providers can accurately diagnose deep fascial edema and guide appropriate management strategies (Pichonnaz et al., 2013). Ultrasound is a valuable tool for real-time imaging and monitoring of edema, providing valuable insights into the extent and severity of the condition and helping to guide treatment decisions (Zhai et al., 2009).

Soft tissue tumors refer to a diverse group of neoplasms that arise from the various nonbony tissues of the body, excluding the viscera and parenchymal organs (Fletcher, 2014). These tumors can originate from mesenchymal tissues such as muscle, fat, connective tissue, nerves, blood vessels, and synovial tissue (Weiss and Goldblum, 2014). Soft tissue tumors encompass a wide range of benign and malignant entities, with varying clinical behaviors and histological features (Clark et al., 2005). Benign soft tissue tumors include lipomas, fibromas, and schwannomas, which typically grow slowly and do not metastasize (Kransdorf, 1995). Malignant soft tissue tumors, known as sarcomas, are less common but can be aggressive and have the potential to metastasize to distant sites (Siegel et al., 2020). Examples of malignant soft tissue tumors include leiomyosarcoma, liposarcoma, and malignant fibrous histiocytoma (Brennan et al., 2014).

The deep fascia serves as a potential pathway for the spread of soft tissue tumors due to its extensive interconnected network throughout the body (Enneking et al., 1980). Tumors originating within or adjacent to the deep fascia can infiltrate this dense connective tissue layer, allowing for local extension along fascial planes and potential dissemination to distant sites via lymphatic or vascular channels. For example, soft tissue sarcomas arising from the musculature adjacent to deep fascial structures may infiltrate the fascia and spread along its pathways, leading to metastasis to regional lymph nodes or distant organs (Gronchi et al., 2005). In addition, tumors originating from neurovascular structures, such as peripheral nerves or blood vessels, can breach the deep fascia and propagate along its pathways, facilitating regional or systemic dissemination (Zagars et al., 2003). The interconnected nature of the deep fascia underscores its role as a conduit for tumor spread and highlights the importance of thorough evaluation and management of soft tissue tumors to prevent local recurrence and distant metastasis (Trovik et al., 2000).

Ultrasound is a valuable imaging modality for evaluating soft tissue tumors due to its ability to provide real-time, high-resolution images without exposure to ionizing radiation (Paltiel et al., 2000). In ultrasound imaging of soft tissue tumors, these lesions typically appear as focal masses within the soft tissue, demonstrating variable echogenicity (brightness) compared to surrounding tissues (Lakkaraju et al., 2009). The characteristics observed on ultrasound can help differentiate between benign and malignant tumors (Bureau et al., 2004). Benign tumors often present as well-defined, homogeneous masses with regular margins and internal echoes similar to surrounding tissue (Inampudi et al., 2004). Examples include lipomas, which appear as hyperechoic (bright) masses, and cysts, which appear as anechoic (black) structures with posterior acoustic enhancement (Hung et al., 2014). Malignant tumors, on the other hand, may exhibit irregular margins, heterogeneous echogenicity, and internal vascularity on Doppler imaging (Fornage, 1999). In addition, features such as infiltration into adjacent structures, necrosis, and calcifications may raise suspicion for malignancy (Chiou et al., 2009). Ultrasound-guided biopsy can be performed to obtain tissue samples for histopathological analysis, aiding in the diagnosis and further management of soft tissue tumors (Torriani et al., 2002). Overall, ultrasound is a valuable tool for initial evaluation, characterization, and guidance of interventions in patients with suspected soft tissue tumors (Kang et al., 2004).

Ultrasound is widely employed for evaluating anterior abdominal wall anatomy due to its noninvasive nature and real-time imaging capabilities (Vlachos et al., 2005). It is particularly useful for visualizing the layers of the abdominal wall, assessing the integrity of muscles, and identifying abnormalities such as hernias or fluid collections. With high-frequency sound waves, ultrasound allows for detailed imaging of the skin, subcutaneous tissue, and muscles, including the rectus abdominis and its tendinous intersections. In addition, ultrasound aids in the diagnosis of hernias by revealing disruptions or weaknesses in the abdominal wall layers (Jamadar et al., 2006). This imaging modality plays a crucial role in guiding procedures such as abdominal wall injections, nerve blocks, and the placement of catheters, enhancing precision and safety in clinical interventions related to the anterior abdominal wall (Tran et al., 2009).

Anatomically, deep fascia is said to be absent in the anterior abdominal wall. Chakravarthi et al. reported a well-defined, thick, bilateral deep fascia in the anterior abdominal wall in their study (Chakravarthi et al., 2013). Such unusual and rare presence of deep fascia in the anterior abdominal wall may prevent or interfere with the bulging of the abdomen, which may lead to elevated intraabdominal pressure (Intraabdominal hypertension) (De Waele et al., 2015). Awareness of these anatomical variations is essential not only to anatomists but also to radiologists, embryologists, and surgeons operating on groin hernias, as it can increase the success of diagnostic evaluation and surgical approaches to the region.

6 Joints

6.1 Introduction

Joints, also known as articulations, are the connections between bones that allow for movement and provide stability to the skeletal system (Gray and Standring, 2016). There are various types of joints, each with its own structure and function (White and Folkens, 2005). Synovial joints, the most common type, have a synovial cavity filled with synovial fluid that lubricates and cushions the joint surfaces, allowing for smooth movement (Tortora and Derrickson, 2017). Examples include the knee, elbow, and shoulder joints (Palastanga et al., 2012). Cartilaginous joints are connected by cartilage and provide limited movement, such as the intervertebral discs between the vertebrae (Moore et al., 2017). Fibrous joints are held together by fibrous connective tissue and have minimal to no movement, like the sutures between the bones of the skull. Joints are essential for mobility, posture, and supporting the body's weight, making them integral to everyday activities and overall health. Ultrasound imaging provides real-time visualization of joint anatomy and pathology, aiding in the diagnosis and management of joint-related conditions (Jacobson and van Holsbeeck, 1998). In ultrasound imaging, joints appear as hypoechoic (darker) spaces surrounded by brighter structures such as bones and tendons (Martinoli, 2010). The synovial fluid within the joint cavity appears as a hypoechoic region, providing lubrication and allowing for smooth movement between the articulating surfaces of bones (McNally, 2008). The joint capsule, a fibrous structure surrounding the joint, may appear as a thin, echogenic (brighter) line surrounding the joint space (Backhaus et al., 2001). Cartilage, which covers the ends of bones within the joint, is not directly visible on ultrasound due to its low echogenicity, but its integrity can indirectly be assessed by evaluating the joint space and the smoothness of the bone surfaces (D'Agostino et al., 2005).

6.1.1 Joint fractures

Joint fractures encompass a variety of injuries involving the articulating surfaces between bones, often resulting from trauma or excessive force. One common example is a distal radius fracture, where the end of the radius bone near the wrist joint breaks, typically occurring due to a fall onto an outstretched hand. Another example is an intra-articular fracture of the ankle, such as a pilon fracture, which involves the tibial plafond and often occurs because of high-energy trauma, like a car accident. These fractures can disrupt the smooth surfaces of the joint, leading to pain, swelling, instability, and decreased range of motion. Treatment varies depending on the severity and location of the fracture but may involve immobilization with splints or casts, closed reduction and fixation with screws or plates, or surgical intervention for complex fractures to restore joint alignment and stability. Early diagnosis and appropriate management are essential to prevent long-term complications and optimize outcomes for patients with joint fractures (Boraiah et al., 2010). Ultrasound plays a crucial role in assessing joint fractures by providing real-time imaging of the affected area, aiding in the diagnosis and management of these injuries. For instance, in evaluating fractures of the wrist, ultrasound can visualize disruptions in the continuity of the bone and surrounding soft tissues, facilitating prompt diagnosis and appropriate treatment planning (Williamson et al., 2000). Similarly, in pediatric patients with suspected growth plate fractures, ultrasound can accurately assess the integrity of the growth plate and surrounding structures, guiding decisions regarding conservative management or surgical intervention (Moritz et al., 2008). In addition, ultrasound can help identify associated soft tissue injuries, such as tendon or ligament tears, which may impact treatment outcomes (Hendrix et al., 2018). Overall, ultrasound offers a noninvasive, readily available modality for assessing joint fractures, enabling timely interventions and optimizing patient care (Poonai et al., 2017).

6.1.2 *Joint disorders*

Joint disorders encompass a wide range of conditions affecting the joints, which are the connections between bones that allow for movement. Joint disorders encompass a spectrum of conditions affecting the articulations between bones, leading to pain, inflammation, and impaired function (Lawrence et al., 2008). Osteoarthritis, the most common type, involves the gradual deterioration of cartilage, often affecting weight-bearing joints like the knees and hips. Rheumatoid arthritis is an autoimmune disorder characterized by chronic inflammation of multiple joints, potentially leading to joint deformity and disability (Scott et al., 2010). Gout, caused by the accumulation of uric acid crystals in the joints, typically affects the big toe but can involve other joints as well (Dalbeth et al., 2016). Other examples include bursitis, an inflammation of the fluid-filled sacs that cushion joints; tendonitis, inflammation of tendons typically caused by overuse; and ligament injuries, such as sprains or tears, resulting from trauma or excessive stretching (Oliveria et al., 1995). These disorders vary in etiology, presentation, and treatment, highlighting the importance of accurate diagnosis and targeted management strategies tailored to everyone's needs (Hunter and Felson, 2006). Ultrasound plays a pivotal role in assessing joint disorders by providing real-time imaging of affected joints and surrounding structures. For instance, in rheumatoid arthritis, ultrasound can detect synovial hypertrophy, joint effusion, and erosions, aiding in disease diagnosis and monitoring (Wakefield et al., 2005). In tendonitis, ultrasound can visualize tendon thickening, tears, and calcifications, guiding treatment decisions such as corticosteroid injections or physical therapy. In addition, ultrasound is valuable in diagnosing traumatic joint injuries by assessing ligament integrity, detecting hemarthrosis, and evaluating joint effusion or fractures. Its versatility extends to conditions like bursitis, where ultrasound can identify inflamed bursae and differentiate between infectious and noninfectious causes, facilitating appropriate management strategies. Overall, ultrasound serves as a versatile and effective tool in the assessment of various joint disorders, offering real-time imaging and guiding targeted interventions for optimal patient care.

7 Discussion

The role of ultrasound in assessing repetitive motion injuries of the musculofascial system has evolved significantly, establishing itself as an indispensable diagnostic modality in contemporary musculoskeletal medicine. This comprehensive imaging approach offers several distinct advantages over traditional imaging methods, particularly in the evaluation of soft tissue pathology associated with repetitive strain injuries.

7.1 Diagnostic superiority and clinical applications

Ultrasound's real-time imaging capabilities provide dynamic assessment opportunities that static imaging modalities cannot match. The ability to perform functional evaluation during movement allows clinicians to observe structural changes and pathological processes as they occur, offering insights into the biomechanical factors contributing to repetitive motion injuries. This dynamic functionality is particularly valuable in assessing muscle tears, tendon pathology, and nerve entrapment syndromes, where symptoms may be position-dependent or movement-related. The high-resolution imaging capabilities of modern ultrasound systems enable detailed visualization of musculofascial structures, facilitating accurate grading of muscle injuries from Grade I strains to complete Grade III ruptures. The ability to differentiate between various pathological processes, including muscle contusions, hematomas, and edema, provides essential information for treatment planning and prognosis determination.

7.2 Nerve entrapment syndromes: A specialized application

The assessment of nerve entrapment syndromes represents one of ultrasound's most significant contributions to the evaluation of repetitive motion injuries. Conditions such as CTS, cubital tunnel syndrome, and other peripheral nerve entrapments benefit tremendously from ultrasound's ability to visualize nerve morphology, cross-sectional area, and dynamic behavior. The capacity to perform comparative studies with the contralateral extremity enhances diagnostic accuracy and provides objective measurements for monitoring treatment response. Dynamic ultrasound evaluation during provocative maneuvers can reveal nerve compression that may not be apparent on static imaging, making it particularly valuable in cases where clinical symptoms are suggestive but conventional nerve conduction studies are normal or equivocal.

7.3 Tendon and fascial pathology assessment

The hierarchical organization of tendons and their covering structures requires sophisticated imaging approaches for comprehensive evaluation. Ultrasound's ability to visualize the epitenon, paratenon, endotenon, and peritenon provides detailed information about tendon health and pathology. This detailed anatomical visualization is crucial for understanding the progression from acute tendinitis to chronic tendinopathy and for guiding targeted therapeutic interventions. The assessment of fascial abnormalities, including plantar fasciitis and ITBS, demonstrates ultrasound's versatility in evaluating diverse repetitive motion injuries. The ability to measure fascial thickness, assess echogenicity changes, and identify structural abnormalities such as tears or calcifications provides comprehensive diagnostic information (Chakravarthi et al., 2017; McMillan et al., 2009).

7.4 Anatomical variations and clinical implications

Recognition of anatomical variations, such as the presence of deep fascia in the anterior abdominal wall, as reported by Chakravarthi et al., highlights the importance of comprehensive anatomical knowledge in ultrasound interpretation. These variations can significantly impact surgical approaches and may contribute to unusual clinical presentations or complications (Chakravarthi, 2022). The identification of muscular variations and anomalies through ultrasound imaging contributes to personalized medicine approaches, where treatment strategies can be tailored based on individual anatomical characteristics rather than standardized protocols alone.

7.5 Limitations and future directions

Despite its numerous advantages, ultrasound imaging has inherent limitations that must be acknowledged. Operator dependency remains a significant concern, requiring extensive training and experience for optimal diagnostic accuracy. The inability to penetrate deeply into certain anatomical regions and the limited visualization of structures obscured by bone or gas represent technical limitations that may necessitate complementary imaging modalities. Future developments in ultrasound technology, including artificial intelligence integration, elastography applications, and enhanced Doppler capabilities, promise to further expand the diagnostic potential of this modality in repetitive motion injury assessment (Ooi et al., 2014).

7.6 Clinical integration and patient care

The integration of ultrasound assessment into comprehensive patient care protocols for repetitive motion injuries requires multidisciplinary collaboration among radiologists, orthopedic surgeons, sports medicine physicians, and rehabilitation specialists. The immediate availability of diagnostic information facilitates prompt treatment initiation and may improve patient outcomes through early intervention strategies. The cost-effectiveness and accessibility of ultrasound imaging make it particularly valuable in both primary care and specialized clinical settings, potentially reducing healthcare costs while maintaining diagnostic accuracy.

8 Conclusion

Ultrasound imaging has established itself as a cornerstone modality in the assessment of repetitive motion injuries affecting the musculofascial system. Its unique combination of real-time imaging, dynamic assessment capabilities, and detailed soft tissue visualization provides comprehensive diagnostic information essential for optimal patient care. Continued technological advancements and expanded clinical applications promise to further enhance its role in musculoskeletal medicine.

References

Albanese, C.T., Rowe, M.I., 1995. Necrotizing enterocolitis. Semin. Pediatr. Surg. 4 (4), 200–206.

Alfen, Pillen, S., Faber, C.G., Laak, Engelen, 2007. Muscle ultrasound for monitoring disease progression in inclusion body myositis: a pilot study. Muscle Nerve 35 (2), 235–239.

Backhaus, M., Burmester, G.-R., Gerber, T., Grassi, W., Machold, K.P., Swen, W.A., Wakefield, R.J., Manger, B., 2001. Guidelines for musculoskeletal ultrasound in rheumatology. Ann. Rheum. Dis. 00034967. 60 (7), 641–649. doi:10.1136/ard.60.7.641.

Blann, A.D., Woywodt, A., Bertolini, F., Bull, T.M., Buyon, J.P., Clancy, R.M., Haubitz, M., Hebbel, R.P., Lip, G.Y.H., Mancuso, P., Sampol, J., Solovey, A., Dignat-George, F., 2005. Circulating endothelial cells. Biomarker of vascular disease. Thromb. Haemost. 93 (2), 228–235. doi:10.1160/TH04-09-0578.

Bodner, G., Buchberger, W., Schocke, M., Bale, R., Huber, B., Harpf, C., Gassner, E., Jaschke, W., 2001. Radial nerve palsy associated with humeral shaft fracture: evaluation with US—initial experience. Radiology. 0033-8419. 219 (3), 811–816. doi:10.1148/radiology.219.3.r01jn09811.

Boraiah, S., Kemp, T.J., Erwteman, A., Lucas, P.A., Asprinio, D.E., 2010. Outcome following open reduction and internal fixation of open pilon fractures. J. Bone Joint Surg. 15351386. 92 (2), 346–352. Available from: http://www.ejbjs.org/cgi/reprint/92/2/346. doi:10.2106/JBJS.H.01678.

Bordoni, B., Mahabadi, N., Varacallo, M., 2023. Anatomy, Fascia. StatPearls Publishing, Treasure Island (FL).

Brennan, M.F., Antonescu, C.R., Moraco, N., Singer, S., 2014. Lessons learned from the study of 10,000 patients with soft tissue sarcoma. Ann. Surg. 15281140. 260 (3), 416–422. http://journals.lww.com/annalsofsurgery/pages/default.aspx. doi:https://doi.org/10.1097/sla.0000000000000869.

Buchbinder, R., 2004. Clinical practice. Plantar fasciitis. N. Engl. J. Med. 15334406. 350 (21), 2159–2166.

Bureau, N.J., Ali, S.S., Chhem, R.K., Cardinal, E., 2004. Ultrasound of soft-tissue masses: from pattern recognition to diagnosis. AJR Am. J. Roentgenol. 183 (6), 1805–1809.

Chakravarthi, K.K., 2022. Anatomical variations of Poupart's ligament in relation to deep fascia of anterior abdomen wall and its clinical implications - a cadaveric study. Medpulse Int. J. Anat. 25507621. 20 (2), 22–25. doi:10.26611/10012022.

Chakravarthi, K.K., 2024. Unilateral accessory muscles of forearm-and its clinical implications. Asian J. Med. Sci. 8 (2), 94–96. doi:10.71152/ajms.v8i2.3996.

Chakravarthi, K.K., Gaddam, T.R., 2026. Congenital agenesis of the obturator foramen in an Asian population and its clinical significance. Ann. Afr. Med. doi:10.4103/aam.aam_763_25.

Chakravarthi, K.K., Nelluri, V., Reghunadhan, D., Sugavasi, R., 2024. Anatomical and radiological findings of abnormal calcified structures within and around the foramen transversarium and its clinical implications. Ann. Afr. Med. 23 (4), 641–648. doi:10.4103/aam.aam_73_24.

Chakravarthi, K.K., Nelluri, V., Sugavasi, R., Reghunadhan, D., 2025. Anatomical study and clinical significance of atlanto-occipital and atlantoaxial assimilation anomaly in Asian population. Ann. Afr. Med. 24 (2), 415–420. doi:10.4103/aam.aam_291_24.

Chakravarthi, K.K., Sarvepalli, A., Siddegowda, S.K., Nelluri, V., 2023. Accessory heads of biceps brachii muscle its embryogenesis and clinical implications: cadaveric study. Ann. Afr. Med. 22 (3), 359–364. doi:10.4103/aam.aam_93_22.

Chakravarthi, K.K., Shetty, P., 2025. Morphological variations of foot lumbricals and their clinical significance: a cadaveric study. Ann. Afr. Med. 24(2), 456–460. doi:10.4103/aam.aam_303_24.

Chakravarthi, K.K., Siddaraju, K.S., Venumadhav, N., Bali, S.A., 2017. Abnormal ossified structures around the hip joint and its clinical implications. Int. J. Res. Med. Sci. 5 (12), 5391. doi:10.18203/2320-6012.ijrms20175461.

Chakravarthi, K.K., Srinivasa, S., Venkateshu, K.V., 2013. Anatomical variations of anterior abdominal wall. J. Clin. Diagn. Res. 7 (8), 1672–1674.

Chiou, H.J., Chou, Y.H., Chiu, S.Y., Wang, H.K., Chen, W.M., Chen, T.H., Chang, C.Y., 2009. Differentiation of benign and malignant superficial soft-tissue masses using grayscale and color Doppler ultrasonography. J. Chin. Med. Assoc. 17287731. 72 (6), 307–315. doi:10.1016/S1726-4901(09)70377-6.

Cho, S., Atwood, J.E., 2002. Peripheral edema. Am. J. Med. 113 (7), 580–586. doi:10.1016/S0002-9343(02)01322-0.

Clark, M.A., Fisher, C., Judson, I., Thomas, J.M., 2005. Soft-tissue sarcomas in adults. N. Engl. J. Med. 0028-4793. 353 (7), 701–711. doi:10.1056/nejmra041866.

D'Agostino, M.A., Conaghan, P., Le Bars, M., Baron, G., Grassi, W., Martin-Mola, E., Wakefield, R., Brasseur, J.L., So, A., Backhaus, M., Malaise, M., Burmester, G., Schmidely, N., Ravaud, P., Dougados, M., Emery, P., 2005. EULAR report on the use of ultrasonography in painful knee osteoarthritis. Part 1: prevalence of inflammation in osteoarthritis. Ann. Rheum. Dis. 00034967. 64 (12), 1703–1709. doi:10.1136/ard.2005.037994.

Dalbeth, N., Merriman, T.R., Stamp, L.K., 2016. Gout. Lancet. 1474547X. 388 (10055), 2039–2052. Available from: http://www.journals.elsevier.com/the-lancet/. doi:10.1016/S0140-6736(16)00346-9.

De Waele, J.J., Malbrain, M.L.N.G., Kirkpatrick, A.W., 2015. The abdominal compartment syndrome: evolving concepts and future directions. Crit. Care. 1466609X. 19 (1). Available from: http://ccforum.com/content/17. doi:10.1186/s13054-015-0879-8.

Diamond, M.P., Freeman, M.L., 2001. Clinical implications of postsurgical adhesions. Hum. Reprod. Update. 13554786. 7 (6), 567–576. doi:10.1093/humupd/7.6.567.

Eaton, S.B., Eaton, S.B., Konner, M.J., 1999. Paleolithic Nutrition Revisited. Oxford University Press (OUP), pp. 313–332. doi:10.1093/oso/9780195103557.003.0013.

Ely, J.W., Osheroff, J.A., Chambliss, M.L., Ebell, M.H., 2006. Approach to leg edema of unclear etiology. J. Am. Board Fam. Med. 15587118. 19 (2), 148–160. Available from: http://www.jabfm.org/cgi/reprint/19/2/148. doi:10.3122/jabfm.19.2.148.

Enneking, W.F., Spanier, S.S., Goodman, M.A., 1980. A system for the surgical staging of musculoskeletal sarcoma. Clin. Orthop. Relat. Res. 0009-921X. 153 (153), 106–120. doi:10.1097/00003086-198011000-00013.

Fairclough, J., Hayashi, K., Toumi, H., Lyons, K., Bydder, G., Phillips, N., Best, T.M., Benjamin, M., 2006. The functional anatomy of the iliotibial band during flexion and extension of the knee: implications for understanding iliotibial band syndrome. J. Anat. 0021-8782. 208 (3), 309–316. doi:10.1111/j.1469-7580.2006.00531.x.

Fede, C., Angelini, A., Stern, R., Macchi, V., Porzionato, A., Ruggieri, P., De Caro, R., Stecco, C., 2018. Quantification of hyaluronan in human fasciae: variations with function and anatomical site. J. Anat. 0021-8782. 233 (4), 552–556. doi:10.1111/joa.12866.

Fletcher, C.D.M., 2014. The evolving classification of soft tissue tumours – an update based on the new 2013 WHO classification. Histopathology. 0309-0167. 64 (1), 2–11. doi:10.1111/his.12267.

Fornage, B.D., 1999. Soft tissue masses: the underutilization of sonography. Semin. Musculoskelet. Radiol. 1098898X. 3 (2), 115–133. Available from: http://www.thieme-connect.com/ejournals/toc/smr. doi:10.1055/s-2008-1080056.

Fredericson, M., Cookingham, C.L., Chaudhari, A.M., Dowdell, B.C., Oestreicher, N., Sahrmann, S.A., 2000. Hip abductor weakness in distance runners with iliotibial band syndrome. Clin. J. Sport Med. 1050642X. 10 (3), 169–175. Available from: http://www.cjsportmed.com/. doi:10.1097/00042752-200007000-00004.

Gray, H., Standring, S., 2016. Gray's Anatomy: The Anatomical Basis of Clinical Practice. Elsevier Limited.

Gronchi, A., Casali, P.G., Mariani, L., Miceli, R., Fiore, M., Vullo, S.L., Bertulli, R., Collini, P., Lozza, L., Olmi, P., Rosai, J., 2005. Status of surgical margins and prognosis in adult soft tissue sarcomas of the extremities: a series of patients treated at a single institution. J. Clin. Oncol. 0732-183X. 23 (1), 96–104. doi:10.1200/jco.2005.04.160.

Hendrix, N., Hauret, K., Burkhart, T., Dowd, 2018. Evaluating pediatric fractures with point-of-care ultrasound in the emergency department. Pediatr. Emerg. Care 34 (1), 43–47.

Holmdahl, L., Risberg, B., Beck, D.E., Burns, J.W., Chegini, N., DiZerega, G.S., Ellis, H., 1997. Adhesions: pathogenesis and prevention - panel discussion and summary. Eur. J. Surg. Acta Chir. Suppl. 577, 56–62,163.

Hung, E.H.Y., Griffith, J.F., Ng, A.W.H., Lee, R.K.L., Lau, D.T.Y., Leung, J.C.S., 2014. Ultrasound of musculoskeletal soft-tissue tumors superficial to the investing fascia. Am. J. Roentgenol. 15463141. 202 (6), W532. Available from: http://www.ajronline.org/doi/pdf/10.2214/AJR.13.11457. doi:10.2214/AJR.13.11457.

Hunter, D.J., Felson, D.T., 2006. Osteoarthritis. BMJ. 0959-8138. 332 (7542), 639–642. doi:10.1136/bmj.332.7542.639.

Hurt, G., Baker, C.L., 2003. Calcific tendinitis of the shoulder. Orthop. Clin. North Am. 00305898. 34 (4), 567–575. doi:10.1016/s0030-5898(03)00089-0.

Inampudi, P., Jacobson, J.A., Fessell, D.P., Carlos, R.C., Patel, S.V., Delaney-Sathy, L.O., van Holsbeeck, M.T., 2004. Soft-tissue lipomas: accuracy of sonography in diagnosis with pathologic correlation. Radiology. 00338419. 233 (3), 763–767. doi:10.1148/radiol.2333031410.

Irving, D.B., Cook, J.L., Young, M.A., Menz, H.B., 2007. Obesity and pronated foot type may increase the risk of chronic plantar heel pain: a matched case-control study. BMC Musculoskelet. Disord. 14712474. 8. doi:10.1186/1471-2474-8-41.

Jacobson, J.A., van Holsbeeck, M.T., 1998. Musculoskeletal ultrasonography. Orthop. Clin. North Am. 00305898. 29 (1), 135–167. doi:10.1016/s0030-5898(05)70010-9.

Jamadar, D.A., Jacobson, J.A., Morag, Y., Girish, G., Ebrahim, F., Gest, T., Franz, M., 2006. Sonography of inguinal region hernias. Am. J. Roentgenol. 0361803X. 187 (1), 185–190. doi:10.2214/AJR.05.1813.

Kang, T., Horton, K.M., Abrams, R.A., Fishman, E.K., Liposarcoma, 2004. CT and MR imaging findings. Radiographics 24 (5), 1353–1365.

Komi, P.V., Fukashiro, S., Järvinen, M., 1992. Biomechanical loading of achilles tendon during normal locomotion. Clin. Sports Med. 02785919. 11 (3), 521–531. doi:10.1016/s0278-5919(20)30506-8.

Koppel, B.S., Tuchman, A.J., Mangiardi, J.R., Daras, M., Weitzner, I., 1988. Epidural spinal infection in intravenous drug abusers. Arch. Neurol. 15383687. 45 (12), 1331–1337. doi:10.1001/archneur.1988.00520360049011.

Kosuri, K.C., Venumadhav, N., Ks, S., 2014. Ossified ligamentum longitudinale anterius in adult human dry vertebrae. J. Clin. Diagn. Res. 8 (8), AC04–AC06. doi:10.7860/JCDR/2014/8691.4702.

Kransdorf, M.J., 1995. Benign soft-tissue tumors in a large referral population: distribution of specific diagnoses by age, sex, and location. Am. J. Roentgenol. 0361803X. 164 (2), 395–402. Available from: http://www.ajronline.org/. doi:10.2214/ajr.164.2.7839977.

Krause, F., Wilke, J., Vogt, L., Banzer, W., 2016. Intermuscular force transmission along myofascial chains: a systematic review. J. Anat. 14697580. 228 (6), 910–918. Available from: http://onlinelibrary.wiley.com/journal/10.1111/(ISSN)1469-7580. doi:10.1111/joa.12464.

Lakkaraju, A., Sinha, R., Garikipati, R., Edward, S., Robinson, P., 2009. Ultrasound for initial evaluation and triage of clinically suspicious soft-tissue masses. Clin. Radiol. 00099260. 64 (6), 615–621. doi:10.1016/j.crad.2009.01.012.

Lawrence, R.C., Felson, D.T., Helmick, C.G., Arnold, L.M., Choi, H., Deyo, R.A., Gabriel, S., Hirsch, R., Hochberg, M.C., Hunder, G.G., Jordan, J.M., Katz, J.N., Kremers, H.M., Wolfe, F., 2008. Estimates of the prevalence of arthritis and other rheumatic conditions in the United States. Part II. Arthritis Rheum. 00043591. 58 (1), 26–35. doi:10.1002/art.23176.

Levy, A.D., Hobbs, C.M., 2004. From the archives of the AFIP. Meckel diverticulum: radiologic features with pathologic correlation. Radiographics. 02715333. 24 (2), 565–587. Available from: http://radiographics.rsnajnls.org/. doi:10.1148/rg.242035187.

Liakakos, T., Thomakos, N., Fine, P.M., Dervenis, C., Young, R.L., 2001. Peritoneal adhesions: etiology, pathophysiology, and clinical significance - recent advances in prevention and management. Digestive Surgery. 02534886. 18 (4), 260–273. doi:10.1159/000050149.

Martinoli, C., 2010. Musculoskeletal ultrasound: technical guidelines. Insights Imaging. 1869-4101. 1 (3), 99–141. doi:10.1007/s13244-010-0032-9.

Maurits, N.M., Bollen, A.E., Windhausen, A., De Jager, A.E.J., Van Der Hoeven, J.H., 2003. Muscle ultrasound analysis: normal values and differentiation between myopathies and neuropathies. Ultrasound Med. Biol. 03015629. 29 (2), 215–225. Available from: https://www.elsevier.com/locate/ultrasmedbio. doi:10.1016/S0301-5629(02)00758-5.

McDonough, J.A., Elia, T., 2008. Emergency musculoskeletal ultrasound. In: Clinical Emergency Radiology. Cambridge University Press, United States, pp. 347–357. Available from doi:10.1017/CBO9780511551734. doi:10.1017/CBO9780511551734.025.

McMillan, A.M., Landorf, K.B., Barrett, J.T., Menz, H.B., Bird, A.R., 2009. Diagnostic imaging for chronic plantar heel pain: a systematic review and meta-analysis. J Foot Ankle Res. 17571146. 2 (1). Available from: http://www.jfootankleres.com/content/2/1/32. doi:10.1186/1757-1146-2-32Australia.

McNally, E.G., 2008. Ultrasound of the small joints of the hands and feet: current status. Skeletal Radiol. 0364-2348. 37 (2), 99–113. doi:10.1007/s00256-007-0356-9.

Miller, T., White, K., Ross, D., 2012. The diagnosis and management of piriformis syndrome: myths and facts. Can. J. Neurol. Sci. 03171671. 39 (5), 577–583. Available from: http://journals.cambridge.org/action/displayBackIssues?jid=CJN. doi:10.1017/S0317167100015298.

Moore, K.L., Dalley, A.F., Agur, A.M., 2017. Clinically Oriented Anatomy, Vol. 8. Wolters Kluwer.

Moritz, J., Berthold, L., Soenksen, S., Alzen, G., 2008. Ultrasound in diagnosis of fractures in children: unnecessary harassment or useful addition to X-ray? Ultraschall Med. Eur. J. Ultrasound. 0172-4614. 29 (03), 267–274. doi:10.1055/s-2008-1027329.

Myers, T.W., 2014. Anatomy Trains: Myofascial Meridians for Manual and Movement Therapists. Churchill Livingstone Elsevier.

O'Brien, J.G., Chennubhotla, S.A., Chennubhotla, R.V., 2005. Treatment of edema. Am. Fam. Physician. 0002838X. 71 (11), 2111–2118.

Oliveria, S.A., Felson, D.T., Reed, J.I., Cirillo, P.A., Walker, A.M., 1995. Incidence of symptomatic hand, Hip, and knee osteoarthritis among patients in a health maintenance organization. Arthritis Rheum. 15290131. 38 (8), 1134–1141. doi:10.1002/art.1780380817.

Ooi, C.C., Malliaras, P., Schneider, M.E., Connell, D.A., 2014. "Soft, hard, or just right?" Applications and limitations of axial-strain sonoelastography and shear-wave elastography in the assessment of tendon injuries. Skeletal Radiol. 14322161. 43 (1), 1–12. doi:10.1007/s00256-013-1695-3.

Orchard, J.W., Fricker, P.A., Abud, A.T., Mason, B.R., 1996. Biomechanics of iliotibial band friction syndrome in runners. Am. J. Sports Med. 03635465. 24 (3), 375–379. Available from: http://ajs.sagepub.com/archive/2008.dtl. doi:10.1177/036354659602400321.

Palastanga, Field, Soames, R., 2012. Anatomy and Human Movement: Structure and Function. Churchill Livingstone Elsevier.

Paltiel, H.J., Burrows, P.E., Kozakewich, H.P.W., Zurakowski, D., Mulliken, J.B., 2000. Soft-tissue vascular anomalies: utility of US for diagnosis. Radiology. 00338419. 214 (3), 747–754. Available from: http://radiology.rsnajnls.org. doi:10.1148/radiology.214.3.r00mr21747.

Pichonnaz, C., Bassin, J.P., Currat, D., Martin, E., Jolles, B.M., 2013. Bioimpedance for oedema evaluation after total knee arthroplasty. Physiother. Res. Int. 15206483. 18 (3), 140–147. doi:10.1002/pri.1540.

Pillen, S., Tak, R.O., Zwarts, M.J., Lammens, M.M.Y., Verrijp, K.N., Arts, I.M.P., van der Laak, J.A., Hoogerbrugge, P.M., van Engelen, B.G.M., Verrips, A., 2009. Skeletal muscle ultrasound: correlation between fibrous tissue and echo intensity. Ultrasound Med. Biol. 03015629. 35 (3), 443–446. doi:10.1016/j.ultrasmedbio.2008.09.016.

Pillen, S., van Keimpema, M., Nievelstein, R.A.J., Verrips, A., van Kruijsbergen-raijmann, W., Zwarts, M.J., 2006. Skeletal muscle ultrasonography: visual versus quantitative evaluation. Ultrasound Med. Biol. 03015629. 32 (9), 1315–1321. doi:10.1016/j.ultrasmedbio.2006.05.028.

Poonai, N., Myslik, F., Joubert, G., Fan, J., Misir, A., Istasy, V., Columbus, M., Soegtrop, R., Goldfarb, A., Thompson, D., Dubrovsky, A.S., Theodoro, D.L., 2017. Point-of-care ultrasound for nonangulated distal forearm fractures in children: test performance characteristics and patient-centered outcomes. Acad. Emerg. Med. 1069-6563. 24 (5), 607–616. doi:10.1111/acem.13146.

Rayan, G.M., Tomasek, J.J., 1994. Generation of contractile force by cultured Dupuytren's disease and normal palmar fibroblasts. Tissue Cell. 00408166. 26 (5), 747–756. doi:10.1016/0040-8166(94)90057-4.

Romo 3rd, T., Sclafani, A.P., Yao, J.F., 2005. Hair restoration surgery: challenges and solutions. Facial Plast Surg 21 (2), 115–122.

Scott, D.L., Wolfe, F., Huizinga, T.W.J., 2010. Rheumatoid arthritis. Lancet. 01406736. 376 (9746), 1094–1108. doi:10.1016/s0140-6736(10)60826-4.

Seror, P., Seror, R., 2006. Meralgia paresthetica: clinical and electrophysiological diagnosis in 120 cases. Muscle Nerve. 0148-639X. 33 (5), 650–654. doi:10.1002/mus.20507.

Siegel, R.L., Miller, K.D., Jemal, A., 2020. Cancer statistics, 2020. CA Cancer J. Clin. 15424863. 70 (1), 7–30. Available from: http://onlinelibrary.wiley.com/journal/10.3322/(ISSN)1542-4863. doi:10.3322/caac.21590.

Standring, S., 2016. Gray's Anatomy: The Anatomical Basis of Clinical Practice. Elsevier Churchill Livingstone.

Stark, H.H., Boyes, Johnson, L., Ashworth, C.R., 1977. The biomechanical properties of the trapezium: application to reconstructive procedures of the thumb. J. Hand Surg. Am. 2 (5), 345–357.

Stecco, A., Masiero, S., Macchi, V., Stecco, C., Porzionato, A., De Caro, R., 2009a. The pectoral fascia: anatomical and histological study. J. Bodyw. Mov. Ther. 13608592. 13 (3), 255–261. doi:10.1016/j.jbmt.2008.04.036.

Stecco, A., Stern, R., Fantoni, I., De Caro, R., Stecco, C., 2016. Fascial disorders: implications for treatment. PM&R. 1934-1482. 8 (2), 161–168. doi:10.1016/j.pmrj.2015.06.006.

Stecco, C., Gagey, O., Belloni, A., Pozzuoli, A., Porzionato, A., Macchi, V., Aldegheri, R., De Caro, R., Delmas, V., 2007. Anatomy of the deep fascia of the upper limb. Second part: study of innervation. Morphologie. 12860115. 91 (292), 38–43. doi:10.1016/j.morpho.2007.05.002.

Stecco, C., Macchi, V., Porzionato, A., Morra, A., Parenti, A., Stecco, A., Delmas, V., De Caro, R., 2010. The ankle retinacula: morphological evidence of the proprioceptive role of the fascial system. Cells Tissues Organs. 1422-6405. 192 (3), 200–210. doi:10.1159/000290225.

Stecco, C., Pirri, C., Fede, C., Fan, C., Giordani, F., Stecco, L., Foti, C., De Caro, R., 2019. Dermatome and fasciatome. Clin. Anat. 0897-3806. 32 (7), 896–902. doi:10.1002/ca.23408.

Terry, G.C., Chopp, T.M., 2000. Functional anatomy of the shoulder. J. Athl. Train. 35 (3), 248–255.

Thomas, J.L., Christensen, J.C., Kravitz, S.R., Mendicino, R.W., Schuberth, J.M., Vanore, J.V., Weil, L.S., Zlotoff, H.J., Bouché, R., Baker, J., 2010. The diagnosis and treatment of heel pain: a clinical practice guideline-revision 2010. J. Foot Ankle Surg. 10672516. 49 (3), S1. doi:10.1053/j.jfas.2010.01.001.

Torriani, M., Etchebehere, M., Amstalden, E.M.I., 2002. Sonographically guided core needle biopsy of bone and soft tissue tumors. J. Ultrasound Med. 02784297. 21 (3), 275–281. doi:10.7863/jum.2002.21.3.275.

Tortora, G.J., Derrickson, B.H., 2017. Principles of Anatomy and Physiology, Vol. 15. John Wiley & Sons.

Tran, T.M.N., Ivanusic, J.J., Hebbard, P., Barrington, M.J., 2009. Determination of spread of injectate after ultrasound-guided transversus abdominis plane block: a cadaveric study. Br. J. Anaesth. 00070912. 102 (1), 123–127. doi:10.1093/bja/aen344.

Trayes, K.P., Studdiford, J.S., Pickle, S., Tully, A.S., 2013. Edema: diagnosis and management. Am. Fam. Physician 15320650. 88 (2), 102–110. Available from http://www.aafp.org/afp/2013/0715/p102.pdf.

Trovik, C.S., Bauer, H.C.F., Alvegård, T.A., Anderson, H., Blomqvist, C., Berlin, O., Gustafson, P., Sæter, G., Wallöe, A., 2000. Surgical margins, local recurrence and metastasis in soft tissue sarcomas: 559 surgically-treated patients from the Scandinavian Sarcoma Group Register. Eur. J. Cancer. 09598049. 36 (6), 710–716. Available from: http://www.journals.elsevier.com/european-journal-of-cancer/. doi:10.1016/S0959-8049(99)00287-7.

Van Der Worp, M.P., Van Der Horst, N., De Wijer, A., Backx, F.J.G., Van Der Sanden, M.W.G.N., 2012. Iliotibial band syndrome in runners: a systematic review. Sports Med. 11792035. 42 (11), 969–992. doi:10.2165/11635400-000000000-00000.

van Goor, H., 2007. Consequences and complications of peritoneal adhesions. Colorectal Disease. 14631318. 9 (2), 25–34. doi:10.1111/j.1463-1318.2007.01358.x.

Vattikuti, R., Towler, D.A., 2004. Osteogenic regulation of vascular calcification: an early perspective. Am. J. Physiol. Endocrinol. Metab. 286 (5), E686. doi:10.1152/ajpendo.00552.2003.

Verdan, C.E., 1975. Primary and Secondary Repair of Flexor and Extensor Tendon Injuries. Williams & Wilkins, pp. 220–275.

Visnes, H., Bahr, R., 2007. The evolution of eccentric training as treatment for patellar tendinopathy (jumper's knee): a critical review of exercise programmes. Br. J. Sports Med. 0306-3674. 41 (4), 217–223. doi:10.1136/bjsm.2006.032417.

Vlachos, K., Archontovasilis, F., Falidas, E., 2005. Ultrasound-guided percutaneous drainage of abdominal abscesses. Eur. J. Radiol. 53 (2), 206–210.

Volpe, A., Rossato, G., Bottanelli, M., Marchetta, A., Caramaschi, P., Bambara, L.M., Bianconi, C., Arcaro, G., Grassi, W., 2009. Ultrasound evaluation of ulnar neuropathy at the elbow: correlation with electrophysiological studies. Rheumatology. 1462-0324. 48 (9), 1098–1101. doi:10.1093/rheumatology/kep167.

Wakefield, R.J., Balint, P.V., Szkudlarek, M., Filippucci, E., Backhaus, M., D'Agostino, M.A., Sanchez, E.N., Iagnocco, A., Schmidt, W.A., Bruyn, G., Kane, D., O'Connor, P.J., Manger, B., Joshua, F., Koski, J., Grassi, W., Lassere, M.N.D., Swen, N., Kainberger, F., Klauser, A., Ostergaard, M., Brown, A.K., Machold, K.P., Conaghan, P.G., 2005. Musculoskeletal ultrasound including definitions for ultrasonographic pathology. J. Rheumatol. 0315162X. 12 (32), 2485–2487. https://www.jrheum.com.

Walker, F.O., 2004. Neuromuscular ultrasound. Neurologic Clinics 22 (3), 563–590. doi:10.1016/j.ncl.2004.03.004.

Walker, F.O., Cartwright, M.S., Wiesler, E.R., Caress, J., 2004. Ultrasound of nerve and muscle. Clin. Neurophysiol. 13882457. 115 (3), 495–507. Available from: https://www.elsevier.com/inca/publications/store/6/0/1/5/2/8. doi:10.1016/j.clinph.2003.10.022.

Wearing, S.C., Smeathers, J.E., Urry, S.R., Hennig, E.M., Hills, A.P., 2006. The pathomechanics of plantar fasciitis. Sports Med. 01121642. 36 (7), 585–611. Available from: http://sportsmedicine.adisonline.com/pt/re/spo/pdfhandler.00007256-200636070-00004.pdf;jsessionid=GhZY7XkLTkrJBgSgsvgKlp41TSy943nRWMJ6HnjjQTMrWNypTf1v!-1734750035!-949856144!8091!-1. doi:10.2165/00007256-200636070-00004Australia.

Weiss, Goldblum, J.R., 2014. Enzinger and Weiss's Soft Tissue Tumors. Mosby Elsevier.

White, T., Folkens, P., 2005. The Human Bone Manual. Elsevier Inc., United States, Available from http://www.sciencedirect.com/science/book/9780120884674. doi:10.1016/C2009-0-00102-0.

Williamson, D., Watura, R., Cobby, M., 2000. Ultrasound imaging of forearm fractures in children: a viable alternative? J. Accid. Emerg. Med. 13510622. 17 (1), 22–24. Available from: https://www.bmjpg.com/data/aem.htm. doi:10.1136/emj.17.1.22.

Woo, S.L.Y., Hollis, J.M., Adams, D.J., Lyon, R.M., Takai, S., 1991. Tensile properties of the human femur-anterior cruciate ligament-tibia complex: the effects of specimen age and orientation. Am. J. Sports Med. 15523365. 19 (3), 217–225. doi:10.1177/036354659101900303.

Zagars, G.K., Ballo, M.T., Pisters, P.W.T., Pollock, R.E., Patel, S.R., Benjamin, R.S., Evans, H.L., 2003. Prognostic factors for patients with localized soft-tissue sarcoma treated with conservation surgery and radiation therapy: an analysis of 1225 patients. Cancer 97 (10), 2530–2543. doi:10.1002/cncr.11365.

Zaidman, C.M., Wu, J.S., Kapur, K., Pasternak, A., Madabusi, L., Yim, S., Pacheck, A., Szelag, H., Harrington, T., Darras, B.T., Rutkove, S.B., 2017. Quantitative muscle ultrasound detects disease progression in Duchenne muscular dystrophy. Ann. Neurol. 15318249. 81 (5), 633–640. Available from: http://onlinelibrary.wiley.com/journal/10.1002/(ISSN)1531-8249. doi:10.1002/ana.24904.

Zhai, G., Valdes, A.M., Cherkas, L., Clement, G., Strachan, T.D., 2009. Spector, natural history and risk factors for bone marrow lesions in women: longitudinal evaluation by MRI over four years. Arthritis Rheum. 60 (11), 3349–3355.

Further reading

Zhai, A., Hungerford, M.W., Nascimento, T.A., 2016. Accuracy of ultrasound in diagnosing lateral femoral cutaneous nerve entrapment: an anatomical study. Reg. Anesth. Pain Med. 41 (4), 523–527.

Zhou, X.T., Ma, Y., Guo, Y., Si, Y.H., Zong, Y.G., Zhong, C., 2019. Application status and mechanism of musculoskeletal ultrasound in orthopedic diagnosis. Chin. J. Tissue Eng. Res. 20954344. 23 (16), 2573–2578. Available from: http://www.crter.org/CN/article/showOldVolumn.do. doi:10.3969/j.issn.2095-4344.1167.

Part V

Chronic disease, yoga, and cognitive well-being

Chapter 18

Muscle imaging in chronic obstructive pulmonary disease

Anup Bhat[a], K. Vaishali[a], Mukesh Kumar Sinha[a], Leena R. David[b,c], Sneha Ravichandran[c], Suresh Sukumar[c], Rajagopal Kadavigere[d], Dilber Uzun Ozsahin[b,e,f], Ilker Ozsahin[f], and Berna Uzun[f]

[a]*Department of Physiotherapy, Manipal College of Health Professions, Manipal Academy of Higher Education, Manipal, India*

[b]*Medical Diagnostic Imaging Department, College of Health Sciences, University of Sharjah, Sharjah, United Arab Emirates*

[c]*Department of Medical Imaging Technology, Manipal College of Health Professions, Manipal Academy of Higher Education, Manipal, Karnataka, India*

[d]*Department of Radiodiagnosis and Imaging, Kasturba Medical College, Manipal Academy of Higher Education, Manipal, India*

[e]*Research Institute for Medical and Health Sciences, University of Sharjah, Sharjah, United Arab Emirates*

[f]*Irfan Suat Gunsel Operational Research Institute, Near East University, Nicosia, Turkish Republic of Northern Cyprus (TRNC), Turkey*

1 Introduction

1.1 Definition and overview of chronic obstructive pulmonary disease

The Global Initiative for Chronic Obstructive Lung Disease, in its 2023 report, defines chronic obstructive pulmonary disease (COPD) as a "heterogeneous lung condition characterized by chronic respiratory symptoms (dyspnea, cough, sputum production, and/or exacerbations) due to abnormalities of the airways (bronchitis, bronchiolitis) and/or alveoli (emphysema) that cause persistent, often progressive, airflow obstruction" (Agustí et al., 2023). Globally, COPD was the third most common cause of death in 2019. According to the Global Burden of Disease 2019 study, 212.3 million cases of COPD were prevalent. The global disability-adjusted life years for COPD were 74.4 million (Safiri et al., 2019). The causes and risk factors of COPD include tobacco smoking, abnormal lung development, infections, occupational exposures, indoor and outdoor pollution, genetics, lung aging, alpha-1 antitrypsin deficiency, and multimorbidity (MacNee, 2006; Viegi et al., 2007). Patients with COPD present with dyspnea, wheezing, cough with or without expectoration, chest tightness, activity limitations, exercise intolerance, muscle weakness, balance impairments, early fatigue, and an overall poor quality of life (Riley and Sciurba, 2019; Park et al., 2020; Singh et al., 2019).

1.2 Significance of muscle involvement in COPD

Patients with COPD present with both ventilatory and nonventilatory skeletal muscle dysfunction (Jaitovich and Barreiro, 2018). Muscle dysfunction includes reduced strength, reduced endurance, and fatigue. Systemic inflammatory responses may trigger cascades of events that lead to muscle atrophy in these patients. As inflammation is not always associated with stable COPD, and muscle changes can be noticed even before pulmonary changes, this suggests that extrapulmonary changes occur simultaneously as the pulmonary system undergoes alterations (Gea et al., 1985). Muscle dysfunction is associated with poor exercise capacity, poor health status, poor quality of life, and increased healthcare utilization. Peripheral muscle dysfunction appears to impact mortality in patients with COPD (Swallow et al., 2007; Marquis et al., 2002).

1.3 Purpose and scope of the chapter

The purpose of this chapter is to provide an overview of how muscle imaging technology can be used to understand muscle function impairment in COPD. This chapter discusses COPD-associated muscle dysfunction, its effects on patients with COPD, various established muscle imaging technologies, and their findings.

Advances in Medical Imaging. https://doi.org/10.1016/B978-0-443-28967-5.00035-6

2 Overview of respiratory muscle dysfunction in patients with COPD

2.1 Overview of respiratory muscles and their role in breathing

Ventilation is a complex process that requires interaction between inspiratory muscles, the chest wall, neural centers and expiratory muscles. The impulses from the pontomedullary centers coordinate to activate the inspiratory pump, which increases the volume of the thorax by acting on the chest wall. The diaphragm, external intercostal muscles, and the parasternal part of the internal intercostal muscles contract to elevate the ribs and decrease the intrathoracic pressure. The muscles of the inspiratory pump that are always active during inspiration are the primary muscles of inspiration. The muscles of the inspiratory pump that become active during forceful inspiration are the accessory muscles of inspiration. These include the sternocleidomastoid, pectoralis, scalene, and trapezius muscles. Expiration is a passive process and is accomplished by the elastic recoil of the lungs and the thoracic cage. The internal intercostal muscles, rectus abdominis, external obliques, internal obliques, and transverse abdominis contract to pull the ribs downward and push the diaphragm upward, thereby reducing the intrathoracic volume. Hence, these muscles are considered accessory muscles of expiration.

2.2 Structure of respiratory muscle

Skeletal muscles are broadly divided into fast-twitch and slow-twitch muscles based on the speed of contraction. This classification is determined by histochemical staining for myofibrillar adenosine triphosphatase (ATPase), as ATPase activity is linked to the speed of contraction. Based on metabolism, fast-twitch muscle fibers (Type II fibers) are further divided into fast oxidative-glycolytic—Type IIa (FTa), which rely on aerobic or oxidative metabolism, and fast glycolytic—Type IIb or IIx (FTb/FTx), which rely more on glycolytic or anaerobic metabolism. Slow-twitch fibers contract slowly but are fatigue-resistant. FTa fibers are fast in contraction and are relatively fatigue-resistant, whereas FTb fibers are very fast but easily fatigable. Based on myosin isoforms, muscle fibers are divided into Type 1, Type 2A, and Type 2X. Type 1 and Type 2A are oxidative, whereas Type 2X relies on glycolytic metabolism (Vikne et al., 2020).

The respiratory muscles have a combination of slow and fast-twitch muscle fibers. The diaphragm contains equal proportions of slow-twitch and fast-twitch muscle fibers. During quiet breathing, slow-twitch, fatigue-resistant muscle fibers are recruited, and during heavy breathing, the recruitment of fast-twitch muscle fibers increases. The diaphragm muscle has high capillary density and high aerobic metabolism, which are necessary for the continuous activity of the diaphragm in breathing. The structure and function of the muscle fibers in the respiratory muscles are not fixed and adapt to training or pathological conditions. The adult diaphragm consists of approximately 55%, 21%, and 24% of slow fibers, fast oxidative fibers, and fast glycolytic muscle fibers, respectively. Intercostal muscles contain about 60% slow-twitch muscle fibers (Polla et al., 2004).

2.3 Respiratory muscle dysfunction in COPD

COPD is characterized by chronic hyperinflation (increased lung compliance) due to airflow limitations. The chronic increase in lung compliance alters the position of the diaphragm; specifically, the dome of the diaphragm is lowered, leading to changes in the length-tension relationship of the diaphragm during inspiration. Alterations in the diaphragm's length-tension relationship reduce the force-generating capacity and mobility of the diaphragm muscle. The flattened position of the diaphragm, caused by a reduced zone of apposition, forces the diaphragm to work harder to overcome the geometrical dysfunction of the chest wall (Bordoni et al., 2016, 2023). Ogan et al. demonstrated that diaphragm thickness has not been altered in people with COPD. Hence, the reason for diaphragm dysfunction is more inclined to structural changes in the chest wall rather than the thickness of the diaphragm (Ogan et al., 2019). The inability to exhale completely due to narrow airways while breathing at higher frequencies (as in exercise and infections) can worsen hyperinflation (O'Donnell et al., 2020; Charususin et al., 2018). Furthermore, hypoxemia, gas exchange abnormalities, or reduced delivery of oxygen cause respiratory muscle dysfunction in COPD. Risk factors for respiratory dysfunction include poor nutrition, medications, comorbidities, exacerbations, smoking, and reduced physical activity.

Morphologically, atrophy of all muscle fiber types is observed. The proportion of oxidative, fatigue-resistant type 1 muscle fibers increases as the proportion of other muscle fibers decreases. This shift in muscle fiber type reduces maximal force development and results in a reduction in muscle power during muscle contraction (17). The increased protease activity in the muscle causes degradative alterations and induces structural injury in the diaphragm. In addition, increased reactive oxygen species cause oxidation of muscle components, such as proteins, and result in structural damage. All the factors mentioned above, combined with malnutrition, comorbidities, and age, contribute to respiratory muscle dysfunction in COPD (Orozco-Levi, 2003; Meyer et al., 2013).

3 Overview of peripheral muscle dysfunction in patients with COPD

3.1 Causes and pathophysiology of peripheral muscle dysfunction

Multiple etiological factors contribute to peripheral muscle dysfunction, such as exacerbation, genetics, cigarette smoke, chronic inflammatory state, hypoxia causing impaired oxygen delivery, acidosis and glycolytic metabolism, malnutrition, comorbidities, age, reduced defense against oxidative stress, disuse atrophy, testosterone deficiencies, and the use of glucocorticoids (Barreiro and Gea, 2016). However, detailed mechanisms of peripheral muscle dysfunction are beyond the scope of this chapter.

3.1.1 Inflammation and malnutrition

Markers of inflammation, such as C-reactive protein, fibrinogen, interleukin-6, interleukin-8, and Tumor necrosis factor—alpha, are elevated in patients with stable COPD. Chronic inflammation and the proinflammatory state in patients with COPD increase the activity of the ubiquitin-proteasome pathway, thereby promoting protein degradation. In addition to this, reduced defense against oxidative stress further damages muscle proteins. To further complicate this situation, many patients with COPD find it difficult to consume nutritious food for various reasons. Muscle proteins are utilized as a source of energy due to the hypermetabolic states present in patients with COPD, causing catabolism. This, coupled with poor nutritional intake, accelerates muscle breakdown (MacIntyre, 2006; Rawal and Yadav, 2015).

3.1.2 Corticosteroids

The drugs that are used in the treatment of COPD exacerbation include corticosteroids. Corticosteroids elicit severe skeletal muscle dysfunction due to the inhibition of pathways that promote protein synthesis and fuel protein degradation. However, corticosteroids do not have effects on muscle function in stable patients (MacIntyre, 2006; Kim et al., 2008).

3.1.3 Disuse and physical inactivity

Patients with COPD often avoid activities and remain sedentary due to the fear of breathlessness. It is a well-known fact that disuse of muscles can cause skeletal muscle adaptations. These adaptations include muscle fiber atrophy, changes in muscle metabolism, reduction in muscle capillary density, and changes in muscle fiber type. These factors are collectively known as deconditioning and affect both the strength and endurance of the muscle. Muscle deconditioning is highly dependent on the degree of inactivity. However, it is worth noting that physical inactivity alone is not the sole cause of muscular dysfunction in COPD participants (Kim et al., 2008).

3.1.4 Advanced age

Even though COPD can be diagnosed in young adults, nearly 95% of the patients are above the age of 50 years. The muscles of older people are smaller, containing more fat and connective tissues than those of younger people. Muscle strength declines with aging, and this factor of aging must be considered in people with COPD (Kim et al., 2008).

3.1.5 Cigarette smoking

Cigarette smoking has been shown to cause muscle dysfunction in both animal studies and humans. The components of cigarette smoke that circulate in the blood trigger the activation of proteolysis and the inhibition of protein synthesis. Blood flow to muscles and oxygen delivery to mitochondria may be impaired in smokers, resulting in muscle fatigue. Smokers exhibit lower proportions of type I muscle fibers compared to nonsmokers (Degens et al., 2015; Kim et al., 2008).

3.1.6 Hypoxemia

Hypobaric oxygen exposure is associated with muscle dysfunction, such as a reduction in strength and endurance. Patients with COPD experience chronic hypoxemia or repeated episodes of hypoxemia. Tissue hypoxia triggers inflammatory mediators and increases oxidative stress in patients with COPD. Quadriceps endurance was reduced in patients with chronic hypoxemia (Kim et al., 2008; Koechlin et al., 2005; Pitsiou et al., 2002).

3.1.7 Acidosis and glycolytic metabolism

In animal models, hypercapnia and acidosis demonstrated an inhibition in the pathways of protein synthesis. Furthermore, hypercapnia has demonstrated increased catabolism. Myotubes exposed to high carbon dioxide showed a reduction in muscle fiber diameter and anabolic capacity (27). Metabolic homeostasis is important in regulating cell proliferation and differentiation. Muscle mitochondrial dysfunction is evident in COPD, causing a loss of oxidative capacity in the muscle. This decreased oxidative capacity leads to glycolytic metabolism, thereby increasing lactic acid production (Couillard and Prefaut, 2005; Puente-Maestu et al., 2013).

3.1.8 Comorbidities

Heart disease, cancer, diabetes, and chronic kidney diseases are the most commonly associated comorbidities in patients with COPD. The prevalence of comorbidities is high in patients with COPD. All these diseases are known to cause muscular dysfunction and/or nutritional impairments (Gea et al., 2015; Poberezhets et al., 2020). The skeletal muscle dysfunction in chronic heart failure and kidney diseases is similar to that of COPD (Dubé and Laveneziana, 2018).

3.1.9 Genetics

Certain gene types are associated with muscular dysfunction and sarcopenia. This explains the variations in muscular dysfunction among patients with similar disease stages and treatments. The muscular dysfunction appears to worsen with hypoxia in the selected gene variants (Attaway et al., 2023).

3.2 Characteristics of muscle dysfunction in COPD patients

Most of the studies that assessed muscle dysfunction included the quadriceps muscle. Quadriceps weakness is one of the most evident features seen in all stages of COPD. Slow-to-fast twitch transformation, i.e., a shift of muscle fibers from type I to type IIx, increased glycolytic capacity, reduced oxidative capacity, fiber atrophy, muscle mass attrition, and reduced capillary density are observed in the studied muscle. Muscular dysfunction is reflected by decreased muscle strength and endurance, as well as increased muscle fatigability. Although muscular dysfunction is frequently studied in quadriceps muscle groups (vastus lateralis, rectus femoris), muscle dysfunction is also evident in the deltoid, biceps brachii, triceps brachii, adductor pollicis, pectoralis major, latissimus dorsi, erector spinae, and abdominal muscle groups. The extent of muscle dysfunction does not appear to be uniformly affected across all muscles.

3.3 Impact of peripheral muscle dysfunction in patients with COPD

Muscle dysfunction, breathlessness, reduced exercise capacity, exercise intolerance, fatigue, and reduced physical activity are interlinked impairments in patients with COPD. Increased lactate production during muscle contraction exacerbates fatigue and exercise intolerance. Fear avoidance leads to a spiral of disability, which further increases muscle dysfunction, breathlessness, reduced exercise capacity, and fatigue. Muscle dysfunction, combined with chronic hypoxia to the brain, causes postural balance impairments in these patients. The above-mentioned impairments contribute to the inability to perform daily tasks and increased dependency, leading to poor health status and quality of life. Muscle dysfunction and exercise intolerance are associated with higher health care utilization and increased mortality (Ansari et al., 2012; Sin and Man, 2006).

4 Imaging techniques for assessing respiratory and peripheral muscle changes in COPD

4.1 Ultrasonography

4.1.1 Respiratory muscle assessment

In 1969, the use of ultrasound to assess diaphragm motion and contour was first described. In 1989, an ultrasonography technique was developed to measure diaphragm thickness. Ultrasound of the diaphragm is considered reliable and reproducible when performed by skilled professionals. The intercostal and subcostal approaches are used to obtain information on diaphragm function.

Intercostal approach: Requires a high-frequency linear array transducer with the probe placed between the anterior and middle axillary lines on the skin, perpendicular to the chest wall. The diaphragm movement can be scanned at the eighth to 11th intercostal spaces. The scan demonstrates three structures: pleura, diaphragm, and peritoneum. The pleura and peritoneum are hyperechoic, and the diaphragm is hypoechoic. Resting thickness and the thickening fraction can be observed. Thickening fraction refers to the difference between end-inspiratory thickness and end-expiratory thickness, divided by end-expiratory thickness. Although various factors, such as body position and lung volume, can affect diaphragm thickness, the lower limit of normal is 1.5 mm, and the thickening ratio is 1.2–1.3.

Subcostal approach: A low-frequency convex probe is placed on the skin at the mid-clavicular line under the rib edge and tilted along the long axis of the intercostal space. The movement of the diaphragm can be observed in M-mode, and the contour of the diaphragm can be visualized through B-mode ultrasound. The movement of the diaphragm is called diaphragm excursion. Normal values during quiet breathing in males are 1.8 ± 0.3 cm and 1.8 ± 0.4 cm for the left and right sides, respectively. Similarly, normal values of diaphragm excursion for females during quiet breathing are 1.6 ± 0.3 cm and 1.6 ± 0.4 cm on the left and right sides, respectively. This increases manyfold during deep breathing (Jia and Zhang, 2022; Laghi et al., 2021).

4.1.2 Peripheral muscle assessment

Ultrasonography of the lower limb, particularly the thigh, is often carried out in COPD. The quadriceps femoris (rectus femoris, vastus intermedius) was the most commonly studied muscle. Linear 3–15 MHz probes are used with bright mode or B-mode ultrasound. The probe is placed either at 1/3 thigh, 3/5 thigh, or mid-thigh between the anterior superior iliac spine and the superior border of the patella. Individuals are placed in a supine position with the rested leg in extension. Measurements are made by placing the transducer perpendicular to the long axis of the thigh. Cross-sectional area of the muscle, quadriceps thickness, echo intensity, and quadriceps contractile index are usually measured parameters. Rectus femoris thickness was significantly lower in individuals with COPD than in healthy, normal individuals. Ye et al. found a significant difference in echo intensity between COPD and age-matched non-COPD individuals. The echo intensity differed significantly among different disease severity grades (Casey et al., 2022; Ramachandran et al., 2020; Ye et al., 2017).

4.2 Computed tomography scans

4.2.1 Respiratory muscle assessment

In 1984, Hayhurst et al. described the use of computed tomography (CT) scans in the diagnosis of emphysema using Hounsfield units (HU) or Electric and Musical Industries (EMI) units (40). In 1997, using the techniques of Pettiaux et al., Cassart et al. demonstrated a reduced total surface area of the diaphragm and its zone of apposition at functional residual capacity. Using a helical scanner, diaphragm morphology has been assessed in COPD (Cassart et al., 1997; Pettiaux et al., 1997). The product of the CT axial slice thickness and the number of axial slices between the inferior-most axial image of the diaphragm and the topmost axial image of the central tendon of the diaphragm is considered the diaphragm dome height. This dome height appears to be associated with COPD severity and exercise capacity. The lower the height, the greater the COPD severity and the poorer the exercise capacity. The diaphragm volume is categorically classified as "normal muscle," "abnormal muscle," and "intermuscular adipose" using HU. The diaphragm volume of individuals with COPD was similar to that of age- and weight-matched control individuals (Donovan et al., 2021).

4.2.2 Peripheral muscle assessment

Computerized tomography has been used to evaluate skeletal muscle cross-sectional area. Most studies have utilized thoracic CT to assess paraspinal muscles, pectoral muscles, psoas/abdominal muscles, and intercostal muscles. The cross-sectional area of muscles had a low-to-moderate association with exercise capacity and body mass index (Nicholson et al., 2022). Pectoralis muscle area was lower in patients with COPD, and there was a significant difference in the pectoralis muscle area (PMA) across all severity levels of COPD (Wilson et al., 2022). In addition, psoas muscle density measured using axial low-dose chest CT was lower in patients with COPD than in controls and is associated with long-term mortality in patients with COPD (Ezponda et al., 2021).

Fig. 18.1 shows the CT scan of the lungs of a 30-year-old male patient with a history of heavy smoking (Mudgal, 2015).

4.3 Magnetic resonance imaging

4.3.1 Respiratory muscle assessment

Magnetic resonance imaging (MRI) can be used to evaluate the shape, position, thickness, and surface area of the diaphragm. The patient lies in a supine position in the MRI apparatus, and using the cine dynamic sequence with a 1.5T MRI, images are obtained during maximal inspiration and maximal expiration. The images are processed using computer software to obtain diaphragm movement and height measurements.

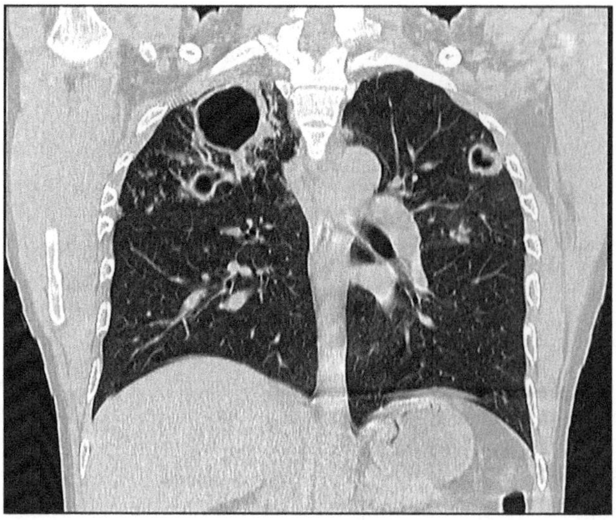

FIG. 18.1 CT chest coronal view showing multiple thick-walled cavitary lesions in the bilateral upper lobes, along with patchy ground-glass nodules and branching centrilobular nodules (tree-in-bud), cystic bronchiectatic changes, and subpleural emphysematous bullae. These findings are suggestive of an infective etiology, most likely pulmonary tuberculosis with chronic obstructive disease changes. *(From Mudgal, P., 2015. Pulmonary tuberculosis with COPD. Radiopaedia.org. doi:10.53347/rID-40830).*

Diaphragm height is measured as the point on the diaphragm contour most distant from a line intersecting the endpoints of a reference diaphragm. Diaphragm movement is assessed by measuring the area swept by the diaphragm in the subsequent images in pixels, which is later converted to cm^2 (Iwasawa et al., 2002).

Dynamic breathing MRI has been used to assess respiratory mechanics, including diaphragm function. During the procedure, subjects are instructed to take 6–8 breaths while images are captured. Using a turbo-fast low-angle shot (FLASH) sequence, scans are obtained at a consistent mid-coronal plane passing through the trachea, as well as at two fixed mid-sagittal planes through the center of each lung. In most patients with emphysema, the diaphragm appears flattened or shows a loss of normal curvature. In addition, an increased vertical (cephalocaudal) movement of the posterior portion of the diaphragm is observed on the mid-sagittal images (Suga et al., 1999). This technique is time-consuming and expensive.

4.3.2 Peripheral muscle assessment

MRI has been used to assess peripheral muscle thickness in patients with COPD. Thigh, calf, forearm, and arm musculature thickness have often been assessed using 1.5T MRI. The thigh is the most common region selected for muscle imaging using MRI. For the thigh region, 5 mm-thick contiguous axial slices from the anterior superior iliac spine to the tibial plateau are acquired, and the images are T1-weighted. These images are later processed using image processing software to analyze the muscle cross-sectional area and other compositions, such as fat and bone. Muscle volume and muscle cross-sectional area are analyzed. In addition, adipose tissues are identified by some researchers (Rozenberg et al., 2017).

4.4 Chest X-ray techniques

4.4.1 Chest X-ray

In 1964, the use of chest X-ray to assess diaphragm movements, level, and contour was described in the criteria for the radiographic diagnosis of emphysema by George Simon. One radiograph is taken during inspiration and another during expiration without moving the X-ray cassette, performed by a trained radiographer. The position of the diaphragm is classified as low, flat, or normal. The position of the diaphragm is considered low "if the top of the right dome is below the level of the inferior margin of the anterior end of the sixth rib". To describe the diaphragm as flat, a line is drawn at the highest point of the curve of each dome, and a parallel line is drawn from the cardiophrenic to the costophrenic angle. If the distance between these two lines is less than 1.5 cm, the diaphragm is classified as flat. Diaphragm movement is assessed by measuring the difference between the dome of the diaphragm in inspiration and expiration films (Nairn et al., 1969; Simon, 1964). Reich et al. described the diaphragm level on chest radiographs by measuring the apex of the diaphragm in reference to the anterior and posterior ribs. In addition, the diaphragmatic arc on the right side in the frontal film was described as "measurement of a line normal to the curve at its apex and normal to a line drawn between the lateral costophrenic angle

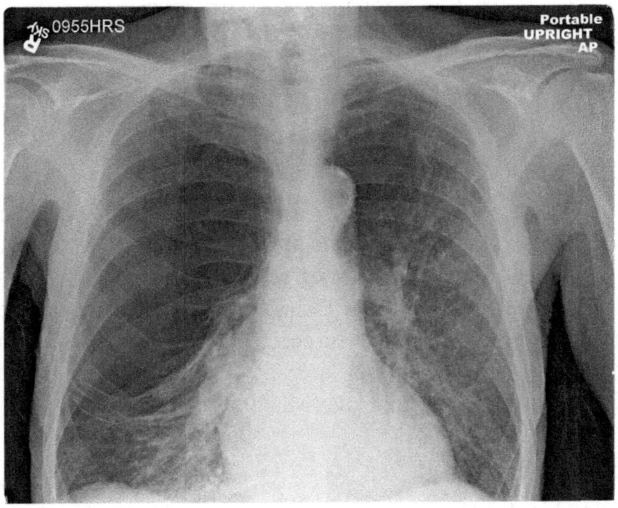

FIG. 18.2 Chest AP X-ray of an active smoker patient with COPD, showing hyperinflated lungs with diffuse reticular changes. *(From Mark, A., 2017. Very large bulla on background of COPD. Radiopaedia.org. doi:10.53347/rID-48598).*

and the cardiophrenic angle" (Reich et al., 1985). Fig. 18.2 shows the chest X-ray of an active smoker, 75 years of age (Mark, 2017).

4.4.2 Dynamic chest X-ray

To detect diaphragmatic movements, a flat-panel X-ray detector can be used to obtain dynamic radiographs. While obtaining a dynamic chest X-ray, patients are asked to inhale completely and exhale completely, and the X-ray is taken in a standing position. The emitting pulse is set at 15 frames per second. These images are then processed using computer software, which automatically detects the edges of the diaphragm on the inspiratory film and expiratory film with inputs from radiologists. The software detects the movements (motion trail). The diaphragm excursion, duration, and peak velocities of diaphragmatic movements can be obtained (Jianghong Chen et al., 2022a).

4.5 Fluoroscopy

X-ray fluoroscopy can serve as an initial assessment tool for respiratory muscle function; however, its images often lack sufficient clarity and expose subjects to radiation hazards. Despite these limitations, fluoroscopy has been used to measure diaphragmatic movements in patients with COPD in interventional studies, such as those evaluating the effects of posture correction or pulmonary rehabilitation.

For imaging analysis, a baseline chest X-ray is obtained to establish reference points. Markers are placed on the rib cage and vertebral bodies of each patient, serving as reference points on the posteroanterior chest radiographs. Patients are instructed to breathe deeply while standing, and fluoroscopy is performed during three deep breaths. Three consecutive chest radiographs are captured at both maximum inspiration and expiration phases using guided fluoroscopy.

On the radiographs, the origin of the diaphragm is marked at the costophrenic angle. The contour of the dome of the diaphragm is traced as a line. Located at the level of the vertebral column where the ribs insert, transverse lines are drawn from the spinal processes to the rib insertion points, forming a horizontal reference line. The displacement of the diaphragm motion is then quantified within this defined area by the vertical line along the vertebral column, the horizontal line, and the shadow of the diaphragm dome on the radiograph (Yanjun Chen et al., 2022b; Jung et al., 2021).

4.6 Positron emission tomography/computed tomography scans

The Fluorine-18 (18F) fluorodeoxyglucose (FDG) (18F-FDG) uptake is used to detect increased usage of respiratory muscles. The procedure for a positron emission tomography (PET)/CT scan includes patients fasting for at least 4 hours before the PET scan. Patients receive 18F-FDG as an IV injection. They need to remain quiet in the injection room for 60 minutes following the IV injection during the 18F-FDG uptake phase. Patients are required to breathe normally without altering their breathing pattern. The CT scan is performed after 60 minutes. The images obtained from the scan are reconstructed and analyzed using software by nuclear medicine physicians. The 18F-FDG uptake was higher in the diaphragm, intercostal, and scalene muscles (Osman et al., 2011).

4.7 Other techniques

4.7.1 Dual-energy X-ray absorptiometry

Dual-energy X-ray absorptiometry (DXA) is predominantly used for body composition assessment of peripheral muscles. The lean body mass, bone-free lean mass, fat mass, and bone mineral density can be evaluated using a DXA scan. Total and segmental lean mass can also be evaluated using DXA scans. Subjects are made to lie in a supine position, and photons from two different X-ray sources are used to perform the scan. Bone and soft tissue are differentiated based on photon attenuation. The differential absorption of the soft tissue is measured. The ratio of the absorption of two energy-level photons is related to the percentage of fat. Muscle mass is then measured by subtracting the sum of fat and bone mass. Lean muscle mass is lower in patients with COPD compared to controls (Malaguti et al., 2006).

4.7.2 Bioelectrical impedance analysis

Body composition assessment using bioelectrical impedance analysis is the most clinically feasible bedside technique for peripheral muscle evaluation. The accuracy of fat-free mass obtained from the bioelectrical impedance method is dictated by the choice of prediction equation. Body composition measurement using bioelectrical impedance underestimated the fat-free mass when compared to the fat-free mass assessed by a DXA scan. Steiner et al. provide a sex-based prediction equation for fat-free mass from bioelectrical impedance. Steiner's formula appears to be more accurate than other equations available to calculate fat-free mass (Rutten et al., 2010; Steiner et al., 2002). A decrease in muscle mass among underweight individuals and an increase in fat mass over three years are associated with a higher risk of deterioration in health status (Rutten et al., 2013).

5 Muscle imaging findings and prognostic value in COPD

5.1 Muscle mass and atrophy

Ultrasound scanning showed no difference in diaphragm thickness between patients with stable COPD and healthy individuals during relaxed breathing; however, upon deep breathing, diaphragm thickness is reduced in patients with stable COPD compared to healthy individuals. This suggests diaphragmatic dysfunction in patients with stable COPD (Jia and Zhang, 2022; Laghi et al., 2021; Hua-Rong et al., 2022; Gorman et al., 2002). Muscle mass is reduced in COPD, and the prevalence ranges from 4% to 35%. Low muscle mass is an independent predictor of mortality, and the cross-sectional area of the thigh is a strong predictor of mortality in patients with COPD of moderate to severe severity (Nyberg et al., 2015).

5.2 Body composition

Fat-free mass index is reduced in a quarter of COPD patients, even with a normal body mass index. Hence, body mass index alone may not be an appropriate measure to assess body composition. A low fat-free mass index is a strong predictor of mortality. A higher lean-to-fat ratio is associated with better functional performance (Eisner et al., 2007; Maltais et al., 2014).

5.3 Muscle function and performance

In patients with COPD, diaphragm excursion or movement was found to be significantly lower when compared with healthy individuals. The diaphragm movement was higher in the mild to moderate severity COPD group when compared with the severe group (Jia and Zhang, 2022; Laghi et al., 2021; Hua-Rong et al., 2022; Gorman et al., 2002). Limb muscle weakness ranges from 20% to 40%, respectively, in patients with moderate and severe COPD. Functional capacity and physical activity levels of a patient with COPD are determined by limb muscle strength. Muscle weakness is associated with increased health care costs, morbidity, and mortality. Quadriceps muscle fatigue is present in the majority of patients with COPD and is unrelated to disease severity (Nyberg et al., 2015). The degree of impairment of muscle dysfunction reflects exercise capacity and the severity of exercise intolerance (Bui et al., 2019).

6 Clinical implications of muscle imaging in COPD

6.1 Guiding treatment strategies

Body composition evaluation must become a standard practice to assess cachexia and muscle dysfunction in patients with COPD. A simple bioelectrical impedance analyzer, CT, or DXA can be used to assess body composition. Identifying patients with modifiable risk factors for muscle dysfunction and ensuring timely referral to pulmonary rehabilitation could provide both physiological and psychological benefits. A multimodal therapy inclusive of aerobic and resistance exercise, along with nutritional supplementation, must be used to stabilize or reverse cachexia. If required, pharmacological treatment, such as testosterone supplementation, may be used (De Brandt et al., 2022).

6.2 Rehabilitation and exercise interventions

Nutritional supplementation, exercise training, and anabolic steroids have been shown to improve muscle function, thereby enhancing the quality of life and reducing symptoms. Aerobic training alone, strength training alone, or a combination of both have been used in the management of muscle dysfunction. Furthermore, these training methods have been found to improve muscle strength and endurance, resulting in an improved quality of life.

Aerobic training: Continuous training can be used in patients with fewer symptoms during exercise, and interval training can be used in patients with severe dyspnea. Aerobic training is offered for a minimum of 3 days per week for at least 20 minutes over a period of 8–12 weeks. At least 20 minutes per session of exercise, with 2–3 minutes of high-intensity and 1–2 minutes of low-intensity exercise or rest, is provided in interval training.

Strength training: Various protocols are used for strength training. Strength training must be performed 2–3 days per week with at least 48 hours between two sessions. Intensity can range from 30% to 80% of one repetition maximum, and repetitions can range from 15 to 25 per set. Higher resistance with lower repetitions improves strength, while lower resistance with higher repetitions improves muscle endurance (Bui et al., 2019).

Respiratory muscle training: Resistive loading, breathing against threshold loading, and normocapnic hyperpnea are various devices used to train respiratory muscle strength. Even though respiratory muscle training can be beneficial in patients with COPD with preserved strength, it is particularly helpful in patients with respiratory muscle weakness, with maximal inspiratory pressure <60 cm of H_2O. Recommended respiratory muscle training is as follows: Intensity: > at 30% of maximal inspiratory pressure; frequency of 1–2 times per day for 3–5 days per week; duration: cumulative duration per day of 30 minutes, and for a total duration of 6 weeks (Crisafulli et al., 2007).

7 Future directions in muscle imaging research

7.1 Emerging imaging technologies

31-Phosphorus magnetic resonance spectroscopy (31P MRS) is an expensive, noninvasive technique that allows in vivo evaluation of skeletal muscle's phosphocreatine (PCr) recovery (Jakobsson and Jorfeldt, 1992).

Near-infrared spectroscopy (NIRS) has recently been used to measure skeletal muscle oxygen delivery and utilization. Muscle oximetry can provide information regarding the oxygenation of muscle tissue using infrared light. Although NIRS has been used to assess local muscle oxidative metabolism since the 1980s, the utilization of NIRS during exercise to monitor muscle metabolism was established in the late 2010s (Szucs et al., 2021).

7.2 Personalized medicine approaches

Sarcopenia seen in COPD is one of the extrapulmonary treatable traits in COPD. Appendicular skeletal muscle mass index of 5.4 $kg \cdot m^{-2}$ is considered as sarcopenia. Nutritional counseling with a high-protein diet and exercise training must be offered to these patients. Pharmacological interventions using anabolic steroids may be considered in patients where exercise training and nutritional supplementation are not feasible. It is worth noting that the strategy of treatable traits is not fully tested or understood.

8 Conclusion

COPD is a progressive, multifactorial condition not confined to the pulmonary system alone but extending its detrimental effects across various organ systems, notably skeletal muscles. A growing body of literature highlights the significance of both respiratory and peripheral muscle dysfunction in the pathophysiology and progression of COPD. These muscular alterations contribute substantially to the exercise intolerance, fatigue, breathlessness, and reduced quality of life commonly experienced by patients, ultimately increasing the risk of morbidity, healthcare utilization, and mortality.

The chapter outlines the intricate mechanisms behind muscle dysfunction, including systemic inflammation, malnutrition, disuse atrophy, corticosteroid use, oxidative stress, aging, comorbidities, and genetic susceptibility. Respiratory muscle dysfunction is compounded by hyperinflation-induced diaphragm flattening, resulting in altered mechanics and compromised inspiratory function. In parallel, peripheral muscle dysfunction is characterized by fiber-type shifting from oxidative to glycolytic, atrophy, and reduced oxidative capacity, particularly in the quadriceps. These muscular impairments are not uniformly distributed but affect several key muscle groups with varying severity.

Advancements in imaging technologies have paved the way for noninvasive, precise evaluation of muscle structure and function in patients with COPD. Techniques such as ultrasound, MRI, CT, dynamic X-rays, and PET/CT scans allow for detailed visualization of muscle mass, excursion, and metabolic activity. These imaging modalities not only aid in diagnosis and monitoring but also hold prognostic significance, offering insights into disease severity and therapeutic responsiveness.

Importantly, muscle imaging plays a crucial role in guiding treatment strategies. By identifying sarcopenia or muscle dysfunction early, clinicians can tailor interventions such as nutritional support, resistance and aerobic training, and respiratory muscle training. These interventions have shown promise in improving muscle strength, endurance, and overall functional capacity, thereby enhancing patient outcomes.

Future directions emphasize the integration of emerging technologies like 31P-MRS and NIRS for in vivo metabolic assessments, as well as the exploration of personalized medicine approaches targeting sarcopenia as a treatable trait. Continued research and clinical application of muscle imaging will be pivotal in shifting COPD management from a solely pulmonary-centered model to a more holistic, patient-specific paradigm that addresses extrapulmonary complications.

References

Agustí, A., Celli, B.R., Criner, G.J., Halpin, D., Anzueto, A., Barnes, P., Bourbeau, J., Han, M.L.K., Martinez, F.J., de Oca, M.M., Mortimer, K., Papi, A., Pavord, I., Roche, N., Salvi, S., Sin, D.D., Singh, D., Stockley, R., Varela, M.V.L., Wedzicha, J.A., Vogelmeier, C.F., 2023. Global initiative for chronic obstructive lung disease 2023 report: GOLD executive summary. Respirology. 14401843. 28 (4), 316–338. Available from: http://onlinelibrary.wiley.com/journal/10.1111/(ISSN)1440-1843. doi:10.1111/resp.14486.

Ansari, K., Keaney, N., Taylor, I., Burns, G., Farrow, M., 2012. Muscle weakness, health status and frequency of exacerbations in chronic obstructive pulmonary disease. Postgrad. Med. J. 88 (1041), 372–376. doi:10.1136/postgradmedj-2011-130293.

Attaway, A.H., Bellar, A., Welch, N., Sekar, J., Kumar, A., Mishra, S., Hatipoğlu, U., McDonald, M.L., Regan, E.A., Smith, J.D., Washko, G., Estépar, R.S.J., Bazeley, P., Zein, J., Dasarathy, S., 2023. Gene polymorphisms associated with heterogeneity and senescence characteristics of sarcopenia in chronic obstructive pulmonary disease. J. Cachexia Sarcopenia Muscle. 21906009. 14 (2), 1083–1095. Available from: http://onlinelibrary.wiley.com/journal/10.1007/13539.2190-6009. doi:10.1002/jcsm.13198.

Barreiro, E., Gea, J., 2016. Molecular and biological pathways of skeletal muscle dysfunction in chronic obstructive pulmonary disease. Chron. Respir. Dis. 1479-9731. 13 (3), 297–311. doi:10.1177/1479972316642366.

Bordoni, B., Escher, A., Compalati, E., Mapelli, L., Toccafondi, A., 2023. The importance of the diaphragm in neuromotor function in the patient with chronic obstructive pulmonary disease. Int. J. Chron. Obstruct. Pulmon. Dis. 1178-2005. 18, 837–848. doi:10.2147/copd.s404190.

Bordoni, B., Marelli, F., Morabito, B., Sacconi, B., 2016. Manual evaluation of the diaphragm muscle. Int. J. Chron. Obstruct. Pulmon. Dis. 1178-2005. 11 (1), 1949–1956. doi:10.2147/copd.s111634.

Bui, K.L., Nyberg, A., Rabinovich, R., Saey, D., Maltais, F., 2019. The relevance of limb muscle dysfunction in chronic obstructive pulmonary disease: a review for clinicians. Clin. Chest Med. 15578216. 40 (2), 367–383. Available from: http://www.elsevier.com/inca/publications/store/6/2/3/3/5/6/index.htt. doi:10.1016/j.ccm.2019.02.013.

Casey, P., Alasmar, M., McLaughlin, J., Ang, Y., McPhee, J., Heire, P., Sultan, J., 2022. The current use of ultrasound to measure skeletal muscle and its ability to predict clinical outcomes: a systematic review. J. Cachexia Sarcopenia Muscle. 2190-5991. 13 (5), 2298–2309. doi:10.1002/jcsm.13041.

Cassart, M., Pettiaux, N., Gevenois, P.A., Paiva, M., Estenne, M., 1997. Effect of chronic hyperinflation on diaphragm length and surface area. Am. J. Respir. Crit. Care Med. 1073449X. 156 (2 I), 504–508. Available from: http://ajrccm.atsjournals.org/. doi:10.1164/ajrccm.156.2.9612089.

Charususin, N., Dacha, S., Gosselink, R., Decramer, M., Leupoldt, A.V., Reijnders, T., Louvaris, Z., Langer, D., 2018. Respiratory muscle function and exercise limitation in patients with chronic obstructive pulmonary disease: a review. Expert Rev. Respir. Med. 1747-6348. 12 (1), 67–79. doi:10.1080/17476348.2018.1398084.

Chen, J., Zhong, Z., Wang, W., Yu, G., Zhang, T., Wang, Z., 2022a. Quantitative evaluation of diaphragmatic motion during forced breathing in chronic obstructive pulmonary disease patients using dynamic chest radiography. Front. Integr. Neurosci. 1662-5145. 16. doi:10.3389/fnint.2022.842404.

Chen, Y., Li, P., Wang, J., Wu, W., Liu, X., 2022b. Assessments and targeted rehabilitation therapies for diaphragmatic dysfunction in patients with chronic obstructive pulmonary disease: a narrative review. Int. J. Chron. Obstruct. Pulmon. Dis. 1178-2005. 17, 457–473. doi:10.2147/copd.s338583.

Couillard, A., Prefaut, C., 2005. From muscle disuse to myopathy in COPD: potential contribution of oxidative stress. Eur. Respir. J. 0903-1936. 26 (4), 703–719. doi:10.1183/09031936.05.00139904.

Crisafulli, E., Costi, S., Fabbri, L.M., Clini, E.M., 2007. Respiratory muscles training in COPD patients. Int. J. Chron. Obstruct. Pulmon. Dis. 11769106. 2 (1), 19–25. doi:10.2147/copd.2007.2.1.19.

De Brandt, J., Beijers, R.J.H.C.G., Chiles, J., Maddocks, M., McDonald, M.-L.N., Schols, A.M.W.J., Nyberg, A., 2022. Update on the etiology, assessment, and management of COPD cachexia: considerations for the clinician. Int. J. Chron. Obstruct. Pulmon. Dis. 1178-2005. 17, 2957–2976. doi:10.2147/copd.s334228.

Degens, H., Gayan-Ramirez, G., Van Hees, H.W.H., 2015. Smoking-induced skeletal muscle dysfunction: from evidence to mechanisms. Am. J. Respir. Crit. Care Med. 15354970. 191 (6), 620–625. Available from: http://www.atsjournals.org/doi/pdf/10.1164/rccm.201410-1830PP. doi:10.1164/rccm.201410-1830PP.

Donovan, A.A., Johnston, G., Moore, M., Jensen, D., Benedetti, A., Coxson, H.O., Gottfried, S.B., Petrof, B.J., Bourbeau, J., Smith, B.M., 2021. Diaphragm morphology assessed by computed tomography in chronic obstructive pulmonary disease. Ann. Am. Thorac. Soc. 23256621. 18 (6), 955–962. Available from: https://www.atsjournals.org/doi/pdf/10.1513/AnnalsATS.202007-865OC. doi:10.1513/AnnalsATS.202007-865OC.

Dubé, B.P., Laveneziana, P., 2018. Effects of aging and comorbidities on nutritional status and muscle dysfunction in patients with COPD. J. Thorac. Dis. 20776624. 10, S1355. Available from: http://jtd.amegroups.com/article/download/19477/pdf. doi:10.21037/jtd.2018.02.20.

Eisner, M.D., Blanc, P.D., Sidney, S., Yelin, E.H., Lathon, P.V., Katz, P.P., Tolstykh, I., Ackerson, L., Iribarren, C., 2007. Body composition and functional limitation in COPD. Respir. Res. 1465993X. 8. Available from: http://respiratory-research.com/content/8/1/7. doi:10.1186/1465-9921-8-7.

Ezponda, A., Casanova, C., Cabrera, C., Martin-Palmero, Á., Marin-Oto, M., Marín, J.M., Pinto-Plata, V., Divo, M., Celli, B.R., Zulueta, J.J., Bastarrika, G., de-Torres, J.P., 2021. Psoas muscle density evaluated by chest CT and long-term mortality in COPD patients. Arch. Bronconeumol. 15792129. 57 (8), 533–539. Available from: https://www.sciencedirect.com/science/journal/03002896. doi:10.1016/j.arbres.2021.04.012.

Gea, J., Agustí, A., Roca, J., 1985. Pathophysiology of muscle dysfunction in COPD. J. Appl. Physiol. 114 (9), 1222–1234. doi:10.1152/japplphysiol.00981.2012.

Gea, J., Pascual, S., Casadevall, C., Orozco-Levi, M., Barreiro, E., 2015. Muscle dysfunction in chronic obstructive pulmonary disease: update on causes and biological findings. J. Thorac. Dis. 20776624. 7 (10), E418. Available from: http://www.jthoracdis.com/article/download/5125/5222. doi:10.3978/j.issn.2072-1439.2015.08.04.

Gorman, R.B., McKenzie, D.K., Pride, N.B., Tolman, J.F., Gandevia, S.C., 2002. Diaphragm length during tidal breathing in patients with chronic obstructive pulmonary disease. Am. J. Respir. Crit. Care Med. 166 (11), 1461–1469. doi:10.1164/rccm.200111-087OC.

Hua-Rong, Z., Liang, C., Rong, L., Yi-Fan, T., Dou-Zi, S., Yue, C., Zu-Lin, L., 2022. Ultrasonographic evaluation of diaphragm function in patients with chronic obstructive pulmonary disease: a systematic review and meta-analysis. Medicine. 1536-5964. 101 (51), e32560. doi:10.1097/md.0000000000032560.

Iwasawa, T., Kagei, S., Gotoh, T., Yoshiike, Y., Matsushita, K., Kurihara, H., Saito, K., Matsubara, S., 2002. Magnetic resonance analysis of abnormal diaphragmatic motion in patients with emphysema. Eur. Respir. J. 0903-1936. 19 (2), 225–231. doi:10.1183/09031936.02.00044602.

Jaitovich, A., Barreiro, E., 2018. Skeletal muscle dysfunction in chronic obstructive pulmonary disease. what we know and can do for our patients. Am. J. Respir. Crit. Care Med. 1073-449X. 198 (2), 175–186. doi:10.1164/rccm.201710-2140CI.

Jakobsson, P., Jorfeldt, L., 1992. Nuclear magnetic resonance spectroscopy: a tool for skeletal muscle metabolic research. Eur. Respir. J. 0903-1936. 5 (2), 151–154. doi:10.1183/09031936.93.05020151.

Jia, Y., Zhang, Q., 2022. Research progress on diaphragm ultrasound in chronic obstructive pulmonary disease: a narrative review. Ultrasound Med. Biol. 03015629. 48 (4), 587–597. doi:10.1016/j.ultrasmedbio.2021.10.019.

Jung, G., Park, C., Hwang, J., You, J.H., Yi, C., Choi, W.J., 2021. Effects of core-postural stabilisation on fluoroscopy diaphragmatic measurement and dyspnea in chronic obstructive pulmonary disease: a randomized single-blinded clinical trial. Technol. Health Care. 09287329. 29 (S1), 359–366. doi:10.3233/thc-218034.

Kim, H.C., Mofarrahi, M., Hussain, S.N.A., 2008. Skeletal muscle dysfunction in patients with chronic obstructive pulmonary disease. Int. J. Chron. Obstruct. Pulmon. Dis. 11769106. 3 (4), 637–658.

Koechlin, C., Maltais, F., Saey, D., Michaud, A., LeBlanc, P., Hayot, M., Préfaut, C., 2005. Hypoxaemia enhances peripheral muscle oxidative stress in chronic obstructive pulmonary disease. Thorax. 00406376. 60 (10), 834–841. doi:10.1136/thx.2004.037531.

Laghi, F.A., Saad, M., Shaikh, H., 2021. Ultrasound and non-ultrasound imaging techniques in the assessment of diaphragmatic dysfunction. BMC Pulm. Med. 14712466. 21 (1). Available from: http://www.biomedcentral.com/bmcpulmmed/. doi:10.1186/s12890-021-01441-6.

MacIntyre, N.R., 2006. Muscle dysfunction associated with chronic obstructive pulmonary disease. Respir. Care. 00201324. 51 (8), 840–848. http://www.rcjournal.com/contents/08.06/08.06.0840.pdf.

MacNee, W., 2006. Pathology, pathogenesis, and pathophysiology. Clin. Res. 332 (7551), 1202–1204.

Malaguti, C., Nery, L.E., Dal Corso, S., Nápolis, L., De Fuccio, M.B., Marise Castro, J.A.N., 2006. Scaling skeletal muscle function to mass in patients with moderate-to-severe COPD. Eur. J. Appl. Physiol. 1439-6319. 98 (5), 482–488. doi:10.1007/s00421-006-0292-8.

Maltais, F., Decramer, M., Casaburi, R., Barreiro, Burelle, Y., Debigare, R., et al., 2014. statement: update on limb muscle dysfunction in chronic obstructive pulmonary disease. Am. J. Respir. Crit. Care Med. 189 (9), 15–62.

Mark, A., 2017. Very large bulla on background of COPD. Radiopaedia.org. doi:10.53347/rID-48598.

Marquis, K., Debigaré, R., Lacasse, Y., LeBlanc, P., Jobin, J., Carrier, G., Maltais, F., 2002. Midthigh muscle cross-sectional area is a better predictor of mortality than body mass index in patients with chronic obstructive pulmonary disease. Am. J. Respir. Crit. Care Med. 1073-449X. 166 (6), 809–813. doi:10.1164/rccm.2107031.

Meyer, A., Zoll, J., Charles, A.L., Charloux, A., de Blay, F., Diemunsch, P., Sibilia, J., Piquard, F., Geny, B., 2013. Skeletal muscle mitochondrial dysfunction during chronic obstructive pulmonary disease: central actor and therapeutic target. Exp. Physiol. 1469445X. 98 (6), 1063–1078. Available from: http://onlinelibrary.wiley.com/journal/10.1111/(ISSN)1469-445X. doi:10.1113/expphysiol.2012.069468.

Mudgal, P., 2015. Pulmonary tuberculosis with COPD. Radiopaedia.org. doi:10.53347/rID-40830.

Nairn, J.R., Prime, F.J., Simon, G., 1969. Association between radiological findings and total and regional function in emphysema. Thorax. 00406376. 24 (2), 218–227. doi:10.1136/thx.24.2.218.

Nicholson, J.M., Orsso, C.E., Nourouzpour, S., Elangeswaran, B., Chohan, K., Orchanian-Cheff, A., Fidler, L., Mathur, S., Rozenberg, D., 2022. Computed tomography-based body composition measures in COPD and their association with clinical outcomes: a systematic review. Chron. Respir. Dis. 14799731. 19. Available from: https://journals.sagepub.com/home/CRD. doi:10.1177/14799731221133387.

Nyberg, A., Saey, D., Maltais, F., 2015. Why and how limb muscle mass and function should be measured in patients with chronic obstructive pulmonary disease. Ann. Am. Thorac. Soc. 2329-6933. 12 (9), 1269–1277. doi:10.1513/annalsats.201505-278ps.

O'Donnell, D.E., Milne, K.M., James, M.D., de Torres, J.P., Neder, J.A., 2020. Dyspnea in COPD: new mechanistic insights and management implications. Adv. Ther. 18658652. 37 (1), 41–60. Available from: http://www.springer.com/springer+healthcare/journal/12325. doi:10.1007/s12325-019-01128-9.

Ogan, N., Aydemir, Y., Evrin, T., Ataç, G.K., Baha, A., Katipoğlu, B., Süzen, B., Akpinar, E.E., 2019. Diaphragmatic thickness in chronic obstructive lung disease and relationship with clinical severity parameters. Turk. J. Med. Sci. 13036165. 49 (4), 1073–1078. Available from: http://journals.tubitak.gov.tr/medical/issues/sag-19-49-4/sag-49-4-15-1901-164.pdf. doi:10.3906/sag-1901-164.

Orozco-Levi, M., 2003. Structure and function of the respiratory muscles in patients with COPD: impairment or adaptation? Eur. Respir. J. 0903-1936. 22 (46 suppl), 41s. doi:10.1183/09031936.03.00004607.

Osman, Tran, Muzaffar, Parkar, Sachdeva, Ruppel, 2011. Does ^{18}F-FDG uptake by respiratory muscles on PET/CT correlate with chronic obstructive pulmonary disease? J. Nucl. Med. Technol. 39 (4), 252–257.

Park, J.K., Deutz, N.E.P., Cruthirds, C.L., Kirschner, S.K., Park, H., Madigan, M.L., Engelen, M.P.K.J., 2020. Risk factors for postural and functional balance impairment in patients with chronic obstructive pulmonary disease. J. Clin. Med. 20770383. 9 (2). Available from: https://www.mdpi.com/2077-0383/9/2/609/pdf. doi:10.3390/jcm9020609.

Pettiaux, N., Cassart, M., Paiva, M., Estenne, M., 1997. Three-dimensional reconstruction of human diaphragm with the use of spiral computed tomography. J. Appl. Physiol. 8750-7587. 82 (3), 998–1002. doi:10.1152/jappl.1997.82.3.998.

Pitsiou, G., Kyriazis, G., Hatzizisi, O., Argyropoulou, P., Mavrofridis, E., Patakas, D., 2002. Tumor necrosis factor-alpha serum levels, weight loss and tissue oxygenation in chronic obstructive pulmonary disease. Respir. Med. 96 (8), 594–598.

Poberezhets, V., Skoczyński, S., Demchuk, A., Oraczewska, A., Tobiczyk, E., Mostovoy, Y., Barczyk, A., 2020. Impact of comorbidities on the deterioration of skeletal muscle dysfunction among COPD patients. Eur. Respir. J. 56, 700. doi:10.1183/13993003.congress-2020.700.

Polla, B., D'Antona, G., Bottinelli, R., Reggiani, C., 2004. Respiratory muscle fibres: specialisation and plasticity. Thorax 59 (9), 808–817. doi:10.1136/thx.2003.009894.

Puente-Maestu, L., Lázaro, A., Humanes, B., 2013. Metabolic derangements in COPD muscle dysfunction. J. Appl. Physiol. 8750-7587. 114 (9), 1282–1290. doi:10.1152/japplphysiol.00815.2012.

Ramachandran, P., Devaraj, U., Patrick, B., Saxena, D., Venkatnarayan, K., Louis, V., Krishnaswamy, U., D'souza, G., 2020. Ultrasonographic assessment of skeletal muscle mass and diaphragm function in patients with chronic obstructive pulmonary disease: a case–control study. Lung India. 0970-2113. 37 (3), 220. doi:10.4103/lungindia.lungindia_103_19.

Rawal, G., Yadav, S., 2015. Nutrition in chronic obstructive pulmonary disease: a review. J. Transl. Intern. Med. 2224-4018. 3 (4), 151–154. doi:10.1515/jtim-2015-0021.

Reich, S.B., Weinshelbaum, A., Yee, J., 1985. Correlation of radiographic measurements and pulmonary function tests in chronic obstructive pulmonary disease. Am. J. Roentgenol. 0361-803X. 144 (4), 695–699. doi:10.2214/ajr.144.4.695.

Riley, C.M., Sciurba, F.C., 2019. Diagnosis and outpatient management of chronic obstructive pulmonary disease: a review. JAMA. 15383598. 321 (8), 745–746. Available from: http://jama.jamanetwork.com/journal.aspx. doi:10.1001/jama.2019.0131.

Rozenberg, D., Martelli, V., Vieira, L., Orchanian-Cheff, A., Keshwani, N., Singer, L.G., Mathur, S., 2017. Utilization of non-invasive imaging tools for assessment of peripheral skeletal muscle size and composition in chronic lung disease: a systematic review. Respir. Med. 15323064. 131, 125–134. Available from: http://www.harcourt-international.com/journals/rmed/. doi:10.1016/j.rmed.2017.08.007.

Rutten, E.P.A., Calverley, P.M.A., Casaburi, R., Agusti, A., Bakke, P., Celli, B., Coxson, H.O., Crim, C., Lomas, D.A., MacNee, W., Miller, B.E., Rennard, S.I., Scanlon, P.D., Silverman, E.K., Tal-Singer, R., Vestbo, J., Watkins, M.L., Wouters, E.F.M., 2013. Changes in body composition in patients with chronic obstructive pulmonary disease: do they influence patient-related outcomes? Ann. Nutr. Metab. 0250-6807. 63 (3), 239–247. doi:10.1159/000353211.

Rutten, E.P.A., Spruit, M.A., Wouters, E.F.M., 2010. Critical view on diagnosing muscle wasting by single-frequency bio-electrical impedance in COPD. Respir. Med. 09546111. 104 (1), 91–98. doi:10.1016/j.rmed.2009.07.004.

Safiri S., Carson-Chahhoud, K., Noori, M., Nejadghaderi S.A., Sullman, M.J.M., Heris, J.A., et al., 2019. Burden of chronic obstructive pulmonary disease and its attributable risk factors in 204 countries and territories, 1990-2019: results from the Global Burden of Disease Study. BMJ 378, e069679. doi:10.1136/bmj-2021-069679.

Simon, G., 1964. Radiology and emphysema. Clin. Radiol. 00099260. 15 (4), 293–306. doi:10.1016/s0009-9260(64)80001-5.

Sin, D.D., Man, S.F.P., 2006. Skeletal muscle weakness, reduced exercise tolerance, and COPD: is systemic inflammation the missing link? Thorax. 00406376. 61 (1), 1–3. doi:10.1136/thx.2005.044941.

Singh, S.S., Bhat, A., Mohapatra, A.K., Manu, M.K., Vaishali, K., 2019. Comparison of reaction time and functional balance in chronic obstructive pulmonary disease and healthy participants. Heart Lung. 15273288. 48 (6), 570–573. Available from: http://www.elsevier.com/inca/publications/store/6/2/3/0/8/9/index.htt. doi:10.1016/j.hrtlng.2019.09.006.

Steiner, M.C., Barton, R.L., Singh, S.J., Morgan, M.D.L., 2002. Bedside methods versus dual energy X-ray absorptiometry for body composition measurement in COPD. Eur. Respir. J. 0903-1936. 19 (4), 626–631. doi:10.1183/09031936.02.00279602.

Suga, K., Tsukuda, T., Awaya, H., Takano, K., Koike, S., Matsunaga, N., Sugi, K., Esato, K., 1999. Impaired respiratory mechanics in pulmonary emphysema: evaluation with dynamic breathing MRI. J. Magn. Reson. Imaging. 10531807. 10 (4), 510–520. doi:10.1002/(SICI)1522-2586(199910)10:4<510::AID-JMRI3>3.3.CO;2-7.

Swallow, E.B., Reyes, D., Hopkinson, N.S., Man, W.D.C., Porcher, R., Cetti, E.J., Moore, A.J., Moxham, J., Polkey, M.I., 2007. Quadriceps strength predicts mortality in patients with moderate to severe chronic obstructive pulmonary disease. Thorax. 00406376. 62 (2), 115–120. doi:10.1136/thx.2006.062026.

Szucs, B., Petrekanits, M., Fekete, M., Varga, J.T., 2021. The use of near-infrared spectroscopy for the evaluation of a 4-week rehabilitation program in patients with COPD. Physiol. Int. 2498602X. 108 (4), 427–439. Available from: http://www.akademiai.com/toc/2060/current. doi:10.1556/2060.2021.00185.

Viegi, G., Pistelli, F., Sherrill, D.L., Maio, S., Baldacci, S., Carrozzi, L., 2007. Definition, epidemiology and natural history of COPD. Eur. Respir. J. 09031936. 30 (5), 993–1013. Available from: http://erj.ersjournals.com/cgi/reprint/30/5/993. doi:10.1183/09031936.00082507Italy.

Vikne, H., Strøm, V., Pripp, A.H., Gjøvaag, T., 2020. Human skeletal muscle fiber type percentage and area after reduced muscle use: a systematic review and meta-analysis. Scand. J. Med. Sci. Sports. 0905-7188. 30 (8), 1298–1317. doi:10.1111/sms.13675.

Wilson, A.C., Bon, J.M., Mason, S., Diaz, A.A., Lutz, S.M., Estepar, R.S.J., Kinney, G.L., Hokanson, J.E., Rennard, S.I., Casaburi, R., Bhatt, S.P., Irvin, M.R., Hersh, C.P., Dransfield, M.T., Washko, G.R., Regan, E.A., McDonald, M.L., 2022. Increased chest CT derived bone and muscle measures capture markers of improved morbidity and mortality in COPD. Respir. Res. 1465993X. 23 (1). Available from: https://respiratory-research.biomedcentral.com/. doi:10.1186/s12931-022-02237-w.

Ye, X., Wang, M., Xiao, H., 2017. Echo intensity of the rectus femoris in stable COPD patients. Int. J. Chron. Obstruct. Pulmon. Dis. 1178-2005. 12, 3007–3015. doi:10.2147/copd.s143645.

Chapter 19

Yoga module to improve the central and peripheral vascular function and cognitive functions among the desk-based workers

Poovitha Shruthi Paramashiva[a], K. Annapoorna[a], K. Vaishali[b], Leena R. David[c,d], Baskaran Chandrasekaran[e], Suresh Sukumar[d], K.N. Shivshankar[f], Sneha Ravichandran[d], Dilber Uzun Ozsahin[c,g,h], Berna Uzun[h], Dilip Shettigar[d], Ilker Ozsahin[h], and Rajagopal Kadavigere[i]

[a]*Division of Yoga, Manipal Academy of Higher Education, Centre for Integrative Medicine and Research, Manipal, India*

[b]*Department of Physiotherapy, Manipal Academy of Higher Education, Manipal College of Health Professions, Manipal, India*

[c]*Medical Diagnostic Imaging Department, College of Health Sciences, University of Sharjah, Sharjah, United Arab Emirates*

[d]*Department of Medical Imaging Technology, Manipal College of Health Professions, Manipal Academy of Higher Education, Manipal, Karnataka, India*

[e]*Symbiosis School of Sports Sciences, Symbiosis International (Deemed University), Pune, India*

[f]*Department of Medicine, Manipal Academy of Higher Education, Kasturba Medical College, Manipal, India*

[g]*Research Institute for Medical and Health Sciences, University of Sharjah, Sharjah, United Arab Emirates*

[h]*Irfan Suat Gunsel Operational Research Institute, Near East University, Nicosia, Turkish Republic of Northern Cyprus (TRNC), Turkey*

[i]*Department of Radiodiagnosis and Imaging, Kasturba Medical College, Manipal, India*

1 Introduction

The growing popularity of work-from-home and desk-based jobs has demonstrated that not all professions require intense physical labor or significant exertion. With the increasing integration of computers and the digital world into daily life, work cultures have gradually shifted toward a more comfortable and convenient lifestyle. While this transition has led to greater flexibility and productivity in many cases, it has also introduced several health concerns, particularly regarding the sedentary nature of desk-based occupations. The desire for a stable and lucrative career has compelled a substantial portion of the working population to adopt prolonged sitting hours, significantly reducing their engagement in physical activities. This shift has, in turn, contributed to a range of physiological and cognitive health risks, particularly those associated with vascular function and cognitive decline among desk-based workers (Young et al., 2016; Biswas et al., 2015).

1.1 The evolution of sedentary work culture

The transformation of the global workforce toward desk-based occupations represents one of the most significant shifts in human labor patterns in recent history. This trend has been accelerating since the late 20th century but reached unprecedented levels following the COVID-19 pandemic, which forced millions of workers worldwide to adopt remote work arrangements. According to recent research, many American workers spend over 7 hours a day at work in primarily sedentary office roles, with office workers sitting for 66% of their working days and only 8% succeeding in interrupting their prolonged periods of sitting within the first 55 minutes (Hadgraft et al., 2016). In addition, more than 80% of jobs in the United States involve mostly sedentary activities (Church et al., 2011), marking a profound departure from the activity patterns that characterized human existence for thousands of years.

Advances in Medical Imaging. https://doi.org/10.1016/B978-0-443-28967-5.00022-8

Historically, human physiology evolved in environments that demanded regular physical movement. Hunter-gatherer societies, agricultural communities, and even early industrial workers maintained relatively high levels of physical activity throughout their daily lives. The human body developed complex physiological systems optimized for regular movement, with cardiovascular, musculoskeletal, and metabolic processes all dependent on consistent physical activity. The abrupt transition to prolonged sedentary behavior represents a fundamental mismatch between our evolutionary design and modern occupational demands (Booth et al., 2017).

The technological revolution has further accelerated this trend, with digital tools enabling greater work efficiency while simultaneously reducing the need for physical movement. Video conferencing, cloud computing, and collaborative digital platforms have eliminated many reasons for workers to leave their desks, creating environments where employees can remain seated for extended periods without interruption. While these technological innovations have delivered substantial productivity benefits, they have inadvertently contributed to increasingly sedentary work patterns that pose significant health risks (Koohsari et al., 2015).

1.2 Physiological impact of prolonged sitting

The human body is not designed for extended periods of immobility. Prolonged sitting has been linked to various adverse health effects, including obesity, cardiovascular disease, metabolic disorders, and musculoskeletal discomfort (Liang et al., 2022; Patterson et al., 2018). Moreover, recent studies have indicated that excessive time spent in the sitting position and the ensuing reduction in leg blood flow-induced shear stress cause endothelial cell dysfunction, a key predisposing factor to peripheral artery disease (Thosar et al., 2015). These vascular changes can contribute to an increased risk of hypertension, atherosclerosis, and other cardiovascular complications over time (Chomistek et al., 2013; Pandey et al., 2016).

The mechanisms through which prolonged sitting affects vascular health are multifaceted. Lower leg and forearm blood flow are both markedly reduced over the course of uninterrupted sitting, thus exposing the vasculature to low levels of shear stress (Thosar et al., 2014). This reduction in shear stress triggers a cascade of negative physiological responses, including decreased nitric oxide production, endothelial dysfunction, and increased oxidative stress within the vascular system. Prolonged, uninterrupted sitting negatively impacts markers of peripheral vascular health, particularly the vasodilatory function of leg arteries (Morishima et al., 2016).

Beyond these immediate effects, chronic sitting behavior leads to more persistent vascular adaptations. Regular exposure to low shear stress conditions promotes arterial stiffening, a process characterized by reduced elasticity in blood vessels that increases cardiac workload and elevates the risk of cardiovascular events (Vlachopoulos et al., 2010). Research using pulse wave velocity measurements—a gold standard for assessing arterial stiffness—has shown that individuals with sedentary occupations exhibit significantly higher arterial stiffness compared to more active populations, even when controlling for traditional cardiovascular risk factors (Celis-Morales et al., 2017).

The impact of sedentary behavior extends beyond the large vessels to the microcirculation as well. Capillary density and function are compromised in chronically sedentary individuals, reducing tissue perfusion and limiting the delivery of oxygen and nutrients to working muscles and organs. This microvascular dysfunction has been implicated in the development of insulin resistance and metabolic syndrome, conditions that further compound cardiovascular risk (Hamburg et al., 2007). The combination of macrovascular and microvascular impairments creates a comprehensive disruption of circulatory function that accelerates cardiovascular aging and increases the risk of cardiovascular mortality.

Musculoskeletal complications represent another significant physiological consequence of prolonged sitting. The seated position places disproportionate stress on the lumbar spine, particularly when poor posture is maintained over extended periods. This mechanical stress leads to accelerated disc degeneration, muscle imbalances, and chronic pain syndromes that affect quality of life and work productivity (Lis et al., 2007). In addition, prolonged immobility reduces joint mobility and flexibility, contributing to a progressive decline in functional capacity that can persist even beyond working hours.

1.3 Cognitive implications of sedentary behavior

In addition to its physiological impact, prolonged sitting also has substantial implications for cognitive health. High sedentary behavior is associated with adverse effects on central vascular function and cognitive function (Colley et al., 2011). Studies suggest that a sedentary lifestyle may negatively affect the brain's structural integrity, particularly in regions responsible for memory and cognition, such as the hippocampus and the posterior cingulate cortex (Voss et al., 2014). A decline in these brain structures has been associated with cognitive impairment, decreased mental sharpness, and an elevated risk of neurodegenerative disorders such as Alzheimer's disease and dementia (Falck et al., 2017). These findings highlight the need for effective interventions to mitigate the negative effects of prolonged sitting on both vascular and cognitive functions.

The relationship between sedentary behavior and cognitive health appears to be mediated through multiple pathways. First, the vascular impairments associated with prolonged sitting directly affect cerebral blood flow and perfusion. The brain, despite representing only 2% of body weight, consumes approximately 20% of cardiac output and oxygen supply, making it particularly vulnerable to circulatory compromise. Reduced cerebral blood flow has been linked to decreased cognitive performance, particularly in domains requiring sustained attention, processing speed, and executive function (Angevaren et al., 2008).

Second, sedentary behavior influences brain health through metabolic pathways. Extended periods of sitting contribute to insulin resistance and glucose dysregulation, conditions that have been repeatedly associated with cognitive decline. Chronically elevated blood glucose levels promote neuroinflammation and oxidative stress, accelerating neuronal damage and compromising synaptic plasticity—the neurological basis for learning and memory formation. Neuroimaging studies have demonstrated that individuals with prolonged sedentary behavior patterns show reduced gray matter volume in brain regions critical for cognitive control and memory consolidation (Sexton et al., 2016).

Third, prolonged sitting affects the production and circulation of brain-derived neurotrophic factor (BDNF), a protein essential for neuronal growth, differentiation, and survival. Physical activity normally stimulates BDNF production, promoting neurogenesis and enhancing synaptic connectivity. Conversely, sedentary behavior suppresses BDNF levels, reducing the brain's capacity for neuroplasticity and adaptation (Voss et al., 2013). This diminished neuroplasticity makes the brain more vulnerable to age-related cognitive decline and neurodegenerative processes.

Finally, sedentary work environments often create conditions of mental fatigue without adequate physical counterbalance. The continuous cognitive demands of desk-based work, combined with physical immobility, create a state of mental exhaustion that impairs cognitive performance over time. Without the rejuvenating effects of physical movement, cognitive resources become progressively depleted throughout the workday, reducing productivity and increasing error rates (Mehta and Agnew, 2012).

1.4 Current interventions for sedentary work environments

A potential solution to counteract the adverse effects of prolonged sitting is the integration of physical activity into daily work routines. A growing body of research suggests that regular movement breaks, ergonomic adjustments, and exercise interventions can significantly improve vascular and cognitive function among desk-based workers (Dempsey et al., 2016). In healthy and overweight/obese adults, interrupting prolonged sitting with activity bouts mitigates impairments in vascular function (Larsen et al., 2014). However, despite the growing awareness of these risks, implementing practical and sustainable physical activity interventions in work environments remains a challenge. Many desk-based workers struggle to find the time or motivation to engage in structured exercise programs, and workplace constraints may limit opportunities for regular movement.

The implementation of standing desks represents one of the most widely adopted interventions to combat sedentary behavior in office environments. These adjustable workstations allow employees to alternate between sitting and standing positions throughout the day, reducing total sitting time and promoting postural variation. Research examining the effectiveness of standing desks has shown that sit-stand desk interventions reduce sitting time at work by 21% (95% CI 18%–25%) and sedentary time by 4.8 min/work-hr (95% CI 4.1–5.4 min/work-hr), which translates into the replacement of 8 hours of sitting time with standing for a 40-hour work week (Pronk et al., 2012). Previous studies that reviewed the effectiveness of introducing sit–stand desks in the workplace reported an average reduction in sitting time of 100 minutes per day (Shrestha et al., 2018).

Beyond energy expenditure, standing workstations appear to offer significant vascular benefits. Studies measuring circulation during standing work sessions have demonstrated improvements in lower limb circulation compared to seated positions. This enhanced circulation helps maintain endothelial function and reduces the risk of venous pooling and thrombosis (Torbeyns et al., 2014). However, prolonged standing presents its own set of challenges, including increased discomfort in the lower back and legs, potential exacerbation of varicose veins, and reduced fine motor precision for certain tasks.

Movement break protocols represent another intervention strategy, encouraging workers to interrupt sitting periods with brief bouts of physical activity. These structured breaks typically recommend 2–5 minutes of movement for every 30–60 minutes of sitting, with activities ranging from simple stretching to more vigorous exercises like stair climbing or desk-based calisthenics. Studies evaluating these protocols have shown that even brief movement breaks can effectively restore vascular function and reduce the metabolic consequences of prolonged sitting (Dunstan et al., 2012). However, compliance remains a significant challenge, with many workers reporting that workflow demands and organizational culture create barriers to regular movement breaks.

More comprehensive workplace wellness programs often incorporate structured exercise sessions within the workday, such as lunchtime fitness classes or dedicated exercise facilities within office buildings. These programs aim to create opportunities for more substantial physical activity while addressing common barriers like time constraints and access to exercise equipment. While these initiatives demonstrate positive outcomes for participants, they typically reach only a fraction of the workforce, with participation rates rarely exceeding 30%–40% of eligible employees (Conn et al., 2009).

Remote work environments present unique challenges and opportunities for addressing sedentary behavior. While home-based workers face fewer social constraints on movement, they also lack the physical cues and infrastructure that might promote activity in traditional offices. In addition, the absence of commuting and workplace walking further reduces incidental physical activity. Interventions targeting remote workers must account for these contextual differences, emphasizing home environment modifications and digital tools that promote regular movement.

Despite the variety of available interventions, their implementation remains inconsistent, and their long-term effectiveness is limited by several factors. First, many approaches treat physical activity as an interruption to work rather than as an integrated element of productive work processes. Second, few interventions address the cognitive dimensions of sedentary behavior, focusing primarily on physical rather than mental health outcomes. Finally, many current strategies require significant behavioral change without adequately addressing the psychological barriers to sustained adoption (Prince et al., 2014).

1.5 Yoga as a comprehensive intervention strategy

Yoga, with its deep historical roots and extensive research backing, presents itself as a viable intervention for addressing these concerns. Dating back to Vedic times (circa 1500–1000 BCE), yoga has been widely practiced for its multifaceted benefits, encompassing physical, mental, and emotional well-being (Iyengar, 1979). Yoga offers a holistic approach that integrates controlled breathing, physical postures, and mindfulness techniques, making it a unique and accessible form of exercise suitable for diverse populations. Unlike conventional aerobic exercises, which may require extensive equipment or high-intensity movement, yoga can be performed in a confined workspace with minimal disruption to workflow.

The philosophical foundation of yoga lies in the integration of mind and body, an approach that directly addresses the dual impact of sedentary behavior on physical and cognitive health. Traditional yogic texts, including the Patanjali Yoga Sutras (compiled around 400 CE), emphasize the concept of "sthira sukham asanam," describing the ideal physical position as both stable (sthira) and comfortable (sukham). This principle acknowledges the importance of balanced physical positioning—neither completely immobile nor excessively active—for optimal physiological and mental functioning. Modern desk-based work environments, with their emphasis on prolonged sitting, represent an extreme deviation from this balanced approach to bodily positioning.

In contrast to conventional exercise interventions that often focus exclusively on physical movement, yoga incorporates three interconnected components: asana (physical postures), pranayama (breath control), and dhyana (meditation or mindfulness). This comprehensive approach offers unique advantages for addressing the multifaceted health challenges faced by desk-based workers. Asanas improve musculoskeletal flexibility, strength, and alignment, directly counteracting the physical strain of prolonged sitting. Pranayama enhances respiratory efficiency and autonomic nervous system function, potentially improving oxygen delivery to tissues and promoting parasympathetic activation. Dhyana cultivates attentional control and mental clarity, addressing the cognitive fatigue associated with desk-based work (Pascoe et al., 2017).

Recent studies have shown promising results regarding yoga's impact on vascular function. A metaanalysis showed an increase in brachial flow-mediated dilation (FMD) through yoga practice (mean difference = −1.23%; 95% confidence interval), indicating improved endothelial function (Cramer et al., 2014). In addition, yoga interventions showed mixed results, with some improvement in endothelial function in older adults but inconsistent effects on arterial stiffness (Chu et al., 2016). These findings indicate that yoga could serve as a practical and effective solution to combat the negative effects of prolonged sitting among desk-based workers.

The vascular benefits of yoga appear to be mediated through multiple physiological mechanisms. Regular practice of yoga asanas promotes muscle contraction and relaxation cycles that enhance blood flow and stimulate the production of nitric oxide, a potent vasodilator that improves endothelial function (Hunter et al., 2013). FMD is widely believed to reflect endothelium-dependent and largely nitric oxide-mediated arterial function and has been used as a surrogate marker of vascular health (Thijssen et al., 2011). Studies using FMD techniques have demonstrated that consistent yoga practice can improve endothelial function in previously sedentary individuals.

Pranayama practices, particularly slow-paced breathing techniques like anuloma viloma (alternate nostril breathing), have been shown to modulate autonomic nervous system activity, promoting parasympathetic dominance and reducing sympathetic overdrive. This autonomic rebalancing contributes to reduced arterial stiffness and improved baroreflex sensitivity—physiological changes that enhance cardiovascular efficiency and reduce the risk of hypertension (Jerath et al., 2006). The rhythmic nature of pranayama practice also creates cyclical patterns of intrathoracic pressure that may enhance venous return and cardiac output, further improving systemic circulation.

From a cognitive perspective, yoga practice engages multiple brain regions and neural networks associated with attention, executive function, and emotional regulation. Neuroimaging studies have revealed that regular yoga practitioners exhibit increased gray matter volume in the hippocampus, prefrontal cortex, and insular cortex—brain regions critical for memory formation, decision-making, and interoceptive awareness (Luders et al., 2009; Froeliger et al., 2012). These structural changes are accompanied by functional improvements in attention networks, with yoga practitioners demonstrating enhanced performance on tasks requiring sustained concentration and cognitive flexibility. Acute studies have also demonstrated a significant overall effect of yoga on cognition, with the effect being strongest for memory, followed by attention, processing speed measures, and executive functions (Gothe et al., 2019).

The meditative component of yoga appears particularly beneficial for counteracting the mental fatigue associated with desk-based work. By training attentional control and promoting present-moment awareness, meditation practices help restore cognitive resources depleted during extended periods of screen-based work (Goyal et al., 2014). In addition, the emphasis on mindful body awareness helps practitioners develop greater sensitivity to physical discomfort, potentially encouraging more frequent postural adjustments and movement breaks during prolonged sitting sessions.

Yoga has been reported to increase brain-wave activity, creating neuroplastic effects with subsequent improvement in cognition, memory, mood, and anxiety (Desai et al., 2015). The practice promotes neuroplasticity through multiple mechanisms, with increased gray matter volume correlating positively with years of yoga practice and negatively with the self-reported number of cognitive failures, suggesting that yoga practice may promote neuroplastic changes in neural systems that support executive functioning (Villemure et al., 2014).

1.6 Research gaps and study rationale

Despite the promising evidence supporting yoga's benefits, there remains a need to develop and validate structured yoga modules tailored specifically to desk-based workers. Many existing yoga programs cater to general fitness or stress management without addressing the specific physiological and cognitive challenges posed by prolonged sitting (Sharma et al., 2017). Therefore, our study aims to develop and validate a comprehensive yoga module designed to improve central and peripheral vascular function and cognitive functions among desk-based workers. By implementing a well-structured yoga intervention, we seek to offer an accessible, scientifically validated, and sustainable approach to enhancing health outcomes in sedentary work environments.

Several factors contribute to the challenge of developing effective yoga interventions for workplace settings. First, conventional yoga sequences often require substantial time commitments (typically 60–90 minutes per session), making them impractical for integration into standard workdays. Second, traditional practices may include postures that are challenging for novice practitioners or inappropriate for office attire and environments. Third, many existing programs lack specific targeting of the physiological systems most affected by sedentary behavior, particularly the vascular and cognitive domains. Finally, the dosage and frequency parameters of yoga practice needed to counteract the effects of prolonged sitting remain poorly defined in the scientific literature (Ross and Thomas, 2010).

The selection of appropriate yoga practices for desk-based workers must balance several considerations. Postures should address common musculoskeletal imbalances resulting from prolonged sitting, such as thoracic kyphosis, cervical strain, and hip flexor tightness, without requiring extensive space or equipment. Breathing techniques should be accessible to novice practitioners while still providing meaningful physiological benefits. Meditation practices should be brief enough to fit into work breaks while offering cognitive restoration. In addition, the complete module must be structured to provide progressive advancement that accommodates varying levels of flexibility, strength, and prior yoga experience.

Previous research on yoga interventions for workplace settings has produced promising but inconsistent results, with methodological limitations hampering definitive conclusions. Many studies have relied on self-reported outcomes without objective physiological measurements, limiting the understanding of the mechanisms underlying observed benefits (Büssing et al., 2012a). Sample sizes have typically been small, and few studies have directly compared yoga interventions to other forms of physical activity or workplace wellness programs. In addition, the heterogeneity of yoga practices included in different studies makes it difficult to identify the specific components responsible for the observed benefits.

This study will explore various aspects of yoga, including specific postures (asanas), breathing techniques (pranayama), and mindfulness practices (meditation), to determine their efficacy in counteracting the negative effects of prolonged sitting. The findings from this research will contribute valuable insights into the development of workplace wellness programs and provide evidence-based recommendations for incorporating yoga into corporate and home-office settings. Given the increasing prevalence of sedentary work cultures, this study has the potential to make a significant impact on public health by promoting healthier, more active work environments.

Through this research, we hope to bridge the gap between traditional yoga practices and modern occupational health needs, ultimately fostering a culture of movement, mindfulness, and well-being in professional settings. The results of this study may not only benefit individual employees but also have broader implications for employers, healthcare professionals, and policymakers striving to create healthier workplaces in an increasingly sedentary world.

2 Methods

2.1 Ethical statement

The Ethical Committee approval was obtained from the Institutional Ethical Board at Kasturba Hospital, Manipal (IEC 260:2021). The study was approved by the Clinical Trials Registry - India (CTRI/2021/03731910/). Written informed consent was obtained from all participants.

2.2 Preparation of yoga module

The inclusion criteria for the yoga practices were to use only those yoga practices that are known to improve vascular and cognitive functions. Data extraction was conducted from June 2021 to August 2021 using Rayyan data extraction software, which is available for free at http://rayyan.qcri.org and is fully funded by Qatar Foundation, a nonprofit organization in the State of Qatar. We extracted the .bib file containing the details of all the articles matching the MeSH terms from the online databases. These were uploaded to the Rayyan software, and the articles were screened by the authors. An extensive literature search was carried out from the yogic literatures: the principal Upanishads (The Yoga Upanishads, 1938), Hathayoga Pradipika (Hatha Yoga Pradipika, n.d.), and Gheranda Samhita (Manninson and Bouchard, 2004) in addition to recent books on yoga-like practices such as "*Light on Yoga*" (Iyengar, 1979) and Asana Pranayama Mudra Bandha (Asana Pranayama Mudra Bandha, 2009). In addition, we also conducted an online search in scientific databases including Scopus, Web of Science, Ovid, Science Direct, and PubMed. The keywords "yoga", "cognitive functions", "vascular changes", and "desk-based jobs" were used in this search. Other than the keyword matches, the main inclusion criteria for the article search were the possible vascular, cognitive, and endothelial improvements using yoga practices as the focus for the final article screening.

2.3 Validation of the manual

In our study, we considered a sample of 10 experts to validate the yoga manual. Trained and certified yoga professionals with a minimum of five years' experience in teaching and clinical yoga were included for validating this manual. Experts without teaching and clinical experience were excluded from the study. Experts were identified through an online search based on academic research papers and educational qualifications. The experts were contacted through email. The detailed graphs are shown in Figs. 19.1 and 19.2. All the experts identified were given validation forms along with an information form explaining the aim of the study (Sukumar, 2023) which was based on the institutional ethics committee's information forms. After the expert's consent, the content validation form and the yoga module were emailed to the experts (Fig. 19.3). The questionnaire's validation was rated across three categories: "essential", "useful, but not necessary", and "not necessary" in the preform's different sections (Sukumar, 2023). Yoga postures that were recommended as "essential" were given a score of one, while the yoga postures marked as "useful, but not necessary" and "not necessary" were given a score of zero. Printed forms were completed by hand by the experts and then scanned and returned to the research team via email.

2.4 Analysis

To evaluate the module's content validity, the content validity ratio (CVR) was calculated according to Lawshe's method. The Lawshe and Davis technique is highly regarded as the favored method for assessing content validity due to its meticulous methodology in evaluating the validity of individual scale items as well as the overall scale (Lawshe, 1975; Ayre and Scally, 2017). According to Lawshe, if more than half of the participants in a study indicate that an item is essential, that item has at least some content validity. The CVR was validated using the following formula. After collecting the printed forms from the experts with their opinions on the yoga practices mentioned, the scores were assigned as 1 point for "essential", 0 points for "useful but not essential", and 0 points for "not necessary". The values were scored, and the final score of each yoga practice was formulated using the formula in Eq. (1).

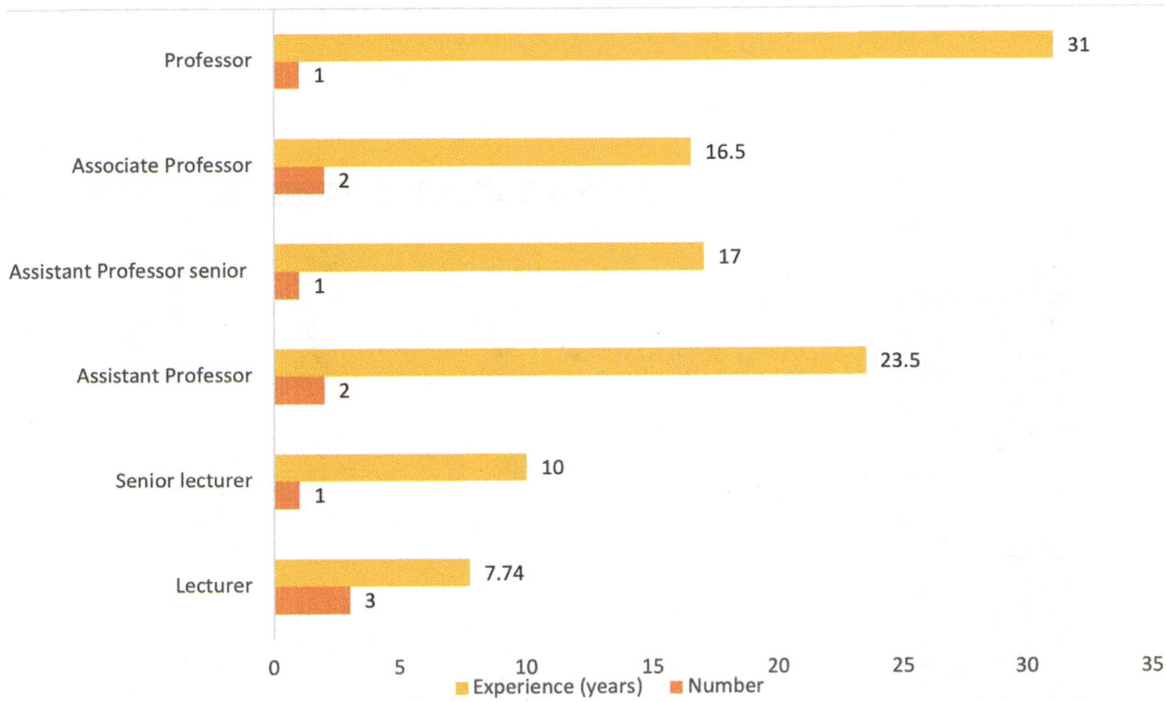

FIG. 19.1 Experience of the experts.

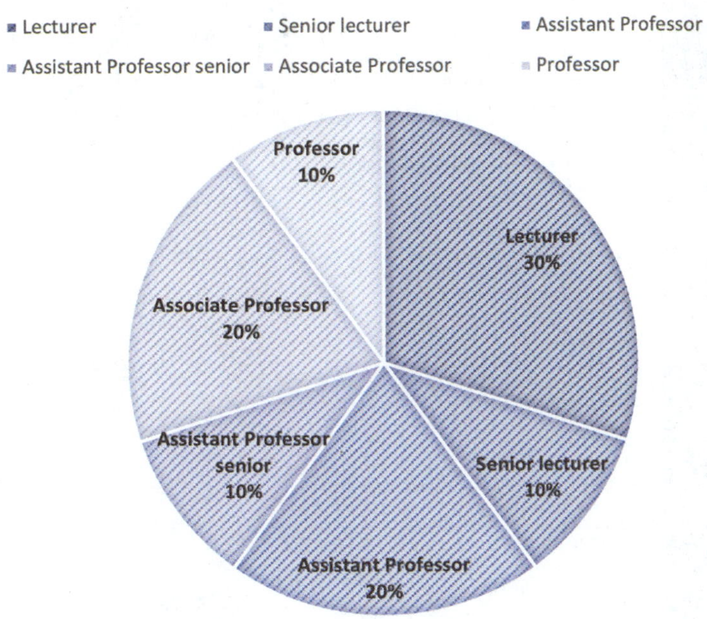

FIG. 19.2 Designation of the experts.

Lawshe's formula:

$$CVR = \frac{Ne - [N/2]}{N/2} \quad (1)$$

CVR—Content validation ratio
Ne—A number of experts who declare an item of importance.
N—The total number of experts' table

Name of the Posture	Steps to Perform	Pictorial Representation
Surya Namaskara	Samasthiti: Stand straight with feet together, arms relaxed by the sides. Vision straight. Fig: A Step 1: Inhaling lift both hands upward and join the palms over the head, look up. Fig: B Step 2: Exhaling bend forward and keep both palms on sides of the both feet, bend the neck and nose should touch the knees. Fig: C Step 3: Inhaling raise the head and look forward. Fig: D Step 4: Exhaling jump backward in down straightly on the ground by the help of palms & toes, parallel to the ground. Fig: E Step 5: Inhaling move the body forward move the front part of the body upward, raise the head and look forward. Fig: F Step 6: Exhaling lift the whole body upward, both feet should touch the floor in a foot gap and move the head so that vision should be on naval region. Breathe for 5 times. Fig: G Step 7: Inhaling look upward and jump forward so that keep the feet between the palms. Fig: H Step 8: Exhaling bring the head backward so that nose should touch the knees. Fig: I Step 9: Inhaling raise the head and slowly stand up and join the palms over the head and look upward. Fig: J Step 10: Exhaling bring down the hands on either side of the body and look front to form samasthiti. Fig: K	
Swastikasana	Samasthiti: Sit erect on the floor with legs stretched straight. Palms on the floor by the sides of the hips. Look front Fig: A Step 1: Inhale Step 2: Exhaling bend the left leg. Keep the left foot below the right thigh. Fig: B Step 3: Inhale Step 4: Exhaling bend the right leg, keep the foot on the left leg in between thigh and calf. Hold the chin mudra. Close the eyes and breathe 5 times Fig: C Step 5: Inhaling stretch right leg Fig: D Step 6: Exhale Step 7: Inhaling stretch the left leg Come back to samasthiti Fig: E	

FIG. 19.3 Yoga postures and its descriptions.

Pavanamuktasana	Samasthiti: Lie flat on the back on the ground. Fig: A Step 1: Inhale Step 2: Exhaling bend the right leg, bring it on the chests, hold the leg by both hands. Fig: B Step 3: Inhaling lift the head up, touch nose or forehead to the knee. Close the eyes. Breathe 5 times. Fig: C Step 4: Exhaling keep the head down on the ground. Fig: D Step 5: Inhaling extend the right leg. Repeat the same steps on left leg & both legs also. Fig: E	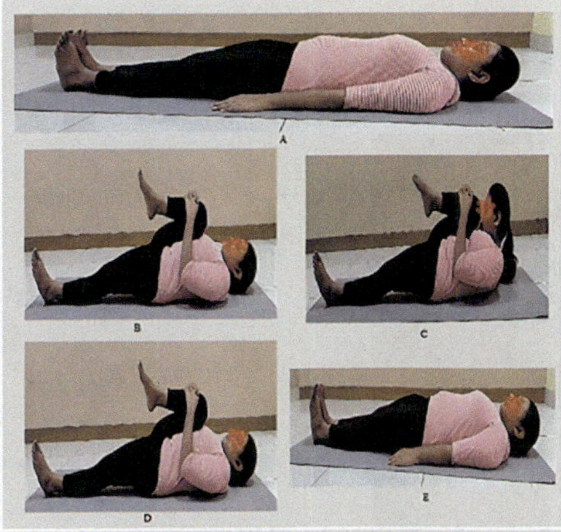
Makarasana	Procedure: Lie on the abdomen, keep the palms one over the other, rest your head on the palms to one side. Keep the legs 3 feet apart. Close the eyes. Relax the mind. Continue this for 1 minutes.	
Dhanurasana	Samasthiti: Lie down on the floor on the stomach face downwards. Fig: A Step 1: Inhale Step 2: Exhaling bend the knees. Stretch the arms back and hold the ankle with respective hands. Fig: B Step 3: Inhaling lift the legs, and head and chest up. Look up. Breathe 5 times. Fig: C Step 4: Exhaling come down. Inhaling release the hands and legs, come back to samasthiti Fig: D	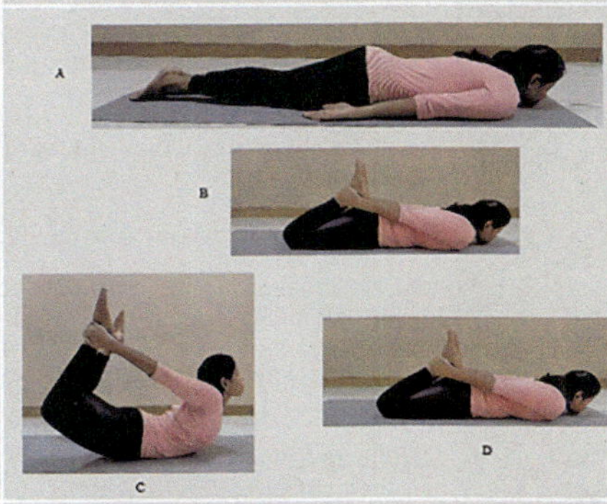
Uttanapadasana	Samasthiti: Lie flat on the back. Fig: A Step 1: Keep the palms on either side of the head. (Figure 15.1) Inhaling lift the head and chest arch it by stretching the neck and moving the head back until the crown of the head rests on the floor. Fig: B Step 2: Exhaling bring the hands forward and join the palms Step 3: Inhaling lift both the legs up until 45° to 50° from floor (parallel to the hands). Look back. Breathe for 5 times. Fig: C Step 4: Exhaling bring down the legs Step 5: Inhaling keep the palms on the floor on either side of the head. Fig: D Exhaling keep the head on the floor and release the hand. Fig: E Back to samasthiti	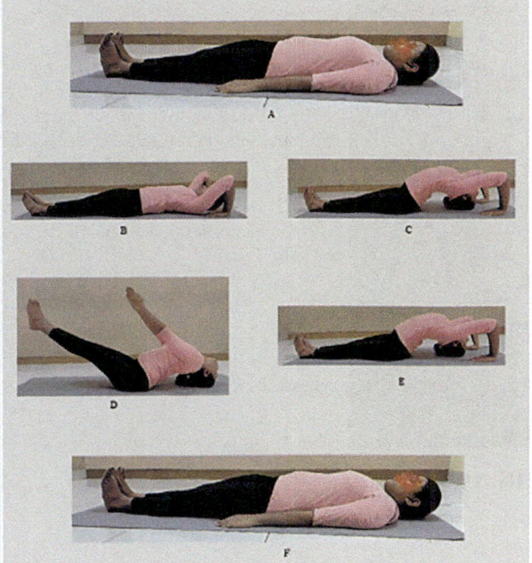

FIG. 19.3 cont'd

Ardha Matsyendrasana	Samasthiti: Sit on the floor with both legs stretched straight, palms on the floor by sides of the body. Look front. Fig: A Step 1: Inhale. Step 2: Exhaling insert the left foot below the right thigh. Fig: B Step 3: Inhaling fold the right leg knee facing up. Fig: C Step 4: Exhaling keep the right foot outside the left knee so that ankle and knee should be in the same line. Fig: D Step 5: Inhale Step 6: Exhaling bring the left hand outside the right knee and hold the right big toe by left hand. Take the right hand back and touch the left thigh. Look back from right side. Breathe 5 times. Fig: E Step 7: Inhaling turn to front and release both the hands and keep the right leg to right side. Fig: F Step 8: Exhaling stretch the right leg Fig: G Step 9: Inhaling stretch the left leg. Fig: H Step 10: Exhaling insert the right leg below the left thigh. Fig: I Step 11: Inhaling fold the left leg knee facing upward. Fig: J Step 12: Exhaling keep the left foot outside the right knee. Fig: K Step 13: Inhale Step 14: Exhaling bring the right hand outside the left knee and hold the left big toe by right hand, take the left hand back and touch the right thigh. Look back from left side. Breathe 5 times. Fig: L Step 15: Inhaling turn to front and release both the hands and keep the left leg to the left side. Fig: M Step 16: Exhaling stretch the right leg. Fig: N Inhaling stretch the right leg. Back to samasthiti. Fig: O	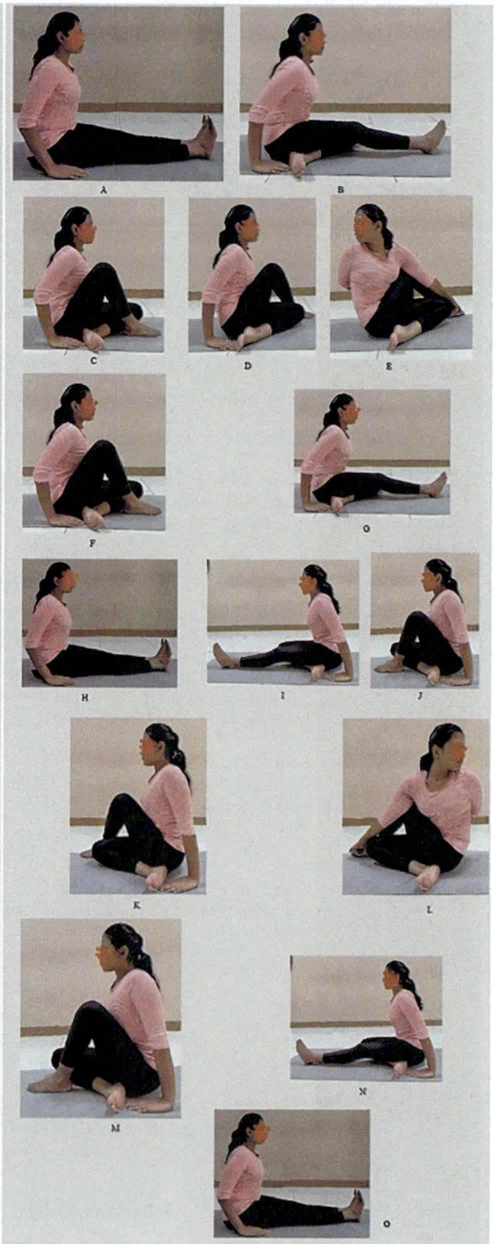

FIG. 19.3 cont'd

Janushirshasana	Samasthiti: Sit on the floor with both legs stretched straight, palms on the floor by sides of the body. Look front. Fig: A Step 1: Inhale Step 2: Exhaling bend the left leg and keep the heel touching the perineum. Fig: B Step 3: Inhale. Step 4: Exhaling bend forward, catch the right big toe with both hand's thumb, index, and middle finger. Keep the forehead on the right knee. Close the eyes. Breathe 5 times. Fig: C Step 5: Inhaling raise the head, release the hands and right leg. Fig: D Step 6: Exhaling fold the right leg and keep the heel touching perineum. Fig: E Step 7: Inhale. Step 8: Exhaling bend forward and hold the left big toe with both hand's thumb, index, and middle finger. Keep the forehead on the left knee. Close the eyes. Breathe for 5 times. Fig: F Step 9: Inhaling come up, release the hands, stretch the right leg and come back to samasthiti	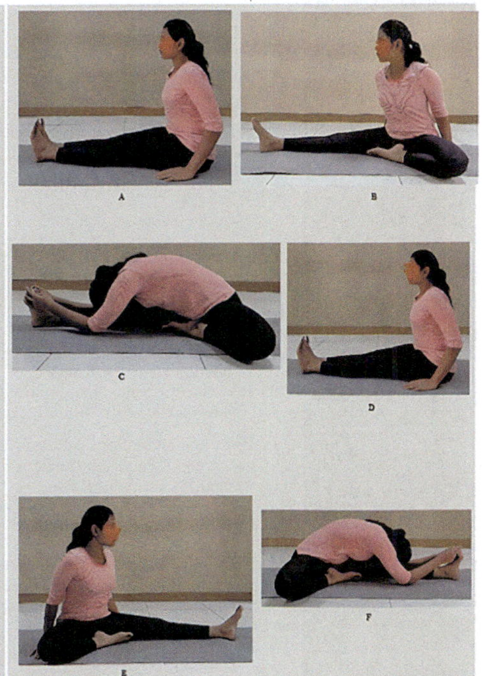
Navasana	Samasthiti: Sit on the floor with both legs stretched straight, palms on the floor by sides of the body. Look front. Fig: A Step 1: Inhaling lift both legs up balancing on the buttocks. Keep the toes in line with the head and hands parallel to the ground. Look front. Breathe for 5 times. Fig: B Exhaling bring legs down to form samasthiti.	

FIG. 19.3 cont'd

Prasarita Padottanasana	Samasthiti: Stand straight with feet together, arms relaxed by the sides. Vision straight. Fig: A Step 1: Inhaling spread the right leg towards the right side to one leg distance. Place the hands on the waist. Fig: B Step 2: Exhaling bend forward. Keep the palms on the floor in line with the shoulders between the feet. Fig: C Step 3: Inhale and simultaneously spread the legs as far as possible. Keep the legs straight. Look front. Fig: D Step 4: Exhaling bend forward, bend the elbow and rest the crown portion of the head on the floor. Both feet, both palms and the head should be in a straight line. Closing the eyes breathe for 5 times. Fig: E Step 5: Inhaling raise the head, straighten the arms and the elbow. Look front. Fig: F Step 6: Exhaling bring the legs nearer to one leg distance. Fig: G Step 7: Inhaling come up to step 1. Exhaling relax the hands, join the right leg to left come back to samasthiti. Fig: H	
Paschimottanasana	Samasthiti: Sit on the floor with legs stretched straight. Palms on the floor by the sides of the hips. Look front. Fig: A Step 1: Inhale. Step 2: Exhaling bend forward and catch the big toes with respective hands(thumb, index, and middle finger) and keep the face on the knees. This is the Asanasthiti. Breathe 5 times. Fig: B Step 3: Inhaling come up, release the hands to form samasthiti.	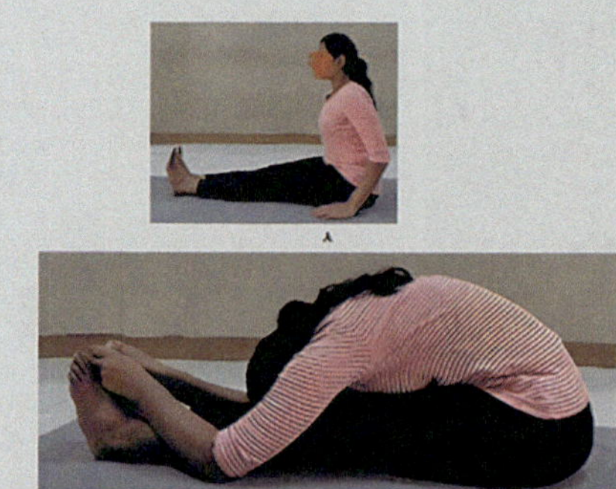
Purvottanasana	Samasthiti: Sit on the floor with legs stretched straight. Palms on the floor by the sides of the hips. Look front. Fig: A Step 1: Inhaling keep the palms on the floor on the back side a foot away from the hips. Fig: B Step 2: Exhale. Step 3: Inhaling lift the body off the floor. Keep the soles on the floor and look back. Breathe 5 times. Fig: C Step 4: Exhaling come down. Bring the hands to the sides of the body to form samasthiti. Fig: D	

FIG. 19.3 cont'd

Vajrasana	Samasthiti: Sit erect on the floor with legs stretched straight. Palms on the floor by the sides of the hips. Look front. Fig: A Step 1: Inhale Step 2: Exhaling sit on the heels by taking both the legs back side in one direction. Keep the palm on the thigh. Keep the back, neck, and head straight. Look front. Breathe 5 times. Fig: B Inhaling bring the legs forward. Back to samasthiti.	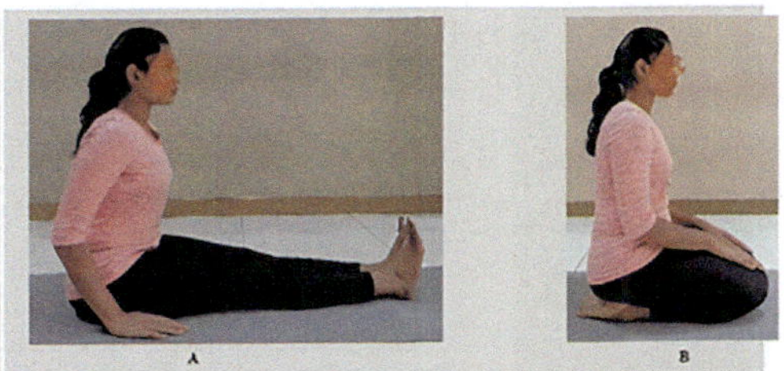
Padangustanasa	Samasthiti: Stand straight with feet together, arms relaxed by the sides. Vision straight. Fig: A Step 1: Inhaling keep the right leg one foot towards right. Fig: B Step 2: Exhaling bend forward, hold the toes by both hands thumb, index, and middle finger, touch the face to the knees. Close the eyes. Breathe for 5 times. Fig: C Step 3: Inhaling come up, stand erect and look front. Fig: D Exhaling join right leg to left to form samasthiti.	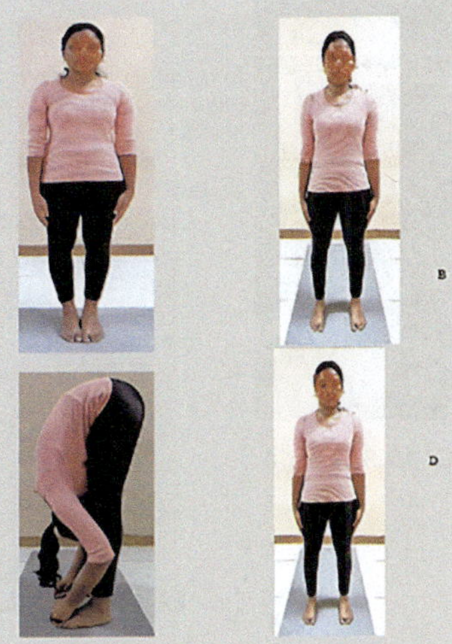
Trikonasana	Samasthiti: Stand straight with feet together, arms relaxed by the sides. Vision straight. Fig: A Step 1: Inhaling spread the right leg to right side about one leg length and simultaneously spread the hands to both sides up to shoulder level. Fig: B Step 2: Exhaling turn right foot to right side, bend to right side and catch the right big toe by 3 fingers of right hand. Look at the left hand which has moved straight upward. Breathe 5 times. Fig: C Step 3: Inhaling come up, turn right foot to front and keep the hands on both sides up to shoulder level and look front. Fig: D Step 4: Exhaling turn left foot to left side, bend to left side and catch the left big toe by 3 fingers of left hand. Move the right hand straight upward. Look at the right hand. Breathe 5 times.Fig: E Step 5: Inhaling come up, turn left foot to front and keep the hands on both sides up to shoulder level and look front . Exhaling join the right leg to left to for samasthiti. Fig: F	

FIG. 19.3 cont'd

Bhramari Pranayama	Preparation: Sit in Swastikasana, keep the palms in chin mudra on the respective knees, close the eyes. Method: Slowly inhale and while exhaling produce honey bee sound ('m' sound). Repeat this for 10 times	
Shavasana	Procedure: Lie down flat on the ground, hand on the floor palm facing up, and keep the legs about 3 feet apart, close the eyes. Breathe normally and keep the mind relaxed. Continue this for 5 mins	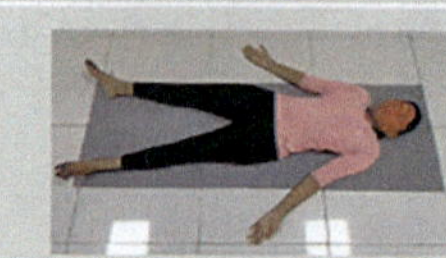
Meditation	Preparation: Sit in Swastikasana, keep the palms in chin mudra on the respective knees Method: • Close the eyes and breathe normally. • Inhaling mentally repeat "so" and while exhaling repeat "ham". • Keep the concentration throughout the practice. • Continue the practice for 5 minutes.	

FIG. 19.3 cont'd

TABLE 19.1 Lawshe's table for minimum values of CVR.

Number of panelists	Minimum value
5	0.99
6	0.99
7	0.99
8	0.75
9	0.78
10	0.62
11	0.59
12	0.56
13	0.54
14	0.51
15	0.49
20	0.42
25	0.37
30	0.33
35	0.31
40	0.29

The criteria for including individual yoga practices were fixed at CVR scores of 0.62 and above to be retained for this study (Kodidala et al., 2022). The filled content validation forms received from the experts were compiled and analyzed. Experts were also requested to provide comments or suggestions on individual yoga practices on the physical validation forms sent to them. Based on Lawshe's method, we gathered the values from 10 experts, which gave us 0.62 as the cutoff value for the acceptance of the content. The detailed values are mentioned in Table 19.1. The details of the yoga practices are added in the online dataverse (Sukumar, 2023).

3 Results

10 experts performed the validation with an average of 16.80 years of experience and a standard deviation of 09.44. The content validity index (CVI) for the entire module was 0.93. Lawshe's method was used to calculate the CVR for each yoga practice (Lawshe, 1975; Ayre and Scally, 2017). Out of 21 practices, we included 19, which are as follows: 1. Surya Namaskara, 2. Swastikasana, 3. Vajrasana, 4. Supta Padangusthasana, 5. Trikonasana, 6. Prasarita Padottanasana, 7. Paschimottasana, 8. Purvottanasana, 9. Janu Sirsasana, 10. Navasana, 11. Ardha Matsyendrasana, 12. Pavanamuktasana, 13. Makarasana, 14. Dhanurasana, 15. Uttana Padasana, 16. Anulomaviloma Pranayama, 17. Bhramari Pranayama, 18. Shavasana, and 19. Meditation, all with a CVR score of 0.62 and above. The outcome of the CVR for each yoga posture is shown in Table 19.2 and Fig. 19.4. Yoga practices such as Tadasana, Suryabedhana Pranayama, and Sarvangasana were excluded due to CVR scores of 0.6, 0.6, and 0.4, respectively.

4 Discussion

Based on the CVR results, experts agreed that the long-term yoga postures validated in the manual may significantly improve central and peripheral vascular function, as well as cognitive functions, among desk-based workers. The high CVI of 0.93 demonstrates strong expert consensus on the selected yoga practices, indicating their potential effectiveness as an intervention for the specific target population.

TABLE 19.2 Yoga manual with scoring criteria results.

Postures	Essential	Useful, but not essential	Not necessary	CVR
Surya namaskara	10	0	0	1
Swastikasana	9	1	0	0.8
Vajrasana	10	0	0	1
Tadasana	8	1	1	0.6
Supta Padangusthasana	9	0	1	0.8
Trikonasana	10	0	0	1
Prasarita Padottanasana	9	1	0	0.8
Paschimottasana	10	0	0	1
Purvottanasana	9	1	0	0.8
Janu Sirsasana	10	0	0	1
Navasana	10	0	0	1
Ardha Matsyendrasana	9	1	0	0.8
Pavanamuktasana	10	0	0	1
Makarasana	10	0	0	1
Dhanurasana	10	0	0	1
Sarvangasana	7	2	1	0.4
Uttana padasana	9	0	0	0.8
Anulomaviloma Pranayama	10	0	0	1
Suryabedhana Pranayama	8	1	1	0.6
Bhramari Pranayam	10	0	0	1
Shavasana	10	0	0	1
Meditation	10	0	0	1

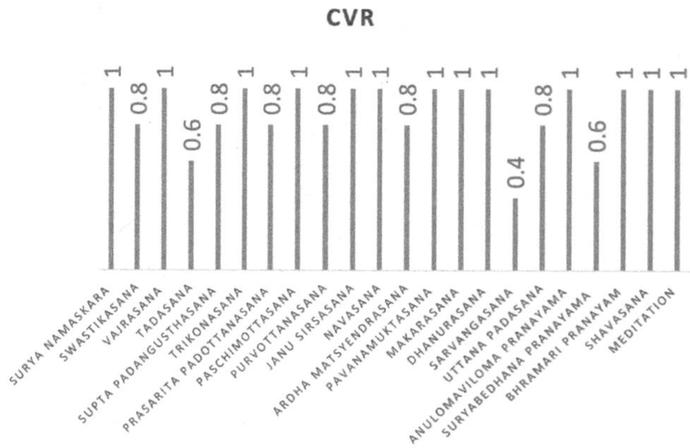

FIG. 19.4 Content validity ratio (CVR) value of each yoga posture.

4.1 Yoga practices and their physiological benefits

Research has observed that cognitive functions like attention, orientation, memory, fluency, and visuospatial ability are improved by the regular practice of Suryanamaskara, which has been the prime component of our manual (Kodidala and Pradhan, 2021; Parajuli et al., 2019). This comprehensive sequence of twelve postures not only serves as a complete body workout but also integrates breath control and mindfulness, potentially activating multiple neural pathways that enhance cognitive processing (Mooventhan and Nivethitha, 2014). The rhythmic nature of Suryanamaskara, which involves coordinated movement with breath, may facilitate improved blood flow to the brain, thereby enhancing cognitive function through increased oxygenation and nutrient delivery to neural tissues (Bhavanani et al., 2003).

Suryanamaskara has been shown to improve focus, memory, mindfulness, attention, orientation, fluency, visuospatial skills, and overall cognitive function by increasing the flow of oxygen-rich blood to the brain (Parajuli et al., 2022). Studies have demonstrated that regular practice of Suryanamaskara improves self-control and mindfulness in participants, with significant improvements observed in stress reduction and emotional intelligence (Sharma et al., 2008; Sengupta et al., 2013). The intensive nature of this yoga exercise routine has been found to successfully treat stress and anxiety, as well as improve psychological well-being (Stec et al., 2023).

Asanas like Trikonasana, Shavasana, and several others have been proven to improve self-confidence, alertness, and productivity in adults (Field, 2011; Vancampfort et al., 2015). Trikonasana (triangle pose), for instance, engages the entire body, promoting improved circulation while simultaneously requiring focus and balance, potentially enhancing both vascular health and cognitive engagement (Hagins et al., 2007). The pose helps in strengthening bones and muscles, thereby boosting immunity and improving overall physical conditioning (Ray et al., 2001). Shavasana (corpse pose), despite its apparent simplicity, plays a crucial role in stress reduction and autonomic nervous system regulation (Kamei et al., 2000). By reducing sympathetic nervous system activity and promoting parasympathetic dominance, Shavasana may contribute to improved vascular health by lowering blood pressure and heart rate, while concurrently enhancing cognitive function through stress reduction and improved attention regulation (Vempati and Telles, 2002).

Compared to physical activities like walking, cycling, and aerobic routines, the yoga postures included in the validated manual have been reported to improve pulse wave velocity of blood vessels (Dhindsa et al., 2008; Hunter et al., 2013; "Yoga vs. aerobic activity: effects on endothelial function and arterial stiffness," 2014). This is particularly significant, as pulse wave velocity serves as a reliable indicator of arterial stiffness, with lower values generally indicating better vascular health (Laurent et al., 2006). Research has shown that yoga programs are more effective than brisk walking in reducing arterial stiffness along with blood pressure in elderly individuals with increased pulse pressure (Patil et al., 2015). The improvement in pulse wave velocity suggests that these yoga practices may effectively counteract the vascular rigidity and endothelial dysfunction that often result from prolonged sitting.

Studies have demonstrated that an 8-month yoga intervention significantly improves arterial compliance in premenopausal women, indicating positive effects on cardiovascular health. The influence of yoga practice on central arterial stiffness has been documented, with regular practitioners showing reduced arterial stiffness compared to sedentary individuals ("Yoga practice improves arterial compliance in premenopausal women," 2014). Yoga has also been shown to significantly reduce sympathetic activity and improve endothelial function, with an enhancement in the bioavailability of nitric oxide (Patil, 2014).

4.2 Specific benefits for desk-based workers

The validated manual addresses several specific concerns related to desk-based work. For instance, Vajrasana has been proven to decrease lower back pain and improve sitting posture in adults (Tekur et al., 2012; Nambi et al., 2014). Since lower back pain represents one of the most common complaints among desk-based workers, the inclusion of Vajrasana offers a targeted intervention for this population-specific issue. Research involving patients with lower back pain found that Vajrasana and other yoga asanas helped reduce discomfort significantly (Sherman et al., 2005). The posture helps strengthen core muscles while relieving pressure on the lumbar spine, potentially improving overall posture during extended sitting periods (Sorosky et al., 2008).

Clinical evaluations of selected yogic procedures have demonstrated their effectiveness in individuals with low back pain, making them particularly relevant for desk-based workers who commonly experience this condition (Yadav and Mehta, 2011). Workplace yoga interventions have been shown to reduce perceived stress and back pain while improving psychological well-being (Hartfiel et al., 2011). Research indicates that yoga can help strengthen back muscles and alleviate low back pain, making it a safe and effective treatment option (Wieland et al., 2017).

Similarly, practices like Paschimottanasana and Janu Sirsasana may help counteract the anterior pelvic tilt commonly observed in individuals who sit for extended periods. These forward bends stretch tight hip flexors and strengthen core muscles, potentially restoring optimal pelvic alignment and reducing the risk of related musculoskeletal disorders (Holtzman and Beggs, 2013). In addition, Ardha Matsyendrasana provides a gentle spinal twist that may help alleviate spinal compression resulting from prolonged sitting (Williams et al., 2009).

From a vascular perspective, the combination of dynamic movements (like Suryanamaskara), static holds (like Trikonasana and Navasana), and targeted breathing practices (like Anuloma Viloma and Bhramari Pranayama) may provide a comprehensive approach to counteracting the negative vascular effects of sedentary behavior (Cramer et al., 2013). Dynamic movements increase blood flow and cardiac output, while static holds may improve vessel elasticity and endothelial function through sustained, moderate pressure changes (Yadav et al., 2019). Breathing practices, particularly those involving breath retention and control, have been shown to influence autonomic nervous system activity, potentially improving vascular tone and function (Jerath et al., 2006).

4.3 Neurophysiological mechanisms

The cognitive benefits observed with the validated yoga practices may stem from multiple neurophysiological mechanisms. Practices like meditation and Bhramari Pranayama have been associated with increased activity in the prefrontal cortex, a brain region crucial for executive functions, attention, and working memory (Froeliger et al., 2012; Luders et al., 2009). Regular engagement in these practices may lead to structural and functional neuroplasticity in this region, potentially enhancing cognitive performance over time (Gothe et al., 2019).

Furthermore, the stress-reducing effects of yoga practices may protect against cognitive decline by mitigating the adverse effects of chronic stress on hippocampal structure and function (Mcewen and Gianaros, 2010). Chronic stress and elevated cortisol levels have been linked to hippocampal atrophy and impaired memory function. By reducing stress and cortisol levels, yoga practices included in the manual may help preserve hippocampal integrity and function, particularly important for desk-based workers who often experience high levels of occupational stress (Goyal et al., 2014).

The inclusion of pranayama practices, like Anuloma Viloma, may also enhance cognitive function through improved oxygen delivery to the brain and regulation of the autonomic nervous system (Sarang and Telles, 2008). Alternating nostril breathing has been shown to balance sympathetic and parasympathetic activity, potentially creating an optimal physiological state for cognitive performance (Sharma et al., 2013). Research has demonstrated that pranayama practices can significantly improve cognitive functions, including attention, memory, and executive functioning (Pradhan and Nagendra, 2008).

Neuroimaging studies have revealed that yoga practice leads to increased gray matter volume in brain regions critical for cognitive control, including the hippocampus, prefrontal cortex, and cingulate cortex (Gotink et al., 2016; Villemure et al., 2014). These structural changes correlate with improved cognitive performance and may represent the neurobiological basis for the cognitive benefits observed with regular yoga practice (Gard et al., 2014).

4.4 Practical implementation considerations

With all these aspects supporting our manual, we have prepared this routine specifically for adult desk-based workers who sit for long hours, providing them with tools to improve cognitive and vascular health compromised by career-based sedentary behavior (Shariat et al., 2018). The comprehensive nature of the validated module, incorporating asanas, pranayama, and meditation, ensures a holistic approach that addresses multiple dimensions of health affected by prolonged sitting (Ross and Thomas, 2010).

The practical implementation of this yoga module in workplace settings deserves careful consideration. Unlike traditional yoga classes that might require dedicated space and extended time, this module has been designed with workplace constraints in mind (Wolever et al., 2012). Many of the validated practices can be performed in limited space, with minimal equipment, and during short breaks throughout the workday. For instance, practices like Vajrasana or seated pranayama could be performed at one's desk, while more dynamic postures like Suryanamaskara might be reserved for longer breaks or dedicated wellness sessions (de Bruin et al., 2017).

The progressive nature of the validated practices also allows for adaptation to various fitness levels and physical limitations, making the module accessible to a diverse workforce (Galantino et al., 2004). Beginning with simpler practices like Swastikasana and gradually progressing to more challenging postures like Dhanurasana ensures that participants can build strength and flexibility over time, reducing the risk of injury while maximizing benefits (Büssing et al., 2012b).

4.5 Long-term adherence and sustainability

For any workplace intervention to be effective, long-term adherence is crucial. The validated yoga module offers several features that may promote sustained engagement (Lazar et al., 2005). The variety of practices included prevents monotony, while the combination of physical postures, breathing techniques, and meditation addresses multiple dimensions of well-being, potentially providing diverse and noticeable benefits that reinforce continued practice (Park et al., 2020).

Moreover, the yoga practices validated in this study can be performed independently after initial instruction, empowering employees to take control of their health without relying on ongoing external support (Sherman et al., 2005). This self-sufficiency may promote a sense of autonomy and competence, key factors in maintaining long-term behavior change according to self-determination theory (Deci and Ryan, 2000). Research has shown that workplace yoga interventions have high acceptability and feasibility, with participants reporting positive experiences and a willingness to continue practice (Brakenridge et al., 2016).

4.6 Organizational implications

Beyond individual health benefits, the implementation of this yoga module has broader organizational implications (Baicker et al., 2010). Improved cognitive function among desk-based workers may translate to enhanced productivity, creativity, and decision-making capacity (Berger et al., 2010). Better vascular health may reduce absenteeism due to related health conditions, increase overall energy levels, and boost work engagement (Aldana, 2001).

Furthermore, offering such wellness initiatives signals an organizational commitment to employee well-being, potentially improving recruitment, retention, and overall workplace satisfaction (Berry et al., 2010). As companies increasingly recognize the importance of workplace wellness in attracting and retaining talent, evidence-based interventions like the validated yoga module may provide a competitive advantage in human resource management (Parks and Steelman, 2008). Studies have shown that workplace wellness programs can reduce healthcare costs while improving employee satisfaction and productivity (Goetzel et al., 2014).

4.7 Limitations and future directions

While the expert validation process provides strong support for the potential efficacy of the yoga module, empirical testing of its effects on vascular and cognitive outcomes among desk-based workers is necessary (Craig et al., 2008). Further studies should implement the validated module in real workplace settings and measure pre–post changes in various vascular parameters (e.g., pulse wave velocity, endothelial function) and cognitive domains (e.g., attention, memory, executive function) (Thijssen et al., 2011).

In addition, investigating the dose-response relationship between yoga practice and health outcomes would provide valuable insights for developing specific recommendations regarding the frequency, intensity, and duration of practice (Garber et al., 2011). Longitudinal studies examining the sustained effects of regular practice over extended periods would also contribute to our understanding of the long-term benefits of this intervention (Kessler et al., 2012).

Future research could also explore the potential synergistic effects of combining the validated yoga module with other workplace interventions, such as ergonomic modifications, nutritional guidance, or stress management programs (Pronk et al., 2012). Such integrated approaches might yield more comprehensive improvements in employee health and well-being than single-modality interventions (Conn et al., 2009). Research comparing the effectiveness of yoga interventions to other workplace wellness programs would also provide valuable insights for organizational decision-making (Shrestha et al., 2018).

5 Conclusion

This study was conducted to validate a manual for desk-based workers to observe the impact of a yoga routine on vascular and cognitive functions. The validation process resulted in a comprehensive yoga module that includes 18 practices encompassing asanas, pranayama, and meditation, all with a CVR score of 0.62 or above. The high average CVI of 0.93 indicates strong expert consensus on the appropriateness and potential efficacy of the included practices. The validated yoga module offers a promising intervention for counteracting the negative effects of prolonged sitting on vascular and cognitive health among desk-based workers. By addressing multiple dimensions of health through a variety of practices, this module provides a holistic approach to workplace wellness that is both scientifically grounded and practically feasible. Further empirical testing of this module in real-world settings will enhance our understanding of its effectiveness and guide refinements for maximal impact on employee health and organizational outcomes.

References

Aldana, S.G., 2001. Financial impact of health promotion programs: a comprehensive review of the literature. Am. J. Health Promot. 08901171. 15 (5), 296–320. Available from: https://journals.sagepub.com/home/AHP. doi:10.4278/0890-1171-15.5.296.

Angevaren, M., Aufdemkampe, G., Verhaar, H.J.J., Aleman, A., Vanhees, L., 2008. Physical activity and enhanced fitness to improve cognitive function in older people without known cognitive impairment. Cochrane Database Syst. Rev. 1469493X. 2. Available from: http://www.mrw.interscience.wiley.com/cochrane/clsysrev/articles/CD005381/pdf_fs.html. doi:10.1002/14651858.CD005381.pub2Netherlands.

Asana Pranayama Mudra Bandha, 2009. Yoga Publications Trust, Munger, Bihar. Available from: http://chrome-extension://efaidnbmnnnibpcajpcglclefindmkaj/https://www.yogkulam.org/books/Asana.pdf.

Ayre, C., Scally, A.J., 2017. Critical values for Lawshe's content validity ratio. Meas. Eval. Couns. Dev. 0748-1756. 47 (1), 79–86. doi:10.1177/0748175613513808.

Baicker, K., Cutler, D., Song, Z., 2010. Workplace wellness programs can generate savings. Health Aff. 0278-2715. 29 (2), 304–311. doi:10.1377/hlthaff.2009.0626.

Berger, A.M., Wielgus, K., Hertzog, M., Fischer, P., Farr, L., 2010. Patterns of circadian activity rhythms and their relationships with fatigue and anxiety/depression in women treated with breast cancer adjuvant chemotherapy. Support Care Cancer. 09414355. 18 (1), 105–114. doi:10.1007/s00520-009-0636-0.

Berry, L.L., Mirabito, A.M., Baun, W.B., 2010. What's the hard return on employee wellness programs? Harv. Bus. Rev. 00178012. 88 (12). Available from: http://hbr.org/2010/12/whats-the-hard-return-on-employee-wellness-programs/ar/1.

Bhavanani, A.B., Madanmohan, Udupa, K., 2003. Acute effect of mukh bhastrika (a yogic bellows type breathing) on reaction time. Indian J. Physiol. Pharmacol. 47 (3), 297–300.

Biswas, A., Oh, P.I., Faulkner, G.E., Bajaj, R.R., Silver, M.A., Mitchell, M.S., Alter, D.A., 2015. Sedentary time and its association with risk for disease incidence, mortality, and hospitalization in adults a systematic review and meta-analysis. Ann. Intern. Med. 15393704. 162 (2), 123–132. Available from: http://annals.org/data/Journals/AIM/931930/0000605-201501200-00007.pdf. doi:10.7326/M14-1651.

Booth, F.W., Roberts, C.K., Thyfault, J.P., Ruegsegger, G.N., Toedebusch, R.G., 2017. Role of inactivity in chronic diseases: evolutionary insight and pathophysiological mechanisms. Physiol. Rev. 15221210. 97 (4), 1351–1402. Available from: http://physrev.physiology.org/content/97/4/1351.full.pdf. doi:10.1152/physrev.00019.2016.

Brakenridge, C.L., Fjeldsoe, B.S., Young, D.C., Winkler, E.A.H., Dunstan, D.W., Straker, L.M., Healy, G.N., 2016. Evaluating the effectiveness of organisational-level strategies with or without an activity tracker to reduce office workers' sitting time: a cluster-randomised trial. Int. J. Behav. Nutr. Phys. Act. 14795868. 13 (1). Available from: http://www.ijbnpa.org/home/. doi:10.1186/s12966-016-0441-3.

Büssing, A., Michalsen, A., Khalsa, S.B.S., Telles, S., Sherman, K.J., 2012a. Effects of yoga on mental and physical health: a short summary of reviews. Evid. Based Complement. Altern. Med. 17414288. 2012. doi:10.1155/2012/165410.

Büssing, A., Michalsen, A., Khalsa, S.B.S., Telles, S., Sherman, K.J., 2012b. Effects of yoga on mental and physical health: a short summary of reviews. Evid. Based Complement. Altern. Med. 1741-427X. 2012, 1–7. doi:10.1155/2012/165410.

Celis-Morales, C.A., Lyall, D.M., Anderson, J., Iliodromiti, S., Fan, Y., Ntuk, U.E., Mackay, D.F., Pell, J.P., Sattar, N., Gill, J.M.R., 2017. The association between physical activity and risk of mortality is modulated by grip strength and cardiorespiratory fitness: evidence from 498 135 UK-Biobank participants. Eur. Heart J. 15229645. 38 (2), 116–122. Available from: http://eurheartj.oxfordjournals.org/. doi:10.1093/eurheartj/ehw249.

Chomistek, A.K., Manson, J.E., Stefanick, M.L., Lu, B., Sands-Lincoln, M., Going, S.B., Garcia, L., Allison, M.A., Sims, S.T., Lamonte, M.J., Johnson, K.C., Eaton, C.B., 2013. Relationship of sedentary behavior and physical activity to incident cardiovascular disease: results from the women's health initiative. J. Am. Coll. Cardiol. 15583597. 61 (23), 2346–2354. doi:10.1016/j.jacc.2013.03.031.

Chu, P., Gotink, R.A., Yeh, G.Y., Goldie, S.J., Hunink, M.G.M., 2016. The effectiveness of yoga in modifying risk factors for cardiovascular disease and metabolic syndrome: a systematic review and meta-analysis of randomized controlled trials. Eur. J. Prev. Cardiol. 20474881. 23 (3), 291–307. Available from: http://cpr.sagepub.com/content/by/year. doi:10.1177/2047487314562741.

Church, T.S., Thomas, D.M., Tudor-Locke, C., Katzmarzyk, P.T., Earnest, C.P., Rodarte, R.Q., Martin, C.K., Blair, S.N., Bouchard, C., Lucia, A., 2011. Trends over 5 decades in U.S. occupation-related physical activity and their associations with obesity. PLoS ONE. 1932-6203. 6 (5), e19657. doi:10.1371/journal.pone.0019657.

Colley, Garriguet, Janssen, Craig, Clarke, Tremblay, 2011. Physical activity of Canadian adults: accelerometer results from the 2007 to 2009 Canadian Health Measures Survey. Health Rep. 22 (1), 7–14.

Conn, V.S., Hafdahl, A.R., Cooper, P.S., Brown, L.M., Lusk, S.L., 2009. Meta-analysis of workplace physical activity interventions. Am. J. Prev. Med. 07493797. 37 (4), 330–339. doi:10.1016/j.amepre.2009.06.008.

Craig, P., Dieppe, P., Macintyre, S., Michie, S., Nazareth, I., Petticrew, M., 2008. Developing and evaluating complex interventions: the new Medical Research Council guidance. BMJ. 1756-1833. 337 (7676), a1655. doi:10.1136/bmj.a1655.

Cramer, H., Lauche, R., Haller, H., Dobos, G., 2013. A systematic review and meta-analysis of yoga for low back pain. Clin. J. Pain. 0749-8047. 29 (5), 450–460. doi:10.1097/ajp.0b013e31825e1492.

Cramer, H., Lauche, R., Haller, H., Steckhan, N., Michalsen, A., Dobos, G., 2014. Effects of yoga on cardiovascular disease risk factors: a systematic review and meta-analysis. Int. J. Cardiol. 01675273. 173 (2), 170–183. doi:10.1016/j.ijcard.2014.02.017.

de Bruin, E.I., Formsma, A.R., Frijstein, G., Bögels, S.M., 2017. Mindful2Work: effects of combined physical exercise, yoga, and mindfulness meditations for stress relieve in employees. a proof of concept study. Mindfulness. 18688535. 8 (1), 204–217. Available from: http://www.springer.com/psychology/cognitive+psychology/journal/12671. doi:10.1007/s12671-016-0593-x.

Deci, E.L., Ryan, R.M., 2000. The "what" and "why" of goal pursuits: human needs and the self-determination of behavior. Psychol. Inq. 1047840X. 11 (4), 227–268. Available from: http://www.tandfonline.com/toc/hpli20/current. doi:10.1207/S15327965PLI1104_01.

Dempsey, P.C., Sacre, J.W., Larsen, R.N., Straznicky, N.E., Sethi, P., Cohen, N.D., Cerin, E., Lambert, G.W., Owen, N., Kingwell, B.A., Dunstan, D.W., 2016. Interrupting prolonged sitting with brief bouts of light walking or simple resistance activities reduces resting blood pressure and plasma noradrenaline in type 2 diabetes. J. Hypertens. 14735598. 34 (12), 2376–2382. Available from: http://journals.lww.com/jhypertension. doi:10.1097/HJH.0000000000001101.

Desai, R., Tailor, A., Bhatt, T., 2015. Effects of yoga on brain waves and structural activation: a review. Complement. Ther. Clin. Pract. 17443881. 21 (2), 112–118. doi:10.1016/j.ctcp.2015.02.002.

Dhindsa, M., Sommerlad, S.M., DeVan, A.E., Barnes, J.N., Sugawara, J., Ley, O., Tanaka, H., 2008. Interrelationships among noninvasive measures of postischemic macro- and microvascular reactivity. J. Appl. Physiol. 15221601. 105 (2), 427–432. Available from: http://jap.physiology.org/cgi/reprint/105/2/427. doi:10.1152/japplphysiol.90431.2008.

Dunstan, D.W., Kingwell, B.A., Larsen, R., Healy, G.N., Cerin, E., Hamilton, M.T., Shaw, J.E., Bertovic, D.A., Zimmet, P.Z., Salmon, J., Owen, N., 2012. Breaking up prolonged sitting reduces postprandial glucose and insulin responses. Diabetes Care. 19355548. 35 (5), 976–983. Available from: http://care.diabetesjournals.org/content/35/5/976.full.pdf+html. doi:10.2337/dc11-1931Australia.

Falck, R.S., Davis, J.C., Liu-Ambrose, T., 2017. What is the association between sedentary behaviour and cognitive function? A systematic review. Br. J. Sports Med. 14730480. 51 (10), 800–811. Available from: http://bjsm.bmjjournals.com/. doi:10.1136/bjsports-2015-095551.

Field, T., 2011. Yoga clinical research review. Complement. Ther. Clin. Pract. 17443881. 17 (1), 1–8. doi:10.1016/j.ctcp.2010.09.007.

Froeliger, B., Garland, E.L., Kozink, R.V., Modlin, L.A., Chen, N.K., McClernon, F.J., Greeson, J.M., Sobin, P., 2012. Meditation-state functional connectivity (msFC): strengthening of the dorsal attention network and beyond. Evid. Based Complement. Altern. Med. 17414288. 2012. doi:10.1155/2012/680407.

Galantino, M.L., Bzdewka, T.M., Eissler-Russo, J.L., Holbrook, M.L., Mogck, E.P., Geigle, P., Farrar, J.T., 2004. The impact of modified hatha yoga on chronic low back pain: a pilot study. Altern. Ther. Health Med. 10786791. 10 (2), 56–59.

Garber, C.E., Blissmer, B., Deschenes, M.R., Franklin, B.A., Lamonte, M.J., Lee, I.M., Nieman, D.C., Swain, D.P., 2011. Quantity and quality of exercise for developing and maintaining cardiorespiratory, musculoskeletal, and neuromotor fitness in apparently healthy adults: guidance for prescribing exercise. Med. Sci. Sports Exerc. 15300315. 43 (7), 1334–1359. doi:10.1249/MSS.0b013e318213fefb.

Gard, T., Hölzel, B.K., Lazar, S.W., 2014. The potential effects of meditation on age-related cognitive decline: a systematic review. Ann. N.Y. Acad. Sci. 17496632. 1307(1), 89–103. Available from: http://onlinelibrary.wiley.com/journal/10.1111/(ISSN)1749-6632. doi:10.1111/nyas.12348.

Goetzel, R.Z., Henke, R.M., Tabrizi, M., Pelletier, K.R., Loeppke, R., Ballard, D.W., Grossmeier, J., Anderson, D.R., Yach, D., Kelly, R.K., McCalister, T., Serxner, S., Selecky, C., Shallenberger, L.G., Fries, J.F., Baase, C., Isaac, F., Crighton, K.A., Wald, P., Exum, E., Shurney, D., Metz, R.D., 2014. Do workplace health promotion (wellness) programs work? J. Occup. Environ. Med. 15365948. 56 (9), 927–934. Available from: http://journals.lww.com/joem. doi:10.1097/JOM.0000000000000276.

Gothe, N.P., Khan, I., Hayes, J., Erlenbach, E., Damoiseaux, J.S., 2019. Yoga effects on brain health: a systematic review of the current literature. Brain Plast. 2213-6304. 5 (1), 105–122. doi:10.3233/bpl-190084.

Gotink, R.A., Meijboom, R., Vernooij, M.W., Smits, M., Hunink, M.G.M., 2016. 8-week mindfulness based stress reduction induces brain changes similar to traditional long-term meditation practice – a systematic review. Brain Cogn. 10902147. 108, 32–41. Available from: http://www.elsevier.com/inca/publications/store/6/2/2/7/9/8/index.htt. doi:10.1016/j.bandc.2016.07.001.

Goyal, M., Singh, S., Sibinga, E.M.S., Gould, N.F., Rowland-Seymour, A., Sharma, R., Berger, Z., Sleicher, D., Maron, D.D., Shihab, H.M., Ranasinghe, P.D., Linn, S., Saha, S., Bass, E.B., Haythornthwaite, J.A., 2014. Meditation programs for psychological stress and well-being: a systematic review and meta-analysis. JAMA Intern. Med. 21686106. 174 (3), 357–368. Available from: http://archinte.jamanetwork.com/article.aspx?articleid=1809754. doi:10.1001/jamainternmed.2013.13018.

Hadgraft, N.T., Brakenridge, C.L., Lamontagne, A.D., Fjeldsoe, B.S., Lynch, B.M., Dunstan, D.W., Owen, N., Healy, G.N., Lawler, S.P., 2016. Feasibility and acceptability of reducing workplace sitting time: a qualitative study with Australian office workers. BMC Public Health. 14712458. 16 (1). Available from: http://www.biomedcentral.com/bmcpublichealth. doi:10.1186/s12889-016-3611-y.

Hagins, M., Moore, W., Rundle, A., 2007. Does practicing hatha yoga satisfy recommendations for intensity of physical activity which improves and maintains health and cardiovascular fitness? BMC Complement Altern. Med. 1472-6882. 7 (1). doi:10.1186/1472-6882-7-40.

Hamburg, N.M., McMackin, C.J., Huang, A.L., Shenouda, S.M., Widlansky, M.E., Schulz, E., Gokce, N., Ruderman, N.B., Keaney, J.F., Vita, J.A., 2007. Physical inactivity rapidly induces insulin resistance and microvascular dysfunction in healthy volunteers. Arteriosc. Thromb. Vasc. Biol. 10795642. 27 (12), 2650–2656. doi:10.1161/ATVBAHA.107.153288.

Hartfiel, N., Havenhand, J., Khalsa, S.B., Clarke, G., Krayer, A., 2011. The effectiveness of yoga for the improvement of well-being and resilience to stress in the workplace. Scand. J. Work Environ. Health. 1795990X. 37 (1), 70–76. Available from: http://www.sjweh.fi/sjwehfile/files/year2011/j_11_1/70_hartfiel.pdf. doi:10.5271/sjweh.2916.

Hatha Yoga Pradipika. Available from: http://chrome-extension://efaidnbmnnnibpcajpcglclefindmkaj/https://ia601301.us.archive.org/33/items/HathaYogaPradipika-SanskritTextWithEnglishTranslatlionAndNotes/HathaYogaPradipika-SanskritTextWithEnglishTranslatlionAndNotes.pdf.

Holtzman, S., Beggs, R.T., 2013. Yoga for chronic low back pain: analysis of randomized controlled trials. Pain Res. Manag. 12036765. 18 (5), 267–272. Available from: http://www.pulsus.com/journals/Login.jsp?sCurrPg=Log%20in&theAction=. doi:10.1155/2013/105919.

Hunter, S.D., Dhindsa, M.S., Cunningham, E., Tarumi, T., Alkatan, M., Nualnim, N., Tanaka, H., 2013. The effect of bikram yoga on arterial stiffness in young and older adults. J. Altern. Complement. Med. 15577708. 19 (12), 930–934. doi:10.1089/acm.2012.0709.

Iyengar, B.K.S., 1979. Light on Yoga. Schocken Books, New York.

Jerath, R., Edry, J.W., Barnes, V.A., Jerath, V., 2006. Physiology of long pranayamic breathing: neural respiratory elements may provide a mechanism that explains how slow deep breathing shifts the autonomic nervous system. Med Hypotheses. 03069877. 67 (3), 566–571. doi:10.1016/j.mehy.2006.02.042.

Kamei, T., Toriumi, Y., Kimura, H., Kumano, H., Ohno, S., Kimura, K., 2000. Decrease in serum cortisol during yoga exercise is correlated with alpha wave activation. Percept. Mot. Skills. 0031-5125. 90 (3), 1027–1032. doi:10.2466/pms.2000.90.3.1027.

Kessler, H.S., Sisson, S.B., Short, K.R., 2012. The potential for high-intensity interval training to reduce cardiometabolic disease risk. Sports Med. 11792035. 42 (6), 489–509. doi:10.2165/11630910-000000000-00000.

Kodidala, S.R., Hans, P.S.R., Sorout, J., Soni, H., 2022. Impact of suryanamaskar training on cardiovascular, respiratory, and cognitive functions among medical students. Avicenna J. Neuro Psycho Physiol. 1–4. doi:10.32592/ajnpp.2022.9.1.100.

Kodidala, S.R., Pradhan, B., 2021. Effect of Surya Namaskara on cognitive functions in school children: a randomized controlled trial. Complement. Ther. Med. 56.

Koohsari, M.J., Sugiyama, T., Sahlqvist, S., Mavoa, S., Hadgraft, N., Owen, N., 2015. Neighborhood environmental attributes and adults' sedentary behaviors: review and research agenda. Prev. Med. 00917435. 77, 141–149. doi:10.1016/j.ypmed.2015.05.027.

Larsen, R.N., Kingwell, B.A., Sethi, P., Cerin, E., Owen, N., Dunstan, D.W., 2014. Breaking up prolonged sitting reduces resting blood pressure in overweight/obese adults. Nutr. Metab. Cardiovasc. Dis. 09394753. 24 (9), 976–982. doi:10.1016/j.numecd.2014.04.011.

Laurent, S., Cockcroft, J., Van Bortel, L., Boutouyrie, P., Giannattasio, C., Hayoz, D., Pannier, B., Vlachopoulos, C., Wilkinson, I., Struijker-Boudier, H., 2006. Expert consensus document on arterial stiffness: methodological issues and clinical applications. Eur. Heart J. 0195-668X. 27 (21), 2588–2605. doi:10.1093/eurheartj/ehl254.

Lawshe, C.H., 1975. A quantitative approach to content validity 1. Pers. Psychol. 0031-5826. 28 (4), 563–575. doi:10.1111/j.1744-6570.1975.tb01393.x.

Lazar, S.W., Kerr, C.E., Wasserman, R.H., Gray, J.R., Greve, D.N., Treadway, M.T., McGarvey, M., Quinn, B.T., Dusek, J.A., Benson, H., Rauch, S.L., Moore, C.I., Fischl, B., 2005. Meditation experience is associated with increased cortical thickness. NeuroReport. 09594965. 16 (17), 1893–1897. doi:10.1097/01.wnr.0000186598.66243.19.

Liang, Z.-D., Zhang, M., Wang, C.-Z., Yuan, Y., Liang, J.-H., 2022. Association between sedentary behavior, physical activity, and cardiovascular disease-related outcomes in adults—A meta-analysis and systematic review. Front. Public Health. 2296-2565. 10. doi:10.3389/fpubh.2022.1018460.

Lis, A.M., Black, K.M., Korn, H., Nordin, M., 2007. Association between sitting and occupational LBP. Eur. Spine J. 09406719. 16 (2), 283–298. doi:10.1007/s00586-006-0143-7.

Luders, E., Toga, A.W., Lepore, N., Gaser, C., 2009. The underlying anatomical correlates of long-term meditation: larger hippocampal and frontal volumes of gray matter. NeuroImage. 10538119. 45 (3), 672–678. doi:10.1016/j.neuroimage.2008.12.061.

Manninson, J., Bouchard, S.V., 2004. The Gherada Samhita. Yoga Vidya. ISBN: 0-971644-2-7.

Mcewen, B.S., Gianaros, P.J., 2010. Central role of the brain in stress and adaptation: links to socioeconomic status, health, and disease. Ann. N.Y. Acad. Sci. 17496632. 1186, 190–222. Available from: http://onlinelibrary.wiley.com/journal/10.1111/(ISSN)1749-6632. doi:10.1111/j.1749-6632.2009.05331.x.

Mehta, R.K., Agnew, M.J., 2012. Influence of mental workload on muscle endurance, fatigue, and recovery during intermittent static work. Eur. J. Appl. Physiol. 14396319. 112 (8), 2891–2902. doi:10.1007/s00421-011-2264-x.

Mooventhan, A., Nivethitha, L., 2014. Scientific evidence-based effects of hydrotherapy on various systems of the body. N. Am. J. Med. Sci. 19472714. 6 (5), 199–209. Available from: http://www.najms.org/temp/NorthAmJMedSci65199-1214094_032220.pdf. doi:10.4103/1947-2714.132935.

Morishima, T., Restaino, R.M., Walsh, L.K., Kanaley, J.A., Fadel, P.J., Padilla, J., 2016. Prolonged sitting-induced leg endothelial dysfunction is prevented by fidgeting. Am. J. Physiol. Heart Circ. Physiol. 0363-6135. 311 (1), H177. doi:10.1152/ajpheart.00297.2016.

Nambi, G.S., Shanmugananth, Devi, S., Inbasekaran, D., Jagannathan, K., Khuman, R., 2014. Changes in pain intensity and health related quality of life with Iyengar yoga in nonspecific chronic low back pain: a randomized controlled study. Int. J. Yoga. 0973-6131. 7 (1), 48. doi:10.4103/0973-6131.123481.

Pandey, A., Salahuddin, U., Garg, S., Ayers, C., Kulinski, J., Anand, V., Mayo, H., Kumbhani, D.J., De Lemos, J., Berry, J.D., 2016. Continuous dose-response association between sedentary time and risk for cardiovascular disease a meta-analysis. JAMA Cardiol. 23806591. 1 (5), 575–583. Available from: http://jamanetwork.com/journals/jamacardiology. doi:10.1001/jamacardio.2016.1567.

Parajuli, N., Pradhan, B., Bapat, S., 2022. Effect of yoga on cognitive functions and anxietyamong female school children with low academic performance: a randomized control trial. Complement. Ther. Clin. Pract. 48, 101614. doi:10.1016/j.ctcp.2022.101614.

Parajuli, N., Shindhe, M., Pradhan, B., 2019. Effect of Surya Namaskara on mental health, self-control and mindfulness of adolescent school children. Res Sq. doi:10.21203/rs.3.rs-955629/v1.

Park, C.L., Quinker, D., Dobos, G., 2020. Motivations for adopting and maintaining a yoga practice: a comparison of beginning and experienced practitioners. Mindfulness (NY) 11 (5), 1090–1096.

Parks, K.M., Steelman, L.A., 2008. Organizational wellness programs: a meta-analysis. J. Occup. Health Psychol. 10768998. 13 (1), 58–68. doi:10.1037/1076-8998.13.1.58.

Pascoe, M.C., Thompson, D.R., Ski, C.F., 2017. Yoga, mindfulness-based stress reduction and stress-related physiological measures: a meta-analysis. Psychoneuroendocrinology. 18733360. 86, 152–168. Available from: www.elsevier.com/locate/psyneuen. doi:10.1016/j.psyneuen.2017.08.008.

Patil, S.G., 2014. Effect of yoga on oxidative stress in elderly with grade-I hypertension: a randomized controlled study. J. Clin. Diagn. Res. 2249782X. 8 (7). doi:10.7860/JCDR/2014/9498.4586.

Patil, S.G., Aithala, M.R., Das, K.K., 2015. Effect of yoga on arterial stiffness in elderly subjects with increased pulse pressure: a randomized controlled study. Complement. Ther. Med. 18736963. 23 (4), 562–569. Available from: http://www.elsevier.com/wps/find/journaldescription.cws_home/623020/authorinstructions. doi:10.1016/j.ctim.2015.06.002.

Patterson, R., McNamara, E., Tainio, M., de Sá, T.H., Smith, A.D., Sharp, S.J., Edwards, P., Woodcock, J., Brage, S., Wijndaele, K., 2018. Sedentary behaviour and risk of all-cause, cardiovascular and cancer mortality, and incident type 2 diabetes: a systematic review and dose response meta-analysis. Eur. J. Epidemiol. 0393-2990. 33 (9), 811–829. doi:10.1007/s10654-018-0380-1.

Pradhan, B., Nagendra, H.R., 2008. Immediate effect of two breathing techniques on cognitive functions in healthy individuals. Indian J. Physiol. Pharmacol. 52 (4), 393–396.

Prince, S.A., Saunders, T.J., Gresty, K., Reid, R.D., 2014. A comparison of the effectiveness of physical activity and sedentary behaviour interventions in reducing sedentary time in adults: a systematic review and meta-analysis of controlled trials. Obes Rev. 1467789X. 15 (11), 905–919. Available from: http://onlinelibrary.wiley.com/journal/10.1111/(ISSN)1467-789X. doi:10.1111/obr.12215Canada.

Pronk, N.P., Katz, A.S., Lowry, M., Payfer, J.R., 2012. Reducing occupational sitting time and improving worker health: the take-a-stand project, 2011. Prev. Chronic Dis. 21665435. 9 (10). Available from: http://www.cdc.gov/pcd/issues/2012/pdf/11_0323.pdf. doi:10.5888/pcd9.110323.

Ray, U.S., Mukhopadhyaya, S., Purkayastha, S.S., Asnani, V., Tomer, O.S., Prashad, R., Thakur, L., Selvamurthy, W., 2001. Effect of yogic exercises on physical and mental health of young fellowship course trainees. Indian J. Physiol. Pharmacol. 45 (1), 37–53.

Ross, A., Thomas, S., 2010. The health benefits of yoga and exercise: a review of comparison studies. J. Altern. Complement. Med. 1075-5535. 16 (1), 3–12. doi:10.1089/acm.2009.0044.

Sarang, S.P., Telles, S., 2008. Immediate effect of two yoga-based relaxation techniques on attention in children. Int. J. Yoga 1 (2), 77–82.

Sengupta, P., Chaudhuri, P., Bhattacharya, K., 2013. Effect of Suryanamaskar on stress and physical fitness in college students. Int. J. Yoga 6 (1), 64–67.

Sexton, C.E., Betts, J.F., Demnitz, N., Dawes, H., Ebmeier, K.P., Johansen-Berg, H., 2016. A systematic review of MRI studies examining the relationship between physical fitness and activity and the white matter of the ageing brain. NeuroImage. 10959572. 131, 81–90. Available from: http://www.elsevier.com/inca/publications/store/6/2/2/9/2/5/index.htt. doi:10.1016/j.neuroimage.2015.09.071.

Shariat, A., Cleland, J.A., Danaee, M., Kargarfard, M., Sangelaji, B., Tamrin, S.B.M., 2018. Effects of stretching exercise training and ergonomic modifications on musculoskeletal discomforts of office workers: a randomized controlled trial. Braz. J. Phys. Ther. 18099246. 22 (2), 144–153. Available from: https://www.journals.elsevier.com/brazilian-journal-of-physical-therapy. doi:10.1016/j.bjpt.2017.09.003.

Sharma, A., Barrett, M.S., Cucchiara, A.J., Gooneratne, N.S., Thase, M.E., 2017. A breathing-based meditation intervention for patients with major depressive disorder following inadequate response to antidepressants: a randomized pilot study. J. Clin. Psychiatry. 01606689. 78 (1), e59. Available from: http://www.psychiatrist.com/JCP/article/_layouts/ppp.psych.controls/BinaryViewer.ashx?Article=/JCP/article/Pages/2017/v78n01/v78n0107.aspx&Type=Article. doi:10.4088/JCP.16m10819.

Sharma, R., Gupta, N., Bijlani, R.L., 2008. Effect of yoga based lifestyle intervention on subjective well-being. Indian J. Physiol. Pharmacol. 00195499. 52 (2), 123–131.

Sharma, V., Trakroo, M., Subramaniam, V., Sahai, A., AnandB Bhavanani, M.R., 2013. Effect of fast and slow pranayama on perceived stress and cardiovascular parameters in young health-care students. Int. J. Yoga. 0973-6131. 6 (2), 104. doi:10.4103/0973-6131.113400.

Sherman, K.J., Cherkin, D.C., Erro, J., Miglioretti, D.L., Deyo, R.A., 2005. Comparing yoga, exercise, and a self-care book for chronic low back pain: a randomized, controlled trial. Ann. Intern. Med. 00034819. 143 (12), 849–856. Available from: http://annals.org/issues.aspx. doi:10.7326/0003-4819-143-12-200512200-00003.

Shrestha, N., Kukkonen-Harjula, K.T., Verbeek, J.H., Ijaz, S., Hermans, V., Pedisic, Z., 2018. Workplace interventions for reducing sitting at work. Cochrane Database Syst. Rev. 14651858. 2018 (6). Available from: https://www.cochranelibrary.com/cdsr/table-of-contents. doi:10.1002/14651858.CD010912.pub4.

Sorosky, S., Stilp, S., Akuthota, V., 2008. Yoga and pilates in the management of low back pain. Curr. Rev. Musculoskelet Med. 1935-9748. 1 (1), 39–47. doi:10.1007/s12178-007-9004-1.

Stec, K., Kruszewski, M., Ciechanowski, L., 2023. Effects of Suryanamaskar, an intensive yoga exercise routine, on the stress levels and emotional intelligence of Indian students. Int. J. Environ. Res. Public Health 20 (4), 2845. doi:10.3390/ijerph20042845.

Sukumar, S., 2023. Development and Validation of Yoga Module to Improve the Central and Peripheral Vascular Function and Cognitive Functions among the Desk-based Workers. Harvard Dataverse. doi:10.7910/DVN/TH1YZJ.

Tekur, P., Nagarathna, R., Chametcha, S., Alex Hankey, H.R.N., 2012. A comprehensive yoga programs improves pain, anxiety and depression in chronic low back pain patients more than exercise: an RCT. Complement. Ther. Med. 09652299. 20 (3), 107–118. doi:10.1016/j.ctim.2011.12.009.

The Yoga Upanishads, 1938. The Adyar Library, Madras.

Thijssen, D.H.J., Black, M.A., Pyke, K.E., Padilla, J., Atkinson, G., Harris, R.A., Parker, B., Widlansky, M.E., Tschakovsky, M.E., Green, D.J., 2011. Assessment of flow-mediated dilation in humans: a methodological and physiological guideline. Am. J. Physiol. Heart Circ. Physiol. 15221539. 300 (1), H2. Available from: http://ajpheart.physiology.org/content/300/1/H2.full.pdf+html. doi:10.1152/ajpheart.00471.2010.

Thosar, S.S., Bielko, S.L., Mather, K.J., Johnston, J.D., Wallace, J.P., 2015. Effect of prolonged sitting and breaks in sitting time on endothelial function. Med. Sci. Sports Exerc. 15300315. 47 (4), 843–849. Available from: http://www.lww.com/product/?0195-9131. doi:10.1249/MSS.0000000000000479.

Thosar, S.S., Bielko, S.L., Wiggins, C.C., Wallace, J.P., 2014. Differences in brachial and femoral artery responses to prolonged sitting. Cardiovasc. Ultrasound. 14767120. 12 (1). Available from: http://www.cardiovascularultrasound.com/home/. doi:10.1186/1476-7120-12-50.

Torbeyns, T., Bailey, S., Bos, I., Meeusen, R., 2014. Active workstations to fight sedentary behaviour. Sports Med. 0112-1642. 44 (9), 1261–1273. doi:10.1007/s40279-014-0202-x.

Vancampfort, D., Stubbs, B., Sienaert, P., 2015. What are the factors associated with physical activity participation in individuals with depression? A systematic review of physical activity correlates. J. Affect Disord 176, 139–148.

Vempati, R.P., Telles, S., 2002. Yoga-based guided relaxation reduces sympathetic activity judged from baseline levels. Psychol. Rep. 0033-2941. 90 (2), 487–494. doi:10.2466/pr0.2002.90.2.487.

Villemure, C., Čeko, M., Cotton, V.A., Bushnell, M.C., 2014. Insular cortex mediates increased pain tolerance in yoga practitioners. Cereb Cortex. 14602199. 24 (10), 2732–2740. Available from: http://cercor.oxfordjournals.org/. doi:10.1093/cercor/bht124.

Vlachopoulos, C., Aznaouridis, K., Stefanadis, C., 2010. Prediction of cardiovascular events and all-cause mortality with arterial stiffness. a systematic review and meta-analysis. J. Am. Coll. Cardiol. 07351097. 55 (13), 1318–1327. doi:10.1016/j.jacc.2009.10.061.

Voss, M.W., Carr, L.J., Clark, R., Weng, T., 2014. Revenge of the "sit" II: does lifestyle impact neuronal and cognitive health through distinct mechanisms associated with sedentary behavior and physical activity? Ment. Health Phys. Act. 18780199. 7 (1), 9–24. Available from: http://www.elsevier.com. doi:10.1016/j.mhpa.2014.01.001.

Voss, M.W., Heo, S., Prakash, R.S., Erickson, K.I., Alves, H., Chaddock, L., Szabo, A.N., Mailey, E.L., Wójcicki, T.R., White, S.M., Gothe, N., Mcauley, E., Sutton, B.P., Kramer, A.F., 2013. The influence of aerobic fitness on cerebral white matter integrity and cognitive function in older adults: results of a one-year exercise intervention. Hum. Brain Mapp. 10970193. 34 (11), 2972–2985. doi:10.1002/hbm.22119.

Wieland, L.S., Skoetz, N., Pilkington, K., Vempati, R., D'Adamo, C.R., Berman, B.M., 2017. Yoga treatment for chronic non-specific low back pain. Cochrane Database Syst. Rev. 14651858. 2017 (1). Available from: https://www.cochranelibrary.com/cdsr/table-of-contents. doi:10.1002/14651858.CD010671.pub2.

Williams, K., Abildso, C., Steinberg, L., Doyle, E., Epstein, B., Smith, D., Hobbs, G., Gross, R., Kelley, G., Cooper, L., 2009. Evaluation of the effectiveness and efficacy of iyengar yoga therapy on chronic low back pain. Spine. 0362-2436. 34 (19), 2066–2076. doi:10.1097/brs.0b013e3181b315cc.

Wolever, R.Q., Bobinet, K.J., McCabe, K., Mackenzie, E.R., Fekete, E., Kusnick, C.A., Baime, M., 2012. Effective and viable mind-body stress reduction in the workplace: a randomized controlled trial. J. Occup. Health Psychol. 19391307. 17 (2), 246–258. doi:10.1037/a0027278.

Yadav, R.K., Mehta, M.D., 2011. Clinical evaluation of selected Yogic procedures in individuals with low back pain. Ancient Sci. Life 31 (1), 12–18.

Yadav, R., Yadav, R.K., Khadgawat, R., 2019. Beneficial effects of yoga practice in diabetes mellitus: a systematic review. Diabetes Metab. J. 43 (2), 173–184.

Young, D.R., Hivert, M.F., Alhassan, S., Camhi, S.M., Ferguson, J.F., Katzmarzyk, P.T., Lewis, C.E., Owen, N., Perry, C.K., Siddique, J., Yong, C.M., 2016. Sedentary behavior and cardiovascular morbidity and mortality: a science advisory from the American Heart Association. Circulation. 15244539. 134 (13), e262. Available from: http://circ.ahajournals.org. doi:10.1161/CIR.0000000000000440.

Chapter 20

Impact of Y-break yoga and pranayama on central and peripheral vascular and cognitive performance among the desk-based workers using ultrasound

Koustubh Kamath[a], Poovitha Shruthi Paramashiva[b], Leena R. David[a,c], Rajagopal Kadavigere[d], K. Vaishali[e], Berna Uzun[f], Dilber Uzun Ozsahin[c,f,g], Baskaran Chandrasekaran[h], Sneha Ravichandran[a], Suresh Sukumar[a], Ilker Ozsahin[f], and Dilip Shettigar[a]

[a]*Department of Medical Imaging Technology, Manipal College of Health Professions, Manipal Academy of Higher Education, Manipal, Karnataka, India*

[b]*Division of Yoga, Manipal Academy of Higher Education, Centre for Integrative Medicine and Research, Manipal, India*

[c]*Medical Diagnostic Imaging Department, College of Health Sciences, University of Sharjah, Sharjah, United Arab Emirates*

[d]*Department of Radiodiagnosis and Imaging, Kasturba Medical College, Manipal, India*

[e]*Department of Physiotherapy, Manipal Academy of Higher Education, Manipal College of Health Professions, Manipal, India*

[f]*Irfan Suat Gunsel Operational Research Institute, Near East University, Nicosia, Turkish Republic of Northern Cyprus (TRNC), Turkey*

[g]*Research Institute for Medical and Health Sciences, University of Sharjah, Sharjah, United Arab Emirates*

[h]*Symbiosis School of Sports Sciences, Symbiosis International (Deemed University), Pune, India*

1 Introduction

A variety of noninvasive medical imaging techniques, such as X-ray radiography, Doppler ultrasound, computed tomography, and magnetic resonance imaging, are widely used to visualize internal anatomical structures and assess physiological functions. These modalities play a crucial role in diagnosing and monitoring a broad spectrum of diseases by providing detailed structural and functional information. Among them, Doppler ultrasound has gained prominence in recent years for evaluating vascular health, including both central and peripheral circulation, due to its safety, accessibility, and real-time imaging capabilities. A Doppler ultrasound is a noninvasive test that can be used to estimate the blood flow through your blood vessels by bouncing high-frequency sound waves (ultrasound) off circulating red blood cells. Doppler scanning of the carotid arteries is a standard diagnostic test for assessing the severity of carotid arterial occlusive disease. The high accuracy of this noninvasive test, compared with digital subtraction angiography and magnetic resonance angiography, has made duplex scanning the main imaging method for screening, preoperative assessment, and follow-up. Diagnosis is based on one or a combination of measurements or observations in the region of a stenosis, including peak systolic velocity (PSV), end-diastolic velocity, the ratio of PSV in the internal carotid artery to common carotid peak velocity, detection of disturbed flow (spectral broadening), and severity of plaque as assessed on the B-mode image (Alexandrov et al., 1997; Carpenter et al., 1996; Moneta et al., 1995).

1.1 Sedentary behavior and vascular function

Excessive sitting in the workplace is associated with increased obesity, altered blood glucose levels, and cardiac disease risk. In recent years, office workers have spent one-third of their time sitting as a result of industrialization and the subsequent computerization of labor. In school and office environments, uninterrupted sitting is linked to increased sedentary time,

Advances in Medical Imaging. https://doi.org/10.1016/B978-0-443-28967-5.00019-8

cardiovascular disease risk, and poor cognitive function (Huang et al., 2019; Wang et al., 2018). Persuasive technology interventions designed to reduce sedentary behavior during occupational sitting have shown promise for improving workplace health culture. Based on a theoretical framework of self-determination and reasoned action, persuasive computer-based applications can prompt and provide real-time feedback for the targeted behavior (Spagnolli et al., 2016). Though convincing evidence behind persuasive computer prompts on sedentary behavior exists, their influence on vascular function and cognitive function associated with desk-based work is still lacking (Carter et al., 2018). Recent evidence suggests that uninterrupted prolonged sitting reduces carotid and superficial femoral artery (SFA) blood flow, which may reduce blood supply to the brain's temporal and frontal areas that are important for memory and cognitive functions (Thosar et al., 2015). In sedentary people, it is widely known that moderate to vigorous physical activity enhances endothelial and vascular function. Though periodic movement breaks can improve the above effects of prolonged sitting, the evidence regarding the intensity of these breaks is still under debate (Pearson and Smart, 2017). Yoga may be as beneficial as or better than aerobic exercise in improving a range of health-related outcome measures in both healthy and ill populations, according to Ross and Thomas et al. Although yoga has been widely employed as an adjunct therapy for ill patients and physical symptoms, little is known about its immediate and long-term effects on cognitive and vascular function (Ross and Thomas, 2010).

1.2 Yoga and vascular imaging

Yoga is an ancient spiritual practice that has been followed in the Eastern regions of the world for over 5000 years and is increasingly gaining popularity in the United States as a means to improve overall health (Thosar et al., 2015). However, limited research has examined its effects on vascular function. Evidence suggests that 12 weeks of stretching can increase carotid artery compliance in previously sedentary individuals (Pearson and Smart, 2017). Furthermore, the isometric contractions associated with yoga postures may affect vascular function. It has been discovered that isometric exercise training lowers blood pressure and improves endothelial function (Ross and Thomas, 2010; Barnes, Bloom and Nahin, 2009).

Meditation is also a component of yoga that has been demonstrated to be a stress-reduction technique (Cortez-Cooper et al., 2008) that can improve vascular function (McGowan et al., 2007), most likely by lowering sympathetic vasoconstrictor tone. Yoga, when taken as a whole, appears to be an excellent intervention technique for improving vascular function in the sedentary population. However, there is no experimental evidence to back up this theory.

In light of these gaps, this study seeks to evaluate the short-term impact of Y-Break yoga on both central and peripheral vascular function, as well as cognitive performance among desk-based workers, using Doppler ultrasound as a primary assessment tool.

2 Methods

This study was a within-subjects, prospective, single-center trial conducted in the Department of Radio-Diagnosis and Imaging, Kasturba Medical Hospital, Manipal, India, from September 2022 to January 2023. Ethical approval was obtained from the Institutional Ethics Committee—(IEC1: 108/2022), and the study was registered with the Clinical Trial Registry of India (CTRI/2022/9/045628). Informed consent was obtained from all participants prior to enrollment. A total of 17 healthy adult desk-based workers (nine males and eight females; mean age: 25 years, age range: 25–38) were recruited. All participants were free from cardiovascular or metabolic diseases, were not taking any medications that could influence vascular function, and none of the female participants were menstruating during the study period. Physical activity levels were assessed using the International Physical Activity Questionnaire (IPAQ), which evaluates time spent on vigorous and moderate activities, walking, and sitting over the past seven days. The sample included participants with moderate ($n = 11$), high ($n = 5$), and low ($n = 1$) levels of physical activity. The IPAQ data can be seen in Table 20.1.

3 Study procedure

Approval for the study was obtained from the Institutional Research Committee of Manipal College of Health Professions and the Institutional Ethics Committee of Kasturba Hospital, Manipal. The study involved the prospective collection of participant data, with informed consent obtained from each individual prior to participation.

Participants were included based on their engagement in office-based, desk-bound occupations and were familiarized with the study protocol through a simulated desk-work session. A counterbalanced crossover design was used, in which each participant underwent three different interventions (Sitting, Pranayama, and Yoga) on separate days, each lasting four hours. A standardized breakfast was provided to all participants to control for potential dietary effects, and all sessions were conducted in the morning between 7:00 a.m. and 9:00 a.m. to reduce circadian variability.

TABLE 20.1 IPAQ results and BMI of the participants.

Height (cm)	Weight (kg)	BMI	IPAQ
167.64	60	21.41	Moderate
172.00	75	25.35	High
179.83	70	21.65	Moderate
167.64	76	27.04	High
188.98	92	25.76	High
189.00	75	21.00	Moderate
187.90	70	19.83	Moderate
189.00	75	21.00	Moderate
167.64	50	17.84	Low
172.00	60	20.28	Moderate
167.64	50	17.93	Moderate
161.00	75	28.93	High
167.64	58	20.70	Moderate
172.00	70	23.66	Moderate
169.00	72	25.21	High
170.00	60	20.76	Moderate
175.00	75	24.49	Moderate

3.1 Interventions

The Sitting (control) condition required participants to continue their desk work without any structured breaks, except for restroom use. In the Pranayama condition, participants performed specific breathing techniques (Nadisodhana, Bhramari Pranayama, and Dhyana) for approximately 5–10 minutes every hour. In the Yoga condition, participants engaged in a series of physical postures (Tadasana, Urdhva Mandukasana, Skandha Chakra, Uttana Mandukasana, Ardha Chakrasana, Kati Chakrasana, and others), combined with deep breathing, also repeated every hour for four hours. All interventions were administered under the supervision of experts. Fig. 20.1 shows a schematic representation of the study.

3.2 Doppler imaging

Physiological assessments were conducted using Doppler ultrasound imaging. Measurements of the common carotid artery (CCA) and SFA were obtained at baseline 0 h, after 2 h, and after 4 h during each intervention. The parameters measured included CCA diameter (in cm), CCA PSV (in cm/s), SFA diameter (in cm), and SFA PSV (in cm/s). In addition to diameter and velocity, shear rate and shear stress of the CCA were also calculated using standard hemodynamic equations. Shear rate was calculated as four times the mean velocity divided by the vessel diameter, and shear stress was estimated by multiplying shear rate by blood viscosity (assumed constant across participants). All ultrasound scans were performed by a trained radiologist using a high-frequency linear transducer, with participants in a relaxed, seated position.

3.3 Executive functions

Executive functions will be measured using the Eriksen Flanker test and the N-Back memory test. Data will be collected at 0, 1, 2, 3, and 4 hours. Testing will be administered through the Deary-Liewald software. The Eriksen Flanker test will be used to measure reaction time and accuracy for congruent and incongruent stimuli. The 1-back and 2-back memory tests will require participants to identify the last one or two words or numbers displayed. Reaction times and accuracy for all stimuli will be recorded.

3.3.1 The modified Eriksen Flanker task

The modified Eriksen Flanker task, based on the visual oddball paradigm, is designed to measure response inhibition and reaction time. The task involves 600 randomly ordered trials, each presenting an array of five letters for 500 ms

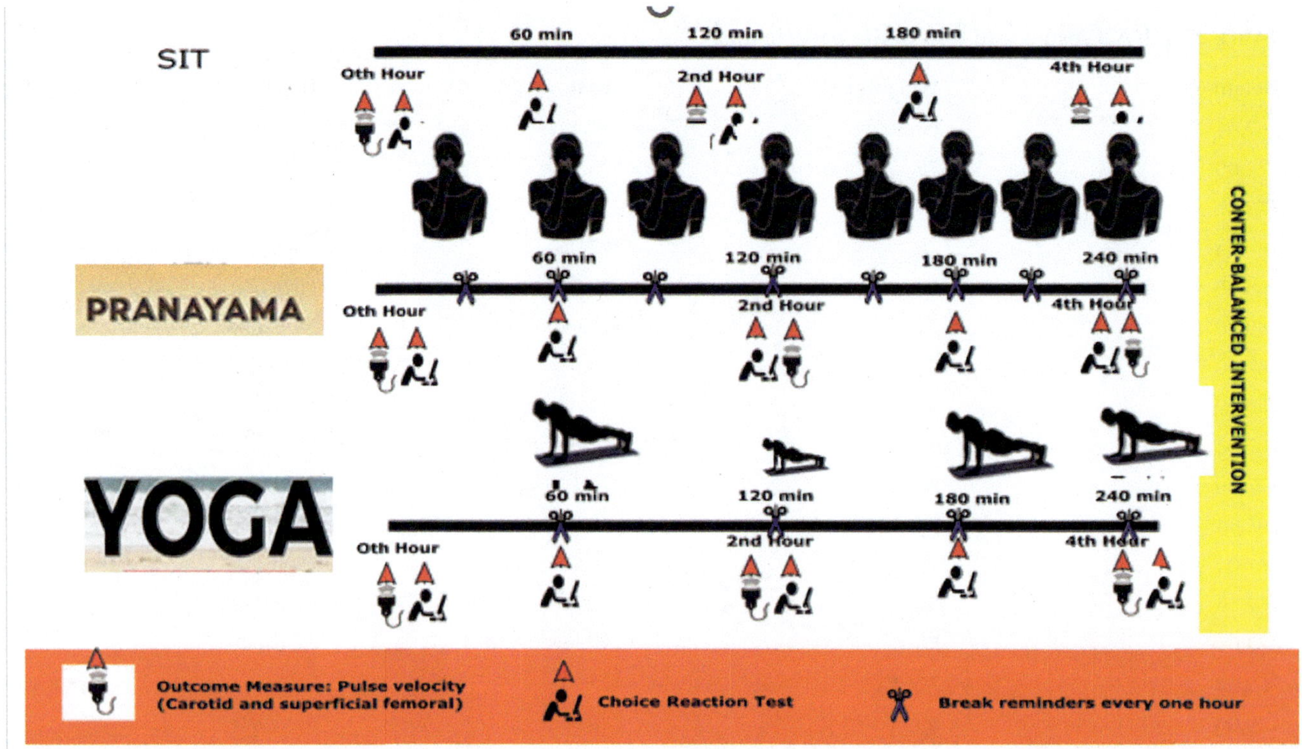

FIG. 20.1 Schematic representation of the study protocol illustrating the sequence and timing of three intervention conditions: sitting, pranayama, and yoga over a four-hour period. Time points indicate measurement intervals for pulse wave velocity (carotid and superficial femoral artery) and choice reaction tests, with hourly break reminders incorporated as part of the counterbalanced intervention design.

H H H H H H H → congruent

K K K H K K K → response compatible

S S S H S S S → response incompatible

FIG. 20.2 Stimulus conditions in the modified Eriksen Flanker task, showing congruent, response-compatible, and response-incompatible trials. *(From Classic experiments explained, n.d. https://support.gorilla.sc/support/educational-resources/classic-psychology-tasks/Flanker-task.)*

(milliseconds), with an interstimulus interval of 2000 ms. As illustrated in Fig. 20.2, the central target letter can appear in three conditions: congruent (e.g., HHHHH), response-compatible (e.g., KKHKK), or response-incompatible (e.g., SS H SS). Two sets of randomized trials are administered under equiprobable congruent and incongruent conditions. Stimuli consist of 3-inch-tall white letters presented against a black background on a 14-inch HP computer monitor (Windows 2016).

Participants will be seated approximately 2 feet from the monitor, with their feet placed flat on the ground. They will be instructed to focus on the central target letter while ignoring the surrounding flanker letters. Responses will be made by pressing joystick keys (up or down) according to the identity of the central target letter.

Throughout the experiment, reaction times and amplitudes for correct and incorrect responses will be recorded. Inference scores will be calculated from correct responses under both congruent and incongruent conditions, providing measures of cognitive control, selective attention, and response inhibition.

3.3.2 N-back test

The N-back test, introduced by Wayne Kirchner in 1958, serves as a reliable measure of working memory capacity, often employed in cognitive neuroscience and psychological research to assess an individual's ability to retain and manipulate

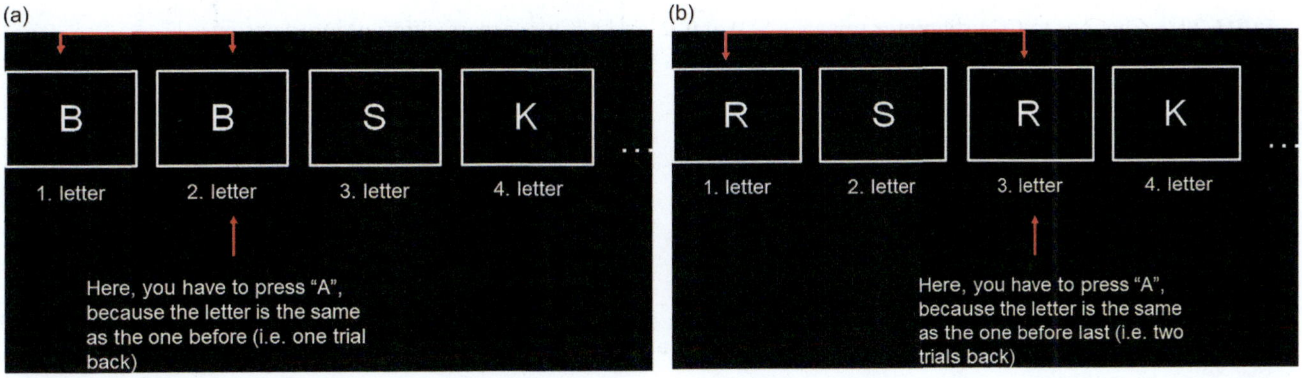

FIG. 20.3 Illustration of the N-back task: (A) 1-back, (B) 2-back. *(From N-BACK TASK, Millisecond Software, 2012. https://www.millisecond.com/download/library/nback.)*

information over short periods. In this task, stimuli are presented sequentially, and participants must determine whether the current stimulus matches one that appeared *N* steps earlier in the sequence. The test is commonly conducted in 1-back and 2-back conditions.

In the 1-back test, participants are required to indicate when the current stimulus matches the one presented immediately before it in the sequence, as illustrated in Fig. 20.3. In contrast, the 2-back test requires participants to respond when the current stimulus matches the one presented two steps earlier in the sequence. This increasing level of difficulty helps assess memory retention, attention, and cognitive flexibility.

The structure of the 1-back and 2-back tests is visually represented, illustrating how participants process and respond to sequential stimuli while engaging their working memory.

All outcome measures, including CCA and SFA diameter and velocity, CCA shear rate and shear stress, as well as cognitive performance indicators such as reaction time and accuracy from the Flanker and N-back tasks, were evaluated to assess the acute physiological and cognitive effects of sitting, pranayama, and yoga interventions among desk-based workers.

4 Results

The study included 17 participants aged between 25 and 38 years. Participants comprised nine males (mean age 25, age range 25–38) and eight females (mean age 25, age range 25–38). The participants had no history of cardiovascular or metabolic disorders; the females were not menstruating, none had depression during the study period, and none were taking medication that might alter vascular function. Self-reported physical activity was assessed using the IPAQ. The questionnaire assesses time (frequency and duration) spent on vigorous and moderate-intensity activities, walking activities, and sitting time in hours for the past seven days.

4.1 Common carotid artery diameter

In CCA, the results show a reduction in the diameter of the artery at each intervention over time, as mentioned in Table 20.2. In comparison over time, the first and second hours exhibit statistical significance with a *P* value of (*P* value = <.001), whereas the third and fourth hours do not exhibit statistical significance (*P* value = .020). Although the Yoga intervention shows significance, it does not show statistical significance when compared to pranayama and group prolonged sitting (*P* value = .014). In addition, the Yoga intervention also does not show statistical significance when compared to the pranayama intervention and prolonged sitting (*P* value = .017).

4.2 CCA velocity

In the CCA, a reduction in velocity was observed following the pranayama and yoga intervention, indicating a potential enhancement in vascular function and blood flow regulation. However, during prolonged sitting, there was a decline in velocity between the 0th and 2nd hour, followed by a subsequent increase in velocity during the final hour, as detailed in Table 20.3. Despite these observed variations, statistical analysis suggests that neither time nor velocity changes in the

TABLE 20.2 Changes in CCA diameter (mean ± SD) at 0, 2, and 4 hours across different interventions.

CCA diameter			
Intervention	0 h (Mean) ± SD	2 h (Mean) ± SD	4 h (Mean) ± SD
Sitting intervention	0.670 ± 0.062	0.579 ± 0.059	0.537 ± 0.067
Pranayama intervention	0.636 ± 0.077	0.585 ± 0.059	0.577 ± 0.054
Yoga intervention	0.606 ± 0.067	0.639 ± 0.115	0.726 ± 0.111

TABLE 20.3 Changes in CCA blood velocity (mean ± SD) at baseline, 2 h, and 4 h across different interventions.

CCA blood velocity			
Intervention	0 h (Mean) ± SD	2 h (Mean) ± SD	4 h (Mean) ± SD
Sitting intervention	116.365 ± 19.713	123.176 ± 11.485	130.206 ± 21.310
Pranayama intervention	124.941 ± 13.797	109.953 ± 25.810	107.482 ± 25.806
Yoga intervention	127.294 ± 28.458	115.524 ± 25.495	107.506 ± 17.020

CCA are significant, indicating that these fluctuations may not hold strong clinical relevance or may be influenced by individual variability rather than a consistent physiological effect.

4.3 SFA diameter

In the SFA, a decrease in diameter was observed during each hour of prolonged sitting and the pranayama intervention, indicating a consistent reduction in arterial size over time. However, during the yoga intervention, there was a reduction in diameter at the 0th and 2nd hour, followed by an increase in diameter at the 4th hour, suggesting a delayed vascular response. Despite these variations, statistical analysis, as referenced in Table 20.4, indicates that neither time nor group differences reached significance, implying that the observed changes may not be substantial enough to draw definitive conclusions regarding their physiological impact.

4.4 SFA velocity

In the SFA, velocity showed a decrease across all interventions, as reported in Table 20.5. When comparing time points, a significant difference was observed between the 0th and 2nd hours ($P = .026$), and another significant change was noted between the 0th and 4th hours ($P = .004$). However, despite these time-based differences, the overall analysis indicates that the group did not exhibit a substantial response to any specific intervention, suggesting that while short-term fluctuations in velocity occurred, they were not strong enough to indicate a meaningful physiological impact at the group level.

4.5 CCA shear and stress

In the CCA, the shear rate increased progressively during prolonged sitting, whereas it decreased during both pranayama and yoga interventions, indicating potential differences in blood flow dynamics between sedentary behavior and active relaxation techniques. In addition, all interventions resulted in a reduction in stress in the CCA, suggesting a general decrease in vascular strain regardless of the approach. However, statistical analysis, as presented in Table 20.6, reveals that neither shear rate nor stress exhibited significant differences across time or between groups, indicating that while trends were observed, they were not substantial enough to be considered significant.

In the Flanker's test, reaction times progressively slowed over the hours, indicating a decline in response speed with time. This trend was observed across all interventions, where reaction times continued to slow hour by hour. A time-based comparison revealed statistical significance between the 0th and 2nd hours, as well as between the 0th and 4th hours, suggesting a notable decline in performance over these periods. However, no statistical significance was observed between the 2nd and 4th hours, indicating a plateau in reaction time reduction after the second hour. Furthermore, when comparing reaction times across different interventions, no significant differences were found, suggesting that none of the interventions had a substantial effect on counteracting the slowing response times. These findings are detailed in Table 20.7.

TABLE 20.4 Changes in SFA diameter (mean ± SD) at baseline, 2 h, and 4 h across different interventions.

SFA diameter			
Intervention	0 h (Mean) ± SD	2 h (Mean) ± SD	4 h (Mean) ± SD
Sitting intervention	6.833 ± 25.555	6.339 ± 23.621	5.796 ± 21.570
Pranayama intervention	0.635 ± 0.051	0.591 ± 0.073	0.589 ± 0.072
Yoga intervention	0.626 ± 0.049	0.650 ± 0.058	0.677 ± 0.076

TABLE 20.5 Changes in SFA velocity (mean ± SD) at baseline, 2 h, and 4 h across different interventions.

SFA velocity			
Intervention	0 h (Mean) ± SD	2 h (Mean) ± SD	4 h (Mean) ± SD
Sitting intervention	100.836 ± 42.410	94.120 ± 39.232	97.401 ± 46.217
Pranayama intervention	98.859 ± 25.082	88.276 ± 23.905	85.288 ± 15.356
Yoga intervention	110.847 ± 24.281	106.024 ± 25.739	98.224 ± 21.270

TABLE 20.6 Changes in CCA shear and stress (mean ± SD) at baseline, 2 h, and 4 h across different interventions.

CCA shear				CCA stress		
Intervention	0 h (Mean) ± SD	2 h (Mean) ± SD	4 h (Mean) ± SD	0 h (Mean) ± SD	2 h (Mean) ± SD	4 h (Mean) ± SD
Sitting intervention	643.633 ± 235.083	642.379 ± 258.007	698.348 ± 277.646	700.154 ± 137.83	861.576 ± 127.689	987.954 ± 208.107
Pranayama intervention	630.400 ± 183.076	612.781 ± 197.767	591.701 ± 145.139	794.020 ± 201.092	763.876 ± 211.299	757.151 ± 214.785
Yoga intervention	713.502 ± 172.524	664.485 ± 194.142	596.131 ± 184.105	860.253 ± 268.289	750.484 ± 233.039	605.912 ± 133.750

TABLE 20.7 Changes in reaction time (ms) on the Flanker task at baseline, 2 hours, and 4 hours across different interventions.

Reaction time with Flanker test			
Intervention	0 h (Mean) ± SD	2 h (Mean) ± SD	4 h (Mean) ± SD
Sitting intervention	433.055 ± 182.599	292.144 ± 206.259	242.618 ± 174.330
Pranayama intervention	299.010 ± 208.574	225.925 ± 183.694	211.445 ± 173.799
Yoga intervention	291.792 ± 209.959	207.889 ± 153.933	190.727 ± 111.468

The accuracy of the Flanker's test showed a progressive decline with each passing hour and across all interventions, indicating a general reduction in cognitive performance over time. However, when comparing accuracy between different time points and intervention groups, no statistically significant differences were observed. This suggests that while accuracy deteriorated over time, the rate of decline remained consistent across interventions, with no specific approach proving more effective in maintaining accuracy. These findings are presented in Table 20.8.

The reaction time in the N-back memory test showed a slower response over time, indicating a general decline in cognitive processing speed. However, statistical analysis revealed no significant differences between groups or across different time points, suggesting that the slowing reaction time was consistent across all interventions and did not indicate a meaningful effect of any specific approach. These findings are detailed in Table 20.9.

TABLE 20.8 Accuracy on the Flanker task at baseline, 2 hours, and 4 hours across different interventions.

Flankers test for accuracy			
Intervention	0 h (Mean) ± SD	2 h (Mean) ± SD	4 h (Mean) ± SD
Sitting intervention	99.621 ± 0.179	99.602 ± 0.198	99.577 ± 0.128
Pranayama intervention	99.608 ± 0.153	99.568 ± 0.145	95.672 ± 16.027
Yoga intervention	99.596 ± 0.116	99.539 ± 0.102	99.530 ± 0.064

TABLE 20.9 Reaction time on the N-back working memory task at baseline, 2 hours, and 4 hours across different interventions.

Reaction time in memory test			
Intervention	0 h (mean) ± SD	2 h (Mean) ± SD	4 h (Mean) ± SD
Sitting intervention	543.548 ± 132.182	546.955 ± 189.407	533.641 ± 161.915
Pranayama intervention	534.130 ± 138.291	542.651 ± 222.897	510.003 ± 188.519
Yoga intervention	502.413 ± 169.203	575.556 ± 242.671	441.075 ± 185.880

TABLE 20.10 Accuracy on the N-back working memory task at baseline, 2 hours, and 4 hours across different interventions.

Accuracy in memory test			
Intervention	0 h (Mean) ± SD	2 h (Mean) ± SD	4 h (Mean) ± SD
Sitting intervention	81.112 ± 8.869	71.892 ± 17.269	73.280 ± 14.459
Pranayama intervention	77.500 ± 15.230	1098.703 ± 4224.139	67.219 ± 36.227
Yoga intervention	76.750 ± 12.320	64.818 ± 33.659	66.121 ± 19.380

The accuracy of the memory tests showed a gradual decline over time and across all interventions, suggesting a reduction in cognitive performance as the study progressed. However, when comparing accuracy across different time points and intervention groups, statistical analysis revealed no significant differences, indicating that the decline was consistent regardless of the intervention. These findings are detailed in Table 20.10.

5 Discussion

The aim of our study is to investigate the effects of yoga on cognitive and vascular function in desk-based workers. Regular exercise plays a crucial role in maintaining and enhancing both physical and mental well-being. Yoga, in particular, has the potential to support overall health by regulating the sympathetic nervous system and the hypothalamic–pituitary–adrenal axis. Through this mechanism, yoga may help improve cardiovascular function while also promoting mental well-being, reducing stress, and enhancing cognitive performance in individuals with sedentary work lifestyles (Moneta et al., 1995).

5.1 Yoga intervention

In our study, Yoga intervention improves the vascular function of the carotid and SFA diameter. Comparing time, the first and second hours exhibit statistical significance with a P-value of $<.001$. Compared to the group, Yoga intervention shows effectiveness but no statistical significance in the carotid artery. In the SFA, the time at the 0th and 2nd hours indicates significance (P-value $= .026$). Still, it is not statistically significant, whereas the 0th and 2nd hours also show significance with a P-value (.004*), but it is also not statistically significant. A study by Ross A found that both Yoga and exercise significantly reduced fasting blood glucose levels at 3 and 6 months (29.48% & 27.43%, $P < .0001$), with Yoga showing better results in social and occupational functioning compared to the exercise group. The study also found that both Yoga and exercise improved total serum cholesterol (<.0001) and low-density protein levels ($P = .030$) (Ross and Thomas, 2010). So, when we compare this to our study, it proves that Yoga shows significant changes.

In the study conducted by (Pearson and Smart, 2017), participants underwent an aerobic exercise intervention lasting between 4 weeks and 6 months. The frequency of exercise sessions varied, with participants engaging in aerobic activity anywhere from twice a week to daily. Each session lasted between 10 and 60 minutes. The study ultimately concluded that aerobic exercise training significantly improved endothelial function, demonstrating its effectiveness in enhancing vascular health.

In comparison, our study focused on a Yoga intervention, which not only contributed to improved endothelial function but also led to a measurable increase in the diameters of both the carotid and superficial femoral arteries. This suggests that Yoga may have a broader impact on vascular structure and function beyond endothelial improvement alone (Pearson and Smart, 2017).

The study conducted by Cortez-Cooper et al. (2008) investigated the effects of different types of exercise interventions on cardiovascular health. Participants were divided into three groups: a strength training group, a combination group that engaged in both strength training and aerobic exercise, and a stretching group. The results demonstrated that both the strength training and combination groups exhibited significant improvements in cardiovascular health markers, including reductions in carotid artery pulse pressure. In addition, the study observed a slight decrease in carotid artery compliance in the strength training and combination groups, whereas the stretching group showed an increase in carotid artery compliance. However, no significant changes were found in carotid-femoral pulse wave velocity in the strength training and combination groups, indicating that these interventions did not impact arterial stiffness in this region.

Overall, the study suggested that combining strength training with aerobic exercise might be more effective in enhancing cardiovascular health compared to stretching alone. In comparison, our study, which implemented a yoga intervention, demonstrated significant improvements across multiple cardiovascular parameters. Unlike the study by Cortez-Cooper, where the impact on arterial compliance varied by intervention type, the yoga intervention in our study resulted in more comprehensive vascular benefits, indicating its potential as an effective modality for cardiovascular health improvement (Cortez-Cooper et al., 2008).

5.2 Prolonged sitting

The study by (Thosar et al., 2015) involved 12 men participating in two randomized 3-hour sitting trials. The first trial was a sitting trial where the subjects were seated in one position without interruption. The second trial was a sit-walk trial where the subjects walked on treadmills for 5 minutes at 2 mph at 30 min, 1.5 hour, and 2.5 hour intervals during the sitting period. The study assessed the flow-mediated dilation (FMD) of the SFA at baseline, 1 hours, 2 hours, and 3 hours in each trial. The results showed that there was no significant decline in the FMD during the sitting trial at each hour from baseline. The mean shear rate also showed a substantial decrease from baseline across time. The sit-walk trial, however, showed a significant difference in the FMD and mean shear rate compared to the sitting trial, indicating that walking during prolonged sitting may benefit arterial health (Thosar et al., 2015). As compared to our study, both CA and SFA were measured. In our study, all interventions aside from prolonged sitting showed a minor decrease in CA velocity over time, which rose in the final four hours of the sitting intervention. Compared to time, the first and second hours exhibited statistical significance with a P-value of <.001, whereas the third and fourth hours exhibited significance but did not exhibit statistical significance (P-value = .020). There was no significance in SFA. However, in our study, yoga showed effectiveness in both CA and SFA.

When people sit for lengthy periods, their CA shear increases, whereas their pranayama and yoga interventions show decreases in CA shear for each hour. In CA shear, neither time nor group will show any significance.

Long periods of sitting have been shown to reduce stress in SFA, and yoga and pranayama interventions have also been shown to reduce stress over time. Both time and group do not exhibit any significant differences when compared.

The study by Carter et al. (2018) recruited 15 office workers (10 male). They had participants sit for hours with 2 minutes of light-intensity treadmill walking breaks every 30 minutes and 4 hours of sitting with 8-minute light-intensity treadmill walking breaks every 120 minutes. Middle cerebral artery (MCA) measures were taken, and it was found that there was no significance in cardiorespiratory and hemodynamic measures, but a significant main effect was observed in cerebral blood flow. Post hoc analysis revealed a greater change in MCA during sitting compared to walking, with a considerable reduction in MCA observed in both the sitting and walking conditions (Carter et al., 2018). Compared to our study, the diameter significantly decreases during the sitting intervention, but velocity increases. Similarly, in the SFA diameter and velocity decrease during the sitting intervention. Over time, the first and second hours exhibit statistical significance with a P-value of <.001, whereas the third and fourth hours demonstrate significance but do not exhibit statistical significance (P-value = .020). The yoga intervention shows significance but not statistical significance when compared to pranayama and group prolonged sitting (P-value = .014). In contrast, the yoga intervention shows significance but not statistical significance when compared to the group in the pranayama intervention and prolonged sitting (P-value = .017). Meanwhile, in CA velocity, neither time nor velocity is significant.

Yoga intervention demonstrates improvement in SFA diameter; however, there is no difference between time and group when compared side by side. Each hour of intervention results in a drop in velocity. The time at the 0th and 2nd hours indicates significance (P-value = .026); still, it is not statistically significant. Similarly, the 0th and 2nd hours also show significance with a P-value (P-value = .004), but it is also not statistically significant. The group has not demonstrated any substantial response to any intervention. Hence, it proves that yoga improves vascular function.

The study conducted by Blumenthal et al. (1989) found that older adults who participated in 16 weeks of aerobic exercise training demonstrated a significant increase in aerobic capacity (cardiorespiratory fitness). However, there was no improvement in their performance on reaction-time tests of attention and memory retrieval compared to a group of young adults. It is suggested that changes in cognitive performance in older adults may require extended periods of training or may be due to differences between physically active and sedentary individuals (Madden et al., 1989). When compared to our study using the Flanker test, reaction times are observed to get slower over the course of an hour. Even during each intervention, the response time tends to slow down as the hour progresses. The comparison of the 0th hour to the 2nd hour exhibits statistical significance, as does the comparison between the 0th and 4th hours, while the 2nd and 4th hours do not exhibit any statistical significance. In addition, no significant differences are demonstrated in any of the interventions when compared to one another.

The accuracy of the Flanker's test decreases with each hour and each intervention. There will be no difference between time and group when compared either.

The reaction time will be slower in the N-back memory test. There will be no indication of any significance in either the group or the moment. The accuracy of memory tests also indicates a decline over the course of an hour and with each intervention. Both time and group comparisons will not demonstrate any significance.

The study by Oken et al. (2004) recruited 135 healthy men and women aged 65–68, who were randomized to 6 months of Hatha yoga classes, walking exercise classes, or a wait-list control. The results showed that there were no effects on cognitive function in the yoga or exercise groups compared to the wait-list control group. In addition,17 participants did not finish the 6-month intervention. The yoga intervention did produce improvements in physical measures as well as quality-of-life measures related to a sense of well-being, energy, and fatigue compared to the control group (Oken et al., 2006). compared to our study, the Flanker test reaction time showed a reduction by an hour, and in each intervention, the response time was reduced. The comparison of the time from the 0th hour to the 2nd hour exhibits statistical significance. However, the accuracy of the Flanker test did not show any changes across both time and group. In contrast, the memory test reaction time was slower, and there were no significant changes in either time or group. In addition, the accuracy in the memory test did not show any statistical significance.

A study conducted by Sharma et al. (2006) investigated the impact of Sahaj Yoga Meditation (SYM) on cognitive function in patients with major depressive disorder. The study involved 30 patients who were randomly divided into two groups: one receiving only conventional antidepressant medication and the other practicing SYM in addition to taking antidepressants. Using a neurocognitive test battery, the study assessed various cognitive functions, including attention span, visuo-motor speed, short-term memory, working memory, and executive functions. The results indicated that patients who practiced SYM showed significantly greater improvements in these cognitive domains compared to those who received only medication. In particular, the meditation group demonstrated enhanced attention span, faster visuo-motor speed, and better memory functions, suggesting that meditation could provide additional cognitive benefits beyond standard pharmacological treatment. These findings highlight the potential role of SYM as a complementary approach to improving cognitive function in individuals with depression (Sharma et al., 2006). When compared to our study, the Flanker test should not have shown any significant changes following any intervention. However, the results indicate that the Flanker task test revealed significant changes after both the first and second hours, suggesting an unexpected effect of the intervention on cognitive control or attention. On the other hand, the N-back memory test did not demonstrate any significant effects in the investigation, indicating that working memory performance remained largely unchanged despite the intervention. These findings contrast with our expectations and warrant further exploration to understand the underlying mechanisms influencing cognitive function in this context.

A study conducted by Chattha et al. (2008) examined the effects of a structured yoga practice on symptoms commonly experienced by perimenopausal women, such as hot flushes, night sweats, and sleep disturbances. The study involved 108 perimenopausal women, who were randomly divided into two groups: one practicing a yoga module that included breathing exercises (pranayama), sun salutations (surya namaskar), and meditation, while the control group engaged in simple physical exercises.

The findings revealed that women in the yoga group experienced a significant reduction in the frequency and intensity of hot flushes and night sweats, along with improved sleep quality. In contrast, the control group, which performed general physical exercises, did not show any notable changes in these symptoms. Furthermore, the study assessed cognitive function using a series of neuropsychological tests and found that the women who practiced yoga demonstrated improvements in various cognitive domains, including memory, attention, and executive functions, compared to the control group.

These results suggest that an integrated yoga-based approach can be beneficial for perimenopausal women, not only in alleviating vasomotor and sleep-related symptoms but also in enhancing cognitive function. The study highlights the potential of yoga therapy as a complementary intervention for managing perimenopausal symptoms, offering a holistic approach that addresses both physical and cognitive well-being (Chattha et al., 2008). In comparison to our study, significant changes are expected to be observed in the Flanker test between the 0th and 2nd hour. However, the accuracy of the Flanker test is not anticipated to show any significant variations across both time and groups. When compared to the memory test, the reaction time is expected to be slower, yet no significant changes will be evident in either time or group. In addition, the accuracy in the memory test does not demonstrate any statistical significance, indicating that the intervention does not have a notable impact on memory performance.

6 Conclusion

In conclusion, the findings of this study could have significant implications for the health and well-being of desk-based workers. If Y-break yoga is proven to enhance cardiovascular health, cognitive function, and overall physical activity, it could serve as a simple, cost-effective, and accessible intervention to mitigate the risks associated with prolonged sitting. By improving vascular function, boosting cognitive performance, and promoting regular movement, Y-break yoga may not only reduce the likelihood of cardiovascular disease but also enhance workplace productivity and overall quality of life for desk-based employees.

7 Limitation

This study may face several limitations that could affect the validity and generalizability of its findings. A small sample size may limit the applicability of results to the broader population of desk-based workers, while participant characteristics, such as age and health status, could further restrict generalizability. In addition, self-selection bias may arise if individuals who are already interested in yoga or perceive themselves as healthier are more likely to participate. The inability to control for external factors, such as diet, exercise habits, and sleep quality, could make it difficult to isolate the effects of Y-break yoga on vascular and cognitive function. Furthermore, reliance on self-reported or subjective assessments may introduce bias and inaccuracies, emphasizing the need for objective measures, such as physiological monitoring and cognitive testing, to strengthen the study's reliability.

8 Future research

Future research on Y-break yoga should explore its long-term effects to determine whether its benefits on vascular and cognitive function are sustained over time. Comparative studies could assess how Y-break yoga measures up against other interventions, such as traditional exercise programs or mindfulness-based stress reduction, to establish its relative effectiveness. In addition, investigating the underlying mechanisms, such as changes in cardiovascular, metabolic, or brain function, could provide deeper insights into how Y-break yoga influences health. Expanding research to include a more diverse range of participants, such as older adults or individuals with chronic conditions, would enhance generalizability. Lastly, examining Y-break yoga as a workplace intervention in real-world settings could help determine its feasibility and practicality in improving employee health and well-being.

References

Alexandrov, A.V., Brodie, D.S., McLean, A., Hamilton, P., Murphy, J., Burns, P.N., 1997. Correlation of peak systolic velocity and angiographic measurement of carotid stenosis revisited. Stroke. 00392499. 28 (2), 339–342. Available from: http://stroke.ahajournals.org/. doi:10.1161/01.STR.28.2.339.

Barnes, P.M., Bloom, B., Nahin, R.L., 2009. Complementary and alternative medicine use among adults and children: United States, 2007. Natl. Health Stat. Rep. 12, 1–23.

Blumenthal, J.A., Emery, C.F., Madden, D.J., George, L.K., Coleman, R.E., Riddle, M.W., McKee, D.C., Reasoner, J., Williams, R.S., 1989. Cardiovascular and behavioral effects of aerobic exercise training in healthy older men and women. J. Gerontol. 00221422. 44 (5), M147. doi:10.1093/geronj/44.5.M147.

Carpenter, J.P., Lexa, F.J., Davis, J.T., 1996. Determination of duplex Doppler ultrasound criteria appropriate to the North American symptomatic carotid endarterectomy trial. Stroke. 00392499. 27 (4), 695–699. Available from: http://stroke.ahajournals.org/. doi:10.1161/01.STR.27.4.695.

Carter, S.E., Draijer, R., Holder, S.M., Brown, L., Thijssen, D.H.J., Hopkins, N.D., 2018. Regular walking breaks prevent the decline in cerebral blood flow associated with prolonged sitting. J. Appl. Physiol. 15221601. 125 (3), 790–798. Available from: https://www.physiology.org/doi/pdf/10.1152/japplphysiol.00310.2018. doi:10.1152/japplphysiol.00310.2018.

Chattha, R., Nagarathna, R., Padmalatha, V., Nagendra, H.R., 2008. Effect of yoga on cognitive functions in climacteric syndrome: a randomised control study. BJOG. 14700328. 115 (8), 991–1000. doi:10.1111/j.1471-0528.2008.01749.x.

Cortez-Cooper, M.Y., Anton, M.M., Devan, A.E., Neidre, D.B., Cook, J.N., Tanaka, H., 2008. The effects of strength training on central arterial compliance in middle-aged and older adults. Eur. J. Prev. Cardiol. 20474881. 15 (2), 149–155. doi:10.1097/HJR.0b013e3282f02fe2.

Huang, Y., Benford, S., Blake, H., 2019. Digital interventions to reduce sedentary behaviors of office workers: scoping review. J. Med. Internet Res. 1438-8871. 21 (2), e11079. doi:10.2196/11079.

Madden, D.J., Blumenthal, J.A., Allen, P.A., Emery, C.F., 1989. Improving aerobic capacity in healthy older adults does not necessarily lead to improved cognitive performance. Psychol. Aging. 08827974. 4 (3), 307–320. doi:10.1037/0882-7974.4.3.307.

McGowan, C.L., Visocchi, A., Faulkner, M., Verduyn, R., Rakobowchuk, M., Levy, A.S., McCartney, N., MacDonald, M.J., 2007. Isometric handgrip training improves local flow-mediated dilation in medicated hypertensives. Eur. J. Appl. Physiol. 14396319. 99 (3), 227–234. doi:10.1007/s00421-006-0337-z.

Moneta, G.L., Edwards, J.M., Papanicolaou, G., Hatsukami, T., Taylor, L.M., Strandness, D.E., Porter, J.M., 1995. Screening for asymptomatic internal carotid artery stenosis: duplex criteria for discriminating 60% to 99% stenosis. J. Vasc. Surg. 07415214. 21 (6), 989–994. doi:10.1016/S0741-5214(95)70228-8.

Oken, B.S., Kishiyama, S., Zajdel, D., Bourdette, D., Carlsen, J., Haas, M., Hugos, C., Kraemer, D.F., Lawrence, J., Mass, M., 2004. Randomized controlled trial of yoga and exercise in multiple sclerosis. Neurology. 0028-3878. 62 (11), 2058–2064. doi:10.1212/01.WNL.0000129534.88602.5C.

Oken, B.S., Zajdel, D., Kishiyama, S., Flegal, K., Dehen, C., Haas, M., Kraemer, D.F., Lawrence, J., Leyva, J., 2006. Randomized, controlled, six-month trial of yoga in healthy seniors: effects on cognition and quality of life. Altern. Ther. Health Med. 10786791. 12 (1), 40–47.

Pearson, M.J., Smart, N.A., 2017. Aerobic training intensity for improved endothelial function in heart failure patients: a systematic review and meta-analysis. Cardiol. Res. Pract. 20900597. 2017. Available from: http://www.hindawi.com/journals/crp/. doi:10.1155/2017/2450202.

Ross, A., Thomas, S., 2010. The health benefits of yoga and exercise: a review of comparison studies. J. Altern. Complement. Med. 1075-5535. 16 (1), 3–12. doi:10.1089/acm.2009.0044.

Sharma, V.K., Das, S., Mondal, S., Goswami, U., Gandhi, A., 2006. Effect of Sahaj Yoga on neuro-cognitive functions in patients suffering from major depression. Indian J. Physiol. Pharmacol. 00195499. 50 (4), 375–383.

Spagnolli, A., Chittaro, L., Gamberini, L., 2016. Interactive persuasive systems: a perspective on theory and evaluation. Int. J. Hum. Comput. Interact. 1044-7318. 32 (3), 177–189. doi:10.1080/10447318.2016.1142798.

Thosar, S.S., Bielko, S.L., Mather, K.J., Johnston, J.D., Wallace, J.P., 2015. Effect of prolonged sitting and breaks in sitting time on endothelial function. Med. Sci. Sports Exerc. 15300315. 47 (4), 843–849. Available from: http://www.lww.com/product/?0195-9131. doi:10.1249/MSS.0000000000000479.

Wang, Y., Wu, L., Lange, J.P., Fadhil, A., Reiterer, H., 2018. Persuasive technology in reducing prolonged sedentary behavior at work: a systematic review. Smart Health. 23526483. 7-8, 19–30. Available from: https://www.journals.elsevier.com/smart-health. doi:10.1016/j.smhl.2018.05.002.

Part VI

Bone health, meditation, and stroke recovery

Chapter 21

Dual-energy X-ray absorptiometry in assessing the effectiveness of yoga and physical activity in improving the bone mineral density in perimenopausal and postmenopausal women

Leena R. David[a,b], Rajagopal Kadavigere[c], Sneha Ravichandran[b], Baskaran Chandrasekaran[d], Berna Uzun[e], Dilber Uzun Ozsahin[a,f,e], Poovitha Shruthi Paramashiva[g], Suresh Sukumar[b], Dilip Shettigar[b], Ilker Ozsahin[e], and Sathya Sabina Muthu[b]

[a]*Medical Diagnostic Imaging Department, College of Health Sciences, University of Sharjah, Sharjah, United Arab Emirates*

[b]*Department of Medical Imaging Technology, Manipal College of Health Professions, Manipal Academy of Higher Education, Manipal, Karnataka, India*

[c]*Department of Radiodiagnosis and Imaging, Kasturba Medical College, Manipal, India*

[d]*Symbiosis School Of Sports Sciences, Symbiosis International (Deemed University), Pune, India*

[e]*Irfan Suat Gunsel Operational Research Institute, Near East University, Nicosia, Turkish Republic of Northern Cyprus (TRNC), Turkey*

[f]*Research Institute for Medical and Health Sciences, University of Sharjah, Sharjah, United Arab Emirates*

[g]*Manipal Academy of Higher Education, Division of Yoga, Centre for Integrative Medicine and Research, Manipal, India*

1 Introduction

Osteoporosis, often referred to as a "silent disease," is a major health concern, particularly for postmenopausal women, who are at a heightened risk due to hormonal changes associated with menopause. The progressive loss of bone density in this population can lead to an increased risk of fractures and reduced mobility, impacting quality of life and independence. While pharmacological interventions are commonly used to manage osteoporosis, nonpharmacological strategies, such as yoga and physical activity, have gained attention for their potential to improve bone health without the side effects often associated with medications.

This comprehensive chapter seeks to investigate the effectiveness of nonpharmacological interventions specifically yoga and physical activity in improving bone mineral density (BMD) in postmenopausal working women. The study aims to explore how these interventions influence bone health markers, with a particular focus on BMD, in a population that often faces time constraints and stress due to work and family responsibilities.

This chapter will also provide an in-depth analysis of the research project, including a review of the background literature, the objectives of the study, the methodology employed, the anticipated outcomes, and the broader significance of the research. By understanding how yoga and physical activity contribute to bone health in postmenopausal women, this study aims to inform evidence-based recommendations for improving osteoporosis management and prevention in this vulnerable population.

1.1 The silent threat of osteoporosis

Osteoporosis and osteopenia are conditions that significantly impact bone health, particularly in postmenopausal women, who experience hormonal changes that accelerate bone loss. Osteoporosis, the more severe of the two conditions, leads to a notable reduction in bone density, resulting in bones that become porous, fragile, and more susceptible to fractures.

Advances in Medical Imaging. https://doi.org/10.1016/B978-0-443-28967-5.00012-5

Osteopenia, on the other hand, is characterized by lower-than-normal bone density and is considered a precursor to osteoporosis, indicating an increased risk of developing the more severe condition.

These conditions pose a substantial health risk, particularly in postmenopausal women. Osteoporosis can lead to falls and fractures, which, in turn, may result in serious consequences such as depression, functional impairment, and a significant decline in quality of life. In some cases, these complications may even lead to mortality. The burden of osteoporosis is especially significant in India, where the incidence is alarmingly high. A study conducted in Pune reported a prevalence of 44.3% of osteoporosis among postmenopausal women, highlighting the urgent need for effective preventive and therapeutic strategies to address this growing health concern (Kadam et al., 2018). The dramatic hormonal changes during menopause, coupled with physical inactivity, contribute to poor calcium absorption and production in this demographic (Kadam et al., 2018).

The impact of osteoporosis extends far beyond individual well-being, with the condition significantly increasing the risk of musculoskeletal discomfort and fractures, which, in turn, impose a substantial economic and social burden. The high incidence of falls and fractures in individuals with osteoporosis leads to increased healthcare costs, long-term disability, and even early mortality. In addition to these physical consequences, osteoporosis is associated with sickness-related absenteeism and reduced productivity in the workplace. As such, addressing this issue becomes not only a matter of personal health but also an urgent societal and economic concern.

For modern working women, the challenge is compounded by the prevalence of sedentary behavior, particularly in desk-based jobs, which may further exacerbate the risk of osteoporosis. Prolonged sitting, combined with the demands of a busy work life, often leads to insufficient physical activity, which plays a crucial role in maintaining bone health. Given this, it is essential to explore the potential benefits of physical activity or yoga on bone health in working women. In addition, understanding the relationship between occupational sitting time and BMD is an area that remains largely unexplored but holds great potential for uncovering strategies to mitigate osteoporosis risk among this population.

Postmenopausal women are strongly associated with the risk of osteoporosis and osteopenia (Nilakanthan et al., 2016; Kerr et al., 2017; Geyer, 2017). Osteoporosis leads to an increased risk of falls and fractures, which often result in depression, functional impairment, and even mortality. More than 55% of the elderly population suffers from osteoporosis, with many unable to afford calcium phosphonates, which are commonly prescribed for bone health. Even when these medications are available, they can lead to side effects such as gastrointestinal disorders and muscular cramps, complicating their use.

The hormonal changes associated with estrogen loss during menopause, combined with physical inactivity, contribute to the risk of diminished calcium absorption and poor calcium production, which further exacerbate the condition. Sedentarism has been identified as an independent risk factor for both osteoporosis and osteopenia in the elderly. These conditions not only affect bone health but also significantly increase the risk of musculoskeletal discomfort and fractures, which impose a major economic burden and disrupt the social functioning of women. In addition, sickness-related absenteeism and reduced work productivity are frequently associated with osteoporosis and osteopenia, making them a pressing issue for both individual health and societal well-being (Sjøgaard et al., 2016).

Working women in modern times are found to experience increased sedentarism at their desks due to computerization. Modern working women spend at least 65% of their waking time sedentary, with 6–7 hours in occupational sitting (Shrestha et al., 2018). To our knowledge, occupational sitting time has not been associated with BMD, and the effects of physical activity or yoga on working women's bone health are largely unknown.

Physical activity is said to improve bone health (Kerr et al., 2017). Breaking up sitting time and engaging in moderate to high-impact exercises like walking, jumping, or dancing may improve the musculoskeletal health of women. Globally, guidelines illustrate that one must accumulate at least 150 minutes of walking every week. Recently, an average of 10,000 steps per day has been associated with improvements in cardiovascular and musculoskeletal health (Hind et al., 2020). Pedometer-based walking interventions have been found to reduce sedentary levels in the working population (Chu et al., 2016; Freak-Poli et al., 2013). Body-supported resistance exercises, particularly curlups, pushups, pullups, squats, lunges, and calf raises, have been found to improve bone health in postmenopausal women. Combined aerobic and resistance exercises may benefit postmenopausal women for effective bone health.

For a decade, yoga has been administered to reverse bone loss and maintain bone mass (McArthur et al., 2016). However, the beneficial effects of yoga are questioned due to the adverse events associated with it (Smith and Boser, 2013). Yoga, an ancient Indian physical activity, is practiced rarely by Indian women due to a lack of awareness. Moreover, the comparative effects of physical activity and yoga on BMD remain largely unknown (Santos et al., 2017; Mansi et al., 2014).

Osteoporosis is a prevalent and concerning condition among perimenopausal and postmenopausal Indian women, often going undiagnosed and untreated. The lack of early detection and intervention exacerbates the risk of fractures and other complications. To address this growing health issue, the present study aims to explore the potential of nonpharmacological interventions, specifically focusing on the impact of yoga on BMD in this population. While there is currently a gap in research examining the effectiveness of yoga in improving BMD among peri and postmenopausal women, this study seeks to fill that gap. If proven effective, incorporating yoga into national health policies could offer an accessible and cost-effective strategy for improving osteoporosis management and prevention, particularly for working women in this age group. Such an approach could complement existing medical treatments and promote holistic well-being (Pharmacological interventions [Internet]. 2023, 2024; Fernández-Rodríguez et al., 2021).

2 Study objectives

2.1 Primary

- To compare the effects of a twelve-month intervention involving yoga and physical activity with those of a usual work routine on improving BMD in postmenopausal working women.

2.2 Secondary

- To assess the effects of two nonpharmacological interventions—yoga and structured physical activityon physical activity levels in postmenopausal working women over a twelve-month period.
- To evaluate changes in muscle strength and key biochemical markers, including serum Vitamin D, thyroid-stimulating hormone, and parathyroid hormone (PTH), associated with long-term engagement in yoga or physical activity among individuals.

3 Methodology

This triple arm, blinded, randomized controlled trial was conducted at a reputable multispecialty teaching hospital, which is equipped with a National Accreditation Board for Hospitals & Healthcare Providers-accredited radiology laboratory, ensuring high-quality diagnostic support and clinical infrastructure. The study commenced only after obtaining formal ethical clearance from the Institutional Ethics Committee of Kasturba Medical College and Hospital. The trial protocol was designed in accordance with internationally accepted standards for clinical research, with its implementation adhering strictly to the principles of good clinical practice and relevant regulatory guidelines.

All research procedures involving human participants were conducted in line with the ethical standards of both institutional and national research committees and complied with the ethical principles outlined in the Declaration of Helsinki. Participants were provided with comprehensive information regarding the nature, purpose, and procedures of the study, ensuring they were fully informed. Written informed consent was obtained from each participant prior to enrollment, safeguarding their rights and ensuring that ethical obligations related to autonomy, safety, confidentiality, and voluntary participation were upheld throughout the study.

3.1 Participants

The study will include postmenopausal women aged 45 years or older, specifically those working in desk-based office jobs. Participants must have a self-reported history of natural menopause, meaning they have not undergone surgical procedures such as hysterectomy. The study requires participants to be willing to wear a triaxial accelerometer on their thigh for seven consecutive days to monitor their physical activity levels objectively. In addition, participants should be willing to undergo dual-energy X-ray absorptiometry (DEXA) scans for BMD measurements. Women with a history of osteoporosis, osteopenia, or fractures within the previous year are excluded, as the study focuses on individuals with normal bone health. Those who are not taking medications or supplements for osteoporosis and have no cardiovascular or metabolic disorders preventing physical activity will be eligible. Furthermore, participants with no underlying bone diseases, such as osteomalacia, or neuromuscular diseases that could hinder their ability to engage in physical activity or yoga will be included. Another requirement is that participants must not anticipate a job shift for the next six months, as this could disrupt their ability to participate in the intervention consistently.

Exclusion criteria include individuals who are unable to comply with at least 60% of the prescribed intervention. Any adverse events that are unrelated to the intervention, or participants who are unwilling to undergo randomization or who express dissatisfaction with the intervention, will also be excluded. If a participant decides to withdraw from the study, their data will not be included in the final analysis. However, those who choose to withdraw will still be given their preferred treatment during the intervention period, though their data will not contribute to the study results.

3.2 Randomization

Participants were randomly assigned to one of three intervention arms using an online randomization tool (www.randomizer.org), ensuring a blinded and unbiased allocation process. A blinded third-party randomizer, who had no involvement in participant recruitment or assessment, conducted the randomization procedure to maintain allocation concealment.

Block randomization was employed to ensure equal distribution across the three study groups. Pregenerated random numbers were placed in sealed, opaque, and sequentially numbered envelopes, which were used to assign participants to their respective groups upon enrollment. The envelopes were opened only after a participant had been deemed eligible and had provided informed consent.

This approach ensured both the randomness of group assignment and the concealment of allocation, minimizing the potential for selection bias and preserving the integrity of the randomized controlled trial design.

3.3 Blinding

This study employed a triple-blinded design to minimize bias and enhance the credibility of the results. Participants, intervention facilitators, and outcome assessors were all blinded to the group allocations throughout the duration of the trial.

The randomization process and assignment to intervention groups were conducted by an independent researcher who was not involved in data collection or analysis. To further ensure blinding, all outcome assessments were carried out by a trained assessor who was not affiliated with the research team and remained unaware of group assignments throughout the study.

All outcome measurement forms and data collection tools were secured in sealed, opaque envelopes, which were accessed only by the blinded assessor after the completion of the intervention period. This process ensured that data collection remained unbiased and that the results reflected genuine differences due to the interventions rather than observer or participant expectations.

3.4 Baseline measurements

The primary baseline measure is physical activity assessed over 7 days using an accelerometer. Baseline measures will include anthropometry, seven-day dietary assessments, DEXA scans (with T-values less than -1.0), thyroid and PTH levels, erythrocyte sedimentation rate (ESR), complete blood count, and vitamin D levels.

3.5 Previous physical activity measurement

To objectively assess physical activity levels, participants will wear a triaxial accelerometer, specifically the activPAL device (PAL Technologies, Glasgow, UK), for a continuous seven-day monitoring period. The device will be affixed to the right thigh of each participant using hypoallergenic adhesive, and participants will be instructed to wear the device continuously, with the exception of bathing and sleeping hours.

The activPAL device is validated for distinguishing between different postural transitions and activity intensities. It will record data in 100-second epochs, capturing a range of metrics, including:

Sit-to-stand transitions and timing
Total duration of sitting, standing, and stepping
Time spent in moderate to vigorous physical activity
Estimated energy expenditure in kilocalories

Following the data collection phase, the recorded data will be downloaded and processed using PAL software. The raw data will be converted into PAL files for analysis. To improve data integrity and eliminate potential noise, a high-pass filter will be applied to remove artifacts. In addition, any period of inactivity exceeding 600 consecutive seconds will be categorized as nonwear time and excluded from analysis.

Color coded visualizations will be employed to analyze and interpret sit-to-stand transitions and related time-stamped metrics, allowing for a detailed examination of activity patterns over the seven-day period. This methodology ensures a robust, objective evaluation of participants' habitual physical activity and sedentary behavior, critical for establishing accurate baseline measures.

3.6 Procedure

Participants were recruited from a large multidisciplinary university campus through a structured outreach campaign. Postmenopausal women who met the predefined inclusion criteria were identified via internal advertisements placed on the university intranet, departmental bulletin boards, and community notice boards. Prior to initiating recruitment, official permission was obtained from the university administration to conduct the study and use campus facilities for outreach and intervention.

Interested individuals were screened for eligibility, and those meeting the inclusion criteria were invited to an orientation session, where they were briefed on the study objectives, procedures, and expectations, including how to properly use the triaxial accelerometer. Written informed consent was obtained from all participants in accordance with ethical guidelines and institutional review board approval.

Following baseline assessments, eligible participants were randomized into one of three intervention arms using a computer-generated randomization process:

Physical Activity Group (PHY)
Yoga Group (YOG)
Waitlist Control Group (CONT)

Randomization ensured equitable distribution across the groups, allowing for comparative analysis of intervention effects on physical activity, musculoskeletal health, and related physiological parameters in postmenopausal working women.

3.7 Intervention

3.7.1 Physical activity

For the PHY group, a structured pedometer-based physical activity intervention will be implemented. Each participant will be provided with a solar-powered pedometer, which will be securely attached around the waist using a belt for ease of use and accurate step-counting. The primary goal of the intervention is to progressively increase daily step count, promoting an active lifestyle among postmenopausal working women.

The intervention protocol consists of the following steps:

Step Count Progression: Participants will be instructed to gradually increase their baseline steps by 1000 steps per day every week. Over the course of 7–8 weeks, participants are expected to achieve a goal of 10,000 steps per day.
Duration of intervention: The total duration of this pedometer-based physical activity intervention will be 8 weeks.

In addition to the pedometer-based walking program, participants in the PHY group will receive supplementary body- supported resistance exercises to further enhance physical fitness and musculoskeletal strength. These exercises will be performed twice weekly and will be based on a pamphlet provided to participants. Each session will last for approximately 30 minutes, with a focus on functional body-weight exercises that can be easily done at home without additional equipment.

To ensure compliance and track progress, participants will be provided with a log monitor to document their daily physical activity. At least 60% attendance for the exercise sessions is mandatory for continued inclusion in the study, ensuring that participants remain actively engaged in the intervention and are following the prescribed exercise regimen.

3.8 The yoga program

It is designed based on a previous pilot study conducted by Fishman et al. in 2009. It consists of a series of yoga poses, including Trikonasana (triangle pose), Adho Mukha Svanasana and Urdhva Mukha Svanasana (upward and downward dog poses), Setu Bandhasana and Urdhva Dhanurasana (bridge and rainbow poses), Janu Sirsasana and Paschimottanasana (one-legged and two-legged forward bends with a focus on maintaining a straight back and preventing kyphosis), Navasana (boat pose), Supta Padangusthasana, and three twisting poses with instructions to keep the back straight (Marichyasana, Matsyendrasana, and Jathara Parivartanasana), as depicted in Fig. 21.1. As part of the structured yoga intervention in this study, participants will be provided with personalized instruction by qualified yoga therapists to ensure the safe and effective practice of each pose. Recognizing that postmenopausal women may present with varying levels of flexibility, strength, and balance, instructors will offer individualized modifications and simplified variations of each asana when necessary. This personalized approach is essential to accommodate participants' physical capabilities, prevent injury, and enhance engagement with the practice.

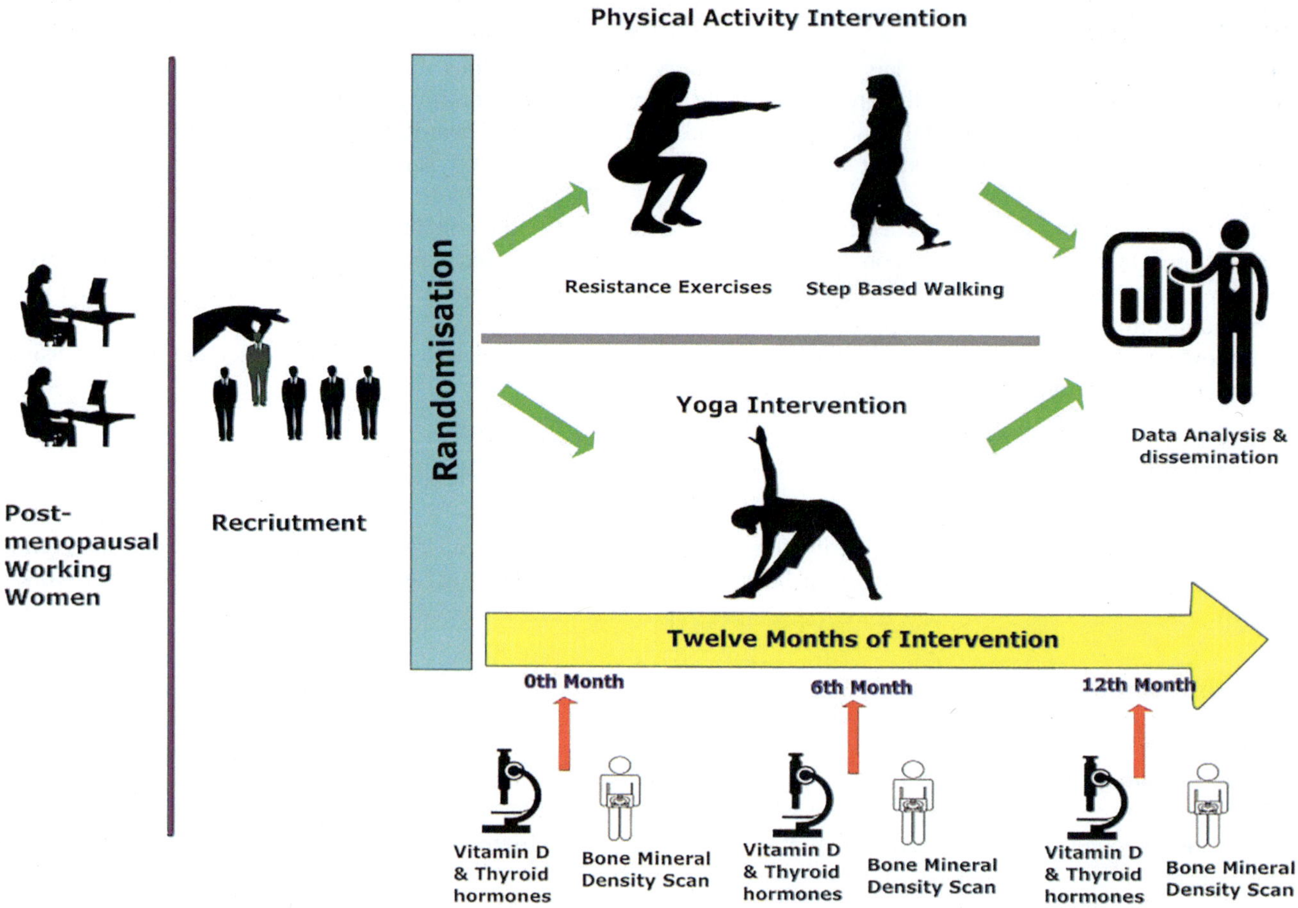

FIG. 21.1 Flow diagram depicting the procedure of the research. A diagram of a physical activity Description automatically generated.

During the sessions, participants will be encouraged to hold each pose for 20 to 30 seconds, allowing sufficient time for musculoskeletal engagement and postural awareness. This duration is intended to activate the muscles and bones through sustained isometric contraction, which may contribute to improved BMD over time. After completing the full set of poses which will take approximately 8 to 10 minutes participants will be guided to take a rest period of 5 to 10 minutes. This relaxation phase, often in a supine position such as *Savasana*, is important for physical recovery and mental relaxation, and helps restore breathing patterns after exertion.

The frequency of practice is set at three times per week, spanning a 12-week intervention period. This regularity is chosen based on existing literature, which suggests that consistent engagement over a sustained period is necessary to yield physiological adaptations, including improvements in BMD, muscle strength, and postural control.

To foster adherence and maintain motivation throughout the study, participants will receive weekly reminders via WhatsApp, a widely used and accessible communication platform. These reminders will serve as prompts to stay on track with their yoga routine. In addition, instructional pamphlets will be shared through WhatsApp, containing visual aids, step-by-step guidance for each pose, and safety tips. These resources are designed to support independent practice, reinforce correct techniques learned during supervised sessions, and enhance participants' confidence.

4 Outcome measurement

4.1 Days physical activity measurement

Physical activity was objectively assessed over a 7-day period using a triaxial accelerometer (PAMSys™, BioSensics LLC, Watertown, MA, USA), which was worn by participants around the chest throughout their daily activities, excluding times spent sleeping or bathing. This accelerometer was designed to capture detailed movement data, providing accurate measurements of both body translations and postural changes.

The accelerometer recorded data at a sampling frequency of 50 Hz, capturing measurements in 100 ms epochs to ensure high temporal resolution of physical activity events. The data collected were then stored in .csv format and transferred to a computer, where analysis was performed using the MATLAB algorithm integrated into PAMware software (BioSensics LLC, Watertown, MA, USA). This software enabled detailed posture and translation analysis by processing the movement data and extracting key activity parameters.

Several key metrics were calculated from the accelerometer data:

Step count: The total number of steps taken each day.
Sitting time: The total time spent seated during the monitoring period.
Active time: The duration of time spent engaging in physical activity.
Calories expended: The total energy expenditure associated with each activity.
Metabolic equivalent levels (METs): Activity intensity levels are based on energy expenditure.
Heart rate: The participant's heart rate during each activity provides insights into the intensity of the physical activity.

The PAMSys accelerometer has demonstrated high validity, with an(ICC) of 0.94 when compared to the double-labeled water technique, a gold standard method for assessing energy expenditure. This high degree of reliability supports the use of PAMSys as a robust tool for accurately monitoring physical activity in the study population.

- **Thyroid and PTH**: Fasting blood samples were collected by a lab technician at the worksite and analyzed for PTH and 25-hydroxy vitamin D in the NABL-accredited Central Laboratory.

DEXA Scan: The assessment of BMD was conducted using DEXA, a highly reliable and noninvasive method that is widely considered the gold standard for measuring bone health. The DEXA scan was performed with the GE Lunar Prodigy Scanner (USA), a device known for its precision and low radiation exposure, making it ideal for research in postmenopausal women.

Participants were positioned either supine (lying on their back) or on their side during the scan to ensure accurate imaging. The DXA machine used dual X-ray beams at two different energy levels to differentiate between bone and soft tissue, allowing for precise measurement of bone density at targeted skeletal sites. The specific sites scanned included the calcaneus (heel bone), tibial tuberosity (on the tibia), femoral neck, and the lumbar vertebrae (L2, L3). Each of these sites was selected for its relevance in understanding the impact of bone density changes, particularly in postmenopausal women who are at a higher risk of fractures, especially in the femoral neck and lumbar spine.

To ensure the reliability of the results, daily quality control procedures were conducted using an anthropomorphic phantom. This phantom mimics the bone composition of a human body, which helps calibrate the DXA machine accurately and ensures that the measurements taken are consistent over time. This method allows for the detection of any potential variations in the equipment's performance, ensuring the precision of the BMD assessments across all participants.

Overall, the use of DXA in this study allowed for a comprehensive and accurate measurement of BMD, providing valuable insights into the bone health of postmenopausal women.

5 Statistical analysis

5.1 Sample size

To accurately assess the impact of yoga and physical activity on BMD in postmenopausal women, the study aimed to detect a minimal change of 1.8% in BMD T-scores, with an effect size of 0.63. This effect size reflects the expected magnitude of the difference between the intervention groups (yoga, physical activity, and the waitlist control group) in terms of changes in BMD. The sample size calculation, using these parameters, revealed that 64 participants would be required, with each group containing 21 women. This number ensures that the study would have sufficient statistical power to identify meaningful differences in BMD between the groups, assuming the effect size is consistent with what is expected based on prior research or clinical estimates.

However, it was anticipated that some participants might drop out of the study or become noncompliant over the course of the trial, which could lead to a reduction in the overall sample size and impact the study's ability to detect significant differences. To mitigate this risk, an additional 20% was added to the calculated sample size to account for potential participant attrition. This resulted in a revised total sample size of 75 participants, which would be divided into three groups: 25 women in the Yoga group, 25 in the Physical Activity group, and 25 in the Waitlist control group.

The addition of participants ensures that the study remains sufficiently powered, even if a certain proportion of the participants withdraw, do not fully comply with the intervention, or drop out for other reasons. The final sample size of

75 also helps to maintain the study's robustness and reliability, ensuring that the results are statistically significant and the conclusions drawn are valid.

The study's sample size calculation is based on a 5% significance level ($\alpha = 0.05$), meaning that there is a 5% chance of making a Type I error, where the study incorrectly concludes that there is a difference when, in fact, there is none. In addition, the study is designed with 80% statistical power, which means there is an 80% chance of detecting a true effect if it exists. This high level of power reduces the likelihood of a Type II error, where the study fails to detect a true difference when there is one. Therefore, the recruitment of 75 participants maximizes the likelihood of obtaining reliable results while accounting for participant attrition.

By ensuring an adequately powered sample, this study is positioned to make a meaningful contribution to understanding the potential effects of yoga and physical activity on bone health in postmenopausal women, a population at higher risk for osteoporosis and other bone-related conditions. The results of this study could provide valuable insights into whether nonpharmacological interventions, such as yoga, can be an effective strategy for improving BMD and reducing the risk of osteoporosis in this population.

5.2 Data analysis

In this study, we included all postmenopausal women who were initially assigned to the intervention groups through randomization, adhering to the intention to treat principle. This means that all participants who were randomized at the beginning were included in the analysis, regardless of whether they completed the intervention or dropped out during the study. For a participant to be included in the analysis, they were required to complete the baseline assessments and at least one follow-up assessment for all the outcome measures.

To visualize the progression of participants throughout the study, we created a flowchart that tracked the enrollment process, randomization, and adherence to the interventions. This provided clarity on the movement of participants at each stage, ensuring transparency in the methodology.

The statistical analysis involved repeated measures analysis of variance (ANOVA) to assess continuous variables, such as BMD (T-scores), vitamin D levels, parathyroid and thyroid hormone levels, uninterrupted sitting time exceeding 30 minutes, step counts, and METs across baseline, 6-month, and 12-month postintervention time points. This analysis allowed us to examine how the intervention groups (yoga, physical activity, and waitlist control) changed over time, considering the within-subject correlation of repeated measures from the same participants.

Before conducting the ANOVA, we confirmed data normality using the Shapiro Wilk test. This test is crucial to ensure that the data meets the necessary assumptions for parametric testing. If the data were not normally distributed, we would have considered nonparametric methods or data transformations. By taking these steps, we aimed to ensure that the statistical conclusions are valid and reliable.

Through these rigorous procedures, we sought to evaluate the impact of yoga and physical activity interventions on various health outcomes in postmenopausal women, with a focus on BMD, physical activity levels, and associated physiological markers like vitamin D and hormones.

To estimate the treatment effect size, we employed Cohen's 'd' using the following formula:

Cohen's d = (M2 – M1) / SD pooled, where SD pooled is calculated as √((SD12 + SD22) / 2), and M2, M1, SD1, and SD2 represent the mean differences and standard deviations among the intervention groups, respectively. We considered results statistically significant if ($p < 0.050$).

Anticipated outcome: In this study, our primary hypothesis is that yoga will lead to a more significant improvement in BMD compared to physical activity in postmenopausal working women. This hypothesis stems from the fact that yoga combines flexibility, strength, and weight-bearing exercises, which are crucial for improving muscle strength, posture, and balance each of which is vital for maintaining bone health, especially after menopause. Yoga's comprehensive approach to mobility and strength may, therefore, have a more profound effect on BMD than typical physical activity alone.

In addition, we expect both yoga and physical activity to result in improvements in other key markers related to bone health, including serum vitamin D levels, thyroid hormone levels, and PTH levels. These hormonal markers are crucial in regulating bone metabolism, and we hypothesize that structured physical activity will positively influence these biomarkers when compared to the usual work routine of the control group. The control group, which does not participate in any intervention, may show little to no change in these parameters, making the comparison between the intervention and control groups a key aspect of this study.

Ultimately, the study seeks to determine the relative benefits of these nonpharmacological interventions (yoga and physical activity) in improving BMD and related biomarkers in postmenopausal women. We anticipate that participants in either intervention group will experience measurable improvements in bone health compared to those in the control group, who maintain their usual work routine without engaging in structured physical activity. This research will contribute valuable insights into nonpharmacological strategies for osteoporosis prevention and management in postmenopausal women.

6 Conclusion

In conclusion, this study addresses the critical issue of osteoporosis in perimenopausal and postmenopausal working women by exploring the efficacy of two nonpharmacological interventions: yoga and physical activity. Osteoporosis, a disease that significantly impacts bone density and strength, is particularly prevalent in postmenopausal women, leading to numerous health complications, fractures, and an increased economic burden. As such, identifying effective, noninvasive interventions to address this issue is of paramount importance.

This study employs a randomized controlled trial design to ensure scientific rigor and minimize biases, providing robust evidence to support the findings. Postmenopausal women who work in office-based jobs and meet the inclusion criteria were randomly assigned to one of three groups: yoga, physical activity, or a control group. Over a span of one year, the interventions were applied, and participants were regularly assessed for changes in BMD, physical activity levels, muscle strength, and the levels of serum vitamin D, thyroid, and PTHs.

The anticipated outcomes of the study are significant. We hypothesize that both yoga and physical activity will positively impact BMD, with yoga potentially offering superior benefits due to its combination of weight bearing and flexibility-enhancing postures. In addition, we expect to see improvements in serum vitamin D and the regulation of thyroid and PTHs, which play a crucial role in bone metabolism. These improvements will likely be more pronounced in the intervention groups compared to the control group, which does not engage in structured exercise.

The results from this research could have substantial implications for public health policy, particularly in India, where the incidence of osteoporosis is rising. The insights gained from comparing yoga and physical activity as treatments for osteoporosis may influence national strategies for osteoporosis prevention and management in perimenopausal and postmenopausal women. Moreover, the findings could lead to the development of evidence-based recommendations for nonpharmacological interventions that improve bone health, enhance quality of life, and reduce the healthcare costs associated with osteoporosis-related fractures and complications.

Ultimately, this study aims to improve the physical health and well-being of postmenopausal working women, reduce the economic burden of osteoporosis, and contribute to the advancement of preventive healthcare strategies. By offering nonpharmacological approaches such as yoga and physical activity, we hope to empower women to take charge of their bone health and maintain an active, healthy lifestyle well into their later years.

References

Chu, A.H.Y., Ng, S.H.X., Tan, C.S., Win, A.M., Koh, D., Müller-Riemenschneider, F., 2016. A systematic review and meta-analysis of workplace intervention strategies to reduce sedentary time in white-collar workers. Obes. Rev. 1467-789X. 17 (5), 467–481. Available from: http://onlinelibrary.wiley.com/journal/10.1111/(ISSN)1467-789X. doi:10.1111/obr.12388.

Fernández-Rodríguez, R., Alvarez-Bueno, C., Reina-Gutiérrez, S., Torres-Costoso, A., de Arenas-arroyo, S.N., Martínez-Vizcaíno, V., Moran, J.M., 2021. Effectiveness of Pilates and Yoga to improve bone density in adult women: a systematic review and meta-analysis. PLoS ONE. 1932-6203. 16 (5), e0251391. doi:10.1371/journal.pone.0251391.

Freak-Poli, R.L.A., Cumpston, M., Peeters, A., Clemes, S.A., 2013. Workplace pedometer interventions for increasing physical activity. Cochrane Database Syst. Rev. 1465-1858. 4 (7). Available from: https://www.cochranelibrary.com/cdsr/table-of-contents. doi:10.1002/14651858.CD009209.pub2.

Geyer, C., 2017. Osteoporosis. Am. J. Lifestyle Med. 11 (2), 125–128.

Hind, K., Hayes, L., Basterfield, L., Pearce, M.S., Birrell, F., 2020. Objectively-measured sedentary time, habitual physical activity and bone strength in adults aged 62 years: the newcastle thousand families study. J. Public Health. 1741-3842. 42 (2), 325–332. doi:10.1093/pubmed/fdz029.

Kadam, N.S., Chiplonkar, S.A., Khadilkar, A.V., Khadilkar, V.V., 2018. Prevalence of osteoporosis in apparently healthy adults above 40 years of age in Pune City, India. Indian J. Endocrinol. Metab. 2230-9500. 22 (1), 67–73. Available from: www.ijem.in/. doi:10.4103/ijem.IJEM_438_17.

Kerr, C., Bottomley, C., Shingler, S., Giangregorio, L., de Freitas, H.M., Patel, C., Randall, S., Gold, D.T., 2017. The importance of physical function to people with osteoporosis. Osteoporos. Int. 1433-2965. 28 (5), 1597–1607. Available from: link.springer.de/link/service/journals/00198/index.htm. doi:10.1007/s00198-017-3911-9.

Mansi, S., Milosavljevic, S., Baxter, G.D., Tumilty, S., Hendrick, P., 2014. A systematic review of studies using pedometers as an intervention for musculoskeletal diseases. BMC Musculoskelet Disord. 1471-2474. 15 (1). doi:10.1186/1471-2474-15-231.

McArthur, C., Laprade, J., Giangregorio, L.M., 2016. Suggestions for adapting yoga to the needs of older adults with osteoporosis. J. Altern. Complement Med. 1557-7708. 22 (3), 223–226. Available from: www.liebertonline.com/acm. doi:10.1089/acm.2014.0397.

Nilakanthan, S., Metri, K., Raghuram, N., Hongasandra, N., 2016. Effect of 6 months intense Yoga practice on lipid profile, thyroxine medication and serum TSH level in women suffering from hypothyroidism: a pilot study. J. Complement Integr. Med. 1553-3840. 13 (2), 189–193. Available from: www.degruyter.com/view/j/jcim. doi:10.1515/jcim-2014-0079.

Pharmacological interventions [Internet]. 2023 British Menopause Society, British Menopause Society, (2024)

Santos, L., Elliott-Sale, K.J., Sale, C., 2017. Exercise and bone health across the lifespan. Biogerontology. 1573-6768. 18 (6), 931–946. Available from: www.wkap.nl/journalhome.htm/1389-5729. doi:10.1007/s10522-017-9732-6.

Shrestha, N., Kukkonen-Harjula, K.T., Verbeek, J.H., Ijaz, S., Hermans, V., Pedisic, Z., 2018. Workplace interventions for reducing sitting at work. Database Syst. Rev. 1465-1858. 2018 (6). Available from: https://www.cochranelibrary.com/cdsr/table-of-contents. doi:10.1002/14651858.CD010912.pub4.

Sjøgaard, G., Christensen, J.R., Justesen, J.B., Murray, M., Dalager, T., Fredslund, G.H., Søgaard, K., 2016. Exercise is more than medicine: the working age population's well-being and productivity. J. Sport Health Sci. 2213-2961. 5 (2), 159–165. Available from: http://www.elsevier.com/wps/find/journaldescription.cws_home/728044/description. doi:10.1016/j.jshs.2016.04.004.

Smith, E.N., Boser, A., 2013. Yoga, vertebral fractures, and osteoporosis: research and recommendations. Int. J. Yoga Therap. 23 (1), 17–23.

Chapter 22

Meditation-induced changes in the default mode network during word repetition and retrieval: A combined functional magnetic resonance imaging and electroencephalogram study

Leena R. David[a,b], Rajagopal Kadavigere[c], Poovitha Shruthi Paramashiva[d], Hari Prakash Palaniswamy[e], M.G. Ramesh Babu[f], Berna Uzun[g], Dilber Uzun Ozsahin[a,g,h], Sneha Ravichandran[b], Suresh Sukumar[b], Ilker Ozsahin[g], and Dilip Shettigar[b]

[a]*Medical Diagnostic Imaging Department, College of Health Sciences, University of Sharjah, Sharjah, United Arab Emirates*

[b]*Department of Medical Imaging Technology, Manipal College of Health Professions, Manipal Academy of Higher Education, Manipal, Karnataka, India*

[c]*Department of Radiodiagnosis and Imaging, Kasturba Medical College, Manipal, India*

[d]*Division of Yoga, Manipal Academy of Higher Education, Centre for Integrative Medicine and Research, Manipal, India*

[e]*Department of Speech and Hearing, Manipal Academy of Higher Education, Manipal College of Health Professions, Manipal, India*

[f]*Division of Physiology, Department of Basic Medical Sciences, Manipal Academy of Higher Education, Manipal, India*

[g]*Irfan Suat Gunsel Operational Research Institute, Near East University, Nicosia, Turkish Republic of Northern Cyprus (TRNC), Turkey*

[h]*Research Institute for Medical and Health Sciences, University of Sharjah, Sharjah, United Arab Emirates*

1 Introduction

Routine speech tasks recruit various networks in the brain depending on their complexity. Routine speech is considered a coordinated activity of both ascending and descending sensory-motor pathways, with its pivotal center being Broca's area (left frontal lobe), Wernicke's area (posterior superior temporal lobe), and the Angular Gyrus. On the contrary, the repetition of a certain word alone could result in anecdotal evidence, wherein it reportedly induces calmness in individuals and is referred to as the "mantra effect", and the words used are called "mantras."

Mantra in Sanskrit can be decoded as manas—mind; tra—tools or instruments, and hence it means "an instrument of thought." This type of meditation practice is followed worldwide (Benson et al., 1975). A mantra is a sound, a word, a group of words, or a short sequence of syllables. It is generally referred to as a sacred phrase or poem (e.g., "om"). Sometimes, even a nonsemantic repetition of words, such as "Table" or "chair," is said to improve cognitive functions such as attention (Benson et al., 1975; Carrington et al., 1980; Baer, 2003; Davanger et al., 2010; Travis and Shear, 2010; Ospina et al., 2007).

Mantras are used in meditation, which, in turn, has been reported to be an excellent choice to attain an altered state of awareness in both religious and nonreligious practitioners. Recently, there has been an increasing trend in using different meditation techniques to alleviate symptoms such as anxiety, depression, and pain, even in Western countries (Ospina et al., 2007).

During mantra meditation, one retrieves the stored information in long-term memory. The process of such spontaneous retrieval of words has always been an enigma for researchers. Verbal fluency, or controlled oral word association, is one of the tasks that test this ability of word retrieval. It involves the rapid generation of words specific to a particular cue within a limited period of sixty seconds (Lezak, 1995). These tasks have been employed for assessing numerous linguistic

Advances in Medical Imaging. https://doi.org/10.1016/B978-0-443-28967-5.00008-3

and cognitive functions, including spontaneous naming, extent of vocabulary, mental organization, search strategies, retrieval mechanisms, endogenous flexibility, lexicon or semantic storage, response speed, and ease of verbal production. Furthermore, it has also been used in rule monitoring, inhibition of inappropriate responses, attention, and working memory (Henry and Crawford, 2004a, 2004b; Raboutet et al., 2010; Riva et al., 2000).

Traditionally, it is believed that there is a calming effect associated with repeating certain words of spiritual importance, such as mantras (Davanger et al., 2010; Ospina et al., 2007). However, it is interesting to note that the repetition of words used in day-to-day life or even nonsense words has also been shown to have similar effects and holds implications for the cognitive functions of the brain (Berkovich-Ohana et al., 2015).

However, the above-mentioned studies suffer from crucial methodological constraints, which include a small sample size, with a maximum sample size of twenty-three (Berkovich-Ohana et al., 2015), and the use of a fixed sequence design, which suffers from serious concerns regarding carryover effects, where the effect of the previous task might have been carried over to the next task (Berkovich-Ohana et al., 2015).

Further, there is considerable variability in the population on which these effects have been tested. For instance, Davanger et al. (2010) reported an increase in frontal activity in a group of experienced and advanced meditators, whereas Berkovich-Ohana et al. (2015) reported that, with reference to the default mode network (DMN), there was a decrease in activation without any counter increase in a group of adults without any experience in meditation, in just 21 seconds of the repetition task.

Thus, to the extent of current understanding, it is unknown whether the repetition of an emotionally significant word would only evoke a calming effect in the brain or if any other word repeated similarly would also result in the same effect. Secondly, it is also unknown whether such an effect is observable only in individuals exposed to meditation training or can also be evidenced in others. Finally, to elucidate the effect of word repetition and the mantra effect, it is essential to use a cross-over design between the groups and within the group.

2 Objective

To elucidate the effect of repeating "OM" and the repetition of semantically relevant and pseudo-words in experienced meditators and nonmeditators.

3 Literature review

A number of studies have investigated brain processing involved in meditation using neurophysiological and neuroimaging methods. The brain activity during meditation, measured using electroencephalogram (EEG), is characterized by an increase in theta and alpha activations and is often associated with proficiency in the meditative technique (Cahn and Polich, 2006). In addition, frontal and midline activities generated in sites of the anterior cingulate cortex, medial prefrontal cortex, and dorsolateral prefrontal cortex during the process of meditation are also said to index the ability to sustain attention (Ishii et al., 1999; Asada et al., 1999).

Neuroimaging methods, including functional magnetic resonance imaging (fMRI), positron emission tomography, and single-photon emission computed tomography (SPECT), have also been used to study brain activation during meditation. These investigations have reported an increase in activation of the inferior frontal gyrus, medial prefrontal cortex, anterior cingulate cortex, posterior cingulate cortex, basal ganglia, supplementary motor area, precentral gyrus, midbrain, inferior parietal lobule, precuneus, and hippocampus (Engström et al., 2010; Bærentsen et al., 2010; O'Sullivan et al., 2010). At the same time, during the meditation process, deactivation of certain cortical regions has also been observed, especially in regions of primary (VI) and secondary (VII) visual cortices, prefrontal cortex, precuneus, insula, anterior cingulate cortex, posterior cingulate cortex, and occipito-parieto-temporal area (Bærentsen et al., 2010; Ives-Deliperi et al., 2011). SPECT results showed that regional cerebral blood flow was significantly increased in the cingulate gyrus, inferior and orbital frontal cortex, dorsolateral prefrontal cortex (DLPFC), and thalamus (Newberg et al., 2001).

Mantra is one of the most universal mental practices in human culture. However, the underlying neuronal mechanisms that may explain its powerful emotional and cognitive impact are not clearly understood. The cognitive impact of regular speech in the brain differs from the prolonged, repetitive use of the same word or group of words, which may either be a traditional mantra (e.g., "om") or a word (e.g., "chair") and is referred to as the mantra effect.

Davanger et al. (2010) suggested that bilateral areas of the inferior frontal gyrus were significantly more activated during the repetition of a meditation sound than during concentrative meditation, similar to cognitive tasks. These meditation-specific brain activations did not habituate over time but increased in strength with continuous meditation bouts. These observations suggest that meditation with a relaxed focus of attention may activate distinct areas of the prefrontal cortex, with implications for the understanding of the neurobiological correlates of meditation.

Engström et al. (2010) reported that for subjects with less than two years of meditation experience, silent mantra meditation activated the bilateral hippocampi. The study was designed as an on–off block design with two alternating blocks: meditate (on), where the subjects were instructed to meditate (using a mantra/method-word as a key), and word (off), where the subjects were instructed to silently (in the same general way as with the mantra/method-word) repeat the short phrase "'table and chairs'".

Wang et al. (2011) observed strong correlations between the depth of meditation and neural activity in the left inferior forebrain areas, including the insula, inferior frontal cortex, and temporal pole. There were persistent changes in the left anterior insula and the precentral gyrus even after meditation had stopped. This study revealed changes in the brain during two different meditation practices in the same individuals, and these changes correlated with the subjective experiences of the practitioners.

Berkovich-Ohana et al. (2015) reported that repetitive speech (repetitive speech word—"one"), an easy cognitive task, is sufficient to deactivate the DMN during the task. Compared to the resting state, there is a reduction in the blood-oxygen-level-dependent (BOLD) activity in cortical regions, including the bilateral posterior and anterior cingulate cortex, precuneus, as well as the right inferior parietal lobule, medial frontal gyrus, and insular cortex.

Nilkamal Singh (2014) reported that there is significant activation in the right middle temporal cortex, right inferior frontal cortex, and left lateral orbital gyrus in an experienced meditator (on thoughts of "Om") as compared to random thinking, nonmeditative focused thinking, and meditative focused thinking.

To the knowledge of the authors, there has been no systematic study on the effect of word repetition in the context of emotionally significant words and pseudo-words among experienced meditators and nonmeditators.

4 Methods

4.1 Study design

This study followed a cross-sectional design, in which data were collected and analyzed at a single time point to assess the characteristics and relationships among the study variables.

4.2 Sample size

A total of 60 participants will be recruited and assigned to two groups, namely the control group ($n = 30$) and the meditators group ($n = 30$).

4.3 Participants

All the participants recruited in the study will have having normal sensory system, including hearing and vision. The subject should be within the age range of 20 to 50 years to isolate the influence of ageing on the cognitive tasks. The participants in the meditation group should have at least 300 h of meditation experience calculated using the formula in hours $= [A \times 52 \times B \times C] + [D \times 52 \times E \times F] + R$ (Hasenkamp and Barsalou, 2012). In each group, all the participants will be matched for their age, sex and handedness.

Ethical committee approval will be obtained from the Institutional Ethical Board, Kasturba Hospital, Manipal. Prior to the administration of any procedure, the participant will be informed about it, and written informed consent will be obtained from them. The obtained data will be monitored by the same committee from time to time.

4.4 Instrumentation

1. Structural and fMRI data will be acquired for all the participants using a Philips Achieva medical imaging 1.5 T scanner from the Radiology Department, KMC Manipal.
2. Functional EEG data will be obtained at the same time as MR using an MR-compatible 64-channel EEG system from BrainProducts.

4.5 Procedure

4.5.1 *Patient positioning*

The patient should be supine, head first into the gantry, with the head in the head-holder. Center the table height such that the external auditory meatus is at the center of the gantry. Head movement will be minimized with foam padding around

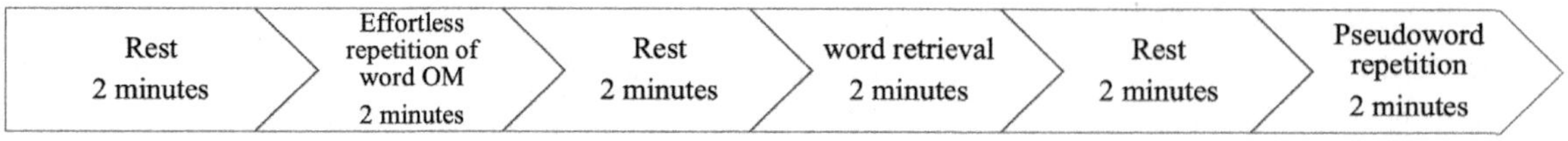

FIG. 22.1 Pictorial representation of the block design.

the head. The fMRI sequence will be recorded for a total of 12 minutes, which is divided into six blocks, each of 2 minutes' duration. The pictorial representation of the block design is shown in Fig. 22.1.

4.5.2 Structural magnetic resonance imaging acquisition parameters

- Three-dimensional sets of data of the whole brain will be collected using a T1-weighted Turbo Fast Echo (TFE) sequence, consisting of sagittal partitions with $1 \times 1 \times 1$ voxel dimensions. This parameter will provide high-resolution images covering the whole brain in 175 slices.
- Structural magnetic resonance imaging of all subjects will be obtained to coregister with fMRI.

4.5.3 Functional MRI acquisition parameters

Functional images will be obtained using a T2* weighted gradient-echo pulse, echo planar imaging (EPI) sequence. TR = 2500 ms, TE = 40 ms, flip angle = 90°, FOV = 196 × 196 mm, 64*64 matrix, slice thickness 4 mm, voxel dimensions = 3 mm*3 mm*4 mm, which will provide whole-brain coverage in 34 slices.

4.5.4 Electroencephalogram acquisition parameters

EEG data will be obtained using an MR-compatible 64-channel EEG system with a sampling rate of 1000 Hz and a band-pass filter set between 0.1 and 100 Hz. The duration of the acquisition of the EEG will be synchronized with the timing parameters as described in the fMRI acquisition protocol.

4.5.5 Analysis of fMRI

- fMRI analysis will be conducted using the functional connectivity (CONN) toolbox software version 15 (cf. http://www.nitrc.org) installed in MATLAB software version R2014a.
- fMRI analysis will be performed in the following steps:

1. Setup: Basic experiment information, data locations, regions of interest (seeds), temporal covariates, and second-level models are defined.
2. Preprocessing steps: Define, explore, and remove possible confounds in the BOLD signal.
3. First-level analyses: The seeds of interest/regions of interest (ROIs) will be defined, and the functional connectivity of different sources will be explored separately for each subject.
4. Second-level analyses: Within and between-subjects contrasts will be defined and explored in this step, which will provide the final result.

- A comparative analysis will be performed between the control and migraine groups.
- A comparative analysis will be performed between meditative and nonmeditative states within the group and between groups.

4.5.6 Analysis of EEG data

The data will be analyzed offline using freely available open-source software, EEGLAB, which runs on the MATLAB platform. EEGLAB version 13.3b will be used with MATLAB (version 2009b). The flow of data analysis is shown as a flow chart in the Figure below. The continuous EEG data will be imported into EEGLAB using the import wizard. Electrode positions will be loaded and configured, and the continuous data will be band-pass filtered between 1 and 30 Hz using an FIR filter with a smoothing factor of 10. Then, all the continuous data will be visually inspected for irregularities (temporary interferences like increased muscular activity, temporary increases in impedance, etc.) and bad channels (nonfunctional channels). If bad channels are present, they will be spline interpolated using the command "eeg = interp [*'channel number'*]" in the MATLAB command window.

In the next step, the data will be decomposed using Independent Component Analysis (ICA) with the "runica" command, which provides 64 components along with topographic plots and their corresponding waveforms. Both blink and horizontal-eye-movement-related components will be identified and then removed. Data across all the participants and sessions will be loaded in to *EEGLAB STUDY* function for analysis. In *"cluster analysis,"* components within a session across all the participants will be classified. Finally, the component activation matrix will be compared across all the tasks to substantiate the fMRI findings.

5 Importance of the proposed project in the context of the current status

To date, little is known about the effect of the repetition of words and their impact on the cortex. Recent studies have advanced slightly in this area, decoding these effects using modern techniques like fMRI. Preliminary evidence suggests that repetitive speech, an easy cognitive task, might be sufficient to induce a widespread unidirectional reduction in activation in the human cortex. This effect is observed even outside the context and training of the commonly practiced mantra. Further ventures in this area, along with the isolation of the repetition effect from meditation and other confounders, could improve our understanding of the neuronal correlates of simple repetitive speech. In addition, identifying ways to reduce the demand on the cortical network in the simplest manner may lead to new areas of relaxation therapy and evidence-based practice.

6 Suggested plan of action for the utilization of research outcomes expected from the project

The findings of the study provide important and new insights into possible nonlinear gating mechanisms of cognitive processes. In addition, it will help improve the current understanding of cortical networks and their modulation during basic tasks such as emotional, nonemotional, and pseudo word repetition. Furthermore, if evidence for a less demanding network exists, it could lead to potentially simpler and novel methods of relaxation. This can be applied to a variety of populations, including those with attention-deficit/hyperactivity disorder (ADHD), geriatric individuals, and other conditions associated with attention-related cognitive functions.

7 Discussion

The proposed protocol addresses several critical gaps in our understanding of the neurobiological mechanisms underlying mantra meditation and repetitive speech effects. The systematic crossover design, involving both experienced meditators and novices, represents a significant methodological advancement over previous studies, which have been limited by small sample sizes and fixed-sequence designs that introduce potential carryover effects (Berkovich-Ohana et al., 2015).

The integration of fMRI and EEG methodologies offers complementary perspectives on brain dynamics during meditation. While fMRI provides excellent spatial resolution for identifying specific brain regions involved in the "mantra effect," EEG offers superior temporal resolution to capture the frequency modulations associated with meditative states. Previous EEG studies have consistently reported increases in theta and alpha band activity during meditation, which correlate with meditative proficiency (Cahn and Polich, 2006). The simultaneous recording approach proposed here will enable direct correlation between BOLD signal changes and electrophysiological markers, potentially revealing the temporal dynamics of meditation-induced neural changes.

The choice of experimental conditions—'OM' repetition, word retrieval, and pseudoword repetition—allows for the systematic isolation of meditation-specific effects from general cognitive processes. This design addresses a fundamental question raised by Berkovich-Ohana et al. (2015), who demonstrated that the repetitive speech of any word could deactivate the DMN. The current protocol will help determine whether traditional mantras like "OM" produce neural effects distinct from simple repetitive speech or word retrieval tasks.

The focus on the DMN is particularly relevant, given its central role in self-referential processing and mind-wandering—cognitive processes that meditation specifically aims to modulate (Davanger et al., 2010). Previous neuroimaging studies have shown both activation and deactivation patterns in various brain regions during meditation, including the prefrontal cortex, anterior cingulate cortex, and posterior cingulate cortex (Bærentsen et al., 2010; Engström et al., 2010). The proposed functional connectivity analysis will provide insights into how these regions interact as networks rather than as isolated brain areas.

One potential limitation of the study is the relatively modest sample size of 30 participants per group, though this represents an improvement over many previous studies in the field. The 300-hour meditation experience criterion for the experienced group provides a reasonable threshold based on established formulas (Hasenkamp and Barsalou, 2012), though the heterogeneity in meditation practices among participants may introduce variability in the results.

The clinical implications of this research extend beyond basic neuroscience. Understanding the minimal cognitive requirements for achieving meditation-like neural states could inform the development of accessible therapeutic interventions for conditions involving attention deficits, anxiety, and stress-related disorders. If simple word repetition produces similar neural benefits to traditional meditation practices, this could democratize access to meditation-based therapies for populations who might otherwise find traditional meditation practices challenging or culturally inappropriate.

The proposed analysis using the CONN toolbox for fMRI data and EEGLAB for electrophysiological data represents current best practices in neuroimaging analysis. The planned removal of artifacts through ICA and the use of cluster analysis across participants will ensure robust and reproducible results.

8 Conclusion

This protocol presents a comprehensive approach to investigating the neural mechanisms underlying the "mantra effect" through a well-controlled experimental design that addresses key methodological limitations of previous studies. By employing both fMRI and EEG methodologies in a crossover design with experienced and novice meditators, the study is positioned to provide definitive evidence regarding whether traditional mantras produce unique neural signatures or whether the observed effects are attributable to general repetitive speech processes.

The findings from this research have the potential to advance our fundamental understanding of how simple cognitive tasks can modulate large-scale brain networks, particularly the DMN. This knowledge could inform the development of evidence-based interventions for mental health conditions and contribute to the growing field of contemplative neuroscience.

Furthermore, the study's focus on both experienced and novice practitioners will provide insights into the role of training and expertise in meditation-related neural changes, addressing whether the benefits of mantra repetition are accessible to individuals without extensive meditation backgrounds. This has important implications for the scalability and clinical applicability of meditation-based interventions.

The integration of multiple neuroimaging modalities and careful experimental controls positions this study to make significant contributions to our understanding of the neurobiological basis of meditative practices and their potential therapeutic applications in promoting mental well-being.

References

Asada, H., Fukuda, Y., Tsunoda, S., Yamaguchi, M., Tonoike, M., 1999. Frontal midline theta rhythms reflect alternative activation of prefrontal cortex and anterior cingulate cortex in humans. Neurosci. Lett. 0304-3940. 274 (1), 29–32. doi:10.1016/s0304-3940(99)00679-5.

Baer, R.A., 2003. Mindfulness training as a clinical intervention: a conceptual and empirical review. Clin. Psychol. Sci. Pract. 10 (2), 125–143. doi:10.1093/clipsy/bpg015.

Bærentsen, K.B., Stødkilde-Jørgensen, H., Sommerlund, B., Hartmann, T., Damsgaard-Madsen, J., Fosnæs, M., Green, A.C., 2010. An investigation of brain processes supporting meditation. Cogn. Process. 1612-4790. 11 (1), 57–84. doi:10.1007/s10339-009-0342-3.

Benson, H., Greenwood, M.M., Klemchuk, H., 1975. The relaxation response: psychophysiologic aspects and clinical applications. Psychiatry Med. 0033-278X. 6 (1–2), 87–98. doi:10.2190/376w-e4mt-qm6q-h0um.

Berkovich-Ohana, A., Wilf, M., Kahana, R., Arieli, A., Malach, R., 2015. Repetitive speech elicits widespread deactivation in the human cortex: the "M antra" effect? Brain Behav. 2162-3279. 5 (7). doi:10.1002/brb3.346.

Cahn, B.R., Polich, J., 2006. Meditation states and traits: EEG, ERP, and neuroimaging studies. Psychol. Bull. 0033-2909. 132 (2), 180–211. doi:10.1037/0033-2909.132.2.180.

Carrington, P., Collings, Benson, H., Robinson, H., Wood, W., Lehrer, P.M., Cole, W., 1980. The use of meditation–relaxation techniques for the management of stress in a working population. J. Occup. Med. 22 (4), 221–231.

Davanger, S., Ellingsen, Ø., Holen, A., Hugdahl, K., 2010. Meditation-specific prefrontal cortical activation during acem meditation: an FMRI study. Percept. Mot. Skills. 0031-5125. 111 (1), 291–306. doi:10.2466/02.04.22.pms.111.4.291-306.

Engström, M., Pihlsgård, J., Lundberg, P., Söderfeldt, B., 2010. Functional magnetic resonance imaging of hippocampal activation during silent mantra meditation. J. Altern. Complement. Med. 1075-5535. 16 (12), 1253–1258. doi:10.1089/acm.2009.0706.

Hasenkamp, W., Barsalou, L.W., 2012. Effects of meditation experience on functional connectivity of distributed brain networks. Front. Hum. Neurosci. 1662-5161. 1–14. Available from: http://www.frontiersin.org/Journal/DownloadFile.ashx?pdf=1&FileId=%2085209&articleId=%2017270&Version=%201&ContentTypeId=21&FileName=%20fnhum-06-00070.pdf. doi:10.3389/fnhum.2012.00038.

Henry, J.D., Crawford, J.R., 2004a. A meta-analytic review of verbal fluency performance following focal cortical lesions. Neuropsychology. 18 (2), 284–295. doi:10.1037/0894-4105.18.2.284.

Henry, J.D., Crawford, J.R., 2004b. Verbal fluency deficits in Parkinson's disease: a meta-analysis. J. Int. Neuropsychol. Soc. 1355-6177. 10 (4), 608–622. Available from: http://journals.cambridge.org/action/displayJournal?jid=INS. doi:10.1017/S1355617704104141.

Ishii, R., Shinosaki, K., Ukai, S., Inouye, T., Ishihara, T., Yoshimine, T., Hirabuki, N., Asada, H., Kihara, T., Robinson, S.E., Takeda, M., 1999. Medial prefrontal cortex generates frontal midline theta rhythm. NeuroReport. 0959-4965. 10 (4), 675–679. doi:10.1097/00001756-199903170-00003.

Ives-Deliperi, V.L., Solms, M., Meintjes, E.M., 2011. The neural substrates of mindfulness: an fMRI investigation. Soc. Neurosci. 1747-0927. 6 (3), 231–242. doi:10.1080/17470919.2010.513495.

Lezak, M.D., 1995. Neuropsychological Assessment, Oxford University Press.

Newberg, A., Alavi, A., Baime, M., Pourdehnad, M., Santanna, J., D'Aquili, E., 2001. The measurement of regional cerebral blood flow during the complex cognitive task of meditation: a preliminary SPECT study. Psychiat. Res. Neuroim. 0925-4927. 106 (2), 113–122. doi:10.1016/S0925-4927(01)00074-9.

Nilkamal Singh, S.T., 2014. A fMRI study of stages of yoga meditation described in traditional texts. J. Psychol. Psychother. 2161-0487. 05 (03). doi:10.4172/2161-0487.1000185.

O'Sullivan, J., O'Connor, O.J., O'Regan, K., Clarke, B., Burgoyne, L.N., Ryan, M.F., Maher, M.M., 2010. An assessment of medical students' awareness of radiation exposures associated with diagnostic imaging investigations. Insights Imaging. 1869-4101. 1 (2), 86–92. doi:10.1007/s13244-010-0009-8.

Ospina, M.B., Bond, K., Karkhaneh, M., Tjosvold, L., Vandermeer, B., Liang, Y., Bialy, L., Hooton, N., Buscemi, N., Dryden, D.M., Klassen, T.P., 2007. Meditation practices for health: state of the research. Evid. Rep. Technol. Assess. 1530-4396. (155), 1–263.

Raboutet, C., Sauzéon, H., Corsini, M.M., Rodrigues, J., Langevin, S., N'Kaoua, B., 2010. Performance on a semantic verbal fluency task across time: dissociation between clustering, switching, and categorical exploitation processes. J. Clin. Exp. Neuropsychol. 1744-411X. 32 (3), 268–280. doi:10.1080/13803390902984464.

Riva, D., Nichelli, F., Devoti, M., 2000. Developmental aspects of verbal fluency and confrontation naming in children. Brain Lang. 0093-934X. 71 (2), 267–284. doi:10.1006/brln.1999.2166.

Travis, F., Shear, J., 2010. Focused attention, open monitoring and automatic self-transcending: categories to organize meditations from Vedic, Buddhist and Chinese traditions. Conscious. Cogn. 1053-8100. 19 (4), 1110–1118. doi:10.1016/j.concog.2010.01.007.

Wang, D.J.J., Rao, H., Korczykowski, M., Wintering, N., Pluta, J., Khalsa, D.S., Newberg, A.B., 2011. Cerebral blood flow changes associated with different meditation practices and perceived depth of meditation. Psychiat. Res. Neuroim. 0925-4927. 191 (1), 60–67. doi:10.1016/j.pscychresns.2010.09.011.

Chapter 23

Role of medical imaging in stroke rehabilitation

Shovan Saha[a], Guruprasad Vijayasarathi[b], Leena R. David[c,d], Sneha Ravichandran[d], Berna Uzun[e], Suresh Sukumar[d], Dilber Uzun Ozsahin[c,e,f], Rajagopal Kadavigere[g], Dilip Shettigar[d], Ilker Ozsahin[e], and Banumathe Karuppaya[h]

[a]*Department of Occupational Therapy, Manipal Academy of Higher Education, Manipal College of Health Professions, Manipal, India*

[b]*Occupational Therapy Program, School of Rehabilitation and Medical Sciences, College of Health Sciences, University of Nizwa, Nizwa, Sultanate of Oman*

[c]*Medical Diagnostic Imaging Department, College of Health Sciences, University of Sharjah, Sharjah, United Arab Emirates*

[d]*Department of Medical Imaging Technology, Manipal College of Health Professions, Manipal Academy of Higher Education, Manipal, Karnataka, India*

[e]*Irfan Suat Gunsel Operational Research Institute, Near East University, Nicosia, Turkish Republic of Northern Cyprus (TRNC), Turkey*

[f]*Research Institute for Medical and Health Sciences, University of Sharjah, Sharjah, United Arab Emirates*

[g]*Department of Radiodiagnosis and Imaging, Kasturba Medical College, Manipal, India*

[h]*Occupational Therapy Program, School of Rehabilitation and Medical Sciences, College of Health Sciences, University of Nizwa, Nizwa, Sultanate of Oman*

1 Introduction to stroke and recovery

1.1 Definition and types of stroke

A stroke, also known as a cerebrovascular accident, occurs when there is an interruption in the blood supply to the brain, resulting in damage to brain tissue. The two primary types of stroke are ischemic stroke and hemorrhagic stroke, each with distinct mechanisms and implications for recovery. Ischemic stroke is the most common type, accounting for about 87% of all stroke cases (Benjamin et al., 2019). It occurs due to the obstruction of blood flow by a thrombus or embolus, leading to a deprivation of oxygen and nutrients in the affected brain regions (Powers, 2018). The result is cell death and loss of function in the areas of the brain served by the blocked artery. Hemorrhagic stroke, which constitutes about 13% of strokes, occurs when a blood vessel in the brain ruptures, leading to intracerebral or subarachnoid hemorrhage (Go et al., 2014). This type of stroke can result from conditions like hypertension, arteriovenous malformations, or aneurysms, causing damage through both direct pressure on brain tissue and disrupted blood flow.

Both types of stroke can lead to significant neurological deficits; the medical management and rehabilitation approaches vary according to the stroke mechanism.

1.2 Epidemiology and impact on global health

Worldwide, stroke is a leading cause of morbidity and mortality, with significant disparities in its distribution across regions. Globally, it is estimated that 12.2 million people experience a stroke each year, resulting in 6.55 million deaths, making stroke the second leading cause of death and the third leading cause of disability (Feigin, 2021). The global burden of stroke has shifted significantly, with low- and middle-income countries now accounting for over 70% of stroke-related deaths and disability-adjusted life years (GBD 2019 Stroke Collaborators, 2021). The incidence of stroke is even more alarming, with approximately 16.9 million people experiencing a stroke each year, translating to a global incidence of 258 per 100,000 people annually (Béjot et al., 2016). Furthermore, the prevalence of stroke survivors has nearly doubled from 1990 to 2010, reaching 33 million, and is projected to rise to 77 million by 2030. This increase in prevalence can

Advances in Medical Imaging. https://doi.org/10.1016/B978-0-443-28967-5.00005-8

be attributed to demographic shifts, aging populations, and improvements in acute stroke management that have led to increased survival rates (Béjot et al., 2016; Béjot, 2022).

Ischemic stroke, which accounts for the majority of cases, is responsible for a considerable proportion of the global stroke burden, particularly in high-income countries, where it is often linked to lifestyle factors such as hypertension, diabetes, and obesity (Feigin, 2021). In contrast, hemorrhagic stroke, though less common, is more fatal and accounts for a significant percentage of stroke-related deaths, particularly in low-income countries, where access to timely medical care is limited (Feigin, 2021). In addition, stroke incidence varies significantly between high- and low-income countries, with higher age-adjusted incidence rates observed in men compared to women (Béjot et al., 2016). This disparity is further exacerbated by limited access to quality healthcare services, a lack of preventive measures, and socioeconomic inequalities.

The economic burden of stroke is considerable, particularly in low- and middle-income countries, where the incidence and mortality rates are rising (Mukherjee and Patil, 2011). The direct medical costs associated with acute stroke care, rehabilitation, and long-term disability management impose significant financial strain on healthcare systems and individuals. Furthermore, the indirect costs of stroke, including lost productivity, long-term caregiving, and social dependency, add to the economic impact. Stroke survivors often face long-term disabilities, including motor impairment, speech difficulties, and cognitive decline, which require ongoing rehabilitation and assistance, further burdening families and healthcare systems (Katan and Luft, 2018). In Asia, poststroke cognitive impairment is a significant concern, affecting stroke survivors' ability to regain independence and function in daily life (Turana et al., 2021).

Several major modifiable risk factors contribute to stroke, including hypertension, diabetes, dyslipidemias, physical inactivity, smoking, poor diet, and excessive alcohol consumption (Boden-Albala et al., 2018; Kuklina et al., 2012). Among these, hypertension is the most prevalent and significant risk factor for stroke, particularly in Asia, where it is a leading contributor to the region's stroke burden (Turana et al., 2021). Effective blood pressure management through lifestyle modifications and pharmacological interventions is crucial in reducing stroke incidence. In addition, atrial fibrillation is a significant risk factor, particularly among the elderly, and its prevalence is expected to increase as populations age, further contributing to the stroke burden (Andrew et al., 2013; Lippi et al., 2021). The identification and management of atrial fibrillation through anticoagulation therapy can substantially reduce the risk of stroke and its associated complications.

As populations continue to age and demographic transitions occur, the global burden of stroke is expected to increase dramatically (Béjot et al., 2016; Béjot, 2022). This anticipated rise underscores the urgent need for comprehensive stroke prevention strategies, improved acute care management, and enhanced rehabilitation services. Primary prevention efforts should focus on controlling risk factors such as hypertension, diabetes, and obesity through public health initiatives, education, and policy changes. Secondary prevention measures, including early detection and treatment of atrial fibrillation and hyperlipidemia, are essential in reducing recurrent strokes and associated disabilities.

The disparity in stroke care between high-income and low-income countries highlights the importance of global health initiatives aimed at improving access to stroke prevention, treatment, and rehabilitation services. Investments in healthcare infrastructure, telemedicine, and community-based interventions can help bridge the gap in stroke care and improve outcomes for patients worldwide. In addition, stroke awareness campaigns and public health programs play a vital role in educating communities about risk factors, early warning signs, and the importance of seeking timely medical attention.

Stroke remains a major global health challenge, with increasing incidence and prevalence driven by aging populations, lifestyle changes, and socioeconomic disparities. The economic and social impacts of stroke are profound, necessitating a multifaceted approach that includes prevention, early intervention, and long-term rehabilitation. Addressing modifiable risk factors, improving access to quality healthcare, and investing in stroke research and innovation are critical steps toward reducing the global stroke burden. Collaborative efforts between governments, healthcare organizations, and researchers are essential in mitigating the impact of stroke and improving the quality of life for stroke survivors worldwide.

1.3 Phases of stroke recovery

Stroke recovery occurs in three different phases acute, subacute, and chronic phases, each characterized by different physiological processes and rehabilitation requirements.

1. **Acute phase**: The acute phase begins immediately after the onset of a stroke. During this period, the primary focus is on medical stabilization and minimizing the extent of brain damage. Acute interventions, such as thrombolysis for ischemic stroke and surgical interventions for hemorrhagic stroke, are critical to preventing further injury and supporting initial recovery. Patients in this phase require close monitoring in a specialized stroke unit to manage complications such as cerebral edema, infections, and deep vein thrombosis. Early mobilization and rehabilitation involving various healthcare professionals, such as physical therapists, occupational therapists, speech therapists, and psychologists, are

now recognized as important components in enhancing functional outcomes during this phase. The effectiveness of early rehabilitation efforts, including passive range-of-motion exercises and positioning strategies, can influence long-term recovery outcomes and reduce complications such as contractures and pressure ulcers.

2. **Subacute phase**: This phase is marked by spontaneous biological recovery, driven by neuroplasticity and repair mechanisms. The subacute period is considered the most responsive to rehabilitation interventions, as the brain's capacity to reorganize and recover function is at its peak. Intensive therapy during this phase can significantly improve motor recovery and cognitive function, with the goal of maximizing independence and functional outcomes. Rehabilitation in this phase typically involves task-specific training, neuromuscular re-education, and activities that promote motor learning. Multidisciplinary teams work together to create individualized rehabilitation programs that address motor deficits, speech impairments, and cognitive dysfunction. Family involvement is crucial at this stage to ensure that patients receive adequate support and motivation during the recovery process.
3. **Chronic phase**: In the chronic phase, recovery tends to slow down, and further progress may rely more on compensatory strategies rather than spontaneous recovery. Rehabilitation in this phase focuses on long-term management of residual deficits, secondary prevention, and adaptive strategies to improve quality of life. Patients may benefit from continued physical therapy, occupational therapy, and speech therapy to maintain gains achieved in earlier stages. In addition, technology-assisted rehabilitation, such as robotic therapy, virtual reality (VR)-based exercises, and electrical stimulation, has shown promise in enhancing recovery during this phase. Although functional improvements can still occur, they tend to be slower and more dependent on continued long-term therapy and support systems. Community reintegration programs, social support groups, and vocational rehabilitation can help stroke survivors regain independence and improve psychological well-being. Addressing mental health concerns, such as poststroke depression and anxiety, is also essential in ensuring holistic recovery and enhancing overall quality of life.

2 Neuroplasticity and stroke recovery

2.1 Mechanisms of neuroplasticity

Neuroplasticity refers to the brain's ability to reorganize itself by forming new neural connections, particularly in response to injuries such as strokes. Several mechanisms underlie neuroplasticity, facilitating functional recovery after strokes.

1. **Axonal sprouting and synaptogenesis**: Following a stroke, surviving neurons can sprout new axons and form new synaptic connections to compensate for lost functions. This process is particularly active in peri-infarct areas, where neurons attempt to reestablish communication pathways disrupted by the stroke (Carmichael, 2016). Axonal sprouting allows for the rewiring of neural circuits, which can lead to partial restoration of motor and sensory functions. The effectiveness of this process can be enhanced by rehabilitative interventions, such as constraint-induced movement therapy (CIMT), task-specific training, and electrical stimulation.
2. **Cortical reorganization**: Neuroplasticity involves the reorganization of cortical maps, wherein intact regions of the brain take over functions previously managed by the damaged areas (Nudo, 2013). Motor functions may be reassigned to adjacent regions of the motor cortex, leading to functional recovery, particularly with rehabilitation. This process is influenced by use-dependent plasticity, meaning that repeated practice of specific movements can strengthen neural pathways and improve motor function. Functional MRI (fMRI) studies have demonstrated shifts in brain activity patterns following a stroke, suggesting that cortical reorganization plays a critical role in regaining lost abilities.
3. **Neurogenesis**: Although more limited in the adult brain, neurogenesis, the formation of new neurons occurs in specific areas such as the hippocampus and can contribute to functional recovery after a stroke. This process is enhanced by rehabilitation interventions and environmental enrichment. Factors such as physical exercise, cognitive stimulation, and pharmacological agents (e.g., neurotrophic factors) have been shown to promote neurogenesis and enhance functional outcomes. In addition, endogenous stem cell migration from neurogenic niches, such as the subventricular zone, to the injured area may support tissue repair and recovery.
4. **Glial cell modulation**: Glial cells, including astrocytes and microglia, play a significant role in neuroplasticity by modulating inflammation, supporting neuronal survival, and facilitating synaptic remodeling. Following a stroke, activated microglia can release neurotrophic factors that promote neuronal repair, while astrocytes contribute to the formation of new synapses and help maintain homeostasis in the damaged brain tissue. Targeting glial cell function through therapeutic interventions may provide novel strategies for enhancing recovery after a stroke.
5. **Hebbian plasticity and long-term potentiation (LTP)**: Hebbian plasticity, often summarized by the phrase "neurons that fire together, wire together," refers to the strengthening of synaptic connections in response to repeated stimulation.

LTP is a key mechanism underlying this process, allowing synaptic connections to become more efficient over time. LTP plays a crucial role in motor learning and functional recovery, particularly when rehabilitation exercises are designed to reinforce synaptic pathways involved in movement and coordination.

2.2 Role of rehabilitation in enhancing neuroplasticity

Neuroplasticity is the brain's ability to reorganize itself by forming new neural connections throughout life. This process is crucial for learning, memory, and recovery from brain injuries such as stroke or traumatic brain injury (Dobkin, 2021; Joshua, 2022). It involves structural and functional changes, including synaptic remodeling, neurogenesis, and the release of neurotrophic factors. Understanding and harnessing neuroplasticity is essential for developing effective rehabilitation strategies aimed at restoring lost functions and improving overall quality of life.

Neuroplasticity occurs at multiple levels, including molecular, cellular, and network levels. At the molecular level, the release of neurotrophic factors, like brain-derived neurotrophic factor and nerve growth factor, supports neuronal survival and synaptic plasticity. Cellular mechanisms involve neurogenesis in specific brain regions, such as the hippocampus, where new neurons are formed and integrated into existing networks. At the network level, neural circuits undergo remodeling through synaptic strengthening (LTP) and weakening (long-term depression), adapting to new experiences and injuries.

Environmental factors, cognitive engagement, and physical activity significantly influence neuroplasticity. Enriched environments with sensory, cognitive, and motor stimulation promote synaptic connectivity and functional recovery following neurological injuries. Moreover, lifestyle factors such as diet, sleep, and stress management play a crucial role in sustaining neuroplastic potential over time.

Rehabilitation leverages neuroplasticity to restore function and improve quality of life following neurological impairments. Various strategies are employed to promote adaptive neuroplasticity, facilitating significant functional recovery (Katz and Dwyer, 2021; Lynskey et al., 2008). These strategies include targeted physical exercises, noninvasive brain stimulation, and task-specific training.

1. **Exercise-based interventions:** Physical exercise is one of the most effective ways to enhance neuroplasticity. Aerobic exercise, resistance training, and mind-body practices such as yoga and tai chi have been shown to enhance neuroplasticity by increasing neurotrophic factors, improving synaptic connectivity, and promoting neurogenesis (Ben Ezzdine et al., 2025; Mang et al., 2013; Penna et al., 2021). Exercise-induced neuroplasticity is primarily mediated by increased cerebral blood flow, metabolic demand, and the release of growth factors that support neural repair and adaptation. High-intensity interval training (HIIT) is particularly effective poststroke, as it enhances cerebral blood flow, mitochondrial function, and neurotrophic factor release, thereby promoting neuroplasticity and cognitive functions (Hugues et al., 2021; Rodrigues et al., 2022). HIIT involves short bursts of intense physical activity followed by recovery periods, optimizing cardiovascular and neuromuscular adaptations. Research indicates that HIIT enhances executive function, working memory, and motor recovery in individuals recovering from neurological impairments. Strength training also contributes to neuroplasticity by improving motor unit recruitment, muscle coordination, and proprioceptive feedback. Regular resistance training has been associated with increased cortical thickness and enhanced motor learning in aging populations and stroke survivors. In addition, dual-task training, which combines physical activity with cognitive tasks, has been found to improve neurocognitive function and motor control simultaneously.
2. **Rehabilitative motor training:** Rehabilitative motor training, including both passive and active exercises, enhances sensorimotor recovery by driving structural and functional plasticity (Lynskey et al., 2008; Torres-Espín et al., 2018). Techniques such as CIMT and robotic-assisted rehabilitation provide intensive, task-specific practice that is critical for inducing neuroplastic changes (Straudi and Basaglia, 2017). CIMT, for instance, forces the use of an impaired limb by restricting the movement of the unaffected one, thereby strengthening neural pathways associated with motor control. Robotic-assisted rehabilitation delivers precise and repetitive motor training, promoting recovery through guided movement patterns. Task-specific training focuses on improving functional movements by engaging individuals in repetitive and meaningful tasks. Activities such as grasping, reaching, and gait training help rewire motor circuits, enabling individuals to regain movement capabilities lost due to neurological damage. VR-based rehabilitation is an emerging field that enhances motor learning by providing interactive, immersive environments that promote active engagement and feedback-driven adjustments. Mirror therapy, another effective intervention, utilizes visual feedback to stimulate motor recovery. By observing the reflection of a functional limb performing movements, individuals can activate mirror neurons and reinforce neural pathways involved in motor control. This technique has been particularly beneficial for stroke patients and individuals with phantom limb pain following amputation.

3. **Noninvasive brain stimulation:** Noninvasive neuromodulation techniques, such as transcranial magnetic stimulation (TMS) and transcranial direct current stimulation (tDCS), can enhance cortical excitability and improve the effects of conventional therapies. These techniques facilitate neuroplasticity by modulating neural network activity, ultimately improving motor and cognitive functions in stroke rehabilitation (Levin, 2006; Lima et al., 2023). TMS uses electromagnetic pulses to stimulate specific brain regions, modulating excitatory and inhibitory circuits to promote motor and cognitive recovery. Repetitive TMS has shown promising results in improving upper limb function, gait, and speech recovery in stroke survivors. Similarly, tDCS delivers low-intensity electrical currents to enhance synaptic plasticity and cortical reorganization. Studies indicate that combining tDCS with rehabilitative exercises leads to greater functional improvements than exercises alone.
4. **Cognitive and psychological rehabilitation:** Cognitive rehabilitation focuses on enhancing neuroplasticity through mental exercises, memory training, and problem-solving tasks. Cognitive stimulation therapy, which involves structured activities to improve attention, reasoning, and executive function, is widely used for patients with neurological conditions such as stroke, traumatic brain injury, and neurodegenerative diseases. Mindfulness-based interventions, including meditation and stress reduction techniques, also contribute to neuroplasticity by modulating neural networks involved in emotional regulation and cognitive processing. Practices such as mindfulness meditation have been associated with increased gray matter density in brain regions responsible for attention and self-regulation.
5. **Pharmacological and nutritional approaches:** Pharmacological interventions aim to support neuroplasticity by targeting neurotransmitter systems and promoting neural repair. Drugs such as selective serotonin reuptake inhibitors and dopamine agonists have been investigated for their potential to enhance neuroplasticity in poststroke rehabilitation. Emerging research suggests that certain nootropic compounds, such as citicoline and piracetam, may facilitate cognitive recovery by modulating neurochemical pathways. Dietary factors also play a crucial role in neuroplasticity. Omega-3 fatty acids, antioxidants, and polyphenols have been shown to support synaptic plasticity and reduce neuroinflammation. A Mediterranean-style diet rich in fruits, vegetables, fish, and healthy fats has been linked to improved cognitive function and neuroprotection.
6. Rehabilitation strategies for stroke patients focus on reducing physical and cognitive impairments, increasing functional independence, and improving quality of life. These approaches are informed by neuroscientific insights into learning mechanisms and training-induced neuroplasticity, ensuring that interventions maximize recovery potential (Dobkin, 2021; Wu et al., 2020). As the field continues to evolve, integrating multidisciplinary approaches, emerging technologies, and personalized interventions will further enhance the effectiveness of rehabilitation and improve patient outcomes.

2.3 Time windows for neuroplastic changes poststroke

Neuroplasticity refers to the brain's ability to reorganize itself by forming new neural connections in response to injury, learning, or experience. Following a stroke, neuroplastic changes occur most prominently within specific time windows, during which the brain is particularly responsive to rehabilitation interventions. Understanding these time frames can help optimize stroke recovery by guiding rehabilitation strategies.

2.3.1 Early poststroke period (0–3 months)

The first three months after a stroke represent the most critical period for neuroplasticity, often referred to as the spontaneous recovery phase. During this time, the brain undergoes heightened plasticity, with several neurobiological processes contributing to recovery. These include synaptogenesis (formation of new synaptic connections), axonal sprouting (growth of new nerve fibers), and cortical remapping (reorganization of brain regions to compensate for lost function) (Zeiler and Krakauer, 2013).

During this phase, the brain experiences a temporary state of increased excitability, allowing for rapid adaptation. The structural and functional reorganization occurring at this stage is largely influenced by both intrinsic mechanisms and external stimuli, including rehabilitation interventions. Intensive rehabilitation during this window can significantly improve functional outcomes, as the brain is more capable of reorganizing itself in response to therapeutic activities. Patients who engage in early and consistent rehabilitation, such as physical therapy, occupational therapy, and speech therapy, tend to achieve better recovery compared to those who delay intervention. However, excessive early training might have adverse effects in some cases, emphasizing the need for a carefully tailored rehabilitation program.

2.3.2 Subacute phase (3–6 months)

Following the early poststroke period, neuroplasticity remains active during the subacute phase, although it begins to taper off compared to the initial three months. This phase still presents a crucial window for meaningful recovery, particularly when rehabilitation is task-specific and repetitive (Krakauer et al., 2012). Patients may continue to see improvements in motor function, cognitive abilities, and speech, though the rate of recovery tends to slow.

Research suggests that the subacute phase is characterized by continued reorganization of neural networks. The brain still exhibits some degree of heightened plasticity, making it possible to refine motor skills and improve functional independence. Rehabilitation efforts should focus on task-oriented exercises, gait training, and activities that encourage bilateral limb use. Engaging patients in real-world tasks, VR therapy, and robotics-assisted rehabilitation can further enhance progress.

Continued rehabilitation during this phase is essential for consolidating the gains made in the early poststroke period. While spontaneous recovery may slow, individuals can still experience substantial benefits with consistent and structured therapy. Moreover, psychological support is crucial during this phase, as many stroke survivors begin to recognize the long-term impact of their condition, which can affect motivation and mental health.

2.3.3 Chronic phase (>6 months)

Neuroplasticity continues into the chronic phase, though the rate of spontaneous recovery slows significantly. Despite this, it remains possible to achieve functional improvements with targeted, high-intensity interventions. Research has shown that rehabilitation efforts, even in the chronic stage, can still lead to neurological adaptations, albeit at a reduced rate compared to earlier phases.

One of the most effective approaches during this stage is CIMT, which involves the intensive, repetitive use of the affected limb while restricting the unaffected limb. This technique encourages cortical reorganization and strengthens neural pathways that support motor recovery. Other interventions, such as TMS and electrical stimulation, have also shown promise in enhancing neuroplastic changes in chronic stroke survivors.

While neuroplastic changes in the chronic phase may be less pronounced, rehabilitation can lead to compensatory strategies that improve daily function. Stroke survivors often learn to adapt their movements and develop new ways to perform tasks, which can significantly enhance their quality of life. Moreover, ongoing cognitive training and social engagement play a crucial role in maintaining brain health and preventing further decline.

Understanding the time windows for neuroplastic changes poststroke is essential for optimizing rehabilitation efforts. The early poststroke period offers the highest potential for recovery, making timely intervention critical. The subacute phase allows for further improvements, reinforcing the gains made earlier. Even in the chronic phase, targeted therapies can still promote neuroplasticity and enhance function. By tailoring rehabilitation strategies to each phase, healthcare professionals can maximize recovery potential and improve long-term outcomes for stroke survivors.

3 Functional imaging in stroke rehabilitation

3.1 Functional MRI: Tracking brain activity and functional recovery

fMRI is a powerful tool for tracking brain activity and mapping functional recovery after a stroke by measuring blood oxygen level-dependent (BOLD) signals. These signals reflect changes in regional brain activity, allowing researchers and clinicians to visualize which areas of the brain are engaged during specific tasks or at rest (Logothetis, 2008). In the context of stroke recovery, fMRI is used to observe cortical reorganization and neuroplastic changes in both affected and unaffected brain areas. For example, during motor recovery, fMRI studies have shown shifts in activation from the damaged motor cortex to adjacent regions or even contralateral motor areas, indicating compensatory mechanisms (Calautti and Baron, 2003). Longitudinal fMRI can track these changes over time, correlating them with improvements in motor and cognitive function, thus providing insights into the effectiveness of rehabilitation interventions. This makes fMRI a valuable tool for both prognosis and the development of personalized rehabilitation strategies.

3.2 Positron emission tomography: Monitoring brain metabolism poststroke

Positron emission tomography (PET) is an imaging technique that enables the visualization and quantification of brain metabolism by tracking radiolabeled tracers, such as fluorodeoxyglucose. PET can assess cerebral glucose metabolism, which is crucial for understanding brain function and recovery after a stroke.

After a stroke, affected brain regions often show hypometabolism, characterized by reduced glucose uptake, even in areas that appear structurally intact. PET imaging is instrumental in identifying these metabolic deficits, which may extend beyond the ischemic core into the surrounding tissue, including the peri-infarct zone, where recovery potential may exist. Monitoring these metabolic changes over time helps to track functional recovery and evaluate the effectiveness of rehabilitation interventions. PET has also been used to explore neuroinflammation and monitor neuroplastic processes, providing a comprehensive picture of poststroke brain activity.

3.3 Transcranial Doppler: Assessment of cerebral blood flow

Transcranial Doppler (TCD) is a noninvasive ultrasound technique used to measure cerebral blood flow velocity in major intracranial arteries. By assessing the speed and direction of blood flow, TCD provides real-time information on cerebral hemodynamics, which is critical in stroke management. TCD is particularly useful for detecting vasospasm following subarachnoid hemorrhage, a common complication that can lead to delayed cerebral ischemia. It also helps in monitoring collateral circulation in cases of large vessel occlusion, assessing emboli in stroke risk stratification, and evaluating cerebral autoregulation (Markus, 2000). Its portability and bedside availability make TCD a valuable tool in both acute stroke settings and long-term care.

3.4 Diffusion tensor imaging: Evaluating white matter tract integrity

Diffusion tensor imaging (DTI) is an advanced MRI technique that measures the diffusion of water molecules in brain tissue, providing detailed information about the integrity of white matter tracts. DTI is particularly valuable in stroke rehabilitation for evaluating corticospinal tract damage, which is critical for motor function recovery.

By quantifying metrics such as fractional anisotropy (FA) and mean diffusivity, DTI allows clinicians to assess the degree of white matter disruption. Reduced FA, for instance, indicates damaged axonal fibers or myelin, which correlates with motor impairment and poor recovery outcomes (Puig et al., 2011). DTI is especially useful for predicting recovery potential by identifying which pathways are structurally intact and likely to support functional improvement, thus helping to tailor rehabilitation interventions such as intensive motor training or neuromodulation.

Through mapping the brain's connectivity, DTI contributes to a better understanding of how white matter integrity influences neuroplasticity and rehabilitation, particularly in guiding personalized therapies.

4 Integration of imaging with rehabilitation protocols

4.1 Use of imaging to personalize rehabilitation approaches

Advanced neuroimaging techniques are increasingly used to personalize rehabilitation strategies by providing detailed insights into brain structure, function, and recovery potential after a stroke. Techniques such as fMRI, DTI, and PET allow clinicians to tailor rehabilitation interventions based on individual patterns of brain damage, neuroplasticity, and recovery trajectories (Cramer, 2018).

For example, fMRI can identify cortical reorganization and activation patterns that reveal which brain areas are compensating for lost function. This information helps clinicians customize therapy to enhance specific motor or cognitive circuits. DTI, which maps white matter integrity, can predict motor recovery by identifying disruptions in corticospinal tracts, guiding more effective motor rehabilitation interventions. PET imaging, by monitoring cerebral metabolism, can further identify regions with residual function or metabolic activity, informing personalized approaches such as neurostimulation or task-specific training to maximize recovery.

These imaging-guided strategies enable the development of rehabilitation programs that align with the unique neurobiological profiles of stroke patients, thereby improving the efficacy of rehabilitation.

4.2 Monitoring progress with imaging tools

Imaging tools such as fMRI, DTI, and PET are crucial for monitoring stroke recovery by providing objective measures of brain activity, structural integrity, and metabolic changes over time. These modalities help track the effectiveness of rehabilitation interventions and guide adjustments to personalized therapy plans.

1. **fMRI:** fMRI tracks brain activation patterns during task performance, allowing clinicians to observe cortical reorganization and compensatory mechanisms as patients recover. Increases in activation in motor areas after motor training, for example, correlate with functional improvements, providing insights into the success of rehabilitation strategies.

2. **DTI:** DTI is used to monitor changes in white matter integrity over time, specifically in pathways like the corticospinal tract. Improvements in FA values are associated with functional recovery, especially in motor skills, indicating neuroplasticity in response to therapy.
3. **PET:** PET imaging assesses changes in cerebral metabolism and neuroinflammation during stroke recovery. Improvements in glucose uptake or reductions in neuroinflammatory markers suggest positive responses to rehabilitation, making PET useful for tracking metabolic recovery and long-term prognosis.

By regularly employing these imaging tools, clinicians can objectively assess progress, refine rehabilitation protocols, and predict long-term outcomes more accurately.

5 Challenges and future directions

5.1 Limitations of current imaging technologies in stroke recovery

While neuroimaging tools such as fMRI, DTI, and PET are valuable for assessing stroke recovery, they have significant limitations that affect their utility in clinical practice.

1. **fMRI:** Despite its ability to map functional changes in the brain, fMRI is limited by its indirect measurement of neural activity, as it relies on BOLD signals rather than actual neuronal firing. In addition, fMRI is susceptible to movement artifacts, making it challenging for stroke patients, particularly those with motor impairments, to remain still during scans (Birn, 2012).
2. **DTI:** While DTI provides insights into white matter integrity, it is limited by its inability to differentiate between causes of reduced FA, such as axonal damage, demyelination, or edema. Furthermore, DTI has lower spatial resolution compared to other imaging methods, making it difficult to map smaller white matter tracts precisely.
3. **PET:** PET imaging is highly sensitive but comes with the drawback of radiation exposure, which limits its repeated use in long-term monitoring. In addition, PET scans have relatively low spatial resolution, making them less effective in pinpointing small or complex regions of the brain. Its high cost and the need for specialized radiotracers further restrict its widespread clinical use.

These limitations highlight the need for further advancements in imaging technology to provide more precise, accessible, and safe monitoring of stroke recovery.

5.2 Future research directions in stroke recovery and medical imaging

1. **Integration of multimodal imaging:** Combining various imaging modalities (e.g., fMRI, DTI, PET) to provide a comprehensive view of brain structure, function, and metabolism could enhance the understanding of stroke recovery processes and improve personalized treatment plans. Research into integrative techniques that fuse data from multiple imaging sources could yield more accurate and holistic assessments of brain health and recovery potential.
2. **Advanced imaging techniques:** Developing novel imaging technologies, such as high-resolution diffusion MRI and advanced neuroimaging biomarkers, could improve the precision of detecting subtle changes in white matter integrity and cortical activation. Research into these advanced techniques aims to overcome current limitations in spatial and temporal resolution, potentially leading to earlier and more accurate detection of recovery patterns.
3. **Neuroimaging biomarkers for predicting outcomes:** Identifying and validating neuroimaging biomarkers that can predict long-term functional outcomes and recovery trajectories will be crucial for personalizing rehabilitation approaches. Future studies could focus on integrating imaging biomarkers with clinical and genetic data to create predictive models for individual recovery prospects.
4. **Real-time and portable imaging:** Developing portable and real-time imaging technologies could revolutionize stroke rehabilitation by allowing for continuous monitoring of brain activity and recovery progress in various settings, including home-based environments. This direction includes advances in wearable imaging technologies and mobile imaging systems.
5. **Neuroplasticity and rehabilitation monitoring:** Research into how specific rehabilitation interventions influence neuroplasticity, as observed through advanced imaging techniques, could lead to more effective therapies. Exploring the relationship between imaging markers of neuroplasticity and functional recovery can refine rehabilitation protocols and enhance therapeutic outcomes.

6 Conclusion

Medical imaging plays a crucial role in stroke rehabilitation by providing detailed insights into brain structure, function, and recovery progress. Advanced imaging techniques, such as MRI, computed tomography (CT), and PET scans, enable clinicians to assess neural damage, monitor treatment efficacy, and guide personalized rehabilitation strategies. By integrating imaging with rehabilitation protocols, healthcare providers can optimize patient outcomes, enhance functional recovery, and improve the quality of life for stroke survivors. Further research and technological advancements will continue to refine imaging applications, making stroke rehabilitation more effective and individualized.

References

Andrew, N.E., Thrift, A.G., Cadilhac, D.A., 2013. The prevalence, impact and economic implications of atrial fibrillation in stroke: what progress has been made? Neuroepidemiology. 40 (4), 229–239. Available from: https://www.scopus.com/inward/record.uri?eid=2-s2.0-84873054619&doi=10.1159%2f000343667&partnerID=40&md5=632c02d33b21b19feb57fb6af4be1b44. doi:10.1159/000343667.

Béjot, Y., 2022. Forty years of descriptive epidemiology of stroke. Neuroepidemiology. 56 (3), 157–162. Available from: https://www.scopus.com/inward/record.uri?eid=2-s2.0-85135431093&doi=10.1159%2f000525220&partnerID=40&md5=9cc046a03cf9a79b12ad76a7532682ff. doi:10.1159/000525220.

Béjot, Y., Daubail, B., Giroud, M., 2016. Epidemiology of stroke and transient ischemic attacks: current knowledge and perspectives. Rev. Neurol. 172 (1), 59–68. Available from: https://www.scopus.com/inward/record.uri?eid=2-s2.0-84950336659&doi=10.1016%2fj.neurol.2015.07.013&partnerID=40&md5=54d4a3926485d9461af2865b60574658. doi:10.1016/j.neurol.2015.07.013.

Ben Ezzdine, L., Dhahbi, W., Dergaa, I., Ceylan, H.İ., Guelmami, N., Ben Saad, H., Chamari, K., Stefanica, V., El Omri, A., 2025. Physical activity and neuroplasticity in neurodegenerative disorders: a comprehensive review of exercise interventions, cognitive training, and AI applications. Front. Neurosci. 19. Available from: https://www.scopus.com/inward/record.uri?eid=2-s2.0-105000349429&doi=10.3389%2ffnins.2025.1502417&partnerID=40&md5=cecbfa57b8a7cc35a29c1a542568fcdd. doi:10.3389/fnins.2025.1502417.

Benjamin, E.J., Muntner, P., Alonso, A., Bittencourt, M.S., Callaway, C.W., Carson, A.P., Chamberlain, A.M., Chang, A.R., Cheng, S., Das, S.R., Delling, F.N., Djousse, L., Elkind, M.S.V., Ferguson, J.F., Fornage, M., Jordan, L.C., Khan, S.S., Kissela, B.M., Knutson, K.L., Kwan, T.W., Lackland, D.T., Lewis, T.T., Lichtman, J.H., Longenecker, C.T., Loop, M.S., Lutsey, P.L., Martin, S.S., Matsushita, K., Moran, A.E., Mussolino, M.E., O'Flaherty, M., Pandey, A., Perak, A.M., Rosamond, W.D., Roth, G.A., Sampson, U.K.A., Satou, G.M., Schroeder, E.B., Shah, S.H., Spartano, N.L., Stokes, A., Tirschwell, D.L., Tsao, C.W., Turakhia, M.P., VanWagner, L.B., Wilkins, J.T., Wong, S.S., Virani, S.S., 2019. Heart disease and stroke statistics-2019 update: a report from the American Heart Association. Circulation. 1524-4539. 139 (10), e56. doi:10.1161/CIR.0000000000000659.

Birn, R.M., 2012. The role of physiological noise in resting-state functional connectivity. NeuroImage. 1095-9572. 62 (2), 864–870. doi:10.1016/j.neuroimage.2012.01.016.

Boden-Albala, B., Appleton, N., Schram, B., 2018. Stroke epidemiology and prevention. In: Wilson, R., Raghavan, P. (Eds.), Stroke Rehabilitation. Elsevier, pp. 1–21. Available from: https://www.scopus.com/inward/record.uri?eid=2-s2.0-85123656466&doi=10.1016%2fB978-0-323-55381-0.00001-9&partnerID=40&md5=65ddbb968a0f118541fa0585491e9e4a. doi:10.1016/B978-0-323-55381-0.00001-9.

Calautti, C., Baron, J.C., 2003. Functional neuroimaging studies of motor recovery after stroke in adults: a review. Stroke. 34 (6), 1553–1566. doi:10.1161/01.STR.0000071761.36075.A6.

Carmichael, S.T., 2016. The 3 rs of stroke biology: radial, relayed, and regenerative. Neurotherapeutics. 1878-7479. 13 (2), 348–359. Available from: http://www.springerlink.com/content/1933-7213/. doi:10.1007/s13311-015-0408-0.

Cramer, S.C., 2018. Treatments to promote neural repair after stroke. J. Stroke. 2287-6405. 20 (1), 57–70. Available from: http://j-stroke.org/upload/pdf/jos-2017-02796.pdf. doi:10.5853/jos.2017.02796.

Dobkin, B.H., 2021. Rehabilitation and recovery of the patient with stroke. In: Grotta, J.C., Albers, G.W., Broderick, J.P., Kasner, M., Sacco, R.L., Schwamm, L.H. (Eds.), Stroke: Pathophysiology, Diagnosis, and Management. Elsevier, pp. 879–887.e2. Available from: https://www.scopus.com/inward/record.uri?eid=2-s2.0-85158958955&doi=10.1016%2fB978-0-323-69424-7.00060-0&partnerID=40&md5=9a51b598148fcd78abb4a35568ce9f1a. doi: 10.1016/B978-0-323-69424-7.00060-0.

Feigin, V.L., et al., 2021. Global, regional, and national burden of stroke and its risk factors, 1990–2019: a systematic analysis for the Global Burden of Disease Study 2019. Lancet Neurol 20 (10), 795–820. doi:10.1016/S1474-4422(21)00252-0.

GBD 2019 Stroke Collaborators, 2021. Global, regional, and national burden of stroke, 1990–2019: a systematic analysis for the Global Burden of Disease Study. Lancet Neurol 20 (10), 795–820. doi:10.1016/S1474-4422(21)00252-0.

Go, A.S., Mozaffarian, D., Roger, V.L., Benjamin, E.J., Berry, J.D., Blaha, M.J., Dai, S., Ford, E.S., Fox, C.S., Franco, S., Fullerton, H.J., Gillespie, C., Hailpern, S.M., Heit, J.A., Howard, V.J., Huffman, M.D., Judd, S.E., Kissela, B.M., Kittner, S.J., Lackland, D.T., Lichtman, J.H., Lisabeth, L.D., Mackey, R.H., Magid, D.J., Marcus, G.M., Marelli, A., Matchar, D.B., McGuire, D.K., Mohler, E.R., Moy, C.S., Mussolino, M.E., Neumar, R.W., Nichol, G., Pandey, D.K., Paynter, N.P., Reeves, M.J., Sorlie, P.D., Stein, J., Towfighi, A., Turan, T.N., Virani, S.S., Wong, N.D., Woo, D., Turner, M.B., 2014. Heart disease and stroke statistics—2014 update: a report from the American Heart Association. Circulation. 1524-4539. 129 (3), e28. doi:10.1161/01.cir.0000441139.02102.80.

Hugues, N., Pellegrino, C., Rivera, C., Berton, E., Pin-Barre, C., Laurin, J., 2021. Is high-intensity interval training suitable to promote neuroplasticity and cognitive functions after stroke? Int. J. Mol. Sci. 22 (6), 1–19. Available from: https://www.scopus.com/inward/record.uri?eid=2-s2.0-85102509752&doi=10.3390%2fijms22063003&partnerID=40&md5=4b31fb4a7885661a89c11a872ca08856. doi:10.3390/ijms22063003.

Joshua, A.M., 2022. Neuroplasticity. In: Joshua, A.M. (Ed.), Physiotherapy for Adult Neurological Conditions. Springer Nature, Singapore, pp. 1–30. Available from: https://www.scopus.com/inward/record.uri?eid=2-s2.0-85171520856&doi=10.1007%2f978-981-19-0209-3_1&partnerID=40&md5=656b17d1d9e4c9b83ba2cafe9a97e2e4. doi: 10.1007/978-981-19-0209-3_1.

Katan, M., Luft, A., 2018. Global burden of stroke. Semin. Neurol. 0271-8235. 38 (02), 208–211. doi:10.1055/s-0038-1649503.

Katz, D.I., Dwyer, B., 2021. Clinical neurorehabilitation: using principles of neurological diagnosis, prognosis, and neuroplasticity in assessment and treatment planning. Semin. Neurol. 41 (2), 111–123. Available from: https://www.scopus.com/inward/record.uri?eid=2-s2.0-85102189381&doi=10.1055%2fs-0041-1725132&partnerID=40&md5=4ac91178d9b01754aa619585f10b9bee. doi:10.1055/s-0041-1725132.

Krakauer, J.W., Carmichael, S.T., Corbett, D., Wittenberg, G.F., 2012. Getting neurorehabilitation right: what can be learned from animal models? Neurorehabilitation and Neural Repair 26 (8), 923–931. doi:10.1177/1545968312440745.

Kuklina, E.V., Tong, X., George, M.G., Bansil, P., 2012. Epidemiology and prevention of stroke: a worldwide perspective. Expert Rev. Neurother. 12 (2), 199–208. Available from: https://www.scopus.com/inward/record.uri?eid=2-s2.0-84856605923&doi=10.1586%2fern.11.99&partnerID=40&md5=b9fd5e917489ed4c4169c6f4e344e2c4. doi:10.1586/ern.11.99.

Levin, H.S., 2006. Neuroplasticity and brain imaging research: implications for rehabilitation. Arch. Phys. Med. Rehabil. 87 (12 SUPPL.), 1. Available from: https://www.scopus.com/inward/record.uri?eid=2-s2.0-33751418945&doi=10.1016%2fj.apmr.2006.09.010&partnerID=40&md5=6dbfebba0e51a79b3eedb8879abef5ce. doi:10.1016/j.apmr.2006.09.010.

Lima, J.P.S., Silva, L.A., Delisle-Rodriguez, D., Cardoso, V.F., Nakamura-Palacios, E.M., Bastos-Filho, T.F., 2023. Unraveling transformative effects after tDCS and BCI intervention in chronic post-stroke patient rehabilitation—an alternative treatment design study. Sensors. 23 (23). Available from: https://www.scopus.com/inward/record.uri?eid=2-s2.0-85179140587&doi=10.3390%2fs23239302&partnerID=40&md5=5c789b7089497687b215d89c6609f33b. doi:10.3390/s23239302.

Lippi, G., Sanchis-Gomar, F., Cervellin, G., 2021. Global epidemiology of atrial fibrillation: an increasing epidemic and public health challenge. Int. J. Stroke. 16 (2), 217–221. Available from: https://www.scopus.com/inward/record.uri?eid=2-s2.0-85078204617&doi=10.1177%2f1747493019897870&partnerID=40&md5=877adb997a5bf0d46189a648dba013ab. doi:10.1177/1747493019897870.

Logothetis, N.K., 2008. What we can do and what we cannot do with fMRI. Nature 453 (7197), 869–878. doi:10.1038/nature06976.

Lynskey, J.V., Belanger, A., Jung, R., 2008. Activity-dependent plasticity in spinal cord injury. J. Rehabil. Res. Dev. 45 (2), 229–240. Available from: https://www.scopus.com/inward/record.uri?eid=2-s2.0-57649203308&doi=10.1682%2fJRRD.2007.03.0047&partnerID=40&md5=e1414b707769d9176cd50548c0c4c2e4. doi:10.1682/JRRD.2007.03.0047.

Mang, C.S., Campbell, K.L., Ross, C.J.D., Boyd, L.A., 2013. Promoting neuroplasticity for motor rehabilitation after stroke: considering the effects of aerobic exercise and genetic variation on brain-derived neurotrophic factor. Phys. Ther. 93 (12), 1707–1716. Available from: https://www.scopus.com/inward/record.uri?eid=2-s2.0-84888992142&doi=10.2522%2fptj.20130053&partnerID=40&md5=49344d97e2292e0d0563dfb4b601ef76. doi:10.2522/ptj.20130053.

Markus, H.S., 2000. Transcranial doppler ultrasound. Br Med Bull. 56 (2), 378–388. doi:10.1258/0007142001903021.

Mukherjee, D., Patil, C.G., 2011. Epidemiology and the global burden of stroke. World Neurosurg. 76 (6 SUPPL.), S85. Available from: https://www.scopus.com/inward/record.uri?eid=2-s2.0-83755170759&doi=10.1016%2fj.wneu.2011.07.023&partnerID=40&md5=2779636aef573e904849adaf40858000. doi:10.1016/j.wneu.2011.07.023.

Nudo, R.J., 2013. Recovery after brain injury: mechanisms and principles. Front. Hum. Neurosci. 1662-5161. 7. Available from: http://www.frontiersin.org/Journal/DownloadFile.ashx?pdf=1&FileId=436107&articleId=74575&Version=1&ContentTypeId=21&FileName=fnhum-07-00887.pdf. doi:10.3389/fnhum.2013.00887.

Penna, L.G., Pinheiro, J.P., Ramalho, S.H.R., Ribeiro, C.F., 2021. Effects of aerobic physical exercise on neuroplasticity after stroke: systematic review. Arq. Neuro-Psiquiatr. 79 (9), 832–843. Available from: https://www.scopus.com/inward/record.uri?eid=2-s2.0-85117373145&doi=10.1590%2f0004-282X-ANP-2020-0551&partnerID=40&md5=5c45d6a7da52abb95d27c9f24705b507. doi:10.1590/0004-282X-ANP-2020-0551.

Powers, W.J., 2018. Guidelines for the early management of patients with acute ischemic stroke: a guideline for healthcare professionals from the American Heart Association/American Stroke Association. Stroke 50 (12), 344–418.

Puig, J., Pedraza, S., Blasco, G., Daunis-i-Estadella, J., Prados, F., Remollo, S., Prats-Galino, A., Soria, G., Boada, I., Castellanos, M., Serena, J., 2011. Acute damage to the posterior limb of the internal capsule on diffusion tensor tractography as an early imaging predictor of motor outcome after stroke. Am. J. Neuroradiol. 0195-6108. 32 (5), 857–863. doi:10.3174/ajnr.A2400.

Rodrigues, L., Moncion, K., Eng, J.J., Noguchi, K.S., Wiley, E., de Las Heras, B., Sweet, S.N., Fung, J., MacKay-Lyons, M., Nelson, A.J., Medeiros, D., Crozier, J., Thiel, A., Tang, A., Roig, M., 2022. Intensity matters: protocol for a randomized controlled trial exercise intervention for individuals with chronic stroke. Trials. 23 (1). Available from: https://www.scopus.com/inward/record.uri?eid=2-s2.0-85130738665&doi=10.1186%2fs13063-022-06359-w&partnerID=40&md5=47b96377d9cbc6c7e2655051392145e5. doi:10.1186/s13063-022-06359-w.

Straudi, S., Basaglia, N., 2017. Neuroplasticity-based technologies and interventions for restoring motor functions in multiple sclerosis. Adv. Exp. Med. Biol. 958, 171–185. Available from: https://www.scopus.com/inward/record.uri?eid=2-s2.0-85009944224&doi=10.1007%2f978-3-319-47861-6_11&partnerID=40&md5=fca7734670fdb5e59ef946c4d990277a. doi:10.1007/978-3-319-47861-6_11.

Torres-Espín, A., Beaudry, E., Fenrich, K., Fouad, K., 2018. Rehabilitative training in animal models of spinal cord injury. J. Neurotrauma. 35 (16), 1970–1985. Available from: https://www.scopus.com/inward/record.uri?eid=2-s2.0-85051170169&doi=10.1089%2fneu.2018.5906&partnerID=40&md5=45273c92265429d2aa52fcfef2631d13. doi:10.1089/neu.2018.5906.

Turana, Y., Tengkawan, J., Chia, Y.C., Nathaniel, M., Wang, J.-G., Sukonthasarn, A., Chen, C.-H., Minh, H.V., Buranakitjaroen, P., Shin, J., Siddique, S., Nailes, J.M., Park, S., Teo, B.W., Sison, J., Soenarta, A.A., Hoshide, S., Tay, J.C., Sogunuru, G.P., Zhang, Y., Verma, N., Wang, T.-D., Kario, K., 2021. Hypertension and stroke in Asia: a comprehensive review from HOPE Asia. J. Clin. Hypertens. 23 (3), 513–521. Available from: https://www.scopus.com/inward/record.uri?eid=2-s2.0-85096667710&doi=10.1111%2fjch.14099&partnerID=40&md5=85c747eb49b66c56895a7650e9a82984. doi:10.1111/jch.14099.

Wu, C.W., Lin, S.-H.N., Hsu, L.-M., Yeh, S.-C., Guu, S.-F., Lee, S.-H., Chen, -C.-C., 2020. Synchrony Between default-mode and sensorimotor networks facilitates motor function in stroke rehabilitation: a pilot fMRI study. Front. Neurosci. 14. Available from: https://www.scopus.com/inward/record.uri?eid=2-s2.0-85087112759&doi=10.3389%2ffnins.2020.00548&partnerID=40&md5=c0ab4a02d6fe4859ad51d6b52983fe95. doi:10.3389/fnins.2020.00548.

Zeiler, S.R., Krakauer, J.W., 2013. The interaction between training and plasticity in the poststroke brain. Curr. Opin. Neurol. 26 (6), 609–616. Available from: https://www.scopus.com/inward/record.uri?eid=2-s2.0-84892373185&doi=10.1097%2fWCO.0000000000000025&partnerID=40&md5=86c400e4ea5462b4e3eab7be2d756e7d. doi:10.1097/WCO.0000000000000025.

Part VII

Traumatic brain injury and auditory neuroscience

Chapter 24

Rehabilitation and medical imaging strategies for coma recovery following traumatic brain injuries

Guruprasad Vijayasarathi[a], Shovan Saha[b], Leena R. David[c,d], Dilip Shettigar[d], Sneha Ravichandran[d], Dilber Uzun Ozsahin[c,e,f], Berna Uzun[f], Suresh Sukumar[d], Rajagopal Kadavigere[g], Ilker Ozsahin[f], and Banumathe Karuppaya[a]

[a]*Occupational Therapy Program, School of Rehabilitation and Medical Sciences, College of Health Sciences, University of Nizwa, Nizwa, Sultanate of Oman*

[b]*Department of Occupational Therapy, Manipal Academy of Higher Education, Manipal College of Health Professions, Manipal, India*

[c]*Medical Diagnostic Imaging Department, College of Health Sciences, University of Sharjah, Sharjah, United Arab Emirates*

[d]*Department of Medical Imaging Technology, Manipal College of Health Professions, Manipal Academy of Higher Education, Manipal, Karnataka, India*

[e]*Research Institute for Medical and Health Sciences, University of Sharjah, Sharjah, United Arab Emirates*

[f]*Irfan Suat Gunsel Operational Research Institute, Near East University, Nicosia, Turkish Republic of Northern Cyprus (TRNC), Turkey*

[g]*Department of Radiodiagnosis and Imaging, Kasturba Medical College, Manipal, India*

1 Overview of traumatic brain injury

1.1 Definition and classification of TBI

1.1.1 Definition

Traumatic brain injury (TBI) is defined as “an injury to the brain resulting from an external force, which can cause significant impairment in cognitive, physical, and emotional functioning. It often results from blunt trauma, such as a fall or a motor vehicle accident, or from penetrating injuries, such as those caused by a gunshot or stab wound” (Lumba-Brown et al., 2018). TBI can have a wide range of effects, from temporary confusion to long-term disability or death. The severity of the injury depends on factors such as the mechanism of injury, the force of impact, and the area of the brain affected.

1.1.2 Classification

Severity of injury

- **Mild TBI (concussion)**: This is the most common form of TBI and is often referred to as a concussion. It is characterized by a brief loss of consciousness (if any) or disorientation lasting under 30 minutes. Symptoms of mild TBI include headache, dizziness, nausea, confusion, difficulty concentrating, and sensitivity to light or noise (McCrory et al., 2017). While most individuals recover fully within days or weeks, some may experience persistent postconcussive symptoms, including memory problems, mood disturbances, and sleep disturbances.
- **Moderate TBI**: This type of injury involves a loss of consciousness lasting between 30 minutes and 24 hours. Posttraumatic amnesia (PTA) can persist from 24 hours to 7 days (Milders et al., 2003). Individuals with moderate TBI may experience significant cognitive, behavioral, and motor impairments that require rehabilitation. They may also suffer from lingering symptoms such as headaches, dizziness, and memory deficits.
- **Severe TBI**: Severe TBI is characterized by a loss of consciousness lasting more than 24 hours and PTA extending beyond 7 days (Masel and DeWitt, 2010). Patients with severe TBI are at risk of long-term disability, including motor deficits, speech impairments, and profound cognitive and emotional dysfunction. Many require extensive rehabilitation and may suffer from permanent impairments that affect their quality of life.

Advances in Medical Imaging. https://doi.org/10.1016/B978-0-443-28967-5.00026-5

Type of injury

- **Closed head injury**: A closed head injury occurs when external forces damage the brain without causing a break in the skull. This can result from rapid acceleration or deceleration forces, such as those experienced in car accidents or falls. Common complications include diffuse axonal injury (DAI), brain contusions, and intracranial bleeding. DAI occurs when the brain shifts inside the skull, causing shearing of neuronal connections, which can lead to widespread brain dysfunction (Giza and Hovda, 2001).
- **Open head injury**: Open head injuries occur when an object penetrates the skull, exposing the brain to external elements. These injuries often involve direct trauma to specific brain regions, leading to focal neurological deficits. Because the brain is exposed, there is a high risk of infection, including meningitis and brain abscess formation (Franjić, 2020). Surgical intervention is often required to remove foreign objects and minimize further damage.

Mechanism of injury

- **Blunt trauma**: This occurs when an external force causes the brain to move within the skull, leading to contusions, intracranial hemorrhages, and DAI. Common causes include falls, motor vehicle accidents, and sports-related injuries. Blunt trauma can result in coup-contrecoup injuries, where the brain is injured both at the site of impact (coup) and on the opposite side due to rebounding forces (contrecoup) (Giza and Hovda, 2001).
- **Penetrating trauma**: This occurs when an object, such as a bullet, knife, or shrapnel, penetrates the skull and directly damages brain tissue. These injuries typically lead to focal damage but can also cause extensive bleeding, brain swelling, and infection. The severity of penetrating trauma depends on the location and depth of penetration, as well as secondary complications such as brain edema and elevated intracranial pressure (Franjić, 2020).

2 Glasgow coma scale

The GCS is a widely used tool for assessing the severity of TBI based on a patient's level of consciousness. It evaluates three domains:

- **Eye-opening response** (scored from 1 to 4)
- **Verbal response** (scored from 1 to 5)
- **Motor response** (scored from 1 to 6)

Scores range from 3 (deep coma) to 15 (fully conscious). The severity of TBI is categorized based on GCS scores:

- Mild TBI: GCS score of 13–15
- Moderate TBI: GCS score of 9–12
- Severe TBI: GCS score of 3–8 (Teasdale and Jennett, 1974)

A lower GCS score indicates a higher level of impairment and a greater need for medical intervention. The scale is commonly used in emergency settings to guide treatment decisions and predict patient outcomes.

3 Posttraumatic amnesia

PTA refers to the period following a TBI during which a person is unable to form new memories. The duration of PTA is used as an indicator of injury severity and recovery prognosis:

- **<24 h**: Mild TBI
- **1 to 7 days**: Moderate TBI
- **>7 days**: Severe TBI (Milders et al., 2003)

PTA is associated with confusion, disorientation, and difficulty retaining new information. The longer the PTA, the greater the likelihood of long-term cognitive deficits.

4 Management and treatment of TBI

The treatment of TBI varies based on severity:

- **Mild TBI (concussion)**: Treatment focuses on symptom management, including rest, pain relief, and a gradual return to activities. Patients are advised to avoid activities that could result in further head injury until they have fully recovered.
- **Moderate to severe TBI**: Treatment involves emergency care to stabilize vital signs, prevent secondary injuries, and manage complications. Surgical interventions may be required to relieve pressure on the brain or remove foreign objects. Rehabilitation programs, including physical therapy, occupational therapy, and cognitive rehabilitation, play a crucial role in recovery.

5 Long-term effects and prognosis

While many individuals with mild TBI recover fully, those with moderate to severe TBI may experience lasting effects, including:

- **Cognitive impairments**: Memory loss, difficulty concentrating, and impaired problem-solving skills
- **Emotional and behavioral changes**: Depression, anxiety, mood swings, and aggression
- **Physical disabilities**: Weakness, coordination problems, and chronic pain
- **Increased risk of neurodegenerative disorders**: Studies suggest a link between TBI and conditions such as Alzheimer's disease and chronic traumatic encephalopathy (CTE)

The prognosis for TBI depends on the severity of the injury, the timeliness of medical intervention, and individual patient factors. Early and intensive rehabilitation can improve outcomes and enhance the quality of life.

6 Epidemiology and impact on functionality of TBI

6.1 Epidemiology

6.1.1 Prevalence and incidence

TBI is a significant public health concern worldwide, impacting millions of individuals annually and leading to substantial economic and social burdens. In the United States, the Centers for Disease Control and Prevention reported that approximately 2.8 million TBIs occur each year, including emergency department visits, hospital admissions, and fatalities (Lumba-Brown et al., 2018). The burden of TBI extends beyond the immediate healthcare system, affecting long-term rehabilitation services, productivity, and quality of life for survivors.

Globally, TBI is a leading cause of disability and mortality, with prevalence and incidence rates varying across different regions. These variations can be attributed to multiple factors, including differences in traffic laws, enforcement of safety regulations, accessibility and quality of healthcare systems, socioeconomic disparities, and public awareness about head injury prevention (Molla and Alemu, 2025). High-income countries often report lower mortality rates due to better emergency response systems and advanced medical care, while low- and middle-income countries struggle with limited healthcare resources, leading to higher TBI-related deaths and long-term disabilities.

Epidemiological studies indicate that TBIs are not uniformly distributed across populations. The rates of occurrence fluctuate based on external factors such as war, natural disasters, and occupational hazards. For instance, military personnel and athletes in contact sports are at a heightened risk of sustaining TBIs due to the nature of their activities. Furthermore, developing nations report a growing incidence of TBIs due to increased motor vehicle accidents, unsafe working conditions, and limited adherence to safety measures such as helmet use and seatbelts.

The demographic profile of TBI victims reveals consistent patterns across studies. Males consistently demonstrate higher rates of TBI than females, with a ratio typically ranging from 1.5:1 to 3:1. This gender disparity is particularly pronounced in adolescence and young adulthood, reflecting greater engagement in high-risk behaviors and occupations. The age distribution of TBI follows a trimodal pattern, with peaks in early childhood, late adolescence/early adulthood, and among the elderly (over 65 years). While younger individuals typically sustain TBIs through sports, violence, or motor vehicle accidents, older adults more commonly experience brain injuries through falls.

The economic impact of TBI is staggering, with direct and indirect costs estimated at $76.5 billion annually in the United States alone. Direct medical costs include emergency services, acute hospitalization, rehabilitation, and long-term care facilities. Indirect costs encompass lost productivity, caregiver burden, and reduced quality of life. The lifetime economic burden for an individual with severe TBI can exceed $4 million, creating substantial financial strain on healthcare systems, insurance providers, families, and society at large.

Prevention strategies have shown promising results in reducing TBI incidence. Legislative measures, such as mandatory helmet laws for motorcyclists and bicyclists, graduated driver licensing programs, and stringent drunk driving penalties, have contributed to decreasing rates of transportation-related TBIs. Technological advancements in vehicle safety features, including airbags, collision avoidance systems, and improved road design, further mitigate injury severity. In sports, rule changes limiting dangerous play, improved protective equipment, and standardized concussion protocols have begun addressing sport-related TBIs, though implementation remains inconsistent across different levels of competition.

The global burden of TBI necessitates coordinated international efforts for prevention, early intervention, and comprehensive rehabilitation. Public health campaigns raising awareness about TBI prevention strategies, improved data collection systems for accurate epidemiological tracking, and enhanced accessibility to rehabilitation services represent critical components of a comprehensive approach to addressing this pervasive health challenge. As urbanization and motorization continue to increase worldwide, particularly in developing regions, proactive measures to combat rising TBI rates will be essential for mitigating the profound personal, social, and economic consequences of these devastating injuries.

6.1.2 Demographics

TBI affects individuals of all age groups, but certain populations are at greater risk. Among children aged 0–4 years, TBIs are predominantly caused by falls, which are common during early childhood as mobility and coordination skills develop (Faul and Coronado, 2015). In this age group, TBIs can result from incidents such as falling from beds, staircases, playground equipment, or accidental impacts with household objects. Due to the vulnerability of the developing brain, early childhood TBIs can lead to cognitive and behavioral impairments that may persist into adulthood.

Similarly, older adults, particularly those aged 75 and above, exhibit high rates of TBI, often as a result of falls and vehicular accidents. Age-related declines in balance, vision, and reflexes contribute to an increased susceptibility to falls, while osteoporosis and frailty make head injuries more severe among the elderly population. TBI in older adults is associated with prolonged hospital stays, higher mortality rates, and an increased risk of neurodegenerative diseases such as Alzheimer's and Parkinson's disease.

In terms of gender disparities, males are disproportionately affected by TBIs compared to females, with incidence rates consistently higher across all age groups (Coronado et al., 2011). This discrepancy may be explained by higher rates of engagement in high-risk behaviors, such as extreme sports, physical altercations, and occupational hazards in industries such as construction, transportation, and law enforcement. In addition, military personnel—predominantly male—are frequently exposed to blast injuries and combat-related TBIs, further contributing to the gender gap. However, some research suggests that females may experience different outcomes post-TBI, with evidence indicating that they may have a greater risk of long-term symptoms, such as post-concussion syndrome and emotional disturbances.

The demographic landscape of TBI is further complicated by socioeconomic factors that create disparities in both incidence and outcomes. Individuals from lower socioeconomic backgrounds face elevated risks of TBI due to environmental hazards, unsafe housing conditions, limited access to protective equipment, and employment in high-risk occupations with inadequate safety protocols. These populations often receive delayed medical attention following injury and have restricted access to specialized rehabilitation services, resulting in poorer long-term functional outcomes and higher disability rates.

Adolescents and young adults between the ages of 15–24 represent another vulnerable demographic, accounting for a significant proportion of TBI cases worldwide. This age group's elevated risk stems from developmental factors, including incomplete frontal lobe maturation, which leads to impulsivity, sensation-seeking behaviors, and susceptibility to peer influence. These characteristics manifest in risky driving behaviors, substance use while operating vehicles, participation in contact sports without proper protective equipment, and engagement in physical altercations. Recent epidemiological data indicate that while motor vehicle accidents remain the leading cause of TBI in this population, sports-related concussions are increasingly recognized as significant contributors, particularly in high-income countries with robust athletic programs.

Racial and ethnic disparities in TBI incidence and outcomes are increasingly documented, though underlying mechanisms remain incompletely understood. Studies in the United States have found that Black and Hispanic populations experience higher rates of TBI-related hospitalizations and mortality compared to non-Hispanic Whites. These disparities likely reflect complex interactions between socioeconomic status, healthcare access, residential segregation patterns that affect exposure to environmental hazards, and potential biases in injury assessment and treatment protocols. Native American and Alaska Native populations also show disproportionately high TBI rates, particularly in rural areas where emergency medical services may be limited and transportation to trauma centers delayed.

Military veterans represent a demographic with unique TBI considerations. Beyond combat-related injuries, service members face elevated TBI risks during training exercises, parachute operations, and even recreational activities on military installations. The signature injuries of recent conflicts in Iraq and Afghanistan include blast-related TBIs from improvised explosive devices, which can produce complex patterns of neural damage through multiple injury mechanisms, including pressure waves, acceleration-deceleration forces, and penetrating fragments. Veterans with TBI often present with comorbid conditions, including post-traumatic stress disorder, chronic pain, and substance use disorders, creating treatment challenges that require integrated care approaches across multiple specialties.

Understanding these demographic patterns is essential for developing targeted prevention strategies, allocating healthcare resources efficiently, and designing rehabilitation programs that address the specific needs of different populations. The heterogeneity of TBI across age groups, genders, socioeconomic strata, and occupational categories underscores the need for customized approaches to both prevention and treatment rather than one-size-fits-all solutions.

6.1.3 Risk factors

Several key risk factors contribute to the likelihood of sustaining a TBI, with motor vehicle accidents, falls, and violent incidents ranking among the leading causes.

1. **Motor vehicle accidents:** Traffic-related accidents account for a significant proportion of TBIs globally, particularly in regions with poor road safety measures, inadequate law enforcement, and a lack of public awareness about seatbelt and helmet usage. High-speed collisions, pedestrian accidents, and cycling incidents all contribute to TBI incidence. In many developing countries, a lack of proper infrastructure, such as pedestrian walkways and dedicated cycling lanes, exacerbates the problem. Young adults, particularly those aged 15–24, are at higher risk of sustaining TBIs from motor vehicle accidents due to reckless driving, speeding, and driving under the influence of alcohol or drugs.
2. **Falls:** Falls represent the leading cause of TBIs in both young children and older adults. In young children, a lack of supervision, unsafe play environments, and household hazards contribute to frequent falls. Among the elderly, conditions such as osteoporosis, arthritis, impaired vision, and medication side effects that cause dizziness increase the risk of falls. Prevention strategies, such as home modifications (e.g., handrails, slip-resistant flooring), physical therapy to improve balance, and regular vision screenings, are crucial in reducing fall-related TBIs in older adults.
3. **Violence and assaults:** Physical assaults, domestic violence, and gunshot wounds also contribute significantly to TBI cases. Interpersonal violence, particularly among marginalized communities and individuals experiencing homelessness, is a notable contributor to TBI rates. In some cases, repeated exposure to head trauma—such as in cases of intimate partner violence—can lead to CTE, a neurodegenerative condition associated with cognitive decline, mood disorders, and motor impairments.
4. **Alcohol and substance use:** Alcohol consumption and drug use are widely recognized as risk factors for TBI. Intoxication impairs judgment, coordination, and reaction time, increasing the likelihood of accidents, falls, and violent encounters. Studies indicate that a substantial proportion of TBI cases involve individuals who were under the influence of alcohol or other substances at the time of injury (Thurman et al., 1999). Public health interventions focusing on responsible alcohol use, addiction treatment, and awareness campaigns may help mitigate this risk.
5. **Occupational and recreational activities:** Certain professions and recreational activities inherently pose a higher risk of TBI. Construction workers, athletes in contact sports (e.g., football, boxing, hockey), and military personnel are particularly vulnerable. Occupational safety measures, including the enforcement of helmet regulations, proper training, and workplace hazard assessments, are essential in reducing TBI incidence in these high-risk groups.

TBI remains a pressing public health issue worldwide, with significant variations in prevalence based on geographical, demographic, and socioeconomic factors. While substantial progress has been made in TBI prevention and treatment, ongoing efforts are needed to address disparities in healthcare access, enforce safety regulations, and implement educational initiatives aimed at reducing the risk factors associated with TBIs. Future research should continue to explore strategies for improving early diagnosis, rehabilitation outcomes, and long-term care for individuals affected by traumatic brain injuries.

6.1.4 Impact on functionality

1. **Physical functioning:** TBIs result in a wide range of impairments in physical functioning, including coordination problems, motor deficits, and difficulties in balance and gait. Severe TBIs often lead to long-term physical disabilities that significantly affect daily living activities and overall quality of life. Many individuals require rehabilitation services, such as physical therapy, occupational therapy, and assistive devices, to regain mobility and independence (Maas et al., 2017).

2. **Cognitive functioning:** TBIs significantly impact cognitive abilities, often leading to deficits in memory, attention, executive functioning, and processing speed. These cognitive challenges make it difficult for individuals to return to work, engage in social activities, and perform everyday tasks. Cognitive rehabilitation programs, including neuropsychological interventions and cognitive-behavioral therapy, play a crucial role in helping individuals adapt to these challenges (Barman et al., 2016).
3. **Emotional and behavioral impact:** Depression, anxiety, aggression, and personality changes are common among individuals with TBIs. Emotional and behavioral impairments complicate rehabilitation efforts by affecting interpersonal relationships and social integration. Psychological interventions, counseling, and support groups are often necessary to help individuals cope with these changes (Yeo, 2021).
4. **Long-term outcomes:** A significant proportion of TBI survivors experience long-term or even permanent disabilities. Studies indicate that approximately 15%–30% of individuals with TBI face persistent impairments that impact their ability to perform daily activities and maintain employment. The long-term consequences of TBI highlight the need for continued support, vocational rehabilitation, and community-based programs to enhance the quality of life for survivors.

7 Coma stimulation therapy

7.1 Definition and rationale

Coma Stimulation Therapy (CST) is a therapeutic approach that aims to stimulate the senses of individuals in a coma or minimally conscious state, thus helping to promote the recovery of consciousness and facilitating neurorehabilitation. CST is based on the principle that sensory input can help activate the neural pathways and improve neural plasticity, providing outcome improvements in patients with severe brain injuries (Padilla and Domina, 2016; Alashram et al., 2020).

7.2 Techniques and modalities

7.2.1 Sensory stimulation

1. Visual stimulation: Visual stimuli, such as bright lights or moving objects, are used. Studies show that brain activity is enhanced in patients with impaired consciousness through visual stimulation (Wijnen et al., 2014).
2. Auditory stimulation: sound and music exposure are the key components. In comatose patients, exposure to familiar voices or music therapy markedly improves responses (Jones et al., 1994).
3. Tactile stimulation: physical contact, such as light touch, gentle massage, or even the texture of different materials, induces tactile stimulation. In patients with severe brain injury, tactile stimulation positively reduces agitation and improves awareness (YekeFallah et al., 2018; Salehi et al., 2019).
4. Olfactory stimulation: This involves the use of scents and odors. Even though very rarely used, olfactory stimulation can help in arousal and memory retrieval in some patients (Glachet et al., 2018; Gao et al., 2019).

7.2.2 Motor stimulation

1. Passive movement: movements provided to the patient without their active participation. Passive movements maintain joint flexibility, prevent muscle atrophy, and provide sensory input to the brain (Petre et al., 2014; Pittaccio et al., 2013).
2. Active movement: movements that are meant to be performed by the patient themselves are possible. Making the patient perform movements actively enhances motor function and improves the overall rehabilitation outcome (Ciccarelli et al., 2005; Iqbal and Mughal, 2007).

7.2.3 Multimodal stimulation

Combination of sensory inputs: Multimodal stimulation involves a combination of visual, auditory, tactile, and olfactory stimuli. According to some studies, the multimodal approach is more effective in enhancing sensory processing and promoting cognitive recovery than single-modal interventions (Megha et al., 2013; Deshpande et al., 2019).

8 Integrating medical imaging and CST

8.1 Role of imaging in guiding therapy

8.1.1 Personalizing stimulation protocols

Medical imaging plays a crucial role in guiding CST by providing detailed insights into the nature and extent of brain injuries. The selection of appropriate stimulation therapies is significantly enhanced by advanced imaging technologies such as magnetic resonance imaging (MRI) and computed tomography (CT). These imaging modalities help identify specific areas of brain damage that may benefit from targeted sensory input, thus enabling clinicians to develop tailored protocols suited to the individual needs of each patient (Giza and Hovda, 2001; Maas et al., 2017).

MRI, particularly diffusion-weighted imaging and susceptibility-weighted imaging, can detect microstructural damage and hemorrhagic lesions that might not be evident on conventional CT scans. CT scans, on the other hand, provide rapid and essential information on acute intracranial pathologies, such as hemorrhage, edema, and skull fractures, which are critical for the initial assessment of patients in comatose states. By using these imaging findings, clinicians can determine the most appropriate coma stimulation interventions, such as auditory, visual, tactile, or olfactory stimulation, depending on the preserved and affected neural pathways.

Furthermore, functional imaging techniques, including functional MRI (fMRI) and positron emission tomography (PET), allow for a dynamic assessment of brain activity and metabolic function. These modalities provide valuable information regarding neural responsiveness and cortical activation patterns, which can indicate the potential for recovery. By monitoring how the brain responds to stimulation over time, clinicians can make real-time adjustments to the therapeutic approach, optimizing the timing, type, and intensity of stimuli to maximize therapeutic benefits (Pouratian and Bookheimer, 2006; Bandettini, 2009).

For instance, in cases where fMRI reveals intact but underactive neural circuits, repetitive stimulation may be applied to those areas to promote neuroplasticity and cortical reorganization. Similarly, PET scans, which measure glucose metabolism and cerebral blood flow, can highlight areas with preserved metabolic activity, guiding targeted interventions aimed at enhancing functional connectivity. These advanced imaging tools ensure that CST is not applied uniformly but is instead personalized to each patient's unique neurological profile, thus improving therapeutic efficacy and recovery prospects.

8.1.2 Monitoring recovery progress

Objective assessment of neurological improvements

Tracking changes in brain function and structure over time is critical in assessing the effectiveness of CST. Imaging studies provide objective and quantifiable measures of recovery progress, helping clinicians determine whether therapeutic interventions are leading to meaningful neurological improvements.

One of the most widely used imaging techniques for monitoring brain function in coma patients is fMRI, which detects changes in blood oxygenation levels associated with neuronal activity. An increase in brain activation patterns on fMRI scans suggests a positive response to CST, indicating that the patient's brain is gradually regaining its functional capabilities (Edlow et al., 2013). This information can guide clinicians in fine-tuning stimulation protocols, such as increasing stimulus intensity or diversifying the types of sensory inputs provided.

Diffusion tensor imaging (DTI), an advanced MRI technique, can be used to assess the integrity of white matter tracts, offering valuable insights into the extent of axonal injury and the potential for neuronal recovery. Serial DTI scans allow clinicians to track white matter changes over time, helping them determine whether the neural pathways involved in sensory processing and consciousness are improving in response to therapy.

Similarly, PET imaging, which measures cerebral glucose metabolism, can help predict long-term outcomes by identifying areas of the brain that retain metabolic activity despite severe injury. Patients with higher metabolic activity in key brain regions, such as the thalamus and prefrontal cortex, have a better prognosis and may benefit more from prolonged and intensive stimulation therapy. Repeated PET scans can also detect shifts in metabolic activity over time, signifying neuroplastic changes and gradual functional restoration.

Electroencephalography (EEG) and magnetoencephalography are additional tools that provide valuable information about neural responsiveness. EEG, which records electrical activity in the brain, is particularly useful for assessing residual cognitive function and detecting emerging signs of awareness. Event-related potentials, a subset of EEG analysis, can measure the brain's response to auditory or visual stimuli, helping to determine whether a patient can process external sensory input, even in the absence of overt behavioral responses.

8.1.3 Clinical implications of imaging-guided therapy adjustments

By leveraging these imaging technologies, clinicians can make evidence-based decisions regarding the continuation, modification, or cessation of CST. For example, if imaging studies indicate minimal neural activity or progressive atrophy in critical brain regions, alternative therapeutic strategies may be explored, including pharmacological interventions or neuromodulation techniques such as transcranial magnetic stimulation or deep brain stimulation. Conversely, if imaging reveals promising signs of neuroplasticity and increasing metabolic activity, the intensity and frequency of CST may be escalated to further enhance recovery outcomes.

Ultimately, the integration of medical imaging into CST represents a paradigm shift in the management of patients with disorders of consciousness. By providing a window into the brain's structural and functional status, imaging enables the development of precision-based therapeutic strategies that maximize each patient's potential for recovery. As imaging technology continues to advance, future innovations—such as artificial intelligence-driven image analysis and machine learning algorithms for predicting recovery trajectories—hold the promise of further revolutionizing the field of neurorehabilitation.

Medical imaging serves as an indispensable tool in the optimization of CST, facilitating personalized treatment approaches and objective monitoring of recovery progress. By harnessing the power of MRI, CT, fMRI, PET, and EEG, clinicians can refine therapeutic interventions to align with the unique neurological needs of each patient. This integration not only enhances the effectiveness of CST but also paves the way for improved patient outcomes and a deeper understanding of the mechanisms underlying brain recovery. As research in this field continues to evolve, the synergy between medical imaging and coma rehabilitation will remain a cornerstone of neurocritical care and consciousness restoration.

9 Future directions and research in CST

9.1 Advances in imaging technologies

9.1.1 High-resolution imaging techniques

Advanced MRI modalities: High-field MRI and DTI are innovative and advanced modalities that provide us with detailed views of the structure and connectivity of the brain. Better assessment of brain injury and recovery is made with these advanced techniques, which facilitate clinicians in providing more precise and individualized coma stimulation therapies (Jones and Leemans, 2011).

Optical imaging: Near-infrared spectroscopy and optical coherence tomography are becoming more reliable tools for monitoring cerebral oxygenation and blood flow, thus emerging as essential tools for assessing the response to coma therapies (Yang et al., 2025).

9.1.2 Real-time functional imaging

Real-time fMRI: During stimulation therapy, real-time fMRI allows dynamic monitoring of brain activity. With the feedback the real-time fMRI provides about brain responses, clinicians can refine stimulation protocols and improve therapeutic outcomes (Birn et al., 2013).

PET and single-photon emission CT (SPECT) advances: Improvements in PET and SPECT techniques have enhanced the ability to visualize brain metabolism and neurotransmitter activity, thus improving the effectiveness of coma stimulation interventions (Brooks, 2005).

9.2 Innovations in CST

9.2.1 Personalized stimulation approaches

Customized sensory protocols: Sensory stimulation protocols must be tailor-made based on an individual's brain injury profile and response patterns, which is currently a major aspect of research. The efficacy of CST is enhanced through personalized approaches that target specific neural pathways and functions (Lombardi et al., 2002).

Adaptive stimulation systems: Adaptive stimulation systems that adjust in real-time based on patient responses and imaging data have been well, leading to improved therapy outcomes. Integration of feedback within these systems through various imaging modalities helps optimize stimulation (Frączek et al., 2021).

9.2.2 Neurotechnological innovations

Brain–computer interfaces (BCIs): BCIs offer new ways to engage patients in coma stimulation therapies through neural signals to control external devices. BCIs are under research, which mainly aims to develop systems that can provide real-time neural feedback through targeted stimulation (Lebedev and Nicolelis, 2006).

Neuroprosthetics: Advancements in neuroprosthetic devices that directly interact with neural tissue could provide new avenues for coma stimulation. Precise control over sensory and motor inputs through these devices enhances their therapeutic efficacy (Hochberg et al., 2006).

9.3 Interdisciplinary approaches

9.3.1 Collaboration between disciplines

Integration of neurology, rehabilitation, and imaging: Neurologists, rehabilitation therapists, and imaging specialists put in a collaborative effort to advance coma stimulation therapies. Comprehensive treatment plans are developed by interdisciplinary teams in order to optimize patient outcomes (Giza and Hovda, 2001).

Research and clinical practice integration: Research findings and clinical practice should always be interlinked for transforming innovative therapies into effective treatments. Implementing new technologies and approaches is always a result of collaboration between researchers and clinicians (McMillan et al., 2009).

10 Conclusion

TBI remains a significant public health concern due to its high prevalence, long-term consequences, and impact on individuals and society. The complexity of TBI, ranging from mild to severe cases, necessitates comprehensive management strategies, including immediate medical intervention, rehabilitation, and long-term care. Advances in imaging techniques, treatment modalities, and neurorehabilitation have improved outcomes for patients, but challenges persist in early diagnosis, effective treatment, and the reduction of secondary brain injury.

Preventive measures, such as helmet use, seatbelt enforcement, and public awareness campaigns, are crucial in minimizing the occurrence of TBI. Continued research is needed to further understand the pathophysiology of TBI, improve treatment protocols, and enhance the quality of life for survivors. A multidisciplinary approach, involving healthcare professionals, caregivers, and policymakers, is essential in mitigating the burden of TBI and ensuring better patient outcomes.

Despite progress in acute care protocols, significant gaps remain in our understanding of postacute recovery trajectories and effective interventions for cognitive, behavioral, and emotional sequelae. The heterogeneity of TBI presentations complicates treatment standardization, while emerging evidence suggests that even mild TBI may trigger progressive neurodegenerative processes in vulnerable individuals. Biomarker discovery and validation represent promising frontiers for improving diagnostic precision and prognostication. Implementation of evidence-based prevention strategies remains inconsistent across regions and populations, particularly in resource-limited settings where TBI incidence continues to rise with increasing motorization. Addressing these challenges requires sustained investment in translational research, healthcare infrastructure, and public policy initiatives that prioritize both prevention and comprehensive care for those affected by this pervasive and often devastating condition.

References

Alashram, A.R., Annino, G., Aldajah, S., Hamad, S.B., Aliswed, B., Padua, E., 2020. Effects of sensory stimulation on level of consciousness in comatose patients after traumatic brain injury: a systematic review. Physiother. Pract. Res. 2213-0691. 41 (2), 143–153. Available from: http://iospress.metapress.com/content/122560/?sortorder=asc. doi:10.3233/PPR-200428.

Bandettini, P.A., 2009. Seven topics in functional magnetic resonance imaging. J. Integr. Neurosci. 0219-6352. 8 (3), 371–403. doi:10.1142/S0219635209002186.

Barman, A., Chatterjee, A., Bhide, R., 2016. Cognitive impairment and rehabilitation strategies after traumatic brain injury. Indian J. Psychol. Med. 0975-1564. 38 (3), 172–181. Available from: http://www.ijpm.info/. doi:10.4103/0253-7176.183086.

Birn, R.M., Molloy, E.K., Patriat, R., Parker, T., Meier, T.B., Kirk, G.R., Nair, V.A., Meyerand, M.E., Prabhakaran, V., 2013. The effect of scan length on the reliability of resting-state fMRI connectivity estimates. NeuroImage. 1095-9572. 83, 550–558. doi:10.1016/j.neuroimage.2013.05.099.

Brooks, D.J., 2005. Positron emission tomography and single-photon emission computed tomography in central nervous system drug development. NeuroRx. 1545-5351. 2 (2), 226–236. Available from: http://www.elsevierhealth.com. doi:10.1602/neurorx.2.2.226.

Ciccarelli, O., Toosy, A.T., Marsden, J.F., Wheeler-Kingshott, C.M., Sahyoun, C., Matthews, P.M., Miller, D.H., Thompson, A.J., 2005. Identifying brain regions for integrative sensorimotor processing with ankle movements. Exp. Brain Res. 0014-4819. 166 (1), 31–42. doi:10.1007/s00221-005-2335-5.

Coronado, V.G., Xu, L., Basavaraju, S.V., McGuire, L.C., Wald, M.M., Faul, M.D., Guzman, B.R., Hemphill, J.D., 2011. Surveillance for traumatic brain injury-related deaths—United States, 1997–2007. MMWR Morb. Mortal. Wkly. Rep. 1545-861X. 60 (5), 1–36. Available from: http://www.cdc.gov/mmwr/pdf/ss/ss6005.pdf.

Deshpande, S.D., Varadharajulu, G., Kanase, S., 2019. Effect of choice-based sensory stimulation as a coma stimulation technique on traumatic brain injury. Indian J. Public Health Res. Dev. 0976-5506. 10 (7), 94–99. Available from: http://indianjournals.com/ijor.aspx?target=ijor:ijphrd&volume=10&issue=7&article=018&type=pdf. doi:10.5958/0976-5506.2019.01543.2.

Edlow, B.L., Giacino, J.T., Wu, O., 2013. Functional mri and outcome in traumatic coma. Curr. Neurol. Neurosci. Rep. 1534-6293. 13 (9). Available from: www.springer.com. doi:10.1007/s11910-013-0375-y.

Faul, M., Coronado, V., 2015. Epidemiology of traumatic brain injury. In: Grafman, J.H., Salazar, A.M. (Eds.), Handbook of Clinical Neurology, vol. 127, Elsevier BV, pp. 3–13. doi:10.1016/b978-0-444-52892-6.00001-5.

Frączek, T.M., Ferleger, B.I., Brown, T.E., Thompson, M.C., Haddock, A.J., Houston, B.C., Ojemann, J.G., Ko, A.L., Herron, J.A., Chizeck, H.J., 2021. Closing the loop with cortical sensing: the development of adaptive deep brain stimulation for essential tremor using the activa PC+S. Front. Neurosci. 1662-453X. 15. Available from: https://www.frontiersin.org/journals/neuroscience#. doi:10.3389/fnins.2021.749705.

Franjić, S., 2020. Head injuries, a general approach. Iberoam. J. Med. 2695-5075. 2 (1), 19–23. doi:10.53986/ibjm.2020.0005.

Gao, Z., Wu, R., Chen, C., Wen, B., Liu, Y., Lu, W., Chen, N., Feng, J., Fan, R., Wang, D., Cui, S., Wang, J.-H., 2019. Coactivations of barrel and piriform cortices induce their mutual synapse innervations and recruit associative memory cells. Brain Res. 0006-8993. 1721, 146333. doi:10.1016/j.brainres.2019.146333.

Giza, C.C., Hovda, D.A., 2001. The neurometabolic cascade of concussion. J. Athl. Train. 36 (3), 228–235.

Glachet, O., Gandolphe, M.-C., Gallouj, K., Antoine, P., El Haj, M., 2018. Effects of olfactory stimulation on autobiographical memory in Alzheimer's disease. Gériatr. Psychol. Neur.. 1760-1703. 16 (3), 311–320. doi:10.1684/pnv.2018.0748.

Hochberg, L.R., Serruya, M.D., Friehs, G.M., Mukand, J.A., Saleh, M., Caplan, A.H., Branner, A., Chen, D., Penn, R.D., Donoghue, J.P., 2006. Neuronal ensemble control of prosthetic devices by a human with tetraplegia. Nature. 14764679. 442 (7099), 164–171. doi:10.1038/nature04970.

Iqbal, K., Mughal, A.M. 2007. Active control vs. passive stiffness in posture and movement coordination. In: Conference Proceedings—IEEE International Conference on Systems, Man and Cybernetics. United States. pp. 3367–3372. 1062922X. doi:10.1109/ICSMC.2007.4414105

Jones, D.K., Leemans, A., 2011. Diffusion tensor imaging. Methods Mol. Biol. 711, 127–144. Available from: https://www.springer.com/series/7651. doi:10.1007/978-1-61737-992-5_6.

Jones, R., Hux, K., Anne Morton-Anderson, K., Knepper, L., 1994. Auditory stimulation effect on a comatose survivor of traumatic brain injury. Arch. Phys. Med. Rehabil. 00039993. 75 (2), 164–171. doi:10.1016/0003-9993(94)90390-5.

Lebedev, M.A., Nicolelis, M.A.L., 2006. Brain-machine interfaces: past, present and future. Trends Neurosci. 0166-2236. 29 (9), 536–546. doi:10.1016/j.tins.2006.07.004.

Lombardi, F.F.L., Taricco, M., De Tanti, A., Telaro, E., Liberati, A., 2002. Sensory stimulation for brain injured individuals in coma or vegetative state. Cochrane Database Syst. Rev. 1465-1858. (2). doi:10.1002/14651858.cd001427.

Lumba-Brown, A., Yeates, K.O., Sarmiento, K., Breiding, M.J., Haegerich, T.M., Gioia, G.A., Turner, M., Benzel, E.C., Suskauer, S.J., Giza, C.C., Joseph, M., Broomand, C., Weissman, B., Gordon, W., Wright, D.W., Moser, R.S., McAvoy, K., Ewing-Cobbs, L., Duhaime, A.C., Putukian, M., Holshouser, B., Paulk, D., Wade, S.L., Herring, S.A., Halstead, M., Keenan, H.T., Choe, M., Christian, C.W., Guskiewicz, K., Raksin, P.B., Gregory, A., Mucha, A., Taylor, H.G., Callahan, J.M., Dewitt, J., Collins, M.W., Kirkwood, M.W., Ragheb, J., Ellenbogen, R.G., Spinks, T.J., Ganiats, T.G., Sabelhaus, L.J., Altenhofen, K., Hoffman, R., Getchius, T., Gronseth, G., Donnell, Z., O'Connor, R.E., Timmons, S.D., 2018. Centers for disease control and prevention guideline on the diagnosis and management of mild traumatic brain injury among children. JAMA Pediatr. 2168-6203. 172 (11). Available from: http://archpedi.jamanetwork.com/issues.aspx. doi:10.1001/jamapediatrics.2018.2853.

Maas, A.I.R., Menon, D.K., David Adelson, P.D., Andelic, N., Bell, M.J., Belli, A., Bragge, P., Brazinova, A., Büki, A., Chesnut, R.M., Citerio, G., Coburn, M., Jamie Cooper, D., Tamara Crowder, A., Czeiter, E., Czosnyka, M., Diaz-Arrastia, R., Dreier, J.P., Duhaime, A.C., Ercole, A., van Essen, T.A., Feigin, V.L., Gao, G., Giacino, J., Gonzalez-Lara, L.E., Gruen, R.L., Gupta, D., Hartings, J.A., Hill, S., Jiang, J.Y., Ketharanathan, N., Kompanje, E.J.O., Lanyon, L., Laureys, S., Lecky, F., Levin, H., Lingsma, H.F., Maegele, M., Majdan, M., Manley, G., Marsteller, J., Mascia, L., McFadyen, C., Mondello, S., Newcombe, V., Palotie, A., Parizel, P.M., Peul, W., Piercy, J., Polinder, S., Puybasset, L., Rasmussen, T.E., Rossaint, R., Smielewski, P., Söderberg, J., Stanworth, S.J., Stein, M.B., von Steinbüchel, N., Stewart, W., Steyerberg, E.W., Stocchetti, N., Synnot, A., Te Ao, B., Tenovuo, O., Theadom, A., Tibboel, D., Videtta, W., Wang, K.K.W., Huw Williams, W., Wilson, L., Yaffe, K., Adams, H., Allanson, J., Coles, J., Hutchinson, P.J., Kolias, A.G., Sahakian, B.J., Stamatakis, E., Williams, G., Agnoletti, V., Martino, C., Masala, A., Teodorani, G., Zumbo, F., Amrein, K., Ezer, E., Kolumbán, B., Kovács, N., Melegh, B., Nyirádi, J., Sorinola, A., Vámos, Z., Andaluz, N., Anke, A., Frisvold, S.K., Antoni, A., van As, A.B., Figaji, A., Audibert, G., Azaševac, A., Dilvesi, D., Golubović, J., Jelača, B., Karan, M., Kolundžija, K., Negru, A., Vulekovic, P., Azouvi, P., Azzolini, M.L., Beretta, L., Baciu, C., Beqiri, V., Chevallard, G., Chieregato, A., Sacchi, M., Badenes, R., Belda, F.J., Bilotta, F., Lozano, A., Barlow, K.M., Schneider, K.J., Bartels, R., den Boogert, H., Hoedemaekers, C., Sir, Ö., Bauerfeind, U., Lefering, R., Schäfer, N., Beauchamp, M., Gravel, J., Beer, D., Beer, R., Helbok, R., Höfer, S., Bellander, B.M., Nelson, D., Bellier, R., Benard, T., Carise, E., Dahyot-Fizelier, C., Giraud, B., Benali, H., Bernard, F., Bertolini, G., Masson, S., Blaabjerg, M., Rosenlund, C., Schou, R.F., Boutis, K., Bouzat, P., Francony, G., Manhes, P., Payen, J.F., Brooks, B., Dewey, D., Emery, C.A., Freedman, S., Kramer, A., Brorsson, C., Koskinen, L.O., Sundström, N., Bullinger, M., Burns, E., Calappi, E., Ortolano, F., Cameron, P., Castaño-León, A.M., Gómez López, P.A., Lagares, A., Causin, F., Freo, U., Persona, P., Rossi, S., Christie, B., Cnossen, M., Dippel, D., Foks, K., Haagsma, J.A., Haitsma, I., Huijben, J.A., van der Jagt, M., Nieboer, D., Volovici, V., Voormolen, D.C., Collett, J., Dawes, H., Esser, P., van Heugten, C., Della Corte, F., Grossi, F., Craig, W., Csato, G., Csomos, A., Curry, N., Dematteo, C., Meade, M., Depreitere, B., van Dijck, J., de Ruiter, G.C.W., Vleggeert-Lankamp, C., Dizdarevic, K., Donoghue, E., Gantner, D., Murray, L., Trapani, T., Vallance, S., Duek, O., Lazar, I., Dulière, G.L., Maréchal, H., Dzeko, A., Eapen, G., Jankowski, S., English, S., Fergusson, D., Osmond, M., Fabricius, M., Kondziella, D., Feng, J., Hui, J., Fleming, J., Latini, R., Gagnon, I., Ptito, A., Galanaud, D., Glocker, B., Kamnitsas, K., Ledig, C., Rueckert, D., Gordon, W.A., Gradisek, P., Griesdale, D., Håberg, A.K., van Hecke, W., Smeets, D., Verheyden, J., Vyvere, T.V., Helseth, E., Røe, C., Røise, O., Horton, L., Jacobs, B., van der Naalt, J., Janssens, K., De Keyser, V., Menovsky, T., Van Praag, D., Jones, K.M., Kapš, R., Katila, A., Posti, J., Takala, R., Kaukonen, K.M., Kivisaari, R., Piippo-Karjalainen, A., Raj, R., Tanskanen, P., Kutsogiannis, D., Kyprianou, T., Lamontagne, F., Lauzier, F., Moore, L., Turgeon, A., Legrand, V., Levi, L., Zaaroor, M., Lightfoot, R., Macdonald, S., Major, S., Vajkoczy, P., Wessels, L., Winkler, M.K.L., Wolf, S., Manara, A., Thomas, M., Mattern, J., Sakowitz, O., Vogt, L., Younsi, A., McFadyen, B., McMahon, C., Correia, M.M., Morganti-Kossmann, M.C., Rosenfeld, J.V., Muehlan, H.,

Schmidt, S., Mukherjee, P., Noirhomme, Q., Oddo, M., Okonkwo, D.O., Oldenbeuving, A.W., Roks, G., Schoonman, G.G., Perlbarg, V., Pichon, N., Pili-Floury, S., Pirinen, M., Pleş, H., Poca, M.A., Radoi, A., Sahuquillo, J., Ragauskas, A., Rocka, S., Real, R.G.L., Telgmann, R., Reed, N., Rhodes, J., Robertson, C., Rosand, J., Rosenthal, G., Salvato, G., Sánchez-Porras, R., Sándor, J., Sangha, G., Schnyer, D., Schöhl, H., Skandsen, T., Stevanovic, A., van Waesberghe, J.V., Stevens, R.D., Taccone, F.S., Taylor, M.S., Zelinkova, V., Temkin, N., Tolias, C.M., Valadka, A.B., Valeinis, E., Vargiolu, A., Vega, E., Vik, A., Vilcinis, R., Wildschut, E., Wood, G., Xirouchaki, N., Zemek, R., 2017. Traumatic brain injury: integrated approaches to improve prevention, clinical care, and research. Lancet Neurol. 1474-4465. 16 (12), 987–1048. Available from: http://www.journals.elsevier.com/the-lancet-neurology/. doi:10.1016/S1474-4422(17)30371-X.

Masel, B.E., DeWitt, D.S., 2010. Traumatic brain injury: a disease process, not an event. J. Neurotrauma. 0897-7151. 27 (8), 1529–1540. doi:10.1089/neu.2010.1358.

McCrory, P., Feddermann-Demont, N., Dvoøák, J., Cassidy, J.D., McIntosh, A., Vos, P.E., Echemendia, R.J., Meeuwisse, W., Tarnutzer, A.A., 2017. What is the definition of sports-related concussion: a systematic review. Br. J. Sports Med. 1473-0480. 51 (11), 877–887. Available from: http://bjsm.bmj-journals.com/. doi:10.1136/bjsports-2016-097393.

McMillan, A.M., Landorf, K.B., Barrett, J.T., Menz, H.B., Bird, A.R., 2009. Diagnostic imaging for chronic plantar heel pain: a systematic review and meta-analysis. J. Foot Ankle Res. 1757-1146. 2 (1). Available from: http://www.jfootankleres.com/content/2/1/32. doi:10.1186/1757-1146-2-32.

Megha, M., Harpreet, S., Nayeem, Z., 2013. Effect of frequency of multimodal coma stimulation on the consciousness levels of traumatic brain injury comatose patients. Brain Inj. 0269-9052. 27 (5), 570–577. doi:10.3109/02699052.2013.767937.

Milders, M., Fuchs, S., Crawford, J.R., 2003. Neuropsychological impairments and changes in emotional and social behaviour following severe traumatic brain injury. J. Clin. Exp. Neuropsychol. 25 (2), 157–172. doi:10.1076/jcen.25.2.157.13642.

Molla, Y.D., Alemu, H.T., 2025. Epidemiology of traumatic brain injury in Ethiopia: a systematic review and meta-analysis of prevalence, mechanisms, and outcomes. PLOS ONE 19326203. 20 (5). Available from: https://journals.plos.org/plosone/article?id=10.1371/journal.pone.0322641. doi:10.1371/journal.pone.0322641.

Padilla, R., Domina, A., 2016. Effectiveness of sensory stimulation to improve arousal and alertness of people in a coma or persistent vegetative state after traumatic brain injury: a systematic review. Am. J. Occup. Ther. 0272-9490. 70 (3). Available from: http://ajot.aota.org/article.aspx?articleid=2512800. doi:10.5014/ajot.2016.021022.

Petre, I., Deaconescu, A., Rogozea, L., Deaconescu, T.I., 2014. Orthopaedic rehabilitation device actuated with pneumatic muscles. Int. J. Adv. Robot. Syst. 1729-8814. 11 (1). Available from: http://cdn.intechopen.com/pdfs/47258.pdf. doi:10.5772/58693.

Pittaccio, S., Zappasodi, F., Tamburro, G., Viscuso, S., Marzetti, L., Garavaglia, L., Tecchio, F., Pizzella, V. 2013. Passive ankle dorsiflexion by an automated device and the reactivity of the motor cortical network. In: Proceedings of the Annual International Conference of the IEEE Engineering in Medicine and Biology Society, EMBS. Italy, pp. 6353–6356. 1557170X. doi:10.1109/EMBC.2013.6611007.

Pouratian, N., Bookheimer, S.Y., 2006. Clinical challenges of fMRI. In: Faro, S.H., Mohamed, F.B. (Eds.), Functional MRI: Basic Principles and Clinical Applications.Springer New York, United States, pp. 99–124. Available from: http://www.springerlink.com/openurl.asp?genre=book&isbn=978-0-387-23046-7. doi:10.1007/0-387-34665-1_5.

Salehi, B., Azimian, J., Yekefallah, L., Mafi, M., 2019. Research paper: effects of foot reflexology and tactile stimulation on consciousness and physiological indicators of traumatic patients. Caspian J. Neurol. Sci. 2423-4818. 5 (3), 133–139. Available from: http://cjns.gums.ac.ir. doi:10.32598/CJNS.5.18.133.

Teasdale, G., Jennett, B., 1974. Assessment of coma and impaired consciousness. A practical scale. Lancet. 0140-6736. 304 (7872), 81–84. doi:10.1016/S0140-6736(74)91639-0.

Thurman, D.J., Alverson, C., Dunn, K.A., Guerrero, J., Sniezek, J.E., 1999. Traumatic brain injury in the United States: a public health perspective. J. Head Trauma Rehabil. 0885-9701. 14 (6), 602–615. Available from: http://journals.lww.com/headtraumarehab. doi:10.1097/00001199-199912000-00009.

Wijnen, V.J.M., Eilander, H.J., de Gelder, B., van Boxtel, G.J.M., 2014. Visual processing during recovery from vegetative state to consciousness: comparing behavioral indices to brain responses. Neurophysiol. Clin. 1769-7131. 44 (5), 457–469. Available from: http://www.elsevier.com/wps/find/journaldescription.cws_home/505818/description#description. doi:10.1016/j.neucli.2014.08.008.

Yang, L., Chen, P., Wen, X., Zhao, Q., 2025. Optical coherence tomography (OCT) and OCT angiography: technological development and applications in brain science. Theranostics. 1838-7640. 15 (1), 122–140. doi:10.7150/thno.97192.

YekeFallah, L., Aghae, F., Azimian, J., Heidari, M.A., Hasandoost, F., 2018. Tactile stimulation improves consciousness and vital signs in patients with traumatic brain injury. Nurs. Crit. Care 1558-447X. 13 (6), 18–22. Available from: http://journals.lww.com/nursingcriticalcare. doi:10.1097/01.CCN.0000546310.00168.30.

Yeo, J., 2021. Emotional disturbances after traumatic brain injury: prevalence, assessment, and treatment. Curr. Phys. Med. Rehabil. Rep. 2167-4833. 9 (3), 85–92. doi:10.1007/s40141-021-00311-w.

Chapter 25

Role of magnetic resonance imaging in observing the structure of amygdala and hippocampus to assess the stress related behavioral changes and the impact on learning and memory

Sudarshan Surendran[a], Venu Madhav Nelluri[b], Suresh Sukumar[c], Leena R. David[c,d], Sneha Ravichandran[c], Berna Uzun[e], Dilber Uzun Ozsahin[d,e,f], M.G. Ramesh Babu[g], Rajagopal Kadavigere[h], Dilip Shettigar[c], and Ilker Ozsahin[e]

[a]*Department of Anatomy, American University of Antigua College of Medicine, Antigua, Antigua and Barbuda*

[b]*Department of Anatomy, Manipal Univeristy College Malaysia, Melaka, Malaysia*

[c]*Department of Medical Imaging Technology, Manipal College of Health Professions, Manipal Academy of Higher Education, Manipal, Karnataka, India*

[d]*Medical Diagnostic Imaging Department, College of Health Sciences, University of Sharjah, Sharjah, United Arab Emirates*

[e]*Irfan Suat Gunsel Operational Research Institute, Near East University, Nicosia, Turkish Republic of Northern Cyprus (TRNC), Turkey*

[f]*Research Institute for Medical and Health Sciences, University of Sharjah, Sharjah, United Arab Emirates*

[g]*Division of Physiology, Department of Basic Medical Sciences, Manipal Academy of Higher Education, Manipal, India*

[h]*Department of Radiodiagnosis and Imaging, Kasturba Medical College, Manipal, India*

1 Introduction

The human brain is a complex organ and is responsible for a multitude of functions. One of the most intricate processes is its capability to manage the body's response to stress. Stress, by itself, is an adaptive and natural reaction to challenging and threatening situations. However, in situations involving exposure to chronic stress or excessive stress, it can have profound consequences on the physical and mental health of an individual. To understand the stress response, it is crucial to explore the specific regions of the brain that play a primary role in this complex process.

The neurobiological basis of the stress response relies on a sophisticated interplay between various brain structures and neuroendocrine systems. At the molecular level, stress activates multiple signaling cascades that ultimately influence gene expression, protein synthesis, and cellular function (Arnsten, 2015). Understanding these mechanisms is not merely an academic pursuit but has profound implications for developing effective treatments for stress-related disorders.

In modern society, chronic stressors have largely replaced the acute physical threats our ancestors faced. Psychological stressors, such as work pressure, financial concerns, and social conflicts, have become predominant. Despite this shift in the nature of stressors, our brain's response mechanisms remain largely unchanged, potentially leading to maladaptive physiological and psychological outcomes when these responses are chronically activated (McEwen, 2017). This mismatch between our evolutionarily conserved stress response systems and contemporary stressors contributes significantly to the rising prevalence of stress-related disorders.

Advances in Medical Imaging. https://doi.org/10.1016/B978-0-443-28967-5.00031-9

1.1 Brain's emotional center, the amygdala

The amygdala forms the core of our stress response system in the human body. It is a small, almond-shaped part of the nervous system located deep within the brain's temporal lobe. It is often mentioned as the brain's emotional center owing to its vital role in processing emotional information, primarily related to fear and anxiety. The amygdala rapidly senses and assesses the situation when a potential threat is detected. Then, the body's fight-or-flight response is triggered, releasing stress hormones such as cortisol and adrenaline to prepare the body for a reaction against the perceived danger.

The amygdala's role in emotional processing extends far beyond simply detecting threats. This complex structure consists of several nuclei, each with distinct functions and connections. The basolateral complex serves as the primary input region, receiving sensory information from thalamic and cortical pathways. The central nucleus, on the other hand, acts as the primary output, projecting to brainstem areas that control autonomic responses (Janak and Tye, 2015). This anatomical organization allows the amygdala to integrate diverse sensory inputs and coordinate appropriate emotional responses.

Recent advances in high-resolution magnetic resonance imaging (MRI) techniques have enabled researchers to visualize these amygdalar subdivisions with unprecedented detail in living human subjects. A study by Saygin et al. (2017) used ultra-high-field 7-Tesla MRI to map the distinct connectivity profiles of amygdalar nuclei, revealing how these subdivisions form networks with other brain regions to process different aspects of emotional stimuli.

1.2 Amygdala in fear conditioning: Mechanisms and clinical relevance

The amygdala plays a critical role in fear conditioning, a form of associative learning where neutral stimuli become associated with aversive outcomes. This process involves several mechanisms, including N-methyl-D-aspartate (NMDA) receptor-dependent synaptic plasticity in the lateral amygdala, which strengthens connections between sensory inputs and amygdala neurons (Maren and Holmes, 2016). These cellular mechanisms provide the substrate for emotional memories that can persist for years, sometimes manifesting as pathological fears in conditions like phobias and posttraumatic stress disorder (PTSD).

In clinical settings, functional MRI (fMRI) studies have demonstrated hyperactivity of the amygdala in patients with anxiety disorders during exposure to fear-relevant stimuli. For instance, individuals with specific phobias show exaggerated amygdala responses when viewing images of their feared objects compared to nonphobic controls (Etkin and Wager, 2007). Similarly, patients with PTSD exhibit heightened amygdala reactivity to trauma-related cues, suggesting a neural basis for their persistent fear responses.

1.3 Case study: Amygdala function in combat veterans with PTSD

A landmark study by Hayes et al. (2012) examined amygdala structure and function in combat veterans with and without PTSD. Using a combination of structural and fMRI, the researchers found that veterans with PTSD showed both reduced amygdala volume and heightened amygdala reactivity to combat-related imagery compared to trauma-exposed veterans without PTSD. These findings suggest that structural alterations in the amygdala may contribute to its functional hyperreactivity in PTSD, potentially explaining the persistent fear responses characteristic of this disorder.

1.4 Memory and stress regulation center, the hippocampus

The hippocampus is located adjacent to the amygdala and forms a part of the limbic system. This seahorse-shaped structure is involved in memory formation and regulation. While the amygdala triggers the rapid response system against stress, the hippocampus dampens this effect. The hippocampus contextualizes the stressor and modulates the body's reaction through the feedback it provides to the amygdala. The hippocampus can be affected in chronic stress conditions, impacting learning and memory functions.

The hippocampus contains an exceptionally high density of glucocorticoid receptors, making it particularly sensitive to the effects of stress hormones (De Kloet et al., 2005). Under normal conditions, these receptors help regulate the hypothalamic–pituitary–adrenal (HPA) axis through negative feedback mechanisms, effectively turning off the stress response once the threat has passed. However, prolonged exposure to elevated glucocorticoid levels can lead to downregulation of these receptors, impairing the hippocampus's ability to effectively regulate the stress response.

1.4.1 Hippocampal subregions and differential vulnerability to stress

Modern neuroimaging techniques have revealed that different subregions of the hippocampus—including the CA1, CA3, dentate gyrus, and subiculum—have distinct functions and varying vulnerabilities to stress. For example, the CA3 region appears particularly susceptible to the effects of chronic stress, with animal studies showing dendritic retraction and reduced neurogenesis in this area following prolonged stress exposure (McEwen et al., 2016).

High-resolution MRI studies in humans have confirmed these subregional effects, with a study by Treadway et al. (2015) demonstrating selective volume reductions in the CA3/dentate gyrus regions in individuals with major depression, a disorder strongly linked to chronic stress. This specificity highlights the importance of examining hippocampal subfields rather than treating the hippocampus as a homogeneous structure when investigating stress-related brain changes.

1.4.2 Clinical relevance: Hippocampal changes in stress-related disorders

Multiple metaanalyses have consistently demonstrated reduced hippocampal volume in patients with major depressive disorder, PTSD, and chronic stress exposure (Videbech and Ravnkilde, 2004). These structural changes correlate with cognitive deficits, particularly in memory domains, suggesting a direct relationship between hippocampal integrity and cognitive function in these populations.

1.4.3 Case study: Cognitive rehabilitation in stress-induced memory impairment

A clinical case report by Bremner and Marmar (2020) described a 45-year-old executive who developed significant memory problems following a period of intense work-related stress. MRI scanning revealed reduced hippocampal volume compared to age-matched controls. The patient underwent a comprehensive treatment program, including stress management techniques, cognitive behavioral therapy, and cognitive rehabilitation exercises specifically targeting memory functions.

1.5 Execution control and stress regulation center, the prefrontal cortex

This region of the brain is located in the front and is responsible for higher-order cognitive functions. These include decision-making, problem-solving, and impulse control. This region of the brain plays a critical role in regulating the stress response. It is also believed to connect to the amygdala and other brain regions. Through these connections, the top-down control over emotional reactions can be exerted and is involved in managing stress.

The prefrontal cortex (PFC) can be divided into several subregions, each with distinct roles in stress regulation. The medial PFC has direct projections to the amygdala and can either enhance or suppress emotional responses, depending on the specific subregion and context. The dorsolateral PFC, meanwhile, contributes to cognitive reappraisal—the ability to reinterpret emotional situations in less threatening ways—which is a key strategy for emotional regulation (Etkin et al., 2015).

fMRI studies have revealed dynamic patterns of connectivity between the PFC and limbic structures during stress regulation. For instance, effective downregulation of negative emotions is associated with increased activity in prefrontal regions, coupled with decreased activity in the amygdala, suggesting top-down inhibitory control (Ochsner et al., 2012). Disruptions in this regulatory circuit have been observed in various stress-related disorders, potentially explaining the emotional dysregulation characteristic of these conditions.

1.5.1 Executive function impairment in chronic stress: Mechanisms and implications

Chronic stress has been consistently associated with deficits in executive functions mediated by the PFC, including working memory, cognitive flexibility, and inhibitory control (Arnsten, 2009). At the cellular level, stress-induced catecholamine release impairs prefrontal function by weakening the connections between pyramidal neurons that maintain task-relevant information.

These stress-induced executive function impairments have significant real-world implications. For example, a study by Liston et al. (2009) examined medical residents during periods of intensive work stress and found both reduced prefrontal activation during attention-switching tasks and corresponding performance deficits. These cognitive effects correlated with measures of perceived stress and reversed after a period of reduced work hours, highlighting both the impact of chronic stress on prefrontal function and the potential for recovery with stress reduction.

1.6 Hormone regulation center, the hypothalamus, and pituitary gland

The hippocampus and the pituitary gland are located deep within the brain and work in tandem to regulate the body's response to stress. The corticotropin-releasing hormone (CRH) from the hypothalamus signals the pituitary gland to release the adrenocorticotropic hormone (ACTH). The ACTH then stimulates the production of cortisol from the adrenal glands, a primary hormone related to stress. This intricate cascade of hormones results in mobilizing the body's resources to deal with stress.

The HPA axis represents a critical interface between the central nervous system and the endocrine system, translating neural signals into hormonal responses. This system operates through a sophisticated negative feedback mechanism, where elevated cortisol levels normally inhibit further CRH and ACTH release, preventing excessive or prolonged stress responses (Herman et al., 2016). Dysfunction in this feedback system has been implicated in numerous stress-related disorders.

1.6.1 Imaging the hypothalamic–pituitary–adrenal axis: Technical challenges and innovations

Directly visualizing the hypothalamus and pituitary using conventional MRI techniques presents significant challenges due to their small size and location. However, advances in neuroimaging have enabled more detailed examination of these structures. Ultra-high-field MRI at 7 T provides the spatial resolution necessary to distinguish hypothalamic nuclei, while specialized sequences can enhance contrast between these densely packed neural structures (Satpute et al., 2019).

Functional imaging of the HPA axis involves additional complexities, as hormonal responses occur on a different timescale than neural activity. Researchers have addressed this challenge through combined imaging and endocrine sampling protocols. For example, a study by Dedovic et al. (2009) used a psychosocial stress paradigm during fMRI scanning while simultaneously collecting saliva samples for cortisol measurement.

1.6.2 Clinical applications: HPA axis dysregulation in psychiatric disorders

Abnormalities in HPA axis function have been documented across numerous psychiatric conditions. In major depression, elevated cortisol levels and impaired negative feedback are common findings, potentially contributing to the hippocampal volume reductions observed in this disorder (Pariante and Lightman, 2008). Conversely, some PTSD patients exhibit hypocortisolism, suggesting a distinct pattern of HPA axis dysregulation characterized by enhanced negative feedback sensitivity (Yehuda and Seckl, 2011).

1.7 What happens when an external disruptive force is introduced—Stress?

In this intensely accelerating world, it is an unwelcome event that leaves a considerable mark on the structures responsible for managing stress and affects our emotions, as well as the areas responsible for recording our experiences. As we dive deep into this fascinating world of neuroimaging, the profound changes in the amygdala and hippocampus become clear to understand under such conditions. MRI provides the advantage of precision and depth for studying these structures, and our access to understanding them in such conditions is increased multifold (McEwen, 2012).

Stress can be conceptualized along several dimensions, including duration (acute versus chronic), predictability (expected versus unexpected), controllability, and intensity. These dimensions influence not only the psychological experience of stress but also its neurobiological impact. For example, uncontrollable stressors typically produce more pronounced physiological and neural responses than controllable ones, even when the objective stressor is identical (Maier and Seligman, 2016).

1.7.1 The allostatic load model: Cumulative effects of stress

The concept of allostatic load, introduced by Bruce McEwen, provides a framework for understanding how repeated or chronic stress exposure affects brain structure and function over time. Allostatic load refers to the cumulative wear and tear on physiological systems resulting from repeated cycles of adaptation to stress (McEwen, 1998). As allostatic load increases, the efficiency of the stress response system diminishes, leading to dysregulation across multiple physiological domains.

Neuroimaging studies have begun to capture the brain's manifestation of allostatic load, with recent research examining the relationship between cumulative life stress and brain structure. A longitudinal study by Gianaros (2015) tracked individuals over a 20-year period, measuring stressful life events and obtaining brain scans at multiple time points. The researchers found that cumulative stress exposure predicted accelerated brain aging, particularly in regions involved in stress regulation, such as the PFC and hippocampus.

1.7.2 Technological advances in stress-related neuroimaging

Recent technological developments have significantly enhanced our ability to visualize stress-related brain changes. Multimodal imaging approaches, combining structural, functional, and molecular techniques, provide complementary information about how stress affects different aspects of brain organization. For example, integrating MRI with positron emission tomography (PET) allows researchers to examine both the structure of stress-sensitive brain regions and the activity of specific neurotransmitter systems within these regions (Krystal, 2013).

Machine learning approaches are increasingly being applied to neuroimaging data to identify patterns associated with stress exposure and stress-related disorders. A study by Admon et al. (2020) used support vector machine learning to analyze multimodal MRI data from trauma-exposed individuals, successfully identifying neural signatures that predicted the subsequent development of PTSD with over 80% accuracy.

1.8 Adult brain plasticity

Initially, the brain was thought to be a static and unchanging organ, apart from its electrophysiological responsiveness, such as long-term potentiation. With the advancement of science and the instrumentation contributing to this, the rewiring mechanisms after brain damage have gained considerable interest (Bliss and Lømo, 1973). This plasticity seen in the brain has led to the investigation of the alterations within the brain's structures in various pathological and neurological conditions.

The discovery of adult neurogenesis—the formation of new neurons in the adult brain—revolutionized our understanding of brain plasticity. While once thought to occur only during development, we now know that neurogenesis continues throughout life in specific brain regions, particularly the dentate gyrus of the hippocampus (Eriksson et al., 1998). This process is highly sensitive to environmental factors, with stress being a potent suppressor of neurogenesis and enriched environments or exercise enhancing it (Kempermann et al., 2018).

The integration of newly born neurons into existing hippocampal circuits has significant functional implications for learning, memory, and stress regulation. Animal studies have shown that ablating adult neurogenesis impairs pattern separation—the ability to distinguish between similar experiences—and enhances anxiety-like behaviors (Anacker and Hen, 2017). Conversely, interventions that promote neurogenesis, such as exercise or antidepressant treatment, can ameliorate stress-induced cognitive and emotional impairments.

1.8.1 Neuroplasticity-based interventions for stress-related disorders

Our expanding knowledge of brain plasticity has inspired novel therapeutic approaches targeting neuroplastic mechanisms to treat stress-related disorders. Cognitive training programs designed to engage and strengthen prefrontal regulatory circuits have shown promise in reducing symptoms of anxiety and depression (Course-Choi et al., 2017). Similarly, mindfulness-based interventions appear to enhance connectivity between prefrontal and limbic regions, potentially strengthening top-down emotional regulation capacity (Tang et al., 2015).

Pharmacological agents that promote neuroplasticity are also being investigated as treatments for stress-related disorders. Ketamine, which enhances synaptic plasticity through the modulation of glutamatergic signaling, has demonstrated rapid antidepressant effects and is now approved for treatment-resistant depression (Duman et al., 2016).

1.9 Stress, a trigger

Several neuropsychiatric disorders have stress as a trigger, and alterations in such conditions are observed in the structural and functional components of the brain regions, along with their networks involved. The prelimbic, cingulate, somatosensory, motor, auditory, and entorhinal cortices, the hippocampus, and the nucleus accumbens are some of the key regions involved in this (Magalhães et al., 2018). Experiments have demonstrated the involvement of multiple regions and related structural atrophy in stressful conditions.

The relationship between stress and psychiatric disorders is bidirectional and complex. While stress can trigger or exacerbate mental health conditions, having a psychiatric disorder can also increase vulnerability to stress and reduce coping capacity. This creates the potential for vicious cycles where stress and psychiatric symptoms mutually reinforce each other (Hammen, 2005).

1.9.1 Developmental timing and stress vulnerability

The timing of stress exposure across the lifespan significantly influences its impact on brain development and function. Early-life stress, particularly during critical developmental windows, can have lasting effects on stress-regulatory systems and increase vulnerability to later psychopathology (Tottenham and Sheridan, 2010). For example, childhood maltreatment is associated with altered development of key stress-regulatory brain regions, including reduced hippocampal and prefrontal volumes, as well as enhanced amygdala reactivity.

Neuroimaging studies have documented how early-life stress affects the structural and functional development of stress-regulatory neural circuits. A longitudinal study by Barch (2016) followed children exposed to early adversity from ages 3 to 15, collecting multiple brain imaging assessments. The researchers found that early poverty and stress predicted altered trajectories of amygdala-prefrontal connectivity development, which, in turn, predicted emotional regulation difficulties in adolescence.

1.9.2 Gene-environment interactions in stress vulnerability

Not everyone exposed to stress develops psychiatric disorders, highlighting the importance of individual differences in stress vulnerability. Genetic factors play a crucial role in determining stress susceptibility, with certain genetic variants conferring resilience or vulnerability to stress-related outcomes. For example, polymorphisms in the serotonin transporter gene (5-HTTLPR) have been associated with differential amygdala reactivity to emotional stimuli and varying risks for depression following stress exposure (Caspi et al., 2010).

Advanced neuroimaging techniques, combined with genetic analyses, are providing new insights into the mechanisms underlying these gene-environment interactions. A study by Bogdan et al. (2017) used fMRI to examine how specific genetic variants influenced the neural response to stress in individuals with varying exposure to childhood adversity.

1.10 Structural changes in the amygdala and hippocampus

MRI studies demonstrate alterations in the amygdala and hippocampus. Chronic stress is shown to reduce the volume of the hippocampus, leading to impaired memory and cognitive functions, whereas there is an increase in the volume of the amygdala in conditions of chronic stress, which is often associated with heightened emotional responses and anxiety. Though stress is an essential part of everyday living, managing this situation has its own heterogeneity in responses (Magalhães et al., 2018).

The volumetric changes observed in these structures are not merely correlational but appear to have causal relationships with stress-related symptoms. Animal studies utilizing optogenetics and chemogenetics have demonstrated that experimentally inducing hyperactivity in the amygdala or impairing hippocampal function can produce anxiety-like behaviors and cognitive deficits similar to those observed in chronic stress conditions (Felix-Ortiz et al., 2013; Anacker, 2018).

1.10.1 Sex differences in stress-induced brain changes

An important consideration in stress research is the presence of sex differences in stress responses and associated brain changes. Men and women show different patterns of stress reactivity at both the hormonal and neural levels, potentially contributing to the well-documented sex differences in prevalence rates for stress-related disorders such as depression and PTSD (Bangasser and Valentino, 2014).

Neuroimaging studies have revealed sex-specific patterns of stress-induced brain changes. Similarly, functional imaging studies have identified sex-specific patterns of neural activation during stress, with women typically showing greater limbic activation and men demonstrating more prefrontal engagement (Wang et al., 2007).

1.10.2 Longitudinal imaging studies: Tracking stress effects over time

While cross-sectional studies have documented associations between stress exposure and brain structure, longitudinal designs provide more compelling evidence for causal relationships. A pioneering study by Lupien et al. (1998) tracked cortisol levels and hippocampal volumes over a four-year period in older adults, finding that individuals with chronically elevated cortisol showed progressive hippocampal atrophy and associated memory impairments compared to those with normal cortisol trajectories.

More recent longitudinal neuroimaging studies have examined how specific stressors affect brain structure over time. For example, Gianaros (2014) followed medical students before, during, and after their intensive exam periods, documenting transient reductions in hippocampal volume during high-stress periods that normalized after stress resolution. Similarly, Papagni et al. (2011) tracked soldiers before and after combat deployment, finding that combat stress exposure predicted decreases in hippocampal volume and increases in amygdala volume, which correlated with PTSD symptom severity.

1.11 Functional changes in the amygdala and hippocampus

The advancement in medical imaging technologies permits new ways of understanding the structural modifications and the functional connectivity of different parts of the brain. One of the approaches used is the voxel-based morphometric study, in which the volume of various regions, such as the somatosensory cortex, motor cortex, auditory cortex, entorhinal cortex, prelimbic cortex, cingulate cortex, retrosplenial cortex, nucleus accumbens, thalamus, and hippocampus, to name a few, can be measured.

At longer durations, the connectivity is strengthening around the limbic regions that form a part of the regions involved in the response to stress. This increase in connectivity strength is seen between the prelimbic cortices, amygdala, piriform cortex, hippocampus, and septum (Castro, 2012; Sandi et al., 2008).

1.11.1 *Advanced functional MRI techniques in stress research*

Recent advances in functional neuroimaging have enhanced our understanding of stress-related brain function. Resting-state functional connectivity analyses measure intrinsic patterns of correlated activity between brain regions in the absence of specific tasks, providing insight into the organization of functional brain networks. Studies using this approach have identified altered connectivity patterns in stress-sensitive circuits in individuals with stress-related disorders (van Oort et al., 2017).

Real-time fMRI neurofeedback represents another innovative approach in stress research. This technique allows individuals to visualize their own brain activity in real-time and learn to modulate it through various cognitive strategies. A study by Young (2017) demonstrated that participants could learn to downregulate amygdala activity through neurofeedback training, resulting in reduced subjective stress responses to emotional stimuli.

Multivariate pattern analysis techniques have also been applied to identify distributed patterns of brain activity associated with different emotional states. Using this approach, Saarimäki et al. (2016) identified distinct neural signatures for different emotional states, including stress-related emotions such as fear and anxiety.

1.11.2 *The role of deep learning and AI in analyzing stress-related neuroimaging data*

The exponential growth in neuroimaging data has created both opportunities and challenges for researchers studying the effects of stress on the brain. Traditional analytical approaches may not fully capture the complex, multivariate patterns of brain changes associated with stress exposure. Artificial intelligence and deep learning methods offer powerful tools to address these challenges.

Convolutional neural networks (CNNs), a class of deep learning algorithms particularly suited to image analysis, have been applied to neuroimaging data to classify stress-related disorders and predict treatment outcomes. A study by Schnyer et al. (2017) used CNNs to analyze structural MRI data from patients with major depression, successfully distinguishing responders from nonresponders to antidepressant treatment with 75% accuracy. Similarly, Drysdale et al. (2017) applied machine learning to resting-state fMRI data and identified four distinct neurophysiological subtypes of depression with different clinical features and treatment response profiles.

1.11.3 *Technological advances in MRI for studying stress-related brain changes*

Advances in MRI technology continue to enhance our ability to visualize stress-sensitive brain structures and their connections. Ultra-high field MRI at 7 T and beyond offers significantly improved spatial resolution compared to conventional 1.5 or 3 T systems, enabling visualization of small brain structures relevant to stress processing. A study by Satpute et al. (2019) used 7 T MRI to distinguish the central and basolateral amygdala in humans, finding differential functional connectivity patterns for these subdivisions.

Simultaneous multimodal imaging approaches combining structural, functional, and metabolic measurements provide complementary information about the effects of stress on the brain. Hybrid PET-MRI systems allow for the simultaneous measurement of brain structure, function, and neurotransmitter activity. A study by Underwood (2018) used this approach to examine the relationships between cortisol levels, serotonin receptor binding, and functional connectivity in individuals with chronic stress exposure.

1.12 Other regions of the brain involved in this condition

The component of fear is a valuable tool for the assessment of neural connections between the regions of the brain, and this connects them functionally in conditions such as PTSD. The cingulate cortex has been shown to be altered during fearful conditions. In symptoms like PTSD, one of the significant risk factors is sleep disturbance, which is linked to the activation of fear extinction. This was studied in 63 PTSD patients, who showed a strong association of changes within the anterior regions of the brain's cortex, involving both the subcortical and cortical regions.

The involvement of multiple regions of the brain was demonstrated in a study on PTSD's effects on the cerebral cortex. The emotion-related prefrontal, insular, and limbic regions were studied. In an extensive study on multiple regions of the brain involved in various neurological conditions, the links were presented where the involvement of prenatal development and postnatal functioning was discussed for a detailed understanding of these conditions (Patel et al., 2021).

1.12.1 *The insula and stress processing*

The insula, a cortical region buried within the lateral sulcus, has emerged as a critical hub in stress processing. This region integrates interoceptive information about the body's physiological state with higher-order cognitive processes, making it uniquely positioned to translate bodily stress responses into conscious awareness and emotional experience (Craig, 2009).

A comprehensive metaanalysis by Kogler et al. (2015) examined neuroimaging studies using various laboratory stress induction protocols. They found that, regardless of the specific stressor used, the anterior insula showed reliable activation, suggesting its involvement in a common neural pathway for stress processing.

The insula's extensive connections with both limbic structures and prefrontal regulatory regions make it a potential target for stress-management interventions. Mindfulness meditation, for example, has been shown to enhance interoceptive awareness through increased insula gray matter volume and functional connectivity with prefrontal regions (Tang et al., 2015).

1.12.2 *The cerebellum: Beyond motor coordination*

Although traditionally viewed primarily as a motor coordination center, the cerebellum is increasingly recognized for its contributions to emotional processing and stress regulation. The cerebellum has extensive connections with limbic and prefrontal regions through cerebello–thalamo–cortical circuits, positioning it to influence both emotional and cognitive aspects of stress processing (Schmahmann et al., 2019).

Neuroimaging studies have documented cerebellar activation during various emotional and stressful tasks, with particular involvement of the posterior cerebellar vermis. A study by Sacchetti et al. (2009) demonstrated that lesions to this region impaired fear conditioning in animal models, suggesting a role in emotional learning.

1.12.3 *The extended amygdala: Bed nucleus of the stria terminalis and beyond*

The extended amygdala, including the bed nucleus of the stria terminalis (BNST), central amygdala, and substantia innominata, forms a functional continuum involved in threat processing and anxiety. While the amygdala proper is primarily implicated in phasic fear responses to immediate threats, the BNST appears particularly important for sustained anxiety responses to uncertain or prolonged threats (Davis et al., 2010).

High-resolution fMRI studies have begun to distinguish the contributions of these closely situated but functionally distinct regions to stress processing. A study by Avery et al. (2016) used 7 T MRI to examine BNST activation during anticipatory anxiety in humans, finding that BNST activity predicted self-reported anxiety levels during threat anticipation but not during acute threat exposure.

The extended amygdala receives dense noradrenergic and CRH inputs, making it highly responsive to stress signals. Pharmacological interventions targeting these systems have shown promise in reducing anxiety-like behaviors in animal models and are being explored as treatments for stress-related disorders in humans (Grillon et al., 2015).

1.13 Clinical applications of neuroimaging in stress-related disorders

The insights gained from neuroimaging studies of stress effects on the brain have significant implications for clinical practice. As our understanding of the neural mechanisms underlying stress-related disorders advances, neuroimaging approaches are increasingly being applied to improve diagnosis, treatment selection, and monitoring of clinical outcomes.

1.13.1 *Neuroimaging biomarkers for diagnosis and treatment planning*

Traditional psychiatric diagnosis relies heavily on subjective symptom reports and clinical observations, which can be imprecise and fail to capture the neurobiological heterogeneity within diagnostic categories. Neuroimaging biomarkers offer potential for more objective, biologically-based classification of stress-related disorders.

A landmark study by Williams et al. (2015) examined whether neuroimaging measures could distinguish between different anxiety disorders better than clinical features alone. The researchers found that patterns of amygdala, insula, and prefrontal activation during emotion processing tasks differentiated panic disorder, social anxiety disorder, and specific phobia with greater accuracy than symptom-based assessments. Similarly, Drysdale et al. (2017) identified four neurophysiologically distinct subtypes of depression based on functional connectivity patterns, each with different clinical features and treatment response profiles.

Neuroimaging measures are also being explored as predictors of treatment response. A study by Whitfield-Gabrieli et al. (2016) found that pretreatment resting-state functional connectivity between the PFC and the default mode network predicted response to cognitive behavioral therapy for depression.

1.13.2 Novel therapeutic approaches based on neuroimaging findings

Understanding the neural circuits affected by stress has inspired the development of targeted interventions aimed at normalizing function in these circuits. Transcranial magnetic stimulation (TMS), which uses magnetic pulses to modulate neural activity in superficial cortical regions, has been approved for treatment-resistant depression. The targeting of TMS has been refined based on neuroimaging findings (Dunlop et al., 2017).

Deep brain stimulation (DBS), involving the surgical implantation of electrodes to modulate activity in specific brain structures, represents another circuit-based approach informed by neuroimaging findings. Initial trials of DBS targeting the subgenual cingulate cortex have demonstrated promising results in severely ill patients (Holtzheimer et al., 2017).

Less invasive approaches based on neuroimaging findings include neurofeedback, where individuals learn to self-regulate activity in specific brain regions using real-time fMRI displays. Both approaches have shown promise in preliminary studies for reducing symptoms in stress-related disorders (Young, 2017; Brunoni, 2017).

1.14 Future directions in neuroimaging research on stress

As neuroimaging technology and analytical methods continue to advance, several promising directions are emerging for future research on the effects of stress on the brain.

1.14.1 Developmental trajectories and early interventions

Longitudinal studies spanning critical developmental periods can provide crucial insights into how stress affects the developing brain and identify sensitive periods when stress exposure may have particularly profound or lasting effects. Such research can inform early intervention strategies aimed at preventing or mitigating adverse outcomes in stress-exposed youth.

The Adolescent Brain Cognitive Development study, a landmark longitudinal project following over 10,000 children from ages 9–10 through early adulthood, includes comprehensive assessments of stress exposure alongside multimodal brain imaging. Early findings from this study have demonstrated associations between early-life stress and accelerated brain maturation in regions involved in emotion processing (Casey et al., 2018).

1.14.2 Precision medicine approaches to stress-related disorders

The substantial heterogeneity within stress-related disorders suggests that personalized, precision medicine approaches may be more effective than one-size-fits-all treatments. Neuroimaging biomarkers hold promise for identifying which individuals are likely to benefit from specific interventions based on their unique neural signatures.

Future research employing large samples and machine learning approaches may identify reproducible neuroimaging-based subtypes within stress-related disorders, similar to the depression biotypes identified by Drysdale et al. (2017). Clinical trials can then test whether matching treatments to these biotypes improves outcomes compared to standard approaches.

2 Conclusion

The brain's response to stress is a highly intricate and dynamic process that involves multiple regions, including the amygdala, hippocampus, PFC, and HPA axis. These structures work in concert to manage and regulate stress, but prolonged exposure to stress can lead to structural and functional changes, with potential long-term consequences on emotional and cognitive well-being.

Advanced neuroimaging techniques have revolutionized our understanding of how stress affects the brain, revealing both structural and functional alterations in key stress-regulatory regions. Structural MRI studies have documented volumetric changes in the hippocampus and amygdala associated with chronic stress exposure, while fMRI has illuminated the dynamic interactions between these regions and their connections with prefrontal regulatory circuits.

The applications of these neuroimaging findings extend far beyond academic interest, with significant implications for clinical practice. Neuroimaging biomarkers hold promise for improving diagnosis, treatment selection, and monitoring of stress-related disorders by moving beyond symptom-based classifications toward a biologically-based, circuit-level understanding of these conditions.

Looking ahead, integrating neuroimaging with other research modalities—including genetics, immunology, and real-world monitoring—will provide an increasingly comprehensive understanding of how stress affects the brain-body system across development and in different environmental contexts. Precision medicine approaches based on neuroimaging biomarkers may enable more personalized, effective interventions for stress-related disorders by matching treatments to the specific neural circuits affected in individual patients.

Understanding these mechanisms is crucial for developing targeted interventions and treatments for stress-related disorders. Future research should focus on identifying specific biomarkers of stress-induced brain changes and exploring the role of neuroplasticity in recovery. By enhancing our knowledge of stress and its impact on the brain, we can work toward improving mental health and overall well-being in individuals affected by chronic stress.

References

Admon, R., Milad, M.R., Hendler, T., 2020. A causal model of post-traumatic stress disorder: disentangling predisposed from acquired neural abnormalities. Trends Cogn. Sci. 24 (10), 826–839.

Anacker, C., et al., 2018. Hippocampal neurogenesis confers stress resilience by inhibiting the ventral dentate gyrus. Nature 559 (7712), 98–102.

Anacker, C., Hen, R., 2017. Adult hippocampal neurogenesis and cognitive flexibility — linking memory and mood. Nat. Rev. Neurosci. 18 (6), 335–346.

Arnsten, A.F.T., 2009. Stress signalling pathways that impair prefrontal cortex structure and function. Nat. Rev. Neurosci. 10 (6), 410–422.

Arnsten, A.F.T., 2015. Stress weakens prefrontal networks: molecular insults to higher cognition. Nat. Neurosci. 18 (10), 1376–1385.

Avery, S.N., Clauss, J.A., Blackford, J.U., 2016. The human BNST: functional role in anxiety and addiction. Neuropsychopharmacology 41 (1), 126–141.

Bangasser, D.A., Valentino, R.J., 2014. Sex differences in stress-related psychiatric disorders: neurobiological perspectives. Front. Neuroendocrinol. 35 (3), 303–319.

Barch, D.M., 2016. Early childhood adverse experiences, inferior frontal gyrus connectivity, and the trajectory of externalizing psychopathology. J. Am. Acad. Child Adolesc. Psychiatry 55 (12), 1033–1041.

Bliss, T.V.P., Lømo, T., 1973. Long-lasting potentiation of synaptic transmission in the dentate area of the anaesthetized rabbit following stimulation of the perforant path. J. Physiol. 232 (2), 331–356.

Bogdan, R., et al., 2017. Imaging genetics and genomics in psychiatry: a critical review of progress and potential. Biol. Psychiatry 82 (3), 165–175.

Bremner, J.D., Marmar, C.R., 2020. Trauma, Memory, and Dissociation, American Psychiatric Association Publishing.

Brunoni, A.R., 2017. Transcranial direct current stimulation for acute major depressive episodes: meta-analysis of individual patient data. Br. J. Psychiatry 210 (5), 322–328.

Casey, B.J., et al., 2018. The adolescent brain cognitive development (ABCD) study: imaging acquisition across 21 sites. Dev. Cogn. Neurosci. 32, 43–54.

Caspi, A., et al., 2010. Genetic sensitivity to the environment: the case of the serotonin transporter gene and its implications for studying complex diseases and traits. Am. J. Psychiatry 167 (5), 509–527.

Castro, J.E., 2012. Role of the amygdala in antidepressant effects on hippocampal cell proliferation and survival and on depression-like behavior in the rat. PloS One 5 (1), 1–11.

Course-Choi, J., Saville, H., Derakshan, N., 2017. The effects of adaptive working memory training and mindfulness meditation training on processing efficiency and worry in high worriers. Behav. Res. Ther. 89, 1–13.

Craig, A.D., 2009. How do you feel - now? The anterior insula and human awareness. Nat. Rev. Neurosci. 10 (1), 59–70.

Davis, M., et al., 2010. Phasic vs sustained fear in rats and humans: role of the extended amygdala in fear vs anxiety. Neuropsychopharmacology 35 (1), 105–135.

De Kloet, E.R., Joëls, M., Holsboer, F., 2005. Stress and the brain: from adaptation to disease. Nat. Rev. Neurosci. 6 (6), 463–475.

Dedovic, K., et al., 2009. The brain and the stress axis: the neural correlates of cortisol regulation in response to stress. NeuroImage 47 (3), 864–871.

Drysdale, A.T., et al., 2017. Resting-state connectivity biomarkers define neurophysiological subtypes of depression. Nat. Med. 23 (1), 28–38.

Duman, R.S., et al., 2016. Synaptic plasticity and depression: new insights from stress and rapid-acting antidepressants. Nat. Med. 22 (3), 238–249.

Dunlop, K., Hanlon, C.A., Downar, J., 2017. Noninvasive brain stimulation treatments for addiction and major depression. Ann. N. Y. Acad. Sci. 1394 (1), 31–54.

Eriksson, P.S., et al., 1998. Neurogenesis in the adult human hippocampus. Nat. Med. 4 (11), 1313–1317.

Etkin, A., Büchel, C., Gross, J.J., 2015. The neural bases of emotion regulation. Nat. Rev. Neurosci. 16 (11), 693–700.

Etkin, A., Wager, T.D., 2007. Functional neuroimaging of anxiety: a meta-analysis of emotional processing in PTSD, social anxiety disorder, and specific phobia. Am. J. Psychiatry 164 (10), 1476–1488.

Felix-Ortiz, A.C., et al., 2013. BLA to vHPC inputs modulate anxiety-related behaviors. Neuron 79 (4), 658–664.

Gianaros, P.J., 2014. Community socioeconomic disadvantage in midlife relates to cortical morphology via neuroendocrine and subclinical cardiovascular pathways. Cereb. Cortex 24 (8), 2139–2151.

Gianaros, P.J., 2015. Inflammatory pathways link socioeconomic inequalities to white matter architecture. Cereb. Cortex 25 (5), 1240–1250.

Grillon, C., et al., 2015. The CRH 1 antagonist GSK561679 increases human fear but not anxiety as assessed by startle. Neuropsychopharmacology 40, 1064–1071.

Hammen, C., 2005. Stress and depression. Annu. Rev. Clin. Psychol. 1 (1), 293–319.

Hayes, J.P., Hayes, S.M., Mikedis, A.M., 2012. Quantitative meta-analysis of neural activity in posttraumatic stress disorder. Biol. Mood Anxiety Disord. 2 (9), 1–13.

Herman, J.P., et al., 2016. Regulation of the hypothalamic-pituitary-adrenocortical stress response. Compr. Physiol. 6 (2), 603–621.

Holtzheimer, P.E., et al., 2017. Subcallosal cingulate deep brain stimulation for treatment-resistant depression: a multisite, randomised, sham-controlled trial. Lancet Psychiatry 4 (11), 839–849.

Janak, P.H., Tye, K.M., 2015. From circuits to behaviour in the amygdala. Nature 517 (7534), 284–292.

Kempermann, G., et al., 2018. Human adult neurogenesis: evidence and remaining questions. Cell Stem Cell 23 (1), 25–30.

Kogler, L., et al., 2015. Psychosocial versus physiological stress - meta-analyses on deactivations and activations of the neural correlates of stress reactions. NeuroImage 119, 235–251.

Krystal, J.H., 2013. Computational approaches to psychiatric nosology: a selective review. Schizophr. Bull. 43 (1), 78–85.

Liston, C., McEwen, B.S., Casey, B.J., 2009. Psychosocial stress reversibly disrupts prefrontal processing and attentional control. Proc. Natl. Acad. Sci. U. S. A. 106 (3), 912–917.

Lupien, S.J., et al., 1998. Cortisol levels during human aging predict hippocampal atrophy and memory deficits. Nat. Neurosci. 1 (1), 69–73.

Magalhães, R., et al., 2018. The dynamics of stress: a longitudinal MRI study of rat brain structure and connectome. Mol. Psychiatry 23 (10), 1998–2006.

Maier, S.F., Seligman, M.E.P., 2016. Learned helplessness at fifty: insights from neuroscience. Psychol. Rev. 123 (4), 349–367.

Maren, S., Holmes, A., 2016. Stress and fear extinction. Neuropsychopharmacology 41 (1), 58–79.

McEwen, B.S., 1998. Stress, adaptation, and disease: allostasis and allostatic load. Ann. N. Y. Acad. Sci. 840 (1), 33–44.

McEwen, B.S., 2012. Brain on stress: how the social environment gets under the skin. Proc. Natl. Acad. Sci. U. S. A. 109 (suppl_2), 17180–17185.

McEwen, B.S., 2017. Neurobiological and systemic effects of chronic stress. Chronic Stress 1, 1–11.

McEwen, B.S., Nasca, C., Gray, J.D., 2016. Stress effects on neuronal structure: hippocampus, amygdala, and prefrontal cortex. Neuropsychopharmacology 41 (1), 3–23.

Ochsner, K.N., Silvers, J.A., Buhle, J.T., 2012. Functional imaging studies of emotion regulation: a synthetic review and evolving model of the cognitive control of emotion. Ann. N. Y. Acad. Sci. 1251 (24), 1–35.

Papagni, S.A., et al., 2011. Effects of stressful life events on human brain structure: a longitudinal voxel-based morphometry study. Stress 14 (2), 227–232.

Pariante, C.M., Lightman, S.L., 2008. The HPA axis in major depression: classical theories and new developments. Trends Neurosci. 31 (9), 464–468.

Patel, Y., et al., 2021. Virtual histology of cortical thickness and shared neurobiology in 6 psychiatric disorders. JAMA Psychiatry 78 (1), 47–63.

Saarimäki, H., et al., 2016. Discrete neural signatures of basic emotions. Cereb. Cortex 26 (6), 2563–2573.

Sacchetti, B., Scelfo, B., Strata, P., 2009. Cerebellum and emotional behavior. Neuroscience 162 (3), 756–762.

Sandi, C., et al., 2008. Chronic stress-induced alterations in amygdala responsiveness and behavior - Modulation by trait anxiety and corticotropin-releasing factor systems. Eur. J. Neurosci. 28 (9), 1836–1848.

Satpute, A.B., et al., 2019. Deconstructing arousal into wakeful, autonomic and affective varieties. Neurosci. Lett. 693, 19–28.

Saygin, Z.M., et al., 2017. High-resolution magnetic resonance imaging reveals nuclei of the human amygdala: manual segmentation to automatic atlas. NeuroImage 155, 370–382.

Schmahmann, J.D., et al., 2019. The theory and neuroscience of cerebellar cognition. Annu. Rev. Neurosci. 42, 337–364.

Schnyer, D.M., et al., 2017. Evaluating the diagnostic utility of applying a machine learning algorithm to diffusion tensor MRI measures in individuals with major depressive disorder. Psychiatry Res. Neuroimaging 264, 1–9.

Tang, Y.Y., Hölzel, B.K., Posner, M.I., 2015. The neuroscience of mindfulness meditation. Nat. Rev. Neurosci. 16 (4), 213–225.

Tottenham, N., Sheridan, M.A., 2010. A review of adversity, the amygdala and the hippocampus: a consideration of developmental timing. Front. Hum. Neurosci. 3, 1–18.

Treadway, M.T., et al., 2015. Illness progression, recent stress, and morphometry of hippocampal subfields and medial prefrontal cortex in major depression. Biol. Psychiatry 77 (3), 285–294.

Underwood, M.D., et al., 2018. Serotonin receptors and suicide, major depression, alcohol use disorder and reported early life adversity. Transl. Psychiatry 8 (1), 279.

van Oort, J., et al., 2017. How the brain connects in response to acute stress: a review at the human brain systems level. Neurosci. Biobehav. Rev. 83, 281–297.

Videbech, P., Ravnkilde, B., 2004. Hippocampal volume and depression: a meta-analysis of MRI studies. Am. J. Psychiatry 161 (11), 1957–1966.

Wang, J., et al., 2007. Gender difference in neural response to psychological stress. Soc. Cogn. Affect. Neurosci. 2 (3), 227–239.

Whitfield-Gabrieli, S., et al., 2016. Brain connectomics predict response to treatment in social anxiety disorder. Mol. Psychiatry 21 (5), 680–685.

Williams, L.M., et al., 2015. Amygdala reactivity to emotional faces in the prediction of general and medication-specific responses to antidepressant treatment in the randomized iSPOT-D trial. Neuropsychopharmacology 40 (10), 2398–2408.

Yehuda, R., Seckl, J., 2011. Minireview: stress-related psychiatric disorders with low cortisol levels: a metabolic hypothesis. Endocrinology 152 (12), 4496–4503.

Young, K.D., et al., 2017. Randomized clinical trial of real-time fMRI amygdala neurofeedback for major depressive disorder: effects on symptoms and autobiographical memory recall. Am. J. Psychiatry. 174 (8), 748–755.

Chapter 26

Imaging in auditory neuroscience

Pradeep Yuvaraj[a], Aravind Kumar Rajasekaran[a], Rajagopal Kadavigere[b], Leena R. David[c,d], Sneha Ravichandran[d], Dilip Shettigar[d], Suresh Sukumar[d], Berna Uzun[e], Dilber Uzun Ozsahin[c,e,f], and Ilker Ozsahin[e]

[a]*Department of Speech Pathology and Audiology, National Institute of Mental Health and Neurosciences, Bengaluru, India*
[b]*Department of Radiodiagnosis and Imaging, Kasturba Medical College, Kasturba Medical College, Manipal, India*
[c]*Medical Diagnostic Imaging Department, College of Health Sciences, University of Sharjah, Sharjah, United Arab Emirates*
[d]*Department of Medical Imaging Technology, Manipal College of Health Professions, Manipal Academy of Higher Education, Manipal, Karnataka, India*
[e]*Irfan Suat Gunsel Operational Research Institute, Near East University, Nicosia, Turkish Republic of Northern Cyprus (TRNC), Turkey*
[f]*Research Institute for Medical and Health Sciences, University of Sharjah, Sharjah, United Arab Emirates*

1 Introduction

Auditory neuroscience seeks to elucidate the complex neural mechanisms by which the human brain processes sound, from basic acoustic features to sophisticated percepts like speech and music. Historically, psychophysical techniques dominated the field, offering indirect insights into auditory processing. However, the advent of neuroimaging technologies has fundamentally shifted this paradigm by providing direct access to the structural and functional characteristics of the auditory system in vivo (Poldrack et al., 2011).

This chapter introduces and critically examines six major imaging techniques that have propelled auditory neuroscience: structural magnetic resonance imaging (MRI), diffusion tensor imaging (DTI), functional MRI (fMRI), magnetic resonance spectroscopy (MRS), positron emission tomography (PET), and functional near-infrared spectroscopy (fNIRS). These modalities offer complementary perspectives on the brain's auditory networks, encompassing morphometric, microstructural, hemodynamic, and metabolic data. MRI-based volumetry and morphometry have revealed cortical plasticity due to music training (Schneider et al., 2023) and anatomical alterations in hearing loss (Qu et al., 2020; Wagner et al., 2022). DTI allows the mapping of white matter tracts in auditory pathways and has been instrumental in understanding pediatric auditory development and in predicting cochlear implant outcomes (Alvand et al., 2023; Huang et al., 2024).

Functional imaging through fMRI has been pivotal in decoding cortical representations of pitch, tonotopy, and contextual loudness (Allen et al., 2022; Behler and Uppenkamp, 2021). Advanced techniques such as arterial spin labeling and laminar fMRI have expanded spatial resolution, while resting-state fMRI has uncovered intrinsic connectivity relevant to auditory learning (Lumaca et al., 2019). MRS complements these approaches by quantifying neurotransmitter concentrations, notably gamma-aminobutyric acid (GABA) and glutamate, which play central roles in auditory signal modulation and pathology, including tinnitus (Wójcik et al., 2023; Sedley et al., 2015b).

PET remains invaluable for studying metabolic and regional cerebral blood flow differences, particularly in clinical contexts such as cochlear implantation (CI) and asymmetrical hearing loss (Olds et al., 2016; Karoui et al., 2023). Meanwhile, fNIRS offers a silent, mobile imaging alternative, particularly suited for pediatric populations, enabling research in ecologically valid environments (Bell et al., 2020; Yaman et al., 2023).

Together, these imaging modalities provide a multidimensional framework for understanding auditory processing in both healthy individuals and clinical populations. By integrating findings across these technologies, researchers and clinicians can develop targeted interventions, evaluate rehabilitation outcomes, and refine theoretical models of auditory cognition and neuroplasticity.

Advances in Medical Imaging. https://doi.org/10.1016/B978-0-443-28967-5.00004-6

2 Auditory system

2.1 Development and organization of the central auditory system

The human auditory system represents one of nature's most remarkable signal-processing networks, handling an extraordinary range of frequencies, intensities, and temporal patterns (Moore, 2012). While the peripheral components of hearing develop relatively early, the central auditory system undergoes a protracted developmental trajectory that extends well beyond childhood (Eggermont and Moore, 2011).

2.2.1 Developmental timeline

The maturation of central auditory pathways follows a carefully orchestrated sequence. Although basic hearing mechanisms are functional at birth, the sophisticated processing capabilities of the auditory cortex continue to develop throughout childhood and adolescence (Kral and Sharma, 2012). This extended developmental window reflects the increasing complexity of auditory processing demands as humans mature.

Research by Ponton et al. (2000) demonstrated that cortical auditory evoked potentials show significant changes in morphology and latency throughout the first two decades of life. These neurophysiological changes correspond to behavioral improvements in complex auditory tasks such as speech perception in noise, auditory scene analysis, and the ability to extract meaningful patterns from rapid acoustic transitions (Werner et al., 2012; Leibold and Buss, 2013).

2.2.2 Hierarchical organization

The central auditory pathway exhibits a hierarchically organized processing stream that begins at the cochlear nucleus in the brainstem and ascends through the superior olivary complex, lateral lemniscus, inferior colliculus, and medial geniculate body before reaching the auditory cortex (Pickles, 2013; Winer et al., 2005). Each successive stage adds layers of processing complexity:

1. Brainstem nuclei: Focus on extracting basic acoustic features such as timing, intensity, and frequency information (Young, 2008)
2. Midbrain (inferior colliculus): Integrates these features and begins more complex pattern recognition (Malmierca and Hackett, 2010)
3. Thalamus (medial geniculate body): Acts as a gateway, filtering and routing information to appropriate cortical regions (Winer et al., 2005)
4. Primary auditory cortex (PAC): Performs detailed spectrotemporal analysis (Hackett, 2011)
5. Secondary auditory regions: Integrate features into coherent auditory objects (Bizley and Cohen, 2013)

2.2.3 Parallel processing streams

As mentioned, auditory information processed at the cortical level follows two primary processing streams (Rauschecker and Scott, 2009):

- The dorsal "where" pathway extends from the PAC through posterior temporal and parietal regions, specializing in spatial localization and motion detection (Lomber and Malhotra, 2008). This pathway is crucial for determining the sound source location and tracking moving sound sources in the environment.
- The ventral "what" pathway projects from the PAC through anterior temporal regions, focusing on sound identification and categorization (Scott and Johnsrude, 2003). This stream enables the recognition of speech sounds, environmental sounds, music, and the identity of speakers.
- This dual-stream organization bears a remarkable similarity to the visual processing system, suggesting evolutionary conservation of sensory processing principles across modalities (Mishkin et al., 1983; Kaas and Hackett, 2000).

2.2.4 Cross-modal integration

The mature auditory system does not operate in isolation. Extensive connections exist between auditory processing regions and other sensory systems, particularly visual and somatosensory networks (Calvert et al., 1998; Schroeder and Foxe, 2005). This multisensory integration provides significant advantages:

- Enhanced detection of events that produce both auditory and visual stimuli (Senkowski et al., 2011)
- Improved localization accuracy through audiovisual integration (Alais and Burr, 2004)

- Disambiguation of unclear signals through contextual information (McGurk and MacDonald, 1976)
- Support for speech perception through visual cues (lip-reading) (Sumby and Pollack, 1954)
- Neuroimaging studies have revealed that even supposedly "pure" auditory tasks activate visual and motor regions, highlighting the extensively interconnected nature of sensory processing networks (Zatorre et al., 2002; Hickok and Poeppel, 2007)

2.2.5 Plasticity and experience-dependent development

The extended developmental timeline of central auditory pathways provides an extended window for experience-dependent plasticity (Kral, 2013). Environmental sound exposure, language learning, and musical training during childhood and adolescence significantly shape the functional architecture of auditory cortical regions (Trainor, 2005; Kuhl, 2010).

This plasticity serves an adaptive function, allowing the auditory system to specialize in the specific acoustic demands of an individual's environment and culture. For instance, tonal language speakers develop enhanced sensitivity to pitch variations compared to nontonal language speakers (Wong et al., 2007), and musicians show enhanced neural representations of acoustic features relevant to musical performance (Herholz and Zatorre, 2012; Zatorre et al., 2012, 2002).

Understanding the complex organization and protracted development of the central auditory system provides crucial insights for educational approaches, clinical interventions for auditory processing disorders (APDs), and rehabilitation strategies following hearing loss or neurological injury (Moore and Hunter, 2013; Sharma et al., 2002).

3 Central auditory processing

The primary function of the auditory system is to perceive incoming sound. The central auditory process involves conduction and various computational processes performed by the auditory neural mechanism in response to the incoming sound. Central auditory processing is considered dysfunctional when there is poor performance in the following skills: sound localization and lateralization, auditory discrimination, auditory pattern recognition, temporal aspects of audition, auditory performance in competing acoustic signals, and auditory performance with degraded acoustic signals (American Speech-Language-Hearing Association, 2005). It is also worth noting that the psychoacoustical parameters of frequency, intensity, and timbre correspond to pitch, loudness, and quality, respectively (Fastl and Zwicker, 2007). Various processes occur throughout the auditory pathway, ultimately enabling the cortex to decode sound with high fidelity.

Central auditory processing encompasses a sophisticated sequence of neural computations that transform acoustic signals into meaningful perceptual experiences (Griffiths and Warren, 2004). When sound waves enter the ear, they initiate a cascade of physiological events that extend far beyond simple signal transmission. The central auditory pathway—spanning from the cochlear nucleus through the brainstem, midbrain, thalamus, and ultimately to the auditory cortex—performs increasingly complex operations on the incoming acoustic information at each processing stage (Pickles, 2013).

Sound localization and lateralization, fundamental abilities identified by American Speech Language Hearing Association (ASHA) as central auditory processes, rely on precise neural comparisons of interaural time and level differences (Grothe et al., 2010). The superior olivary complex in the lower brainstem contains specialized neurons that detect microsecond timing differences between the ears, allowing listeners to determine sound direction with remarkable accuracy (Joris et al., 1998). This spatial hearing capacity proves crucial for navigating environments, focusing attention, and separating sound sources in complex acoustic scenes (Middlebrooks and Green, 1991).

Auditory discrimination, another critical function, enables listeners to differentiate between similar but distinct sounds (Moore, 2012). This ability depends on the preservation of spectral and temporal features throughout the auditory pathway and reaches its highest refinement in the PAC, where neurons demonstrate exquisite selectivity for specific acoustic patterns (Wang et al., 2005). The discrimination of speech sounds, essential for language processing, represents a specialized form of this capability, requiring the detection of rapid acoustic transitions that distinguish phonemes (Liberman and Mattingly, 1985; Hickok and Poeppel, 2007).

Temporal processing—the analysis of sounds over time—stands as one of the auditory system's most remarkable achievements (Phillips, 2008). The system can resolve temporal gaps as brief as 2–3 ms, track amplitude modulations that create rhythm perception, and detect frequency modulations that contribute to melody perception (Moore, 2003). These temporal abilities underpin the perception of speech prosody, musical structures, and environmental sound patterns. Specialized neural circuits throughout the auditory pathway, particularly in the inferior colliculus and auditory cortex, appear dedicated to extracting different aspects of temporal information (Schreiner and Langner, 1997; Bendor and Wang, 2005).

The ability to function in challenging acoustic environments demonstrates the sophisticated nature of central auditory processing (Bronkhorst, 2000). When faced with competing signals (such as understanding speech in a noisy restaurant) or degraded signals (like a poorly connected phone call), the auditory system employs sophisticated computational strategies, including spectral-temporal integration, harmonic grouping, spatial filtering, and attentional mechanisms, to enhance signals of interest while suppressing irrelevant information (Bregman, 1990; Cherry, 1953). These processes involve extensive cortical networks that interact with subcortical auditory structures through robust feedback connections (Winer, 2006; Suga, 2008).

The perceptual qualities of sound—pitch, loudness, and timbre—emerge from the neural coding of acoustic features (Plack and Oxenham, 2005). Pitch perception, corresponding to the physical parameter of frequency, relies on both temporal coding mechanisms (phase-locking to waveform periodicity) and place coding (tonotopic organization) that begin in the cochlea and are maintained throughout the auditory pathway (Oxenham, 2012). Loudness perception, while initiated by the intensity of sound waves, undergoes substantial recalibration through processes like dynamic range adaptation and gain control at multiple levels of the auditory system, allowing listeners to perceive meaningful loudness relationships across widely varying sound levels (Moore et al., 1999; Planck and Oxenham, 2005). Timbre, the quality that distinguishes sounds of similar pitch and loudness, emerges from the integration of spectral content, temporal envelope, and fine temporal structure, requiring sophisticated pattern recognition mechanisms primarily implemented in auditory cortical regions (McAdams, 2019; Caclin et al., 2007).

The central auditory system does not operate in isolation but maintains extensive connections with attentional networks, memory systems, emotional centers, and other sensory modalities (Fritz et al., 2003; LeDoux, 2000). These connections enable context-dependent processing, where prior experience, current goals, and cross-modal information can significantly modify auditory perception (Weinberger, 2004). This contextual influence explains phenomena such as the cocktail party effect (selectively attending to one voice among many) and perceptual learning (improved recognition of initially unfamiliar speech accents or sound patterns with exposure) (Shinn-Cunningham, 2008; Samuel and Kraljic, 2009).

Dysfunction in central auditory processing, as defined by ASHA, can result from developmental anomalies, neurological disorders, auditory deprivation, or aging (Musiek et al., 2005; Bellis, 2003). These impairments often manifest as a disproportionate difficulty understanding speech in challenging listening environments despite normal hearing sensitivity (Chermak and Musiek, 1997). The complexity of central auditory processing explains why conventional audiometric testing may fail to identify these disorders, necessitating specialized behavioral and electrophysiological assessments that target specific auditory functions such as temporal processing, binaural integration, and pattern recognition (Musiek and Chermak, 2007; Billiet and Bellis, 2011).

Understanding the intricate mechanisms of central auditory processing provides crucial insights for clinical interventions, educational approaches for individuals with APDs, and the development of technologies that enhance communication in challenging acoustic environments (Dillon, 2013; Ferguson and Henshaw, 2015). The remarkable fidelity with which the auditory system decodes complex acoustic signals represents one of the brain's most sophisticated computational achievements (Wang, 2013).

4 Magnetic resonance imaging-volumetry/morphometry

The key insights about the structural properties of auditory regions are captured through quantitative analysis of brain structure, specifically MRI volumetry. This information helps to elucidate the effect of structural alterations on the observed auditory functions (Ratnanather, 2020). The plasticity of the brain to auditory stimuli, such as music, is an interesting phenomenon. Through MRI-based morphometry, the effect of positive stimulation, like music, on brain structures can be studied. Similarly, the lack of auditory stimulation, such aging and hearing loss, has a negative impact on the brain.

Identifying and anatomically segmenting the auditory cortex into its sub-regions has been a challenge and is best accomplished using histological methods. However, researchers have employed MRI methods to achieve this. Meyer et al. (2014) utilized volumetric 3-dimensional T1-weighted MRI to assess the cortical regions related to auditory function for their volume, cortical thickness (CT), and cortical surface area. The authors reported hemispherical asymmetry in the tested parameters. Zoellner et al. (2019) attempted to identify the PAC by analyzing CT. A 3 T(3 Tesla) T1-weighted 3D MRI of the brain was performed, and the results were correlated with MEG-based functional activity. In the right hemisphere, it could be mapped to the medial portion of Heschl's gyrus (HG), whereas in the left hemisphere, it was located more laterally. Furthermore, CTH reflected individual variability among the subjects tested, making it robust enough to be expanded as an in-vivo marker to identify the PAC.

Hearing loss is known to cause changes in cortical structure, and MRI can be employed to study its effects in detail. Qu et al. (2020) performed T1-weighted MRI on children with profound hearing loss. Surface-based morphometry (SBM) and subcortical analyses were conducted for the whole-brain. Decreased CT, an increase in the local gyrification index, and gray-matter volume changes in many brain regions were observed in children with hearing loss. The sub-cortical regions exhibited regional contractions and decreased volumes. In a study using automated volumetric segmentation, Wagner et al. (2022) evaluated younger (7–12 months) and older (13 months–18 years) children with gap junction protein beta 2 GJB2)-related hearing loss. They calculated the relative volume of preselected brain structures, such as HG relation to the whole brain volume. The results revealed smaller ratios in certain parameters in the older group and differences in other parameters in the younger group. The significance of these observations is that brain volumetric assessment could be employed to study the impact of hearing loss and its age-related changes. Music appears to have an impact on the auditory cortex at a cortical macrostructural level (Schneider et al., 2023). The researchers conducted a longitudinal MRI study over 12 years in children undergoing musical training. They correlated structural and functional parameters. The results revealed maturational and learning-induced plasticity.

5 Diffusion tensor imaging

DTI has added a new dimension to auditory neuroscience. It provides information on the structural connectivity and microstructural integrity of auditory pathways. Auditory processing involves complex interconnections between the auditory centers and other nonauditory regions along the auditory pathway. DTI helps in studying the white-matter pathways associated with auditory processing. Its application in studying auditory processing in children needs to be emphasized, especially during their developmental period.

It was found that DTI could reveal microstructural alterations in children with APD. Findings, such as increased average path length, correlated with listening difficulties (Alvand et al., 2023). In a recent study, (Madzime et al., 2024) examined the white matter integrity of the auditory pathway in an 11 year-old child with perinatally acquired HIV. They included regions at the level of the brainstem (lateral lemniscus) to higher auditory centers (acoustic radiation) and nonauditory regions. In addition, they employed graph theory-based analyses, such as nodal strength and efficiency of the central auditory system (CAS) and other brain regions in the structural brain network. The test results revealed abnormalities in the form of higher mean diffusivity (MD), higher axial diffusivity, and higher radial diffusivity in the auditory and nonauditory regions. However, nodal properties showed no significant difference between the patient and the control.

Brain stem implants are performed to habilitate patients with bilateral hearing loss, most commonly caused by Neurofibromatosis Type 2. In brain stem implants, the cochlea and the auditory nerve are bypassed, and the stimulating electrodes are placed over the cochlear nuclei (CN). However, pinpointing the CN is a challenge for surgeons. High-resolution DTI enables surgeons and scientists to visualize and study the CN structure (Epprecht et al., 2020). The authors employed a 7 T (T) MRI to perform DTI at a high resolution of about 50 μm. The structure (lamellar) and orientation (orthogonal to the fiber tracts of the inferior cerebellar peduncle) could be elucidated. This methodology could be incrementally refined and protocolized for studying the auditory nuclei and implanting devices (Epprecht et al., 2020).

It is vital to predict the outcomes of any intervention programs, more so if the patients undergo invasive procedures like cochlear implants. DTI equips us with a unique tool to study the white matter microstructure in this context (Huang et al., 2015). By employing the DTI measures, the researchers attempted to predict the outcome of CI in patients with congenital hearing loss. Fractional anisotropy (FA) and MD of the auditory pathway (trapezoid body to HG) were considered. FA appears to be a better predictor of postsurgical auditory performance (Huang et al., 2015).

6 Functional magnetic resonance imaging

fMRI has been widely used in auditory neuroscience; however, it comes with some challenges. The presentation of the sound stimulus itself is a challenge in fMRI due to the magnetic properties of regular transducers. However, with the advent of MR-compatible headphones, this has become possible. Even this arrangement restricts the study of auditory processing in subjects using hearing aids, cochlear implants, or brainstem implants. Another constraint is the noise generated during fMRI, which, without appropriate ear protection, is sure to cause a threshold shift and, sometimes, even tinnitus. Again, active noise cancellation headphones are of some help in this aspect. However, unintended activation of auditory regions must be kept in mind, even if the ambient noise is minimal.

Loudness is the psychophysical aspect of sound intensity. In an earlier study, Röhl and Uppenkamp (2012) examined the relationship between behavioral measures (categorical loudness scaling) and the fMRI parameter (percent signal change). They reported a linear relationship between behavioral measures and cortical activity. Loudness perception in the real world is relative and operates in the context of other sounds. It is natural to assume that the location of a louder sound source is closer compared to a fainter one. This assumption aids in the perception of both the location and the movement of the sound source. Behler and Uppenkamp (2016) investigated the blood-oxygen-level-dependent (BOLD) response to different loudness levels and varying loudness ramps (increasing and decreasing inputs). Although both types of stimuli elicited similar responses in the auditory regions, higher responses were observed for the loudness ramps in nonauditory regions. Furthermore, there was a difference between the ramp-up and ramp-down presentations. These findings indicated that contextual loudness evaluation at the auditory cortex involves cognitive processes. Behler and Uppenkamp (2016) also studied the effect of unilateral stimulation on loudness perception. They compared the BOLD responses during unilateral, diotic, and silent phases. The sound intensity revealed differential responses in the auditory regions, with the highest sensitivity to intensity observed in the PAC. The inferior colliculus showed the strongest response to contralateral stimuli, which became less pronounced at the cortical level, possibly due to the interhemispheric transfer function. Cortical regions demonstrated a linear relationship with categorical loudness, albeit at different levels across the auditory pathway, from the inferior colliculus to the cortex.

The capacity to discern pitch is paramount for a multitude of human experiences, including the appreciation of music, perceiving the emotional aspects of speech, and understanding speech in challenging acoustic environments, such as speech in noise. Pitch is the psychoacoustic parameter of frequency. The frequency sensitivity of the cochlear basilar membrane is arranged such that high frequencies are represented at the basal end and low frequencies at the apical end of the cochlea. This distinctive separation or spatial arrangement of frequencies along the auditory pathway, like that of the cochlea, is called tonotopicity. Allen et al. (2022) performed high-resolution (7 T) fMRI on normal-hearing young adults. They aimed to understand the difference in brain activation (spatial organization) between the tonotopic organization of sounds and pitch perception. They used pure tones, complex pitch, and complex timbre sounds as stimuli. The tonotopic map displayed a clear frequency gradient, with lower frequencies represented more anteriorly and higher frequencies more posteriorly within the PAC. Pitch, however, had some overlapping areas, and the anterior belt regions of the auditory cortex showed distinct activation. Furthermore, pitch processing appears to be a higher-order function, involving complex neural computations beyond simple frequency mapping. The findings have significant implications for understanding how the brain processes speech and music, where both tonotopic and pitch-related information play crucial roles. The distinct processing of tonotopy and pitch might explain how humans can recognize the pitch of complex sounds, such as voices and musical instruments, even in noisy environments where pure frequency information might be ambiguous or missing. Understanding these distinct neural mechanisms could inform treatments for APDs, where either tonotopy or pitch perception might be impaired.

There are other fMRI methods to study tonotopy/pitch. Arterial spin labeling (ASL), which observes blood flow to an area of interest, is one such method (Gardumi et al., 2017). Using ASL, Gardumi et al. (2017) attempted to create a cortical tonotopic map. The PAC (A1) revealed a clear tonotopic layout, with low frequencies represented more anteriorly and high frequencies more posteriorly. Furthermore, the nonprimary auditory areas (belt and parabelt) exhibited complex frequency representations. ASL provides a direct measure of blood flow, unlike BOLD. ASL may provide unique data and complement BOLD fMRI studies. Vascular space occupancy (VASO) is a CBV measure with high spatial specificity and can be combined with BOLD acquisition, which has high detection sensitivity (Lu et al., 2013). Faes et al. (2023) proposed a VASO protocol to perform laminar fMRI investigations of the auditory cortices. They employed 7 T fMRI to acquire concurrent gradient-echo BOLD (GE-BOLD) and VASO data and recommended a similar methodology for gaining better insights. GE-BOLD fMRI is used to collect submillimeter data. GE-BOLD fMRI stands out for its ability to capture the dynamic functional organization and connectivity of the auditory cortex. This technique leverages the BOLD signal, which reflects changes in blood flow associated with neural activity, to provide insights into the spatial and temporal patterns of brain activation. GE-BOLD methods could achieve even better results when a combined phase regression is performed (Pais-Roldán et al., 2022). The connectivity between brain regions should be hardwired for well-established behavior. Unlike task-based fMRI, resting-state fMRI (rs-fMRI) evaluates the connections between the brain's auditory networks during rest. Lumaca et al. (2019) reported that rs-FC within fronto-temporal auditory networks predicted the social learning and transmission of music. The researchers used both task-based and rs-fMRI. Other methods, like gradient-echo and spin-echo (GRASE) methods, can also be used to study cortical auditory processing (De Martino et al., 2015). A novel MRI sequence, "presaturated echo planar imaging (EPI) using Multiple delays in Steady-state", enhances auditory fMRI by allowing variable intervolume delays while maintaining signal stability, which is crucial for flexible stimulus design in auditory studies (Shrestha et al., 2021). This advancement facilitates more accurate assessments of auditory processing by improving the quality of the fMRI data collected.

The stimulus presentation can be modified to achieve finer details in auditory processing. fMRI adaptation is a method used to study frequency tuning (Besle et al., 2022). The method works on the principle that repeated stimulation of neurons leads to suppression of neural activity at that frequency. It helps achieve finer spatial resolution than conventional methods. (Besle et al., 2022) measured how the responses of the auditory cortex to a high-frequency sound (3.8 kHz) were suppressed when preceded by sounds of varying frequencies (from 0.5 to 3.8 kHz). The results revealed that multiple adapters (preceding sounds) are better than a single-time presentation of the adapter sound. The primary and nonprimary auditory regions exhibited variable adaptation tuning, indicating that the adaptation observed is due to a cortical phenomenon and not the effect of subcortical processing.

Advanced analysis methods are used to understand auditory cortical processing effectively (Gu et al., 2019). By applying univariate analyses, multivariate analyses, and a support vector machine classifier, the researchers studied the brain areas associated with auditory imagery and perception. They were able to distinguish and decode brain activation for different sound inputs, identify regions activated for auditory imagery and perception, and identify distinct representations of auditory imagery and perception in the overlapping brain regions. These findings provide new insights into the neural mechanisms involved in complex auditory processing. The scope of the findings could be extended to study clinical conditions like auditory processing deficits and auditory hallucinations.

The use of fMRI with a searchlight analysis technique is based on the principle that spatial voxel positions around BOLD activity are the useful ones (Urbschat et al., 2021). Machine learning algorithms combined with fMRI data can be used to analyze the decoding errors of the brain, thereby achieving new insights about the processing mechanisms (Jung et al., 2018). In this study, the researchers aimed to identify the neural signatures during auditory scene analysis. They performed multivoxel pattern analysis, decoding error analysis, linear mixed-effects modeling, and an exploratory searchlight analysis. It was found that anterior regions are related to behavior, while the posterior regions of the brain search for auditory structure. New processing strategies can be developed for hearing aids, cochlear implants, or other auditory devices, making them more effective in complex environments.

fMRI is successfully employed to study various clinical conditions. Hearing loss and tinnitus are known to distort cortical tonotopy, and fMRI can be employed to study these changes. Ghazaleh et al. (2017) examined the tonotopy of the PAC using high-resolution fMRI in patients with unilateral hearing loss and tinnitus. Patients with unilateral hearing loss exhibited a shift in the representation of frequencies, while patients with tinnitus showed hyperactivity and increased spatial representation for frequencies around the tinnitus frequency. Understanding these changes in the auditory cortex might inform treatment approaches for tinnitus. A more recent study by Berlot et al. (2020) used 7T fMRI to investigate processing in the cortical and subcortical regions in patients with tinnitus. The study revealed reduced frequency tuning in both the subcortical (medial geniculate body) and cortical areas. Furthermore, the study identified poor thalamocortical and cortico-cortical connectivity. Small-scale neural structures and subtle changes in brain activity, which might not be detectable with lower-field MRI, were observed. Higher-resolution data could provide researchers with a better understanding of the underlying neural mechanisms and expand the scope of auditory neuroscience. In the case of tinnitus, insights into the subtle changes in the auditory pathway or the connectivity alterations between structures would help in understanding its pathophysiology more comprehensively and devising an appropriate intervention plan accordingly.

7 Magnetic resonance spectroscopy

Auditory processing and disorders involve metabolic changes in the neural structures. This enables researchers to identify specific biomarkers and the underlying mechanisms linked to APDs. Any changes in neurotransmitter levels, such as glutamate and GABA, which are crucial for auditory signal processing, can be detected through MRS. An integrative approach combining MRS with fMRI/DTI may facilitate a comprehensive understanding of the auditory pathway and its connectivity.

Sedley et al. (2015a) evaluated patients with unilateral tinnitus by employing the Mescher-Garwood proton resolved spectroscopy technique. They explored both the left and right auditory cortices for GABA concentration, which was found to be significantly reduced in the right auditory cortex. The tinnitus side did not affect this finding. Wójcik et al. (2023) hypothesized increased levels of glutamate and glutamine (Glx) in the temporal lobe, excluding the PAC, in patients with tinnitus. They performed single-voxel point-resolved spectroscopy sequence MRS in patients with unilateral and bilateral tinnitus and compared the findings with those of normal controls. There were no significant differences between the groups in the temporal regions. However, the results revealed higher Glx levels for the bilateral tinnitus group in the left prefrontal brain area, and the authors postulated an active frontal-limbic maintenance circuit. Other techniques, like two-dimensional J-resolved spectroscopy (2D-JPRESS), can be utilized to study the metabolics of tinnitus (Isler et al., 2022).

MRS can be utilized to study the effect of tinnitus treatment (Cacace et al., 2018). In the study (Cacace et al., 2018), repetitive transcranial magnetic stimulation was administered over the left auditory cortex in patients with tinnitus. They reported reduced glutamate concentrations in the left auditory cortex and noted that the findings correlated with a reduction in tinnitus perception by the subjects.

In tinnitus, it is a continuous perception of auditory sensation by the neural system. The lack of stimulation, on the other hand, may have a different metabolic basis. Kilicarslan et al. (2014) utilized single-voxel spectroscopy to evaluate the effect of unilateral acoustic neuroma on the metabolites of HG. They compared the N-acetylaspartate, creatine, and choline values between the tumor and the nontumor side HG. The HG on the nontumor side is affected as it lacks stimulation from the contralateral ear (tumor side).

8 Positron emission tomography

PET is employed to perform functional neuroimaging of the auditory system. Brain activation for auditory stimuli appears different even for postlingual cochlear implantees compared to individuals with normal hearing (Wong et al., 1999). CIs showed more activation in the right temporal lobe, whereas for individuals with normal hearing, activation was observed in the left inferior frontal gyrus. Furthermore, CIs alone exhibited activation in the auditory regions in response to multi-talker babble stimuli. These findings indicate differences in auditory processing between the two groups. Berding et al. (2015) aimed to evaluate the factors behind differential performance in individuals with various auditory implants. They performed O-15-water PET while subjects listened to speech or continuous multitone stimuli. The study revealed a correlation between better speech perception and bilateral auditory cortex activation. For continuous multitone stimuli, activation was observed in the unilateral auditory cortex. In addition, loudness adaptation, which is a causative factor for poor speech understanding, led to reduced activation. Supportive suppression activity related to adaptation was observed in the ventral prefrontal cortex. O-15-water PET can also be performed in children who are to be implanted under general anesthesia (Mamach et al., 2018). Researchers were able to identify the auditory and extra-auditory regions involved during auditory activation. Auditory stimulation was provided through the promontory or cochlear implants. In a recent study, Karoui et al. (2023) conducted a similar investigation in patients with asymmetrical hearing loss who were unilaterally implanted with cochlear implants. They explored the relationship between spatial hearing, the side of implantation, and brain activation. PET data collected during the resting period (between stimulations) can be used to understand brain plasticity in patients with CI (Strelnikov et al., 2024).

9 Functional near-infrared spectroscopy

Auditory experiments are best carried out in silence. Unlike probing through the visual modality, auditory modality studies are restricted in fMRI due to high background noise. The fNIRS, on the other hand, employs LED/laser technology and optical neuroimaging, thereby enabling silent experiments suitable for auditory neurosciences. fNIRS is portable and can be easily set up, making it suitable for performing studies in naturalistic settings and in difficult-to-test populations. The real-time data display, in the form of lighting of brain regions related to the stimulation, facilitates the understanding of dynamic auditory processes. It may lead to innovative applications in clinical settings and help develop targeted interventions. In addition, fNIRS can be integrated with other techniques like EEG or fMRI. This allows researchers to access multidimensional information about the brain's auditory centers and their interaction with other brain regions during complex auditory tasks. Unlike fMRI, fNIRS is a new entrant in the area of auditory neuroscience.

The fNIRS has proved to be a boon for studying the effects of training and, thereby, neural plasticity (Mai et al., 2024). A short-term training program for adults with hearing loss in speech-in-noise revealed a significant correlation between the behavioral outcomes and the imaging results. Furthermore, the fNIRS indicated emerging improvements even before behavioral changes were observed. The changes noted were in the form of reduced differences in the left auditory cortex between speech and nonspeech conditions, as well as decreased cross-modal activation of brain regions (Mai et al., 2024).

In studying auditory and language processing in the pediatric population, the applicability of fNIRS holds great potential. Furthermore, various stimuli can be used to perform the experiments (Mushtaq et al., 2019). They performed fNIRS on 6 to 12 year-old children to study the differences in brain activation patterns for various auditory stimuli. In a similar age group, Mushtaq et al. (2019) investigated speech intelligibility under degraded conditions. The left temporal area showed greater activation when stimuli were more intelligible. These findings have significant applications in monitoring children with cochlear implants, brainstem implants, or any APD (Mushtaq et al., 2019).

Yaman et al. (2023), using fNIRS, evaluated auditory processing in patients post-SARS-CoV-2 infection. The regions of interest were the frontal and bilateral temporal regions, and an auditory odd ball paradigm was employed to evaluate auditory processing. Both the patient and control groups had auditory complaints. Findings revealed higher activation in patients compared to normal controls in the prefrontal cortex, as well as the superior and medial temporal gyri. Furthermore, higher activation was observed in the right hemisphere of the patients, indicating compensation in these individuals. A 16 source-detector setting was used to create 45 fNIRS channels covering the frontal and bilateral temporal lobes. These findings support the applicability of fNIRS in evaluating higher-order auditory processing.

10 Limitations and future directions

Despite these advancements, several limitations persist. Imaging artifacts, scanner noise, and constraints on patient compatibility (especially with implants) pose challenges for auditory fMRI studies. The spatial resolution of fNIRS and the metabolic specificity of MRS are still limited compared to other modalities. In addition, interindividual variability in cortical organization and neuroplastic responses complicates the generalization of findings.

Future research should focus on integrating multimodal data—combining high-resolution anatomical imaging with functional and neurochemical markers—to create comprehensive models of auditory processing. Advances in machine learning and computational modeling may facilitate real-time decoding of auditory brain activity and the development of personalized auditory rehabilitation protocols. Longitudinal studies using pediatric and aging populations will be critical for mapping experience-dependent plasticity and for identifying biomarkers of auditory dysfunction across the lifespan. Continued refinement of imaging protocols, particularly for noninvasive and implant-compatible methods, will be essential for expanding the clinical utility of neuroimaging in auditory science.

11 Conclusion

The integration of diverse neuroimaging modalities has significantly enriched our understanding of auditory neuroscience by enabling noninvasive, high-resolution analysis of the brain's structural, functional, and neurochemical dimensions. Structural imaging techniques, such as MRI and DTI, elucidate macro- and microstructural features of auditory pathways, shedding light on developmental plasticity and pathological changes associated with hearing loss and neurodevelopmental disorders. Functional modalities, particularly fMRI and PET, provide dynamic insights into cortical processing of sound features, including pitch, tonotopy, and loudness, while MRS and fNIRS extend this understanding by uncovering neurochemical substrates and allowing silent, ecologically valid assessments. The convergence of findings across these methods offers a multidimensional view of auditory perception and disorders, from basic sensory encoding to higher-order cognitive and clinical phenomena. As imaging technologies continue to evolve, their combined application will play an increasingly pivotal role in refining diagnostic criteria, tailoring rehabilitation strategies, and advancing theoretical models in auditory neuroscience.

References

Alais, D., Burr, D., 2004. The ventriloquist effect results from near-optimal bimodal integration. Curr. Biol. 09609822. 14 (3), 257–262. doi:10.1016/j.cub.2004.01.029.

Allen, E.J., Mesik, J., Kay, K.N., Oxenham, A.J., 2022. Distinct representations of tonotopy and pitch in human auditory cortex. J. Neurosci. 15292401. 42 (3), 416–434. Available from: https://www.jneurosci.org/content/jneuro/42/3/416.full.pdf. doi:10.1523/JNEUROSCI.0960-21.2021.

Alvand, A., Kuruvilla-Mathew, A., Roberts, R.P., Pedersen, M., Kirk, I.J., Purdy, S.C., 2023. Altered structural connectome of children with auditory processing disorder: a diffusion MRI study. Cereb. Cortex 14602199. 33 (12), 7727–7740. Available from: http://cercor.oxfordjournals.org/. doi:10.1093/cercor/bhad075.

American Speech-Language-Hearing Association, 2005. (Central) auditory processing disorders [Technical report].

Behler, O., Uppenkamp, S., 2016. Auditory fMRI of sound intensity and loudness for unilateral stimulation. Adv. Exp. Med. Biol. 22148019. 894, 165–174. Available from: http://www.springer.com/series/5584. doi:10.1007/978-3-319-25474-6_18.

Behler, O., Uppenkamp, S., 2021. Contextual effects on loudness judgments for sounds with continuous changes of intensity are reflected in nonauditory areas. Hum. Brain Mapp. 1065-9471. 42 (6), 1742–1757. doi:10.1002/hbm.25325.

Bell, L., Peng, Z. E., Pausch, F., Reindl, V., Neuschaefer-Rube, C., Fels, J., Konrad, K., 2020. fNIRS assessment of speech comprehension in children with normal hearing and children with hearing aids in virtual acoustic environments: pilot data and practical recommendations. Children. 2227-9067. 7 (11), 219. doi:10.3390/children7110219.

Bellis, T.J., 2003. Assessment and Management of Central Auditory Processing Disorders in the Educational Setting: From Science to Practice, Thomson Delmar Learning.

Bendor, D., Wang, X., 2005. The neuronal representation of pitch in primate auditory cortex. Nature. 0028-0836. 436 (7054), 1161–1165. doi:10.1038/nature03867.

Berding, G., Wilke, F., Rode, T., Haense, C., Joseph, G., Meyer, G.J., Mamach, M., Lenarz, M., Geworski, L., Bengel, F.M., Lenarz, T., Lim, H.H., 2015. Positron emission tomography imaging reveals auditory and frontal cortical regions involved with speech perception and loudness adaptation. PLoS ONE 19326203. 10 (6). Available from: http://www.plosone.org/article/fetchObject.action?uri=info:doi/10.1371/journal.pone.0128743&representation=PDF. doi:10.1371/journal.pone.0128743.

Berlot, E., Arts, R., Smit, J., George, E., Gulban, O.F., Moerel, M., Stokroos, R., Formisano, E., De Martino, F., 2020. A 7 Tesla fMRI investigation of human tinnitus percept in cortical and subcortical auditory areas. Neuroimage Clin. 22131582. 25, 102166. doi:10.1016/j.nicl.2020.102166.

Besle, J., Sánchez-Panchuelo, R.M., Francis, S., Krumbholz, K., 2022. Can single-neuron frequency tuning in human auditory cortex be quantified through fMRI adaptation? bioRxiv 26928205. Available from: https://www.biorxiv.org. doi:10.1101/2022.01.06.475208.

Billiet, C.R., Bellis, T.J., 2011. The relationship between brainstem temporal processing and performance on tests of central auditory function in children with reading disorders. J. Speech Lang. Hear. Res. 10924388. 54 (1), 228–242. Available from: http://jslhr.asha.org/cgi/reprint/54/1/228. doi:10.1044/1092-4388(2010/09-0239).

Bizley, J.K., Cohen, Y.E., 2013. The what, where and how of auditory-object perception. Nat. Rev. Neurosci. 14710048. 14 (10), 693–707. doi:10.1038/nrn3565.

Bregman, A.S., 1990. Auditory Scene Analysis: The Perceptual Organization of Sound. The MIT Press. doi:10.7551/mitpress/1486.001.0001.

Bronkhorst, A.W., 2000. The cocktail party phenomenon: a review of research on speech intelligibility in multiple-talker conditions. Acustica 86 (1), 117–128.

Cacace, A.T., Hu, J., Romero, S., Xuan, Y., Burkard, R.F., Tyler, R.S., 2018. Glutamate is down-regulated and tinnitus loudness-levels decreased following rTMS over auditory cortex of the left hemisphere: a prospective randomized single-blinded sham-controlled cross-over study. Hear. Res. 18785891. 358, 59–73. Available from: https://www.elsevier.com/locate/heares. doi:10.1016/j.heares.2017.10.017.

Caclin, A., Giard, M.H., Smith, B.K., McAdams, S., 2007. Interactive processing of timbre dimensions: a Garner interference study. Brain Res. 00068993. 1138 (1), 159–170. doi:10.1016/j.brainres.2006.12.065.

Calvert, G.A., Brammer, M.J., Iversen, S.D., 1998. Crossmodal identification. Trends Cogn. Sci. 13646613. 2 (7), 247–253. doi:10.1016/S1364-6613(98)01189-9.

Chermak, G.D., Musiek, F.E., 1997. Central Auditory Processing Disorders: New Perspectives, Singular Publishing Group.

Cherry, E.C., 1953. Some experiments on the recognition of speech, with one and with two ears. J. Acoust. Soc. Am. 00014966. 25 (5), 975–979. doi:10.1121/1.1907229.

De Martino, F., Moerel, M., Ugurbil, K., Goebel, R., Yacoub, E., Formisano, E., 2015. Frequency preference and attention effects across cortical depths in the human primary auditory cortex. Proc. Natl. Acad. Sci. U. S. A. 10916490. 112 (52), 16036–16041. Available from: http://www.pnas.org/content/112/52/16036.full.pdf. doi:10.1073/pnas.1507552112.

Dillon, H., 2013. Hearing Aids. Informa UK Limited, pp. 3631–3641. doi:10.1201/b15118-293.

Eggermont, J.J., Moore, J.K., 2011. Morphological and Functional Development of the Auditory Nervous System. Springer Science and Business Media LLC, pp. 61–105. doi:10.1007/978-1-4614-1421-6_3.

Epprecht, L., Qureshi, A., Kozin, E.D., Vachicouras, N., Huber, A.M., Kikinis, R., Makris, N., Brown, M.C., Reinshagen, K.L., Lee, D.J., 2020. Human cochlear nucleus on 7 tesla diffusion tensor imaging: insights into micro-anatomy and function for auditory brainstem implant surgery. Otol. Neurotol. 15374505. 41 (4), e484. Available from: http://journals.lww.com/otology-neurotology. doi:10.1097/MAO.0000000000002565.

Faes, L.K., De Martino, F., Huber, L., Ahveninen, J., 2023. Cerebral blood volume sensitive layer-fMRI in the human auditory cortex at 7T: challenges and capabilities. PLOS ONE. 1932-6203. 18 (2), e0280855. doi:10.1371/journal.pone.0280855.

Fastl, H., Zwicker, E., 2007. Psychoacoustics: Facts and Models Psychoacoustics: Facts and Models. Springer Berlin Heidelberg, Germany, pp. 1–463. Available from: http://link.springer.com/book/10.1007%2F978-3-540-68888-4. doi:10.1007/978-3-540-68888-4.

Ferguson, M.A., Henshaw, H., 2015. Auditory training can improve working memory, attention, and communication in adverse conditions for adults with hearing loss. Front. Psychol. 1664-1078. 6. doi:10.3389/fpsyg.2015.00556.

Fritz, J., Shamma, S., Elhilali, M., Klein, D., 2003. Rapid task-related plasticity of spectrotemporal receptive fields in primary auditory cortex. Nat. Neurosci. 1097-6256. 6 (11), 1216–1223. doi:10.1038/nn1141.

Gardumi, A., Ivanov, D., Havlicek, M., Formisano, E., Uludağ, K., 2017. Tonotopic maps in human auditory cortex using arterial spin labeling. Hum. Brain Mapp. 1065-9471. 38 (3), 1140–1154. doi:10.1002/hbm.23444.

Ghazaleh, N., van der Zwaag, W., Clarke, S., De Ville, D.V., Maire, R., Saenz, M., 2017. High-resolution fMRI of auditory cortical map changes in unilateral hearing loss and tinnitus. Brain Topogr. 0896-0267. 30 (5), 685–697. doi:10.1007/s10548-017-0547-1.

Griffiths, T.D., Warren, J.D., 2004. What is an auditory object? Nat. Rev. Neurosci. 1471003X. 5 (11), 887–892. doi:10.1038/nrn1538.

Grothe, B., Pecka, M., McAlpine, D., 2010. Mechanisms of sound localization in mammals. Physiol. Rev. 0031-9333. 90 (3), 983–1012. doi:10.1152/physrev.00026.2009.

Gu, J., Zhang, H., Liu, B., Li, X., Wang, P., Wang, B., 2019. An investigation of the neural association between auditory imagery and perception of complex sounds. Brain Struct Funct. 224 (8), 2925–2937. doi:10.1007/s00429-019-01948-z.

Hackett, T.A., 2011. Information flow in the auditory cortical network. Hear. Res. 03785955. 271 (1-2), 133–146. doi:10.1016/j.heares.2010.01.011.

Herholz, S.C., Zatorre, R.J., 2012. Musical training as a framework for brain plasticity: behavior, function, and structure. Neuron. 10974199. 76 (3), 486–502. doi:10.1016/j.neuron.2012.10.011.

Hickok, G., Poeppel, D., 2007. The cortical organization of speech processing. Nat. Rev. Neurosci. 1471-003X. 8 (5), 393–402. doi:10.1038/nrn2113.

Huang, L., Zheng, W., Chunxiao, W., Wei, X., Xianguang, W., Wang, Y., Zheng, H., Zhou, J., 2015. Diffusion tensor imaging of the auditory neural pathway for clinical outcome of cochlear implantation in pediatric congenital sensorineural hearing loss patients. PLOS ONE. 1932-6203. 10 (10), e0140643. doi:10.1371/journal.pone.0140643.

Huang, R., Wang, A., Zhang, Y., Guochao, L., Lin, Y., Xinru, B., Bao, X., Yunxin, L., Zhang, G., 2024. Alterations of the cerebral microstructure in patients with noise-induced hearing loss: a diffusion tensor imaging study. Brain Behav. 2162-3279. 14 (4). doi:10.1002/brb3.3479.

Isler, B., von Burg, N., Kleinjung, T., Meyer, M., Stämpfli, P., Zölch, N., Neff, P., 2022. Lower glutamate and GABA levels in auditory cortex of tinnitus patients: a 2D-JPRESS MR spectroscopy study. Sci. Rep. 2045-2322. 12 (1). doi:10.1038/s41598-022-07835-8.

Joris, P.X., Smith, P.H., Yin, T.C.T., 1998. Coincidence detection in the auditory system: 50 years after Jeffress. Neuron 08966273. 21 (6), 1235–1238. Available from: http://www.cell.com/neuron/home. doi:10.1016/S0896-6273(00)80643-1.

Jung, Y., Larsen, B., Walther, D.B. 2018. Using decoding error patterns to trace the neural signature of auditory scene perception. International Workshop on Pattern Recognition in Neuroimaging, PRNI 2018. Institute of Electrical and Electronics Engineers Inc, Canada, 9781538668597. http://ieeexplore.ieee.org/xpl/mostRecentIssue.jsp?punumber=8412440. doi:10.1109/PRNI.2018.8423950.

Kaas, J.H., Hackett, T.A., 2000. Subdivisions of auditory cortex and processing streams in primates. Proc. Natl. Acad. Sci. U. S. A. 0027-8424. 97 (22), 11793–11799. doi:10.1073/pnas.97.22.11793.

Karoui, C., Strelnikov, K., Payoux, P., Salabert, A.S., James, C.J., Deguine, O., Barone, P., Marx, M., 2023. Auditory cortical plasticity after cochlear implantation in asymmetric hearing loss is related to spatial hearing: a PET H215O study. Cereb. Cortex 14602199. 33 (5), 2229–2244. Available from: http://cercor.oxfordjournals.org/. doi:10.1093/cercor/bhac204.

Kilicarslan, R., Alkan, A., Aralasmak, A., Aksoy, F., Toprak, H., Yetis, H., Ozturan, O., 2014. Magnetic resonance spectroscopy features of heschl's gyri in patients with unilateral acoustic neuroma: preliminary study. Acad. Radiol. 18784046. 21 (12), 1501–1505. Available from: https://www.journals.elsevier.com/academic-radiology. doi:10.1016/j.acra.2014.07.007.

Kral, A., 2013. Auditory critical periods: a review from system's perspective. Neuroscience. 03064522. 247, 117–133. doi:10.1016/j.neuroscience.2013.05.021.

Kral, A., Sharma, A., 2012. Developmental neuroplasticity after cochlear implantation. Trends Neurosci. 01662236. 35 (2), 111–122. doi:10.1016/j.tins.2011.09.004.

Kuhl, P.K., 2010. Brain mechanisms in early language acquisition. Neuron. 08966273. 67 (5), 713–727. doi:10.1016/j.neuron.2010.08.038.

LeDoux, J.E., 2000. Emotion circuits in the brain. Annu. Rev. Neurosci. 0147006X. 23, 155–184. doi:10.1146/annurev.neuro.23.1.155.

Leibold, L.J., Buss, E., 2013. Children's identification of consonants in a speech-shaped noise or a two-talker masker. J. Speech Lang. Hear. Res. 10924388. 56 (4), 1144–1155. Available from: http://jslhr.asha.org/cgi/reprint/56/4/1144. doi:10.1044/1092-4388(2012/12-0011).

Liberman, A.M., Mattingly, I.G., 1985. The motor theory of speech perception revised. Cognition. 00100277. 21 (1), 1–36. doi:10.1016/0010-0277(85)90021-6.

Lomber, S.G., Malhotra, S., 2008. Double dissociation of 'what' and 'where' processing in auditory cortex. Nat. Neurosci. 10976256. 11 (5), 609–616. doi:10.1038/nn.2108.

Lu, H., Hua, J., van Zijl, P.C.M., 2013. Noninvasive functional imaging of cerebral blood volume with vascular-space-occupancy (VASO) MRI. NMR Biomed. 10991492. 26 (8), 932–948. doi:10.1002/nbm.2905.

Lumaca, M., Kleber, B., Brattico, E., Vuust, P., Baggio, G., 2019. Functional connectivity in human auditory networks and the origins of variation in the transmission of musical systems. eLife. 2050-084X. 8. doi:10.7554/elife.48710.

Madzime, J., Jankiewicz, M., Meintjes, E.M., Torre, P., Laughton, B., van der Kouwe, A.J.W., Holmes, M., 2024. Reduced white matter maturation in the central auditory system of children living with HIV. Front. Neuroimaging. 2813-1193. 3. doi:10.3389/fnimg.2024.1341607.

Mai, G., Jiang, Z., Wang, X., Tachtsidis, I., Howell, P., 2024. Neuroplasticity of Speech-in-noise processing in older adults assessed by functional near-infrared spectroscopy (fNIRS). Brain Topogr. 15736792. 37 (6), 1139–1157. Available from: https://www.springer.com/journal/10548. doi:10.1007/s10548-024-01070-2.

Malmierca, M.S., Hackett, T.A., 2010. Structural Organization of the Ascending Auditory Pathway the Oxford Handbook of Auditory Science: The Auditory Brain. Oxford University Press, pp. 9–41.

Mamach, M., Kessler, M., Bankstahl, J.P., Wilke, F., Geworski, L., Bengel, F.M., Kurt, S., Berding, G., Malmierca, M.S., 2018. Visualization of the auditory pathway in rats with 18F-FDG PET activation studies based on different auditory stimuli and reference conditions including cochlea ablation. PLOS ONE. 1932-6203. 13 (10), e0205044. doi:10.1371/journal.pone.0205044.

McAdams, S., 2019. The Perceptual Representation of Timbre. Springer Science and Business Media LLC, pp. 23–57. doi:10.1007/978-3-030-14832-4_2.

McGurk, H., MacDonald, J., 1976. Hearing lips and seeing voices. Nature. 0028-0836. 264 (5588), 746–748. doi:10.1038/264746a0.

Meyer, M., Liem, F., Hirsiger, S., Jäncke, L., Hänggi, J., 2014. Cortical surface area and cortical thickness demonstrate differential structural asymmetry in auditory-related areas of the human cortex. Cereb. Cortex 14602199. 24 (10), 2541–2552. Available from: http://cercor.oxfordjournals.org/. doi:10.1093/cercor/bht094.

Middlebrooks, J.C., Green, D.M., 1991. Sound localization by human listeners. Annu. Rev. Psychol. 00664308. 42 (1), 135–159. Available from: http://arjournals.annualreviews.org/loi/psych. doi:10.1146/annurev.ps.42.020191.001031.

Mishkin, M., Ungerleider, L.G., Macko, K.A., 1983. Object vision and spatial vision: two cortical pathways. Trends Neurosci. 01662236. 6 (C), 414–417. doi:10.1016/0166-2236(83)90190-X.

Moore, B.C., 2012. An Introduction to the Psychology of Hearing, Academic Press.

Moore, B.C.J., Vickers, D.A., Plack, C.J., Oxenham, A.J., 1999. Inter-relationship between different psychoacoustic measures assumed to be related to the cochlear active mechanism. J. Acoust. Soc. Am. 00014966. 106 (5), 2761–2778. Available from: http://scitation.aip.org/content/asa/journal/jasa. doi:10.1121/1.428133.

Moore, B.C.J., 2003. Temporal Processing in the Auditory System an Introduction to the Psychology of Hearing. Academic Press, pp. 175–207.

Moore, D.R., Hunter, L.L., 2013. Auditory processing disorder (APD) in children: a marker of neurodevelopmental syndrome. Hear. Balance Commun. 21695725. 11 (3), 160–167. doi:10.3109/21695717.2013.821756.

Mushtaq, F., Wiggins, I.M., Kitterick, P.T., Anderson, C.A., Hartley, D.E.H., 2019. Evaluating time-reversed speech and signal-correlated noise as auditory baselines for isolating speech-specific processing using fNIRS. PLoS ONE 19326203. 14 (7). Available from: https://journals.plos.org/plosone/article/file?id=10.1371/journal.pone.0219927&type=printable. doi:10.1371/journal.pone.0219927.

Musiek, F.E., Chermak, G.D., 2007. Handbook of (Central) Auditory Processing Disorder, Volume I: Auditory Neuroscience and Diagnosis, Plural Publishing.

Musiek, F.E., Shinn, J.B., Jirsa, R., Bamiou, D.E., Baran, J.A., Zaida, E., 2005. GIN (Gaps-In-Noise) test performance in subjects with confirmed central auditory nervous system involvement. Ear Hear. 01960202. 26 (6), 608–618. doi:10.1097/01.aud.0000188069.80699.41.

Olds, C., Pollonini, L., Abaya, H., Larky, J., Loy, M., Bortfeld, H., Beauchamp, M.S., Oghalai, J.S., 2016. Cortical activation patterns correlate with speech understanding after cochlear implantation. Ear Hear. 15384667. 37 (3), e160. Available from: http://journals.lww.com/ear-hearing/pages/default.aspx. doi:10.1097/AUD.0000000000000258.

Oxenham, A.J., 2012. Pitch perception. J. Neurosci. 15292401. 32 (39), 13335–13338. Available from: http://www.jneurosci.org/content/32/39/13335.full.pdf+html. doi:10.1523/JNEUROSCI.3815-12.2012.

Pais-Roldán, P., Yun, S.D., Shah, N.J., 2022. Pre-processing of Sub-millimeter GE-BOLD fMRI Data for Laminar Applications. Front. Neuroimaging. 2813-1193. 1. doi:10.3389/fnimg.2022.869454.

Phillips, D.P., 2008. Auditory gap detection, perceptual channels, and temporal resolution in speech perception. J. Am. Acad. Audiol. 19 (7), 546–556.

Pickles, J.O., 2013. An Introduction to the Physiology of Hearing, Academic Press.

Plack, C.J., Oxenham, A.J., 2005. The psychophysics of pitch. In: Plack, C.J., Oxenham, A.J., Fay, R.R., Popper, A.N. (Eds.), Pitch: Neural Coding and Perception. Springer, New York, pp. 7–55.

Poldrack, R.A., Mumford, J.A., Nichols, T.E., 2011. Handbook of Functional MRI Data Analysis. Cambridge University Press. doi:10.1017/CBO9780511895029.

Ponton, C.W., Eggermont, J.J., Kwong, B., Don, M., 2000. Maturation of human central auditory system activity: evidence from multi-channel evoked potentials. Clin. Neurophysiol. 13882457. 111 (2), 220–236. doi:10.1016/S1388-2457(99)00236-9.

Qu, H., Tang, H., Pan, J., Zhao, Y., Wang, W., 2020. Alteration of cortical and subcortical structures in children with profound sensorineural hearing loss. Front. Hum. Neurosci. 1662-5161. 14. doi:10.3389/fnhum.2020.565445.

Ratnanather, J.T., 2020. Structural neuroimaging of the altered brain stemming from pediatric and adolescent hearing loss—Scientific and clinical challenges. Wiley Interdiscip. Rev. Syst. Biol. Med. 1939005X. 12 (2). Available from: http://onlinelibrary.wiley.com/journal/10.1002/(ISSN)1939-005X. doi:10.1002/wsbm.1469.

Rauschecker, J.P., Scott, S.K., 2009. Maps and streams in the auditory cortex: nonhuman primates illuminate human speech processing. Nat. Neurosci. 10976256. 12 (6), 718–724. doi:10.1038/nn.2331.

Röhl, M., Uppenkamp, S., 2012. Neural coding of sound intensity and loudness in the human auditory system. J. Assoc. Res. Otolaryngol. 1525-3961. 13 (3), 369–379. doi:10.1007/s10162-012-0315-6.

Samuel, A.G., Kraljic, T., 2009. Perceptual learning for speech. Atten. Percept. Psychophys. 19433921. 71 (6), 1207–1218. Available from: http://app.psychonomic-journals.org/content/71/6/1207.full.pdf. doi:10.3758/APP.71.6.1207.

Schneider, P., Engelmann, D., Groß, C., Bernhofs, V., Hofmann, E., Christiner, M., Benner, J., Bücher, S., Ludwig, A., Serrallach, B.L., Zeidler, B.M., Turker, S., Parncutt, R., Seither-Preisler, A., 2023. Neuroanatomical disposition, natural development, and training-induced plasticity of the human auditory system from childhood to adulthood: a 12-year study in musicians and nonmusicians. J. Neurosci. 0270-6474. 43 (37), 6430–6446. doi:10.1523/jneurosci.0274-23.2023.

Schreiner, C.E., Langner, G., 1997. Laminar fine structure of frequency organization in auditory midbrain. Nature 00280836. 388 (6640), 383–386. Available from: http://www.nature.com/nature/index.html. doi:10.1038/41106.

Schroeder, C.E., Foxe, J., 2005. Multisensory contributions to low-level, 'unisensory' processing. Curr. Opin. Neurobiol. 15 (4), 454–458. doi:10.1016/j.conb.2005.06.008.

Scott, S.K., Johnsrude, I.S., 2003. The neuroanatomical and functional organization of speech perception. Trends Neurosci. 26 (2), 100–107. doi:10.1016/S0166-2236(02)00037-1.

Sedley, W., Gander, P.E., Kumar, S., Oya, H., Kovach, C.K., Nourski, K.V., Kawasaki, H., Howard, M.A., Griffiths, T.D., 2015b. Intracranial mapping of a cortical tinnitus system using residual inhibition. Curr. Biol. 18790445. 25 (9), 1208–1214. Available from: http://www.elsevier.com/journals/current-biology/0960-9822. doi:10.1016/j.cub.2015.02.075.

Sedley, W., Parikh, J., Edden, R.A.E., Tait, V., Blamire, A., Griffiths, T.D., 2015a. Human auditory cortex neurochemistry reflects the presence and severity of tinnitus. J. Neurosci. 15292401. 35 (44), 14822–14828. Available from: http://www.jneurosci.org/content/35/44/14822.full.pdf. doi:10.1523/JNEUROSCI.2695-15.2015.

Senkowski, D., Saint-Amour, D., Höfle, M., Foxe, J.J., 2011. Multisensory interactions in early evoked brain activity follow the principle of inverse effectiveness. NeuroImage 56 (4), 2200–2208. doi:10.1016/j.neuroimage.2011.03.075.

Sharma, A., Dorman, M.F., Spahr, A.J., 2002. A sensitive period for the development of the central auditory system in children with cochlear implants: implications for age of implantation. Ear Hear. 01960202. 23 (6), 532–539. Available from: http://journals.lww.com/ear-hearing/pages/default.aspx. doi:10.1097/00003446-200212000-00004.

Shinn-Cunningham, B.G., 2008. Object-based auditory and visual attention. Trends Cogn. Sci. 13646613. 12 (5), 182–186. doi:10.1016/j.tics.2008.02.003.

Shrestha, M., Lee, H.S., Nöth, U., Deichmann, R., 2021. A novel sequence to improve auditory functional MRI with variable silent delays. Magn. Reson. Med. 15222594. 85 (2), 883–896. Available from: http://onlinelibrary.wiley.com/journal/10.1002/(ISSN)1522-2594. doi:10.1002/mrm.28479.

Strelnikov, K., Karoui, C., Payoux, P., Salabert, A.S., James, C., Deguine, O., Barone, P., Marx, M., 2024. Adaptive Strategies of Single-Sided Deaf Cochlear-Implant Users Revealed Through Resting State Activity: an Auditory PET Brain Imaging Study. Hear. Res. 03785955. 451, 109079. doi:10.1016/j.heares.2024.109079.

Suga, N., 2008. Role of corticofugal feedback in hearing. J. Comp. Physiol. A. 0340-7594. 194 (2), 169–183. doi:10.1007/s00359-007-0274-2.

Sumby, W.H., Pollack, I., 1954. Visual contribution to speech intelligibility in noise. J. Acoust. Soc. Am. 0001-4966. 26 (2), 212–215. doi:10.1121/1.1907309.

Trainor, L.J., 2005. Are there critical periods for musical development? Dev. Psychobiol. 46 (3), 262–278. doi:10.1002/dev.20059.

Urbschat, A., Uppenkamp, S., Anemüller, J., 2021. Searchlight classification informative region mixture model (SCIM): identification of cortical regions showing discriminable BOLD patterns in event-related auditory fMRI data. Front. Neurosci. 1662-453X. 14. doi:10.3389/fnins.2020.616906.

Wagner, M.W., Cushing, S.L., Tshuma, M., Gordon, K.A., Ertl-Wagner, B.B., Vidarsson, L., 2022. Volumetric analysis of hearing-related structures of brain in children with GJB2-related congenital deafness. Children 22279067. 9 (6). Available from: https://www.mdpi.com/2227-9067/9/6/800/pdf?version=1653908109. doi:10.3390/children9060800.

Wang, X., 2013. The harmonic organization of auditory cortex. Front. Syst. Neurosci. 1662-5137. 7. doi:10.3389/fnsys.2013.00114.

Wang, X., Lu, T., Snider, R.K., Liang, L., 2005. Sustained firing in auditory cortex evoked by preferred stimuli. Nature 435 (7040), 341–346. doi:10.1038/nature03565.

Weinberger, N.M., 2004. Specific long-term memory traces in primary auditory cortex. Nat. Rev. Neurosci. 14710048. 5 (4), 279–290. doi:10.1038/nrn1366.

Werner, L.A., Leibold, L.J., Baer, T., 2012. The Auditory System: Anatomy, Physiology, and Clinical Correlates, Lippincott Williams & Wilkins.

Winer, J.A., 2006. Decoding the auditory corticofugal systems. Hear. Res. 03785955. 212 (1-2), 1–8. doi:10.1016/j.heares.2005.06.014.

Winer, J.A., Miller, L.M., Lee, C.C., Schreiner, C.E., 2005. Auditory thalamocortical transformation: structure and function. Trends Neurosci. 01662236. 28 (5), 255–263. Available from: https://www.elsevier.com/locate/tins. doi:10.1016/j.tins.2005.03.009.

Wójcik, J., Kochański, B., Cieśla, K., Lewandowska, M., Karpiesz, L., Niedziałek, I., Raj-Koziak, D., Skarżyński, P.H., Wolak, T., 2023. An MR spectroscopy study of temporal areas excluding primary auditory cortex and frontal regions in subjective bilateral and unilateral tinnitus. Sci. Rep. 20452322. 13 (1). Available from: https://www.nature.com/srep/. doi:10.1038/s41598-023-45024-3.

Wong, D., Miyamoto, R.T., Pisoni, D.B., Sehgal, M., Hutchins, G.D., 1999. PET imaging of cochlear-implant and normal-hearing subjects listening to speech and nonspeech. Hear. Res. 03785955. 132 (1-2), 34–42. Available from: https://www.elsevier.com/locate/heares. doi:10.1016/S0378-5955(99)00028-3.

Wong, P.C.M., Skoe, E., Russo, N.M., Dees, T., Kraus, N., 2007. Musical experience shapes human brainstem encoding of linguistic pitch patterns. Nat. Neurosci. 10976256. 10 (4), 420–422. doi:10.1038/nn1872.

Yaman, H., Yılmaz, O., Hanoğlu, L., Bayazıt, Y., 2023. fNIRS-based evaluation of the impact of SARS-CoV-2 infection central auditory processing. Brain Behav. 21623279. Available from: http://onlinelibrary.wiley.com/journal/10.1002/(ISSN)2157-9032. doi:10.1002/brb3.3303.

Young, E.D., 2008. Neural representation of spectral and temporal information in speech. Philos. Trans. R. Soc. Lond. B Biol. Sci. 09628436. 363 (1493), 923–945. Available from: http://rstb.royalsocietypublishing.org/. doi:10.1098/rstb.2007.2151.

Zatorre, R.J., Belin, P., Penhune, V.B., 2002. Structure and function of auditory cortex: music and speech. Trends Cogn. Sci. 13646613. 6 (1), 37–46. doi:10.1016/S1364-6613(00)01816-7.

Zatorre, R.J., Fields, R.D., Johansen-Berg, H., 2012. Plasticity in gray and white: neuroimaging changes in brain structure during learning. Nat. Neurosci. 15461726. 15 (4), 528–536. doi:10.1038/nn.3045.

Zoellner, S., Benner, J., Zeidler, B., Seither-Preisler, A., Christiner, M., Seitz, A., Goebel, R., Heinecke, A., Wengenroth, M., Blatow, M., Schneider, P., 2019. Reduced cortical thickness in Heschl's gyrus as an in vivo marker for human primary auditory cortex. Hum. Brain Mapp. 1065-9471. 40 (4), 1139–1154. doi:10.1002/hbm.24434.

Chapter 27

Inner ear: Imaging and functional correlation

Pradeep Yuvaraj[a], Aravind Kumar Rajasekaran[a], Sneha Ravichandran[b], Leena R. David[b,c], Rajagopal Kadavigere[d], Dilber Uzun Ozsahin[c,e,f], Dilip Shettigar[b], Suresh Sukumar[b], Ilker Ozsahin[f], and Berna Uzun[f]

[a]*Department of Speech Pathology and Audiology, National Institute of Mental Health and Neurosciences, Bengaluru, India*

[b]*Department of Medical Imaging Technology, Manipal College of Health Professions, Manipal Academy of Higher Education, Manipal, Karnataka, India*

[c]*Medical Diagnostic Imaging Department, College of Health Sciences, University of Sharjah, Sharjah, United Arab Emirates*

[d]*Department of Radiodiagnosis and Imaging, Kasturba Medical College, Manipal, India*

[e]*Research Institute for Medical and Health Sciences, University of Sharjah, Sharjah, United Arab Emirates*

[f]*Irfan Suat Gunsel Operational Research Institute, Near East University, Nicosia, Turkish Republic of Northern Cyprus (TRNC), Turkey*

1 Introduction

The inner ear houses both hearing and balance mechanisms. The gravity of the problem, if these structures are compromised, is high. Hearing loss is the most prevalent among sensory impairments. With the ever-growing global population, the number of people with hearing loss is expected to increase dramatically. According to the World Report on Hearing, by the year 2050, around 2.5 billion people may have hearing loss (Chadha et al., 2021). The primary cause of permanent hearing loss is sensorineural, which involves the cochlea (a part of the inner ear). Adding to the complexity is cochlear hearing loss, which can be caused by genetic, congenital, or acquired factors and can affect individuals across the age range. The impact is enormous for children, as it affects speech and language development. When vestibular structures (parts of the inner ear) are affected, it impacts motor development in children and leads to balance dysfunctions across all age groups. A population-based study reported the prevalence of peripheral vestibular disorders as 6.5%, with a higher incidence in older age groups (Hülse et al., 2019). Due to the enormity and complexity of the problems that inner ear compromise can cause, it is pertinent to have in-depth knowledge about it. Technological advancements in inner ear imaging have facilitated clinical and research endeavors for the benefit of affected patients.

The human inner ear represents one of the most intricately designed sensory organs in the body, comprising the cochlea for auditory function and the vestibular labyrinth for balance perception. This architectural marvel, encased within the densest bone of the human body—the temporal bone—exemplifies nature's precision engineering. The cochlea, a spiral-shaped structure resembling a snail shell, contains approximately 15,000 hair cells arranged in exquisite tonotopic organization, allowing humans to discriminate between thousands of frequency variations (Pickles, 2012). The vestibular system, consisting of three semicircular canals and two otolith organs (utricle and saccule), detects angular and linear accelerations, respectively, providing constant feedback about head position and movement in three-dimensional space (Goldberg et al., 2012).

The global burden of inner ear disorders represents a significant public health challenge that extends beyond the raw statistics. While the World Report on Hearing projects an alarming increase in hearing loss prevalence to 2.5 billion by 2050, these numbers fail to capture the profound impact on quality of life across multiple domains. Hearing impairment disrupts communication, diminishes educational and employment opportunities, increases social isolation, and contributes to cognitive decline, particularly in aging populations (Livingston et al., 2020). Furthermore, the economic burden of unaddressed hearing loss has been estimated at $980 billion annually worldwide, encompassing healthcare costs, productivity losses, and social support expenditures (World Health Organization, 2021).

Sensorineural hearing loss (SNHL), the predominant form of permanent hearing impairment, arises from dysfunction within the cochlea or along the auditory neural pathway. The etiological landscape of SNHL is remarkably diverse, spanning genetic mutations, prenatal infections, ototoxic medications, noise exposure, aging processes, autoimmune disorders,

Advances in Medical Imaging. https://doi.org/10.1016/B978-0-443-28967-5.00020-4

and vascular pathologies (Bowl and Dawson, 2015). Over 150 genetic loci have been implicated in hereditary hearing loss, with mutations in the GJB2 gene (encoding connexin 26) representing the most common cause of congenital hearing impairment in many populations (Kelsell et al., 1997). The inherent vulnerability of cochlear hair cells, which lack regenerative capacity in mammals, means that damage from any cause typically results in permanent impairment, unlike many other cellular systems in the body that can repair or replace damaged components (Forge and Schacht, 2000).

The consequences of cochlear dysfunction in pediatric populations extend far beyond auditory perception. Early hearing is fundamental to the acquisition of spoken language, literacy development, social-emotional regulation, and academic achievement (Yoshinaga-Itano et al., 1998). Children with untreated hearing loss demonstrate vocabulary gaps of 2–3 years compared to peers with normal hearing, and these discrepancies often widen rather than narrow over time (Hart and Risley, 1995). Even mild or unilateral hearing loss can significantly impact classroom performance, with affected children showing higher rates of grade repetition and a need for educational support services (Bess et al., 1998). The critical periods for auditory development and language acquisition underscore why early identification and intervention for hearing loss have become the standard of care in developed healthcare systems, with universal newborn hearing screening programs implemented in many countries (Joint Committee on Infant Hearing, 2019).

Vestibular disorders, though less recognized than hearing impairment, exert equally profound effects on functioning and quality of life. Beyond the 6.5% prevalence of peripheral vestibular disorders reported by Hülse et al. (2019), vestibular dysfunction has been documented in 35% of adults over 40 years old in the United States (Agrawal et al., 2009). The clinical manifestations range from acute, incapacitating vertigo to chronic imbalance and spatial disorientation. In children, vestibular dysfunction can delay motor milestones, impair balance-dependent activities, and affect spatial cognition, a domain increasingly recognized as important for mathematical and abstract reasoning (Rajendran and Mitchell, 2007). Benign paroxysmal positional vertigo, Ménière's disease, vestibular neuritis, and bilateral vestibulopathy represent common peripheral vestibular pathologies, each with distinct pathophysiological mechanisms and clinical presentations (Strupp and Brandt, 2013). The high comorbidity between hearing and balance disorders reflects their shared anatomical and physiological foundations within the inner ear.

Recent technological innovations have revolutionized our ability to visualize and assess inner ear structures and functions. High-resolution computed tomography (HRCT) and magnetic resonance imaging (MRI) with specialized inner ear protocols now permit detailed evaluation of cochlear morphology, semicircular canal anomalies, endolymphatic hydrops (EH), and neural pathway integrity (Casselman et al., 1993). Advanced electrophysiological techniques like electrocochleography, vestibular evoked myogenic potentials (VEMPs), and video head impulse testing (vHIT) provide functional assessment of specific inner ear components (Rosengren et al., 2019). These diagnostic modalities have transformed clinical approaches to conditions like enlarged vestibular aqueduct (EVA) syndrome, superior semicircular canal (SCC) dehiscence, and auditory neuropathy spectrum disorder—entities that were poorly understood or unrecognized just decades ago (Minor et al., 1998; Starr et al., 1996).

The therapeutic landscape for inner ear disorders has similarly expanded. Conventional interventions, such as hearing aids and vestibular rehabilitation therapy, have been complemented by cochlear implants, which bypass damaged hair cells to directly stimulate auditory nerve fibers. More than 700,000 individuals worldwide have received these devices, with expanding candidacy criteria and improving outcomes (Zeng et al., 2008). Emerging regenerative approaches, including gene therapy, stem cell transplantation, and small molecule drugs targeting specific cellular pathways, hold promise for restoring inner ear function (Géléoc and Holt, 2014). Clinical trials investigating ATOH1 gene delivery for hair cell regeneration and FGF delivery for spiral ganglion protection represent the leading edge of translational research in this field (Mizutari et al., 2013).

The interdisciplinary nature of inner ear disorders necessitates collaboration among audiologists, otolaryngologists, neurologists, pediatricians, geneticists, speech-language pathologists, and educators. This comprehensive approach acknowledges the complex interplay between hearing, balance, communication, cognition, and psychosocial well-being. Public health initiatives focusing on prevention, including noise exposure reduction, ototoxicity monitoring, and immunization against infections that cause hearing loss, complement clinical interventions for existing conditions (Nelson et al., 2005). The World Health Organization's "Make Listening Safe" campaign exemplifies efforts to address recreational noise exposure, an increasingly significant cause of hearing loss in younger populations (World Health Organization, 2015).

As our understanding of inner ear function continues to evolve through basic science research, clinical investigations, and epidemiological studies, so too will our approaches to the diagnosis, treatment, and prevention of conditions affecting this remarkable sensory system. The convergence of molecular biology, biomedical engineering, and computational neuroscience offers unprecedented opportunities to address the growing global burden of hearing and balance disorders across the lifespan.

2 Inner ear anatomy

The inner ear is housed within the temporal bone of the skull. Lateral to the inner ear is the middle ear, and medially lies the internal acoustic meatus. The oval window and the round window are membranes that cover the two openings between the middle and inner ear. The inner ear has an outer bony labyrinth and an inner membranous labyrinth. The bony labyrinth protects the delicate structures of the membranous labyrinth. The membranous labyrinth encompasses the cochlear duct, semicircular ducts, the utricle, and the saccule, and is filled with endolymph. The space between the bony and membranous labyrinths is filled with perilymph. The inner ear can be divided into the cochlea, vestibule, and SCCs. The cochlea is related to hearing, while the other two are responsible for balance. The cochlear duct is divided into the scala vestibuli, scala media, and scala tympani. The floor of the scala media contains the vital structure for hearing, the organ of Corti, which houses the outer and inner hair cells. It is from these hair cells that auditory information is transduced to the auditory nerve fibers. The cochlea is a snail-like structure, with its basal end sensitive to high frequencies and its apical end sensitive to low frequencies. The basal turn of the cochlea causes a bony prominence in the middle ear called the promontory. There are three SCCs: the anterior, lateral, and posterior SCC. These canals are positioned at right angles to each other, and the lateral SCC forms a bulge in the middle ear. The SCCs detect the rotational movements of the head. Each SCC has an enlarged region near its end, which houses the sensory organ for balance, called the crista ampullaris. The otolith organs, namely the utricle and saccule, detect linear acceleration. The sensory structure within the otolith organs is the macula.

The inner ear represents one of nature's most extraordinary sensory organs, encapsulated within the petrous portion of the temporal bone, the densest osseous structure in the human body (Schuknecht, 1993). This strategic location provides critical protection for the delicate neurosensory apparatus contained within. The inner ear's positioning between the middle ear laterally and the internal acoustic meatus medially facilitates both its sound transduction function and its neural connections to the central nervous system.

Communication between the middle and inner ear occurs through two specialized membrane-covered apertures: the oval window and the round window. The oval window serves as the input portal where stapes vibrations are transmitted to the perilymph, while the round window functions as a pressure relief valve, allowing fluid displacement to accommodate the incoming sound energy (Merchant and Rosowski, 2008). This complementary mechanical arrangement prevents pressure buildup that would otherwise dampen cochlear responses.

The architectural organization of the inner ear follows a remarkable dual-compartment design. The outer bony labyrinth, carved within the temporal bone, forms a rigid, protective shell around the more delicate membranous labyrinth. The bony labyrinth consists of three interconnected spaces: the cochlea anteriorly, the vestibule centrally, and the SCCs posteriorly. This osseous framework maintains precise spatial relationships critical for both auditory and vestibular functions (Igarashi, 1967).

Within this bony framework floats the membranous labyrinth, a continuous system of interconnected fluid-filled tubes and sacs. The membranous labyrinth is suspended within the bony labyrinth through a network of fine connective tissue strands and is bathed externally by perilymph, a fluid with an ionic composition similar to extracellular fluid (high sodium, low potassium). The membranous structures themselves contain endolymph, which has an unusual ionic profile more resembling intracellular fluid (high potassium, low sodium). This distinctive fluid environment creates the electrochemical gradient essential for sensory transduction (Wangemann, 2006).

The cochlear portion of the membranous labyrinth, known as the cochlear duct or scala media, assumes a triangular cross-sectional configuration when viewed in radial section. This duct divides the bony cochlear canal into three parallel fluid-filled chambers that spiral around the modiolus, the central bony core of the cochlea. Above the cochlear duct lies the scala vestibuli, which receives vibrations from the oval window; below is the scala tympani, which terminates at the round window. These scalae contain perilymph and communicate at the cochlear apex through an opening called the helicotrema, forming a continuous fluid pathway (Békésy, 1960).

The cochlear duct itself is bounded superiorly by Reissner's membrane, laterally by the stria vascularis, and inferiorly by the basilar membrane. The stria vascularis serves as a metabolically active epithelium responsible for maintaining the distinctive ionic composition of endolymph and generating the +80 mV endocochlear potential that drives hair cell transduction (Hibino et al., 2010). The basilar membrane exhibits graduated mechanical properties, being narrow and stiff at the base and progressively wider and more compliant toward the apex. This structural gradient underlies the cochlea's tonotopic organization, with high frequencies preferentially vibrating the basal region and low frequencies the apical region (Greenwood, 1990).

Upon the basilar membrane rests the organ of Corti, the auditory sensory epithelium containing approximately 15,000 hair cells arranged in one row of inner hair cells and three rows of outer hair cells. Despite their numerical predominance, outer hair cells constitute only 5%–10% of afferent innervation; their primary function is the amplification of cochlear vibrations through electromotility, a unique property whereby these cells change length in response to membrane potential changes (Ashmore, 2008). Inner hair cells, conversely, serve as the primary sensory receptors, transducing mechanical displacement into neural signals that are transmitted via type I spiral ganglion neurons to the cochlear nucleus (Fuchs, 2005).

The stereocilia of cochlear hair cells are embedded in the overlying tectorial membrane, a gelatinous structure that pivots with basilar membrane vibrations. This mechanical coupling enables the shearing forces that deflect stereocilia, opening mechanically-gated transduction channels and initiating the receptor potential (Hudspeth, 2008). The intricate spatial arrangement of stereocilia in precisely graded heights, interconnected by tip links, confers directional sensitivity to mechanical stimulation.

The vestibular apparatus, comprising the SCCs and the vestibule housing the otolith organs, detects head movement and position through similar mechanoelectrical transduction mechanisms but with specialized structural adaptations. The three SCCs—anterior (superior), posterior, and lateral (horizontal)—are oriented approximately orthogonally to each other, enabling detection of angular acceleration in all three planes of rotation (Rabbitt et al., 2004). Each SCC forms approximately two-thirds of a circle with a diameter of about 6.5 mm and contains a membranous duct of significantly smaller diameter (0.4 mm) that follows the contour of the bony canal.

Near the junction of each SCC with the vestibule, the membranous duct expands to form the ampulla, housing the crista ampullaris, the sensory epithelium for rotational movement detection. The crista consists of specialized hair cells, whose stereocilia and single kinocilium project into a gelatinous cupula that extends across the ampulla. During rotational head movements, the relative inertia of the endolymph creates a pressure differential that deflects the cupula, bending the hair cell stereocilia and generating receptor potentials (Goldberg and Fernandez, 1971). The orientation of hair cells within each crista allows for bidirectional sensitivity, with depolarization occurring during ampullopetal flow (toward the ampulla) in the horizontal canal and ampullofugal flow (away from the ampulla) in the vertical canals.

Within the central vestibule lie the otolith organs, the utricle and saccule, which detect linear acceleration and head position relative to gravity. Their sensory epithelia, called maculae, contain hair cells whose stereocilia are embedded in an otolithic membrane containing calcium carbonate crystals (otoconia). These dense crystals provide inertial mass that, during linear acceleration or head tilt, create shearing forces across the hair cell stereocilia (Goldberg et al., 2012). The macula utriculi lies predominantly in the horizontal plane, making it most sensitive to horizontal linear acceleration and head tilt, while the macula sacculi assumes a more vertical orientation, responding primarily to vertical linear movement.

The specialized orientation of hair cells within each macula forms a complex pattern radiating from a central curved line called the striola. This arrangement ensures that, for any given head movement, some population of hair cells will be excited (depolarized), while others are inhibited (hyperpolarized), providing the central nervous system with precisely encoded information about the direction and magnitude of linear forces acting on the head (Goldberg et al., 2012).

Both cochlear and vestibular hair cells synapse with primary afferent neurons, whose cell bodies reside in the spiral and vestibular ganglia, respectively. These bipolar neurons project centrally through the internal acoustic meatus as components of the vestibulocochlear nerve (cranial nerve VIII), carrying sensory information to brainstem nuclei for initial processing before ascending to higher centers (Kandel et al., 2013).

This remarkably compact and elegantly designed sensory organ, measuring barely 2 cm across, enables humans to detect sound pressures varying by a millionfold in amplitude and frequencies spanning three orders of magnitude, while simultaneously monitoring head position and movement with exceptional precision. The inner ear's architectural complexity and functional sophistication represent a pinnacle of evolutionary adaptation in sensory system design.

3 Common inner ear abnormalities

The inner ear can present with various structural abnormalities. Complete labyrinthine aplasia (CLA), or Michel Deformity, is a rare and severe inner ear abnormality. Arrested otic placode development seems to be the cause of CLA. It presents as the absence of the cochlea, vestibule, SCCs, vestibular and cochlear aqueducts, involving both the bony and membranous labyrinth. CLA can be subdivided into: (1) CLA with hypoplastic or aplastic petrous bone, (2) CLA without an otic capsule, and (3) CLA with an otic capsule (Sennaroğlu and Bajin, 2017). The rudimentary otocyst is a condition where the development of the otic capsule is rudimentary, with the absence of an internal auditory canal. Sometimes, SCCs may be vaguely present. In cochlear aplasia, the cochlea is absent. It has two types: cochlear aplasia with (1) normal labyrinth and (2) cochlear aplasia with a dilated vestibule. Common cavity is a condition in which the cochlea and vestibule are undifferentiated (Jackler et al., 1987). In cochlear hypoplasia, there is a clear distinction between the cochlea and the vestibule; however, the cochlea is smaller. Incomplete partition of the cochlea is a condition characterized by internal architectural variations, with distinct cochlea and vestibule making up this disorder. There are three types of incomplete partitions

(Sennaroğlu and Bajin, 2017): Type 1, otherwise called cystic cochleovestibular malformation, lacks the entire modiolus and interscalar septa. Type 2 exhibits the triad characteristics of the Mondini deformity, namely a defective apical part of the modiolus, a slightly dilated vestibule, and an EVA. In Type 3, the modiolus is absent, whereas the interscalar septa of the cochlea are present.

In the EVA, the cochlea, vestibule, and SCCs appear normal except for the vestibular aqueduct (Valvassori and Clemis, 1978). The cochlear aperture abnormalities could be hypoplastic (width is <1.4 mm) or aplastic. The above are congenital inner ear conditions. Beyond these, one must be aware of other abnormalities involving the inner ear, for which imaging plays a role. Cochlear otosclerosis is generally an advanced condition of middle ear otosclerosis. In cochlear otosclerosis, the otic capsule deforms (Cureoglu et al., 2010). A temporal bone fracture, which can be longitudinal or transverse, could result from head trauma. Here, the bony structures of the inner ear could be damaged (Dahiya et al., 1999). Inner ear cholesteatoma could result from chronic middle ear disease, which is generally found in the middle ear. Rarely, these cholesteatomas can invade the inner ear structures as well (Palva et al., 1989). Labyrinthitis is the infection or inflammation of the membranous labyrinth of the inner ear. It may be due to an inner ear infection or trauma. It may cause ossification of the membranous labyrinth or the endosteum, a condition called labyrinthitis ossificans (Schuknecht and Donovan, 1986).

Ménière's disease involves both the cochlea and the vestibular system. The presence of an abnormally high quantity of inner ear fluid, called EH, is said to be its cause (Hallpike and Cairns, 1938). The bone over the superior SCC dehiscence is thinned out in the superior SCC (Minor et al., 1998). The thinning could reach such a level that it may create a third window to the inner ear, along with the round and oval windows. Autoimmune and autoinflammatory inner ear diseases (AIED/AID) are conditions where the immune system compromises the inner ear structures, and these could be due to autoimmune conditions such as lupus or rheumatoid arthritis (Cogan, 1945). The well-compartmentalized middle and inner ear are compromised, leading to the leakage of perilymph, and the condition is termed a perilymphatic fistula. The leading cause of this condition is barotrauma (Goodhill, 1971).

4 Computed tomography

The inner ear encompasses intricate bony structures and is deeply located within the temporal bone. Detection of abnormalities in these minute structures requires advanced imaging techniques. HRCT images are vital for identifying inner ear bony labyrinth abnormalities. HRCT can provide thin-slice images (<0.5 mm), aiding in better imaging of inner ear structures, including cochlear turns.

Further, the scope of multiplane image reconstruction is vital. Along with congenital inner ear malformation, HRCT helps in the diagnosis of conditions like superior SCC dehiscence (Waldeck et al., 2022), the effect of otosclerosis on the inner ear (Wichova et al., 2019), Ménière's disease (Xia et al., 2024) and others. Imaging in children is challenging. However, precochlear implantation, it is essential to know about the inner ear structures in detail. This knowledge could be detrimental in selecting the type of implant and the electrode insertion. Multidetector computed tomography can acquire enough data quickly, leading to better 3D reconstruction of the inner ear structures (Maiolo et al., 2013). At times, cone beam computed tomography could be explored to study cochlear implantation, cholesteatomas, and other conditions. This method has relatively less radiation exposure. Advanced analysis methods like adaptive statistical iterative reconstruction and model-based iterative reconstruction facilitate better imaging outcomes.

EVA is commonly seen in children and presents with progressive SNHL (Griffith and Wangemann, 2011). Hearing loss, however, can be stable and fluctuating/nonfluctuating (Ruthberg et al., 2022). The ear ipsilateral to EVA shows a higher degree of hearing loss and could range from moderate to profound. Some patients may even present with normal hearing (Macielak et al., 2019). Many times, progressive hearing loss may have a sudden onset (Wendt and Hashikawa, 2019). The vestibular aqueduct width ($\geq$2.0 mm opercular or $\geq$1 mm at the midpoint), as established by computed tomography (CT), is found to correlate with the progression of hearing loss in EVA (Boston et al., 2007). Recent studies suggest HRCT with 3D reconstruction and VA volumetry for the assessment of EVA (Weiss et al., 2023). Beyond hearing loss, 45% of patients with EVA exhibit vestibular signs and symptoms (Zalewski et al., 2015). Besides PTA, audiovestibular tests like VEMP and acoustically evoked short-latency negative responses are employed for their clinical utility in patients with EVA. These test measures are related to the degree of hearing loss and could be studied in relation to imaging findings. EVA could be associated with syndromic conditions like Pendred syndrome (Pryor et al., 2005). Endolymphatic sac tumors (ELST) arise from endolymphatic sacs (Tang et al., 2023). CT presents a distorted vestibular aqueduct and local bone invasion as a "honeycomb" pattern (Geng et al., 2022). Clinical tests reveal hearing loss and facial palsy in patients with ELST (Bae et al., 2020). These patients may present with vertigo and tinnitus along with hearing loss (Villegas-González et al., 2022). In a recent study, Hansen et al. (2022) performed various audiovestibular tests on five patients with ELST. Patients revealed unilateral abnormal findings on pure tone audiometry (moderate hearing loss), caloric test (unilateral weakness), vHIT (refixation saccades), and VEMP (absence). Furthermore, patients reported tinnitus and dizziness. The authors recommended consideration of imaging upon clinical findings of unilateral hearing loss (Hansen et al., 2022). SCC dysplasia is a condition in which one or all the SCCs are hypoplastic or aplastic. Lateral SCC dysplasia (LSCD) is the most common among them (Kwak et al., 2020). The authors evaluated unilateral and bilateral LSCD patients and reported abnormal caloric responses and VOR gain. It should be noted that only 26.7% of the patients reported dizziness (Kwak et al., 2020). This could be due to central compensation. There seems to be a relationship between the severity of LSCD and the vestibular test findings (Yun et al., 2024). The severity of aplasia (detected by CT/MRI) could be lateral semicircular canal (LSCC) with central bony island (CBI) or vestibule without CBI. The caloric test revealed an abnormality, whereas vHIT findings remained normal. Patients without CBI had poorer findings than patients whose imaging revealed the presence of CBI.

5 Magnetic resonance imaging

Advancements in MRI have opened avenues for understanding inner ear disorders. The anatomical atlas and template created by high-resolution MRI can be applied in individual cases or conditions for better identification and clinical correlation (Ahmadi et al., 2021). However, a 7 T MRI provides ultra-high-resolution images suitable for imaging intricate inner ear structures in research protocols (Van Egmond et al., 2014). It (7 T) can produce more detailed images of the interscalar septum, utricular macula, and nerve bundles (Van Egmond et al., 2014). Some researchers have reported that higher-resolution 7 T imaging may benefit precochlear implant imaging studies (van der Jagt et al., 2015). In regular clinical MRI protocols (1.5 and 3 T), T1 and T2-weighted protocols are employed to image the inner ear.

T1-weighted MRI highlights the differences between tissues, aiding inner ear imaging. Furthermore, T1W imaging with contrast yields clinically relevant information for labyrinthine lesions (Dubrulle et al., 2010). Patients with intralabyrinthine schwannomas, labyrinthitis, and inflammatory granulomas reveal enhancement on T1W (postgadolinium contrast). Precontrast T1-weighted images are recommended to eliminate spontaneous hyperintensities. Intralabyrinthine hemorrhage shows spontaneous T1W hyperintensities. These findings may correlate with hearing loss and vestibular symptoms (Dubrulle et al., 2010). The study reported that the patients had varied hearing loss ranging from mild to severe, and around 85% experienced vertigo. Techniques like the Multipoint Dixon Method facilitate better inner ear imaging (Cakmak et al., 2021). The superior soft tissue contrast in T2-weighted imaging is vital for understanding labyrinthine structures, intra labyrinthine fluid, and the vestibular nerve.

MRI with contrast (gadolinium) helps identify the effects of inner ear inflammation or infection, such as labyrinthitis and vascular/neoplastic disorders. AIED/AID can cause SNHL (Vambutas and Pathak, 2016). At times, imaging helps refine the clinical diagnosis of this condition, especially if there is an associated conductive pathology (Lee et al., 2021). To diagnose AIED/AID, Lee et al. (2021) employed a MRI-based decision tree using classification and regression tree. Patients with AIED/AID exhibited significant inner ear post-T1WI and fluid-attenuated inversion recovery (FLAIR) enhancement. Findings revealed that AIED/AID could be differentiated from conductive pathology with >90% specificity. The EVA is primarily evaluated with HRCT. MRI could detect EVA with a sensitivity of 97% (Clarke et al., 2021). The findings were comparable to those of HRCT. In T2-weighted axial MRI, measures such as vestibular aqueduct midpoint width, endolymphatic sac width, and signal heterogeneity are also effective for assessing EVA (Saeed et al., 2023). In the case of SCD, MRI can be employed to determine the anterior-posterior width and the width of the bone island of LSCC (Alam-Eldeen et al., 2017). This measurement aids in detecting and characterizing LSCD. A 3D inversion recovery with REAL reconstruction (3D-REAL-IR to study the endolymphatic volumetric ratio could be used in studying LSCD (Suárez-Vega et al., 2021).

MRI aids in diagnosing clinical conditions leading to sudden hearing loss, such as intracochlear hemorrhage (Jrad et al., 2021). Jrad et al. (2021) reported higher signal intensity on T1W and T2 FLAIR in the respective cochlea of a patient with intracochlear hemorrhage. Audiological evaluation revealed unilateral profound hearing loss. However, a linear correlation between the MRI findings and the audiological results may not be expected. Labyrinthitis is the inflammation of the inner ear. Audiological test findings of labyrinthitis could indicate profound hearing loss, as clearly demonstrated by pure tone audiometry and the absence of OAE. Postcontrast T1W MRI shows enhancement, and in the T2 FLAIR, signal intensity is significantly brighter in these patients (Benson et al., 2020). Labyrinthitis ossificans is a condition characterized by ossification of the membranous labyrinth. Taxak and Ram (2020) documented postcontrast enhancement of the labyrinth on MRI in a case study. Furthermore, the patient presented with progressive SNHL, tinnitus, and vertigo (Taxak and Ram, 2020).

Labyrinthine schwannomas are tumors arising from the inner ear's membranous labyrinth. MRI identifies and subcategorizes them based on their origin and extension: intracochlear, intravestibular, vestibulocochlear, transmodiolar, transmacular, and transotic. Patients report progressive SNHL, tinnitus, and vertigo (Salzman et al., 2012). ELST arise from endolymphatic sacs (Tang et al., 2023). MRI shows varied intensities (hyper, hypo, iso-, mixed) on T1W and T2W imaging (Geng et al., 2022). T1W MRI may show a mixture of hyper and hypo-intensity within the tumor (Bae et al., 2020). EH is a condition in which the endolymphatic space is enlarged, resulting in the distortion of various inner ear structures, causing Meniere's disease. The distortion of the structures may vary based on their mechanical properties (Salt and Plontke, 2010). The triad of symptoms associated with this disease includes episodic vertigo, fluctuating hearing loss, and tinnitus. Pure tone audiometry may reveal low-frequency hearing loss. As the disease progresses, the symptoms and audiometric patterns change. MRI with contrast can create an inner ear atlas to study the endolymphatic space. Objective evidence of EH can be achieved by comparing this standard atlas (Valerie Kirsch et al., 2019). In individual patients, gadolinium chelate administration may help evaluate those with EH who exhibit reduced perilymphatic space (Nakashima et al., 2007). Imaging with contrast in patients presenting symptoms of Meniere's disease can aid in the differential diagnosis of EH and other incidental associated or unassociated inner ear conditions (Attyé et al., 2017). Using MRI 3D 3D-FLAIR and hybrid imaging, it is now possible to visualize EH by differentiating endolymphatic and perilymphatic spaces. These findings provide definitive and objective evidence for disease diagnosis and treatment. Recent studies have revealed some successful findings (Zanetti et al., 2021). EH can be sub-categorized into cochlear and vestibular hydrops and can be graded (Baráth et al., 2014). In this study, researchers reported abnormal image findings in 22% of the clinically nonaffected sides. Variable findings in audio-vestibular tests may correlate with these imaging results.

6 Positron emission tomography

Studies employing positron emission tomography (PET) in inner ear disorders are limited. However, technical advancements in the future may equip us with newer methods to study the inner ear's functional aspects by recording its metabolic activity, achieved through PET imaging. Integrating the findings of PET with other imaging and clinical assessments can provide a more comprehensive picture of the inner ear's structure and function. Attempts are underway to employ tracers like fluorodeoxyglucose to study cochlear metabolism. A deep understanding of the molecular and cellular mechanisms may shed light on inner ear pathophysiology; thereby, new treatment procedures can be attempted (Rasilla et al., 2018).

7 Optical coherence tomography

Imaging the microstructures of the inner ear is challenging but rewarding. Optical coherence tomography (OCT) allows micron-scale resolution of the inner ear, thereby facilitating imaging of hair cells, which is otherwise impossible. The application of spectral-domain OCT (SD-OCT) and swept-source OCT has been tested for in-depth imaging of the cochlear and vestibular structures. Clinical benefits of OCT have been evaluated in cochlear implantees (Starovoyt et al., 2019). OCT can be performed rapidly, making it an ideal candidate for intraoperative imaging during cochlear implantation. Furthermore, in postoperative scenarios, it can assess the surgical outcome by visualizing electrode position or detecting inflammation in the tissues surrounding the electrodes. OCT can also be further tested to study the progression of cholesteatoma, Ménière's disease, and labyrinthitis.

8 Multimodal imaging of the inner ear

A combination of imaging techniques is always beneficial, especially in the case of complex structures like the inner ear. A sensitivity of >80% can be achieved by employing CT and MRI to diagnose perilymphatic fistula (Venkatasamy et al., 2020). The researchers performed both CT and MRI on 17 patients with surgically confirmed perilymphatic fistulas in round and/or oval windows. The common findings were fluid-filling in the window niches and disorientation of the footplate or pneumolabyrinth. In a case study, Taxak and Ram (2020) reported the CT and MRI findings in a patient with labyrinthitis ossificans. The patient had bilateral hearing loss with sudden progression of hearing loss unilaterally. The CT revealed ossification with sclerosis, and the MRI revealed postcontrast enhancement of the labyrinth.

9 Conclusion

The inner ear, a marvel of biological engineering, plays a critical role in hearing and balance. Its structural intricacy and functional sophistication make it vulnerable to a wide array of congenital and acquired disorders, each with profound clinical implications. Technological advancements in imaging, particularly high-resolution CT, MRI, PET, and OCT, have revolutionized the diagnosis, characterization, and management of inner ear pathologies. These modalities not only reveal anatomical anomalies but also offer functional insights that guide clinical decision-making. A multidisciplinary approach, integrating imaging findings with audiovestibular assessments and clinical presentation, remains essential for optimal diagnosis, treatment, and ongoing research into this complex and vital sensory system.

References

Agrawal, Y., Carey, J.P., Santina, C.C.D., Schubert, M.C., Minor, L.B., 2009. Disorders of balance and vestibular function in US adults: data from the National Health and Nutrition Examination Survey, 2001-2004. Arch. Intern. Med. 15383679. 169 (10), 938–944. Available from: http://archinte.ama-assn.org/cgi/reprint/169/10/938. doi:10.1001/archinternmed.2009.66.

Ahmadi, S.A., Raiser, T.M., Rühl, R.M., Flanagin, V.L., Zu Eulenburg, P., 2021. IE-Map: a novel in-vivo atlas and template of the human inner ear. Sci. Rep. 20452322. 11 (1). Available from: https://www.nature.com/srep/index.html. doi:10.1038/s41598-021-82716-0.

Alam-Eldeen, M., Hasana, N.A., Rashad, U., 2017. Value of magnetic resonance imaging measurement of lateral semicircular canal bone island in sensorineural hearing loss patients. J. Curr. Med. Res. Pract. 2357-0121. 2 (1), 35. doi:10.4103/2357-0121.210306.

Ashmore, J., 2008. Cochlear outer hair cell motility. Physiol. Rev. 0031-9333. 88 (1), 173–210. doi:10.1152/physrev.00044.2006.

Attyé, A., Eliezer, M., Galloux, A., Pietras, J., Tropres, I., Schmerber, S., Dumas, G., Krainik, A., 2017. Endolymphatic hydrops imaging: differential diagnosis in patients with Meniere disease symptoms. Diagn. Interv. Imaging. 22115684. 98 (10), 699–706. doi:10.1016/j.diii.2017.06.002.

Bae, S.H., Kim, S.S., Kwak, S.H., Jung, J.S., Choi, J.Y., Moon, I.S., 2020. Clinical features and treatment of endolymphatic sac tumor. Acta Otolaryngol. 16512251. 140 (6), 433–437. Available from: http://www.tandfonline.com/loi/ioto20. doi:10.1080/00016489.2020.1722855.

Baráth, K., Schuknecht, B., Naldi, A.M., Schrepfer, T., Bockisch, C.J., Hegemann, S.C.A., 2014. Detection and grading of endolymphatic hydrops in Menière disease using MR imaging. Am. J. Neuroradiol. 1936959X. 35 (7), 1387–1392. Available from: http://www.ajnr.org/content/35/7/1387.full.pdf+html. doi:10.3174/ajnr.A3856.

Békésy, 1960. Experiments in Hearing, McGraw-Hill.

Benson, J.C., Carlson, M.L., Lane, J.I., 2020. MRI of the internal auditory canal, labyrinth, and middle ear: how we do it. Radiology 15271315. 297 (2), 252–265. Available from: http://pubs.rsna.org/doi/10.1148/radiol.2020201767. doi:10.1148/RADIOL.2020201767.

Bess, F.H., Dodd-Murphy, J., Parker, R.A., 1998. Children with minimal sensorineural hearing loss: prevalence, educational performance, and functional status. Ear Hear. 01960202. 19 (5), 339–354. doi:10.1097/00003446-199810000-00001.

Boston, M., Halsted, M., Meinzen-Derr, J., Bean, J., Vijayasekaran, S., Arjmand, E., Choo, D., Benton, C., Greinwald, J., 2007. The large vestibular aqueduct: a new definition based on audiologic and computed tomography correlation. Otolaryngol. Head Neck Surg. 0194-5998. 136 (6), 972–977. doi:10.1016/j.otohns.2006.12.011.

Bowl, M.R., Dawson, S.J., 2015. The mouse as a model for age-related hearing loss - A mini-review. Gerontology 14230003. 61 (2), 149–157. Available from: https://www.karger.com/journals/ger/ger_jh.htm. doi:10.1159/000368399.

Cakmak, P., Herek, D., Yagci, A.B., Sagtas, E., Ufuk, F., Çakmak, V., 2021. Efficiency of fat suppression in t1-weighted inner ear magnetic resonance imaging: multipoint dixon method versus hybrid techniques. Curr. Med. Imaging 18756603. 17 (7), 884–888. Available from: https://www.eurekaselect.com/190313/article. doi:10.2174/1573405617666210114141300.

Casselman, J.W., Kuhweide, R., Deimling, M., Ampe, W., Dehaene, I., Meeus, L., 1993. Constructive interference in steady state-3DFT MR imaging of the inner ear and cerebellopontine angle. Am. J. Neuroradiol. 14 (1), 47–57.

Chadha, S., Kamenov, K., Cieza, A., 2021. The world report on hearing, 2021. Bull. World Health Organ. 15640604. 99 (4), 242–242. Available from: https://www.who.int/bulletin/volumes/99/4/21-285643.pdf. doi:10.2471/BLT.21.285643.

Clarke, R.L., Isaacson, B., Kutz, J.W., Xi, Y., Booth, T.N., 2021. MRI evaluation of the normal and abnormal endolymphatic duct in the pediatric population: a comparison with high-resolution CT. Am. J. Neuroradiol. 1936959X. 42 (10), 1865–1869. Available from: http://www.ajnr.org/content/42/10/1865. doi:10.3174/ajnr.A7224.

Cogan, D.G., 1945. Syndrome of nonsyphilitic interstitial keratitis and vestibuloauditory symptoms. Arch. Ophthalmol. 0003-9950. 33 (2), 144–149. doi:10.1001/archopht.1945.00890140064007.

Cureoglu, S., Baylan, M.Y., Paparella, M.M., 2010. Cochlear otosclerosis. Curr. Opin. Otolaryngol. Head Neck Surg. 10689508. 18 (5), 357–362. doi:10.1097/MOO.0b013e32833d11d9.

Dahiya, R., Keller, J.D., Litofsky, N.S., Bankey, P.E., Bonassar, L.J., Megerian, C.A., 1999. Temporal bone fractures: otic capsule sparing versus otic capsule violating clinical and radiographic considerations. J. Trauma Inj. Infect. Crit. Care 15298809. 47 (6), 1079–1083. http://www.jtrauma.com. doi:10.1097/00005373-199912000-00014.

Dubrulle, F., Kohler, R., Vincent, C., Puech, P., Ernst, O., 2010. Differential diagnosis and prognosis of T1-weighted post-gadolinium intralabyrinthine hyperintensities. Eur. Radiol. 0938-7994. 20 (11), 2628–2636. doi:10.1007/s00330-010-1835-2.

Forge, A., Schacht, J., 2000. Aminoglycoside antibiotics. Audiol. Neurotol. 1420-3030. 5 (1), 3–22. doi:10.1159/000013861.

Fuchs, P.A., 2005. Time and intensity coding at the hair cell's ribbon synapse. J. Physiol. 566 (1), 7–12. doi:10.1113/jphysiol.2004.082214.

Géléoc, G.S.G., Holt, J.R., 2014. Sound strategies for hearing restoration. Science 10959203. 344 (6184). Available from: http://www.sciencemag.org/content/344/6184/1241062.full.pdf. doi:10.1126/science.1241062.

Geng, Y., Gu, X., Lin, M., Chen, W., Tang, Z., Sha, Y., Song, Y., 2022. Endolymphatic sac tumor: exploring the role of CT and MRI features in the diagnosis of 22 cases. Clin. Radiol. 77 (8), 592. doi:10.1016/J.CRAD.2022.04.010.

Goldberg, J.M., Fernandez, C., 1971. Physiology of peripheral neurons innervating semicircular canals of the squirrel monkey. I. Resting discharge and response to constant angular accelerations. J. Neurophysiol. 00223077. 34 (4), 635–660. doi:10.1152/jn.1971.34.4.635.

Goldberg, J.M., Wilson, V.J., Cullen, K.E., Angelaki, D.E., Broussard, D.M., Buttner-Ennever, J., Fukushima, K., Minor, L.B., 2012. The Vestibular System: A Sixth Sense The Vestibular System: A Sixth Sense. Oxford University Press, United States, pp. 1–560. Available from: http://www.oxfordscholarship.com/view/10.1093/acprof:oso/9780195167085.001.0001/acprof-9780195167085. doi:10.1093/acprof:oso/9780195167085.001.0001.

Goodhill, V., 1971. Sudden deafness and round window rupture. Laryngoscope. 0023-852X. 81 (9), 1462–1474. doi:10.1288/00005537-197109000-00010.

Greenwood, D.D., 1990. A cochlear frequency-position function for several species—29 years later. J. Acoust. Soc. Am. 00014966. 87 (6), 2592–2605. doi:10.1121/1.399052.

Griffith, A.J., Wangemann, P., 2011. Hearing loss associated with enlargement of the vestibular aqueduct: mechanistic insights from clinical phenotypes, genotypes, and mouse models. Hear. Res. 18785891. 281 (1-2), 11–17. doi:10.1016/j.heares.2011.05.009.

Hallpike, C.S., Cairns, H., 1938. Observations on the pathology of Ménière's syndrome. J. Laryngol. Otol. 0022-2151. 53 (10), 625–655. doi:10.1017/s0022215100003947.

Hansen, J.O., West, N.C., Cayé-Thomasen, P., 2022. Impact of an endolymphatic sac tumor on subjective and objective audiovestibular function. Otol. Neurotol. 15374505. 43 (8), 956–961. Available from: http://journals.lww.com/otology-neurotology. doi:10.1097/MAO.0000000000003601.

Hart, B., Risley, T.R., 1995. Meaningful Differences in the Everyday Experience of Young American Children, Paul H Brookes Publishing.

Hibino, H., Nin, F., Tsuzuki, C., Kurachi, Y., 2010. How is the highly positive endocochlear potential formed? The specific architecture of the stria vascularis and the roles of the ion-transport apparatus. Pflugers Arch. 0031-6768. 459 (4), 521–533. doi:10.1007/s00424-009-0754-z.

Hudspeth, A.J., 2008. Making an effort to listen: mechanical amplification in the ear. Neuron. 08966273. 59 (4), 530–545. doi:10.1016/j.neuron.2008.07.012.

Hülse, R., Biesdorf, A., Hörmann, K., Stuck, B., Erhart, M., Hülse, M., Wenzel, A., 2019. Peripheral vestibular disorders: an epidemiologic survey in 70 million individuals. Otol. Neurotol. 15374505. 40 (1), 88–95. Available from: http://journals.lww.com/otology-neurotology. doi:10.1097/MAO.0000000000002013.

Igarashi, M., 1967. Dimensional study of the vestibular apparatus. Laryngoscope. 0023-852X. 77 (10), 1806–1817. doi:10.1288/00005537-196710000-00003.

Jackler, R.K., Luxford, W.M., House, W.F., 1987. Congenital malformations of the inner ear: a classification based on embryo genesis. Laryngoscope. 15314995. 97 (3), 2–14. doi:10.1002/lary.5540971301.

Joint Committee on Infant Hearing 2019. Position statement: principles and guidelines for early hearing detection and intervention programs. doi:10.15142/fptk-b748.

Jrad, M., Zlitni, H., Boumediene, M., Nasr, A.B., Bouzrara, M., Iezzi, R., 2021. Intracochlear hemorrhage: a rare cause of sudden sensorineural hearing loss. Case Rep. Radiol. 2090-6870. 2021, 1–4. doi:10.1155/2021/1072047.

Kandel, E.R., Schwartz, J.H., Jessell, T.M., Siegelbaum, S.A., Hudspeth, A.J., 2013. Principles of Neural Science, McGraw-Hill.

Kelsell, D.P., Dunlop, J., Stevens, H.P., Lench, N.J., Liang, J.N., Parry, G., Mueller, R.F., Leigh, I.M., 1997. Connexin 26 mutations in hereditary non-syndromic sensorineural deafness. Nature. 0028-0836. 387 (6628), 80–83. doi:10.1038/387080a0.

Kwak, S.H., Kim, M.K., Kim, S.H., Jung, J., 2020. Audiological and vestibular functions in patients with lateral semicircular canal dysplasia and aplasia. Clin. Exp. Otorhinolaryngol. 20050720. 13 (3), 255–260. Available from: http://e-ceo.org/. doi:10.21053/ceo.2019.01053.

Lee, B., Bae, Y.J., Choi, B.Y., Kim, Y.S., Han, J.H., Kim, H., Choi, B.S., Kim, J.H., 2021. Construction of an MRI-based decision tree to differentiate autoimmune and autoinflammatory inner ear disease from chronic otitis media with sensorineural hearing loss. Sci. Rep. 20452322. 11 (1). Available from: https://www.nature.com/srep/index.html. doi:10.1038/s41598-021-98557-w.

Livingston, G., Huntley, J., Sommerlad, A., Ames, D., Ballard, C., Banerjee, S., Brayne, C., Burns, A., Cohen-Mansfield, J., Cooper, C., Costafreda, S.G., Dias, A., Fox, N., Gitlin, L.N., Howard, R., Kales, H.C., Kivimäki, M., Larson, E.B., Ogunniyi, A., Orgeta, V., Ritchie, K., Rockwood, K., Sampson, E.L., Samus, Q., Schneider, L.S., Selbæk, G., Teri, L., Mukadam, N., 2020. Dementia prevention, intervention, and care: 2020 report of the Lancet Commission. Lancet 1474547X. 396 (10248), 413–446. Available from: http://www.journals.elsevier.com/the-lancet/. doi:10.1016/S0140-6736(20)30367-6.

Macielak, R.J., Mattingly, J.K., Findlen, U.M., Moberly, A.C., Malhotra, P.S., Adunka, O.F., 2019. Audiometric findings in children with unilateral enlarged vestibular aqueduct. Int. J. Pediatr. Otorhinolaryngol. 18728464. 120, 25–29. Available from: https://www.elsevier.com/locate/ijporl. doi:10.1016/j.ijporl.2019.01.034.

Maiolo, V., Savastio, G., Modugno, G.C., Barozzi, L., 2013. Relationship between multidetector CT imaging of the vestibular aqueduct and inner ear pathologies. Neuroradiol. J. 19714009. 26 (6), 683–692. Available from: https://journals.sagepub.com/home/NEU. doi:10.1177/197140091302600612.

Manuel Suárez-Vega, V., Domínguez, P., Pérez-Fernández, N., 2021. MRI evaluation of vestibular endolymphatic space in patients with isolated cystic lateral semicircular canal malformation. J. Otorhinolaryngol. Hear. Balanc. Med. 2504-463X. 2 (1), 3. doi:10.3390/ohbm2010003.

Merchant, S.N., Rosowski, J.J., 2008. Conductive hearing loss caused by third-window lesions of the inner ear. Otol. Neurotol. 1531-7129. 29 (3), 282–289. doi:10.1097/mao.0b013e318161ab24.

Minor, L.B., Solomon, D., Zinreich, J.S., Zee, D.S., 1998. Sound- and/or pressure-induced vertigo due to bone dehiscence of the superior semicircular canal. Arch. Otolaryngol. Head Neck Surg. 08864470. 124 (3), 249–258. Available from: http://archotol.jamanetwork.com/journal.aspx. doi:10.1001/archotol.124.3.249.

Mizutari, K., Fujioka, M., Hosoya, M., Bramhall, N., Okano, H.J., Okano, H., Edge, A.S.B., 2013. Notch inhibition induces cochlear hair cell regeneration and recovery of hearing after acoustic trauma. Neuron. 10974199. 77 (1), 58–69. doi:10.1016/j.neuron.2012.10.032.

Nakashima, T., Naganawa, S., Sugiura, M., Teranishi, M., Sone, M., Hayashi, H., Nakata, S., Katayama, N., Ishida, I.M., 2007. Visualization of endolymphatic hydrops in patients with Meniere's disease. Laryngoscope. 0023852X. 117 (3), 415–420. doi:10.1097/MLG.0b013e31802c300c.

Nelson, D.I., Nelson, R.Y., Concha-Barrientos, M., Fingerhut, M., 2005. The global burden of occupational noise-induced hearing loss. Am. J. Ind. Med. 02713586. 48 (6), 446–458. doi:10.1002/ajim.20223.

Palva, T., Kärjä, J., Palva, A., 1989. Opening of the labyrinth during chronic ear surgery. Arch. Otolaryngol. Head Neck Surg. 115 (11), 1333–1337.

Pickles, J.O., 2012. An Introduction to the Physiology of Hearing, Emerald Group Publishing.

Pryor, S.P., Madeo, A.C., Reynolds, J.C., Sarlis, N.J., Arnos, K.S., Nance, W.E., Yang, Y., Zalewski, C.K., Brewer, C.C., Butman, J.A., Griffith, A.J., 2005. SLC26A4/PDS genotype-phenotype correlation in hearing loss with enlargement of the vestibular aqueduct (EVA): evidence that Pendred syndrome and non-syndromic EVA are distinct clinical and genetic entities. J. Med. Genet. 00222593. 42 (2), 159–165. doi:10.1136/jmg.2004.024208.

Rabbitt, R.D., Boyle, R., Highstein, S.M., 2004. Sensory transduction of head velocity and acceleration in the toadfish horizontal semicircular canal. J. Neurophysiol. 72 (2), 891–911.

Rajendran, V., Mitchell, P., 2007. Influence of weight and body composition on balance-related parameters in children with developmental coordination disorder. Res. Dev. Disabil. 28 (1), 35–47.

Rasilla, J.M., Evan, L.O., Villarreal, I., Ramón García-Berrocal, J., 2018. Can positron emission tomography support the characterization of immune-mediated inner ear disease? Rev. Esp. Med. Nucl. Imagen Mol. (English Edition). 22538089. 37 (5), 290–295. doi:10.1016/j.remnie.2018.04.002.

Rosengren, S.M., Colebatch, J.G., Young, A.S., Govender, S., Welgampola, M.S., 2019. Vestibular evoked myogenic potentials in practice: methods, pitfalls and clinical applications. Clin. Neurophysiol. Pract. 2467981X. 4, 47–68. Available from: http://www.journals.elsevier.com/clinical-neurophysiology-practice. doi:10.1016/j.cnp.2019.01.005.

Ruthberg, J.S., Kocharyan, A., Farrokhian, N., Stahl, M.C., Hicks, K., Scarborough, J., Murray, G.S., Wu, S., Manzoor, N., Otteson, T., 2022. Hearing loss patterns in enlarged vestibular aqueduct syndrome: do fluctuations have clinical significance? Int. J. Pediatr. Otorhinolaryngol. 18728464. 156. Available from: https://www.elsevier.com/locate/ijporl. doi:10.1016/j.ijporl.2022.111072.

Saeed, H.S., Rajai, A., Dixon, R., Kapadia, T., Bruce, I.A., Stivaros, S., 2023. Can MRI biomarkers for hearing loss in enlarged vestibular aqueduct be measured reproducibly? Br. J. Radiol. 96 (1147), 20220274. doi:10.1259/BJR.20220274/7469167.

Salt, A.N., Plontke, S.K., 2010. Endolymphatic hydrops: pathophysiology and experimental models. Otolaryngol. Clin. North Am. 00306665. 43 (5), 971–983. Available from: http://www.elsevier.com/inca/publications/store/6/2/3/1/7/0/index.htt. doi:10.1016/j.otc.2010.05.007.

Salzman, K.L., Childs, A.M., Davidson, H.C., Kennedy, R.J., Shelton, C., Harnsberger, H.R., 2012. Intralabyrinthine schwannomas: imaging diagnosis and classification. Am. J. Neuroradiol. 0195-6108. 33 (1), 104–109. doi:10.3174/ajnr.A2712.

Schuknecht, H.F., 1993. Pathology of the Ear, Lea & Febiger.

Schuknecht, H.F., Donovan, E.D., 1986. The pathology of idiopathic sudden sensorineural hearing loss. Arch. Otorhinolaryngol. 0302-9530. 243 (1), 1–15. doi:10.1007/bf00457899.

Sennaroğlu, L., Bajin, M.D., 2017. Classification and current management of inner ear malformations. Balkan Med. J. 21463131. 34 (5), 397–411. Available from: http://balkanmedicaljournal.org/pdf.php?&id=1789. doi:10.4274/balkanmedj.2017.0367.

Starovoyt, A., Putzeys, T., Wouters, J., Verhaert, N., 2019. High-resolution imaging of the human cochlea through the round window by means of optical coherence tomography. Sci. Rep. 2045-2322. 9 (1). doi:10.1038/s41598-019-50727-7.

Starr, A., Picton, T.W., Sininger, Y., Hood, L.J., Berlin, C.I., 1996. Auditory neuropathy. Brain. 00068950. 119 (3), 741–753. doi:10.1093/brain/119.3.741.

Strupp, M., Brandt, T., 2013. Peripheral vestibular disorders. Curr. Opin. Neurol. 1350-7540. 26 (1), 81–89. doi:10.1097/wco.0b013e32835c5fd4.

Tang, J.D., Grady, A.J., Nickel, C.J., Ryan, L.E., Malone, A., Canvasser, L., Boyev, K.P., 2023. Systematic review of endolymphatic Sac tumor treatment and outcomes. Otolaryngol. Head Neck Surg. 0194-5998. 168 (3), 282–290. doi:10.1177/01945998221108313.

Taxak, P., Ram, C., 2020. Labyrinthitis and labyrinthitis ossificans - A case report and review of the literature. J. Radiol. Case Rep. 1943-0922. 14 (5), 1–6. doi:10.3941/jrcr.v14i5.3706.

Valerie Kirsch, F.N., Ahmadi, S.-A., Dieterich, M., Ertl-Wagner, B., 2019. A probabilistic atlas of the human inner ear's bony labyrinth enables reliable atlas-based segmentation of the total fluid space. J. Neurol. 0340-5354. 266 (S1), 52–61. doi:10.1007/s00415-019-09488-6.

Valvassori, G.E., Clemis, J.D., 1978. The large vestibular aqueduct syndrome. Laryngoscope. 0023852X. 88 (5), 723–728. doi:10.1002/lary.1978.88.5.723.

Vambutas, A., Pathak, S., 2016. AAO: autoimmune and autoinflammatory (disease) in otology: what is new in immune-mediated hearing loss. Laryngoscope Investig. Otolaryngol. 2378-8038. 1 (5), 110–115. doi:10.1002/lio2.28.

van der Jagt, M.A., Brink, W.M., Versluis, M.J., Steens, S.C.A., Briaire, J.J., Webb, A.G., Frijns, J.H.M., Verbist, B.M., 2015. Visualization of human inner ear anatomy with high-resolution MR imaging at 7T: initial clinical assessment. Am. J. Neuroradiol. 0195-6108. 36 (2), 378–383. doi:10.3174/ajnr.A4084.

Van Egmond, S.L., Visser, F., Pameijer, F.A., Grolman, W., 2014. Ex vivo and in vivo imaging of the inner ear at 7 Tesla MRI. Otol. Neurotol. 15374505. 35 (4), 725–729. doi:10.1097/MAO.0000000000000276.

Venkatasamy, A., Ohraini, Z.A., Karol, A., Karch-Georges, A., Riehm, S., Rohmer, D., Charpiot, A., Veillon, F., 2020. CT and MRI for the diagnosis of perilymphatic fistula: a study of 17 surgically confirmed patients. Eur. Arch. Otorhinolaryngol. 0937-4477. 277 (4), 1045–1051. doi:10.1007/s00405-020-05820-3.

Villegas-González, M.J., Angel, J.A.M.D., González, J.L.T., Aranda-Garcia, D., Soto-Galindo, G.A., Pérez-Peña, A.M., 2022. Endolymphatic sac tumor as a Ménière-like vertiginous syndrome: a case report. J. Int. Adv. Otol. 21483817. 18 (1), 84–87. Available from: https://www.advancedotology.org/Content/files/sayilar/113/JIAO_20200068_nlm_new_indd.pdf. doi:10.5152/iao.2022.21102.

Waldeck, S., Lanfermann, H., von Falck, C., Froelich, M.F., Chapot, R., Brockmann, M., Overhoff, D., 2022. New classification of superior semicircular canal dehiscence in HRCT. PLoS ONE 19326203. 17 (1). Available from: https://journals.plos.org/plosone/article/file?id=10.1371/journal.pone.0262758&type=printable. doi:10.1371/journal.pone.0262758.

Wangemann, P., 2006. Supporting sensory transduction: cochlear fluid homeostasis and the endocochlear potential. J. Physiol. 0022-3751. 576 (1), 11–21. doi:10.1113/jphysiol.2006.112888.

Weiss, N.M., Breitsprecher, T.M., Pscheidl, A., Bächinger, D., Volkenstein, S., Dazert, S., Mlynski, R., Langner, S., Roland, P., Dhanasingh, A., 2023. Volumetry improves the assessment of the vestibular aqueduct size in inner ear malformation. Eur. Arch. Otorhinolaryngol. 14344726. 280 (5), 2155–2163. Available from: https://www.springer.com/journal/405. doi:10.1007/s00405-022-07681-4.

Wendt, W.J., Hashikawa, A., 2019. Enlarged vestibular aqueduct syndrome: sudden hearing loss in a child with a cerebral shunt. Pediatr. Emerg. Care 15351815. 35 (7), E135. Available from: http://journals.lww.com/pec-online. doi:10.1097/PEC.0000000000001043.

Wichova, H., Alvi, S., Boatright, C., Ledbetter, L., Staecker, H., Lin, J., 2019. High-resolution computed tomography of the inner ear: effect of otosclerosis on cochlear aqueduct dimensions. Ann. Otol. Rhinol. Laryngol. 0003-4894. 128 (8), 749–754. doi:10.1177/0003489419842579.

World Health Organization, 2015. Making Listening Safe [Internet]. WHO, Geneva.

World Health Organization, 2021. World Report on Hearing [Internet]. World Health Organization, Geneva. Available from: https://www.who.int/publications/i/item/world-report-on-hearing.

Xia, K., Lei, P., Liu, Y., Chen, C., Pan, H., Leng, Y., Liu, B., 2024. Comparison of vestibular aqueduct visualization on computed tomography and magnetic resonance imaging in patients with Ménière's disease. BMC Med. Imaging. 1471-2342. 24 (1). doi:10.1186/s12880-024-01275-8.

Yoshinaga-Itano, C., Sedey, A.L., Coulter, D.K., Mehl, A.L., 1998. Language of early- and later-identified children with hearing loss. Pediatrics. 00314005. 102 (5), 1161–1171. doi:10.1542/peds.102.5.1161.

Yun, J.M., Kim, S.H., Bae, S.H., 2024. Vestibular dysfunction in lateral semicircular canal dysplasia. Front. Neurol. 16642295. 15. Available from: http://www.frontiersin.org/Neurology. doi:10.3389/fneur.2024.1341812.

Zalewski, C.K., Chien, W.W., King, K.A., Muskett, J.A., Baron, R.E., Butman, J.A., Griffith, A.J., Brewer, C.C., 2015. Vestibular dysfunction in patients with enlarged vestibular aqueduct. Otolaryngol. Head Neck Surg. 10976817. 153 (2), 257–262. Available from: http://oto.sagepub.com/content/by/year. doi:10.1177/0194599815585098.

Zanetti, D., Conte, G., Scola, E., Casale, S., Lilli, G., Di Berardino, F., 2021. Advanced imaging of the vestibular endolymphatic space in Ménière's disease. Front. Surg. 2296-875X. 8. doi:10.3389/fsurg.2021.700271.

Zeng, F.G., Rebscher, S., Harrison, W., Sun, X., Feng, H., 2008. Cochlear implants: system design, integration, and evaluation. IEEE Rev. Biomed. Eng. 19411189. 1, 115–142. doi:10.1109/RBME.2008.2008250.

Part VIII

Imaging in oncology and systemic diseases

Chapter 28

Retrocochlear condition: Imaging and clinical correlates

Pradeep Yuvaraj[a], Aravind Kumar Rajasekaran[a], Rajagopal Kadavigere[b], Leena R. David[c,d], Berna Uzun[e], Dilber Uzun Ozsahin[c,e,f], Sneha Ravichandran[d], Suresh Sukumar[d], Ilker Ozsahin[e], and Dilip Shettigar[d]

[a]*Department of Speech Pathology and Audiology, National Institute of Mental Health and Neurosciences, Bengaluru, India*

[b]*Department of Radiodiagnosis and Imaging, Kasturba Medical College, Manipal, India*

[c]*Medical Diagnostic Imaging Department, College of Health Sciences, University of Sharjah, Sharjah, United Arab Emirates*

[d]*Department of Medical Imaging Technology, Manipal College of Health Professions, Manipal Academy of Higher Education, Manipal, Karnataka, India*

[e]*Irfan Suat Gunsel Operational Research Institute, Near East University, Nicosia, Turkish Republic of Northern Cyprus (TRNC), Turkey*

[f]*Research Institute for Medical and Health Sciences, University of Sharjah, Sharjah, United Arab Emirates*

1 Introduction

Retrocochlear hearing disorders represent a complex subset of auditory pathology that extends beyond the peripheral cochlear structures to involve the neural pathways responsible for auditory and vestibular processing. Unlike conventional sensorineural hearing loss (SNHL), which primarily affects cochlear hair cells, retrocochlear conditions involve dysfunction at the level of the eighth cranial nerve (vestibulocochlear nerve) and the central auditory-vestibular pathways extending from the brainstem to the cortex (Musiek and Baran, 2007).

The clinical significance of a retrocochlear condition lies not only in its potential for causing profound hearing and balance deficits but also in its association with serious underlying conditions, including space-occupying lesions, vascular events, inflammatory disorders, and neurodegenerative diseases (Sedwick et al., 2001). Early identification of these conditions is crucial, as they may represent treatable causes of hearing loss or herald more extensive neurological involvement requiring urgent intervention.

The diagnostic approach to retrocochlear conditions has evolved significantly with advances in electrophysiological testing, high-resolution imaging, and our understanding of central auditory-vestibular neuroanatomy. Modern assessment combines behavioral audiometry, objective electrophysiological measures, vestibular function testing, and advanced neuroimaging to localize pathology and guide appropriate management strategies.

This chapter provides a comprehensive overview of retrocochlear conditions, integrating anatomical knowledge with clinical presentation patterns, diagnostic findings, and imaging correlates. Understanding these complex relationships is essential for clinicians involved in the evaluation and management of patients with suspected central auditory-vestibular dysfunction.

2 Literature review

The auditory and vestibular systems, responsible for hearing and balance respectively, are intricately intertwined both anatomically and physiologically, forming a unified and expansive audio-vestibular network. This complex system can be broadly categorized into peripheral and central components, both of which play pivotal roles in sensory signal transmission and interpretation. While the central auditory pathway is typically considered to commence at the cochlear nucleus, a more integrated understanding includes the vestibulocochlear nerve, or cranial nerve VIII, which serves as a conduit for both auditory and vestibular signals (Corrêa et al., 2023; Benoudiba et al., 2013).

Cranial nerve VIII comprises two distinct yet closely aligned branches: the auditory nerve, arising from the cochlea, and the vestibular nerve, arising from the semicircular canals, utricle, and saccule. The vestibular nerve itself is subdivided into

Advances in Medical Imaging. https://doi.org/10.1016/B978-0-443-28967-5.00015-0

superior and inferior branches, each innervating specific structures within the vestibular apparatus. Together, these nerve fibers carry critical auditory and balance-related sensory input to the brainstem (Bucchieri et al., 2019; Chen et al., 2021). Once there, they synapse in specialized nuclei: the cochlear nucleus for auditory signals and the vestibular nuclei for balance signals, thus initiating central processing (Corrêa et al., 2023).

The auditory pathway proceeds from the cochlea to the cochlear nucleus, located at the junction of the medulla and pons. From this point, auditory information ascends bilaterally through the superior olivary complex, which plays a central role in the localization of sound. The signals then travel via the lateral lemniscus to the inferior colliculus in the midbrain, continuing on to the medial geniculate body (MGB) of the thalamus. Ultimately, auditory inputs are relayed to the auditory cortex in the temporal lobe via the auditory radiations. Each station along this pathway decodes specific auditory information, such as pitch, intensity, timing, and spatial localization, enabling accurate perception and interpretation of environmental sounds, particularly speech and language (Peterson et al., 2018; Musiek, 1986; Dieterich and Brandt, 2015).

In parallel, the vestibular system begins its central course at the vestibular nuclei complex, located medially in the medulla and adjacent to the fourth ventricle (Musiek, 1986; Yoo and Mihaila, 2022). These nuclei act as primary processing centers for balance-related input and project along several critical pathways. One major route is the vestibulo-thalamocortical pathway, which relays information to the vestibular cortex via the thalamus, allowing for the conscious perception of balance and spatial orientation (Dieterich and Brandt, 2015; Yoo and Mihaila, 2022). Another key pathway is the vestibulo-cerebellar tract, which enters the cerebellum through the inferior cerebellar peduncle, aiding in the coordination of movements and postural stability (Musiek, 1986; Papadopoulou et al., 2022).

In addition, the vestibular nuclei communicate with the extraocular motor nuclei through the vestibulo-ocular reflex (VOR) pathway, which is vital for gaze stabilization during head movement. This pathway travels along the medial longitudinal fasciculus and engages cranial nerves III, IV, and VI, which are responsible for eye muscle control (Dieterich and Brandt, 2015; Hamid and Sheykholeslami, 2006; Papadopoulou et al., 2022). The vestibulospinal tract, originating from the vestibular nuclei, is another critical route divided into medial and lateral components. The medial vestibulospinal tract influences head and neck posture, while the lateral tract extends further to coordinate postural adjustments of the trunk and lower limbs (Baloh and Honrubia, 1979; Dieterich and Brandt, 2015; Yoo and Mihaila, 2022).

Given this intricate anatomical framework, the audio-vestibular system is susceptible to a broad range of central nervous system (CNS) pathologies. Vascular events, inflammatory and demyelinating diseases, neurodegenerative disorders, tumors, and infections can all disrupt these pathways, leading to clinical symptoms such as hearing loss, vertigo, imbalance, and visual disturbances like oscillopsia (Hamid and Sheykholeslami, 2006; Papadopoulou et al., 2022; Baloh and Honrubia, 1979). The wide distribution of this network—from the brainstem and midbrain to the thalamus, cerebellum, and cortex—explains the diverse symptomatology that may arise from lesions at various levels of the neuraxis.

From a functional perspective, the auditory system plays a vital role in decoding and interpreting sound stimuli, with speech being one of the most complex and significant forms. It allows individuals to recognize and differentiate between various sounds and to respond appropriately to auditory cues in their environment. Beyond mere detection, the system converts sound waves into neural signals, which are then processed through ascending relay stations to facilitate the perception, discrimination, and comprehension of language and other auditory inputs (Musiek, 1986; Peterson et al., 2018).

Conversely, the vestibular system is essential for balance control, postural adjustments, and spatial awareness. It provides real-time information about motion, head orientation, and gravitational forces. Through its integration with visual and proprioceptive systems, it ensures the body's ability to maintain equilibrium and navigate the environment smoothly and efficiently. Gaze stabilization, one of its fundamental functions, is achieved through the coordination of eye and head movements, ensuring a stable visual field even during rapid bodily movements (Dieterich and Brandt, 2015; Hamid and Sheykholeslami, 2006; Curthoys, 2017; Baloh and Honrubia, 1979).

When the integrity or function of this interconnected audio-vestibular system is compromised, individuals may experience a wide range of clinical deficits. These may manifest as hearing loss, tinnitus, vertigo, imbalance, disequilibrium, spatial disorientation, or visual instability (Dieterich and Brandt, 2015; Hamid and Sheykholeslami, 2006; Papadopoulou et al., 2022). Such symptoms can profoundly affect quality of life, communication, mobility, and general well-being. To diagnose and localize the site of dysfunction, clinicians rely on a comprehensive array of audiovestibular assessments. These include audiometry to evaluate hearing thresholds, tympanometry to assess middle ear function, otoacoustic emissions (OAEs) for cochlear integrity, and auditory brainstem responses (ABR) for brainstem-level auditory conduction (Musiek, 1986; Peterson et al., 2018; Hamid and Sheykholeslami, 2006; ProQuest, 2025).

Vestibular assessment tools such as videonystagmography (VNG), vestibular evoked myogenic potentials (VEMP), caloric testing, and video head impulse testing (vHIT) provide valuable insights into the functional status of the peripheral and central vestibular systems (Dieterich and Brandt, 2015; Yoo and Mihaila, 2022; Hamid and Sheykholeslami, 2006; Papadopoulou et al., 2022; Baloh and Honrubia, 1979). These diagnostic modalities not only aid in identifying the underlying pathology but also guide treatment planning and rehabilitation strategies tailored to the individual's specific deficits (Dieterich and Brandt, 2015; Papadopoulou et al., 2022; ProQuest, 2025).

3 Retrocochlear disorders

3.1 Auditory neuropathy spectrum disorder

Auditory Neuropathy Spectrum Disorder (ANSD) is a complex and heterogeneous hearing disorder characterized by a disruption in neural signaling along the auditory pathway, despite often normal or near-normal cochlear function. Unlike typical SNHL, where damage affects both the sensory hair cells and the auditory nerve, ANSD involves a functional disconnection between the inner ear and the central auditory system. This results in a unique hearing profile, where individuals may have preserved outer hair cell function but struggle with speech perception, particularly in noisy environments.

The defining features of ANSD include the presence of OAEs or cochlear microphonics, indicating intact outer hair cell function, alongside absent or severely abnormal ABRs. This dissociation confirms that, while the cochlea can detect sound normally, the neural transmission of auditory signals is impaired (Berlin et al., 2010). The underlying pathology can occur at various sites, including the inner hair cell synapses, the auditory nerve itself, or even more central auditory pathways.

One of the most common lesion sites in ANSD is the synapse between the inner hair cells and the auditory nerve. Mutations in genes such as OTOF, which encodes the synaptic protein otoferlin, can disrupt neurotransmitter release, leading to poor neural synchronization (Moser and Starr, 2016). This desynchronization is particularly detrimental to speech perception, as the auditory system relies on precise timing cues to decode rapid changes in speech sounds. In other cases, the disorder may stem from auditory nerve hypoplasia, demyelination, or structural abnormalities such as cochlear nerve deficiency or cochlear malformations.

The clinical presentation of ANSD is highly variable. Some individuals may have normal pure-tone thresholds but significant speech recognition difficulties, while others may exhibit moderate to profound hearing loss (Starr et al., 2000). This variability complicates diagnosis and management, as conventional hearing aids—which simply amplify sound—may not adequately address the neural dyssynchrony. Cochlear implants, which directly stimulate the auditory nerve, can be beneficial in some cases, particularly when the primary deficit is presynaptic (Roush et al., 2011; Breneman et al., 2012). However, outcomes are unpredictable and depend on the integrity of the auditory nerve and central pathways.

Due to its heterogeneity, ANSD requires a comprehensive diagnostic approach, including OAEs, ABRs, electrocochleography (ECochG), and imaging studies such as magnetic resonance imaging (MRI) to assess cochlear and auditory nerve integrity (Manchaiah et al., 2011). Early identification is crucial, as delayed intervention can impact speech and language development in children (Bielecki et al., 2011). Ongoing research aims to better understand the molecular mechanisms, refine diagnostic criteria, and develop targeted therapies, such as gene therapy for synaptic disorders like OTOF-related ANSD (Akil et al., 2012).

3.2 Tumors

Vestibular schwannoma (VS), also known as acoustic neuroma, is the most common intracranial tumor found in the cerebellopontine angle (CPA) and originates predominantly from the vestibular nerve, specifically its inferior division. These benign tumors arise from Schwann cells, which provide the myelin sheath around peripheral nerves. The clinical manifestations, diagnostic findings, and impact of a VS can vary significantly based on its size, precise location, and extent of growth.

The tumor may be confined within the internal auditory canal (IAC), referred to as intracanalicular VS, or it may extend beyond into the CPA, where it can potentially compress surrounding neural structures such as the cochlear nerve (a division of cranial nerve VIII), the facial nerve (cranial nerve VII), and even brainstem structures in advanced cases. This spatial variation directly influences the symptoms experienced by patients. In its early stages, VS typically presents with unilateral SNHL, often gradual and initially affecting high frequencies. Patients may also report tinnitus, imbalance, or vertigo. As the tumor grows, it may lead to facial weakness or numbness due to involvement of the facial nerve, and more severe symptoms such as headaches or ataxia if brainstem compression occurs.

Audiovestibular testing in VS reveals a diverse spectrum of findings. Pure-tone audiometry may show hearing thresholds ranging from normal to profound loss, while speech audiometry often reveals disproportionately poor speech discrimination scores relative to the degree of hearing loss. This discrepancy is an important clinical indicator suggesting a retrocochlear condition (Prendergast et al., 2018). OAEs, which reflect outer hair cell function in the cochlea, may be present or absent depending on whether the tumor compromises the vascular supply to the cochlea. ABRs, another key diagnostic tool, may show absent waveforms or delayed peak latencies, indicating disrupted neural conduction along the auditory pathway (Schmidt et al., 2001; Don et al., 2005).

Vestibular testing also provides insight into nerve involvement. VEMPs can help distinguish between involvement of the inferior and superior branches of the vestibular nerve. An absent cervical VEMP suggests inferior vestibular nerve involvement, while an absent ocular VEMP (oVEMP) may point to superior nerve involvement. vHIT, which assess the VOR, may reveal dysfunction in the semicircular canals, contributing to the clinical picture of vestibular imbalance.

MRI remains the gold standard for diagnosing VS (Cueva, 2004). Contrast-enhanced MRI can accurately delineate the tumor, assess its size, its confinement to the IAC, or its extension into the CPA, and determine whether it causes brainstem compression. These imaging findings are essential for surgical planning or deciding on noninvasive management strategies. Histopathological examination following tumor resection confirms the diagnosis. In addition, computed tomography (CT) can complement MRI findings by showing bony changes, such as IAC widening, which may suggest the presence of a slowly growing intracanalicular tumor.

In summary, VS is a complex condition with a wide array of clinical and diagnostic presentations. Its identification relies on a combination of audiovestibular assessments and high-resolution imaging, with MRI providing the most definitive evaluation. Early recognition and accurate diagnosis are crucial, as they guide appropriate treatment decisions, whether that involves monitoring, radiotherapy, or surgical intervention, with the ultimate aim of preserving neural function and quality of life.

A wide range of space-occupying lesions at the CPA can lead to audiovestibular dysfunction, with VS being the most commonly encountered tumor in this region (Matthies and Samii, 1997). However, following VS, other notable CPA lesions include meningiomas, epidermoid cysts, arachnoid cysts, facial nerve schwannomas, and, more rarely, hemangiomas, all of which can impinge on the auditory and vestibular systems depending on their size, location, and rate of growth (Fukaya et al., 2011).

Meningiomas, which arise from the meninges, are the second most common CPA tumors (Jackler and Whinney, 2001). When they occur near the CPA, they can exert pressure on the vestibulocochlear nerve (CN VIII) and facial nerve (CN VII). The audiovestibular manifestations depend largely on the degree and location of this compression. Clinically, these tumors typically present with unilateral, progressive hearing loss. Audiological evaluations, such as ABR, may reveal delayed wave latencies or even absent waveforms, especially when the auditory nerve is compressed. However, the severity of these findings can vary based on the stage of tumor progression. Interestingly, in early stages or small tumors, patients may maintain normal hearing thresholds, while ABR abnormalities may still be evident. In such cases, VEMPs may remain intact, indicating that the vestibular branch of CN VIII is not yet compromised.

Epidermoid cysts, the third most common CPA lesions, often present with symptoms similar to VS and meningiomas (Berger and Wilson, 1985). However, because these tumors are soft and infiltrative rather than compressive, they may not always produce clear abnormalities in audiological or vestibular tests. This can lead to diagnostic challenges, especially when structural imaging has not yet been performed. Despite this, patients may still report symptoms such as dizziness, hearing changes, or tinnitus.

Arachnoid cysts are rare but noteworthy lesions that may develop in the CPA. When large enough, they can exert pressure on the adjacent auditory and vestibular structures, leading to hearing loss and balance disturbances. Their effects, like those of other CPA tumors, depend on their growth and proximity to key neural elements. Recent literature (Nisson et al., 2024) has emphasized their clinical relevance, particularly when they produce symptoms despite being benign and slow-growing.

Facial nerve schwannomas represent another CPA pathology that may lead to audiovestibular dysfunction, especially when the tumor arises in the IAC. This proximity to the cochlear and vestibular nerves means even relatively small lesions can have significant functional consequences. In such cases, MRI can be particularly helpful in distinguishing facial nerve schwannomas from CN VIII tumors. The presence of tumor growth in the geniculate ganglion or labyrinthine segment of the facial nerve is a distinctive feature. Because the facial nerve is part of the acoustic reflex arc, tumors affecting it can specifically disrupt acoustic reflexes. In addition, electroneuronography can be valuable in evaluating facial nerve integrity in such presentations (Libell et al., 2023).

Finally, hemangiomas of the CPA, though exceedingly rare, are benign vascular tumors that may develop near the cisternal segment of CN VIII. Their vascular nature and location can cause hearing loss, often through compression or compromised blood supply to the cochlea. Case studies, such as those referenced by Yu et al. (2023), underscore their diagnostic significance despite their rarity.

In summary, while VSs dominate CPA pathology, a wide array of other lesions can produce similar or overlapping audiovestibular symptoms. Accurate diagnosis relies heavily on a combination of clinical assessment, electrophysiological testing, and, crucially, high-resolution imaging—especially MRI. Each tumor type carries its own implications for management and prognosis, highlighting the importance of a thorough differential diagnosis in patients presenting with unilateral hearing loss, tinnitus, or balance issues.

Trigeminal schwannomas, though rare, are tumors that arise from the Schwann cells of the fifth cranial nerve (CN V). While primarily affecting the trigeminal nerve, their anatomical proximity to the CPA and the auditory structures can result in secondary involvement of the auditory pathway due to mass effect. In some cases, patients with trigeminal schwannomas may present with symptoms that closely mimic those of VS, including hearing loss, tinnitus, and dizziness, as noted by Ganesan et al., 2015). This overlap can complicate differential diagnosis without detailed imaging. Furthermore, during surgical intervention for trigeminal schwannomas, there is a risk of inadvertent damage to the motor branch of the trigeminal nerve. Such a compromise can impair the function of the tensor veli palatini muscle, which plays a critical role in Eustachian tube function. Consequently, patients may develop Eustachian tube dysfunction, leading to conductive hearing loss, as reported by Ito et al., 2016).

Glomus jugulare tumors are another category of rare, slow-growing neoplasms that originate from the paraganglionic cells within the jugular foramen. Although benign, their location in the jugular fossa gives them the potential to invade adjacent middle ear structures. As these tumors enlarge, they can initially manifest with pulsatile tinnitus—a hallmark symptom—and conductive hearing loss due to direct involvement of the ossicles or tympanic cavity. With further extension, these tumors may involve the inner ear or the auditory nerve, resulting in SNHL. The gradual progression of auditory symptoms underscores the importance of early detection, especially in cases where patients initially present with isolated tinnitus (Baguley et al., 1994).

Gliomas, which originate from glial cells, are a broad group of tumors with subtypes including astrocytomas, oligodendrogliomas, and ependymomas. When gliomas affect brainstem structures such as the cochlear or vestibular nuclei, they may present with SNHL and vestibular dysfunction. Involvement of auditory brainstem centers, such as the cochlear nuclei, is often reflected in abnormal ABR results—specifically, delayed or absent waveforms beyond Wave III. Gliomas involving the cerebellum or its connections may also disrupt the VOR and smooth pursuit eye movements, leading to abnormal findings on VNG, including impaired VOR suppression and saccadic intrusions, as noted by Handelsman (2005). Ruiz-García et al. (2024) have further emphasized the need to consider gliomas in differential diagnoses when SNHL and vestibular abnormalities occur with otherwise unremarkable otologic findings.

Medulloblastoma, the most common malignant brain tumor in children, may also involve the auditory and vestibular systems. Its rapid growth and location in the posterior fossa allow it to compress or infiltrate brainstem nuclei and pathways involved in hearing and balance. As a result, patients may present with progressive SNHL or, in rarer cases, sudden SNHL (Roy et al., 2019; Terakawa et al., 2011). These presentations often accompany other signs of increased intracranial pressure or cerebellar dysfunction, making thorough neurologic and audiological evaluations crucial.

Certain central tumors can also affect higher-order auditory processing centers. A tumor in the region of the third ventricle has been documented to cause pure word deafness—a rare auditory agnosia characterized by preserved hearing thresholds and recognition of environmental sounds, but a complete inability to comprehend spoken language. This has been attributed to bilateral compression of the MGB and internal capsule, as demonstrated in a case by Shivashankar et al. (2001). Such findings illustrate the intricate nature of central auditory processing and the potential for localized brain lesions to produce selective perceptual deficits.

Pineal region tumors, including pineal germinomas, may also cause auditory symptoms, particularly if they grow large enough to compress the brainstem or midbrain auditory pathways. In one case, a patient with bilateral hearing loss due to a pineal tumor experienced full recovery of hearing following surgical removal of the mass, highlighting the compressive rather than destructive nature of the lesion (Mahboob et al., 2023). These tumors may also present with classic signs of obstructive hydrocephalus due to their position near the cerebral aqueduct.

CNS lymphomas, though rare, can produce isolated vestibular symptoms, often without concurrent hearing loss. Patients with lymphoma involving the brainstem or cerebellum may show distinct VNG abnormalities, including gaze-evoked nystagmus, impaired smooth pursuit, and irregular saccades. These findings, as reported by Lee et al. (2024), underscore the importance of central causes in cases where vestibular dysfunction is present but peripheral testing is inconclusive.

Collectively, these varied intracranial tumors illustrate the complex interplay between CNS pathology and audiovestibular function. Whether through direct compression, infiltration, or disruption of neural pathways, such lesions can present with a spectrum of auditory and balance-related symptoms, often requiring a multidisciplinary approach for accurate diagnosis and management. Advanced neuroimaging, alongside electrophysiological and behavioral audiological assessments, remains critical in distinguishing these central pathologies from more common peripheral disorders.

3.3 Vascular lesions

Posterior circulation cerebrovascular stroke is well-documented to manifest with a wide spectrum of audiovestibular symptoms, which can range from subtle balance disturbances to profound hearing deficits. According to Koura et al. (2024), abnormalities in various audiological and vestibular diagnostic parameters were frequently observed in such patients.

The study reported prolonged ABR peak latencies, altered cervical VEMP (cVEMP) latencies and amplitudes, along with abnormal VNG parameters, including reduced slow-phase peak velocity, VOR gain, and altered saccadic movements. These findings underscore the diverse impact of posterior circulation strokes on both auditory and vestibular pathways.

Among the individuals evaluated in this study, the majority exhibited pontine involvement. One of the classic syndromes associated with posterior circulation stroke is Lateral Medullary Syndrome (also known as Wallenberg Syndrome), which results from infarction in the lateral medulla. This is typically due to occlusion of the posterior inferior cerebellar artery or vertebral artery. In such cases, vestibular symptoms—such as vertigo, nausea, imbalance, and spontaneous nystagmus—are more prevalent than hearing loss. However, when the stroke involves more rostral regions of the medulla, vHIT abnormalities become more apparent, as noted by Lee et al., 2020).

Of clinical significance is the observation that posterior circulation strokes can present as isolated vestibular syndromes in up to 25% of patients. This poses a diagnostic challenge, especially when imaging findings are initially negative. In such scenarios, advanced vestibular function testing, including VNG and vHIT, plays a crucial role in identifying central causes of vestibular dysfunction and guiding appropriate clinical decisions.

Beyond the brainstem, thalamic strokes also have implications for auditory function. Sudden bilateral hearing loss has been reported in cases involving the thalamic-midbrain junction. Lee et al. (2016) presented a case where MRI revealed chronic infarction at the right thalamic-midbrain junction and subacute infarction at the left, corresponding with bilateral auditory symptoms. Notably, Agarwal et al. (2016) documented instances where hearing loss was the sole presenting symptom of thalamic infarct. These cases demonstrated normal ABR findings despite significant hearing loss, as shown by pure tone audiometry (PTA). This discrepancy suggests that lesions in the thalamus—an important relay station in the central auditory pathway—may be better detected using mid- or long-latency auditory evoked potentials, such as middle latency responses (MLR) or late latency responses (LLR), which assess cortical and subcortical auditory processing beyond the brainstem.

At the cortical level, cerebrovascular strokes can severely disrupt auditory perception and processing. The temporal lobe, particularly the primary auditory cortex located in Heschl's gyrus, is the final cortical destination for auditory signals transmitted via the acoustic radiations. Stroke-related damage in this region can result in central auditory processing disorders or cortical deafness. If medical intervention is timely, there is potential for considerable functional recovery. For instance, Pihl et al. (2021) described a patient with an acute ischemic lesion in the left temporoparietal region involving the insular cortex who experienced complete recovery from bilateral hearing loss. Objective auditory electrophysiological tools like LLR and mismatch negativity (MMN) can be useful not only in identifying cortical auditory dysfunction but also in tracking prognosis.

Neuroimaging techniques have further contributed to understanding the impact of cortical strokes on hearing. Gwak et al. (2021) employed resting-state functional MRI (rs-fMRI) and diffusion tensor imaging (DTI) to examine connectivity in the auditory pathways. Their findings showed reduced functional connectivity in auditory cortices and disrupted acoustic radiation, the latter being a key white matter tract transmitting auditory signals from the thalamus to the cortex. Similarly, putaminal hemorrhages have demonstrated compromised acoustic radiation pathways on DTI and have been linked to severe hearing deficits (Koyama and Domen, 2016).

Advanced imaging techniques, such as voxel-based morphometry and tractography, can provide detailed visualization of disrupted cortical connections. In cases of cortical deafness, such tools help localize the extent of damage and guide rehabilitation. Semenza et al. (2012) emphasized the role of tractography in revealing the neural disconnections responsible for auditory agnosia and deafness. A striking example of this is reported by Lachowska et al. (2021), where bilateral damage to Heschl's gyri—confirmed by MRI following an ischemic stroke—resulted in a complete loss of sound awareness, despite normal findings in peripheral auditory tests such as OAE, ABR, and acoustic reflexes. However, the absence of MMN in this patient confirmed central auditory processing failure, providing compelling evidence of cortical deafness.

3.4 Other conditions

Multiple sclerosis (MS) is a chronic demyelinating disorder of the CNS marked by relapses and remissions. Though the disease is primarily known for its motor and sensory manifestations, it can also involve the audiovestibular system. In recent years, studies have recognized auditory and vestibular disturbances in MS patients, which may occur at various stages of the disease. According to Gür et al. (2022), a significant number of patients experience these symptoms, and while Hellmann et al. (2011) estimated hearing loss in approximately 4.35% of MS cases, the actual prevalence remains uncertain due to inconsistent reporting across studies. This variability may arise from differences in diagnostic tools, clinical populations, and criteria for defining hearing impairment.

Hearing loss in MS can present as either temporary or permanent, and in some cases, it may even be the first presenting symptom of the disease. PTA and ABR are commonly used to evaluate these deficits, with ABR often showing prolonged interpeak latencies suggestive of delayed neural conduction in the brainstem. Interestingly, treatment with corticosteroids during acute relapses can sometimes lead to partial or complete recovery of hearing function, as observed by Lee et al. (2019). Neuroimaging studies have provided further insight, with Stadio et al. (2018) reporting the presence of white matter hyperintensities in areas such as the upper auditory pathways, the root entry zone of the cochlear nerve, and even the cochlea itself, highlighting the multifocal nature of demyelinating lesions in MS.

Vestibular dysfunction is another prominent symptom in MS and may include vertigo, imbalance, oscillopsia, and dizziness. These symptoms can be traced to demyelinating plaques in regions such as the medulla, cerebellar peduncles, posterior pontine tegmentum, intrapontine eighth nerve fascicle, and midbrain. The presence of such lesions often correlates with oculomotor abnormalities and may manifest in clinical signs like nystagmus or impaired gaze stabilization. Pula et al., 2013) emphasized the importance of these associations, which not only aid in diagnosis but also help in understanding the functional impact of MS on central vestibular pathways.

In addition to MS, other neurological disorders show significant involvement of the audiovestibular system. For instance, in Parkinson's disease, high-frequency SNHL and delayed ABR peaks have been reported, which may be related to dopaminergic dysfunction within the auditory system (de Souza et al., 2025). Similarly, Wernicke encephalopathy, a neurological emergency caused by thiamine deficiency, can present with auditory and vestibular symptoms. MRI findings often reveal symmetrical lesions in the medial thalami, mammillary bodies, tectal plate, and periaqueductal region, although atypical areas may also be affected (Zuccoli et al., 2009). Patients may exhibit bilateral SNHL, as noted by Walker et al. (2014), and vestibular testing can reveal peripheral or central involvement. Lee et al., 2018) reported bilateral dysfunction of the lateral semicircular canals in such patients using vHIT, suggesting a significant central component to the vestibular deficit.

Autoimmune encephalitis, particularly anti-NMDA receptor encephalitis, has also been associated with audiovestibular symptoms. Bilateral hearing loss has been documented, although imaging studies may remain normal in up to 50% of cases. When present, MRI changes are often localized to the hippocampus (Zhang et al., 2018). These findings underscore the possibility of subclinical involvement or functional disruptions not easily captured by conventional imaging and the need for comprehensive electrophysiological evaluations.

Lastly, CNS infections, such as acute bacterial meningitis, remain a known cause of permanent hearing loss. This occurs as a result of direct damage to the cochlea, auditory nerve, or surrounding structures due to the inflammatory process. Jensen et al. (2023) highlighted the long-term audiological consequences of meningitis, which further emphasize the importance of early diagnosis and intervention.

4 Diagnostic methods and imaging

4.1 Electrophysiological testing

A wide array of audiological and vestibular assessments is employed to evaluate retrocochlear conditions, conditions that affect auditory pathways beyond the cochlea, involving the auditory nerve, brainstem, and higher central auditory processing centers. The diagnostic approach to these disorders typically begins with foundational behavioral tests, which help establish a baseline understanding of hearing function and identify any patterns that may point toward central or neural involvement rather than peripheral cochlear pathology.

Among the initial tests, PTA plays a central role by measuring hearing thresholds across a range of frequencies. It offers a general view of the hearing profile but does not directly indicate the site of the lesion. Complementing PTA, speech audiometry evaluates the individual's ability to detect and understand spoken words. In retrocochlear conditions, a hallmark finding may be disproportionately poor speech recognition scores compared to what would be expected from the pure tone thresholds. Such discrepancies suggest that the problem lies beyond the cochlea, possibly involving the auditory nerve or higher centers, where the fidelity of sound processing is compromised (Young and Cornejo, 2023).

To further investigate the neural pathways involved in hearing, electrophysiological tests are indispensable. The ABR is a highly sensitive and objective test that records neural activity along the auditory nerve and brainstem pathways (Jewett and Williston, 1971). The test identifies waveforms (typically labeled I through V), each representing neural activity at specific points along the ascending auditory pathway. Abnormalities such as delayed latencies, reduced amplitudes, or the absence of specific waves can provide strong evidence of a retrocochlear lesion, including VSs, demyelinating diseases like MS, or brainstem infarctions (Robinette et al., 2000; Selters and Brackmann, 1977).

MLR and LLR offer additional layers of diagnostic insight. MLR assesses neural conduction between the brainstem and the primary auditory cortex, reflecting thalamo-cortical function, while LLR provides information about cortical auditory processing. These responses are particularly useful in evaluating auditory processing at higher neurological levels and may also be applied in contexts such as auditory training and rehabilitation in individuals with central auditory processing deficits.

In addition to these objective electrophysiological tests, specific behavioral tests can be used to probe central auditory function in greater detail. Tests such as the dichotic digits test, competing sentences test, and gaps-in-noise test are valuable in assessing the auditory system's ability to handle complex auditory information. These tests challenge the auditory system by introducing competing signals or requiring temporal resolution, and they are especially helpful in identifying deficits in auditory processing that are not evident in basic threshold testing.

Vestibular evaluation is equally critical when retrocochlear involvement is suspected, particularly if the patient presents with symptoms such as vertigo, imbalance, or spatial disorientation. VEMPs are powerful tools for assessing the integrity of the otolithic organs—the saccule and utricle—as well as the reflex pathways they activate. cVEMP evaluates the saccule and vestibulo-collic reflex pathway via the sternocleidomastoid muscle, while oVEMP assesses the utricle and VOR pathway via the inferior oblique muscle. Abnormalities in these tests may indicate either peripheral or central vestibular dysfunction, depending on the pattern of responses (Rosengren et al., 2010).

vHIT is another essential tool, assessing the functional status of the semicircular canals and the corresponding VOR. This test evaluates how well the eyes compensate for rapid head movements, with impaired responses pointing to deficits in the central integration of vestibular input or in the peripheral canal function (Halmagyi et al., 2017).

Video nystagmography (VNG) offers a comprehensive evaluation of both spontaneous and provoked eye movements. It includes various subtests, one of which is the caloric test, uniquely suited to stimulating the horizontal semicircular canals. By irrigating the ear canal with warm or cold air or water, the caloric test induces endolymphatic flow, triggering nystagmus. Abnormal or absent responses can help distinguish between peripheral and central vestibular involvement.

Together, these audiological and vestibular tests provide a layered and integrative assessment approach to diagnosing retrocochlear and central vestibular disorders. Each test contributes a specific dimension to the understanding of the patient's condition, whether it is localizing a lesion to the auditory nerve, evaluating central auditory processing, or assessing the integrity of balance mechanisms. This holistic diagnostic framework enables clinicians to pinpoint the source of dysfunction more precisely, thus guiding targeted intervention strategies and improving patient outcomes.

4.2 Neuroimaging techniques

4.2.1 Magnetic resonance imaging

MRI remains the gold standard for diagnosing retrocochlear conditions, offering superior soft tissue contrast and multiplanar imaging capabilities essential for evaluating the complex anatomy of the auditory and vestibular pathways (Cueva, 2004; Gurgel et al., 2012). High-resolution MRI with gadolinium enhancement can accurately delineate anatomical structures, assess lesion characteristics, and determine the extent of pathology affecting the IAC, CPA, and central auditory pathways.

Modern MRI protocols for retrocochlear evaluation typically employ high-resolution T1-weighted sequences before and after gadolinium administration, T2-weighted sequences, and fluid-attenuated inversion recovery (FLAIR) imaging. The standard slice thickness ranges from 0.6 to 1.5 mm, providing detailed visualization of small structures such as the facial and vestibulocochlear nerves within the IAC (Baird et al., 2019). Three-dimensional constructive interference in steady state or 3D-Fast Imaging Employing Steady-state Acquisition (FIESTA) sequences are particularly valuable for demonstrating the cisternal segments of cranial nerves VII and VIII, allowing differentiation between nerve components and early detection of small intracanalicular lesions.

MRI demonstrates characteristic enhancement patterns that aid in differential diagnosis. VSs typically show homogeneous enhancement with occasional cystic components, while meningiomas demonstrate intense enhancement with the pathognomonic "dural tail" sign (Jackler and Whinney, 2001). Epidermoid cysts characteristically follow cerebrospinal fluid signal intensity on most sequences but demonstrate high signal intensity on FLAIR imaging, distinguishing them from arachnoid cysts (Berger and Wilson, 1985). In inflammatory conditions such as MS, FLAIR sequences reveal characteristic periventricular white matter hyperintensities that may involve auditory pathways (Stadio et al., 2018).

While MRI provides excellent soft tissue contrast, certain limitations exist. Small intracanalicular lesions may be challenging to detect, and motion artifacts can compromise image quality (Arriaga et al., 1995). Contraindications include certain metallic implants, claustrophobia, and gadolinium allergy, though newer gadolinium-based contrast agents have reduced nephrotoxicity concerns.

4.2.2 Computed tomography

CT serves as a complementary imaging modality, particularly valuable for evaluating bony changes associated with retrocochlear conditions. High-resolution CT with thin-section (0.5–1.0 mm) axial and coronal reconstructions can demonstrate IAC widening, temporal bone erosion, and ossicular chain abnormalities that may accompany retrocochlear pathology (Schmidt et al., 2001).

CT imaging is particularly useful in detecting subtle bone changes, such as IAC enlargement in cases of slowly growing intracanalicular tumors. Glomus jugulare tumors demonstrate characteristic bone erosion and may show calcifications not readily visible on MRI. In cases of temporal bone trauma or suspected congenital malformations, CT provides superior bone detail compared to MRI (Don et al., 2005).

4.2.3 Diffusion tensor imaging

DTI represents an advanced MRI technique that provides insights into white matter tract integrity by measuring the directional movement of water molecules along neural fibers. This technique is particularly valuable in evaluating acoustic radiation pathways in stroke patients, demyelinating conditions, and other central auditory pathway disorders (Gwak et al., 2021).

DTI measures fractional anisotropy (FA), mean diffusivity (MD), and other diffusion parameters that reflect the microstructural integrity of white matter tracts. The acoustic radiations, which connect the MGB of the thalamus to the primary auditory cortex, can be specifically traced using tractography algorithms, allowing visualization of pathway disruption in various pathological conditions (Koyama and Domen, 2016).

In stroke patients, DTI demonstrates disrupted acoustic radiation pathways, correlating with auditory processing deficits. Reduced FA values and increased MD indicate a loss of white matter integrity, while tractography can visualize the extent of pathway disruption. In patients with MS, DTI reveals microstructural changes in normal-appearing white matter that may not be visible on conventional MRI sequences, providing insights into subclinical disease involvement (Semenza et al., 2012).

DTI has revealed important insights into the relationship between white matter integrity and auditory function. Studies have demonstrated correlations between acoustic radiation DTI parameters and speech processing abilities, suggesting potential biomarkers for central auditory processing disorders (Lachowska et al., 2021).

4.2.4 Functional MRI

Functional MRI (fMRI) offers assessment of auditory cortex activation patterns and can reveal functional disruptions not visible on structural imaging. This technique measures blood oxygen level-dependent signal changes in response to auditory stimuli, providing insights into central auditory processing at the cortical level (Giroud et al., 2017).

Auditory fMRI presents unique technical challenges due to scanner noise interference with auditory stimuli. Sparse temporal sampling techniques and specialized MRI-compatible audio systems are employed to minimize these artifacts. Block design paradigms using speech, music, or pure tone stimuli allow mapping of different auditory processing regions within the temporal lobe (Pihl et al., 2021).

In patients with cortical deafness following a stroke, fMRI can demonstrate absent or reduced activation in the primary auditory cortex despite preserved peripheral hearing function. Functional connectivity analysis reveals disrupted networks between auditory processing regions, providing insights into the mechanisms of central auditory dysfunction. In cochlear implant candidates, preoperative fMRI assessment of auditory cortex integrity may help predict postimplantation outcomes.

fMRI studies have revealed remarkable plasticity in the auditory system following intervention. Patients recovering from cortical deafness show progressive restoration of auditory cortex activation, correlating with functional improvement. Cross-modal plasticity, where visual or somatosensory cortices are recruited for auditory processing, has been demonstrated in cases of profound deafness.

4.2.5 Nuclear imaging techniques

Positron emission tomography (PET) and Single Photon Emission CT provide metabolic information, complementing structural imaging. 18F-FDG PET can demonstrate altered glucose metabolism in auditory processing regions, while specific radioligands may target neurotransmitter systems involved in auditory function (Terakawa et al., 2011).

Magnetic resonance spectroscopy offers a noninvasive assessment of brain metabolism, measuring concentrations of various metabolites, including *N*-acetylaspartate (NAA), choline, and lactate. Reduced NAA levels may indicate neuronal dysfunction in retrocochlear conditions affecting central auditory pathways.

Ultra-high-field MRI (7 T and above) provides unprecedented spatial resolution and signal-to-noise ratio, enabling detailed visualization of small anatomical structures and subtle pathological changes not detectable at conventional field strengths. These techniques hold promise for the early detection of retrocochlear conditions and for monitoring treatment response (Bharadwaj et al., 2014).

Arterial spin labeling represents a noninvasive perfusion imaging technique that can assess cerebral blood flow in auditory processing regions without the need for contrast administration. This technique may be valuable in evaluating vascular causes of retrocochlear dysfunction and monitoring treatment response in inflammatory conditions.

5 Discussion

The comprehensive evaluation of retrocochlear conditions presents unique challenges that require the integration of clinical expertise, advanced diagnostic testing, and sophisticated neuroimaging techniques. The complexity of the central auditory-vestibular system, with its multiple anatomical levels and interconnected pathways, creates opportunities for diverse pathological processes to manifest with overlapping symptom profiles.

One of the most significant clinical challenges in retrocochlear conditions is the potential for diagnostic delay or misdiagnosis. Many retrocochlear conditions, particularly in their early stages, can present with symptoms that mimic more common peripheral hearing disorders. For instance, the gradual unilateral hearing loss associated with small VSs may initially be attributed to presbycusis o noise-induced hearing loss, particularly in older patients. This highlights the critical importance of maintaining a high index of suspicion for retrocochlear conditions, especially when patients present with asymmetric hearing loss, poor speech discrimination scores disproportionate to pure-tone thresholds, or accompanying vestibular symptoms (Carlson et al., 2015).

The evolution of diagnostic techniques has significantly improved our ability to identify and localize retrocochlear lesions. Electrophysiological testing, particularly ABR, remains a cornerstone of retrocochlear evaluation, offering an objective assessment of neural conduction from the auditory nerve through the brainstem pathways. There are several factors that need to be taken into consideration when interpreting ABR results, including the patient's age, the severity of their hearing loss, and the technical parameters of the test. Recent advances in ABR technology, including chirp-evoked responses and high-frequency ABR, have enhanced sensitivity for detecting subtle retrocochlear abnormalities (Cebulla et al., 2014).

The integration of vestibular function testing has proven invaluable in the comprehensive assessment of retrocochlear conditions. These advancements have revolutionized the diagnosis and management of vestibular disorders, enabling clinicians to pinpoint specific deficits with greater precision. Together, they provide a comprehensive assessment of the vestibular system, enhancing treatment planning and patient outcomes. These tools have not only improved diagnostic accuracy but have also enhanced our understanding of the functional impact of various pathological processes on the vestibular system (Curthoys, 2010).

Neuroimaging, particularly high-resolution MRI with gadolinium enhancement, has become indispensable in the evaluation of suspected retrocochlear conditions. The ability to visualize anatomical structures in detail has transformed the diagnosis and management of conditions such as VS, where imaging findings directly influence treatment decisions. Advanced imaging techniques, including DTI and functional MRI, are beginning to provide insights into the microstructural and functional changes associated with retrocochlear conditions (Gurgel et al., 2012).

The management of retrocochlear conditions has evolved considerably, with treatment approaches becoming increasingly individualized based on specific pathology, patient factors, and functional impact. For VSs, the traditional approach of immediate surgical intervention has given way to more nuanced decision-making that considers tumor size, growth rate, patient age, hearing status, and quality-of-life factors. The introduction of stereotactic radiosurgery has provided an additional treatment option that may be particularly suitable for certain patient populations (Carlson et al., 2015).

The recognition of conditions such as ANSD has highlighted the importance of understanding the heterogeneity within retrocochlear conditions. The variable presentation and outcomes associated with ANSD underscore the need for personalized approaches to diagnosis and management, as well as the importance of genetic counseling and testing in appropriate cases.

An emerging area of interest is the relationship between retrocochlear conditions and cognitive function. There is growing evidence that central auditory processing deficits may have broader implications for cognitive performance, particularly in areas such as working memory and attention. This connection suggests that the impact of retrocochlear conditions may extend beyond traditional audiological measures and may require more comprehensive assessment and rehabilitation approaches (Moore et al., 2014).

The role of rehabilitation in retrocochlear conditions is an area requiring further development. While hearing aids and cochlear implants may provide benefits in some cases, the neural dyssynchrony and central processing deficits characteristic of many retrocochlear conditions often require specialized rehabilitation strategies. Auditory training programs, particularly those targeting temporal processing and speech understanding in noise, demonstrate potential for improving functional outcomes (Cuda et al., 2012).

6 Future directions

The field of retrocochlear conditions is poised for significant advancement across multiple domains, driven by technological innovation, improved understanding of underlying mechanisms, and the development of novel therapeutic approaches.

6.1 Advances in diagnostic technologies

The future of retrocochlear diagnosis lies in the development of more sensitive and specific diagnostic tools. Next-generation electrophysiological techniques, including high-frequency ABR, envelope-following responses, and auditory steady-state responses, promise to enhance our ability to detect subtle neural conduction abnormalities (Bharadwaj et al., 2014). The integration of artificial intelligence and machine learning algorithms into diagnostic interpretation may improve accuracy and reduce interexaminer variability in test interpretation.

Advanced neuroimaging techniques are expected to provide unprecedented insights into retrocochlear conditions. Ultra-high-field MRI (7 T and beyond) offers superior spatial resolution and may enable visualization of microscopic anatomical changes not detectable with current imaging modalities. Functional imaging techniques, including task-based and resting-state fMRI, will likely provide new insights into the functional consequences of retrocochlear lesions and may guide rehabilitation strategies (Giroud et al., 2017).

The development of hybrid imaging approaches, combining structural and functional information, may offer a more comprehensive assessment of retrocochlear conditions. Techniques such as simultaneous PET-MRI may provide insights into both anatomical changes and metabolic activity, potentially improving diagnostic accuracy and treatment monitoring.

6.2 Molecular and genetic approaches

The future of retrocochlear medicine will likely be significantly influenced by advances in molecular biology and genetics. Whole-genome sequencing and transcriptomic analysis are beginning to reveal the genetic basis of various retrocochlear conditions, particularly ANSD. This knowledge is expected to lead to more precise diagnostic approaches and personalized treatment strategies (Shearer et al., 2017).

Gene therapy represents a promising therapeutic frontier for retrocochlear conditions. Early-phase clinical trials for conditions such as otoferlin (OTOF) gene-related ANSD are showing encouraging results, suggesting that direct genetic intervention may become a viable treatment option for specific forms of retrocochlear conditions (Al-Moyed et al., 2019).

The application of clustered regularly interspaced short palindromic repeats-associated protein 9 (CRISPR)-Cas9 and other gene-editing technologies may enable the precise correction of genetic defects underlying certain retrocochlear conditions. While still in early development, these approaches offer the potential for curative treatment of previously untreatable conditions.

6.3 Novel therapeutic approaches

The development of targeted therapies for retrocochlear conditions represents a significant opportunity for improving patient outcomes. Neuroprotective agents, including neurotrophic factors and antiapoptotic compounds, may help preserve neural function in degenerative conditions affecting the auditory nerve and central pathways (Stöver and Lenarz, 2011).

Regenerative medicine approaches, including stem cell therapy and tissue engineering, may offer new treatment options for conditions involving neural degeneration. While still largely experimental, early studies suggest that these approaches may have the potential for restoring neural function in certain retrocochlear conditions.

The integration of neuromodulation techniques, such as transcranial magnetic stimulation and transcranial direct current stimulation, with traditional rehabilitation approaches, may enhance treatment outcomes. These techniques may help optimize neural plasticity and improve the effectiveness of auditory training programs (Yeshurun et al., 2017).

6.4 Personalized medicine and precision diagnostics

The future of retrocochlear medicine will likely be characterized by increasingly personalized approaches to diagnosis and treatment. The integration of genomic data, detailed phenotypic characterization, and advanced biomarker analysis may enable precise prediction of disease progression and treatment response.

The development of liquid biopsy techniques for detecting circulating biomarkers associated with retrocochlear conditions may enable earlier detection and monitoring of disease progression. Such approaches could be particularly valuable for conditions such as VS, where early detection may influence treatment decisions.

6.5 Rehabilitation and quality of life

Future rehabilitation approaches for retrocochlear conditions will likely incorporate advanced technologies such as virtual reality and augmented reality to create more effective and engaging training environments. These technologies may enable more precise control of auditory training parameters and improve patient compliance and outcomes (Kwak et al., 2022).

The development of more sophisticated hearing aid and cochlear implant technologies, specifically designed to address the unique challenges posed by retrocochlear conditions, represents another important area of development. Signal processing strategies that account for neural dyssynchrony and central processing deficits may improve outcomes for patients with retrocochlear conditions.

6.6 Research priorities

Several key research priorities emerge for advancing the field of retrocochlear conditions. These include the development of standardized diagnostic protocols and outcome measures, the investigation of the relationship between retrocochlear conditions and cognitive function, and the establishment of large-scale longitudinal studies to better understand the disease's natural history and treatment outcomes.

The investigation of environmental and lifestyle factors that may influence the development and progression of retrocochlear conditions represents another important research priority. Understanding these factors may lead to preventive strategies and improved patient counseling.

7 Conclusion

Retrocochlear conditions represent a diverse spectrum of disorders affecting the auditory-vestibular system beyond the cochlea, encompassing pathologies from the eighth cranial nerve to the cortical processing centers. The complexity of these conditions demands a comprehensive, multidisciplinary approach that integrates detailed clinical assessment, sophisticated electrophysiological testing, and advanced neuroimaging techniques. Early recognition of retrocochlear conditions is crucial, as these conditions may signal serious underlying neurological disorders requiring prompt intervention. The characteristic clinical presentations—including asymmetric hearing loss, disproportionately poor speech discrimination, and accompanying vestibular symptoms—should prompt thorough evaluation to exclude central pathology. The evolution of diagnostic modalities, from traditional ABR to advanced VEMP testing and high-resolution MRI, has significantly improved our ability to localize lesions and guide appropriate management strategies. However, the heterogeneous nature of retrocochlear conditions continues to present diagnostic challenges, emphasizing the importance of maintaining clinical vigilance and employing systematic diagnostic protocols. As our understanding of the molecular mechanisms underlying retrocochlear conditions advances, the field is moving toward more personalized approaches to diagnosis and treatment. The integration of genetic testing, targeted therapies, and advanced rehabilitation strategies holds promise for improving outcomes in patients with these complex conditions. Future developments in artificial intelligence, ultra-high-field imaging, gene therapy, and regenerative medicine are expected to transform the landscape of retrocochlear disorder management, offering new hope for patients with previously untreatable conditions. The continued collaboration between clinicians, researchers, and technology developers will be essential for translating these advances into improved patient care and quality of life.

References

Agarwal, A., Baruah, U., Gaikwad, S., Gupta, A., Mahore, A., Hardjasudarma, M., 2016. Sudden bilateral hearing loss as the sole presenting symptom of thalamic infarct. J. Clin. Neurosci. 34, 41–43.

Akil, O., Seal, R.P., Burke, K., Wang, C., Alemi, A., During, M., Edwards, R.H., Lustig, L.R., 2012. Restoration of hearing in the VGLUT3 knockout mouse using virally mediated gene therapy. Neuron. 10974199. 75 (2), 283–293. doi:10.1016/j.neuron.2012.05.019.

Al-Moyed, H., Cepeda, A.P., Jung, S.Y., Moser, T., Kügler, S., Reisinger, E., 2019. A dual-AAV approach restores fast exocytosis and partially rescues auditory function in deaf otoferlin knock-out mice. EMBO Mol. Med. 17574684. 11 (1). Available from: http://embomolmed.embopress.org/. doi:10.15252/emmm.201809396.

Arriaga, M.A., Carrier, D., Houston, G.D., 1995. False-positive magnetic resonance imaging of small internal auditory canal tumors: a clinical, radiologic, and pathologic correlation study. Otolaryngol Head Neck Surg. 113 (1), 61–70. doi:10.1016/S0194-5998(95)70146-X.

Baguley, D.M., Moffat, D.A., Hardy, D.G., 1994. What is the effect of translabyrinthine acoustic schwannoma removal upon tinnitus. J. Laryngol. Otol. 108 (11), 914–917.

Baird, S.M., Nguyen, K., Bakhshayesh, H., Psaltis, A.J., 2019. Inner ear and retrocochlear pathology on magnetic resonance imaging for sudden and progressive asymmetrical sensorineural hearing loss. ANZ J. Surg. 89 (7–8), 915919. doi:10.1111/ans.14957.

Baloh, R.W., Honrubia, V., 1979. Clinical neurophysiology of the vestibular system. Contemp. Neurol. Ser. 18, 1–21.

Benoudiba, F., Toulgoat, F., Sarrazin, J.-L., 2013. The vestibulocochlear nerve (VIII). Diagn. Interv. Imaging. 22115684. 94 (10), 1043–1050. doi:10.1016/j.diii.2013.08.015.

Berger, M.S., Wilson, C.B., 1985. Epidermoid cysts of the posterior fossa. J. Neurosurg. 00223085. 62 (2), 214–219. doi:10.3171/jns.1985.62.2.0214.

Berlin, C.I., Hood, L.J., Morlet, T., Wilensky, D., Li, L., Mattingly, K.R., Taylor-Jeanfreau, J., Keats, B.J.B., John, P.S., Montgomery, E., Shallop, J.K., Russell, B.A., Frisch, S.A., 2010. Multi-site diagnosis and management of 260 patients with auditory neuropathy/dys-synchrony (auditory neuropathy spectrum disorder. Int. J. Audiol. 17088186. 49 (1), 30–43. doi:10.3109/14992020903160892.

Bharadwaj, H.M., Verhulst, S., Shaheen, L., Liberman, M.C., Shinn-Cunningham, B.G., 2014. Cochlear neuropathy and the coding of supra-threshold sound. Front. Syst. Neurosci. 16625137. 8. Available from: http://www.frontiersin.org/Journal/DownloadFile.ashx?pdf=1&FileId=115620&articleId=68800&Version=1&ContentTypeId=21&FileName=fnsys-08-00026.pdf. doi:10.3389/fnsys.2014.00026.

Bielecki, I., Horbulewicz, A., Wolan, T., 2011. Risk factors associated with hearing loss in infants: an analysis of 5282 referred neonates. Int. J. Pediatr. Otorhinolaryngol. 01655876. 75 (7), 925–930. doi:10.1016/j.ijporl.2011.04.007.

Breneman, A.I., Gifford, R.H., DeJong, M.D., 2012. Cochlear implantation in children with auditory neuropathy spectrum disorder: long-term outcomes. J. Am. Acad. Audiol. 10500545. 23 (1), 5–17. Available from: http://docserver.ingentaconnect.com/deliver/connect/aaa/10500545/v23n1/s2.pdf?expires=1328075883&id=66964053&titleid=72010016&accname=Elsevier&checksum=B72832447DBA45232FBE333E503800A6. doi:10.3766/jaaa.23.1.2.

Bucchieri, F., Carletti, F., David, S., Cappello, F., Ferraro, G., Sardo, P., 2019. Anatomy and physiology of the peripheral and central auditory system. In: Dispenza, F., Martines, F. (Eds.), Sensorineural Hearing Loss: Pathophysiology, Diagnosis and Treatment. Nova Science Publishers, Inc, pp. 3–23. Available from: https://www.scopus.com/inward/record.uri?eid=2-s2.0-85077641730&partnerID=40&md5=f7023055e99dd4fb04c3fd1612fd32d4.

Carlson, M.L., Habermann, E.B., Wagie, A.E., Driscoll, C.L., Van Gompel, J.J., Jacob, J.T., Link, M.J., 2015. The changing landscape of vestibular schwannoma management in the United States—a shift toward conservatism. Otolaryngol. Head Neck Surg. 10976817. 153 (3), 440–446. http://oto.sagepub.com/content/by/year. doi:10.1177/0194599815590105.

Cebulla, M., Lurz, H., Shehata-Dieler, W., 2014. Evaluation of waveform, latency and amplitude values of chirp ABR in newborns. Int. J. Pediatr. Otorhinolaryngol. 01655876. 78 (4), 631–636. doi:10.1016/j.ijporl.2014.01.020.

Chen, Z.C., Zhang, Y.Z., Zhao, H.D., Wei, X., Murofushi, T., Hu, J., Cheng, Y., Chen, F.Y., Xu, X.D., Xu, M., Zhang, Q., 2021. Retro-labyrinthine lesion site detected by galvanic vestibular stimulation elicited vestibular-evoked myogenic potentials in patients with auditory neuropathy. Curr. Med. Sci. 2523899X. 41 (4), 695–704. Available from: https://rd.springer.com/journal/11596. doi:10.1007/s11596-021-2411-5.

Corrêa, D.G., da Cruz, L.C.H., Freddi, T.D.A.L., 2023. The vestibulocochlear nerve: anatomy and pathology. Semin Ultrasound CT MR. 15585034. 44 (2), 81–94. Available from: http://www.elsevier.com/inca/publications/store/6/2/3/1/9/2/index.htt. doi:10.1053/j.sult.2023.03.007.

Crist, J., Hodge, J.R., Frick, M., Leung, F.P., Hsu, E., Gi, M.T., Venkatesh, S.K., 2017. Magnetic resonance imaging appearance of schwannomas from head to toe: a pictorial review. J. Clin. Imaging Sci. 21565597. 7 (1). Available from: https://clinicalimagingscience.org/. doi:10.4103/jcis.jcis_40_17.

Cuda, D., Mancini, P., Santarelli, R., 2012. Retrocochlear auditory dysfunctions (RADs) and their treatment: a narrative review. Audiol. Res. 16 (1), 5. doi:10.3390/audiolres16010005.

Cueva, R.A., 2004. Auditory brainstem response versus magnetic resonance imaging for the evaluation of asymmetric sensorineural hearing loss. Laryngoscope. 15314995. 114 (10), 1686–1692. doi:10.1097/00005537-200410000-00003.

Curthoys, I.S., 2010. A critical review of the neurophysiological evidence underlying clinical vestibular testing using sound, vibration and galvanic stimuli. Clin. Neurophysiol. 13882457. 121 (2), 132–144. doi:10.1016/j.clinph.2009.09.027.

Curthoys, I.S., 2017. The new vestibular stimuli: sound and vibration—anatomical, physiological and clinical evidence. Exp. Brain Res. 14321106. 235 (4), 957–972. Available from: https://www.link.springer.de/link/service/journals/00221/index.htm. doi:10.1007/s00221-017-4874-y.

de Souza, R.V., Silva, L.A.F., Matas, C.G., 2025. Auditory pathway abnormalities in parkinson's disease. Arq. Neuropsiquiatr. 83 (1), 1–11. doi:10.1055/s-0045-1801844 .

Dieterich, M., Brandt, T., 2015. The bilateral central vestibular system: its pathways, functions, and disorders. Ann. N. Y. Acad. Sci. 0077-8923. 1343 (1), 10–26. doi:10.1111/nyas.12585.

Don, M., Kwong, B., Tanaka, C., Brackmann, D., Nelson, R., 2005. The stacked ABR: a sensitive and specific screening tool for detecting small acoustic tumors. Audiol. Neurotol. 1420-3030. 10 (5), 274–290. doi:10.1159/000086001.

Fukaya, R., Yoshida, K., Ohira, T., Kawase, T., 2011. Trigeminal schwannomas: experience with 57 cases and a review of the literature. Neurosurg. Rev. 14372320. 34 (2), 159–171. Available from: https://www.link.springer.de/link/service/journals/10143/index.htm. doi:10.1007/s10143-010-0289-y.

Ganesan, P., Sankaran, P., Kothandaraman, P.P., 2015. A rare case of hearing impairment due to cerebello-pontine angle lesion: trigeminal schwannoma. J. Int. Adv. Otol. 11 (2), 170–172. doi:10.5152/IAO.2015.252.

Giroud, N., Lemke, U., Reich, P., Matthes, K.L., Meyer, M., 2017. The impact of hearing aids and age-related hearing loss on auditory plasticity across three months—an electrical neuroimaging study. Hear. Res. 18785891. 353, 162–175. Available from: www.elsevier.com/locate/heares. doi:10.1016/j.heares.2017.06.012.

Gür, E., Binkhamis, G., Kluk, K., 2022. Effects of multiple sclerosis on the audio-vestibular system: a systematic review. BMJ Open 12 (8), e060540. doi:10.1136/bmjopen-2021-060540.

Gurgel, R.K., Jackler, R.K., Dobie, R.A., Popelka, G.R., 2012. A new standardized format for reporting hearing outcome in clinical trials. Otolaryngol. Head Neck Surg. 0194-5998. 147 (5), 803–807. doi:10.1177/0194599812458401.

Gwak, D.W., Park, E., Park, J.S., Kim, E., Kang, M.G., Kim, A.R. et al, 2021. Alterations of functional connectivity in auditory and sensorimotor neural networks: a case report in a patient with cortical deafness after bilateral putaminal hemorrhagic stroke. Medicine (Baltimore) 100 (3), e24302. doi:10.1097/MD.0000000000024302.

Halmagyi, G.M., Chen, L., MacDougall, H.G., Weber, K.P., McGarvie, L.A., Curthoys, I.S., 2017. The video head impulse test. Front. Neurol. 16642295. 8. Available from: http://journal.frontiersin.org/article/10.3389/fneur.2017.00258/full. doi:10.3389/fneur.2017.00258.

Hamid, A.M., Sheykholeslami, S., 2006. Clinical Anatomy and Physiology of the Auditory and Vestibular Systems Introduction to Auditory and Vestibular Systems, Thieme Medical Publishers.

Handelsman, J.A., 2005. Evaluating patients with dizziness and unsteadiness: a team approach. ASHA Leader 10 (2), 6–7. doi:10.1044/leader.FTR1.10022005.6.

Hellmann, M.A., Steiner, I., Mosberg-Galili, R., 2011. Sudden sensorineural hearing loss in multiple sclerosis: clinical course and possible pathogenesis. Acta Neurol. Scand. 00016314. 124 (4), 245–249. doi:10.1111/j.1600-0404.2010.01463.x.

Ito, C.J., Malone, A.K., Wong, R.H., van Loveren, H.R. Boyev, K.P., 2016. Denervation of the Eustachian tube and hearing loss following trigeminal schwannoma resection. J. Neurol. Surg. Rep. 77 (1), e56–e59. doi:10.1055/s-0035-1570389.

Jackler, R.K., Whinney, D., 2001. A century of eighth nerve surgery. Otol. Neurotol. 15317129. 22 (3), 401–416. Available from: http://journals.lww.com/otology-neurotology. doi:10.1097/00129492-200105000-00023.

Jensen, E.S., Cayé-Thomasen, P., Bodilsen, J., Nielsen, H., Friis-Hansen, L., Christensen, T. et al, 2023. Hearing loss in bacterial meningitis revisited—evolution and recovery. Open Forum Infect. Dis. 10 (3), ofad056. doi:10.1093/ofid/ofad056.

Jewett, D.L., Williston, J.S., 1971. Auditory-evoked far fields averaged from the scalp of humans. Brain 94 (4), 681–696. doi:10.1093/brain/94.4.681.

Koura, S.S., El-Badry, A., Hussein, M., Ali, A., Radwan, A., oussef, I., 2024. Evaluation of audiovestibular functions in patients with posterior circulation cerebrovascular stroke. Egypt J. Otolaryngol. 40 (1). doi:10.1186/s43163-024-00609-1.

Koyama, T., Domen, K.A., 2016. A case of hearing loss after bilateral putaminal hemorrhage: a diffusion-tensor imaging study. Prog. Rehabil. Med. 1 (1), 20160003. doi:10.2490/prm.20160003.

Kwak, C., Han, W., Bahng, J., 2022. Systematic review and meta-analysis of the application of virtual reality in hearing disorders. Am. J. Audiol. 26 (4), 169–181. doi:10.7874/jao.2022.00234.

Lachowska, M., Pastuszka, A., Sokołowski, J., Szczudlik, P., Niemczyk, K., 2021. Cortical deafness due to ischaemic strokes in both temporal lobes. J. Audiol. Otol. 25 (3), 163–170. doi:10.7874/jao.2020.00269.

Lee, J.H., Park, S.S., Ahn, J.Y., Heo, J.H., 2016. Bilateral sudden hearing difficulty caused by bilateral thalamic infarction. J. Clin. Neurol. 13 (1), 107–108. doi:10.3988/jcn.2017.13.1.107.

Lee, S.H., Kim, J.M., Schuknecht, B., Tarnutzer, A.A., 2020. Vestibular and ocular motor properties in lateral medullary stroke critically depend on the level of the medullary lesion. Front. Neurol. 11, 390. doi:10.3389/fneur.2020.00390.

Lee, S.H., Kim, S.H., Kim, J.M., Tarnutzer, A.A., 2018. Vestibular dysfunction in wernicke's encephalopathy: predominant impairment of the horizontal semicircular canals. Front. Neurol. 9, 141. doi:10.3389/fneur.2018.00141.

Lee, S., Jeon, E.S., Cho, H.H., 2019. Auditory evoked potential inconsistency in sudden unilateral hearing loss with multiple sclerosis. J. Int. Adv. Otol. 15 (1), 160–164. doi:10.5152/iao.2018.5225.

Lee, S.M., Choi, S.Y., Choi, J.H., Oh, E.H., Yoo, D., Lee, S.U. et al, 2024. Ocular motor and vestibular dysfunction in central nervous system lymphoma. Eur. J. Neurol. 31 (5), e16242. doi:10.1111/ene.16242.

Libell, J.L., Balar, A.B., Libell, D.P., Joseph, J.T., Hogg, J.P., Lakhani, D.A. et al, 2023. Facial nerve schwannoma: case report and brief review of the literature. Radiol. Case Rep. 18 (10), 3442–3447. doi:10.1016/j.radcr.2023.06.043.

Mahboob, H., Hamzah, A., Alothman, A., Ashqar, A.A., Samkari, A., Alsinani, T.A. et al, 2023. Pineal region tumors presenting with auditory symptoms: a case series and literature review. Case Rep. Oncol. 16 (1), 837–845. doi:10.1159/000533518.

Manchaiah, V.K.C., Zhao, F., Danesh, A.A., Duprey, R., 2011. The genetic basis of auditory neuropathy spectrum disorder (ANSD). Int. J. Pediatr. Otorhinolaryngol. 01655876. 75 (2), 151–158. doi:10.1016/j.ijporl.2010.11.023.

Matthies, C., Samii, M., 1997. Management of 1000 vestibular schwannomas (acoustic neuromas): clinical presentation. Neurosurgery 40 (1), 1–9. doi:10.1097/00006123-199701000-00001.

Moore, D.R., Ferguson, M.A., Edmondson-Jones, A.M., Ratib, S., Riley, A., 2014. Nature of auditory processing disorder in children. Pediatrics 126 (2), 382.

Moser, T., Starr, A., 2016. Auditory neuropathy—neural and synaptic mechanisms. Nat. Rev. Neurol. 1759-4758. 12 (3), 135–149. doi:10.1038/nrneurol.2016.10.

Musiek, F.E., 1986. Neuroanatomy, neurophysiology, and central auditory assessment part III: corpus callosum and efferent pathways. Ear Hear. 15384667. 7 (6), 349–358. doi:10.1097/00003446-198612000-00001.

Musiek, F.E., Baran, J.A., 2007. The Auditory System: Anatomy, Physiology, and Clinical Correlates, second ed. Plural Publishing, 487.

Nisson, P.L., Quintero-Consuegra, M.D., Lekovic, G.P., 2024. Arachnoid cyst of the cerebellopontine angle: a systematic literature review. Neurosurg. Rev. 182, e675–e691. doi:10.1016/j.wneu.2023.12.018.

Papadopoulou, A.M., Bakogiannis, N., Sofokleous, V., Skrapari, I., Bakoyiannis, C., 2022. The impact of vascular loops in the cerebellopontine angle on audio-vestibular symptoms: a systematic review. Audiol. Neurotol. 14219700. 27 (3), 200–207. Available from: www.karger.ch/journals/aud/aud_jh.htm. doi:10.1159/000521792.

Peterson, D.C., Reddy, V., Hamel, R.N., 2018. Neuroanatomy, Auditory Pathway, StatPearls Publishing.

Pihl, C.E., Back, C.F., Iversen, H.K., Amin, F.M., 2021. Sudden bilateral deafness in a patient with transient ischemic attack: a case report. J. Stroke Cerebrovasc. Dis. 13 (1), 119–124. doi:10.1159/000512403.

Prendergast, G., Tu, W., Guest, H., Millman, R.E., Kluk, K., Couth, S., Munro, K.J., Plack, C.J., 2018. Supra-threshold auditory brainstem response amplitudes in humans: test-retest reliability, electrode montage and noise exposure. Hear. Res. 18785891. 364, 38–47. Available from: www.elsevier.com/locate/heares. doi:10.1016/j.heares.2018.04.002.

ProQuest, 2025. Neuroimaging of the Auditory and Vestibular Systems a Clinician's Guide. Plural Publishing.

Pula, J.H., Newman-Toker, D.E., Kattah, J.C., 2013. Multiple sclerosis as a cause of the acute vestibular syndrome. J. Neurol. 260 (6), 1649–1654. doi:10.1007/s00415-013-6850-1.

Robinette, M.S., Bauch, C.D., Olsen, W.O., Cevette, M.J., 2000. Auditory brainstem response and magnetic resonance imaging for acoustic neuromas: costs by prevalence. Arch. Otolaryngol. Head Neck Surg. 08864470. 126 (8), 963–966. Available from: http://archotol.jamanetwork.com/journal.aspx. doi:10.1001/archotol.126.8.963.

Rosengren, S.M., Welgampola, M.S., Colebatch, J.G., 2010. Vestibular evoked myogenic potentials: past, present and future. Clin. Neurophysiol. 13882457. 121 (5), 636–651. doi:10.1016/j.clinph.2009.10.016.

Roush, P., Frymark, T., Venediktov, R., Wang, B., 2011. Audiologic management of auditory neuropathy spectrum disorder in children: a systematic review of the literature. Am. J. Audiol. 1059-0889. 20 (2), 159–170. doi:10.1044/1059-0889(2011/10-0032).

Roy, C.F., Yang, L., Daniel, S.J., 2019. A pediatric medulloblastoma presenting as isolated sensorineural hearing loss: case report and review of the literature. Int. J. Pediatr. Otorhinolaryngol. 126, 109640. doi:10.1016/j.ijporl.2019.109640.

Ruiz-García, C., Lassaletta, L., López-Larrubia, P., Varela-Nieto, I., Murillo-Cuesta, S., 2024. Tumors of the nervous system and hearing loss: beyond vestibular schwannomas. Hear Res. 447, 109012. doi:10.1016/j.heares.2024.109012.

Schmidt, R.J., Sataloff, R.T., Newman, J., Spiegel, J.R., Myers, D.L., 2001. The sensitivity of auditory brainstem response testing for the diagnosis of acoustic neuromas. Arch. Otolaryngol. Head Neck Surg. 08864470. 127 (1), 19–22. Available from: http://archotol.jamanetwork.com/journal.aspx. doi:10.1001/archotol.127.1.19.

Sedwick, J.D., Gajewski, B.J., Antonelli, P.J., 2001. Magnetic resonance imaging in the search for retrocochlear pathology. Otolaryngol. Head Neck Surg. 124 (6), 622–627. doi:10.1177/019459980112400610.

Selters, W.A., Brackmann, D.E., 1977. Acoustic tumor detection with brain stem electric response audiometry. Arch. Otolaryngol. 1538361X. 103 (4), 181–187. doi:10.1001/archotol.1977.00780210037001.

Semenza, C., Cavinato, M., Rigon, J., Battel, I., Meneghello, F., Venneri, A., 2012. Persistent cortical deafness: a voxel-based morphometry and tractography study. Neuropsychologia 26 (6), 675–683. doi:10.1037/a0029688.

Shearer, A.E., Hildebrand, M.S., Smith, R.J., 2017. Hereditary Hearing Loss and Deafness Overview . GeneReviews®, University of Washington.

Shivashankar, N., Shashikala, H.R., Nagaraja, D., Jayakumar, P.N., 2001. Pure word deafness in two patients with subcortical lesions. Clin. Neurol Neurosurg. 103 (4), 201–205. doi:10.1016/S0303-8467(01)00136-6.

Stadio, A.D., Dipietro, L., Ralli, M., Meneghello, F., Minni, A., Greco, A., 2018. Sudden sensorineural hearing loss in patients with multiple sclerosis: a systematic review. Eur. Rev. Med. Pharmacol. Sci. 22(14), 4611–4624. doi:10.26355/eurrev_201807_15520.

Starr, A., Sininger, Y.S., Pratt, H., 2000. The varieties of auditory neuropathy. J. Basic Clin. Physiol. Pharmacol. 2191-0286. 11 (3), 215–230. doi:10.1515/jbcpp.2000.11.3.215.

Stöver, T., Lenarz, T., 2011. Biomaterials in cochlear implants. GMS Curr. Top. Otorhinolaryngol. Head Neck. Surg. 8, 10. doi:10.3205/cto000062.

Terakawa, Y., Tsuyuguchi, N., Takami, T., Ohata, K., 2011. Medulloblastoma manifesting as sudden sensorineural hearing loss. J. Korean Neurosurg. Soc. 50 (1), 51–53. doi:10.3340/jkns.2011.50.1.51.

Walker, M.A., Zepeda, R., Afari, H.A., Cohen, A.B., 2014. Hearing loss in wernicke encephalopathy. Neurol. Clin. Pract. 4 (6), 511–515. doi:10.1212/CPJ.0000000000000072.

Yeshurun, S., Carrasco, M., Maloney, L.T., 2017. Bias and sensitivity in two-interval forced choice procedures: tests of the difference model. Vis. Res. 143, 29–40.

Yoo, H., Mihaila, D.M., 2022. Neuroanatomy, Vestibular Pathways, StatPearls Publishing.

Young, A., Cornejo, J., 2023. Spinner, Auditory Brainstem Response. StatPearls Publishing, Treasure Island, FL.

Yu, Z., Lu, T., Yu, R., Liang, R., 2023. Cavernous hemangioma of the cisternal segment of the auditory nerve: case report. BMC Neurol. 23 (1), 217. doi:10.1186/s12883-023-03275-7.

Zhang, W., Cui, L., Wang, W., Jiao, Y., Zhang, Y., Jiao, J., 2018. Early identification of anti-NMDA receptor encephalitis presenting cerebral lesions in unconventional locations on magnetic resonance imaging. J. Neuroimmunol 320, 101–106. doi:10.1016/j.jneuroim.2018.04.012.

Zuccoli, G., Gallucci, M., Capellades, J., Regnicolo, L., Tumiati, B., Pasquini, L., Scotti, 2009. Wernicke encephalopathy: MR findings at clinical presentation in twenty-six patients. Am. J. Neuroradiol. 30 (9), 1792–1794.

Chapter 29

The role of imaging in oncology

Winniecia Dkhar[a], Rajagopal Kadavigere[b], S. Sharath[a], Abhimanyu Pradhan[a], Leena R. David[a,c], Berna Uzun[d], Dilber Uzun Ozsahin[c,d,e], Ilker Ozsahin[d], and Neil Abraham Barnes[a]

[a]*Department of Medical Imaging Technology, Manipal College of Health Professions, Manipal Academy of Higher Education, Manipal, Karnataka, India*

[b]*Department of Radiodiagnosis and Imaging, Kasturba Medical College, Manipal, India*

[c]*Medical Diagnostic Imaging Department, College of Health Sciences, University of Sharjah, Sharjah, United Arab Emirates*

[d]*Irfan Suat Gunsel Operational Research Institute, Near East University, Nicosia, Turkish Republic of Northern Cyprus (TRNC), Turkey*

[e]*Research Institute for Medical and Health Sciences, University of Sharjah, Sharjah, United Arab Emirates*

1 Cancer: A global health challenge

Cancer has emerged as one of the most formidable public health challenges of the 21st century, accounting for millions of deaths each year and placing an immense burden on healthcare systems globally. Characterized by the uncontrolled proliferation of cells that acquire the ability to invade surrounding tissues and metastasize to distant organs, cancer is a multifactorial disease influenced by genetic, environmental, and lifestyle factors (de Martel et al., 2020). Its complexity, high treatment costs, and devastating impact on patients and their families underscore the urgent need for comprehensive, globally coordinated responses. According to the World Health Organization (WHO), cancer accounted for nearly 10 million deaths in 2020, making it the second-leading cause of death globally after cardiovascular diseases. As such, cancer not only affects individuals on a biological and emotional level but also disrupts communities, economies, and development agendas worldwide (Anonymous, n.d.a; Wu et al., 2024b).

The biological basis of cancer lies in the disruption of the normal regulatory mechanisms that control cell division and apoptosis. Mutations in oncogenes, tumor suppressor genes, and DNA repair genes lead to cellular transformations, allowing cells to escape normal growth controls (Nenclares and Harrington, n.d.). These aberrant cells accumulate genetic and epigenetic alterations over time, eventually forming malignant tumors. If not detected early or left untreated, these tumors can metastasize through blood or lymphatic systems to vital organs such as the liver, lungs, and brain, making the disease difficult to manage and often fatal (Zhang et al., 2024).

Globally, cancer incidence is on an upward trajectory, and this trend is expected to continue due to a range of converging factors. One of the most significant contributors is the aging population. With advances in medical care, sanitation, and public health infrastructure, people are living longer than ever before. However, aging is associated with increased cellular senescence, accumulated genetic mutations, and declining immune function, all of which heighten cancer susceptibility. It is estimated that more than 60% of new cancer cases and 70% of cancer deaths occur in individuals aged 60 and older. This demographic shift is particularly pronounced in high-income and rapidly developing middle-income countries, where life expectancy has risen sharply over the past few decades (Prathap et al., 2024; Pilleron et al., 2019).

Urbanization and globalization are additional driving forces behind the cancer epidemic. These phenomena bring about rapid changes in lifestyle and environmental exposure. Urban residents are more likely to adopt sedentary behaviors, consume high-calorie, processed foods, and experience chronic stress factors that collectively increase the risk of several types of cancer, including colorectal, breast, and prostate cancers (Hurbain et al., 2024; Ribeiro and Fecht, 2021). Furthermore, the global spread of tobacco use and alcohol consumption has had profound implications for cancer epidemiology. Tobacco alone is responsible for approximately 22% of global cancer deaths, making it the single largest preventable cause of cancer worldwide. Alcohol consumption, especially in excess, has also been linked to cancers of the liver, esophagus, and breast, among others (Jun et al., 2024).

Environmental and occupational exposures represent another significant cluster of risk factors. Industrialization and modernization have led to increased human exposure to various known and suspected carcinogens. Air pollution, for instance, contains particulate matter and volatile organic compounds strongly associated with lung cancer. The International Agency for Research on Cancer has classified outdoor air pollution and particulate matter as Group 1 carcinogens. In agricultural and industrial settings, workers are often exposed to hazardous chemicals such as benzene, asbestos, and pesticides, which pose long-term cancer risks (Ajayi and Ogunjobi, 2024; Dike and McGowan, 2020).

Advances in Medical Imaging. https://doi.org/10.1016/B978-0-443-28967-5.00024-1

Advancements in diagnostic tools and medical imaging technologies have led to earlier and more accurate detection of cancer, contributing, in part, to the apparent increase in incidence rates. Techniques such as magnetic resonance imaging (MRI), computed tomography (CT) scans, mammography, and liquid biopsies have significantly improved our ability to identify malignancies at their nascent stages. Moreover, the implementation of widespread screening programs, particularly in high-income countries, has facilitated the early diagnosis of common cancers such as cervical, colorectal, and breast cancer (Shin et al., 2015; Ueno et al., 2018). While this has improved patient outcomes and survival rates, it also results in a paradoxical increase in reported incidence, as cancers that might have gone undiagnosed in previous decades are now being detected (Firjani Naef, 2014).

Despite these technological advances, vast disparities in cancer care persist across different regions and populations. In many low- and middle-income countries (LMICs), healthcare infrastructure remains underdeveloped, and access to early diagnosis, specialized treatment, and palliative care is severely limited (Bamodu and Chung, 2024). Patients in these regions often present with late-stage disease, when therapeutic options are limited and prognosis is poor. These disparities are fueled by socioeconomic inequities, lack of awareness, cultural stigmas, and financial barriers to care. The shortage of oncologists, diagnostic equipment, and essential medications further compounds the problem. As a result, cancer survival rates in LMICs lag far behind those in high-income nations, underscoring the need for targeted interventions to close these gaps (Krishnan et al., 2023; Chan et al., 2024; Anonymous, n.d.b).

In addition to its medical and epidemiological dimensions, cancer exerts profound social and economic impacts. The financial burden associated with cancer treatment, including surgery, chemotherapy, radiation therapy, and long-term care, is substantial. For many patients and families, out-of-pocket expenditures can lead to catastrophic health spending, driving households into poverty. This is especially true in countries without robust health insurance systems or government subsidies (Alzehr et al., 2022; Schlander et al., 2024). The economic cost of cancer also manifests in lost productivity, as patients are often unable to work during treatment, and caregivers must forgo employment opportunities to provide support. On a national scale, the cumulative economic losses due to cancer-related morbidity and mortality are staggering, affecting national income and development trajectories (Goyanka, 2021).

Addressing the global cancer crisis requires a multifaceted and sustained approach centered around the four pillars of cancer control: prevention, early detection, treatment, and palliation. Prevention strategies are perhaps the most cost-effective and far-reaching (Mussallem, 2022). Public health campaigns that discourage tobacco use, promote healthy eating, encourage physical activity, and limit alcohol consumption have demonstrated significant potential in reducing cancer incidence (Bandini et al., 2023; Anonymous, n.d.c). Vaccination programs against oncogenic viruses, such as the human papillomavirus and hepatitis B virus, are vital in preventing cervical and liver cancers, respectively (Schiller and Lowy, 2010). Regulatory measures to limit exposure to environmental carcinogens, including clean air initiatives, safe occupational practices, and food safety standards, also play a critical role in cancer prevention (Enokida et al., 2021).

Early detection is equally crucial in improving cancer outcomes. Screening programs must be expanded, especially in underserved regions, and adapted to the specific cancer burden of the population (Schlander et al., 2024). Community-based outreach, mobile diagnostic units, and telemedicine can bridge the gap in rural and remote areas. Education and awareness campaigns are essential in empowering individuals to recognize early signs and symptoms of cancer, thereby prompting timely medical consultation (Alzehr et al., 2022; Schlander et al., 2024).

Equally important is the need to strengthen cancer treatment infrastructure. This includes investing in specialized facilities, training healthcare professionals in oncology, and ensuring the availability of essential drugs and technologies. Integrative care models that combine conventional treatment with psychological and social support services can significantly enhance patient outcomes and quality of life. International partnerships and funding mechanisms, such as the Global Fund and Global Alliance for Vaccines and Immunization (GAVI), could be leveraged to support cancer control initiatives in resource-limited settings (Akingbola et al., 2024).

Finally, ongoing investment in cancer research is imperative. Scientific innovation holds the key to developing more effective, less toxic therapies, as well as predictive biomarkers that can tailor treatment to the individual patient. The promise of precision oncology—using genetic, proteomic, and epigenomic information to guide treatment decisions—represents a major advancement in the field. Research into cancer immunotherapy, including checkpoint inhibitors and chimeric antigen receptor T-cell (CAR-T) cell therapy, has already revolutionized treatment for certain malignancies and holds promise for others (Chehelgerdi et al., 2023).

2 Epidemiology and risk factors of cancer

The global cancer burden is expected to rise significantly in the coming decades, with the WHO projecting that cancer incidence will reach 28.4 million cases annually by 2040, marking a nearly 50% increase from 2020 (Sung et al., 2021). Among the most prevalent types of cancer, lung cancer remains the leading cause of cancer-related deaths worldwide, primarily attributed to tobacco smoking, prolonged exposure to air pollution, and occupational hazards such as asbestos and radon gas (Ahmad et al., 2024). Breast cancer, the most commonly diagnosed cancer in women, is influenced by genetic predisposition, hormonal changes, and lifestyle factors such as obesity, alcohol consumption, and physical inactivity. Similarly, colorectal cancer, which is strongly associated with dietary habits, obesity, and a sedentary lifestyle, continues to be a major public health concern, particularly in high-income countries where processed food consumption is prevalent (Mao et al., 2019). Prostate cancer, another frequently diagnosed malignancy in men, is often detected through prostate-specific antigen testing, with risk factors including age, family history, and hormonal imbalances (Rosino Sánchez et al., 2025). In addition, liver and stomach cancers show higher prevalence in regions with elevated hepatitis B and C infection rates, as well as dietary risks such as aflatoxin exposure and excessive salt consumption. As cancer cases continue to rise, understanding these risk factors is crucial for developing targeted prevention strategies, promoting early detection, and implementing effective public health interventions to reduce the global cancer burden (Chen et al., 2024).

3 Diagnostic modalities in cancer detection and management

The detection and management of cancer rely on a diverse range of imaging modalities, each offering distinct advantages in identifying, characterizing, and assessing tumors based on their anatomical, functional, and metabolic properties. These imaging techniques play a pivotal role in early diagnosis, staging, treatment planning, and monitoring therapeutic response, thereby ensuring a comprehensive approach to cancer care (Nussbaum et al., 2022; Prathap et al., 2024).

3.1 X-ray imaging—A primary screening tool for structural abnormalities

X-ray imaging is a primary screening tool for detecting lung tumors and bone metastases. While widely accessible and cost-effective, its lower sensitivity compared to advanced imaging techniques limits its role in early cancer detection (Heindel et al., 2014).

X-ray imaging has long been considered a fundamental tool in medical diagnostics and remains one of the most widely used methods for the initial screening of a variety of conditions, including cancer. In the context of cancer detection, X-rays are particularly valuable for identifying lung tumors and bone metastases, where their role in detecting structural abnormalities can provide early insights into cancer development or spread. Despite its widespread availability, affordability, and ability to offer rapid results, X-ray imaging has certain limitations, especially in the early detection of cancers (Kim and Kim, 2020).

X-rays work by passing a controlled amount of radiation through the body, with the varying densities of tissues and organs absorbing and reflecting the radiation differently. This difference in absorption creates an image on a photographic plate or digital detector. In oncology, X-ray imaging is primarily used for identifying gross abnormalities in the structure of tissues, such as changes in bone density, the appearance of masses or lesions, and signs of tumor spread, especially to the bones (Kim and Kim, 2020; Çallı et al., 2021).

In the case of lung cancer, X-ray imaging, often in the form of a chest X-ray, can detect signs of pulmonary nodules, tumors, or fluid accumulation within the lungs, which may suggest the presence of cancer. Chest X-rays are commonly used as a first-line screening tool for patients at risk of lung cancer, particularly those with a history of smoking or exposure to lung carcinogens. The simplicity and speed of chest X-rays make them a valuable tool for this initial screening, as they can quickly identify potential lung tumors or other abnormalities (Kim and Kim, 2020).

In addition to lung cancer, X-ray imaging is critical in detecting bone metastases, a common site of cancer spread. When cancer cells from primary tumors (such as breast, prostate, or lung cancer) spread to bones, X-ray imaging can reveal osteolytic lesions (areas of bone destruction) or osteoblastic lesions (areas of abnormal bone growth). These abnormalities are significant because they can help in identifying secondary cancer involvement in the bones, which may influence treatment decisions (Heindel et al., 2014).

3.2 Computed tomography—High-resolution imaging for staging and treatment planning

CT scans provide detailed cross-sectional images, making them indispensable for staging malignancies and evaluating tumor size, location, and metastasis. Widely used in the diagnosis of lung, abdominal, and pelvic cancers, CT imaging offers comprehensive insights into lymph node involvement and disease progression, aiding in precise treatment planning (Unadkat et al., 2022).

CT has become a cornerstone of modern cancer care, providing high-resolution cross-sectional images that allow clinicians to visualize and assess tumors in great detail. This imaging modality uses X-rays and advanced computer processing to generate a series of thin, detailed slices of the body, which can then be reconstructed into 3D images (Singh et al., 2024).

CT scans are indispensable in the staging of malignancies, helping clinicians evaluate not only the size and location of a tumor but also its potential spread to nearby tissues and organs, including the involvement of lymph nodes and distant metastases. The ability to obtain these detailed images quickly makes CT a crucial tool for both diagnosis and treatment planning in oncology (Hussain et al., 2022).

CT imaging plays an essential role in the diagnosis and staging of several types of cancer, including lung, abdominal, and pelvic cancers. In lung cancer, CT scans are particularly useful for detecting small pulmonary nodules that may be missed on conventional X-rays (Singh et al., 2024). They also help assess the extent of the tumor and its relationship with surrounding structures, such as major blood vessels and the chest wall. CT is also pivotal in identifying lymph node involvement, which is critical for determining the stage of the disease. Enlarged or abnormally shaped lymph nodes can be a sign that the cancer has spread beyond the primary site, which, in turn, impacts prognosis and treatment decisions (Pesapane et al., 2020).

In abdominal cancers, such as those affecting the liver, pancreas, or kidneys, CT scans provide detailed images that allow for precise tumor localization. The images help determine whether the tumor is confined to its original site or if it has invaded adjacent organs. For instance, in pancreatic cancer, CT scans can provide essential information about the tumor's size, the extent of vascular invasion, and whether there is liver metastasis, all of which are crucial for treatment planning (Rubin, 2015; Henzler et al., 2012).

In pelvic cancers, particularly in colorectal cancer or ovarian cancer, CT scans are used to evaluate the extent of tumor spread to nearby structures such as the rectum, bladder, and uterus, as well as to assess regional lymph node involvement (Kyriazi et al., 2010). CT is instrumental in determining whether the cancer has infiltrated the surrounding tissues or spread to distant organs like the liver or lungs, which significantly influences the choice of treatment and the overall management plan. In these cases, CT can also detect metastatic lesions, providing critical information about the disease's progression (Rubin, 2015; Alcázar et al., 2019).

One of the most significant advantages of CT is its ability to offer detailed images of not only the primary tumor but also the regional lymph nodes and distant metastases (Unadkat et al., 2022). This comprehensive view is crucial for staging the cancer accurately, which, in turn, helps clinicians devise personalized treatment plans. For example, a Stage III cancer diagnosis, indicating regional lymph node involvement, may prompt a more aggressive treatment regimen, including chemotherapy and radiation therapy, in addition to surgery (Santana-Davila and Martins, 2016; Hanson et al., 2021). Conversely, a Stage IV diagnosis, indicating distant metastases, may result in a focus on palliative care or systemic therapies aimed at controlling the disease (Dillon et al., 2021).

Moreover, CT is invaluable for monitoring treatment response. After chemotherapy, radiation, or surgery, repeated CT scans can help determine whether the tumor has shrunk, stabilized, or progressed (Henzler et al., 2012; Albano et al., 2021; Itani et al., 2021). This is particularly important in evaluating the effectiveness of treatment and in making necessary adjustments, such as switching to alternative therapies or altering the radiation dose. CT scans can also identify complications of treatment, such as postsurgical infections or radiation-induced changes in the surrounding tissues (Firjani Naef, 2014).

The use of contrast agents in CT imaging further enhances its ability to differentiate between healthy tissue and malignant areas. Intravenous contrast agents highlight blood vessels and enhance the visibility of tumors, making it easier to detect even small or poorly defined lesions (Yi et al., 2009). This contrast also aids in distinguishing between different tissue types, providing more accurate information about the tumor's vascularity, necrosis, or areas of inflammation. In cases of angiogenesis, which is commonly seen in tumors, CT can reveal these new vessels, which are often a sign of tumor growth and aggressiveness (Hussain et al., 2022; Williams and Probst, 2016; Gupta et al., 2023).

3.3 Magnetic resonance imaging—Superior soft tissue contrast for complex tumors

MRI is the imaging modality of choice for detecting and characterizing tumors in the brain, spinal cord, breast, and prostate, owing to its superior soft tissue contrast. Its high-resolution differentiation between normal and pathological tissues facilitates the identification of small or complex tumors that may not be discernible on CT scans, making it particularly advantageous in neuro-oncology and musculoskeletal cancer diagnostics (Mittal and Tayal, 2021).

MRI is a powerful imaging modality that has revolutionized cancer diagnosis and management, particularly in the detection and characterization of tumors in soft tissues. MRI excels in providing superior soft tissue contrast, making it the imaging technique of choice for a wide range of cancers, especially in the brain, spinal cord, breast, and prostate. Its ability to differentiate between normal and pathological tissues at high resolution allows for the detection of small or complex tumors that may be difficult to visualize with other imaging methods, such as CT scans. This unique capability has made MRI indispensable in neuro-oncology, musculoskeletal cancer diagnostics, and various other oncological applications (Lo Gullo et al., 2024; Hegde et al., 2013; Hu et al., 2023).

One of the most significant advantages of MRI is its ability to differentiate between various types of soft tissues, thanks to its use of magnetic fields and radiofrequency waves to generate detailed images. This ability is crucial in detecting and characterizing tumors in areas where soft tissue structures are intricate and closely interwoven, such as the brain and spinal cord. For brain tumors, MRI provides detailed images of the tumor's size, location, and relationship with critical structures like the cerebral cortex, blood vessels, and cranial nerves. This level of detail is invaluable for planning surgery, determining whether the tumor is operable, and evaluating the extent of tumor infiltration into surrounding tissues. MRI also aids in identifying brain metastases from primary cancers in other organs, such as the lungs, breast, or kidneys, which are common sites of spread to the brain (Hu et al., 2023; Elizabeth et al., 2014; Horvat et al., 2018).

In spinal cord tumors, MRI is indispensable due to its superior ability to differentiate between the soft tissues of the spinal cord and the surrounding bone structures. Spinal cord gliomas or meningiomas, as well as metastatic tumors, can be identified, and MRI helps assess the tumor's relationship to critical spinal structures, including nerve roots and the spinal canal. MRI's ability to evaluate tumor involvement in these regions provides essential information for surgical planning and treatment decisions, including the potential need for radiotherapy or chemotherapy (Neilio and Ginat, 2024; Takamiya et al., 2024).

In breast cancer diagnosis, MRI has become an essential tool, particularly in women with dense breast tissue, where conventional mammography or ultrasound may not provide sufficient detail. MRI is highly sensitive in detecting small breast cancers, particularly those located in dense areas that might otherwise be missed. Dynamic contrast-enhanced MRI is particularly useful for evaluating the vascularity of breast tumors, helping differentiate benign lesions from malignant ones based on their blood supply. MRI is also valuable in breast cancer staging, providing detailed images that allow clinicians to assess the extent of tumor spread to lymph nodes, muscle tissues, and adjacent organs (Lo Gullo et al., 2024; Partridge et al., 2011; Zhang et al., 2020).

For prostate cancer, MRI plays a critical role in both diagnosis and treatment planning. The ability to differentiate between healthy and cancerous tissues in the prostate is one of the reasons MRI is so valuable in this context. Multiparametric MRI, which combines anatomic MRI with functional imaging techniques such as diffusion-weighted imaging and dynamic contrast-enhanced imaging, is particularly effective in detecting prostate tumors, guiding biopsy procedures, and assessing the cancer's local invasion. MRI is also used to evaluate lymph node involvement and extracapsular extension, which are crucial factors in determining the appropriate course of treatment, whether surgical or nonsurgical (Hegde et al., 2013; Maurer and Heverhagen, 2017; Matsukawa et al., 2024; Patel and Oto, 2016).

Beyond these specific cancers, MRI is an essential tool for detecting musculoskeletal tumors, including those found in the bone, muscles, and soft tissues. For bone tumors, MRI provides superior soft tissue contrast, which is essential in identifying the tumor's precise location and its interaction with surrounding soft tissues. This information helps in determining whether the tumor is confined to the bone or has spread to nearby muscle or fat tissue. In soft tissue sarcomas, MRI is invaluable for evaluating the extent of the tumor and for assessing its relationship to surrounding structures, which influences surgical resection and other treatment decisions (Samartzis et al., 2015; Singh et al., 2018; Takamiya et al., 2024).

3.4 Positron emission tomography—Functional imaging for tumor metabolism and staging

Positron emission tomography (PET) is a highly effective imaging technique that provides detailed insights into the metabolic activity and cellular function of tumors, distinguishing it from conventional imaging methods that primarily focus on anatomical structures. PET imaging is particularly invaluable in cancer staging, evaluating treatment response, and detecting recurrence, as it helps identify abnormal metabolic processes that often precede visible anatomical changes. This ability to assess functional activity makes PET an indispensable tool for understanding how tumors are behaving at a cellular level, offering a more nuanced view of cancer progression and its response to therapy (Treglia and Muoio, 2020; Ateş and Akdemir, 2024).

One of the key advantages of PET imaging lies in its ability to detect hypermetabolic areas, where cancer cells tend to exhibit increased glucose metabolism, a hallmark of malignant growth. These areas are captured using radiotracers such as fluorodeoxyglucose, which mimics glucose and is taken up by cells with high metabolic activity, such as those found in tumors. This allows PET scans to identify cancerous regions even before significant structural changes are detectable by other imaging modalities like CT or MRI. As a result, PET is particularly useful in early cancer detection, offering the potential for earlier intervention and more effective treatment outcomes (Ebner et al., 2024; Byrd et al., 2017; Decazes et al., 2020).

In lung cancer, PET is widely utilized to evaluate the extent of disease spread, particularly in identifying nodal involvement or distant metastases, which are crucial for determining the appropriate treatment plan. PET can also differentiate between benign and malignant lung nodules, improving diagnostic accuracy and reducing unnecessary biopsies. Similarly, in the case of lymphoma, PET imaging helps assess the metabolic activity of lymph nodes, enabling accurate staging and monitoring of therapeutic efficacy. PET scans can identify regions of active disease that may be missed with conventional anatomical imaging, providing valuable insights into tumor response to chemotherapy or radiotherapy (Gündooan et al., n.d.; Bashir et al., 2015).

In metastatic disease, PET is essential in detecting secondary tumors that may not be easily visible through conventional imaging. By identifying areas with high metabolic activity, PET can reveal the presence of metastases in organs such as the liver, bones, or brain, allowing for a more comprehensive assessment of disease spread and guiding treatment decisions. PET is also effective in evaluating treatment response, as it can assess changes in tumor metabolism during and after therapy. A decrease in metabolic activity often indicates a positive therapeutic response, while persistent or increasing activity may suggest resistance to treatment or the presence of residual disease (Treglia and Muoio, 2020; Abgral and Couturier, 2024).

In addition, PET imaging plays a crucial role in the detection of recurrence. In cancers that have been treated with surgery, chemotherapy, or radiation, PET scans can help identify any returning tumor cells, even before they are large enough to be detected on anatomical imaging. This early detection of recurrence is critical in initiating timely interventions, such as salvage therapy, and improving overall patient prognosis (Ebner et al., 2024).

PET imaging provides crucial insights into tumor metabolism and cellular activity, making it highly effective for cancer staging, evaluating treatment response, and detecting recurrence. It is particularly useful in lung cancer, lymphoma, and metastatic disease, as it enables the identification of hypermetabolic cancerous regions before significant anatomical changes occur (Wahl and Hicks, 2021).

3.5 Single photon emission computed tomography—Accessible functional imaging for bone and neuroendocrine tumors

Single photon emission CT (SPECT) is a widely accessible functional imaging modality that provides detailed information about the metabolic activity and functional processes of tissues, similar to PET but with broader availability and at a more cost-effective price. SPECT uses gamma rays emitted from radiotracers, such as technetium-99m or iodine-131, which are injected into the patient. These tracers emit single photons, which are detected by the scanner to create detailed 3D images of metabolic activity within the body. Because of its relative accessibility and affordability, SPECT is often used in clinical settings where PET might not be feasible due to cost or availability constraints, especially in areas with limited resources (Xiao and Xin, 2022; Stanton et al., 2016; Farkas et al., 2024).

SPECT is especially valuable in the diagnosis and management of bone metastases, a common complication of many cancers, including breast cancer, prostate cancer, and lung cancer. Bone scans using SPECT provide functional images that reveal areas of abnormal bone metabolism, helping to detect the spread of cancer to the bones even when structural changes may not yet be visible on conventional X-ray or CT imaging. This early detection is crucial for guiding treatment strategies, such as determining the need for radiotherapy, chemotherapy, or bisphosphonate therapy to manage bone metastases. SPECT imaging can also evaluate the extent of bone involvement and monitor the response to treatments aimed at controlling metastatic bone disease (Yang et al., 2018; Gherghe et al., 2022; Frantellizzi et al., 2023).

Beyond bone metastases, SPECT is widely utilized in the assessment of neuroendocrine tumors, including thyroid and adrenal tumors, where it provides critical insights into the functional status of these glands. In thyroid cancer, SPECT can be used to evaluate thyroid function and detect abnormal growths or metastases, particularly when combined with other imaging modalities like iodine-131 scanning. Similarly, for adrenal tumors, SPECT can assess the metabolic activity of the adrenal glands, identifying tumors that may be producing excess hormones, such as catecholamines or steroids, which could lead to conditions like pheochromocytoma or Cushing's syndrome. By accurately identifying functional abnormalities, SPECT contributes significantly to the characterization of these tumors, aiding in the decision-making process for surgical resection or medical management (Wang et al., 2024; Chernov et al., 2016).

Moreover, SPECT plays an essential role in evaluating functional abnormalities and disease progression in various cancers. Its ability to assess metabolic changes and radiotracer uptake provides valuable information that anatomical imaging techniques, like CT or MRI, may not offer. For instance, SPECT is particularly useful in detecting early-stage disease or identifying tumors that are not yet large enough to be seen on other imaging modalities, offering a functional perspective that complements the structural insights provided by conventional imaging (Zhang et al., 2023; Arvola et al., 2022).

Similar to PET imaging but more widely available, SPECT is commonly employed in bone scans and neuroendocrine tumor assessments. It plays a significant role in detecting bone metastases and evaluating functional abnormalities in thyroid and adrenal tumors, thereby contributing to accurate disease characterization and management (Fuccio et al., 2013).

3.6 Ultrasound imaging—Real-time, noninvasive assessment of superficial and abdominal cancers

Ultrasound is a noninvasive, radiation-free imaging modality that is extensively used in diagnosing breast, liver, and gynecological cancers. It is particularly effective in differentiating between benign and malignant masses and, when combined with Doppler ultrasound, allows for the evaluation of tumor vascularity, enhancing cancer characterization and diagnosis (Delorme, 2012).

Each of these imaging techniques contributes uniquely to the comprehensive detection, staging, and management of cancer, facilitating early intervention and optimizing patient outcomes.

3.7 Hybrid imaging techniques

Advanced hybrid imaging technologies integrate anatomical and metabolic data, enhancing diagnostic accuracy. PET/CT and PET/MRI combine the strengths of PET's metabolic insights with CT or MRI's structural details, providing more precise tumor characterization. SPECT/CT enhances functional imaging by integrating SPECT's nuclear imaging capabilities with CT's anatomical resolution, improving diagnostic confidence in oncology cases (Beuthien-Baumann et al., 2021; Queiroz et al., 2018). In addition, contrast-enhanced imaging, widely applied in MRI and mammography, enhances tumor visibility by increasing contrast between normal and malignant tissues, aiding in more accurate lesion detection and characterization. These diagnostic modalities, when used in combination, play a crucial role in the comprehensive evaluation and personalized management of cancer patients (Gelardi et al., 2022; Khalifa and Albadawy, 2024).

4 Advances in cancer imaging technologies

4.1 Artificial intelligence in cancer imaging

Artificial intelligence (AI)-driven image analysis has significantly transformed cancer diagnostics by enhancing accuracy and efficiency. Through advanced machine learning algorithms, AI systems can process and analyze vast amounts of imaging data, detecting subtle abnormalities that may be challenging for the human eye to identify. These systems can uncover complex patterns within medical scans, which help in the early detection of cancer and improve prognostic assessments, ultimately leading to more timely and effective interventions. AI-based automated tumor detection systems have shown immense promise in identifying lung nodules, breast masses, and brain tumors with greater precision, helping reduce false positives and false negatives, thus improving diagnostic reliability and workflow efficiency. This is particularly important in reducing unnecessary procedures or treatments that may arise from incorrect diagnoses (Alsharif, 2024; Saeidnia et al., 2025; Bi et al., 2019).

An innovative AI-based approach known as radiomics has further advanced cancer imaging. Radiomics extracts a wealth of quantitative imaging features from medical scans, such as texture, shape, and intensity patterns, which provide in-depth information about the tumor's characteristics. By evaluating these features, AI can assist in better tumor characterization, determining factors like tumor aggressiveness, heterogeneity, and even potential genetic traits. This allows for a more precise prediction of disease behavior, helping clinicians make more informed decisions about treatment strategies (Bi et al., 2019; Díaz et al., 2024).

In addition, AI-powered predictive analytics models are being integrated into cancer care to continuously monitor disease progression. These models use data from imaging, clinical records, and treatment responses to track changes over time. They enable healthcare providers to assess how well a patient is responding to therapy and make real-time adjustments to the treatment plan based on this information. This approach supports personalized medicine, ensuring that interventions are tailored to each patient's unique tumor characteristics and response to treatment, thereby improving the likelihood of a successful outcome (Koh et al., 2022; Alowais et al., 2023; Khalifa and Albadawy, 2024).

AI-driven image analysis enhances diagnostic precision by detecting subtle abnormalities that the human eye may miss. Machine learning algorithms process vast datasets to identify complex patterns, aiding in early cancer detection and prognosis (Saeidnia et al., 2025). Automated tumor detection helps identify lung nodules, breast masses, and brain tumors with greater efficiency, reducing false positives and improving workflow. Radiomics, an AI-driven approach, extracts quantitative imaging features from medical scans, allowing for better tumor characterization and predicting disease behavior. In addition, predictive analytics models use AI to track cancer progression, enabling personalized treatment planning by assessing response to therapy and adjusting interventions accordingly (Chassagnon et al., 2023).

4.2 Molecular and functional imaging

Advancements in molecular and functional imaging are revolutionizing cancer diagnostics by enabling the visualization of biological processes at the cellular and molecular levels. These techniques provide clinicians with powerful tools to understand tumor behavior beyond structural changes, facilitating early and precise diagnosis, personalized treatment planning, and real-time therapy monitoring. One of the most transformative developments in this domain is theranostics, a combined approach that integrates diagnostic imaging with targeted therapy. This technique allows for the simultaneous identification of cancerous lesions and the delivery of specific treatments, thereby enhancing therapeutic precision while minimizing harm to surrounding healthy tissues (Velpula & Buddolla, 2025; Wu et al., 2024a).

In addition, *liquid biopsy* has emerged as a noninvasive method that detects circulating tumor DNA (ctDNA) and other cancer-related biomarkers in the bloodstream. This technique offers a promising alternative to traditional tissue biopsies, providing early insights into tumor recurrence, metastasis, and genetic mutations without subjecting patients to repeated invasive procedures. Liquid biopsies are particularly valuable in tracking disease progression, assessing treatment response, and guiding dynamic treatment adjustments based on tumor evolution (An et al., 2025; Rowe and Pomper, 2022).

Another innovative frontier in imaging is the integration of *nanotechnology*, which enhances imaging resolution and specificity through the use of nanoparticle-based contrast agents. These agents can be engineered to target specific tumor markers, thereby improving the visualization of cancerous tissues at very early stages. Nanotechnology also holds potential in image-guided drug delivery, where therapeutic agents are delivered directly to tumor sites, enhancing treatment efficacy while reducing systemic toxicity (Bhange and Telange, 2025).

Collectively, these cutting-edge approaches—theranostics, liquid biopsies, and nanotechnology—are reshaping the landscape of cancer imaging. They enable more accurate disease characterization, early detection of recurrence, and personalized intervention strategies. As research and clinical applications continue to evolve, molecular and functional imaging are expected to play an increasingly pivotal role in advancing precision oncology and improving patient outcomes worldwide (Gangadaran et al., 2025).

Emerging imaging techniques provide deeper insights into tumor biology at the molecular and cellular levels, allowing for more precise diagnosis and targeted treatment. Theranostics combine diagnostic imaging with targeted therapies, enabling real-time monitoring of treatment effectiveness while minimizing damage to healthy tissues (Editors, n.d.; Langbein et al., 2019). Liquid biopsies, a noninvasive approach, detect ctDNA and biomarkers in the bloodstream, providing an early indication of cancer recurrence or progression without requiring invasive tissue sampling (Bibikova and Fan, 2023). Furthermore, nanotechnology in imaging enhances contrast and resolution, improving early tumor detection by enabling highly sensitive imaging agents to target cancer cells more precisely. These advancements in imaging technology are transforming oncology by enabling early diagnosis, precise treatment planning, and improved patient outcomes (Umadevi et al., 2024; Liu and Grodzinski, 2021).

5 Conclusion

Cancer continues to be a leading global health burden, and tackling it effectively requires a multifaceted approach involving cutting-edge technology, research, and comprehensive healthcare strategies. Advances in imaging technologies, particularly with the integration of AI, have revolutionized early detection, offering more precise and earlier identification of tumors. Alongside AI, radiomics and molecular diagnostics are further enhancing the understanding of tumor biology, enabling more accurate diagnosis, personalized treatment strategies, and improved monitoring of therapeutic responses. These innovations, when combined with public awareness and better access to healthcare, hold the potential to significantly reduce cancer-related mortality rates. However, there is still much to be done. Future research should prioritize refining AI-driven imaging techniques, enhancing radiomic approaches, and exploring the potential of liquid biopsies to provide noninvasive, real-time insights into tumor dynamics and patient responses. In addition, collaboration between researchers, clinicians, and policymakers will be essential to overcome challenges related to healthcare disparities, ensuring that advances in cancer care are accessible to all populations globally. This collective effort will ultimately contribute to improving both cancer outcomes and patient quality of life, making significant strides in the global fight against cancer. Through continued advancements and coordinated efforts, we can hope to transform cancer care, offering better prognosis and treatment options for patients worldwide.

References

Abgral, R., Couturier, O.F., 2024. Assessment of therapeutic response by FDG-PET in solid tumors (PERCIST 1.0 criteria). Med. Nucl. doi:10.1016/j.mednuc.2024.09.004.

Ahmad, I., Jasim, S.A., Sharma, M.K., Renuka Jyothi, S., Hjazi, A., Mohammed, J.S., et al, 2024. New paradigms to break barriers in early cancer detection for improved prognosis and treatment outcomes. J. Gene Med., 26. doi:10.1002/jgm.3730.

Ajayi, R.O., Ogunjobi, T.T., 2024. Environmental exposures and cancer risk: a comprehensive review. Medinformatics. doi:10.47852/bonview-MEDIN42023598.

Akingbola, A., Adegbesan, A., Ojo, O., Otumara, J.U., Alao, U.H., 2024. Artificial intelligence and cancer care in Africa. J. Med. Surg. Public Health 3, 100132. doi:10.1016/j.glmedi.2024.100132.

Albano, D., Benenati, M., Bruno, A., Bruno, F., Calandri, M., Caruso, D., et al, 2021. Imaging side effects and complications of chemotherapy and radiation therapy: a pictorial review from head to toe. Insights Imaging, 12. doi:10.1186/s13244-021-01017-2.

Alcázar, J.L., Caparros, M., Arraiza, M., Mínguez, J.Á., Guerriero, S., Chiva, L., et al., 2019. Pre-operative assessment of intra-abdominal disease spread in epithelial ovarian cancer: a comparative study between ultrasound and computed tomography. Int. J. Gynecol. Cancer 29, 227–233. doi:10.1136/ijgc-2018-000066.

Alowais, S.A., Alghamdi, S.S., Alsuhebany, N., Alqahtani, T., Alshaya, A.I., Almohareb, S.N., et al, 2023. Revolutionizing healthcare: the role of artificial intelligence in clinical practice. BMC Med. Educ. 23. doi:10.1186/s12909-023-04698-z.

Alsharif, F., 2024. Artificial intelligence in oncology: applications, challenges and future frontiers. Int. J. Pharm. Investig. 14, 647–656. doi:10.5530/ijpi.14.3.76.

Alzehr, A., Hulme, C., Spencer, A., Morgan-Trimmer, S., 2022. The economic impact of cancer diagnosis to individuals and their families: a systematic review. Support. Care Cancer 30, 6385–6404. doi:10.1007/s00520-022-06913-x.

An, L., Liu, Y., Liu, Y., 2025. Advancements in circulating tumor cell detection for early cancer diagnosis: an integration of machine learning algorithms with microfluidic technologies. Biosensors (Basel) 15. doi:10.3390/bios15040220.

Anonymous, Statistics at a glance. n.d.a.

Anonymous, n.d.b. Essential Medicines List in National Cancer Control Plans.

Anonymous, n.d.c. Rationale for Prevention.

Arvola, S., Seppänen, M., Malaspina, S., Mätzke, S., Raiko, J., Timonen, K.L., et al., 2022. Comparison of reprojected bone SPECT/CT and planar bone scintigraphy for the detection of bone metastases in breast and prostate cancer. Nucl. Med. Commun. 43, 510–517. doi:10.1097/MNM.0000000000001536.

Ateş, S.G., Akdemir, Ü.Ö., 2024. PET/MR imaging in neurooncology. Nucl. Med. Semin. 10, 257–273. doi:10.4274/nts.galenos.2024.70288.

Bamodu, O.A., Chung, -C.-C., 2024. Cancer care disparities: overcoming barriers to cancer control in low- and middle-income countries. JCO Glob. Oncol. doi:10.1200/go.23.00439.

Bandini, L., Schatz, A., Hood, V., Clark, N., Hall, M.J., Carlson, R.W., 2023. NCCN Policy Summit: reducing the cancer burden through prevention and early detection. JNCCN J. Natl. Compr. Cancer Netw. 21, 910–914. doi:10.6004/jnccn.2023.7040.

Bashir, U., Mallia, A., Stirling, J., Joemon, J., MacKewn, J., Charles-Edwards, G., et al., 2015. PET/MRI in oncological imaging: state of the art. Diagnostics 5, 333–357. doi:10.3390/diagnostics5030333.

Beuthien-Baumann, B., Sachpekidis, C., Gnirs, R., Sedlaczek, O., 2021. Adapting imaging protocols for PET-CT and PET-MRI for immunotherapy monitoring. Cancers (Basel) 13. doi:10.3390/cancers13236019.

Bhange, M., Telange, D., 2025. Convergence of nanotechnology and artificial intelligence in the fight against liver cancer: a comprehensive review. Discov. Oncol. 16. doi:10.1007/s12672-025-01821-y

Bi, W.L., Hosny, A., Schabath, M.B., Giger, M.L., Birkbak, N.J., Mehrtash, A., et al., 2019. Artificial intelligence in cancer imaging: clinical challenges and applications. CA Cancer J. Clin. 69, 127–157. doi:10.3322/caac.21552.

Bibikova, M., Fan, J., 2023. Liquid biopsy for early detection of lung cancer. Chin. Med. J. Pulm. Crit. Care Med. 1, 200–206. doi:10.1016/j.pccm.2023.08.005.

Byrd, D., Christopfel, R., Buatti, J., Moros, E., Nehmeh, S., Opanowski, A., et al., 2017. Multicenter survey of PET/CT protocol parameters that affect standardized uptake values. J. Med. Imaging 5, 1. doi:10.1117/1.jmi.5.1.011012.

Çallı, E., Sogancioglu, E., van Ginneken, B., van Leeuwen, K.G., Murphy, K., 2021. Deep learning for chest X-ray analysis: a survey. Med. Image Anal. 72. doi:10.1016/j.media.2021.102125.

Chan, A., Eng, L., Jiang, C., Dagsi, M., Ke, Y., Tanay, M., et al, 2024. Global disparities in cancer supportive care: an international survey. Cancer Med., 13. doi:10.1002/cam4.70234.

Chassagnon, G., De Margerie-Mellon, C., Vakalopoulou, M., Marini, R., Hoang-Thi, T.N., Revel, M.P., et al., 2023. Artificial intelligence in lung cancer: current applications and perspectives. Jpn. J. Radiol. 41, 235–244. doi:10.1007/s11604-022-01359-x.

Chehelgerdi, M., Chehelgerdi, M., Allela, O.Q.B., Pecho, R.D.C., Jayasankar, N., Rao, D.P., et al, 2023. Progressing nanotechnology to improve targeted cancer treatment: overcoming hurdles in its clinical implementation. Mol. Cancer, 22. doi:10.1186/s12943-023-01865-0.

Chen, J.G., Zhang, Y.H., Lu, J.H., Kensler, T.W., 2024. Liver cancer etiology: old issues and new perspectives. Curr. Oncol. Rep. doi:10.1007/s11912-024-01605-7.

Chernov VI, M., AA, Z.R.V., Sinilkin, I.G., Stasyuk, E.S., Larionova, L.A., et al., 2016. Radiopharmaceuticals for SPECT cancer detection. IOP Conf. Ser. Mater. Sci. Eng. 135. doi:10.1088/1757-899X/135/1/012008.

de Martel, C., Georges, D., Bray, F., Ferlay, J., Clifford, G.M., 2020. Global burden of cancer attributable to infections in 2018: a worldwide incidence analysis. Lancet Glob. Health 8, e180–90. doi:10.1016/S2214-109X(19)30488-7.

Decazes, P., Thureau, S., Modzelewski, R., Damilleville-Martin, M., Bohn, P., Vera, P., 2020. Benefits of positron emission tomography scans for the evaluation of radiotherapy. Cancer/Radiotherapie 24, 388–397. doi:10.1016/j.canrad.2020.02.007.

Delorme, S., 2012. Ultraschall in der onkologie screening und staging. Internist 53, 271–281. doi:10.1007/s00108-011-2958-5.

Díaz, O., Rodríguez-Ruíz, A., Sechopoulos, I., 2024. Artificial intelligence for breast cancer detection: technology, challenges, and prospects. Eur. J. Radiol. 175. doi:10.1016/j.ejrad.2024.111457.

Dike, S., McGowan, T., 2020. Environmental risk factors: the role of oncology nurses in assessing and reducing the risk for exposure. Clin. J. Oncol. Nurs. 24, 31–38. doi:10.1188/20.CJON.S2.31-38.

Dillon, E.C., Meehan, A., Li, J., Liang, S.Y., Lai, S., Colocci, N., et al., 2021. How, when, and why individuals with stage IV cancer seen in an outpatient setting are referred to palliative care: a mixed methods study. Support. Care Cancer 29, 669–678. doi:10.1007/s00520-020-05492-z.

Ebner, R., Sheikh, G.T., Brendel, M., Ricke, J., Cyran, C.C., 2024. ESR Essentials: staging and restaging with FDG-PET/CT in oncology—practice recommendations by the European Society for Hybrid, Molecular and Translational Imaging. Eur. Radiol. doi:10.1007/s00330-024-11094-8.

Editors, K, n.d. Cancer Drug Discovery and Development Increasing the Therapeutic Ratio of Radiotherapy.

Enokida, T., Moreira, A., Bhardwaj, N., 2021. Vaccines for immunoprevention of cancer. J. Clin. Investig. 131. doi:10.1172/JCI146956.

Farkas, I., Sipka, G., Bakos, A., Maráz, A., Bajory, Z., Mikó, Z., et al, 2024. Diagnostic value of [^{99m}Tc]Tc-PSMA-I&S-SPECT/CT for the primary staging and restaging of prostate cancer. Ther. Adv. Med. Oncol. 16. doi:10.1177/17588359231221342.

Firjani Naef, A., 2014. A Non-invasive Image Based System for Early Diagnosis of Prostate Cancer. University of Louisville. doi:10.18297/etd/1710.

Frantellizzi, V., Ricci, M., Cimini, A., Filippi, L., Conte, M., De Feo, M.S., et al, 2023. The role of PET and SPECT imaging in prostate cancer targeted alpha therapy: when and how?. Appl. Sci. (Switzerland), 13. doi:10.3390/app13031890.

Fuccio, C., Spinapolice, E.G., Chondrogiannis, S., Maffione, A.M., Trifirò, G., Colletti, P.M., et al, 2013. Evolving Role of SPECT/CT in Neuroendocrine Tumors Management Staging, Treatment Response, and Follow-up.

Gangadaran, P., Onkar, A., Rajendran, R.L., Goenka, A., Oh, J.M., Khan, F., et al, 2025. Noninvasive in vivo imaging of macrophages: understanding tumor microenvironments and delivery of therapeutics. Biomark. Res. 13. doi:10.1186/s40364-025-00735-9.

Gelardi, F., Ragaini, E.M., Sollini, M., Bernardi, D., Chiti, A., 2022. Contrast-enhanced mammography versus breast magnetic resonance imaging: a systematic review and meta-analysis. Diagnostics 12. doi:10.3390/diagnostics12081890.

Gherghe, M., Mutuleanu, M.D., Stanciu, A.E., Irimescu, I., Lazar, A., Bacinschi, X., et al, 2022. Quantitative analysis of SPECT-CT data in metastatic breast cancer patients—the clinical significance. Cancers (Basel), 14. doi:10.3390/cancers14020273.

Goyanka, R., 2021. Economic and non-economic burden of cancer: a propensity score matched analysis using household health survey data of India. Cancer Res. Stat. Treat. 4, 29–36. doi:10.4103/crst.crst_6_21.

Gündooan, C., Tatar, G., Erkan, E., Semih Çakr, M., Forouz, A., Gökhan Toktaa, M., et al, n.d. The Impact of IV Contrast Enhanced F-FDG PET/CT on the Staging of Bladder Cancer.

Gupta, D., Roy, I., Gandhi, S., 2023. Metallic nanoparticles for CT-guided imaging of tumors and their therapeutic applications. OpenNano 12. doi:10.1016/j.onano.2023.100146.

Hanson, L.C., Wessell, K.L., Hanspal, J., Lin, F.C., Collichio, F.A., DeWalt, D., et al., 2021. Pre-post evaluation of collaborative oncology palliative care for patients with stage IV cancer. J. Pain Symptom Manage. 62, e56–64. doi:10.1016/j.jpainsymman.2021.02.032.

Hegde, J.V., Mulkern, R.V., Panych, L.P., Fennessy, F.M., Fedorov, A., Maier, S.E., et al., 2013. Multiparametric MRI of prostate cancer: an update on state-of-the-art techniques and their performance in detecting and localizing prostate cancer. J. Magn. Reson. Imaging 37, 1035–1054. doi:10.1002/jmri.23860.

Heindel, W., Gübitz, R., Vieth, V., Weckesser, M., Schober, O., Schäfers, M., 2014. The diagnostic imaging of bone metastases. Dtsch Arztebl. Int. 111, 741–747. doi:10.3238/arztebl.2014.0741.

Henzler, T., Shi, J., Jafarov, H., Schoenberg, S.O., Manegold, C., Fink, C., et al., 2012. Functional CT imaging techniques for the assessment of angiogenesis in lung cancer. Transl. Lung Cancer Res. 1, 78–83. doi:10.3978/j.issn.2218-6751.2012.01.02.

Horvat, J.V., Durando, M., Milans, S., Patil, S., Massler, J., Gibbons, G., et al., 2018. Apparent diffusion coefficient mapping using diffusion-weighted MRI: impact of background parenchymal enhancement, amount of fibroglandular tissue and menopausal status on breast cancer diagnosis. Eur. Radiol. 28, 2516–2524. doi:10.1007/s00330-017-5202-4.

Hu, L., Fu, C., Song, X., Grimm, R., von Busch, H., Benkert, T., et al. Automated deep-learning system in the assessment of MRI-visible prostate cancer: comparison of advanced zoomed diffusion-weighted imaging and conventional technique. Cancer Imaging 2023;23. doi:10.1186/s40644-023-00527-0.

Hurbain, P., Strickland, M.J., Liu, Y., Li, D., 2024. Environmental inequality in estimated cancer risk from airborne toxic exposure across United States communities from 2011 to 2019. Environ. Sci. Technol. doi:10.1021/acs.est.4c02526.

Hussain, S., Mubeen, I., Ullah, N., Shah, S.S.U.D., Khan, B.A., Zahoor, M., et al, 2022. Modern diagnostic imaging technique applications and risk factors in the medical field: a review. Biomed. Res. Int. 2022. doi:10.1155/2022/5164970.

Itani, M., Menias, C.O., Mellnick, V.M., El Zakhem, A., Elsayes, K., Katabathina, V., et al., 2021. Imaging of abdominal and pelvic infections in the cancer patient. Abdom. Radiol. 46, 2920–2941. doi:10.1007/s00261-020-02896-7.

Jun, S., Park, H., Kim, U.J., Lee, H.A., Park, B., Lee, S.Y., et al., 2024. The combined effects of alcohol consumption and smoking on cancer risk by exposure level: a systematic review and meta-analysis. J. Korean Med. Sci. 39, 1–15. doi:10.3346/jkms.2024.39.e185.

Khalifa, M., Albadawy, M., 2024. AI in diagnostic imaging: revolutionising accuracy and efficiency. Comput. Methods Programs Biomed. 5. doi:10.1016/j.cmpbup.2024.100146

Kim, J., Kim, K.H., 2020. Role of chest radiographs in early lung cancer detection. Transl. Lung Cancer Res. 9, 522–531. doi:10.21037/tlcr.2020.04.02.

Koh, D.M., Papanikolaou, N., Bick, U., Illing, R., Kahn, C.E., Kalpathi-Cramer, J., et al, 2022. Artificial intelligence and machine learning in cancer imaging. Commun. Med., 2. doi:10.1038/s43856-022-00199-0.

Krishnan, M., Agarwal, P., Pinninti, R., Rajappa, S., 2023. Global inequalities in availability of systemic therapies for cancer care and strategies to address them. J. Surg. Oncol. 128, 1038–1044. doi:10.1002/jso.27439.

Kyriazi, S., Kaye, S.B., Desouza, N.M., 2010. Imaging ovarian cancer and peritoneal metastases—current and emerging techniques. Nat. Rev. Clin. Oncol. 7, 381–393. doi:10.1038/nrclinonc.2010.47.

Langbein, T., Weber, W.A., Eiber, M., 2019. Future of theranostics: an outlook on precision oncology in nuclear medicine. J. Nucl. Med. 60, 13S–19S. doi:10.2967/jnumed.118.220566.

Liu, C.H., Grodzinski, P., 2021. Nanotechnology for cancer imaging: advances, challenges, and clinical opportunities. Radiol. Imaging Cancer 3. doi:10.1148/rycan.2021200052.

Lo Gullo, R., Marcus, E., Huayanay, J., Eskreis-Winkler, S., Thakur, S., Teuwen, J., et al., 2024. Artificial intelligence-enhanced breast MRI: applications in breast cancer primary treatment response assessment and prediction. Invest. Radiol. 59, 230–242. doi:10.1097/RLI.0000000000001010.

M, E.S., Schopp, J.G., Peacock, S., DeMartini, W.D., Rahbar, H., Lehman, C.D., et al, 2014. Diffusion-weighted MRI: association between patient characteristics and apparent diffusion coefficients of normal breast fibroglandular tissue at 3 T. Am. J. Roentgenol. 202. doi:10.2214/AJR.13.11159.

Mao, J., Shi, J., He, S., Li, N., Shi, J., Yang, F., 2019. Review Article: reproductive factors and risk of breast cancer: a pooled meta-analysis of 55 case-control studies based on different source populations. Int. J. Clin. Exp. Med. 12.

Matsukawa, A., Yanagisawa, T., Bekku, K., Kardoust Parizi, M., Laukhtina, E., Klemm, J., et al, 2024. Comparing the performance of digital rectal examination and prostate-specific antigen as a screening test for prostate cancer: a systematic review and meta-analysis. Eur. Urol. Oncol. doi:10.1016/j.euo.2023.12.005.

Maurer, M.H., Heverhagen, J.T., 2017. Diffusion weighted imaging of the prostate-principles, application, and advances. Transl. Androl. Urol. 6, 490–498. doi:10.21037/tau.2017.05.06.

Mittal, N., Tayal, S., 2021. Advance computer analysis of magnetic resonance imaging (MRI) for early brain tumor detection. Int. J. Neurosci. 131, 555–570. doi:10.1080/00207454.2020.1750390.

Mussallem, D., 2022. Lifestyle for breast cancer risk reduction. Menopause 29, 979–981. doi:10.1097/GME.0000000000002006.

Neilio, J.M., Ginat, D.T., 2024. Emerging head and neck tumor targeting contrast agents for the purpose of CT, MRI, and multimodal diagnostic imaging: a molecular review. Diagnostics 14. doi:10.3390/diagnostics14151666.

Nenclares, P., Harrington, K.J., n.d. The Biology of Cancer.

Nussbaum, S., Shoukry, M., Ashary, M.A., Kasbi, A.A., Baksh, M., Gabriel, E., 2022. Advanced tumor imaging approaches in human tumors. Cancers (Basel) 14. doi:10.3390/cancers14061549.

Partridge, S.C., Singer, L., Sun, R., Wilmes, L.J., Klifa, C.S., Lehman, C.D., et al., 2011. Diffusion-weighted MRI: influence of intravoxel fat signal and breast density on breast tumor conspicuity and apparent diffusion coefficient measurements. Magn. Reson. Imaging 29, 1215–1221. doi:10.1016/j.mri.2011.07.024.

Patel, P., Oto, A., 2016. Magnetic resonance imaging of the prostate, including pre- and postinterventions. Semin. Interv. Radiol. 33, 186–195. doi:10.1055/s-0036-1586144.

Pesapane, F., Downey, K., Rotili, A., Cassano, E., Koh, D.M., 2020. Imaging diagnosis of metastatic breast cancer. Insights Imaging 11. doi:10.1186/s13244-020-00885-4.

Pilleron, S., Sarfati, D., Janssen-Heijnen, M., Vignat, J., Ferlay, J., Bray, F., et al., 2019. Global cancer incidence in older adults, 2012 and 2035: a population-based study. Int. J. Cancer 144, 49–58. doi:10.1002/ijc.31664.

Prathap, R., Kirubha, S., Rajan, A.T., Manoharan, S., Elumalai, K., 2024. The increasing prevalence of cancer in the elderly: an investigation of epidemiological trends. Aging Med. 7, 516–527. doi:10.1002/agm2.12347.

Queiroz, M.A., De Galiza Barbosa, F., Buchpiguel, C.A., Cerri, G.G., 2018. Positron emission tomography/magnetic resonance imaging (PET/MRI): an update and initial experience at HC-FMUSP. Rev. Assoc. Med. Bras. 64, 71–84. doi:10.1590/1806-9282.64.01.71.

Ribeiro, A.I., Fecht, D., 2021. Environmental determinants of the social gradient in cancer incidence. In: Social Environment and Cancer in Europe: Towards an Evidence-Based Public Health Policy. Springer International Publishing, 221–233. doi:10.1007/978-3-030-69329-9_15.

Rosino Sánchez, A., García Torralba, E., Girela Baena, E., Macías Cerrolaza, J.A., Tudela Pallares, J., Zafra Povés, M., et al, 2025. Early diagnosis of prostate cancer in healthy men with germline mutations in DNA damage response (DDR) pathways: a literature review and proposal for a screening strategy. Actas Urol. Esp. doi:10.1016/j.acuro.2025.501686.

Rowe, S.P., Pomper, M.G., 2022. Molecular imaging in oncology: current impact and future directions. CA Cancer J. Clin. 72, 333–352. doi:10.3322/caac.21713.

Rubin, G.D., 2015. Lung nodule and cancer detection in computed tomography screening. J. Thorac. Imaging 30, 130–138. doi:10.1097/RTI.0000000000000140.

Saeidnia, H.R., Firuzpour, F., Kozak, M., Majd, H.S., 2025. Advancing cancer diagnosis and treatment: integrating image analysis and AI algorithms for enhanced clinical practice. Artif. Intell. Rev. 58. doi:10.1007/s10462-025-11117-w.

Samartzis, D., Gillis, C.C., Shih, P., O'Toole, J.E., Fessler, R.G., 2015. Intramedullary spinal cord tumors: part I-epidemiology, pathophysiology, and diagnosis. Glob. Spine J. 5, 425–435. doi:10.1055/s-0035-1549029.

Santana-Davila, R., Martins, R., 2016. Treatment of stage IIIA non-small-cell lung cancer: a concise review for the practicing oncologist. J. Oncol. Pract. 12, 601–606. doi:10.1200/JOP.2016.013052.

Schiller, J.T., Lowy, D.R., 2010. Vaccines to prevent infections by oncoviruses. Annu. Rev. Microbiol. 64, 23–41. doi:10.1146/annurev.micro.112408.134019.

Schlander, M., Pham, P.D., Ubels, J., Hernandez, D., Hernandez-Villafuerte, K., Eckford, M.H.R.R.D. et al, 2024. The socioeconomic impact of cancer on patients and their relatives: organisation of European Cancer Institutes task force consensus recommendations on conceptual framework, taxonomy, and research directions. Lancet Oncol. 25.

Shin, S., Ko, E.S., Kim, R.B., Han, B.K., Nam, S.J., Shin, J.H., et al., 2015. Effect of menstrual cycle and menopausal status on apparent diffusion coefficient values and detectability of invasive ductal carcinoma on diffusion-weighted MRI. Breast Cancer Res. Treat. 149, 751–759. doi:10.1007/s10549-015-3278-6.

Singh, P.K., Gupta, P.K., Gupta, V., Singh, S., Akhtar, S., 2024. Making cancer imaging smarter: emerging techniques and computational outlook to guide precision diagnostics. Open Bioinforma. J. 17. doi:10.2174/0118750362344593241111114723.

Singh, S., Bray, T.J.P., Hall-Craggs, M.A., 2018. Quantifying bone structure, micro-architecture, and pathophysiology with MRI. Clin. Radiol. 73, 221–230. doi:10.1016/j.crad.2017.12.010.

Stanton, S.E., Eary, J.F., Marzbani, E.A., Mankoff, D., Salazar, L.G., Higgins, D., et al, 2016. Concurrent SPECT/PET-CT imaging as a method for tracking adoptively transferred T-cells in vivo. J. Immunother. Cancer, 4. doi:10.1186/s40425-016-0131-3.
Sung, H., Ferlay, J., Siegel, R.L., Laversanne, M., Soerjomataram, I., Jemal, A., et al., 2021. Global Cancer Statistics 2020: GLOBOCAN estimates of incidence and mortality worldwide for 36 cancers in 185 countries. CA Cancer J. Clin. 71, 209–249. doi:10.3322/caac.21660.
Takamiya, S., Malvea, A., Ishaque, A.H., Pedro, K., Fehlings, M.G., 2024. Advances in imaging modalities for spinal tumors. Neurooncol. Adv. doi:10.1093/noajnl/vdae045.
Treglia, G., Muoio, B., 2020. Evidence-based PET for brain tumours. In: Evidence-based Positron Emission Tomography: Summary of Recent Meta-analyses on PET. Springer International Publishing, 25–33. doi:10.1007/978-3-030-47701-1_3.
Ueno, Y.R., Tamada, T., Takahashi, S., Tanaka, U., Sofue, K., Kanda, T., et al., 2018. Computed diffusion-weighted imaging in prostate cancer: basics, advantages, cautions, and future prospects. Korean J. Radiol. 19, 832–837. doi:10.3348/kjr.2018.19.5.832.
Umadevi, K., Sundeep, D., Vighnesh, A.R., Misra, A., Krishna, A.G., 2024. Current trends and advances in nanoplatforms-based imaging for cancer diagnosis. Indian J. Microbiol. doi:10.1007/s12088-024-01373-9.
Unadkat, B., Kashikar, S., Mishra, G., Dhande, R., 2022. Evaluation of preoperative multidetector row computed tomography in colon malignancy with reference to (Local Staging and Lymph Node Status). J. Datta Meghe Inst. Med. Sci. Univ. 17, 1019–1022. doi:10.4103/jdmimsu.jdmimsu_452_22.
Velpula, T., Buddolla, V., 2025. Enhancing detection and monitoring of circulating tumor cells: integrative approaches in liquid biopsy advances. J. Liq. Biopsy 8, 100297. doi:10.1016/j.jlb.2025.100297.
Wahl, R.L., Hicks, R.J., 2021. PET diagnosis and response monitoring in oncology. In: Molecular Imaging: Principles and Practice. Elsevier, 1049–1076. doi:10.1016/B978-0-12-816386-3.00048-X.
Wang, Q., Ketteler, S., Bagheri, S., Ebrahimifard, A., Luster, M., Librizzi, D., et al, 2024. Diagnostic efficacy of [^{99m}Tc]Tc-PSMA SPECT/CT for prostate cancer: a meta-analysis. BMC Cancer, 24. doi:10.1186/s12885-024-12734-4.
Williams, K., Probst, H., 2016. Use of IV contrast media in radiotherapy planning CT scans: a UK audit. Radiography 22, S28–32. doi:10.1016/j.radi.2016.06.006.
Wu, Y., Shang, J., Zhang, X., Li, N., 2024a. Advances in molecular imaging and targeted therapeutics for lymph node metastasis in cancer: a comprehensive review. J. Nanobiotechnol. 22. doi:10.1186/s12951-024-02940-4.
Wu, Z., Xia, F., Lin, R., 2024b. Global burden of cancer and associated risk factors in 204 countries and territories, 1980–2021: a systematic analysis for the GBD 2021. J. Hematol. Oncol. 17. doi:10.1186/s13045-024-01640-8.
Xiao, L., Xin, J., 2022. Advances in clinical oncology research on 99mTc-3PRGD2 SPECT imaging. Front. Oncol. 12. doi:10.3389/fonc.2022.898764.
Yang, C.T., Ghosh, K.K., Padmanabhan, P., Langer, O., Liu, J., Eng, D.N.C., et al., 2018. PET-MR and SPECT-MR multimodality probes: development and challenges. Theranostics 8, 6210–6232. doi:10.7150/thno.26610.
Yi, A., Hak, H.K., Hee, J.S., Mi, O.H., Seung, D.A., Bo, K.S., 2009. Radiation-induced complications after breast cancer radiation therapy: a pictorial review of multimodality imaging findings. Korean J. Radiol. 10, 496–507. doi:10.3348/kjr.2009.10.5.496.
Zhang, J., Chen, Y., Zhang, E., Li, N., Yuan, H., Lang, N., 2020. Use of monoexponential diffusion-weighted imaging and diffusion kurtosis imaging and dynamic contrast-enhanced-MRI for the differentiation of spinal tumors. Eur. Spine J. 29, 1112–1120. doi:10.1007/s00586-020-06330-w.
Zhang, S., Xiao, X., Yi, Y., Wang, X., Zhu, L., Shen, Y., et al, 2024. Tumor initiation and early tumorigenesis: molecular mechanisms and interventional targets. Signal Transduct. Target. Ther. 9. doi:10.1038/s41392-024-01848-7.
Zhang, Y., Shi, Y., Ye, L., Li, T., Wei, Y., Lin, Z., et al, 2023. Improving diagnostic efficacy of primary prostate cancer with combined ^{99m}Tc-PSMA SPECT/CT and multiparametric-MRI and quantitative parameters. Front. Oncol., 13. doi:10.3389/fonc.2023.1193370.

Chapter 30

Recent trends and techniques in breast imaging

S. Sharath[a], Rajagopal Kadavigere[b], Veena Mayya[c], Leena R. David[a,d], Berna Uzun[e], Dilber Uzun Ozsahin[d,e,f], Abhimanyu Pradhan[a], Winniecia Dkhar[a], Ilker Ozsahin[e], and Neil Abraham Barnes[a]

[a]*Department of Medical Imaging Technology, Manipal College of Health Professions, Manipal Academy of Higher Education, Manipal, Karnataka, India*

[b]*Department of Radiodiagnosis and Imaging, Kasturba Medical College, Manipal, India*

[c]*Department of Information & Communication Technology, Manipal Academy of Higher Education, Manipal, India*

[d]*Medical Diagnostic Imaging Department, College of Health Sciences, University of Sharjah, Sharjah, United Arab Emirates*

[e]*Irfan Suat Gunsel Operational Research Institute, Near East University, Nicosia, Turkish Republic of Northern Cyprus (TRNC), Turkey*

[f]*Research Institute for Medical and Health Sciences, University of Sharjah, Sharjah, United Arab Emirates*

1 Introduction

Breast cancer is the most commonly diagnosed malignancy among women globally, accounting for approximately 11.7% of all cancer cases and remaining a leading cause of cancer-related deaths among women. In 2020 alone, over 2.3 million women were diagnosed with breast cancer, and 685,000 succumbed to the disease. With an estimated 7.8 million women living within five years of a breast cancer diagnosis, it is now the most prevalent cancer worldwide, surpassing even lung cancer in terms of overall incidence among women (Breast Cancer Facts & Figures, 2019). Although breast cancer primarily affects women, it is important to note that men can also develop the disease, albeit rarely, comprising less than 1% of all male cancer cases. Male breast cancer, often diagnosed at a later stage due to lack of awareness and screening, carries unique clinical challenges. The global rise in breast cancer incidence is attributed to various factors, including improved screening practices such as mammography and breast self-examination, changes in reproductive behaviors, urbanization, lifestyle modifications, and socioeconomic development (Bray et al., 2024). In high-income countries such as the United States, Canada, Australia, and many parts of Europe, the incidence has shown a modest annual increase of less than 0.5% since 1997, primarily due to effective screening programs, health awareness, and advancements in early diagnosis (Dafni et al., 2019; Lei et al., 2021; Barrios et al., 2018). In contrast, low- and middle-income countries (LMICs), particularly those in Africa, Latin America, and parts of Asia, have seen a more rapid and concerning rise in breast cancer cases, largely driven by limited access to healthcare, delayed diagnosis, cultural barriers, lack of awareness, and changing lifestyle patterns. As urbanization spreads and life expectancy increases, lifestyle-related risk factors such as sedentary behavior, obesity, alcohol consumption, reduced parity, late-age childbirth, shorter duration of breastfeeding, and hormone replacement therapy have become more prominent in LMICs, contributing to a growing burden of breast cancer (Manson and Achel, 2023; Tfayli et al., 2010). This disparity is further exacerbated by the absence of organized screening programs, poor access to diagnostic imaging, and lack of trained healthcare professionals, resulting in a higher proportion of late-stage diagnoses and, consequently, lower survival rates. While developed countries have achieved significant reductions in breast cancer mortality through early detection and effective treatment protocols—including surgery, radiotherapy, chemotherapy, hormonal therapy, and targeted biologics like trastuzumab and CDK4/6 inhibitors—developing countries continue to struggle with limited treatment options and poor health infrastructure (Gao et al., 2024; Newman, 2023; Ataeinia et al., 2021). Breast cancer is a heterogeneous disease, with several molecular subtypes including hormone receptor-positive, HER2-positive, and triple-negative breast cancer (TNBC), each with distinct prognostic and therapeutic implications. TNBC, which is more aggressive and less responsive to hormone therapy, is disproportionately observed in younger women and women of African descent, often contributing to worse outcomes in these populations. Research into the genetic and molecular basis of breast cancer, including the role of BRCA1 and BRCA2 mutations,

Advances in Medical Imaging. https://doi.org/10.1016/B978-0-443-28967-5.00030-7

has enhanced our understanding and enabled more personalized treatment approaches. Meanwhile, screening remains the cornerstone of breast cancer control, with mammography playing a key role in early detection (Jain et al., 2025; Mahtani et al., 2021). Countries with population-based screening programs have demonstrated significant improvements in stage-at-diagnosis and survival outcomes. However, the implementation of such programs in LMICs is hindered by economic constraints, infrastructure limitations, and insufficient public health policy support (Tfayli et al., 2010; Gao et al., 2024). Innovative solutions, including mobile screening units, community-based awareness campaigns, and the application of artificial intelligence (AI) in image interpretation, are being explored to improve access to early detection in resource-constrained settings. Moreover, breast cancer survivorship is an emerging area of focus, with attention shifting toward improving the quality of life for survivors, managing long-term side effects, and addressing psychological, social, and financial burdens (Ahn et al., 2023; Anderson et al., 2003). Survivors often face challenges such as fatigue, lymphedema, pain, anxiety, depression, and fear of recurrence, all of which require comprehensive follow-up care and rehabilitation services. Global health organizations such as the World Health Organization, the Union for International Cancer Control, and various nongovernmental organizations are working collaboratively to reduce the breast cancer burden through policy advocacy, research support, capacity building, and equitable access to screening and treatment. Despite advances in diagnostics and therapeutics, breast cancer continues to reflect broader social, gender, and health inequities. Addressing these challenges requires a coordinated, multisectoral approach that includes strengthening health systems, ensuring universal health coverage, investing in early detection infrastructure, and empowering women through education and health literacy. Ultimately, reducing the global burden of breast cancer will depend not only on scientific breakthroughs but also on political will, health system readiness, and international collaboration to ensure that every individual, regardless of geography or socioeconomic status, has access to timely, effective, and affordable breast cancer care (Baughan et al., 2022; Jeong and Park, 2021; Bitencourt and Pinker, 2023; Stubblefield, 2017).

1.1 Overview of breast imaging modalities

Breast imaging is commonly performed using mammography, ultrasound, and magnetic resonance imaging (MRI). Mammography is the only scientifically proven method for detecting breast cancer in asymptomatic women at average risk. Using ultrasound, solid masses can be distinguished from cystic masses, and biopsies can be guided. In addition to screening high-risk individuals, MRI is used to evaluate the extent of disease and monitor treatment progress (Steyerova and Burgetova, 2022).

1.2 Mammography

Mammography has long served as the cornerstone of breast cancer screening, with its scientifically validated ability to reduce mortality through early detection. As breast cancer remains the most frequently diagnosed malignancy and a leading cause of cancer-related deaths among women globally, mammography's role in timely diagnosis and intervention is crucial. Utilizing low-dose ionizing radiation, mammography produces high-resolution images capable of detecting microcalcifications, small masses, and architectural distortions, which are often invisible during physical examination. It is particularly effective in identifying ductal carcinoma in situ (DCIS), a noninvasive cancer confined to the milk ducts (Liu et al., 2023). The procedure involves compressing the breast to improve image clarity and reduce radiation dose, allowing for detailed visualization of the breast parenchyma. Over the decades, mammography has undergone significant technological advancements—from the introduction of molybdenum targets in 1966 and single-emulsion films in 1972, to the development of dedicated mammographic units in the 1990s equipped with automatic exposure control, microfocus x-ray tubes, high-frequency generators, and antiscatter grids (Lamb et al., 2022). The transition to digital mammography further enhanced diagnostic accuracy by enabling image manipulation, storage, and computer-aided detection. Standard mammographic views include the craniocaudal and mediolateral oblique projections, offering complementary perspectives, while additional views like magnification, spot compression, rolled, and extended views are employed when necessary. Mammography serves both screening and diagnostic functions, targeting asymptomatic women over 40 for early detection, while also evaluating symptomatic patients (Gelardi et al., 2022). However, its sensitivity decreases in women with dense breast tissue—common among younger individuals—where overlapping fibroglandular tissue can obscure malignancies or mimic abnormalities, leading to false-negative or false-positive outcomes. In such cases, adjunct imaging modalities like ultrasound or MRI may be recommended. Despite the low radiation dose (typically 1–3 mGy per image), concerns exist regarding cumulative exposure, particularly in high-risk populations such as BRCA1/2 mutation carriers, for whom MRI may be preferred (Ramos-Pollán et al., 2012). Nonetheless, modern systems prioritize patient safety through dose optimization and adherence to the ALARA (As Low As Reasonably Achievable) principle. Mammography's sensitivity ranges between 68% and 90%, with specificity around 85%–90%, depending on breast density and tumor characteristics. It detects

TABLE 30.1 Features of dedicated mammographic imaging systems.

High voltage generator	High frequency, 5–10 kHz generator
Target/filter	W/60 μm Mo
	Mo/30 μm Mo
	Mo/50 μm Rh
	Rh/50 μm Rh
kVp	20–35 kVp in 1-kVp increments
Compression	Low Z, auto-adjust, and release
Grids ratio	3:1 to 5:1, 30 lines/cm
Exposure control	Automatic to account for tissue thickness, composition, and reciprocity law failure.
Focal spot	0.3 mm/0.1 mm (large/small)
Magnification	≤2
SID	50–80 cm

up to 78.9% of breast cancers through routine screening, offering the benefit of diagnosing lesions at more treatable stages, thereby improving survival and reducing treatment invasiveness. In conclusion, while mammography has limitations, its evolution from film-based to digital systems—and ongoing innovations like digital breast tomosynthesis (DBT), contrast-enhanced mammography (CEM), and AI integration—ensures it remains a central, effective tool in breast cancer detection, enabling more precise, personalized, and timely care for women worldwide (Dembrower et al., 2023) Table 30.1.

1.3 Ultrasonography: A valuable adjunct in breast imaging

Ultrasonography, or breast ultrasound, is a noninvasive imaging technique that utilizes high-frequency sound waves to visualize internal structures of the breast. It operates by transmitting ultrasound waves, typically ranging from 3 to 12 MHz, through a hand-held probe known as a transducer. As these waves travel through different tissues, they encounter varying acoustic impedances—properties unique to each tissue type that influence the reflection of sound waves (Evans et al., 2018). These reflected echoes are captured by the probe and converted into real-time grayscale images by the ultrasound machine. The brightness of the image is directly related to the strength of the reflected signal, allowing for detailed visualization of the breast's anatomical structures (Hooley et al., 2013).

One of the primary clinical utilities of breast ultrasound lies in its ability to differentiate between solid and cystic lesions. Solid masses may be benign (such as fibroadenomas) or malignant, while cysts are usually benign and fluid-filled. This capability is crucial for accurate diagnosis and for guiding patient management. In addition, ultrasound is frequently used in interventional procedures, such as ultrasound-guided needle biopsies, where a fine or core needle is directed into a suspicious area under real-time imaging to obtain tissue samples for histopathological examination (Lee et al., 2024).

Ultrasound plays an especially critical role in evaluating women with dense breast tissue—a group in which mammography often struggles to detect malignancies due to the similar radiographic appearance of dense fibroglandular tissue and tumors (Hooley et al., 2013). In dense breasts, lesions may be obscured on mammograms, leading to false-negative results. When used in conjunction with mammography, whole-breast ultrasound has been shown to significantly enhance cancer detection rates, especially for small, node-negative invasive cancers that may not be visible on mammograms alone. As such, ultrasound is increasingly advocated as a supplemental screening tool for women with dense breasts or those at elevated risk of breast cancer (Evans et al., 2018).

It is important to note, however, that ultrasound is not a substitute for mammography. While it offers advantages in specific contexts, such as improved lesion characterization and better detection in dense tissue, ultrasound is operator-dependent and may not reliably detect microcalcifications—an early sign of breast cancer best visualized with mammography. Therefore, ultrasound is most effective when integrated into a multimodal imaging approach rather than used in isolation (Wang et al., 2022).

Beyond screening and diagnostic evaluation, breast ultrasound is also employed in monitoring therapeutic responses. During treatments like chemotherapy or radiation therapy, ultrasound can track changes in tumor size and vascularity, helping oncologists assess the effectiveness of interventions over time. This dynamic monitoring allows for timely adjustments in treatment plans and facilitates personalized care (Giannakou et al., 2025).

A major advantage of ultrasound over other imaging modalities, like mammography and computed tomography, is its safety profile. Ultrasound does not involve ionizing radiation, making it particularly suitable for repeated use and for younger women, pregnant patients, or those requiring frequent follow-ups. The procedure is painless, widely available, relatively cost-effective, and does not require contrast agents, reducing the risk of allergic reactions or nephrotoxicity (Iacob et al., 2024).

1.4 Magnetic resonance imaging in breast imaging—Expanded paragraph version

MRI of the breast is a sophisticated, noninvasive imaging technique that provides highly detailed cross-sectional images of breast tissues using powerful magnetic fields and radiofrequency pulses, rather than ionizing radiation. Unlike traditional mammography, which utilizes X-rays, breast MRI capitalizes on the principles of nuclear magnetic resonance (Dkhar et al., 2020). In this process, hydrogen protons in the body align with a strong external magnetic field. When exposed to a specific radiofrequency pulse, these protons are temporarily displaced. As they relax back to their original alignment, they emit signals that are captured by the MRI scanner to generate detailed images of the breast in multiple planes. This technique offers exceptional soft tissue contrast, which is especially beneficial in identifying subtle abnormalities in dense breast tissue or in cases where conventional imaging methods are inconclusive (Partridge et al., 2001).

A significant advantage of breast MRI is the use of a contrast agent, typically a gadolinium-based compound, administered intravenously during the scan. This process is referred to as dynamic contrast-enhanced MRI (DCE-MRI). It allows radiologists to analyze how different tissues in the breast absorb and wash out the contrast over time. Malignant tumors usually have increased blood supply and permeability, leading them to enhance rapidly after contrast administration and then wash out quickly. This pattern is distinct from that of benign lesions or normal tissue, aiding in more accurate differentiation and detection of malignancies (Elizabeth S et al., 2014). The use of DCE-MRI significantly enhances the diagnostic accuracy of MRI, offering a high sensitivity rate of around 93% and a specificity of approximately 71%, making it a valuable tool, particularly in the evaluation of high-risk individuals (Partridge et al., 2011).

Breast MRI is especially recommended for women who have a strong family history of breast cancer, carry BRCA1 or BRCA2 gene mutations, or have undergone radiation therapy to the chest at a young age. In such high-risk populations, breast MRI is used as a supplemental screening tool alongside mammography. It is also frequently used when findings from mammograms or ultrasounds are inconclusive or when further evaluation is required to determine the extent of known disease. MRI is highly sensitive in detecting multifocal (multiple tumors in the same breast) and multicentric (tumors in different quadrants of the breast) disease and in identifying contralateral breast involvement in patients already diagnosed with cancer in one breast. This information plays a crucial role in treatment planning, especially when deciding between breast-conserving surgery and mastectomy (Shin et al., 2015).

In addition to anatomical imaging, breast MRI includes advanced techniques such as diffusion-weighted imaging, which evaluates the movement of water molecules within tissues. Malignant lesions often restrict this diffusion, showing up as areas of high signal intensity with low apparent diffusion coefficient values, which can help differentiate cancer from benign lesions. Another innovative method is magnetic resonance (MR) spectroscopy, which evaluates the chemical composition of tissues. Elevated choline levels, commonly found in malignant tissues due to increased cellular turnover, can be detected and used as a metabolic biomarker. MR elastography, though still under investigation, assesses tissue stiffness and could potentially distinguish between soft benign lesions and harder malignant tumors. These functional MRI techniques not only provide anatomical details but also offer insights into the biological behavior of lesions (Dkhar et al., 2020, 2021, 2023).

The safety profile of breast MRI is generally excellent. As it does not use ionizing radiation, it is especially suitable for repeated use in younger women or those requiring long-term surveillance. However, there are considerations regarding the use of gadolinium-based contrast agents, especially in patients with impaired kidney function, as they may be at risk for a rare condition known as nephrogenic systemic fibrosis. Despite this, the overall risk is minimal when proper protocols are followed. MRI is also contraindicated in individuals with certain implanted medical devices, such as some types of pacemakers, though MRI-compatible devices are becoming more common. Some patients may experience claustrophobia due to the enclosed nature of the scanner, but modern wide-bore MRI machines and the use of mild sedation have helped mitigate this concern (Kieturakis et al., 2021; Si et al., 2024; Sardanelli et al., 2020).

Despite its numerous advantages, breast MRI is not without limitations. Its high sensitivity can sometimes lead to false positives, detecting benign abnormalities that may never become clinically significant, potentially resulting in unnecessary biopsies or anxiety. In addition, MRI is less effective in identifying microcalcifications, which are early signs of DCIS and are best visualized with mammography (Ramos-Pollán et al., 2012). MRI is also more expensive and less widely available than mammography and ultrasound, which can limit access, especially in resource-limited settings. Furthermore, interpreting breast MRI requires specialized expertise, and variability in reporting can occur if standard protocols are not adhered to.

2 Recent trends and techniques in breast imaging: Digital breast tomosynthesis

Advancements in breast imaging have significantly transformed the landscape of breast cancer diagnosis and management. These technological innovations have improved the accuracy, sensitivity, and specificity of imaging modalities, particularly in early cancer detection, which is crucial for timely intervention and better prognosis. Among these, DBT has emerged as a revolutionary technique that addresses many limitations of traditional mammography, especially in women with dense breast tissue (Bernardi et al., 2020).

DBT, commonly referred to as three-dimensional (3D) mammography, is a relatively recent imaging technique that creates a 3Dreconstruction of the breast by acquiring multiple low-dose X-ray images from different angles around the breast. These images are then synthesized into thin slices, usually 1 mm thick, allowing radiologists to evaluate the breast tissue layer by layer. This eliminates the superimposition of overlapping tissues, a common limitation in conventional two-dimensional (2D) mammography, which can obscure lesions or mimic abnormalities that may not actually exist (Richman et al., 2021).

One of the major advantages of DBT is its enhanced sensitivity in detecting breast cancers, especially in women with dense breasts, where the fibroglandular tissue may obscure small lesions on standard mammograms. Studies have shown that DBT can detect 15%–30% more breast cancers than 2D mammography. Dense breast tissue is not only common but also associated with a higher risk of developing breast cancer, making DBT an invaluable tool in screening programs (Manda-Mapalo et al., 2020).

Another key benefit of DBT is the reduction in false-positive rates. In conventional mammography, overlapping structures can sometimes appear as suspicious areas, leading to unnecessary callbacks, further imaging, and even biopsies. DBT reduces this issue by allowing radiologists to scroll through the breast tissue slice by slice, improving lesion visibility and characterization. As a result, false positives are reduced by approximately 15%–20%, minimizing unnecessary interventions, reducing patient anxiety, and optimizing resource use in clinical practice (Tejerina Bernal et al., 2012).

Furthermore, DBT has been shown to improve the detection of multifocal and multicentric breast cancers, which can be missed in 2D imaging due to tissue overlap. By improving the visualization of lesion margins, architectural distortion, and asymmetries, DBT provides a more comprehensive assessment of breast pathology and aids in more accurate tumor staging. This contributes significantly to informed decision-making regarding surgical planning and overall treatment strategy (Mariscotti et al., 2016).

Despite its significant diagnostic benefits, DBT is not without limitations. One key concern is its reduced sensitivity in detecting microcalcifications, which are often early indicators of DCIS; these are typically better visualized on high-resolution 2D mammograms. Although some DBT systems incorporate synthesized 2D images to compensate, their sensitivity may still not fully match that of conventional mammography. Another drawback is the slightly increased radiation dose—while each individual projection in DBT uses a lower dose, the cumulative effect from multiple projections results in approximately 8% higher radiation exposure compared to standard digital mammography. Nonetheless, this dose remains within regulatory safety limits and is generally considered acceptable in light of DBT's enhanced diagnostic capabilities. In addition, the high cost of DBT equipment and the need for specialized training limit its availability, particularly in low-resource settings. Interpreting DBT images also requires more time and expertise, posing challenges for radiologists and potentially reducing efficiency in high-volume screening programs (Bernardi et al., 2020; Manda-Mapalo et al., 2020; Wu et al., 2024).

2.1 Elastography in breast imaging

Elastography is a modern imaging technique that enhances conventional modalities, such as ultrasound and MRI, by measuring the mechanical properties of tissues, particularly their stiffness or elasticity. This method is especially valuable in breast imaging, where distinguishing between benign and malignant lesions is crucial for accurate diagnosis and effective treatment planning. Unlike traditional imaging, which primarily provides structural information, elastography adds a functional dimension by evaluating tissue biomechanics (Ricci et al., 2014). The technique is based on the principle that pathological tissues, especially malignant tumors, exhibit increased stiffness compared to surrounding normal or benign tissues. When an external force is applied—such as low-frequency vibrations in ultrasound or MR elastography—affected tissues resist deformation, a property that can be quantified and visualized through imaging. This resistance, or lack of elasticity, is often a marker for malignancy, as tumors typically have a denser cellular structure and an altered extracellular matrix that makes them stiffer (Mariscotti et al., 2016).

The elasticity of tissue is commonly expressed using Young's modulus, which measures the degree of deformation under stress. In healthy breast tissue, elasticity generally ranges from 1 to 100 kilopascals (kPa), while malignant tumors can be up to 15 times stiffer than benign conditions such as fibroadenomas. This significant variation in mechanical properties helps clinicians distinguish between different types of lesions. Elastography is particularly useful in assessing indeterminate breast lesions, evaluating palpable abnormalities not clearly seen on mammography, and characterizing tissue in women with dense breasts, where traditional imaging may fall short. Furthermore, elastography plays a vital role in reducing the number of unnecessary biopsies by providing additional information that helps classify lesions as likely benign or suspicious for malignancy (Mao et al., 2022).

There are two principal types of elastography employed in breast imaging: strain elastography and shear wave elastography. Strain elastography measures tissue deformation, or "strain," in response to manual or automatic compression. It generates a qualitative or semiquantitative color-coded map, allowing radiologists to visually assess which areas of tissue deform more than others. In strain elastography, softer tissue deforms more and appears in warmer colors (e.g., red or yellow), whereas stiffer tissue deforms less and appears in cooler tones (e.g., blue). This method is particularly useful for identifying stiff lesions that may correspond to malignancy. However, its accuracy is somewhat limited by operator dependency and variability in the amount of pressure applied during imaging (McKnight, 2002; Jamshidi et al., 2024).

Shear wave elastography, on the other hand, offers a more quantitative and reproducible approach. It utilizes acoustic radiation forces generated by the ultrasound probe to produce shear waves within the tissue. The velocity of these waves is directly related to tissue stiffness—stiffer tissues conduct shear waves faster than softer ones. By measuring shear wave speed, the system calculates and displays tissue elasticity in either kilopascals or meters per second. This technique allows for real-time assessment of tissue stiffness and provides numerical data that enhances objectivity and reduces interobserver variability. In clinical practice, shear wave elastography has demonstrated superior diagnostic accuracy in differentiating benign from malignant breast lesions and is less influenced by the skill of the operator (Orguc and Açar, 2022; Goddi et al., 2012).

The incorporation of elastography into routine breast imaging significantly enhances diagnostic performance. It improves clinician confidence, especially when conventional imaging results are inconclusive, and helps to reduce patient anxiety by minimizing unnecessary invasive procedures. For example, a lesion that appears suspicious on B-mode ultrasound but shows soft characteristics on elastography may be monitored instead of biopsied, thus reducing overdiagnosis and overtreatment. In addition, elastography has shown promise in evaluating response to treatment. As tumors undergo changes during chemotherapy or radiotherapy, alterations in tissue stiffness can be tracked, providing an early indicator of therapeutic effectiveness and helping to tailor treatment plans accordingly (Faruk et al., 2015).

Despite its many benefits, elastography is not without limitations. Strain elastography is highly dependent on the operator's technique and the consistency of applied pressure, which can affect image reproducibility. Meanwhile, shear wave elastography, while more robust, may face challenges in imaging deep-seated or very small lesions, where accurate shear wave propagation is harder to achieve. Moreover, conditions such as inflammation, fibrosis, or postsurgical changes may also increase tissue stiffness, potentially resulting in false-positive findings. Nevertheless, these limitations are increasingly being addressed with advances in imaging technology, software improvements, and increased clinician training (Ricci et al., 2014; Goddi et al., 2012).

2.2 Contrast-enhanced mammography in breast imaging

CEM represents one of the most promising advancements in breast imaging over the last decade. By integrating the use of intravenous contrast agents with digital mammography, CEM offers a powerful diagnostic approach for visualizing breast lesions, particularly in women with dense breast tissue or inconclusive findings on traditional imaging (Sensakovic et al., 2021). This technique is based on the principle of tumor angiogenesis, the formation of new blood vessels that supply nutrients to growing tumors. Since malignant breast tumors typically exhibit increased vascularity and altered permeability, they can be effectively highlighted using contrast agents. When an iodinated contrast medium is injected intravenously, it accumulates in areas with increased blood supply, which allows radiologists to visualize hypervascular lesions more clearly on mammographic images. As a result, CEM not only provides high-resolution anatomical information but also enhances the visualization of functional characteristics associated with cancer (Neeter et al., 2021).

CEM is particularly valuable in scenarios where conventional mammography might fall short. Traditional digital mammography, although effective, often has limitations in women with dense breast tissue, where overlapping fibroglandular structures can obscure lesions. CEM overcomes this limitation by providing dual-energy images that distinguish between normal and abnormal tissue based on contrast uptake. The technique involves acquiring two sets of images—low-energy and high-energy exposures—before and after contrast injection (Coffey and Jochelson, 2022). The low-energy images are similar to conventional mammograms and provide standard anatomical detail, while the high-energy images are used to isolate and subtract the contrast-enhanced components. The resulting recombined image highlights areas of contrast uptake, which are typically indicative of increased vascularization associated with malignancy (Sogani et al., 2021).

One of the most significant advantages of CEM is its high sensitivity, which is comparable to that of breast MRI. Studies have reported sensitivity rates between 85% and 90% for CEM, making it a reliable alternative, particularly for patients who cannot undergo MRI due to contraindications such as claustrophobia, implantable devices, or impaired renal function. Moreover, CEM offers the added benefit of shorter acquisition times, increased patient comfort, and lower cost compared to MRI, which makes it more accessible in certain clinical settings. In many cases, CEM has also demonstrated fewer false positives than MRI, reducing unnecessary biopsies and follow-up procedures. This balance of high sensitivity and improved specificity enhances diagnostic confidence, particularly when evaluating suspicious findings from screening mammograms or when monitoring known breast abnormalities (Weaver et al., 2023).

In terms of clinical utility, CEM is used for a range of diagnostic purposes. These include the evaluation of abnormal screening findings, assessment of the extent of disease in newly diagnosed breast cancer, monitoring of treatment response, and problem-solving in cases with discordant imaging results. For instance, in women with newly diagnosed breast cancer, CEM can help determine whether the disease is multifocal or multicentric, or if there is contralateral involvement. This information is crucial for surgical planning and treatment decisions. In addition, CEM is often used as an adjunct to ultrasound and standard mammography in complex diagnostic cases, particularly when initial imaging fails to clarify the nature of a lesion (Sensakovic et al., 2021).

Despite its advantages, CEM is not without limitations. One of the primary concerns is its relatively lower specificity compared to its sensitivity, which can result in false positives in certain cases. In addition, while CEM enhances detection of invasive cancers and some in situ lesions, it may be less effective in identifying DCIS, especially those without significant vascularity. This is particularly important because DCIS is a noninvasive form of breast cancer that, if undetected, could progress to a more aggressive form. Furthermore, the use of iodinated contrast agents introduces a potential risk for allergic reactions and nephrotoxicity, although such adverse events are rare. As with any procedure involving contrast media, careful patient selection and prescreening for renal function and contrast allergies are essential to ensure safety (Bennett et al., 2023; Wong et al., 2025).

Availability and accessibility of CEM remain another challenge. While its use is expanding, CEM is not yet universally available across all healthcare settings. The equipment and software required for dual-energy mammography with contrast are specialized, and not all imaging centers are equipped with this technology. In addition, radiologists require specific training to accurately interpret CEM images, as the technique involves analyzing both anatomical and functional data (Wong et al., 2025).

2.3 Scintimammography

Scintimammography is an imaging tool used to detect cancer cells in women with dense breast tissue, scar tissue from surgery, or implants in the breast. A radioisotope is used in scintimammography imaging to visualize breast lesions. Because mammography cannot detect breast cancer in dense breast tissue, it yields a high number of false-positive results (Sree, 2011).

Radiotracers are injected into a patient's bloodstream during scintimammography to detect tumors. Sestamibi, a radiotracer based on technetium-99 m, is typically used (Yang, 2011). A high-resolution, breast-specific gamma camera is used to capture images of the breast tissue taking up the radiotracer, which is absorbed by cancer cells. This imaging technique works well with dense breasts, implants, large and palpable abnormalities, and multiple tumors. Scintimammography is widely known for its high sensitivity in detecting breast cancer, especially in dense breast tissue where mammography may be less effective. This test can also be used to distinguish benign from malignant breast lesions based on the tissue's metabolic activity. Compared to other imaging modalities like mammography, scintimammography has a lower specificity, leading to false-positives. In addition to the radiation dose from the radiotracer, the procedure involves exposure to a small amount of ionizing radiation. However, radiation doses are generally considered safe and low (Iranmakani et al., 2020).

2.4 Positron emission mammography

The positron emission mammography (PEM) produces 3D images. A PET scan for breast cancer is essentially a high-resolution, small-field-of-view, dedicated scan. It detects a pair of rays coming from a radionuclide introduced into the body. Normal cells do not have an increased glucose metabolism, whereas malignant tumors do. Cancerous cells and normal cells can be distinguished well by PET imaging in this way. In addition to identifying the grade and stage of cancer, as well as evaluating treatment response, PET is often used to assess a patient's response to treatment. The procedure is mainly recommended when MRI scans are not indicated or are infeasible (Keshavarz et al., 2020).

A dedicated PEM scanner can offer advantages, such as better resolution, over a standard whole-body PET scanner. With PEM, the breast is scanned using a specialized scanning system that contains a ring of detectors or a set of two gamma radiation detectors. Breast compression is used to spread the breast and reduce its thickness (Sree, 2011). PEM is an emerging modality for imaging breast tissue functionally. Due to its low specifications and a dose equivalent to 30–40 mammograms, PEM is not very common for the early detection of low-risk cases (Iranmakani et al., 2020).

2.5 Optical imaging

In medical imaging, optical imaging uses visible, near-infrared, and infrared light to create images of tissues, organs, and other structures in the body. To generate images, this imaging method utilizes absorption, scattering, reflection, and fluorescence to interact with light and biological tissues (Sree, 2011). There are numerous applications for optical imaging in clinical and research settings, including monitoring physiological processes and diagnosing diseases. It employs light to observe and measure various properties of breast tissue in a noninvasive manner (Aflalo et al., 2024).

A visible light optical probe is primarily used for imaging superficial tissues, such as skin cancers. The use of optical imaging in the living body and in animal and plant tissues is particularly useful for studying cellular and molecular function. The use of optical imaging for research has been primarily restricted to small-animal models, but optical molecular imaging will have a substantial impact on clinical practice in several areas. Diagnosis and management of breast cancer are enhanced by optical imaging, which identifies breast abnormalities with increased hemoglobin, acting as a surrogate for cancer. It is expected that optical imaging will play a major role in breast cancer imaging in the future (Bennett et al., 2023).

A major benefit of optical imaging is its ability to provide high-resolution images, functional information, and noninvasiveness, which are all important considerations in medical research and clinical practice. In order to address the limitations of optical imaging and expand its clinical utility, researchers continue to develop and refine optical imaging techniques (Raghunathan et al., 2024).

3 Conclusions

Recent advancements in breast imaging have significantly enhanced early detection, diagnostic accuracy, and treatment planning, which has created a revolution in breast healthcare. The detection of breast pathologies is now more accurate and earlier, and radiation levels have been reduced due to innovations. Techniques such as DBT, CEM, and elastography, alongside AI integration, offer superior imaging capabilities, particularly for dense breast tissues. Despite these innovations, challenges remain, including accessibility, cost, and standardization of emerging modalities. The advancements in the field of breast imaging have resulted in better patient care, with a reduced mortality rate due to early and more accurate detection. Further improvements in breast imaging can be expected as technology continues to advance, enabling us to diagnose and treat breast cancer more effectively.

Breast imaging techniques have progressed significantly in recent years, creating a revolution in breast healthcare. Breast cancer detection is now more accurate and earlier, and radiation levels have been reduced due to these innovations. In combination, these advancements result in better patient outcomes and lower mortality rates due to earlier and more accurate cancer detection. Further improvements in breast imaging can be expected as technology continues to advance, enabling us to diagnose and treat breast cancer more effectively.

References

Breast Cancer Facts & Figures, 2019. American Cancer Society.

Aflalo, K., Gao, P., Trivedi, V., Sanjeev, A., Zalevsky, Z., 2024. Optical super-resolution imaging: a review and perspective. Opt. Laser Eng. 01438166. 183, 108536. doi:10.1016/j.optlaseng.2024.108536.

Ahn, J.S., Shin, S., Yang, S.A., Park, E.K., Kim, K.H., Cho, S.I., Ock, C.Y., Kim, S., 2023. Artificial intelligence in breast cancer diagnosis and personalized medicine. J. Breast Cancer. 20929900. 26 (5), 405–435. Available from: https://www.ejbc.kr/search.php?where=aview&id=10.4048/jbc.2023.26.e45&code=0096JBC&vmode=FULL. doi:10.4048/jbc.2023.26.e45.

Anderson, B.O., Braun, S., Lim, S., Smith, R.A., Taplin, S., Thomas, D.B., 2003. Early detection of breast cancer in countries with limited resources. Breast J. 9 (2), S51–S59. doi:10.1046/j.1524-4741.9.s2.4.x.

Ataeinia, B., Moghaddam, S.S., Shabani, M., Gohari, K., Sheidaei, A., Rezaei, N., Naderimagham, S., Ghasemi, E., Khalilabad, M.R., Roshani, S., Farzi, Y., Farzadfar, F., 2021. National and subnational incidence, mortality, and years of life lost due to breast cancer in Iran: trends and age-period-cohort analysis since 1990. Front. Oncol. 2234-943X. 11. doi:10.3389/fonc.2021.561376.

Barrios, C.H., Reinert, T., Werutsky, G., 2018. Global breast cancer research: moving forward. Am. Soc. Clin. Oncol. Educ. Book. 1548-8748. 38 (38), 441–450. doi:10.1200/edbk_209183.

Baughan, N., Douglas, L., Giger, M.L., 2022. Past, present, and future of machine learning and artificial intelligence for breast cancer screening. J. Breast Imaging. 26316129. 4 (5), 451–459. Available from: https://academic.oup.com/jbi. doi:10.1093/jbi/wbac052.

Bennett, C., Woodard, S., Zamora, K., 2023. A pictorial guide to artifacts on contrast mammography: how to avoid pitfalls and improve interpretation. Clin. Imaging. 08997071. 101, 215–222. doi:10.1016/j.clinimag.2023.06.019.

Bernardi, D., Gentilini, M.A., De Nisi, M., Pellegrini, M., Fantò, C., Valentini, M., Sabatino, V., Luparia, A., Houssami, N., 2020. Effect of implementing digital breast tomosynthesis (DBT) instead of mammography on population screening outcomes including interval cancer rates: results of the Trento DBT pilot evaluation. Breast. 15323080. 50, 135–140. Available from: http://www.elsevier-international.com/journals/brst/. doi:10.1016/j.breast.2019.09.012.

Bitencourt, A., Pinker, K., 2023. Artificial Intelligence in Breast Imaging Holistic Approach to Breast Disease. Springer Nature, Brazil, pp. 489–500. Available from https://link.springer.com/book/10.1007/978-981-99-0035-0. doi:10.1007/978-981-99-0035-0_36.

Bray, F., Laversanne, M., Sung, H., Ferlay, J., Siegel, R.L., Soerjomataram, I., Jemal, A., 2024. Global cancer statistics 2022: GLOBOCAN estimates of incidence and mortality worldwide for 36 cancers in 185 countries. CA Cancer J. Clin. 15424863. 74 (3), 229–263. Available from: http://onlinelibrary.wiley.com/journal/10.3322/(ISSN)1542-4863. doi:10.3322/caac.21834.

Coffey, K., Jochelson, M.S., 2022. Contrast-enhanced mammography in breast cancer screening. Eur. J. Radiol. 18727727. 156. Available from: www.elsevier.com/locate/ejrad. doi:10.1016/j.ejrad.2022.110513.

Dafni, U., Tsourti, Z., Alatsathianos, I., 2019. Breast cancer statistics in the european union: incidence and survival across european countries. Breast Care. 16613805. 14 (6), 344–353. Available from: www.karger.com/brc. doi:10.1159/000503219.

Dembrower, K., Crippa, A., Colón, E., Eklund, M., Strand, F., 2023. Artificial intelligence for breast cancer detection in screening mammography in Sweden: a prospective, population-based, paired-reader, non-inferiority study. Lancet Digit. Health. 25897500. 5 (10), e703. doi:10.1016/s2589-7500(23)00153-x.

Dkhar, W., Kadavigere, R., Mustaffa, S.P., 2021. Quantitative evaluation for differential diagnosis of breast lesions in diffusion-weighted MR imaging. Health Technol. 21907196. 11 (6), 1269–1275. Available from: http://www.springer.com/engineering/biomedical+eng/journal/12553. doi:10.1007/s12553-021-00604-z.

Dkhar, W., Kadavigere, R., Paruthikunnan, S.M., 2020. Image quality analysis for optimization of b-value in diffusion weighted MRI of breast. Malaysian J. Med. Health Sci. 26369346. 16 (4), 126–131. Available from https://medic.upm.edu.my/upload/dokumen/2020120209374117_MJMHS_0078.pdf.

Dkhar, W., Kadavigere, R., Sukumar, S., Pradhan, A., Sharath, S., 2023. Diagnostic performances of ADC value in diffusion-weighted MR imaging for differential diagnosis of breast lesions in 1.5 T: a systematic review and meta-analysis. J. Med. Biol. Eng. 1609-0985. 43 (5), 497–507. doi:10.1007/s40846-023-00825-9.

Elizabeth S, M., Schopp, J.G., Peacock, S., DeMartini, W.D., Rahbar, H., Lehman, C.D., Partridge, S.C., Diffusion-weighted, M.R.I., 2014. Association between patient characteristics and apparent diffusion coefficients of normal breast fibroglandular tissue at 3 T. Am. J. Roentgenol. 15463141. 202 (5), W496. Available from: http://www.ajronline.org/doi/pdf/10.2214/AJR.13.11159. doi:10.2214/AJR.13.11159.

Evans, A., Trimboli, R.M., Athanasiou, A., Balleyguier, C., Baltzer, P.A., Bick, U., Herrero, J.C., Clauser, P., Colin, C., Cornford, E., Fallenberg, E.M., Fuchsjaeger, M.H., Gilbert, F.J., Helbich, T.H., Kinkel, K., Heywang-Köbrunner, S.H., Kuhl, C.K., Mann, R.M., Martincich, L., Panizza, P., Pediconi, F., Pijnappel, R.M., Pinker, K., Zackrisson, S., Forrai, G., Sardanelli, F., 2018. Breast ultrasound: recommendations for information to women and referring physicians by the European Society of Breast Imaging. Insights Imaging. 18694101. 9 (4), 449–461. Available from: http://www.springer.com/medicine/radiology/journal/13244. doi:10.1007/s13244-018-0636-z.

Faruk, T., Islam, M.K., Arefin, S., Haq, M.Z., 2015. The journey of elastography: background, current status, and future possibilities in breast cancer diagnosis. Clin. Breast Cancer. 19380666. 15 (5), 313–324. Available from: http://www.journals.elsevier.com/clinical-breast-cancer. doi:10.1016/j.clbc.2015.01.002.

Gao, M., Wik, S.L., Yu, Q.Y., Xue, F.Y., Chan, S.C., Chow, S.H., Adebisi, Y.A., Zhong, C.C., Lucero-Prisno, D.E., Wong, M.C.S., Huang, J., Burden, D., 2024. Risk factors, and temporal trends in breast cancer in low- and middle-income countries: a global study. Public Health Chall. 3 (3), 27692450. Available from https://onlinelibrary.wiley.com/journal/27692450. doi:10.1002/puh2.223.

Gelardi, F., Ragaini, E.M., Sollini, M., Bernardi, D., Chiti, A., 2022. Contrast-enhanced mammography versus breast magnetic resonance imaging: a systematic review and meta-analysis. Diagnostics. 20754418. 12 (8). Available from: http://www.mdpi.com/journal/diagnostics/. doi:10.3390/diagnostics12081890.

Giannakou, A., Dagdeviren, C., Ozmen, T., 2025. Conformable ultrasound breast patch—the future of breast cancer screening? Eur. J. Breast Health. 25870831. 21 (1), 90–92. doi:10.4274/ejbh.galenos.2024.2024-11-3.

Goddi, A., Bonardi, M., Alessi, S., 2012. Breast elastography: a literature review. J. Ultrasound. 19713495. 15 (3), 192–198. doi:10.1016/j.jus.2012.06.009.

Hooley, R.J., Scoutt, L.M., Philpotts, L.E., 2013. Breast ultrasonography: state of the art. Radiology. 15271315. 268 (3), 642–659. Available from: http://radiology.rsna.org/content/268/3/642.full.pdf+html. doi:10.1148/radiol.13121606.

Iacob, R., Iacob, E.R., Stoicescu, E.R., Ghenciu, D.M., Cocolea, D.M., Constantinescu, A., Ghenciu, L.A., Manolescu, D.L., 2024. Evaluating the role of breast ultrasound in early detection of breast cancer in low- and middle-income countries: a comprehensive narrative review. Bioengineering. 23065354. 11 (3). Available from: www.mdpi.com/journal/bioengineering. doi:10.3390/bioengineering11030262.

Iranmakani, S., Mortezazadeh, T., Sajadian, F., Ghaziani, M.F., Ghafari, A., Khezerloo, D., Musa, A.E., 2020. A review of various modalities in breast imaging: technical aspects and clinical outcomes. Egypt. J. Radiol. Nucl. Med. 20904762. 51 (1). Available from: https://ejrnm.springeropen.com/. doi:10.1186/s43055-020-00175-5.

Jain, A., Barge, A., Parris, C.N., 2025. Combination strategies with PARP inhibitors in BRCA-mutated triple-negative breast cancer: overcoming resistance mechanisms. Oncogene. 14765594. 44 (4), 193–207. Available from: https://www.nature.com/onc/. doi:10.1038/s41388-024-03227-6.

Jamshidi, M.H., Karami, A., Keshavarz, A., Fatemi, A., Ghanavati, S., 2024. Magnetic resonance elastography for breast cancer diagnosis through the assessment of tissue biomechanical properties. Health Sci. Rep. 2398-8835. 7 (12). doi:10.1002/hsr2.70253.

Jeong, S.M., Park, S.M., 2021. Care for breast cancer survivors. Adv. Exp. Med. Biol. 1187, 511–524. Available from http://www.springer.com/series/5584. doi:10.1007/978-981-32-9620-6_27.

Keshavarz, K., Jafari, M., Lotfi, F., Bastani, P., Salesi, M., Gheisari, F., Hemami, M.R., 2020. Positron Emission Mammography (PEM) in the diagnosis of breast cancer: a systematic review and economic evaluation. Med. J. Islam. Repub. Iran. 10161430. 34. doi:10.47176/mjiri.34.100.

Kieturakis, A.J., Wahab, R.A., Vijapura, C., Mahoney, M.C., 2021. Current recommendations for breast imaging of the pregnant and lactating patient. Am. J. Roentgenol. 15463141. 216 (6), 1462–1475. Available from: https://www.ajronline.org/doi/pdfplus/10.2214/AJR.20.23905. doi:10.2214/AJR.20.23905.

Lamb, L.R., Lehman, C.D., Gastounioti, A., Conant, E.F., Bahl, M., 2022. Artificial intelligence (AI) for screening mammography, from the AJR special series on AI applications. Am. J. Roentgenol. 15463141. 219 (3), 369–381. Available from: https://www.ajronline.org/doi/epdf/10.2214/AJR.21.27071. doi:10.2214/AJR.21.27071.

Lee, C.U., Hesley, G.K., Pierson, T.A., Higgins, R.L., Urban, M.W., 2024. Breast ultrasound knobology and the knobology of twinkling for marker detection. Transl. Breast Cancer Res. 22186778. 5. Available from: https://tbcr.amegroups.org/article/download/92269/68194. doi:10.21037/tbcr-24-30.

Lei, S., Zheng, R., Zhang, S., Wang, S., Chen, R., Sun, K., Zeng, H., Zhou, J., Wei, W., 2021. Global patterns of breast cancer incidence and mortality: a population-based cancer registry data analysis from 2000 to 2020. Cancer Commun. 2523-3548. 41 (11), 1183–1194. doi:10.1002/cac2.12207.

Liu, J., Lei, J., Ou, Y., Zhao, Y., Tuo, X., Zhang, B., Shen, M., 2023. Mammography diagnosis of breast cancer screening through machine learning: a systematic review and meta-analysis. Clin. Exp. Med. 1591-9528. 23 (6), 2341–2356. doi:10.1007/s10238-022-00895-0.

Mahtani, R., Kittaneh, M., Kalinsky, K., Mamounas, E., Badve, S., Vogel, C., Lower, E., Schwartzberg, L., Pegram, M., 2021. Advances in therapeutic approaches for triple-negative breast cancer. Clin. Breast Cancer. 15268209. 21 (5), 383–390. doi:10.1016/j.clbc.2020.12.011.

Manda-Mapalo, M.T., Fine, S.G., Safadi, S., Lee, J.H., Du, R., Sussman, A.L., Mishra, S., Selwyn, R.G., Saline, J.L., Hine, W.L., Brown-Glaberman, U.A., 2020. Breast cancer screening among medically underserved women in New Mexico: potential for lower recall rates with digital breast tomosynthesis. J. Womens Health. 1931843X. 29 (12), 1596–1601. Available from: www.liebertonline.com/jwh. doi:10.1089/jwh.2020.8402.

Manson, E.N., Achel, D.G., 2023. Fighting breast cancer in low-and-middle-income countries – what must we do to get every woman screened on regular basis? Sci. Afr. 24682276. 21. Available from: https://www.journals.elsevier.com/scientific-african. doi:10.1016/j.sciaf.2023.e01848.

Mao, Y.J., Lim, H.J., Ni, M., Yan, W.H., Wong, D.W.C., Cheung, J.C.W., 2022. Breast tumour classification using ultrasound elastography with machine learning: a systematic scoping review. Cancers. 20726694. 14 (2). Available from: https://www.mdpi.com/2072-6694/14/2/367/pdf. doi:10.3390/cancers14020367.

Mariscotti, G., Durando, M., Houssami, N., Zuiani, C., Martincich, L., Londero, V., Caramia, E., Clauser, P., Campanino, P.P., Regini, E., Luparia, A., Castellano, I., Bergamasco, L., Sapino, A., Fonio, P., Bazzocchi, M., Gandini, G., 2016. Digital breast tomosynthesis as an adjunct to digital mammography for detecting and characterising invasive lobular cancers: a multi-reader study. Clin. Radiol. 00099260. 71 (9), 889–895. doi:10.1016/j.crad.2016.04.004.

McKnight, A., 2002. MR elastography of breast cancer: preliminary results. Am. J. Roentgenol. 178 (6), 1411–1417. doi:10.2214/AJR.178.6.1781411.

Neeter, L.M.F.H., Raat, H.P.J., Alcantara, R., Robbe, Q., Smidt, M.L., Wildberger, J.E., Lobbes, M.B.I., 2021. Contrast-enhanced mammography: what the radiologist needs to know. BJR Open. 25139878. 3 (1). Available from: https://academic.oup.com/bjro/advance-articles. doi:10.1259/bjro.20210034.

Newman, L., 2023. Oncologic anthropology: global variations in breast cancer risk, biology, and outcome. J. Surg. Oncol. 0022-4790. 128 (6), 959–966. doi:10.1002/jso.27459.

Orguc, S., Açar, Ç.R., 2022. Correlation of shear-wave elastography and apparent diffusion coefficient values in breast cancer and their relationship with the prognostic factors. Diagnostics. 20754418. 12 (12). Available from: http://www.mdpi.com/journal/diagnostics/. doi:10.3390/diagnostics12123021.

Partridge, S.C., McKinnon, G.C., Henry, R.G., Hylton, N.M., 2001. Menstrual cycle variation of apparent diffusion coefficients measured in the normal breast using MRI. J. Magn. Reson. Imaging. 10531807. 14 (4), 433–438. doi:10.1002/jmri.1204.

Partridge, S.C., Singer, L., Sun, R., Wilmes, L.J., Klifa, C.S., Lehman, C.D., Hylton, N.M., Diffusion-weighted, M.R.I., 2011. Influence of intravoxel fat signal and breast density on breast tumor conspicuity and apparent diffusion coefficient measurements. Magn. Reson. Imaging. 18735894. 29 (9), 1215–1221. doi:10.1016/j.mri.2011.07.024.

Raghunathan, R., Vasquez, M., Zhang, K., Zhao, H., Wong, S.T.C., 2024. Label-free optical imaging for brain cancer assessment. Trends Cancer. 24058033. 10 (6), 557–570. Available from: https://www.sciencedirect.com/science/journal/24058033. doi:10.1016/j.trecan.2024.03.005.

Ramos-Pollán, R., Guevara-López, M.A., Suárez-Ortega, C., Díaz-Herrero, G., Franco-Valiente, J.M., Rubio-Del-Solar, M., González-De-Posada, N., Vaz, M.A.P., Loureiro, J., Ramos, I., 2012. Discovering mammography-based machine learning classifiers for breast cancer diagnosis. J. Med. Syst. 1573689X. 36 (4), 2259–2269. doi:10.1007/s10916-011-9693-2.

Ricci, P., Maggini, E., Mancuso, E., Lodise, P., Cantisani, V., Catalano, C., 2014. Clinical application of breast elastography: state of the art. Eur. J. Radiol. 0720048X. 83 (3), 429–437. doi:10.1016/j.ejrad.2013.05.007.

Richman, I.B., Long, J.B., Hoag, J.R., Upneja, A., Hooley, R., Xu, X., Kunst, N., Aminawung, J.A., Kyanko, K.A., Busch, S.H., Gross, C.P., 2021. Comparative effectiveness of digital breast tomosynthesis for breast cancer screening among women 40–64 years old. J. Natl. Cancer Inst. 14602105. 113 (11), 1515–1522. Available from: http://jnci.oxfordjournals.org/. doi:10.1093/jnci/djab063.

Sardanelli, F., Cozzi, A., Trimboli, R.M., Schiaffino, S., 2020. Gadolinium retention and breast MRI screening: more harm than good? Am. J. Roentgenol. 15463141. 214 (2), 324–327. Available from: https://www.ajronline.org/doi/pdf/10.2214/AJR.19.21988. doi:10.2214/AJR.19.21988.

Sensakovic, W.F., Carnahan, M.B., Czaplicki, C.D., Fahrenholtz, S., Panda, A., Zhou, Y., Pavlicek, W., Patel, B., 2021. Contrast-enhanced mammography: how does it work? Radiographics. 15271323. 41 (3), 829–839. Available from: https://pubs.rsna.org/doi/pdf/10.1148/rg.2021200167. doi:10.1148/rg.2021200167.

Shin, S., Ko, E.S., Kim, R.B., Han, B.K., Nam, S.J., Shin, J.H., Hahn, S.Y., 2015. Effect of menstrual cycle and menopausal status on apparent diffusion coefficient values and detectability of invasive ductal carcinoma on diffusion-weighted MRI. Breast Cancer Res. Treat. 15737217. 149 (3), 751–759. Available from: www.wkap.nl/journalhome.htm/0167-6806. doi:10.1007/s10549-015-3278-6.

Si, G., Du, Y., Tang, P., Ma, G., Jia, Z., Zhou, X., Dan, M., Shen, Y., Lu, Y., Mao, Y., Chen, C., Yan, L., Ning, G., 2024. Unveiling the next generation of MRI contrast agents: current insights and perspectives on ferumoxytol-enhanced MRI. Natl. Sci. Rev. 2095-5138. 11 (5). doi:10.1093/nsr/nwae057.

Sogani, J., Mango, V.L., Keating, D., Sung, J.S., Jochelson, M.S., 2021. Contrast-enhanced mammography: past, present, and future. Clin. Imaging. 18734499. 69, 269–279. Available from: www.elsevier.com/locate/clinimag. doi:10.1016/j.clinimag.2020.09.003.

Sree, S.V., 2011. Breast imaging: a survey. World J. Clin. Oncol. 2218-4333. 2 (4), 171. doi:10.5306/wjco.v2.i4.171.

Steyerova, P., Burgetova, A., 2022. Current imaging techniques and impact on diagnosis and survival—a narrative review. Ann. Breast Surg. 26162776. 6, 25. doi:10.21037/abs-21-22.

Stubblefield, M.D., 2017. The underutilization of rehabilitation to treat physical impairments in breast cancer survivors. PM R. 19341482. 9 (9), S317. Available from: http://www.pmrjournal.org/. doi:10.1016/j.pmrj.2017.05.010.

Tejerina Bernal, A., Bernal, A.T.B., Rabadán Doreste, F., González, A.D.L., Roselló Llerena, J.A., Gómez, A.T., 2012. Breast imaging: how we manage diagnostic technology at a multidisciplinary breast center. J. Oncol. 16878469. doi:10.1155/2012/213421.

Tfayli, A., Temraz, S., Mrad, R.A., Shamseddine, A., 2010. Breast cancer in low- and middle-income countries: an emerging and challenging epidemic. J. Oncol. 1687-8450. 2010, 1–5. doi:10.1155/2010/490631.

Wang, Y., Li, Y., Song, Y., Chen, C., Wang, Z., Li, L., Liu, M., Liu, G., Yali, X., Zhou, Y., Sun, Q., Shen, S., 2022. Comparison of ultrasound and mammography for early diagnosis of breast cancer among Chinese women with suspected breast lesions: a prospective trial. Thorac. Cancer. 1759-7706. 13 (22), 3145–3151. doi:10.1111/1759-7714.14666.

Weaver, O.O., Yang, W.T., Scoggins, M.E., Adrada, B.E., Arribas, E., Moseley, T.W., Esquivel, J., Melgar, Y., Kornecki, A., 2023. Challenging contrast-enhanced mammography–guided biopsies: practical approach using real-time multimodality imaging and a proposed procedural algorithm. Am. J. Roentgenol. 15463141. 220 (4), 512–523. Available from: https://www.ajronline.org/doi/10.2214/AJR.22.28572. doi:10.2214/AJR.22.28572.

Wong, Y.T., Yang, S., Leong, P.W., Chow, D.L.Y., Kwok, K.Y., Lau, K.Y.K., Li, O.C.A., 2025. Artefacts and diagnostic pitfalls in contrast-enhanced mammography: a pictorial review. Clin. Radiol. 00099260. 81, 106763. doi:10.1016/j.crad.2024.106763.

Wu, P.S., Hong, Y.T., Shen, C.H., Lee, C.H., Chou, C.P., 2024. Digital breast tomosynthesis screening improves early breast cancer detection and survival in Taiwan. J. Breast Imaging. 26316129. 6 (6), 601–609. Available from: https://academic.oup.com/jbi. doi:10.1093/jbi/wbae044.

Yang, W.T., 2011. Emerging techniques and molecular imaging in breast cancer. Semin. Ultrasound CT MRI. 15585034. 32 (4), 288–299. doi:10.1053/j.sult.2011.03.003.

Chapter 31

Recent trends and techniques in prostate imaging

Kaushik Nayak[a], Winniecia Dkhar[a], Rajagopal Kadavigere[b], Anand Venugopal[c], N. Praveen[d], Veena Mayya[e], Leena R. David[a,f], Abhimanyu Pradhan[a], Berna Uzun[g], Dilber Uzun Ozsahin[f,g,h], S. Sharath[a], Ilker Ozsahin[g], and Neil Abraham[a]

[a]*Department of Medical Imaging Technology, Manipal College of Health Professions, Manipal Academy of Higher Education, Manipal, Karnataka, India*
[b]*Department of Radiodiagnosis and Imaging, Kasturba Medical College, Manipal, India*
[c]*Manipal Academy of Higher Education, Manipal, India*
[d]*Department of Radiodiagnosis & Imaging, Kasturba Hospital, Manipal, India*
[e]*Department of Information & Communication Technology, Manipal Academy of Higher Education, Manipal, India*
[f]*Medical Diagnostic Imaging Department, College of Health Sciences, University of Sharjah, Sharjah, United Arab Emirates*
[g]*Irfan Suat Gunsel Operational Research Institute, Near East University, Nicosia, Turkish Republic of Northern Cyprus (TRNC), Turkey*
[h]*Research Institute for Medical and Health Sciences, University of Sharjah, Sharjah, United Arab Emirates*

1 Introduction

The prostate gland is a small, walnut-sized organ that plays a vital role in the male reproductive system. It is situated just beneath the urinary bladder and encircles the upper portion of the urethra, the tube that carries urine and semen out of the body through the penis. Structurally and functionally, the prostate is composed of both glandular and muscular elements, and its secretory function is pivotal to male fertility. The gland produces a significant portion of the seminal fluid that, along with sperm cells from the testes and fluids from other glands like the seminal vesicles, forms semen. The prostate's secretions contain enzymes, prostate-specific antigen (PSA), and other substances that help nourish and transport sperm, enhancing their motility and viability for successful fertilization (Henry et al., 2018; Sethi et al., n.d.).

Given the prevalence and potential severity of prostate conditions, accurate and timely diagnosis is essential. This is where prostate imaging comes into play. Medical imaging allows clinicians to visualize the internal structure of the prostate gland, identify abnormalities, and make informed decisions about management and treatment. Over the past few decades, advances in imaging technology have dramatically enhanced our ability to evaluate the prostate noninvasively (Huang et al., 2025; Porter and Frankel, 2020).

Among the available imaging techniques, magnetic resonance imaging (MRI) has emerged as the cornerstone of prostate imaging, especially in the context of prostate cancer. MRI provides excellent soft tissue contrast, making it particularly effective in differentiating normal prostate tissue from abnormal areas. Its ability to provide high-resolution, multiparametric images allows for a detailed assessment of the gland's anatomy and pathology. MRI is unique in that it combines multiple imaging sequences—such as T2-weighted imaging (T2WI), diffusion-weighted imaging (DWI), dynamic contrast-enhanced imaging (DCE), and sometimes magnetic resonance spectroscopy (MRS)—to deliver a comprehensive overview of the prostate's structure and function (Mussi, 2018; Wang et al., 2014).

One of the most important contributions of MRI in prostate care is in cancer detection, localization, and staging. Unlike traditional imaging methods such as ultrasound or computed tomography (CT), which offer limited soft tissue differentiation, MRI can visualize tumors within the prostate and assess their extent with remarkable accuracy. This is crucial for differentiating between clinically significant cancers that require treatment and indolent tumors that may be safely monitored through active surveillance. Multiparametric MRI (mpMRI) has significantly improved the detection of clinically significant prostate cancer, especially in patients with elevated PSA levels but negative biopsy results (Hegde et al., 2013; Morrison et al., 2017; Kaninjing et al., 2018).

Advances in Medical Imaging. https://doi.org/10.1016/B978-0-443-28967-5.00033-2

MRI is also invaluable in guiding targeted prostate biopsies. Traditionally, prostate biopsies have been performed under transrectal ultrasound guidance, which can result in random sampling and missed tumors. With MRI, suspicious lesions can be identified with precision, and biopsies can be targeted to these areas, improving diagnostic yield and reducing the likelihood of underdiagnosis. This approach is referred to as MRI-transrectal ultrasonography (TRUS) fusion biopsy, where MRI findings are fused with real-time ultrasound images during the biopsy procedure (Huang et al., 2025; Porter and Frankel, 2020).

In addition to detection and diagnosis, MRI plays a critical role in treatment planning and posttreatment follow-up. For patients undergoing surgical intervention, such as radical prostatectomy, MRI can help assess the relationship of the tumor to important anatomical structures, like the neurovascular bundles, aiding in nerve-sparing surgical planning. In the context of radiation therapy, MRI helps in delineating the target volumes with precision. After treatment, MRI is used to monitor for residual or recurrent disease, particularly in patients who have undergone radiation therapy or focal treatments, like cryotherapy and high-intensity focused ultrasound (HIFU) (Sethi et al., n.d.; Huang et al., 2025).

MRI has also been instrumental in improving the standardization of prostate imaging. The development of the Prostate Imaging Reporting and Data System (PI-RADS) has facilitated uniform interpretation and reporting of MRI findings. PI-RADS scores lesions from 1 to 5 based on the likelihood of clinically significant cancer, aiding clinicians in decision-making regarding the need for biopsy or further intervention (Turkbey and Purysko, 2023).

2 Anatomy and functions of the prostate

The prostate is a walnut-sized gland of the male reproductive system located just below the bladder and surrounding the upper part of the urethra. It is composed of both glandular and fibromuscular tissue and is anatomically divided into four zones: the peripheral zone, central zone, transition zone, and anterior fibromuscular stroma (Shin et al., 2023; Ali et al., 2022). Given its anatomical location and functional responsibilities, the prostate is prone to several pathological conditions, particularly in aging males. The most common prostate disorders include prostate cancer, benign prostatic hyperplasia (BPH), and prostatitis. These conditions can significantly affect a man's urinary and sexual function and, in some cases, can be life-threatening if not detected and treated early. Prostate cancer is among the most commonly diagnosed cancers in men worldwide and remains a leading cause of cancer-related mortality. It often develops slowly and may initially present with few or no symptoms, making early detection critical. BPH, a noncancerous enlargement of the prostate, is extremely common in older men and can cause troublesome lower urinary tract symptoms (LUTS) such as frequent urination, weak urinary stream, and nocturia. Prostatitis, which refers to inflammation of the prostate gland, can result from bacterial infections or other causes and may present with pelvic pain, urinary difficulties, and general discomfort (Cui et al., 2024; Rani et al., 2018; Dahut et al., 2019). The peripheral zone is the largest and the most common site for prostate cancer, while the transition zone is typically affected in BPH (Holder et al., 2021; Lin et al., 2016). The main function of the prostate is to produce and secrete prostatic fluid, which makes up about 20%–30% of the semen. This fluid contains enzymes such as PSA, citric acid, and zinc, which help nourish and protect sperm and aid in sperm motility by liquefying the ejaculate. The prostate also plays a role in urinary continence by supporting the internal urethral sphincter and in ejaculation by preventing the backward flow of semen into the bladder (Verze et al., 2016; Marcoccia et al., 2025; Risbridger and Taylor, 2006). Its growth and function are regulated by male hormones, particularly dihydrotestosterone (DHT). Clinically, the prostate is significant due to common conditions such as prostatitis, BPH, and prostate cancer (Carson and Rittmaster, 2003).

3 Prostate pathologies

The prostate gland is susceptible to several pathological conditions that vary in severity and clinical impact. One of the most common is BPH, a noncancerous enlargement of the prostate, typically involving the transition zone and associated with age-related hormonal changes, particularly elevated DHT (Holder et al., 2021; Carson and Rittmaster, 2003). BPH often leads to LUTS such as frequent urination, nocturia, weak stream, and incomplete bladder emptying. Another common condition is prostatitis, which refers to inflammation of the prostate and is classified into acute bacterial prostatitis, chronic bacterial prostatitis, chronic pelvic pain syndrome, and asymptomatic inflammatory prostatitis (Tang and Yang, 2009; Jarvis et al., 2014; Casals et al., 2021). Symptoms can range from pelvic pain and painful urination to systemic signs like fever. Prostate cancer, the most serious pathology, usually arises in the peripheral zone and may remain asymptomatic in early stages. It is typically diagnosed through elevated PSA levels and confirmed by biopsy. Risk factors include age, genetics, and ethnicity, and treatment may involve surgery, radiation, or hormone therapy. Prostatic intraepithelial neoplasia, particularly the high-grade type, is considered a precursor to cancer and is often detected incidentally during biopsies

(Tang and Yang, 2009; Roobol et al., 2009; Rani et al., 2018). Prostate abscesses, though less common, can result from untreated bacterial prostatitis and present with severe urinary symptoms and fever, often requiring drainage. Lastly, prostatic calculi, or stones within the prostate, are frequently seen in older men or those with chronic prostatitis and are usually asymptomatic, though they can occasionally contribute to discomfort or infection (Deshpande et al., 2013; Wu et al., 2024).

4 Imaging modalities for prostate evaluation

4.1 Magnetic resonance imaging's importance in prostate imaging

For a number of compelling reasons, MRI is crucial in prostate imaging. *Excellent Soft Tissue Contrast:* MRI's superior soft tissue contrast makes it possible to see the prostate gland and its surroundings in detail. For identifying normal and diseased tissue in the prostate, this is especially useful. *Imaging using multiparametric data:* mpMRI is an advanced imaging technique used for the evaluation and assessment of the prostate gland. It combines several imaging sequences to deliver comprehensive data on the prostate tissue and is very helpful for identifying and classifying prostate cancer (Mussi, 2018; Hu et al., 2023). T2WI, DWI, DCE-MRI, and occasionally MRS are some of the MRI sequences that are combined in mpMRI. The prostate's anatomy, cellularity, perfusion, and metabolic activity can all be fully understood using this multiparametric method in a single examination (Maurer and Heverhagen, 2017).

4.2 T2-weighted imaging

T2WI is an important technique in prostate imaging because it provides valuable information about the anatomy and pathology of the prostate gland. It allows for the differentiation between various tissue types within the prostate, which is essential for distinguishing normal prostate tissue from anomalies such as tumors, cysts, or inflammation. This imaging modality offers excellent anatomical visualization, clearly depicting features like the prostate capsule, seminal vesicles, and adjacent organs, including the bladder and rectum, due to its high contrast resolution (Duenweg et al., 2023). T2-weighted images are particularly effective in detecting malignant tumors, as cancerous tissue typically appears hypointense (darker) compared to the hyperintense (brighter) signal of healthy tissue. This contrast aids in the identification and localization of tumors. In addition, T2WI plays a critical role in evaluating tumor characteristics such as size, location, and extent, which are vital for staging prostate cancer and planning appropriate therapeutic interventions (Nketiah et al., 2021; Bucher et al., 2024). It also enables the assessment of extracapsular extension, helping determine whether cancer has spread beyond the prostate capsule into surrounding tissues, thereby influencing treatment strategies. Furthermore, T2WI is useful in identifying anatomical variations in the prostate's shape, size, and glandular tissue distribution, which is important for personalized surgical and treatment planning. It can also be used for longitudinal monitoring, as serial T2-weighted scans allow clinicians to observe changes in the prostate over time, assess disease progression, and evaluate the effectiveness of treatments. This imaging modality supports interventional procedures such as needle biopsies by guiding the sampling of targeted regions within the prostate. Overall, T2-weighted images are integral to the comprehensive management of prostate conditions, aiding clinicians in making informed decisions about treatment options and follow-up care based on detailed anatomical and pathological insights (Emad-Eldin et al., 2014; Haider et al., 2007; Chatterjee et al., 2019).

4.3 T1-weighted imaging

T1-weighted imaging is a vital and complementary technique in prostate MRI that, when used alongside T2WI, offers a more comprehensive and detailed evaluation of the prostate gland. It plays a significant role in the characterization of prostate tissue, as T1-weighted images provide insights into the inherent signal properties of various tissue types. The differences in T1 relaxation times among normal prostate tissue, malignant tumors, cysts, and hemorrhagic regions result in varying signal intensities on T1-weighted sequences, which aid in differentiating between these structures. This capability is particularly important for identifying abnormalities that may not be as distinctly visualized on other sequences. One of the key advantages of T1-weighted imaging is its sensitivity to hemorrhage within the prostate gland. Hemorrhagic regions often appear hyperintense (brighter) on T1-weighted images, allowing radiologists to detect internal bleeding or vascular abnormalities, which could be associated with recent biopsy procedures or pathological changes. In addition, T1-weighted imaging is useful for distinguishing between fat- and water-containing tissues, helping to identify areas of fatty infiltration or nodules that might be linked to certain benign or malignant conditions (Lovegrove et al., 2018; Giganti et al., 2019).

This imaging technique is also integral to DCE-MRI, where T1-weighted images are acquired before and after the intravenous administration of contrast agents to evaluate vascularity and perfusion within the prostate. Tumors, especially aggressive ones, tend to exhibit early and persistent enhancement due to increased neovascularity, and T1-weighted imaging helps in recognizing these suspicious patterns, thereby improving the accuracy of prostate cancer detection and localization. Moreover, T1-weighted images are commonly used for measuring the volume of the prostate gland, a critical parameter in diagnosing conditions such as BPH and in formulating surgical or interventional plans. It also plays a role in the assessment of posttreatment changes, such as those following radiation therapy, cryotherapy, or focal ablation, by revealing alterations in tissue architecture, density, and enhancement patterns that may indicate fibrosis, residual tumor, or recurrence (Baur et al., 2020; Aydin et al., 2015).

Furthermore, T1-weighted imaging can assist in identifying calcifications within the prostate, which may appear as signal voids and be associated with chronic inflammation, infection, or prior treatment. Although not always the primary imaging sequence for prostate cancer detection, T1-weighted imaging provides crucial supplemental information that enhances diagnostic accuracy when interpreted alongside other sequences. The findings from T1-weighted images, along with other mpMRI components, are synthesized using the PI-RADS. PI-RADS standardizes the interpretation and reporting of prostate MRI findings, ensuring consistency and improving communication among radiologists, urologists, and oncologists. In summary, T1-weighted imaging enriches the overall mpMRI assessment by contributing valuable anatomical, structural, and functional insights, thereby playing a pivotal role in the diagnosis, risk assessment, treatment planning, and posttherapeutic evaluation of prostate diseases (Higaki et al., 2024; Kyk et al., 2016).

4.4 DWI-weighted imaging

Prostate imaging requires the use of DWI, which adds essential data to other MRI sequences, including T1- and T2WI, contributing to a more comprehensive assessment of the prostate gland. DWI functions by measuring the Brownian motion of water molecules within tissues, and this molecular motion reflects tissue cellularity and structural integrity. In the context of prostate cancer, changes such as increased cell proliferation, reduced extracellular space, and architectural disruption lead to restricted diffusion of water molecules. This makes DWI a powerful tool in identifying and characterizing malignant tissue, which typically exhibits lower apparent diffusion coefficient values due to restricted water mobility, in contrast to the relatively higher diffusion observed in normal or benign tissues. High-grade prostate tumors, which are more aggressive and clinically significant, often show more pronounced diffusion restriction because of their dense cellular architecture, making DWI especially helpful in detecting these lesions, many of which may not be readily visible on conventional T1- or T2-weighted sequences (Maurer and Heverhagen, 2017; Emad-Eldin et al., 2014).

DWI also plays a vital role in localizing suspicious lesions within the prostate, particularly when combined with other MRI sequences as part of mpMRI. This precise lesion localization assists in directing targeted biopsies, ensuring that samples are taken from the most diagnostically relevant areas, thus increasing the likelihood of detecting clinically significant cancer. In terms of staging, DWI contributes to the identification of extracapsular extension (ECE) and seminal vesicle invasion (SVI), both of which are critical indicators of locally advanced disease and essential for determining the most appropriate treatment pathway. Since ECE is associated with a higher risk of metastasis and disease progression, its accurate detection significantly influences prognosis and therapeutic decision-making (Tan et al., 2011; Foster et al., 2023).

DWI also provides valuable information in monitoring treatment response, especially following interventions such as radiation therapy, chemotherapy, or focal therapies like cryotherapy and HIFU. Tumors that respond effectively to treatment often demonstrate increased diffusion, whereas persistent restriction may suggest residual disease. In addition, DWI is instrumental in evaluating suspected recurrences, as regions of recurrent prostate cancer often present with restricted diffusion similar to primary tumors, even after therapeutic intervention. This functional aspect of imaging allows for noninvasive and ongoing assessment of disease activity over time (Lovegrove et al., 2018; Tamada et al., 2022; Erena et al., 2020).

In the biopsy setting, DWI helps reduce sampling errors by guiding the needle to the most suspicious regions with constrained diffusion, thereby improving biopsy accuracy and minimizing unnecessary tissue sampling. Furthermore, DWI can aid in differentiating prostate cancer from BPH, a common condition in older men that may mimic cancer on conventional imaging. Due to their unique diffusion profiles, BPH nodules typically show less restriction compared to malignant tumors, allowing DWI to contribute to more accurate diagnosis and treatment selection. This distinction is particularly important in preventing the overtreatment of indolent or slow-growing tumors, which may not require immediate intervention. By identifying patients suitable for active surveillance, DWI helps avoid the adverse effects of unnecessary treatments, thus supporting more personalized patient care (Patel and Oto, 2016; Matsukawa et al., 2024).

MRI, in general, is highly sensitive and specific for prostate cancer detection, often identifying small or early-stage tumors—especially in the peripheral zone, which is the most common site of origin for prostate cancer. The high spatial and contrast resolution of MRI, combined with DWI, enhances the visualization and characterization of these lesions. In terms of risk classification, MRI findings are interpreted using the PI-RADS, which standardizes image assessment and assigns scores that indicate the likelihood of clinically significant cancer. This system supports clinical decision-making by helping differentiate between lesions that require intervention and those that may be monitored (Bucher et al., 2024; Haider et al., 2007; Adams et al., 2022).

Moreover, DWI, along with other MRI sequences, plays a central role in targeted biopsy techniques such as MRI-guided in-bore biopsies and MRI-ultrasound fusion biopsies. These methods improve lesion targeting, reduce procedural complications, and enhance the accuracy of prostate cancer diagnosis. In focal therapy planning, MRI helps define the exact size, location, and margins of tumors, allowing clinicians to treat only the affected portion of the prostate while preserving healthy surrounding tissue, thereby maintaining urinary and sexual function in many cases (Cui et al., 2024; Morlando et al., 2017).

DWI also contributes to functional imaging, offering physiological information about tissue characteristics, such as cell density, perfusion, and viability, which helps differentiate aggressive tumors from benign conditions. Alongside DCE-MRI, DWI enriches the diagnostic framework by providing both structural and functional insights. After treatment, MRI is used to evaluate therapeutic outcomes, assess residual or recurrent disease, and guide further management decisions. It is particularly useful in complex posttreatment anatomy, where conventional imaging may fall short (Nagayama et al., 2011; Lee et al., 2018).

MRI, including DWI, can also be integrated with other imaging modalities such as positron emission tomography (PET), particularly prostate-specific membrane antigen (PSMA) PET, for a more thorough evaluation of disease, especially when metastasis is suspected. This hybrid approach improves the detection of cancer spread and supports more informed treatment strategies. With the ongoing evolution of imaging technologies, including advancements in artificial intelligence (AI), radiomics, and machine learning, DWI and MRI are continually being refined to enhance diagnostic precision, reduce variability in interpretation, and promote early and individualized prostate cancer management. These innovations are helping to shape a future where prostate imaging not only detects disease more accurately but also plays a central role in predicting outcomes and guiding personalized, evidence-based care (Cuocolo et al., 2019).

4.5 Transrectal ultrasonography

TRUS is a widely utilized and essential imaging technique for examining and evaluating the prostate gland, often performed in conjunction with MRI to provide a comprehensive and accurate assessment of prostate pathology. This minimally invasive procedure involves the insertion of a specialized ultrasound probe, approximately the size of a finger, into the rectum, allowing it to be positioned near the prostate for optimal image acquisition. Before the procedure, patients are typically instructed to empty their bladders and bowels to improve image clarity and are positioned on their side with knees bent for ease and comfort during the examination. The probe, once lubricated, emits high-frequency sound waves that reflect off the prostate tissues. These echoes are captured and converted into real-time images displayed on a monitor, enabling the healthcare provider to assess the prostate's shape, size, texture, and internal architecture, as well as identify abnormalities such as nodules, lesions, or areas of calcification that may suggest benign conditions or malignancies like prostate cancer. Once the necessary images are acquired, the probe is carefully withdrawn (Power et al., 2022).

TRUS offers several notable benefits. One of its most significant advantages is the ability to deliver real-time imaging, which is crucial for guiding prostate biopsies and other interventional procedures with precision. It allows urologists to target specific areas within the prostate that appear suspicious, thereby increasing the diagnostic accuracy and effectiveness of biopsy procedures. In prostate cancer diagnosis and grading, TRUS-guided biopsies remain a standard approach due to their ability to access and sample tissue from multiple prostate zones. The technique is also used for calculating prostate volume, which is vital in diagnosing and managing BPH, determining PSA density, and planning surgical or medical interventions. TRUS is integral during therapeutic procedures such as cryotherapy, brachytherapy (radioactive seed implantation), and transrectal biopsies, as it helps ensure that treatments are accurately delivered to the intended regions of the prostate (Liau et al., 2019; Grey et al., 2020).

Furthermore, TRUS can be employed for long-term monitoring of prostate diseases, particularly in patients on active surveillance for prostate cancer. It helps track changes in the gland's size, appearance, or the emergence of new lesions over time. TRUS also aids in localizing intraprostatic lesions, cysts, and calcifications, and in delineating anatomical features like the seminal vesicles and the bladder neck—structures that are important for surgical planning and staging of disease. It is especially useful in postprocedural monitoring, helping to evaluate complications such as bleeding, infection, or abscess formation following biopsies or therapeutic interventions (Calderone et al., 2023).

While TRUS provides a wide range of benefits, it also has certain limitations, particularly in its ability to distinguish between benign and malignant tissue purely based on echotexture. Its soft tissue resolution is inferior to that of MRI, making it less effective in detecting and characterizing early or subtle prostate cancers, especially those located in the anterior or transitional zones. As a result, TRUS is often combined with MRI to overcome these limitations. MRI offers superior soft tissue contrast, multiplanar capabilities, and functional imaging sequences such as DWI and DCE, making it highly sensitive for tumor localization and staging. When used together, TRUS and MRI provide a synergistic approach—TRUS offers real-time procedural guidance, while MRI contributes detailed anatomical and functional insights. This integration enhances diagnostic accuracy, improves biopsy targeting, and supports personalized treatment planning, ultimately leading to better clinical outcomes for patients with prostate diseases (Grey et al., 2020; Lang et al., 2024).

4.6 Positron emission tomography

The prostate can be imaged using PET, which is frequently coupled with other imaging modalities, like CT, to provide more comprehensive data. Because they can help detect areas of elevated metabolic activity, which may be symptomatic of malignant tissue, PET scans are helpful in the examination of prostate cancer (Evangelista and Fanti, 2023).

Metastasis detection: PET imaging, specifically utilizing radiotracers like PSMA, choline, or sodium fluoride (NaF), is highly sensitive in detecting metastatic prostate cancer lesions, even in cases of low PSA levels. It can spot metastases in lymph nodes, bones, and other distant locations. Early recurrence detection: Following initial treatment, such as radical prostatectomy or radiation therapy, PET imaging is useful for identifying early indicators of prostate cancer recurrence. It can detect distant metastases or local recurrence inside the prostate bed, allowing for prompt intervention. Risk Assessment: By assessing PSMA expression in prostate lesions, PET imaging, particularly with PSMA-PET, can assist in stratifying patients' risk levels. The determination of the cancer's aggressiveness and the development of individualized treatments are both aided by this information. Treatment planning: By giving precise information on the location and size of prostate cancer lesions, PET imaging helps treatment decisions. Choosing the best course of treatment, such as surgery, radiation therapy, systemic therapy, or focused therapy, is aided by this. Monitoring treatment response: To track a patient's response to treatment, PET scans are performed (Hoffman and Amiel, 2023; Dondi et al., 2024). On subsequent PET scans, changes in the size and metabolic activity of lesions can show if a treatment is working or whether changes are required. Biopsy guidance: PET imaging, especially when paired with other imaging modalities like MRI, can help guide targeted biopsies. It ensures that biopsies are carried out on locations with elevated PSMA expression or metabolic activity, improving diagnostic precision. Identification of primary lesions: When the primary prostate tumor is hard to find or identify, PET imaging can help. This is especially true for patients with high PSA levels and negative results from earlier imaging tests. Evaluation of extracapsular extension: PET scans can help in evaluating prostate cancer's ECE, which is important for treatment planning and surgical judgment. Combination with MRI: PET-MRI fusion imaging integrates the metabolic data from PET with the structural and functional information from MRI to provide a more thorough examination of prostate cancer. This multimodal strategy improves diagnostic precision (Keidar et al., 2018; Grizzi and Taverna, 2024).

Digital rectal examination (DRE): During this physical exam, a healthcare professional inserts a gloved finger into the rectum to feel for any prostate gland anomalies, such as lumps or enlargement. The first stage in prostate assessment frequently involves DRE. TRUS: This kind of ultrasound is frequently used to assess the prostate. The prostate gland may be imaged in great detail thanks to a tiny ultrasonic probe that is placed into the rectum. TRUS is frequently used as prostate biopsy guidance and aids in determining the size, shape, and texture of the prostate. Prostate MRI: This is a sophisticated imaging technique that offers fine-grained pictures of the prostate gland. Prostate cancer staging and detection are becoming more common. MRI can assist in locating abnormal prostate tissue that may need further examination or biopsy. CT Scan: When prostate cancer has spread to adjacent lymph nodes or structures, CT scans may be employed. In most cases, it is not the main imaging technique used to assess the prostate. PET scan: PET scans can be used to find prostate cancer that has spread to other parts of the body. They are frequently paired with a radiotracer like 18F-choline or 68Ga-PSMA (Maurer et al., 2016; Lawhn-Heath et al., 2021). It is very helpful for determining how far prostate cancer has spread. mpMRI: This specialized MRI method combines different imaging sequences to offer comprehensive details regarding the structure and function within the prostate. For the diagnosis and staging of prostate cancer, it is widely used. PSMA PET/CT: This cutting-edge imaging method uses a radiotracer that targets PSMA, a protein that is frequently overexpressed in prostate cancer cells. PSMA PET/CT is extremely accurate in finding prostate cancer metastases and is capable of detecting small tumors (Franzese et al., 2021).

The patient's symptoms, risk factors, and a prostate problem, whether suspected or diagnosed, all influence the imaging modality decision. Prostate imaging plays a vital role in assisting treatment decisions, such as prostate cancer staging and treatment planning, and is a crucial factor in diagnosing and managing prostate illnesses (Ye et al., 2024).

5 Treatment planning and posttherapy imaging

Treatment planning for prostate pathologies, particularly prostate cancer, requires a multidisciplinary approach that includes clinical staging, histopathological evaluation, biomarker assessment (e.g., PSA levels), and advanced imaging techniques (Lawhn-Heath et al., 2021; McLaughlin et al., 2024). MRI, especially mpMRI, plays a critical role in local staging, tumor localization, and assessing extracapsular extension, SVI, and nodal involvement, which are vital for guiding decisions between active surveillance, surgery (radical prostatectomy), radiation therapy, or hormonal therapy (Adams et al., 2022; Patel and Oto, 2016; Albano et al., 2021). For benign conditions like BPH, imaging (e.g., transrectal ultrasound or MRI) helps assess prostate volume and bladder outlet obstruction before planning medical or surgical treatment (e.g., transurethral resection of the prostate (TURP)). Posttherapy imaging is essential for monitoring treatment response, detecting residual or recurrent disease, and assessing complications. In prostate cancer, postradical prostatectomy imaging (typically mpMRI or PSMA PET/CT) is used to detect local recurrence or metastasis, especially in the presence of rising PSA levels (McLaughlin et al., 2024; Shin et al., 2023; Ali et al., 2022; Holder et al., 2021; Lin et al., 2016; Deshpande et al., 2013; Wu et al., 2024; Verze et al., 2016; Marcoccia et al., 2025; Risbridger and Taylor, 2006; Carson and Rittmaster, 2003; Tang and Yang, 2009; Jarvis et al., 2014; Casals et al., 2021; Roobol et al., 2009; Rani et al., 2018; Lawhn-Heath et al., 2021; McLaughlin et al., 2024; Adams et al., 2022; Patel and Oto, 2016; Albano et al., 2021; Franzese et al., 2021; Porter and Frankel, 2020; Grey et al., 2020; Grizzi and Taverna, 2024; Lang et al., 2024). After radiation therapy, imaging can differentiate between postradiation changes and tumor persistence. In BPH and prostatitis, follow-up imaging helps evaluate therapeutic efficacy, resolution of inflammation, or detect abscess formation. Overall, imaging serves as a cornerstone in both initial treatment planning and long-term disease surveillance, enhancing precision in prostate disease management (Lasocki et al., 2025).

6 Artificial intelligence and radiomics in prostate imaging

AI plays an increasingly transformative role in prostate cancer imaging, particularly in the interpretation and analysis of mpMRI. By utilizing machine learning and deep learning techniques, AI enhances diagnostic accuracy, reduces interobserver variability, and enables faster and more consistent interpretation of prostate imaging. One of its primary contributions is automated prostate gland and lesion segmentation, which saves time and standardizes image analysis across institutions (Tătaru et al., 2021). AI algorithms can also assist in lesion detection and classification by assigning PI-RADS scores and generating probability maps or heatmaps that highlight regions of interest, supporting radiologists in identifying clinically significant cancer. Furthermore, AI can integrate mpMRI data with clinical variables such as PSA levels, Gleason scores, and patient history to stratify cancer risk and distinguish between indolent and aggressive tumors. Radiomics, a subset of AI, extracts advanced quantitative features from MRI data, such as shape, texture, and intensity, that are not visible to the human eye, enabling predictions about tumor behavior, treatment response, and prognosis (Ayyad et al., 2025; Chaddad et al., 2023). AI also supports MRI-ultrasound fusion biopsy by improving the alignment of MRI-identified lesions with real-time ultrasound images, enhancing targeting accuracy, and reducing unnecessary sampling. In treatment planning, AI aids in determining tumor extent and involvement of adjacent structures, which is essential for surgical and radiation therapy strategies. Posttreatment, AI can track subtle changes across serial imaging to detect recurrence or assess therapeutic response. Despite challenges related to data diversity, model generalizability, regulatory approval, and clinical implementation, AI holds enormous potential to revolutionize prostate cancer diagnosis and management. As AI tools become more integrated into clinical workflows and validated through large-scale studies, they are expected to complement radiologists by offering faster, more precise, and individualized care in the diagnosis, treatment planning, and follow-up of prostate cancer (Nair et al., 2025; Sunoqrot et al., 2022; Sun et al., 2024; Esteban et al., 2025).

7 Conclusion

MRI has emerged as a transformative tool in the evaluation and management of prostate diseases, particularly prostate cancer. Its multiparametric capability, integrating T2WI, DWI, DCE, and occasionally MR spectroscopy, enables detailed anatomical visualization, functional assessment, and metabolic characterization of prostate tissue. This comprehensive approach allows for accurate detection, localization, and staging of clinically significant prostate cancer while minimizing unnecessary biopsies and reducing the risk of overdiagnosis and overtreatment.

MRI has also revolutionized the way prostate biopsies are conducted. With the advent of MRI-guided and MRI-ultrasound fusion biopsies, clinicians can now target suspicious lesions with higher precision, improving diagnostic yield and patient outcomes. Moreover, MRI is invaluable in treatment planning—whether for surgery, radiotherapy, or focal therapy—by providing critical insights into tumor size, location, and possible extension beyond the prostatic capsule. Posttreatment, MRI serves as a noninvasive modality to monitor disease recurrence and evaluate therapeutic efficacy.

Looking ahead, advances such as AI, radiomics, and theranostic imaging hold immense potential to further refine prostate MRI. AI algorithms are being developed to enhance image interpretation and reduce variability among radiologists, while radiomics enables the extraction of high-dimensional imaging features that may predict tumor behavior and treatment response. Theranostic imaging, which combines diagnostic and therapeutic capabilities, represents a futuristic approach in personalized prostate cancer care.

References

Adams, L.C., Makowski, M.R., Engel, G., Rattunde, M., Busch, F., Asbach, P., Niehues, S.M., Vinayahalingam, S., van Ginneken, B., Litjens, G., Bressem, K.K., 2022. Prostate158—an expert-annotated 3T MRI dataset and algorithm for prostate cancer detection. Comput. Biol. Med. 18790534. 148. Available from: www.elsevier.com/locate/compbiomed. doi:10.1016/j.compbiomed.2022.105817.

Alabd, M., 2015. Severe urinary tract infection with complicating prostatic abscess. In: Radiopaedia.org. doi:10.53347/rID-41479.

Albano, D., Benenati, M., Bruno, A., Bruno, F., Calandri, M., Caruso, D., Cozzi, D., De Robertis, R., Gentili, F., Grazzini, I., Micci, G., Palmisano, A., Pessina, C., Scalise, P., Vernuccio, F., Barile, A., Miele, V., Grassi, R., Messina, C., Albano, D., Benenati, M., Bruno, A., Bruno, F., Calandri, M., Caruso, D., Cozzi, D., De Robertis, R., Gentili, F., Grazzini, I., Micci, G., Palmisano, A., Pessina, C., Scalise, P., Vernuccio, F., Messina, C., 2021. Imaging side effects and complications of chemotherapy and radiation therapy: a pictorial review from head to toe. Insights Imaging. 1869-4101. 12 (1). doi:10.1186/s13244-021-01017-2.

Ali, A., Feu, A.D., Oliveira, P., Choudhury, A., Bristow, R.G., Baena, E., 2022. Prostate zones and cancer: lost in transition? Nat. Rev. Urol. 1759-4812. 19 (2), 101–115. doi:10.1038/s41585-021-00524-7.

Aydin, H., Kizilgoz, V., Tekin, B.O., 2015. Overview of current multiparametric magnetic, resonance imaging approach in the diagnosis and staging of prostate cancer. Kaohsiung J. Med. Sci. 24108650. 31 (4), 167–178. Available from: http://www.elsevier.com/wps/find/journaldescription.cws_home/708701/description#description. doi:10.1016/j.kjms.2015.01.002.

Ayyad, S.M., Abdel-Hamid, N.B., Ali, H.A., Labib, L.M., 2025. Multimodality imaging in prostate cancer diagnosis using artificial intelligence: basic concepts and current state-of-the-art. Multimed. Tools Appl. 15737721. Available from: https://www.springer.com/journal/11042. doi:10.1007/s11042-025-20786-2.

Baur, A.D.J., Hansen, C.M., Rogasch, J., Posch, H., Elezkurtaj, S., Maxeiner, A., Erb-Eigner, K., Makowski, M.R., 2020. Evaluation of T1 relaxation time in prostate cancer and benign prostate tissue using a Modified Look-Locker inversion recovery sequence. Sci. Rep. 20452322. 10 (1). Available from: www.nature.com/srep/index.html. doi:10.1038/s41598-020-59942-z.

Bucher, A.M., Egger, J., Dietz, J., Strecker, R., Hilbert, T., Frodl, E., Wenzel, M., Penzkofer, T., Hamm, B., Chun, F.K.H., Vogl, T., Kleesiek, J., Beeres, M., 2024. Value of MRI—T2 mapping to differentiate clinically significant prostate cancer. J. Imaging Inform. Med. 2948-2933. 37 (6), 3304–3315. doi:10.1007/s10278-024-01150-6.

Calderone, C.E., Turner, E.M., Hayek, O.E., Summerlin, D., West, J.T., Rais-Bahrami, S., Galgano, S.J., 2023. Contemporary review of multimodality imaging of the prostate gland. Diagnostics. 20754418. 13 (11). Available from: http://www.mdpi.com/journal/diagnostics/. doi:10.3390/diagnostics13111860.

Carson, C., Rittmaster, R., 2003. The role of dihydrotestosterone in benign prostatic hyperplasia. Urology. 00904295. 61 (4), 2–7. doi:10.1016/s0090-4295(03)00045-1.

Casals, R., Badalato, G., Rutman, M., 2021. Role of diet in the development and progression of benign prostatic hyperplasia. In: Molecular Mechanisms of Nutritional Interventions and Supplements for the Management of Sexual Dysfunction and Benign Prostatic Hyperplasia. Elsevier, United States, pp. 81–94. Available from: https://www.sciencedirect.com/book/9780128197653. doi:10.1016/B978-0-12-819765-3.00011-X.

Chaddad, A., Tan, G., Liang, X., Hassan, L., Rathore, S., Desrosiers, C., Katib, Y., Niazi, T., 2023. Advancements in MRI-based radiomics and artificial intelligence for prostate cancer: a comprehensive review and future prospects. Cancers. 2072-6694. 15 (15), 3839. doi:10.3390/cancers15153839.

Chatterjee, A., Devaraj, A., Mathew, M., Szasz, T., Antic, T., Karczmar, G.S., Oto, A., 2019. Performance of T2 maps in the detection of prostate cancer. Acad. Radiol. 18784046. 26 (1), 15–21. doi:10.1016/j.acra.2018.04.005.

Cui, H., Zhang, W., Zhang, L., Yang, Q., Zhengxing, X., Tan, Z., Yan, P., Tang, M., Yang, C., Wang, Y., Chen, L., Xiao, C., Zou, Y., Liu, Y., Zhang, L., Yang, Y., Yao, Y., Jiayuan, L., Liu, Z., Yang, C., Jiang, X., Zhang, B., Chaudhuri, A.A., 2024. Risk factors for prostate cancer: an umbrella review of prospective observational studies and mendelian randomization analyses. PLoS Med. 1549-1676. 21 (3), e1004362. doi:10.1371/journal.pmed.1004362.

Cuocolo, R., Cipullo, M.B., Stanzione, A., Ugga, L., Romeo, V., Radice, L., Brunetti, A., Imbriaco, M., 2019. Machine learning applications in prostate cancer magnetic resonance imaging. Eur. Radiol. Exp. 25099280. 3 (1). Available from: https://link.springer.com/journal/41747. doi:10.1186/s41747-019-0109-2.

Dahut, W.L., Couvillon, A., Pinto, P.A., Turkbey, B., Karzai, F., 2019. Natural history and imaging in men with high genetic risk for developing prostate cancer. Can. J. Urol. 11959479. 26 (5), 7–8.

Deshpande, A., Haleblian, G., Rapose, A., 2013. Prostate abscess: MRSA spreading its influence into Gram-negative territory: case report and literature review. BMJ Case Rep. 1757-790X. 2013, bcr2013009057. doi:10.1136/bcr-2013-009057.

Dondi, F., Antonelli, A., Suardi, N., Treglia, G., Bertagna, F., 2024. The role of PSMA PET imaging in the classification of the risk of prostate cancer patients: a systematic review on the insights to guide an active surveillance approach. Cancers. 2072-6694. 16 (6), 1122. doi:10.3390/cancers16061122.

Duenweg, S.R., Bobholz, S.A., Barrett, M.J., Lowman, A.K., Winiarz, A., Nath, B., Stebbins, M., Bukowy, J., Iczkowski, K.A., Jacobsohn, K.M., Vincent-Sheldon, S., LaViolette, P.S., 2023. T2-weighted MRI radiomic features predict prostate cancer presence and eventual biochemical recurrence. Cancers. 20726694. 15 (18). Available from: http://www.mdpi.com/journal/cancers/. doi:10.3390/cancers15184437.

Emad-Eldin, S., Halim, M., Metwally, L.I.A., Abdel-Aziz, R.M., 2014. Diffusion-weighted MR imaging and ADC measurement in normal prostate, benign prostatic hyperplasia and prostate carcinoma. Egypt. J. Radiol. Nucl. Med. 20904762. 45 (2), 535–542. Available from: http://www.elsevier.com/wps/find/journaldescription.cws_home/724146/description#description. doi:10.1016/j.ejrnm.2014.02.017.

Erena, A.N., Shen, G., Lei, P., 2020. Prostate cancer awareness and screening practice among Kenyan men. Eur. J. Cancer Prev. 14735709. 29 (3), 252–258. Available from: http://journals.lww.com/eurjcancerprev/pages/default.aspx. doi:10.1097/CEJ.0000000000000501.

Esteban, L.M., Borque-Fernando, Á., Escorihuela, M.E., Esteban-Escaño, J., Abascal, J.M., Servian, P., Morote, J., 2025. Integrating radiological and clinical data for clinically significant prostate cancer detection with machine learning techniques. Sci. Rep. 20452322. 15 (1). Available from: https://www.nature.com/srep/. doi:10.1038/s41598-025-88297-6.

Evangelista, L., Fanti, S., 2023. PET/CT in prostate cancer. Cancers. 2072-6694. 15 (15), 3751. doi:10.3390/cancers15153751.

Foster, S., Carvallo, M., Wenske, M., Lee, J., 2023. Damaged masculinity: how honor endorsement can influence prostate cancer screening decision-making and prostate cancer mortality rates. Pers. Soc. Psychol. Bull. 0146-1672. 49 (2), 296–308. doi:10.1177/01461672211065293.

Franzese, E., De Falco, S., Laterza, M.M., Montella, L., Facchini, S., Liguori, C., Coppola, P., Diessa, Y., Berretta, M., Pisconti, S., Trama, F., Ferro, M., Crocetto, F., Fasano, M., Buonerba, C., Facchini, G., 2021. The use of 68Ga prostate-specific membrane antigen PET-CT in prostate cancer: diagnostic challenges and therapeutic opportunities. Future Sci. OA. 20565623. 7 (6). Available from: http://www.future-science.com/loi/fso?open=2015. doi:10.2144/fsoa-2021-0035.

Gaillard, F., 2008. Benign prostatic hyperplasia. In: Radiopaedia.org. doi:10.53347/rID-2586.

Giganti, F., Rosenkrantz, A.B., Villeirs, G., Panebianco, V., Stabile, A., Emberton, M., Moore, C.M., 2019. The evolution of MRI of the prostate: the past, the present, and the future. Am. J. Roentgenol. 15463141. 213 (2), 384–396. Available from: https://www.ajronline.org/doi/pdf/10.2214/AJR.18.20796. doi:10.2214/AJR.18.20796.

Grey, A.D.R., Connor, M.J., Tam, J., Loch, T., 2020. Can transrectal prostate ultrasound compete with multiparametric MRI in the detection of clinically significant prostate cancer? Transl. Androl. Urol. 22234683. 9 (3), 1492–1500. doi:10.21037/tau.2020.02.26.

Grizzi, F., Taverna, G., 2024. Editorial: PET/CT and MRI in prostate cancer. Front. Oncol. 2234-943X. 14. doi:10.3389/fonc.2024.1421542.

Hacking, C., 2020. Prostate gland and seminal vesicles (Gray's illustration) In: Radiopaedia.org. doi:10.53347/rID-81748.

Haider, M.A., Kwast, T.H.V.D., Tanguay, J., Evans, A.J., Hashmi, A.T., Lockwood, G., Trachtenberg, J., 2007. Combined T2-weighted and diffusion-weighted MRI for localization of prostate cancer. Am. J. Roentgenol. 0361803X. 189 (2), 323–328. doi:10.2214/AJR.07.2211.

Hegde, J.V., Mulkern, R.V., Panych, L.P., Fennessy, F.M., Fedorov, A., Maier, S.E., Tempany, C.M.C., 2013. Multiparametric MRI of prostate cancer: an update on state-of-the-art techniques and their performance in detecting and localizing prostate cancer. J. Magn. Reson. Imaging. 15222586. 37 (5), 1035–1054. doi:10.1002/jmri.23860.

Henry, G.H., Malewska, A., Joseph, D.B., Malladi, V.S., Lee, J., Torrealba, J., Mauck, R.J., Gahan, J.C., Raj, G.V., Roehrborn, C.G., Hon, G.C., MacConmara, M.P., Reese, J.C., Hutchinson, R.C., Vezina, C.M., Strand, D.W., 2018. A cellular anatomy of the normal adult human prostate and prostatic urethra. Cell Rep. 22111247. 25 (12), 3530. Available from: http://www.sciencedirect.com/science/journal/22111247. doi:10.1016/j.celrep.2018.11.086.

Higaki, A., Tamada, T., Kido, A., Takeuchi, M., Ono, K., Miyaji, Y., Yoshida, K., Sanai, H., Moriya, K., Yamamoto, A., 2024. Short repetition time diffusion-weighted imaging improves visualization of prostate cancer. Jpn. J. Radiol. 1867-1071. 42 (5), 487–499. doi:10.1007/s11604-023-01519-7.

Hoffman, A., Amiel, G.E., 2023. The impact of PSMA PET/CT on modern prostate cancer management and decision making—the urological perspective. Cancers. 20726694. 15 (13). Available from: http://www.mdpi.com/journal/cancers/. doi:10.3390/cancers15133402.

Holder, K., Galvan, B., Sakya, J., Frost, J., de Riese, W., 2021. Anatomical changes of the peripheral zone depending on benign prostatic hyperplasia size and their potential clinical implications: a review for clinicians. Urol. Pract. 2352-0779. 8 (2), 259–263. doi:10.1097/upj.0000000000000201.

Hu, L., Fu, C., Song, X., Grimm, R., von Busch, H., Benkert, T., Kamen, A., Lou, B., Huisman, H., Tong, A., Penzkofer, T., Choi, M.H., Shabunin, I., Winkel, D., Xing, P., Szolar, D., Coakley, F., Shea, S., Szurowska, E., Guo, J.-Y., Liang, L., Yue-hua, L., Zhao, J.-G., 2023. Automated deep-learning system in the assessment of MRI-visible prostate cancer: comparison of advanced zoomed diffusion-weighted imaging and conventional technique. Cancer Imaging. 1470-7330. 23 (1). doi:10.1186/s40644-023-00527-0.

Huang, X., Ye, H., Hu, Y., Lei, Y., Tian, Y., Huang, X., Zhang, J., Zhang, Y., Gui, B., Liu, Q., Zhang, G., Deng, Q., 2025. Ultrasound super-resolution imaging for non-invasive assessment of microvessel in prostate lesion. Cancer Imaging. 1470-7330. 25 (1). doi:10.1186/s40644-024-00819-z.

Jarvis, T.R., Chughtai, B., Kaplan, S.A., 2014. Bladder outlet obstruction and BPH. Curr. Bladder Dysfunct. Rep. 19317220. 9 (4), 372–378. Available from: http://www.springer.com/medicine/urology/journal/11884. doi:10.1007/s11884-014-0263-6.

Kaninjing, E., Lopez, I., Nguyen, J., Odedina, F., Young, M.E., 2018. Prostate cancer screening perception, beliefs, and practices among men in Bamenda, Cameroon. Am. J. Mens Health. 15579891. 12 (5), 1463–1472. Available from: http://www.sagepub.com. doi:10.1177/1557988318768596.

Keidar, Z., Gill, R., Goshen, E., Israel, O., Davidson, T., Morgulis, M., Pirmisashvili, N., Ben-Haim, S., 2018. 68Ga-PSMA PET/CT in prostate cancer patients – patterns of disease, benign findings and pitfalls. Cancer Imaging. 1470-7330. 18 (1). doi:10.1186/s40644-018-0175-3.

Kyk, W., Cyh, C., Ky, K., Wk, W., Kw, T., 2016. Magnetic resonance imaging of prostate cancer. Hong Kong J. Radiol. 22236619. 19 (1), 64–73. Available from: http://www.hkjr.org/system/files/v19n1_64Magnetic.pdf. doi:10.12809/hkjr1615299.

Lang, J., McClure, T.D., Margolis, D.J.A., 2024. MRI—ultrasound fused approach for prostate biopsy—how it is performed. Cancers. 20726694. 16 (7). Available from: http://www.mdpi.com/journal/cancers/. doi:10.3390/cancers16071424.

Lasocki, A., Sia, J., Stuckey, S.L., 2025. Differentiation between radiation necrosis and true tumour progression after radiotherapy to intracranial metastases. J. Med. Imaging Radiat. Oncol. 17549485. 69 (3), 414–424. Available from: http://onlinelibrary.wiley.com/journal/10.1111/(ISSN)1754-9485. doi:10.1111/1754-9485.13847.

Lawhn-Heath, C., Salavati, A., Behr, S.C., Rowe, S.P., Calais, J., Fendler, W.P., Eiber, M., Emmett, L., Hofman, M.S., Hope, T.A., 2021. Prostate-specific membrane antigen PET in prostate cancer. Radiology. 15271315. 299 (2), 248–260. Available from: http://pubs.rsna.org/doi/10.1148/radiol.2021202771. doi:10.1148/RADIOL.2021202771.

Lee, H., Hwang, S.I., Lee, H.J., Byun, S.S., Lee, S.E., Hong, S.K., 2018. Diagnostic performance of diffusion-weighted imaging for prostate cancer: peripheral zone versus transition zone. PLoS ONE. 19326203. 13 (6). Available from: http://journals.plos.org/plosone/article/file?id=10.1371/journal.pone.0199636&type=printable. doi:10.1371/journal.pone.0199636.

Liau, J., Goldberg, D., Arif-Tiwari, H., 2019. Prostate cancer detection and diagnosis: role of ultrasound with MRI correlates. Curr. Radiol. Rep. 2167-4825. 7 (3). doi:10.1007/s40134-019-0318-8.

Lin, Y.T., Amouyal, G., Correas, J.M., Pereira, H., Pellerin, O., Giudice, C.D., Déan, C., Thiounn, N., Sapoval, M., 2016. Can prostatic arterial embolisation (PAE) reduce the volume of the peripheral zone? MRI evaluation of zonal anatomy and infarction after PAE. Eur. Radiol. 14321084. 26 (10), 3466–3473. Available from: www.link.springer.de/link/service/journals/00330/index.htm. doi:10.1007/s00330-015-4177-2.

Lovegrove, C.E., Matanhelia, M., Randeva, J., Eldred-Evans, D., Tam, H., Miah, S., Winkler, M., Ahmed, H.U., Shah, T.T., 2018. Prostate imaging features that indicate benign or Malignant pathology on biopsy. Transl. Androl. Urol. 22234691. 7, S420. Available from: www.amepc.org/tau/index. doi:10.21037/tau.2018.07.06.

Marcoccia, D., Mollari, M., Galli, F.S., Cuva, C., Tassinari, V., Mantovani, A., 2025. Prostate as a target of endocrine disrupting chemicals: relevance, pathways, assays. Reprod. Toxicol. 18731708. 133. Available from: https://www.sciencedirect.com/science/journal/08906238. doi:10.1016/j.reprotox.2025.108867.

Matsukawa, A., Yanagisawa, T., Bekku, K., Parizi, M.K., Laukhtina, E., Klemm, J., Chiujdea, S., Mori, K., Kimura, S., Fazekas, T., Miszczyk, M., Miki, J., Kimura, T., Karakiewicz, P.I., Rajwa, P., Shariat, S.F., 2024. Comparing the performance of digital rectal examination and prostate-specific antigen as a screening test for prostate cancer: a systematic review and meta-analysis. Eur. Urol. Oncol. 25889311. 7 (4), 697–704. doi:10.1016/j.euo.2023.12.005.

Maurer, M.H., Heverhagen, J.T., 2017. Diffusion weighted imaging of the prostate-principles, application, and advances. Transl. Androl. Urol. 22234691. 6 (3), 490–498. Available from: http://tau.amegroups.com/article/download/14949/15072. doi:10.21037/tau.2017.05.06.

Maurer, T., Eiber, M., Schwaiger, M., Gschwend, J.E., 2016. Current use of PSMA-PET in prostate cancer management. Nat. Rev. Urol. 17594820. 13 (4), 226–235. Available from: http://www.nature.com/nrurol/index.html. doi:10.1038/nrurol.2016.26.

McLaughlin, P.W., Cousins, M.M., Tsodikov, A., Soni, P.D., Crook, J.M., 2024. Mortality reduction and cumulative excess incidence (CEI) in the prostate-specific antigen (PSA) screening era. Sci. Rep. 20452322. 14 (1). Available from: https://www.nature.com/srep/. doi:10.1038/s41598-024-55859-z.

Morlando, M., Pelullo, C.P., Di Giuseppe, G., 2017. Prostate cancer screening: knowledge, attitudes and practices in a sample of men in Italy. A survey. PLoS ONE. 19326203. 12 (10). Available from: http://journals.plos.org/plosone/article/file?id=10.1371/journal.pone.0186332&type=printable. doi:10.1371/journal.pone.0186332.

Morrison, B.F., Aiken, W.D., Mayhew, R., Gordon, Y., Odedina, F.T., 2017. Prostate cancer knowledge, prevention, and screening behaviors in Jamaican Men. J. Cancer Educ. 15430154. 32 (2), 352–356. Available from: http://www.springer.com/biomed/cancer/journal/13187. doi:10.1007/s13187-016-0991-8.

Mussi, T.C., 2018. Current state of multiparameter magnetic resonance imaging of the prostate. Einstein (São Paulo. 2317-6385. 16 (2). doi:10.1590/s1679-45082018md4408.

Nagayama, M., Watanabe, Y., Terai, A., Araki, T., Notohara, K., Okumura, A., Amoh, Y., Ishimori, T., Nakashita, S., Dodo, Y., 2011. Determination of the cutoff level of apparent diffusion coefficient values for detection of prostate cancer. Jpn. J. Radiol. 1867-1071. 29 (7), 488–494. doi:10.1007/s11604-011-0586-6.

Nair, A., Ong, W., Lee, A., Leow, N.W., Makmur, A., Ting, Y.H., Lee, Y.J., Ong, S.J., Tan, J.J.H., Kumar, N., Hallinan, J.T.P.D. 2025. Enhancing radiologist productivity with artificial intelligence in magnetic resonance imaging (MRI): a narrative review. Diagnostics. 2075-4418. 15 (9), 1146. doi:10.3390/diagnostics15091146.

Nketiah, G.A., Elschot, M., Scheenen, T.W., Maas, M.C., Bathen, T.F., Selnæs, K.M., Attenberger, U.I., Baltzer, P.A.T., Bathen, T.F., Fütterer, J.J., Haider, M.A., Helbich, T.H., Kiefer, B., Maas, M.C., Macura, K.J., Margolis, D.J.A., Padhani, A.R., Polanec, S.H., Praet, M., Scheenen, T.W., Schoenberg, S.O., Selnæs, K.M., van der Kwast, T.H., Villeirs, G.M., Viset, T., von Busch, H., 2021. Utility of T2-weighted MRI texture analysis in assessment of peripheral zone prostate cancer aggressiveness: a single-arm, multicenter study. Sci. Rep. 20452322. 11 (1). Available from: www.nature.com/srep/index.html. doi:10.1038/s41598-021-81272-x.

Patel, M., 2011. Prostatic abscess. In: Radiopaedia.org. doi:10.53347/rID-13513.

Patel, P., Oto, A., 2016. Magnetic resonance imaging of the prostate, including pre- and postinterventions. Semin. Interv. Radiol. 0739-9529. 33 (03), 186–195. doi:10.1055/s-0036-1586144.

Porter, C.R., Frankel, J.K., 2020. Ultrasound for Prostate Biopsy Practical Urological Ultrasound, third ed. Springer International Publishing, United States, pp. 199–212. Available from: https://link.springer.com/book/10.1007/978-3-030-52309-1. doi:10.1007/978-3-030-52309-1_11.

Power, J., Murphy, M., Hutchinson, B., Murphy, D., McNicholas, M., O'Malley, K., Murray, J., Cronin, C., 2022. Transperineal ultrasound-guided prostate biopsy: what the radiologist needs to know. Insights Imaging. 1869-4101. 13 (1). doi:10.1186/s13244-022-01210-x.

Rani, P., Singh, K., Arjuna, A., Devi, S., 2018. Detection of prostate cancer: a review. Asian J. Pharm. Clin. Res. 24553891. 11 (6), 25–31. Available from: https://innovareacademics.in/journals/index.php/ajpcr/article/download/24739/14941. doi:10.22159/ajpcr.2018.v11i6.24739.

Risbridger, G., Taylor, R., 2006. Physiology of the Male Accessory Sex Structures: The Prostate Gland, Seminal Vesicles, and Bulbourethral Glands, vol. 1. Elsevier BV, pp. 1149–1172. doi:10.1016/b978-012515400-0/50028-2.

Roobol, M.J., Schröder, F.H., Crawford, E.D., Freedland, S.J., Sartor, A.O., Fleshner, N., Andriole, G.L., 2009. A Framework for the Identification of Men at Increased Risk for Prostate Cancer. J. Urol. 00225347. 182 (5), 2112–2122. doi:10.1016/j.juro.2009.07.018.

Sethi, G., Ashrafizadeh, M., Ebrahimi, n.d. Prostate Cancer: Molecular Events and Therapeutic Modalities.

Shin, C., Sundaram, K., Arora, S., 2023. MRI of the prostate: technique, anatomy, and noncancerous disease processes. In: Magnetic Resonance Imaging of the Pelvis: A Practical Approach. Elsevier, United States, pp. 113–128. Available from: https://www.sciencedirect.com/book/9780323898546. doi:10.1016/B978-0-323-89854-6.00015-6.

Sun, Z., Wang, K., Wu, C., Chen, Y., Kong, Z., She, L., Song, B., Luo, N., Pengsheng, W., Wang, X., Zhang, X., Wang, X., 2024. Using an artificial intelligence model to detect and localize visible clinically significant prostate cancer in prostate magnetic resonance imaging: a multicenter external validation study. Quant. Imaging Med. Surg. 22234292. 14 (1), 43–60. doi:10.21037/qims-23-791.

Sunoqrot, M.R.S., Saha, A., Hosseinzadeh, M., Elschot, M., Huisman, H., 2022. Artificial intelligence for prostate MRI: open datasets, available applications, and grand challenges. Eur. Radiol. Exp. 25099280. 6 (1). Available from: https://link.springer.com/journal/41747. doi:10.1186/s41747-022-00288-8.

Tamada, T., Ueda, Y., Ueno, Y., Kojima, Y., Kido, A., Yamamoto, A., 2022. Diffusion-weighted imaging in prostate cancer. Magn. Reson. Mater. Phys. Biol. Med. 1352-8661. 35 (4), 533–547. doi:10.1007/s10334-021-00957-6.

Tan, C.H., Wang, J., Kundra, V., 2011. Diffusion weighted imaging in prostate cancer. Eur. Radiol. 09387994. 21 (3), 593–603. doi:10.1007/s00330-010-1960-y.

Tang, J., Yang, J., 2009. Etiopathogenesis of benign prostatic hypeprlasia. Indian J. Urol. 19983824. 25 (3), 312–317. doi:10.4103/0970-1591.56179.

Tătaru, O.S., Vartolomei, M.D., Rassweiler, J.J., Virgil, O., Lucarelli, G., Porpiglia, F., Amparore, D., Manfredi, M., Carrieri, G., Falagario, U., Terracciano, D., de Cobelli, O., Busetto, G.M., Giudice, F.D., Ferro, M., 2021. Artificial intelligence and machine learning in prostate cancer patient management—current trends and future perspectives. Diagnostics. 20754418. 11 (2). Available from: https://www.mdpi.com/2075-4418/11/2/354/pdf. doi:10.3390/diagnostics11020354.

Turkbey, B., Purysko, A.S., 2023. PI-RADS: where next? Radiology. 15271315. 307 (5). Available from: https://pubs.rsna.org/doi/10.1148/radiol.223128. doi:10.1148/radiol.223128.

Verze, P., Cai, T., Lorenzetti, S., 2016. The role of the prostate in male fertility, health and disease. Nat. Rev. Urol. 1759-4812. 13 (7), 379–386. doi:10.1038/nrurol.2016.89.

Wang, S., Burtt, K., Turkbey, B., Choyke, P., Summers, R.M., 2014. Computer aided-diagnosis of prostate cancer on multiparametric MRI: a technical review of current research. Biomed. Res. Int. 23146141. 2014. Available from: http://www.hindawi.com/journals/biomed/. doi:10.1155/2014/789561.

Wu, Y., Wang, C., He, X., Li, Y., Yang, Y., Lu, Q., 2024. A rare case report of severe prostate abscess treated with artificial intelligence-assisted mpMRI-TRUS real-time-guided puncture drainage. SAGE Open Med. Case Rep. 2050-313X. 12. doi:10.1177/2050313x241301860.

Ye, Z., Kou, Y., Shen, J., Dang, J., Tan, X., Jiang, X., Wang, X., Hao, L., Chen, S., Cheng, Z., 2024. A comparative study of 18F-PSMA-1007 PET/CT and pelvic MRI in newly diagnosed prostate cancer. BMC Med. Imaging. 1471-2342. 24 (1). doi:10.1186/s12880-024-01376-4.

Chapter 32

Recent trends and techniques in thorax and abdomen imaging

Abhimanyu Pradhan[a], Winniecia Dkhar[a], S. Sharath[a], Rajagopal Kadavigere[b], Anand Venugopal[c], N. Praveen[d], Leena R. David[a,e], Berna Uzun[f], Dilber Uzun Ozsahin[e,f,g], Kaushik Nayak[a], Ilker Ozsahin[f], and Neil Abraham[a]

[a]*Department of Medical Imaging Technology, Manipal College of Health Professions, Manipal Academy of Higher Education, Manipal, Karnataka, India*

[b]*Department of Radiodiagnosis and Imaging, Kasturba Medical College, Manipal, India*

[c]*Manipal Academy of Higher Education, Manipal, India*

[d]*Department of Radiodiagnosis & Imaging, Kasturba Hospital, Manipal, India*

[e]*Medical Diagnostic Imaging Department, College of Health Sciences, University of Sharjah, Sharjah, United Arab Emirates*

[f]*Irfan Suat Gunsel Operational Research Institute, Near East University, Nicosia, Turkish Republic of Northern Cyprus (TRNC), Turkey*

[g]*Research Institute for Medical and Health Sciences, University of Sharjah, Sharjah, United Arab Emirates*

1 Introduction

Medical imaging has undergone a remarkable transformation over the past decades, fundamentally revolutionizing the diagnosis, staging, and management of diseases affecting the thorax and abdomen (Lu et al., 2022; Lewiecki, 2013). The evolution from conventional radiography to sophisticated multimodal imaging approaches has dramatically enhanced our ability to detect, characterize, and monitor pathological processes with unprecedented precision and accuracy. In contemporary clinical practice, imaging modalities serve as the cornerstone of diagnostic workflows, particularly in oncology, where early detection and accurate staging directly impact patient outcomes and therapeutic decision-making (Ni et al., 2022; Moffett et al., 2016).

The thoracic and abdominal regions present unique imaging challenges due to their complex anatomical structures, diverse organ systems, and varying tissue compositions. The thorax encompasses critical structures, including the lungs, mediastinum, pleura, and chest wall, while the abdomen contains vital organs such as the liver, pancreas, kidneys, and gastrointestinal tract (Lin et al., 2019; Lai et al., 2012). Each of these regions requires tailored imaging approaches that can effectively visualize both normal anatomy and pathological changes while providing the necessary diagnostic information for clinical management.

Traditional imaging modalities, including conventional radiography and ultrasonography (USG), continue to play essential roles as initial diagnostic tools due to their accessibility, cost-effectiveness, and rapid acquisition capabilities (Prasad et al., 2016; García-Ortega et al., 2016). However, the limitations of these conventional techniques in detecting early-stage diseases and providing comprehensive tissue characterization have driven the development and widespread adoption of advanced imaging technologies. Cross-sectional imaging modalities, particularly computed tomography (CT) and magnetic resonance imaging (MRI), have emerged as indispensable tools that offer superior anatomical detail, multiplanar reconstruction capabilities, and enhanced tissue contrast resolution (Khoroshun et al., 2024; Stark et al., 1985).

In oncological imaging, the integration of functional and molecular imaging techniques has opened new frontiers in understanding tumor biology and treatment response. The advent of hybrid imaging systems, including positron emission tomography-CT (PET-CT) and the fusion of multiple imaging modalities, has enabled clinicians to combine anatomical and metabolic information, providing a more comprehensive assessment of malignant processes (Hu et al., 2023; Manafi-Farid et al., 2022). These technological advances have not only improved diagnostic accuracy but have also facilitated personalized treatment approaches and precise monitoring of therapeutic efficacy.

Recent technological innovations continue to push the boundaries of what is achievable in thoracic and abdominal imaging. Advanced CT techniques, such as dual-energy CT (DECT) and spectral CT, offer enhanced tissue characterization and material differentiation capabilities (Edmund et al., 2023; Kim et al., 2020; Sodickson et al., 2021). Similarly, functional

Advances in Medical Imaging. https://doi.org/10.1016/B978-0-443-28967-5.00027-7

MRI sequences, including diffusion-weighted imaging (DWI) and dynamic contrast-enhanced MRI (DCE-MRI), provide insights into tissue microstructure and vascular perfusion patterns (Al-Saadi et al., 2015; Grizzi and Taverna, 2024; Si et al., 2024). These developments have particular significance in oncology, where understanding tumor biology and treatment response is crucial for optimal patient management.

The integration of artificial intelligence (AI) and machine learning algorithms into imaging workflows represents another paradigm shift, promising improved image quality, automated detection capabilities, and enhanced diagnostic confidence (Garrido et al., 2017; Feng et al., 2017). These technological advances, combined with sophisticated reconstruction algorithms and dose optimization techniques, address longstanding challenges related to radiation exposure while maintaining or improving diagnostic quality (Lang et al., 2024; Lei et al., 2024).

This chapter provides a comprehensive overview of the current state and recent advances in thoracic and abdominal imaging, examining each major modality's strengths, limitations, and specific applications in clinical practice. By understanding these imaging techniques and their evolving capabilities, healthcare professionals can make informed decisions about appropriate imaging strategies, ultimately improving patient care and clinical outcomes in the diagnosis and management of thoracic and abdominal pathologies (Froeling, 2019; Horvat et al., 2018; Sharma, 2009).

2 X-ray or radiography

X-ray imaging, also known as radiography, is one of the most widely used and accessible diagnostic imaging modalities in clinical practice. It plays a pivotal role in the initial assessment and diagnosis of a broad range of conditions, particularly those affecting the thorax and abdomen. Due to its availability, rapid image acquisition, and relatively low cost, it often serves as the first-line investigation in both emergency and outpatient settings (Lu et al., 2022; Lewiecki, 2013; Prasad et al., 2016).

Chest radiography is among the most frequently performed imaging studies globally. It is an essential tool in evaluating respiratory complaints, cardiovascular abnormalities, and thoracic pathologies (Ni et al., 2022). The standard projections used in chest radiography include the posteroanterior and lateral views. In critically ill or immobile patients, an anteroposterior view is often utilized. These projections help in the visualization of lung fields, heart size and contour, mediastinal structures, pleura, and the bony thoracic cage (Moffett et al., 2016; Lin et al., 2019; Lai et al., 2012).

In the context of lung cancer, chest radiography serves as an initial diagnostic tool. It may detect obvious masses, lung nodules, pleural effusions, or signs of metastatic disease. However, its sensitivity is limited, particularly for small or centrally located tumors that may be obscured by overlapping structures such as the mediastinum, ribs, or diaphragm. Hence, while chest radiographs can raise suspicion of lung cancer, they often necessitate further evaluation through advanced imaging techniques, particularly CT, which offers cross-sectional imaging and better tissue contrast (Khoroshun et al., 2024; Stark et al., 1985).

Similarly, chest radiographs are often the first step in evaluating mediastinal tumors. These tumors may present as a widening of the mediastinum or displacement of adjacent structures. While X-rays can suggest the presence of a mass, they are not specific in characterizing the lesion or determining its exact extent or nature. CT imaging is usually recommended to provide a more detailed and accurate assessment, including delineation of anatomical relationships, involvement of adjacent structures, and tissue characterization (Hall, 2023; Archer et al., 2023; Gaubert et al., 2010).

In abdominal imaging, plain radiography is also used, though to a lesser extent compared to thoracic imaging. It is commonly employed in the evaluation of bowel obstruction, perforation, or the presence of radio-opaque foreign bodies or calculi. When it comes to abdominal tumors, however, the diagnostic yield of X-rays is limited. Tumors may not be directly visualized unless they cause secondary signs such as displacement of bowel loops or calcifications. As with thoracic masses, CT significantly outperforms radiography in the detection and characterization of abdominal tumors. CT provides detailed anatomical resolution and allows for multiplanar reconstructions, which are invaluable for surgical planning and staging (Nenezić and Stević, 2009; Nelms and Kann, 2021; Raut et al., 2020).

Despite its limitations, radiography remains an indispensable initial tool in diagnostic workflows. Its utility lies in its rapid availability, low radiation exposure compared to CT, and its ability to guide further investigations. In many clinical pathways, chest and abdominal radiographs serve as important gatekeepers, helping clinicians decide when more advanced imaging is warranted.

3 Ultrasonography

USG is a well-established imaging modality widely used across various domains of medical imaging due to its noninvasiveness, real-time imaging capability, portability, and absence of ionizing radiation. However, in the context of thoracic oncology, its application is relatively limited compared to other advanced imaging techniques, such as CT and MRI. This is primarily due to the physical limitations of ultrasound in imaging air-filled structures, such as the lungs, which pose significant acoustic impedance to sound waves (García-Ortega et al., 2016; Scarlata et al., 2023).

Despite these limitations, USG plays a critical and valuable role in specific clinical scenarios related to thoracic oncology. One of its most important uses is in image-guided procedures. USG is frequently employed to guide biopsies of peripheral lung masses, pleural-based lesions, or enlarged lymph nodes, particularly those located in accessible areas like the supraclavicular or axillary regions. Ultrasound guidance enhances the safety, precision, and success rate of these procedures, minimizing complications and improving diagnostic yield (Scarlata et al., 2023; Koukaki et al., 2025; Gupta et al., 2025).

Another significant application of USG in thoracic oncology is the assessment and management of pleural effusions. USG is highly sensitive in detecting even small volumes of pleural fluid and can differentiate between simple and complex effusions. It also provides valuable information regarding septations, loculations, or associated pleural thickening. Moreover, ultrasound guidance is routinely used during thoracentesis, a procedure to remove pleural fluid, making it safer and more accurate (Kreuter and Mathis, 2014; Lim et al., 2025; Hafez et al., 2017).

Ultrasound is also useful in the evaluation of thoracic lymphadenopathy. Superficial lymph nodes, especially in the supraclavicular area, can be easily assessed for size, morphology, vascularity, and echogenicity. Suspicious nodes can be sampled under real-time ultrasound guidance, facilitating diagnosis and staging in cases of lung cancer or metastatic disease (Hafez et al., 2017; Schauer et al., 2024).

In addition, USG may be used to monitor treatment response in patients with pleural-based lesions or superficial masses. Changes in lesion size or vascularity can provide insights into the efficacy of ongoing therapies. It can also be utilized for the assessment and drainage of abscesses, empyemas, or posttreatment collections in the thoracic region (Soni et al., 2015).

One of the notable limitations of thoracic USG is its inability to visualize lesions that are deep-seated within aerated lung parenchyma. Sound waves cannot penetrate air effectively, which significantly hinders the visualization of intrapulmonary masses unless they are abutting the pleura or located in lung regions affected by consolidation or collapse. This makes ultrasonography an unsuitable standalone tool for comprehensive lung cancer screening or assessment of central thoracic structures such as the mediastinum (Romero et al., 2024).

Thus, the utility of USG in thoracic oncology imaging is largely dependent on the location and nature of the lesion. It is most beneficial when the pathology involves the pleura, chest wall, superficial lymph nodes, or peripheral lung regions. While it does not replace CT or MRI in thoracic oncology, USG serves as an essential adjunct tool, particularly in procedural guidance and bedside evaluation, offering significant clinical value when used appropriately.

3.1 Guidance for biopsy and fine needle aspiration cytology

Typically, after the proper diagnosis using advanced modalities such as CT and MRI, a USG-guided biopsy or fine needle aspiration cytology (FNAC) is planned. Patient positioning depends on the location of the lesions, usually in a prone or lateral position. The skin puncture area is determined using prior images from either CT or MRI or by performing a survey USG scan. Once the area is confirmed, the puncture site is cleaned and sterilized. Local anesthesia is administered, or in some cases, general anesthesia, especially if the procedure is more invasive. Using a low-frequency curvilinear transducer, a biopsy needle is carefully inserted through the skin, avoiding the ribs, to reach the target lesion. Once the needle is in the correct position (confirmed by the USG image), a tissue sample is obtained. Depending on the type of biopsy, a large sample of tissue is collected using a hollow needle (core biopsy), or a thin, fine needle is used to collect cells or fluid from the lesions. After sample collection, the needle is withdrawn, and the area is imaged again to confirm that the site has been adequately sampled (Yang et al., 2020; Perez-Gutierrez et al., 2024; Islam et al., 2013).

3.2 Evaluation of lymph nodes

Evaluation of lymph nodes is performed using a specialized technique called endobronchial ultrasound-guided transbronchial needle aspiration (EBUS-TBNA). This technique is used to assess lymph nodes within or adjacent to the lungs. The findings of EBUS-TBNA help in staging lung cancer by determining whether the cancer has spread to nearby lymph nodes, which is essential for treatment planning. Prior to the procedure, the patient is instructed to remain nil per oral for at least four hours to avoid aspiration. Local anesthesia is administered to the throat and airway to ensure patient comfort during the procedure. The patient is positioned either in a sitting or lying-down position, depending on the approach chosen. A flexible bronchoscope is first inserted through the mouth or nose into the lungs, followed by the EBUS bronchoscope, which is equipped with a high-frequency transducer probe. The EBUS probe is introduced through the working channel of the bronchoscope and adjusted to visualize the target area as needed. Once the EBUS probe is activated, real-time images are obtained to identify the lymph node of interest. After identifying the target lymph node, an FNAC needle is inserted through the working channel and advanced into the lymph node. A sample is collected once the lymph node is accessed. Multiple passes may be made to collect adequate samples (Lim et al., 2025; Hafez et al., 2017).

3.3 Monitoring treatment response

Monitoring treatment response in the lung using USG is less common than other modalities such as CT scans and MRI scans. However, by using serial USG examinations, one can manage pleural effusion by tracking changes in effusion size, examine changes in the size of tumors and pneumothorax. In the case of a lung abscess, USG can be used to guide drainage procedures. By using USG instead of X-rays or CT scans to monitor the above-mentioned changes, a significant amount of radiation exposure can be avoided, leading to better patient care.

The use of USG in oncology imaging of the abdomen is often employed for various purposes in the assessment and management of abdominal cancers. It is considered a valuable oncology imaging modality. USG is especially utilized for initial screening, detecting the presence of abdominal masses, and assessing their size, location, and characterization. Other applications of USG range from monitoring treatment responses, assessment of lymph nodes, evaluation of ascites, pancreatic imaging, pediatric oncology, assistance in interventional procedures, and guided biopsy and aspiration (García-Ortega et al., 2016; Gupta et al., 2025; Kreuter and Mathis, 2014; Bartolomé-Solanas et al., 2024).

3.4 Screening, detection and characterization

USG of the abdomen is one of the most commonly performed scans. The most important aspect of a USG scan is the preparation. The patient is asked to fast for at least 4 hours and instructed to ensure the bladder is full before the scan. The scan is performed using a curvilinear transducer. Using an ultrasound gel as a coupling medium, the scan is conducted. Several abdominal-related tumors can be screened and characterized using USG. The echogenicity of the USG images helps in the detection and characterization of tumors, making it easier to differentiate tumors from cysts in a USG scan. In addition, the use of Doppler can aid in the characterization of tumors (Sinan et al., 2003).

3.5 Monitoring treatment responses

Performing a series of USG scans can track the changes in tumor size over time. This technique is a safer choice compared to performing a CT scan due to the excessive radiation dose involved (Romero et al., 2024).

4 Recent trends and techniques of USG in oncology imaging

Medical technology is evolving quite quickly. There are several recent trends and techniques in USG that can be used to aid in the diagnosis or characterization of tumors.

4.1 Three-dimensional and four-dimensional ultrasound

In recent years, the use of three-dimensional (3D) and four-dimensional (4D) USG has become increasingly prevalent in medical imaging, including oncology. These advanced ultrasound techniques offer significant improvements in anatomical visualization and diagnostic accuracy compared to traditional two-dimensional (2D) imaging. The evolution of ultrasound from 2D to 3D and 4D represents a significant technological leap, providing clinicians with enhanced capabilities in tumor evaluation and monitoring (Chen and Lai, 2011; Sinan et al., 2003). 3D ultrasound is achieved by acquiring multiple 2D images from different angles and reconstructing them into a volumetric image. This allows for a more comprehensive view of anatomical structures, enabling precise localization and detailed assessment of lesions. In oncology, 3D ultrasound can be particularly beneficial in evaluating the size, shape, and extent of tumors, especially those located in superficial organs or adjacent to the pleura. It also facilitates accurate volume measurements, which are important in tracking tumor growth or response to therapy over time (Umek et al., 2013; Pandey et al., 2018). 4D ultrasound, often referred to as real-time 3D imaging, incorporates the element of time into 3D imaging. This dynamic technique enables the continuous visualization of structures in motion, providing live images of tumors and surrounding tissues. In thoracic oncology, 4D ultrasound can aid in procedures such as real-time needle guidance for biopsies or drain placements, improving safety and accuracy (Alcázar and Jurado, 2011; Alcázar, 2013). The ability of 3D and 4D ultrasound to generate volumetric scans of tumors enhances diagnostic precision and tissue characterization. These modalities allow for better spatial understanding of lesion margins and vascularity, contributing to more informed clinical decision-making. While their application in thoracic oncology is still evolving, these technologies hold great promise in advancing noninvasive, real-time tumor assessment and procedural planning (Deslandes et al., 2022).

4.2 Contrast-enhanced ultrasound

To visualize or assess the vascularization and perfusion in the lesions, contrast-enhanced ultrasound (CEUS) is most valuable. Typically, a microbubble is administered intravenously to improve the visibility of blood vessels, organs, and lesions. Most importantly, CEUS is used in assessing liver lesions such as hepatocellular carcinoma (HCC), liver metastases, and benign liver lesions. It can also be used to characterize renal masses, pancreatic masses, splenic masses, and assess the gastrointestinal tract (Schauer et al., 2024).

4.3 Liver elastography

Elastography is an advanced ultrasound-based imaging technique that measures the stiffness and elasticity of tissues, particularly the liver. It provides a noninvasive, quantitative assessment of tissue mechanical properties, which can offer critical insights into various pathological conditions. In clinical practice, elastography has become a valuable tool for evaluating liver fibrosis, monitoring disease progression, and assessing the therapeutic impact of liver-directed treatments, especially in oncology. The principle behind elastography is that diseased or fibrotic tissue tends to be stiffer than healthy tissue. By applying mechanical vibrations or using acoustic radiation force impulses, elastography generates shear waves through the liver tissue (Fang and Sidhu, 2020; Murad Gutiérrez and Romero Enciso, 2018). The velocity of these waves correlates with tissue stiffness—higher velocities indicate increased stiffness, which often corresponds to advanced fibrosis or cirrhosis. Two common types of elastography techniques include transient elastography and shear wave elastography, both widely used for liver assessment. In the context of oncology, elastography has particular importance (Shimy et al., 2022; Fang et al., 2020). Chronic liver disease and cirrhosis are major risk factors for HCC. By accurately staging liver fibrosis, elastography helps identify patients at higher risk of developing liver cancer, thereby supporting early detection and surveillance strategies. Furthermore, elastography can be employed to evaluate the impact of liver-directed cancer therapies, such as radiofrequency ablation, transarterial chemoembolization, or targeted systemic therapies. Changes in liver stiffness following treatment may reflect treatment response, tissue necrosis, or progression of underlying liver disease (Medyńska-Przęczek et al., 2024; Deslandes et al., 2022; Piscaglia et al., 2014). In addition, elastography may assist in the characterization of focal liver lesions. Malignant tumors are often stiffer than benign lesions or surrounding parenchyma, and elastography can complement conventional ultrasound findings to improve diagnostic accuracy. Overall, elastography is a promising, noninvasive adjunct in liver imaging that enhances the evaluation of fibrosis, supports cancer surveillance, and contributes to more precise management of liver tumors and therapy monitoring (Frulio and Trillaud, 2013; Suffredini et al., 2024).

5 Computed tomography

The use of CT in oncology imaging is crucial. It has a wide range of applications in oncology imaging, including diagnosis, localization, characterization, treatment planning, and monitoring. However, the use of CT should be justified to reduce the risk of biological effects relative to its benefits. Although CT has better sensitivity and specificity for diagnosis, localization, and characterization, for follow-ups and treatment monitoring, other modalities such as USG and MRI can be used to reduce the radiation burden on the patient (Karthik and Moss, 2011; Song et al., 2025).

CT scans are often used as the initial imaging modality to detect suspicious lesions of the thorax. Noncontrast, high-resolution CT (HRCT), or contrast-enhanced CT examination (CECT) is performed for diagnosis. In the case of thorax imaging, an HRCT thorax provides substantial information to diagnose tumors. An HRCT consists of thin-slice, high-resolution images acquired with a high matrix size of at least 728 × 728. The diagnosis of bronchogenic carcinoma and chest wall invasion of tumors can be achieved with high accuracy through HRCT thorax examination (Ginsberg, 2014; Garrido et al., 2017).

In the case of a CECT examination, as per protocol, a noncontrast or HRCT sequence is acquired prior to the contrast-enhanced phase. This can be obtained either by using bolus tracking or timed acquisition. For bolus tracking acquisition, the locator and tracker are placed at the descending aorta at the level of the tracheal bifurcation (Carina). The pressure injector is loaded with 60 ml of iodinated contrast medium and 30 ml of normal saline, and the injection rate is maintained at 4 ml/second. By adjusting the threshold to 150 Hounsfield units (HU) and setting a postthreshold delay of 16 seconds, the scan is acquired. The most important part of the CECT phase is the inclusion of the adrenal glands in the scan volume. This plays a very important role when evaluating metastasis (Johnstone and Moss, 2015; Feng et al., 2017; Kim et al., 2020).

Using a timed phase instead of a bolus tracking phase can reduce the radiation dose to the patient because, in the bolus tracking phase, a series of scans is taken to track the contrast bolus. However, the timed phase requires a proper understanding of timing parameters for the required enhancement. For example, after the initiation of contrast injection, it usually takes an average of 19 seconds (under normal conditions) to achieve the required enhancement of the descending aorta in the bolus tracking phase. This, along with a postthreshold delay of 16 seconds, can give an average timing parameter of 35 seconds. Therefore, 35 seconds can be selected for better results (Feng et al., 2017; Kim et al., 2020).

CECT thorax shows an accuracy of 85% in detecting the presence of malignancy. The diagnosis of small cell lung carcinoma and carcinoid tumors can be performed using CECT thorax examination. However, one needs to understand that CT uses X-rays to produce images, and exposure to ionizing radiation must be justified and optimized to keep the radiation dose as low as reasonably achievable. This can be achieved by reducing the technical parameters for image acquisition. Specifically, the tube voltage can be adjusted to control exposure. Tailoring the tube voltage to the body mass index of the patient or body weight can help achieve this. The increased noise in the images due to the use of low tube voltage can be compensated by recently developed iterative reconstruction techniques or AI-based reconstruction algorithms (Garrido et al., 2017; Johnstone and Moss, 2015; Feng et al., 2017; Kim et al., 2020).

Similar benefits of a CECT examination can be obtained with an abdominal scan. CECT abdomen is a multiphase examination wherein each phase has its own use in the differential diagnosis of tumors. Usually, a noncontrast scan is performed prior to the contrast-enhanced phase, followed by contrast-enhanced multiphase acquisition. This can be achieved either by bolus tracking or timed techniques. In the case of bolus tracking, the locator and tracker are kept at the abdominal aorta (at the level between the 12th thoracic and first lumbar vertebra). The pressure injector is loaded with 80 ml of iodinated contrast medium and 40 ml of normal saline, and the injection rate is maintained at 4 ml/second. Adjusting the threshold to 150 HU, multiphase acquisition is planned, where an 8-second postthreshold delay is used for the arterial phase, 25 seconds for the parenchymal phase, and 45 seconds for the portal-venous phase (Garrido et al., 2017; Johnstone and Moss, 2015; Feng et al., 2017).

The multiphase acquisition includes the combination of arterial and portal-venous, arterial, parenchymal and portal-venous, and arterial, portal-venous and delayed phase (usually 3 minutes). These are the different combinations used depending upon the area of interest. For example, an arterial and portal-venous phase can be acquired for areas of interest other than the liver. The combination of arterial, parenchymal and portal-venous is performed specifically when the pancreas is the area of interest. The combination of arterial, portal-venous and delayed phase is used in cases of HCC (Garrido et al., 2017; Johnstone and Moss, 2015).

In cases of tumors related to the urinary system, CECT urography is performed. In this examination, including noncontrast phase, arterial phase, and portal-venous phase, a nephrography phase is taken at a 90-second postthreshold delay. This is followed by a three-minute delay, an eight-minute delay, a 15-minute delay, and imaging with a full bladder. However, due to the multiple phases of scanning, there is always a concern related to radiation exposure. An optimization technique can be used to reduce exposure. Since all the phases play a very important role in diagnosis, the volume coverage can be optimized by limiting the scan length for the delayed phase. Furthermore, acquiring the arterial and portal-venous phases together can reduce radiation exposure to patients. This, along with reduced tube voltage and the use of iterative reconstruction or AI-based reconstruction to reduce noise caused by reduced tube voltage, can be implemented (Ginsberg, 2014; Garrido et al., 2017).

6 Recent advancement in CT technology for oncology imaging

DECT: DECT is an advanced imaging modality that utilizes two X-ray sources operating at different energy levels (typically low and high kilovoltage) to acquire diagnostic images simultaneously. This dual-energy approach allows for improved tissue characterization and material differentiation beyond the capabilities of conventional single-energy CT. DECT works by analyzing how different tissues and substances within the body attenuate X-rays at varying energy levels. This enables the creation of material-specific maps, such as iodine, calcium, or uric acid maps, and virtual noncontrast images, enhancing diagnostic confidence and reducing the need for additional scans. In oncology, DECT offers significant advantages. It can help differentiate between benign and malignant tissues, improve lesion conspicuity, and more accurately define tumor margins. DECT is especially useful in identifying and quantifying contrast enhancement, which is critical in evaluating tumor vascularity, assessing response to therapy, and detecting recurrence. In addition, it enhances visualization of lymph nodes, vascular invasion, and metastases. DECT also aids in reducing artifacts from metal implants or dense structures, leading to clearer images in postoperative or complex anatomical regions. Overall, DECT is a powerful tool in oncologic imaging, offering enhanced tissue contrast, functional information, and greater diagnostic precision (Sodickson et al., 2021; Edmund et al., 2023).

Spectral CT: Spectral CT represents a sophisticated advancement in CT imaging, building upon the principles of DECT. While DECT uses two energy levels to distinguish materials based on their different attenuation properties, Spectral CT provides an even more refined analysis by acquiring and processing X-ray data across a broader range of energy spectra. This allows for detailed spectral decomposition and the generation of multiple image reconstructions that can highlight specific materials or tissue characteristics with high precision. One of the most powerful features of Spectral CT is its ability to perform material-specific imaging, such as quantifying iodine, calcium, or fat content within tissues. In oncology, this is particularly valuable for measuring iodine concentration within tumors, which serves as a surrogate marker of vascularity. Tumors that exhibit high iodine uptake are generally more vascular and metabolically active, which can correlate with aggressiveness or response to treatment. Conversely, a reduction in iodine content during follow-up imaging may suggest effective therapy or tumor necrosis. Spectral CT also aids in differentiating between malignant and benign lesions, improving lesion detection and characterization by enhancing contrast resolution. It allows for virtual noncontrast imaging, eliminating the need for separate unenhanced scans and thereby reducing radiation exposure. Moreover, Spectral CT can reduce image artifacts and improve visualization in challenging areas, such as those affected by metal implants. Its advanced postprocessing capabilities make it a powerful tool for functional imaging and personalized treatment assessment, positioning it as a promising modality in modern oncologic imaging workflows (Kim et al., 2020).

Cone beam CT (CBCT): CBCT is a specialized imaging technique that plays a critical role in modern radiation therapy, particularly in the context of image-guided radiotherapy. Unlike conventional CT, which uses a fan-shaped X-ray beam and multiple detector rows, CBCT utilizes a cone-shaped X-ray beam and a flat-panel detector to acquire volumetric data. This setup allows for the rapid generation of high-resolution 3D images of a patient's anatomy directly in the treatment position. CBCT is typically integrated into the linear accelerator system used for delivering radiation therapy. Before each treatment session, a CBCT scan is performed to verify the tumor location and patient positioning in real-time. This capability enables clinicians to adjust the patient's alignment and radiation beams with submillimeter precision, ensuring that the radiation dose is accurately targeted to the tumor while sparing surrounding healthy tissues. This technology has significantly enhanced the accuracy and safety of radiotherapy, particularly in tumors located in areas prone to movement, such as the lungs, liver, or prostate. It allows for the detection of anatomical changes over the course of treatment, such as tumor shrinkage or organ shifts, enabling adaptive radiotherapy strategies. CBCT's integration into daily treatment workflows has revolutionized precision cancer therapy, improving clinical outcomes by minimizing radiation exposure to nontarget tissues and maximizing tumor control (Hu et al., 2025; Zhou et al., 2025).

PET-CT/CT-MRI Fusion: The integration of different imaging modalities, such as PET-CT and CT-MRI fusion, has significantly enhanced the diagnostic and therapeutic capabilities of oncologic imaging. These hybrid techniques combine the strengths of anatomical and functional imaging, providing a more comprehensive evaluation of tumors than either modality alone. PET-CT combines metabolic information from PET with detailed anatomical data from CT. In PET imaging, radiotracers like fluorodeoxyglucose are used to highlight areas of increased glucose metabolism, which is often a hallmark of malignant cells. By fusing this data with HRCT images, clinicians can localize metabolically active lesions with precision, aiding in early tumordetection, accurate staging, treatment planning, and response assessment. PET-CT is especially valuable in cancers such as lung, lymphoma, colorectal, and head and neck malignancies, where metabolic activity is a key marker of disease progression or recurrence. Similarly, CT-MRI fusion involves aligning CT images with MRI scans to leverage the superior soft-tissue contrast of MRI along with the bony and spatial detail of CT. This fusion is particularly useful in complex anatomical regions like the brain, pelvis, or spine, where precise delineation of tumor boundaries and involvement of adjacent structures is critical. These fusion imaging techniques facilitate multidimensional tumor assessment, improve diagnostic accuracy, and support personalized treatment strategies. They also play a crucial role in guiding biopsies, radiotherapy planning, and surgical interventions, making them indispensable tools in modern oncology practice (Manafi-Farid et al., 2022; Hu et al., 2023).

7 Magnetic resonance imaging

MRI is an advanced and highly versatile imaging modality that plays a crucial role in modern oncology. Due to its superior soft tissue contrast resolution, multiplanar imaging capabilities, and the absence of ionizing radiation, MRI is particularly valuable in the diagnosis, staging, treatment planning, and monitoring of various types of cancers. These advantages make MRI one of the most critical imaging tools for evaluating tumors, especially those located in complex anatomical regions. One of the key strengths of MRI lies in its ability to differentiate tissues based on their intrinsic properties, such as T1 and T2 relaxation times and proton density. By manipulating extrinsic parameters like Repetition Time and Echo Time, MRI generates contrast in images that highlight differences in water content and tissue structure. This makes MRI highly effective in characterizing tumors and distinguishing between benign and malignant lesions. Malignant tissues often have different signal intensities compared to normal or benign tissues, allowing for early and more accurate detection. MRI can produce high-resolution, multiplanar images that allow clinicians to visualize tumors and their relationship with adjacent structures in exquisite detail. This is particularly useful for surgical planning, as well as for radiation therapy targeting. In addition, since MRI does not involve ionizing radiation, it is safer for repeated imaging, especially in younger patients or those requiring long-term monitoring (Sharma, 2009; Froeling, 2019; Horvat et al., 2018).

While CT remains the primary imaging modality for the detection and staging of lung cancers due to its speed, resolution, and sensitivity to pulmonary lesions, MRI still holds significant value in thoracic oncology in specific contexts. MRI is particularly useful in evaluating the invasion of tumors into the chest wall, mediastinum, diaphragm, or pleura, where soft tissue delineation is paramount. For example, T1-weighted sequences are excellent for assessing pleural and pericardial invasion by tumors, while Short Tau Inversion Recovery and fat-suppressed T2-weighted images are useful for visualizing soft tissue edema and tumor infiltration into the chest wall or vertebrae. MRI may also be used in evaluating superior sulcus (Pancoast) tumors, where the involvement of the brachial plexus, subclavian vessels, and spine needs to be assessed with precision. These scenarios demonstrate MRI's supplemental value in thoracic oncology, especially in areas where CT may be limited by its lower soft tissue contrast (Winder et al., 2023; Bucher et al., 2024; Goehde et al., 2005; Haider et al., 2007).

MRI is particularly well-suited for imaging abdominal and pelvic organs, making it a cornerstone modality in the evaluation of tumors in these regions. It is routinely used for detecting and characterizing liver lesions, such as HCC, cholangiocarcinoma, and metastases from other primary cancers. MRI with liver-specific contrast agents, like gadoxetic acid, allows for functional liver imaging and improves the detection of small or isoattenuating lesions not well seen on CT. In prostate cancer, multiparametric MRI is now considered the gold standard for local staging and active surveillance. It combines anatomical and functional sequences, including T2-weighted imaging, DWI, and DCE imaging, to assess tumor aggressiveness and guide biopsy. MRI is also widely used in gynecologic cancers, such as cervical, endometrial, and ovarian malignancies, as well as in rectal cancer, where it provides detailed evaluation of the primary tumor, mesorectal fascia, and lymph node involvement. In these applications, MRI contributes significantly to treatment decisions, including the need for surgery, chemotherapy, or radiation therapy (Liau et al., 2019; Adams et al., 2022; Grey et al., 2020).

In recent years, the integration of functional imaging sequences into standard MRI protocols has transformed oncology imaging. These include DWI and DCE-MRI. The DWI sequence provides insight into tissue cellularity by measuring the Brownian motion of water molecules. Malignant tumors typically restrict water diffusion due to their dense cellular architecture, which appears as high signal on DWI and low signal on the Apparent Diffusion Coefficient (ADC) map. ADC values are quantitative and can help differentiate between benign and malignant lesions, monitor treatment response, and detect early recurrence. Meanwhile, the DCE-MRI technique assesses tumor vascularity and perfusion by evaluating contrast agent kinetics. Tumors often exhibit abnormal angiogenesis, leading to rapid contrast uptake and washout patterns that can be visualized and quantified using DCE-MRI. This can help in treatment planning, particularly in distinguishing viable tumor tissue from necrosis after therapy. These functional techniques enhance the sensitivity and specificity of MRI, contributing not just to structural imaging but also to tumor biology assessment, which is key in personalized oncology care (Al-Saadi et al., 2015; Grizzi and Taverna, 2024; Si et al., 2024; Partridge et al., 2011; Elizabeth et al., 2014).

While imaging modalities such as CT, PET-CT, and X-ray are indispensable in oncology, they do expose patients to ionizing radiation, which carries a potential risk, especially when repeated scans are needed. Therefore, the use of such modalities must be clinically justified, weighing the diagnostic benefits against the cumulative radiation dose. MRI, being radiation-free, offers a safer alternative for long-termmonitoring, particularly in pediatric patients, young adults, and those requiring frequent follow-ups. When imaging with ionizing radiation is necessary, steps should be taken to optimize exposure factors, such as using low-dose CT protocols or limiting the scan area to the region of interest, to minimize harm while maintaining diagnostic efficacy (Lei et al., 2024; Lang et al., 2024).

MRI is a powerful, noninvasive imaging modality that offers exceptional soft tissue contrast, functional imaging capabilities, and radiation-free evaluation—making it invaluable in oncology. While its role in lung cancer may be more limited compared to CT, its applications in liver, prostate, gynecologic, and pelvic cancers are extensive and continuously expanding. The addition of functional imaging techniques like DWI and DCE-MRI further enhances its utility, allowing for a more comprehensive assessment of tumor biology and treatment response. As imaging continues to evolve, MRI will remain a cornerstone in the safe, precise, and effective management of cancer patients (Neilio and Ginat, 2024; Sunoqrot et al., 2022).

8 Discussion

The landscape of thoracic and abdominal imaging has evolved dramatically, with each imaging modality offering unique advantages that complement one another in comprehensive patient care. The integration of multiple imaging techniques has transformed diagnostic workflows, particularly in oncology, where precise characterization and staging are paramount for optimal treatment planning (Hu et al., 2023; Manafi-Farid et al., 2022).

Conventional radiography, despite its limitations, remains an essential first-line imaging tool due to its universal availability, rapid acquisition, and cost-effectiveness (Lu et al., 2022; Prasad et al., 2016). However, its role has evolved from a primary diagnostic tool to an initial screening modality that guides the need for more advanced imaging. The sensitivity limitations of chest radiography, particularly for small or centrally located tumors, underscore the importance of cross-sectional imaging in comprehensive evaluation (Khoroshun et al., 2024; Stark et al., 1985).

USG has demonstrated remarkable versatility, particularly in image-guided procedures and real-time assessment. The development of advanced ultrasound techniques, including 3D/4D imaging, CEUS, and elastography, has significantly expanded its clinical applications (Chen and Lai, 2011; Fang and Sidhu, 2020; Schauer et al., 2024). These innovations have made USG an indispensable tool for procedural guidance, treatment monitoring, and functional tissue assessment, while maintaining its radiation-free advantage.

CT continues to serve as the backbone of thoracic and abdominal imaging, offering excellent spatial resolution and rapid acquisition capabilities (Johnstone and Moss, 2015; Garrido et al., 2017). The introduction of advanced CT techniques, including dual-energy and spectral CT, has enhanced tissue characterization capabilities beyond conventional imaging (Edmund et al., 2023; Kim et al., 2020). However, the challenge of radiation exposure necessitates careful consideration of risk-benefit ratios and the implementation of dose optimization strategies (Lei et al., 2024; Lang et al., 2024).

MRI has established itself as the premier modality for soft tissue characterization, offering superior contrast resolution without ionizing radiation exposure (Froeling, 2019; Sharma, 2009). The integration of functional imaging sequences, particularly diffusion-weighted imaging and DCE-MRI, has transformed tumor evaluation by providing insights into cellular architecture and vascular perfusion patterns (Al-Saadi et al., 2015; Grizzi and Taverna, 2024). This functional capability has proven particularly valuable in treatment response assessment and the early detection of recurrence.

The emergence of hybrid imaging systems and multimodal fusion techniques represents a paradigm shift toward comprehensive, personalized imaging approaches. PET-CT and CT-MRI fusion combine anatomical and functional information, enabling more accurate staging, treatment planning, and response monitoring (Hu et al., 2023; Manafi-Farid et al., 2022). These integrated approaches exemplify the trend toward precision medicine, where imaging biomarkers guide individualized treatment strategies.

AI and machine learning integration into imaging workflows promise to address several current limitations, including image quality optimization, automated detection, and dose reduction (Feng et al., 2017; Garrido et al., 2017). As these technologies mature, they are expected to enhance diagnostic accuracy while reducing interpretation time and variability.

The selection of appropriate imaging modalities requires careful consideration of multiple factors, including clinical presentation, suspected pathology, patient characteristics, and available resources. A multimodal approach, utilizing the complementary strengths of different techniques, often provides the most comprehensive assessment. For instance, initial chest radiography may identify suspicious findings that warrant CT evaluation, which, in turn, may guide MRI assessment for chest wall invasion or ultrasound-guided biopsy for tissue confirmation.

9 Conclusion

The field of thoracic and abdominal imaging continues to evolve rapidly, driven by technological innovations and the growing demand for precision medicine approaches. Each imaging modality discussed in this chapter contributes unique capabilities to the diagnostic arsenal, and their strategic integration forms the foundation of modern medical imaging practice. From conventional radiography, serving as an accessible first-line tool, to advanced cross-sectional modalities providing detailed anatomical and functional information, the imaging landscape offers comprehensive solutions for complex diagnostic challenges. The development of functional imaging techniques has revolutionized our understanding of tissue biology and treatment response, particularly in oncology, where early and accurate assessment directly impacts patient outcomes. The future of thoracic and abdominal imaging lies in the continued integration of advanced technologies, including AI, machine learning, and novel contrast agents, which promise to enhance diagnostic accuracy while optimizing patient safety. Hybrid imaging systems and multimodal fusion techniques will likely become increasingly sophisticated, providing comprehensive assessments in single examinations. As technology advances, the emphasis on radiation safety and dose optimization remains paramount. Healthcare professionals must stay informed about evolving technologies and their appropriate clinical applications, understanding the strengths and limitations of each modality for optimal diagnostic decision-making. The ultimate goal remains the provision of accurate, timely, and safe imaging that guides effective treatment strategies and improves patient outcomes in the management of thoracic and abdominal pathologies.

References

Adams, L.C., Makowski, M.R., Engel, G., Rattunde, M., Busch, F., Asbach, P., Niehues, S.M., Vinayahalingam, S., van Ginneken, B., Litjens, G., Bressem, K.K., 2022. Prostate158—an expert-annotated 3T MRI dataset and algorithm for prostate cancer detection. Comput. Biol. Med. 18790534. 148. Available from: https://www.elsevier.com/locate/compbiomed. doi:10.1016/j.compbiomed.2022.105817.

Al-Saadi, W.I., Shallab, E.N., Naji, S., 2015. Diffusion weighted MRI in the characterization of solitary breast mass. Egypt. J. Radiol. Nucl. Med. 20904762. 46 (4), 1337–1341. Available from: https://www.elsevier.com/wps/find/journaldescription.cws_home/724146/description#description. doi:10.1016/j.ejrnm.2015.08.007.

Alcázar, J.L., 2013. Three-dimensional static ultrasound and 3D power doppler in gynecologic pelvic tumors. J. Ultrasound Obstet. Gynecol. 09751912. 7 (2), 187–199. Available from: https://www.dsjuog.com/. doi:10.5005/jp-journals-10009-1283.

Alcázar, J.L., Jurado, M., 2011. Three-dimensional ultrasound for assessing women with gynecological cancer: a systematic review. Gynecol. Oncol. 10956859. 120 (3), 340–346. doi:10.1016/j.ygyno.2010.10.023.

Archer, J.M., Ahuja, J., Strange, C.D., Shroff, G.S., Gladish, G.W., Sabloff, B.S., Truong, M.T., 2023. Multimodality imaging of mediastinal masses and mimics. Mediastinum. 25226711. 7. Available from: https://med.amegroups.com/article/view/7698/html. doi:10.21037/med-22-53.

Bartolomé-Solanas, A., Porta-Vilaró, M., Soler-Perromat, J.C., Del Amo, M., García-Diez, A.I., Radalov, I., Cornellas, L., Pomés Lopez, I., Isern-Kebschull, J., Tomás, X., 2024. Narrative review of chest wall ultrasound: a practical approach. Quant. Imaging Med. Surg. 22234306. 14 (11), 7983–8000. Available from: https://qims.amegroups.org/article/view/128841/html. doi:10.21037/qims-24-355.

Bucher, A.M., Egger, J., Dietz, J., Strecker, R., Hilbert, T., Frodl, E., Wenzel, M., Penzkofer, T., Hamm, B., Chun, F.K.H., Vogl, T., Kleesiek, J., Beeres, M., 2024. Value of MRI—T2 mapping to differentiate clinically significant prostate cancer. J. Imaging Inform. Med. 29482933. 37 (6), 3304–3315. Available from: https://www.springer.com/journal/10278. doi:10.1007/s10278-024-01150-6.

Chen, D.R., Lai, H.W., 2011. Three-dimensional ultrasonography for breast malignancy detection. Expert Opin. Med. Diagn. 17530067. 5 (3), 253–261. doi:10.1517/17530059.2011.561314.

Deslandes, A., Croft, A., Panuccio, C., 2022. Three-dimensional gynaecological ultrasound, part one: how, when and why? Sonography. 2202-8323. 9 (1), 7–15. doi:10.1002/sono.12294.

Edmund, J., Rønjom, M.F., van Overeem Felter, M., Maare, C., Dam, A.M.J., Tsaggari, E., Wohlfahrt, P., 2023. Split-filter dual energy computed tomography radiotherapy: from calibration to image guidance. Phys. Imaging Radiat. Oncol. 24056316. 28, 100495. doi:10.1016/j.phro.2023.100495.

Elizabeth, M.S., Schopp, J.G., Peacock, S., DeMartini, W.B., Rahbar, H., Lehman, C.D., Partridge, S.C., 2014. Diffusion-weighted MRI: association between patient characteristics and apparent diffusion coefficients of normal breast fibroglandular tissue at 3 T. Am. J. Roentgenol. 15463141. 202 (5), W496. Available from: https://www.ajronline.org/doi/pdf/10.2214/AJR.13.11159. doi:10.2214/AJR.13.11159.

Fang, C., Lim, A., Sidhu, P.S., 2020. Ultrasound-based liver elastography in the assessment of fibrosis. Clin. Radiol. 00099260. 75 (11), 822–831. doi:10.1016/j.crad.2020.01.005.

Fang, C., Sidhu, P.S., 2020. Ultrasound-based liver elastography: current results and future perspectives. Abdom. Radiol. 23660058. 45 (11), 3463–3472. Available from: https://link.springer.com/journal/261. doi:10.1007/s00261-020-02717-x.

Feng, Z., Li, M., Liu, F., Peng, Y., Ren, W., Xie, H., Peng, Z., 2017. Analysis of giant thoracic neoplasms: correlations between imaging, pathology and surgical management. Thorac. Cancer. 1759-7706. 8 (5), 402–409. doi:10.1111/1759-7714.12448.

Froeling, M., 2019. QMRTools: a mathematica toolbox for quantitative MRI analysis. J. Open Source Softw. 2475-9066. 4 (38), 1204. doi:10.21105/joss.01204.

Frulio, N., Trillaud, H., 2013. Ultrasound elastography in liver. Diagn. Interv. Imaging. 22115684. 94 (5), 515–534. doi:10.1016/j.diii.2013.02.005.

García-Ortega, A., Briones-Gómez, A., Fabregat, S., Martínez-Tomás, R., Martínez-García, M.Á., Cases, E., 2016. Benefit of chest ultrasonography in the diagnosis of peripheral thoracic lesions in an interventional pulmonology unit. Arch. Bronconeumol. 15792129. 52 (5), 244–249. Available from: https://www.elsevier.com/wps/find/journaldescription.cws_home/710346/description#description. doi:10.1016/j.arbr.2016.03.004.

Garrido, P., Sánchez, M., Belda Sanchis, J., Moreno Mata, N., Artal, Á., Gayete, Á., Matilla González, J.M., Galbis Caravajal, J.M., Isla, D., Paz-Ares, L., Seijo, L.M., 2017. Reflections on the implementation of low-dose computed tomography screening in individuals at high risk of lung cancer in spain. Arch. Bronconeumol. 15792129. 53 (10), 568–573. Available from: https://www.elsevier.com/wps/find/journaldescription.cws_home/710346/description#description. doi:10.1016/j.arbr.2017.08.006.

Gaubert, J.-Y., Cohen, F., Vidal, V., Louis, G., Moulin, G., Bartoli, J.-M., Jacquier, A., 2010. Imagerie des tumeurs du médiastin. Rev. Pneumol. Clin. 07618417. 66 (1), 17–27. doi:10.1016/j.pneumo.2009.12.011.

Ginsberg, M.S., 2014. Imaging of thoracic cavity tumors. Surg. Oncol. Clin. N. Am. 15585042. 23 (4), 709–733. Available from: https://www.elsevier.com/inca/publications/store/6/2/3/3/8/3/index.htt. doi:10.1016/j.soc.2014.06.005.

Goehde, S.C., Hunold, P., Vogt, F.M., Ajaj, W., Goyen, M., Herborn, C.U., et al., 2005. Full-body MRI for early disease detection full-body cardiovascular and tumor mri for early detection of disease: feasibility and initial experience in 298 subjects. Am. J. Roentgenol. 184, 598–611. Available from: https://pubmed.ncbi.nlm.nih.gov/15671386/.

Grey, A.D.R., Connor, M.J., Tam, J., Loch, T., 2020. Can transrectal prostate ultrasound compete with multiparametric MRI in the detection of clinically significant prostate cancer? Transl. Androl. Urol. 22234683. 9 (3), 1492–1500. doi:10.21037/tau.2020.02.26.

Grizzi, F., Taverna, G., 2024. Editorial: PET/CT and MRI in prostate cancer. Front. Oncol. 2234-943X. 14. doi:10.3389/fonc.2024.1421542.

Gupta, V., Gowing, S.D., Pandya, R., Tan, L., Liu, R.Y., Srinathan, S.K., Kidane, B., 2025. Surgeon-led point-of-care ultrasound-guided thoracic biopsy: a new paradigm in efficient diagnosis and resource-sparing care. J. Thorac. Cardiovasc. Surg. 1097685X. Available from: https://www.sciencedirect.com/science/journal/00225223. doi:10.1016/j.jtcvs.2024.11.039.

Hafez, M.R., Sobh, E.S.M., Elsawy, S.B., Abo-Elkheir, O.I., 2017. The usefulness of thoracic ultrasonography in diagnosis and staging of bronchogenic carcinoma. Ultrasound. 17431344. 25 (4), 200–212. Available from: https://www.uk.sagepub.com/journals/Journal202202?siteId=sage-uk&prodTypes=any&q=Ultrasound&fs=1. doi:10.1177/1742271X17721264.

Haider, M.A., Van Der Kwast, T.H., Tanguay, J., Evans, A.J., Hashmi, A.T., Lockwood, G., Trachtenberg, J., 2007. Combined T2-weighted and diffusion-weighted MRI for localization of prostate cancer. Am. J. Roentgenol. 0361803X. 189 (2), 323–328. doi:10.2214/AJR.07.2211.

Hall, L.T., 2023. Molecular Imaging and Therapy. Exon Publications, United States, pp. 1–218. Available from: https://exonpublications.com/index.php/exon/issue/view/molecular-imaging. doi:10.36255/molecular-imaging.

Horvat, J.V., Durando, M., Milans, S., Patil, S., Massler, J., Gibbons, G., Giri, D., Pinker, K., Morris, E.A., Thakur, S.B., 2018. Apparent diffusion coefficient mapping using diffusion-weighted mri: impact of background parenchymal enhancement, amount of fibroglandular tissue and menopausal status on breast cancer diagnosis. Eur. Radiol. 14321084. 28(6), 2516–2524. Available from: https://www.link.springer.de/link/service/journals/00330/index.htm. doi:10.1007/s00330-017-5202-4.

Hu, C., Cao, N., Li, X., He, Y., Zhou, H., 2025. CBCT-to-CT synthesis using a hybrid U-Net diffusion model based on transformers and information bottleneck theory. Sci. Rep. 2045-2322. 15 (1), doi:10.1038/s41598-025-92094-6.

Hu, Q., Li, K., Yang, C., Wang, Y., Huang, R., Gu, M., Xiao, Y., Huang, Y., Chen, L., 2023. The role of artificial intelligence based on PET/CT radiomics in NSCLC: disease management, opportunities, and challenges. Front. Oncol. 2234-943X. 13. doi:10.3389/fonc.2023.1133164.

Islam, T., Hossain, F., Rumpa, A.P., Sikder, N.H., Bhuiyan, M.A., Karim, E., Hossain, A., 2013. Ultrasound guided fine needle aspiration cytology: a sensitive diagnostic tool for diagnosis of intra-abdominal lesions. Bangladesh Med. Res. Counc. Bull. 2224-7238. 39 (1), 14–17. doi:10.3329/bmrcb.v39i1.15792.

Johnstone, A., Moss, H., 2015. Thoracic surgical radiographic and CT pathology. Anaesth. Intensive Care Med. 16 (2), 45–53.

Karthik, S., Moss, H., 2011. Thoracic surgical plain radiographic and computed tomography pathology. Anaesth. Intensive Care Med. 12 (11), 507–512.

Khoroshun, E., Makarov, V., Nehoduiko, V., Yasinskyi, O., Sharmazanova, O., Pulyaev, S., 2024. Oblique projections in the analysis of multislice computed tomography data in gunshot wounds. Emerg. Med. (Ukraine). 23071230. 20 (3), 211–216. Available from: https://emergency.zaslavsky.com.ua/index.php/journal/article/view/1694. doi:10.22141/2224-0586.20.3.2024.1694.

Kim, C., Kim, W., Park, S.J., Lee, Y.H., Hwang, S.H., Yong, H.S., Oh, Y.W., Kang, E.Y., Lee, K.Y., 2020. Application of dual-energy spectral computed tomography to thoracic oncology imaging. Korean J. Radiol. 12296929. 21 (7), 838–850. Available from: https://kjronline.org/Synapse/Data/PDFData/0068KJR/kjr-21-838.pdf. doi:10.3348/kjr.2019.0711.

Koukaki, E., Anagnostopoulos, N., Bakiri, A., Zaneli, S., Stratakos, G., 2025. Endobronchial ultrasound access to pulmonary vasculature in thoracic malignancy. Cancers. 2072-6694. 17 (4), 616. doi:10.3390/cancers17040616.

Kreuter, M., Mathis, G., 2014. Emergency ultrasound of the chest. Respiration. 0025-7931. 87 (2), 89–97. doi:10.1159/000357685.

Lai, V., Tsang, W.K., Chan, W.C., Yeung, T.W., 2012. Diagnostic accuracy of mediastinal width measurement on posteroanterior and anteroposterior chest radiographs in the depiction of acute nontraumatic thoracic aortic dissection. Emerg. Radiol. 14381435. 19 (4), 309–315. doi:10.1007/s10140-012-1034-3.

Lang, J., McClure, T.D., Margolis, D.J.A., 2024. MRI–ultrasound fused approach for prostate biopsy—how it is performed. Cancers. 20726694. 16 (7). Available from: https://www.mdpi.com/journal/cancers/. doi:10.3390/cancers16071424.

Lei, Y., Feng, B., Wan, M., Xu, K., Cui, J., Ma, C., Sun, J., Yao, C., Gan, S., Shi, J., Cui, E., 2024. Predicting microvascular invasion in hepatocellular carcinoma with a CT- and MRI-based multimodal deep learning model. Abdom. Radiol. 2366-0058. 49 (5), 1397–1410. doi:10.1007/s00261-024-04202-1.

Lewiecki, E.M., 2013. Imaging technologies for assessment of skeletal health in men. Curr. Osteoporos. Rep. 15442241. 11 (1), 1–10. doi:10.1007/s11914-012-0128-x.

Liau, J., Goldberg, D., Arif-Tiwari, H., 2019. Prostate cancer detection and diagnosis: role of ultrasound with MRI correlates. Curr. Radiol. Rep. 2167-4825. 7 (3), doi:10.1007/s40134-019-0318-8.

Lim, C., Karnam, S., Irving, L., Hammerschlag, G., Taverner, J., 2025. High diagnostic yield of ultrasound-guided percutaneous needle biopsy of peripheral lung and pleural lesions, lymph nodes, and other sites performed by respiratory physicians. J. Bronchol. Interv. Pulmonol. 1948-8270. 32 (2), doi:10.1097/lbr.0000000000001010.

Lin, H.C., Huang, C.Y., Huang, W.M., Jhou, Z.Y., Chen, C.H., Chien, Y.C., Huang, C.C., 2019. Oblique views of chest radiography from a designed rotation angle recommendation increase the contrast ratio between obscured lesions and surrounding structures. Thorac. Cancer. 17597714. 10 (10), 2057–2063. Available from: https://onlinelibrary.wiley.com/journal/10.1111/(ISSN)1759-7714. doi:10.1111/1759-7714.13167.

Lu, S., Wang, S., Wang, G., 2022. Automated universal fractures detection in X-ray images based on deep learning approach. Multimed. Tools Appl. 1380-7501. 81 (30), 44487–44503. doi:10.1007/s11042-022-13287-z.

Manafi-Farid, R., Askari, E., Shiri, I., Pirich, C., Asadi, M., Khateri, M., Zaidi, H., Beheshti, M., 2022. [18F]FDG-PET/CT radiomics and artificial intelligence in lung cancer: technical aspects and potential clinical applications. Semin. Nucl. Med. 00012998. 52 (6), 759–780. doi:10.1053/j.semnuclmed.2022.04.004.

Medyńska-Przęczek, A., Stochel-Gaudyn, A., Wędrychowicz, A., 2024. Liver fibrosis assessment in pediatric population—can ultrasound elastography be an alternative method to liver biopsy? A systematic review. Adv. Med. Sci. 18961126. 69 (1), 8–20. doi:10.1016/j.advms.2023.12.001.

Moffett, B.K., Panchabhai, T.S., Nakamatsu, R., Arnold, F.W., Peyrani, P., Wiemken, T., Guardiola, J., Ramirez, J.A., 2016. Comparing posteroanterior with lateral and anteroposterior chest radiography in the initial detection of parapneumonic effusions. Am. J. Emerg. Med. 15328171. 34 (12), 2402–2407. Available from: https://www.journals.elsevier.com/american-journal-of-emergency-medicine/. doi:10.1016/j.ajem.2016.09.021.

Murad Gutiérrez, V., Romero Enciso, J.A., 2018. Liver elastography: what it is, how it is done, and how it is interpreted. Radiologia. 15781771. 60 (3), 183–189. Available from: https://www.elsevier.com/wps/find/journaldescription.cws_home/718580/description#description. doi:10.1016/j.rx.2017.11.002.

Neilio, J.M., Ginat, D.T., 2024. Emerging head and neck tumor targeting contrast agents for the purpose of CT, MRI, and multimodal diagnostic imaging: a molecular review. Diagnostics. 20754418. 14 (15). Available from: https://www.mdpi.com/journal/diagnostics/. doi:10.3390/diagnostics14151666.

Nelms, D.W., Kann, B.R., 2021. Imaging modalities for evaluation of intestinal obstruction. Clin. Colon Rectal Surg. 15309681. 34 (4), 205–218. Available from: https://www.thieme-connect.com/ejournals/toc/ccrs. doi:10.1055/s-0041-1729737.

Nenezić, D.U., Stević, R., 2009. Computerized tomography of mediastinal tumors. Acta Chir. Iugosl. 0354950X. 56 (4), 69–75. doi:10.2298/ACI0904069N.

Ni, J., Bhuyan, Z., Chen, Q., Sun, X., Wang, D., Cao, Y., Liu, B. 2022. Enhance chest X-ray classification with multi-image fusion and pseudo-3d reconstruction. In: Proceedings of the International Joint Conference on Neural Networks. 9781728186719. Institute of Electrical and Electronics Engineers Inc., United States. doi:10.1109/IJCNN55064.2022.9892095.

Pandey, H., Guruvare, S., Kadavigere, R., Rao, C.R., 2018. Utility of three dimensional (3-D) ultrasound and power Doppler in identification of high risk endometrial cancer at a tertiary care hospital in southern India: a preliminary study. Taiwan. J. Obstet. Gynecol. 18756263. 57 (4), 522–527. Available from: https://www.sciencedirect.com/science/journal/10284559. doi:10.1016/j.tjog.2018.06.007.

Partridge, S.C., Singer, L., Sun, R., Wilmes, L.J., Klifa, C.S., Lehman, C.D., Hylton, N.M., 2011. Diffusion-weighted MRI: influence of intravoxel fat signal and breast density on breast tumor conspicuity and apparent diffusion coefficient measurements. Magn. Reson. Imaging. 18735894. 29 (9), 1215–1221. doi:10.1016/j.mri.2011.07.024.

Perez-Gutierrez, V., Khalid, A., Camere, M., Perwaiz, M., 2024. Ultrasound-guided percutaneous fine-needle aspiration performed by in-training pulmonologist a comparative study with interventional radiologists. J. Bronchol. Interv. Pulmonol. 19488270. 32 (1). Available from: https://journals.lww.com/bronchology/pages/default.aspx. doi:10.1097/LBR.0000000000000998.

Piscaglia, F., Marinelli, S., Bota, S., Serra, C., Venerandi, L., Leoni, S., Salvatore, V., 2014. The role of ultrasound elastographic techniques in chronic liver disease: current status and future perspectives. Eur. J. Radiol. 0720048X. 83 (3), 450–455. doi:10.1016/j.ejrad.2013.06.009.

Prasad, P.M.K., Prasad, D.Y.V., Rao, G.S., 2016. Performance analysis of orthogonal and biorthogonal wavelets for edge detection of X-ray images. Procedia Comput. Sci. 18770509. 87, 116–121. Available from: https://www.sciencedirect.com/science/journal/18770509.

Raut, A.A., Naphade, P.S., Maheshwari, S., 2020. Abdominal radiograph. J. Gastrointest. Abdom. Radiol. 2581-9178. 3 (S 01), S22. doi:10.1055/s-0040-1701327.

Romero, B.R., Vollmer Torrubiano, I., Martín Juan, J., Frades, S.H., Pérez Pallares, J., Ruiz, V.P., Wangüemert Pérez, A., Cristina Ramos, H., Viedma, E.C., 2024. Ultrasound in the study of Thoracic diseases: innovative aspects. Arch. Bronconeumol. 03002896. 60 (1), 33–43. doi:10.1016/j.arbres.2023.10.009.

Scarlata, S., Okoye, C., Zotti, S., Lauretani, F., Nouvenne, A., Cerundolo, N., Bruni, A.A., Torrini, M., Finazzi, A., Mazzarone, T., Lunian, M., Zucchini, I., Maccioni, L., Guarino, D., Della Faggiola, S.F., Capacci, M., Bianco, M.G., Guarona, G., Bellelli, G., Monzani, F., Virdis, A., Incalzi, R.A., Ungar, A., Ticinesi, A., 2023. Advancing healthcare through thoracic ultrasound research in older patients. Aging Clin. Exp. Res. 1720-8319. 35 (12), 2887–2901. doi:10.1007/s40520-023-02590-w.

Schauer, M.I., Jung, E.M., Hofmann, H.-S., da Silva, N.P.B., Akers, M., Ried, M., 2024. Performance of intraoperative contrast-enhanced ultrasound (Io-CEUS) in the diagnosis of primary lung cancer. Diagnostics. 2075-4418. 14 (15), 1597. doi:10.3390/diagnostics14151597.

Sharma, H.A., 2009. MRI physics-basic principles. Acta Neuropsychiatr. 16015215. 21 (4), 200–201. Available from: https://www.wiley.com/bw/journal.asp?ref=0924-2708&site=1. doi:10.1111/j.1601-5215.2009.00404.x.

Shimy, R.M., Ali, A.M., Elsattar, A.A., Hanna, A., Desoukey, S.H.I., Hassan, M.H., Hussein, M.M.M., Morad, K., 2022. Ultrasound elastography role versus liver biopsy in diagnosing and staging of hepatic fibrosis in chronic liver disease patients. Egypt. J. Hosp. Med. 20907125. 89 (2), 6693–6701. Available from: https://ejhm.journals.ekb.eg/. doi:10.21608/EJHM.2022.270760.

Si, G., Du, Y., Tang, P., Ma, G., Jia, Z., Zhou, X., Mu, D., Shen, Y., Lu, Y., Mao, Y., Chen, C., Li, Y., Gu, N., 2024. Unveiling the next generation of MRI contrast agents: current insights and perspectives on ferumoxytol-enhanced MRI. Natl. Sci. Rev. 2095-5138. 11 (5), doi:10.1093/nsr/nwae057.

Sinan, T., Leven, H., Sheikh, M., 2003. Is fasting a necessary preparation for abdominal ultrasound? BMC Med. Imaging. Available from: https://pmc.ncbi.nlm.nih.gov/articles/PMC183866/. doi:10.1186/1471-2342-3-1 .

Sodickson, A.D., Keraliya, A., Czakowski, B., Primak, A., Wortman, J., Uyeda, J.W., 2021. Dual energy CT in clinical routine: how it works and how it adds value n. Emerg. Radiol. doi:10.1007/s10140-020-01785-2/Published.

Song, J., Hwang, E.J., Yoon, S.H., Park, C.M., Goo, J.M., 2025. Emerging trends and innovations in radiologic diagnosis of Thoracic diseases. Investig. Radiol. 15360210. Available from: https://www.journals.lww.com/investigativeradiology/pages/default.aspx. doi:10.1097/RLI.0000000000001179.

Soni, N.J., Franco, R., Velez, M.I., Schnobrich, D., Dancel, R., Restrepo, M.I., Mayo, P.H., 2015. Ultrasound in the diagnosis and management of pleural effusions. J. Hosp. Med. 15535606. 10 (12), 811–816. Available from: https://onlinelibrary.wiley.com/journal/10.1002/(ISSN)1553-5606. doi:10.1002/jhm.2434.

Stark, P., Lester, R.G., Greene, R.E., 1985. Pseudolesion of the chest. A conglomerate shadow on the lateral radiograph. Chest. 00123692. 87 (4), 541–543. doi:10.1378/chest.87.4.541.

Suffredini, G., Gao, W.D., Dodd-o, J.M., 2024. Ultrasound shear wave elastography evaluation of the liver and implications for perioperative medicine. J. Clin. Med. 20770383. 13 (13). Available from: https://www.mdpi.com/journal/jcm. doi:10.3390/jcm13133633.

Sunoqrot, M.R.S., Saha, A., Hosseinzadeh, M., Elschot, M., Huisman, H., 2022. Artificial intelligence for prostate MRI: open datasets, available applications, and grand challenges. Eur. Radiol. Exp. 25099280. 6 (1). Available from: https://link.springer.com/journal/41747. doi:10.1186/s41747-022-00288-8.

Umek, W., Laterza, R.M., Yerlikaya, G., Kölbl, H., 2013. Dreidimensionale sonographie in der urogynäkologie. Gynakologe. 14330393. 46 (7), 448–452. doi:10.1007/s00129-012-3123-5.

Winder, M., Grabowska, S., Hitnarowicz, A., Barczyk-Gutkowska, A., Gruszczyńska, K., Steinhof-Radwańska, K., 2023. The application of abbreviated MRI protocols in malignant liver lesions surveillance. Eur. J. Radiol. 0720048X. 164, 110840. doi:10.1016/j.ejrad.2023.110840.

Yang, X., Zhai, D., Zhang, T., Zhang, S., 2020. Use of strain ultrasound elastography versus fineneedle aspiration cytology for the differential diagnosis of thyroid nodules: a retrospective analysis. Clinics. 18075932. 75, 1–8. Available from: https://www.scielo.br/pdf/clin/v75/1807-5932-clin-75-e1594pdf. doi:10.6061/clinics/2020/e1594.

Zhou, J., Gopal, A., Zhang, B., Xu, H., Chen, S., Yi, B.Y., Lasio, G., 2025. A novel optimization of cone beam CT frequency for lung radiation therapy based on an image-guided radiation therapy protocol and patient classification method. Transl. Lung Cancer Res. 22264477. 14 (1), 81–95. Available from: https://tlcr.amegroups.org/article/view/95648/html. doi:10.21037/tlcr-24-606.

Chapter 33

Future and challenges for the implementation of computational medical imaging in oncology

Veena Mayya[a], Winniecia Dkhar[b], Rajagopal Kadavigere[c], S. Sharath[b], Ashok Rao[a], Abhimanyu Pradhan[b], Anuradha Rao[a], Leena R. David[b,d], Berna Uzun[d,e], Dilber Uzun Ozsahin[e,f], Kaushik Nayak[b], Ilker Ozsahin[e], and Neil Abraham[b]

[a]*Manipal Institute of Technology, Manipal Academy of Higher Education, Manipal, India*

[b]Department of Medical Imaging Technology, Manipal College of Health Professions, Manipal Academy of Higher Education, Manipal, Karnataka, India

[c]*Department of Radiodiagnosis and Imaging, Kasturba Medical College, Manipal, India*

[d]*Medical Diagnostic Imaging Department, College of Health Sciences, University of Sharjah, Sharjah, United Arab Emirates*

[e]*Irfan Suat Gunsel Operational Research Institute, Near East University, Nicosia, Turkish Republic of Northern Cyprus (TRNC), Turkey*

[f]*Research Institute for Medical and Health Sciences, University of Sharjah, Sharjah, United Arab Emirates*

1 Introduction

Computational imaging for oncology stands as a transformative force in modern healthcare, reshaping the landscape of cancer diagnosis, treatment, and management. By employing sophisticated algorithms and computational techniques, this approach delves deeper into the complexities of medical imaging data, unveiling dimensions of analysis that transcend the capabilities of human perception and traditional interpretative methods. As cancer remains one of the leading causes of morbidity and mortality worldwide, there is an increasing demand for diagnostic tools that are not only accurate and efficient but also predictive and personalized. Computational imaging fulfills this demand by revolutionizing how clinicians acquire, process, and interpret imaging data. This interdisciplinary field merges the domains of medicine, computer science, biomedical engineering, and applied mathematics, fostering a collaborative ecosystem that is essential for the development and deployment of intelligent imaging solutions. At its core, computational imaging integrates advanced image processing, machine learning (ML), deep learning (DL), computer vision, and statistical modeling with well-established medical imaging modalities such as magnetic resonance imaging (MRI) (Winder et al., 2023), computed tomography (CT) (Goehde et al., 2005) and positron emission tomography (PET) (Yang et al., 2018). Through this integration, computational imaging enables researchers and clinicians to identify, localize, and characterize tumors with heightened accuracy, reduced human error, and improved reproducibility, ultimately contributing to better outcomes in oncology care (Itani et al., 2021).

One of the most significant contributions of computational imaging lies in its ability to overcome the limitations of manual interpretation. Traditional imaging assessments rely heavily on visual inspection and expert judgment, which, while valuable, are susceptible to variability and subjective bias. By contrast, computational approaches utilize quantitative imaging and pattern recognition to extract high-dimensional data features, often referred to as radiomics, which can capture subtle differences in tumor shape, texture, intensity, and heterogeneity that are imperceptible to the human eye. These features serve as potential biomarkers, offering insights into tumor biology, treatment response, and prognosis (Li and Zhou, 2022; Tabassum et al., 2023).

Artificial intelligence (AI) technologies, including convolutional neural networks (CNNs), recurrent neural networks, and transformers, have shown exceptional capabilities in segmenting tumors, classifying tissue types, and predicting clinical outcomes based on imaging data. For instance, AI-driven algorithms can automate the segmentation of complex anatomical structures, delineate tumor margins with precision, and differentiate between benign and malignant lesions—all in a fraction of the time required by human experts. This capability not only accelerates the diagnostic workflow but also

Advances in Medical Imaging. https://doi.org/10.1016/B978-0-443-28967-5.00001-0

reduces interobserver variability, leading to more consistent and objective evaluation (Saeidnia et al., 2025; Hussain et al., 2024; Bhise et al., 2021; Gonzalez et al., 2024).

Moreover, computational imaging facilitates early cancer detection, often before clinical symptoms emerge or visible lesions develop. In oncology, early diagnosis is closely linked to improved survival rates, particularly in aggressive malignancies such as pancreatic, lung, and ovarian cancers. Through advanced computational models, subtle changes in tissue structure or perfusion patterns can be detected and flagged for further investigation, enabling timely interventions. For example, texture analysis in mammography or multiparametric MRI can identify early-stage breast and prostate cancers with greater specificity than conventional assessments (Matsukawa et al., 2024; Firjani Naef, 2014).

Another major advantage of computational imaging is its role in personalized medicine. By analyzing the spatial and temporal characteristics of tumors, computational models can provide clinicians with detailed maps of intratumoral heterogeneity, a critical factor in determining how different regions of a tumor may respond to therapy. These maps can inform adaptive treatment planning, allowing oncologists to tailor radiotherapy or chemotherapy regimens to the biological behavior of each patient's disease. Personalized treatment strategies not only improve therapeutic efficacy but also minimize toxic side effects, enhancing the overall quality of life for patients (Cobanaj et al., 2024).

Furthermore, the integration of computational imaging with longitudinal patient data supports the real-time monitoring of disease progression and treatment response. Advanced imaging biomarkers can be used to track tumor shrinkage, metabolic changes, or recurrence, providing an objective basis for clinical decision-making. In this context, radiogenomics, the fusion of radiomic features with genomic, transcriptomic, and proteomic data, emerges as a particularly promising approach. Radio genomic studies aim to uncover correlations between imaging phenotypes and underlying molecular profiles, thereby enabling noninvasive tumor characterization and stratification. Such integrative analyses can identify patients who are likely to benefit from specific targeted therapies or immunotherapies, further refining precision oncology (Li and Zhou, 2022; Tabassum et al., 2023).

Despite these advancements, the field of computational imaging faces several challenges that must be addressed to ensure its successful translation into routine clinical practice. Data quality and standardization remain key concerns. The performance of computational models is heavily influenced by the quality, diversity, and consistency of the imaging data used for training and validation. Variations in imaging protocols, scanner types, and acquisition parameters across institutions can limit the generalizability of AI algorithms. Standardization initiatives, such as the Quantitative Imaging Biomarkers Alliance and The Cancer Imaging Archive, are crucial for developing harmonized protocols and datasets that facilitate model development, benchmarking, and regulatory approval (Komura et al., 2025; Hanna et al., 2025a; Panayides et al., 2020).

Another critical issue is the interpretability of AI-driven decisions. Black-box models, while highly accurate, often lack transparency, making it difficult for clinicians to understand the rationale behind their outputs. This lack of explainability can hinder clinical adoption, especially in high-stakes scenarios such as cancer diagnosis or treatment planning. To address this, researchers are developing explainable AI (XAI) frameworks that provide visualizations, feature importance maps, and decision rules to make model predictions more understandable and trustworthy (Panayides et al., 2020; Chan et al., 2020).

Ethical considerations, including data privacy, algorithmic bias, and informed consent, also play a vital role in shaping the future of computational imaging. Ensuring that AI models are trained on diverse, representative populations is essential to avoid disparities in care. Moreover, robust governance frameworks are needed to manage the use of patient data and safeguard against potential misuse or unintended consequences (Hanna et al., 2025b; Pham, 2025).

Looking ahead, the future of computational imaging in oncology is promising. The convergence of multimodal imaging, big data analytics, and cloud computing will pave the way for integrated diagnostic platforms capable of delivering comprehensive, data-driven insights in real time. As computational tools continue to evolve, they will not replace clinicians but rather augment their expertise, offering powerful support in navigating the complexity of cancer care (Radanliev, 2025).

In conclusion, computational imaging represents a paradigm shift in oncological practice. It empowers clinicians and researchers with the ability to decode the hidden language of medical images, offering a richer understanding of cancer biology and enabling more precise, timely, and personalized interventions. By addressing current challenges and embracing collaborative innovation, computational imaging has the potential to redefine the standard of care in oncology, ultimately improving outcomes for patients worldwide.

2 Advancements in computational imaging for oncology: State-of-the-art

Advancements in computational imaging for oncology represent a significant leap forward in cancer diagnosis, treatment planning, and monitoring, marking a transition toward precision medicine and data-driven healthcare. This interdisciplinary field fuses the capabilities of medical imaging, AI, computer vision, data science, and clinical oncology to extract, analyze, and interpret vast amounts of data from imaging modalities such as MRI, CT, PET, and, more recently, hybrid imaging technologies (Hussain et al., 2024; Beuthien-Baumann et al., 2021; Henzler et al., 2012; Hegde et al., 2013). By leveraging ML, DL, and other computational algorithms, clinicians and researchers can uncover patterns and biomarkers in imaging data that are imperceptible to the human eye, thereby facilitating early diagnosis, accurate tumor classification, individualized treatment strategies, and dynamic treatment monitoring. These capabilities are increasingly being integrated into AI-based decision support systems that assist radiologists and oncologists in automating routine tasks and enhancing diagnostic accuracy. For instance, deep CNNs have demonstrated remarkable performance in segmenting tumors, detecting lesions, and classifying cancer stages, often surpassing traditional rule-based or manual interpretation approaches in terms of speed and consistency. Studies have shown the application of computational imaging in various oncological tasks, such as automated tumor detection, lesion characterization, cancer stage classification, segmentation of anatomical structures, and personalized therapy planning (Adams et al., 2022; Shi et al., 2024) Relevant studies often illustrate a generic pipeline where computational imaging feeds into an AI-driven framework to support clinical decision-making. These frameworks typically follow a multistep workflow, including image acquisition, preprocessing, feature extraction (using radiomics or DL models, classification or prediction, and visualization or decision output (Goehde et al., 2005; Hou et al., 2023). Fig. 33.1 represents the generic workflow for computational medical imaging-driven AI-based oncology decision support systems. Notably, most studies rely on single-modality imaging data, although there is a growing push toward multimodal data integration to gain a more comprehensive view of tumor behavior and biology. Advanced imaging technologies such as cone-beam CT and real-time MRI-guided radiotherapy are further enhancing the precision of therapeutic interventions by allowing for image-guided, adaptive treatment planning. These technologies are supported by computational imaging algorithms that enable dynamic updates to radiation plans based on changes in tumor volume, position, or patient anatomy. This is crucial in minimizing damage to healthy tissues and maximizing treatment efficacy (Cobanaj et al., 2024; Van de Wiele et al., 2021; Anon, n.d.). Moreover, computational imaging is integral to the extraction of quantitative biomarkers that serve as objective indicators of disease status and response to therapy. Such biomarkers include tumor heterogeneity, perfusion characteristics, texture patterns, and metabolic signatures, which are derived through radiomic analysis—a process that quantifies features from medical images and correlates them with clinical outcomes. Radiomics, when combined with genomics in an emerging field known as radiogenomics, allows researchers to establish associations between imaging phenotypes and gene expression profiles. This integrative approach has shown promise in predicting tumor behavior, treatment response, and prognosis without the need for invasive biopsies (Li and Zhou, 2022; Tabassum et al., 2023) The fusion of computational imaging and radio genomic data creates a more complete portrait of the tumor microenvironment and its underlying biology. Importantly, the use of AI in computational imaging does not come without challenges. One of the primary issues is the "black-box" nature of DL models, where even experts may not fully understand how the model reaches its conclusions. This lack of interpretability hinders clinical trust and the broader adoption of AI tools in practice. In response, researchers have developed XAI techniques that aim to shed light on the inner workings of these models by identifying the features or image regions most influential to a given prediction (Saeidnia et al., 2025; Uthamacumaran et al., 2023; An et al., 2025). By making AI models more transparent and interpretable, XAI enables clinicians to better validate and trust AI-driven outputs. Another challenge lies in the requirement for large, annotated datasets that are diverse and representative of the broader patient population. Most existing datasets are either institution-specific or limited in terms of demographic variability, which can lead to biases in AI models and reduced generalizability (Koh et al., 2022; Singh et al., 2024). Standardization of imaging protocols, data curation practices, and quality control measures is essential to mitigate this risk. In addition, rigorous validation of AI models through multicenter clinical trials is needed to ensure their reliability and reproducibility across different clinical settings. Ethical considerations, such as patient privacy, data security, and algorithmic accountability, must also be addressed to foster responsible and equitable use of computational imaging

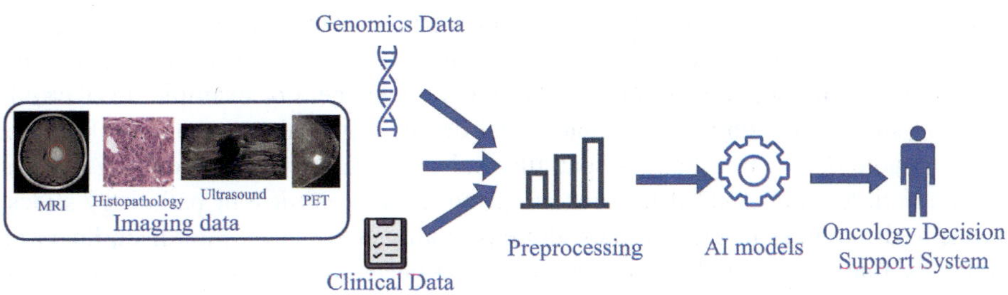

FIG. 33.1 Overview of the pipeline for computational medical imaging-based oncology decision-support systems.

technologies. Federated learning and privacy-preserving AI models are being explored to allow model training across multiple institutions without compromising patient confidentiality. Furthermore, regulatory frameworks for the clinical deployment of AI systems must evolve to accommodate the unique challenges posed by these dynamic and data-intensive tools. Despite these challenges, the future of computational imaging in oncology remains promising. With continuous advancements in hardware (e.g., graphical processing unit (GPU) acceleration), software (e.g., improved model architectures), and data-sharing infrastructures (e.g., cloud-based platforms and open-source repositories), the adoption of computational imaging in routine oncology practice is expected to grow rapidly. Interdisciplinary collaboration among radiologists, oncologists, data scientists, and engineers will be key to unlocking the full potential of this technology (Paudyal et al., 2023; Bi et al., 2019; Eidex et al., 2024). Moreover, patient-centered approaches that involve end-users in the design and implementation of AI tools will ensure that these technologies meet the practical needs of clinicians and enhance, rather than disrupt, existing workflows. As AI and computational imaging continue to evolve, their role will expand beyond diagnosis and treatment planning to encompass areas such as cancer prevention, population health screening, and even health equity analysis by identifying gaps in care delivery (Koh et al., 2022). In summary, computational imaging has emerged as a cornerstone of modern oncology, offering unprecedented insights into tumor biology, enabling more precise and personalized interventions, and improving patient outcomes through data-informed decision-making. As this field matures, it will not only transform clinical practice but also redefine our understanding of cancer itself.

3 Challenges and limitations in implementing computational imaging in oncology

Implementing computational imaging in oncology presents a range of challenges and limitations that need to be carefully addressed to ensure the successful integration of these advanced technologies into routine clinical practice. While computational imaging holds great promise for enhancing diagnostic accuracy, treatment planning, and patient outcomes in oncology, its translation from research settings to clinical environments is complex. A primary enabler of successful implementation is access to large, high-quality, and well-annotated datasets. Without such data, computational models may be inaccurate, biased, or unable to generalize across diverse patient populations (Cobanaj et al., 2024). Below, several core challenges associated with data quality, standardization, privacy, multimodality integration, interpretability, workflow alignment, and regulatory compliance are discussed in detail.

3.1 Data quality and standardization

One of the foundational requirements for computational imaging is high-quality, standardized data. Medical imaging datasets often suffer from variability in acquisition protocols, scanner hardware, patient positioning, resolution, and postprocessing methods. This heterogeneity introduces noise and bias into computational models, limiting their reproducibility and generalizability across institutions. For example, MRI scans from two different hospitals may have different echo times, slice thicknesses, or coil configurations, which can affect the extracted radiomic features and model predictions. Therefore, the development of consistent imaging protocols is imperative. In addition, the establishment of international standards for data acquisition and annotation can facilitate data interoperability and support multicenter research collaborations. Quality control measures must also be embedded throughout the imaging pipeline to detect and correct artifacts, missing data, and other inconsistencies. Establishing central repositories for curated, quality-assured datasets that are diverse and representative of the broader patient population is a critical step toward overcoming these challenges (Ru et al., 2024; Saw et al., 2025).

3.2 Multimodal imaging data integration

Oncology often relies on multiple imaging modalities, such as CT, MRI, PET, ultrasound, and histopathology—as well as genomic and clinical data—to form a comprehensive understanding of tumor behavior and patient status. However, integrating such diverse data types poses technical and conceptual challenges. Each imaging modality differs in its spatial resolution, contrast mechanism, acquisition time, and clinical relevance. For example, PET provides metabolic information, while MRI offers soft tissue contrast and anatomical detail. Combining these modalities requires sophisticated data fusion techniques capable of aligning images spatially and temporally while preserving their unique informational content. In addition, integrating nonimaging data, such as genomics, proteomics, or pathology slides, introduces another layer of complexity due to differences in data formats, dimensionality, and scales. Interdisciplinary approaches involving data scientists, radiologists, oncologists, and bioinformaticians are necessary to develop robust algorithms for multimodal data integration. Advances in AI, particularly in multistream DL architectures, have begun to address this challenge, but many models remain in the experimental phase and require further validation before clinical adoption (Waqas et al., 2024; Lother et al., 2023).

3.3 Data privacy and security

Medical imaging data inherently contains sensitive patient information, raising critical concerns around privacy, security, and ethical use. With the increasing use of cloud-based systems and distributed computing for AI model development, ensuring the confidentiality and security of patient data is paramount. Data breaches, unauthorized access, or improper sharing could have serious legal and ethical consequences. To mitigate these risks, robust security frameworks must be implemented, including encryption of data in storage and transmission, access control mechanisms, anonymization procedures, and compliance with international data protection regulations such as Health Insurance Portability and Accountability Act in the United States and General Data Protection Regulation in the European Union. Furthermore, federated learning—a decentralized approach where AI models are trained locally on institutional data without sharing the raw data—has gained attention as a privacy-preserving solution. In this model, only the trained parameters are shared, ensuring that patient data remains secure and localized. However, federated learning introduces its own challenges, such as handling data heterogeneity, model synchronization, and ensuring robustness against adversarial attacks (Lee et al., 2020; Olakanmi and Odeyemi, 2023).

3.4 Interpretability and explainability of AI models

A key barrier to the clinical adoption of AI-based computational imaging is the "black-box" nature of many ML models, particularly DL architectures. Clinicians often require a clear understanding of how and why an AI system arrives at a particular decision or diagnosis to build trust and accountability. Without interpretability, the clinical utility of AI is limited, as practitioners may be reluctant to rely on systems that cannot justify their outputs. XAI seeks to address this issue by developing techniques that provide human-interpretable insights into model behavior. For instance, heatmaps, saliency maps, and feature attribution methods can highlight which parts of an image most influenced a prediction. Some models incorporate attention mechanisms that mimic human decision-making patterns. Despite these advancements, much work remains in standardizing interpretability metrics and integrating them into user-friendly clinical interfaces. Moreover, explanations must be tailored to the needs and understanding levels of different stakeholders—including radiologists, oncologists, and patients—to be truly effective. As computational imaging tools evolve, enhancing their transparency will be essential to ensure safe, ethical, and confident clinical decision-making (Chen et al., 2022; González-Alday et al., 2023).

3.5 Integration with clinical workflow

For computational imaging to realize its potential, it must be seamlessly integrated into existing clinical workflows. Currently, many AI tools function in isolation or as research prototypes rather than being embedded into electronic health records or radiology information systems. This creates inefficiencies and disrupts clinical routines. Integration challenges may include compatibility with Picture Archiving and Communication Systems, standardization of output formats, and alignment with clinicians' time constraints and diagnostic pathways. Furthermore, clinicians must be adequately trained to use and interpret AI-driven tools. Training programs, certification courses, and continuing medical education modules focused on computational imaging are essential to promote adoption and avoid misuse. Human–computer interaction studies can help design interfaces that are intuitive, responsive, and aligned with clinical logic. Workflow integration also involves developing protocols for when and how computational imaging results should be reviewed, validated, and acted upon. Ideally, AI tools should augment rather than replace human expertise, functioning as intelligent assistants that enhance speed, accuracy, and confidence in diagnosis and treatment planning. Designing AI systems with a clinician-centered approach will be vital to minimizing resistance and ensuring sustainable use (Hawkins et al., 2023; Wang et al., 2016).

3.6 Regulatory approval and clinical validation

Another major hurdle is the rigorous regulatory landscape governing the approval of new medical technologies. Regulatory bodies such as the United States Food and Drug Administration, European Medicines Agency, and other national agencies require evidence of clinical safety, efficacy, and performance before approving clinical use. For computational imaging tools, especially those using AI, demonstrating reproducibility, robustness, and generalizability is particularly challenging. Many AI models are trained and validated on retrospective datasets, which may not accurately reflect real-world clinical variability. Prospective, multicenter clinical trials are needed to establish clinical utility and compliance with regulatory standards. In addition, regulatory frameworks must evolve to keep pace with the rapid development of AI and computational technologies. Unlike traditional medical devices, AI tools may change and improve over time through continuous learning, raising questions about how to regulate dynamic systems. Some agencies are exploring adaptive regulatory pathways and postmarket surveillance mechanisms to monitor AI tools in real-world settings. Nonetheless, navigating the regulatory approval process remains time-consuming, resource-intensive, and a potential bottleneck in innovation (Hanna et al., 2025b; Pham, 2025; Lee et al., 2021).

3.7 Ethical and societal considerations

Beyond technical and regulatory barriers, the implementation of computational imaging in oncology raises broader ethical and societal concerns. Algorithmic bias is a significant risk if AI models are trained on nonrepresentative datasets that underrepresent certain populations based on age, gender, ethnicity, or socioeconomic status. Such bias can lead to disparities in care and undermine public trust. To mitigate this, AI development must prioritize inclusivity and equity by involving diverse patient cohorts and continuously monitoring model performance across subgroups. Ethical frameworks should also address issues of accountability, consent, and the role of AI in decision-making. For example, if an AI system makes an incorrect diagnosis that leads to harm, determining liability can be complex. Moreover, patients and providers must be fully informed about the capabilities and limitations of AI tools to make shared decisions (Hanna et al., 2025b; Pham, 2025; Radanliev, 2025).

4 Conclusions

Advancements in computational imaging techniques have revolutionized oncology by providing more accurate diagnoses, personalized treatment planning, and improved patient outcomes. In this work, we have delved into the current advancements in computational imaging for oncology, shedding light on the innovative techniques and methodologies reshaping the landscape of cancer diagnosis, treatment, and monitoring. We have explored how computational imaging technologies revolutionize oncology by providing more accurate and precise methods for detecting tumors, characterizing their behavior, and guiding treatment decisions.

However, amidst these advancements, we have discussed the significant challenges and obstacles in implementing computational imaging in oncology. These challenges range from technical issues, such as data standardization and interoperability, to clinical concerns regarding the interpretability and trustworthiness of ML models. In addition, regulatory and logistical hurdles, such as obtaining approval and reimbursement for computational imaging procedures, further complicate the integration of these technologies into clinical practice.

By highlighting both the advancements and challenges in computational imaging for oncology, we aim to provide a comprehensive understanding of the field's current state. Continued innovation and interdisciplinary collaboration among researchers, clinicians, industry partners, regulatory agencies, and policymakers are essential to harness the full potential of computational imaging in cancer care.

References

Adams, L.C., Makowski, M.R., Engel, G., Rattunde, M., Busch, F., Asbach, P., et al., 2022. Prostate158—an expert-annotated 3T MRI dataset and algorithm for prostate cancer detection. Comput. Biol. Med. 148. doi:10.1016/j.compbiomed.2022.105817.

An, L., Liu, Y., Liu, Y., 2025. Advancements in circulating tumor cell detection for early cancer diagnosis: an integration of machine learning algorithms with microfluidic technologies. Biosensors (Basel) 15. doi:10.3390/bios15040220.

Anon, n.d. Expanding the scope of oncological understanding via multimodal imaging.

Beuthien-Baumann, B., Sachpekidis, C., Gnirs, R., Sedlaczek, O., 2021. Adapting imaging protocols for PET-CT and PET-MRI for immunotherapy monitoring. Cancers (Basel) 13. doi:10.3390/cancers13236019.

Bhise, S., Bepari, S., Gadekar, S., Kale, D., Singh Gaur, A., Aswale, S., 2021. Breast cancer detection using machine learning techniques. Int. J. Eng. Res. Technol.

Bi, W.L., Hosny, A., Schabath, M.B., Giger, M.L., Birkbak, N.J., Mehrtash, A., et al., 2019. Artificial intelligence in cancer imaging: clinical challenges and applications. CA Cancer J. Clin. 69, 127–157. doi:10.3322/caac.21552.

Chan, H.P., Samala, R.K., Hadjiiski, L.M., Zhou, C., 2020. Deep learning in medical image analysis. In: Advances in Experimental Medicine and Biology, Vol. 1213. Springer, pp. 3–21. doi:10.1007/978-3-030-33128-3_1.

Chen, H., Gomez, C., Huang, C.M., Unberath, M., 2022. Explainable medical imaging AI needs human-centered design: guidelines and evidence from a systematic review. NPJ Digit. Med. 5. doi:10.1038/s41746-022-00699-2.

Cobanaj, M., Corti, C., Dee, E.C., McCullum, L., Boldrini, L., Schlam, I., et al., 2024. Advancing equitable and personalized cancer care: novel applications and priorities of artificial intelligence for fairness and inclusivity in the patient care workflow. Eur. J. Cancer 198, 113504. doi:10.1016/j.ejca.2023.113504.

Eidex, Z., Ding, Y., Wang, J., Abouei, E., Qiu, R.L.J., Liu, T., et al., 2024. Deep learning in MRI-guided radiation therapy: a systematic review. J. Appl. Clin. Med. Phys. 25. doi:10.1002/acm2.14155.

Firjani Naef, A., 2014. A Non-invasive Image Based System for Early Diagnosis of Prostate Cancer. University of Louisville. doi:10.18297/etd/1710.

Goehde, S.C., Hunold, P., Vogt, F.M., Ajaj, W., Goyen, M., Herborn, C.U., et al., 2005. Full-body MRI for early disease detection full-body cardiovascular and tumor MRI for early detection of disease: feasibility and initial experience in 298 subjects. Am. J. Roentgenol.

González-Alday, R., García-Cuesta, E., Kulikowski, C.A., Maojo, V., 2023. A scoping review on the progress, applicability, and future of explainable artificial intelligence in medicine. Appl. Sci. (Switz.) 13. doi:10.3390/app131910778.

Gonzalez, R., Nejat, P., Saha, A., Campbell, C.J.V., Norgan, A.P., Lokker, C., 2024. Performance of externally validated machine learning models based on histopathology images for the diagnosis, classification, prognosis, or treatment outcome prediction in female breast cancer: a systematic review. J. Pathol. Inform. 15. doi:10.1016/j.jpi.2023.100348.

Hanna, M.G., Pantanowitz, L., Dash, R., Harrison, J.H., Deebajah, M., Pantanowitz, J., et al., 2025a. Future of artificial intelligence—machine learning trends in pathology and medicine. Mod. Pathol., 38. doi: 10.1016/j.modpat.2025.100705.

Hanna, M.G., Pantanowitz, L., Jackson, B., Palmer, O., Visweswaran, S., Pantanowitz, J., et al., 2025b. Ethical and bias considerations in artificial intelligence/machine learning. Mod. Pathol. 38, 100686. doi:10.1016/j.modpat.2024.100686.

Hawkins, J.R., Olson, M.P., Harouni, A., Qin, M.M., Hess, C.P., Majumdar, S., et al., 2023. Implementation and prospective real-time evaluation of a generalized system for in-clinic deployment and validation of machine learning models in radiology. PLOS Digit. Health 2. doi:10.1371/journal.pdig.0000227.

Hegde, J.V., Mulkern R, V., Panych, L.P., Fennessy, F.M., Fedorov, A., Maier, S.E., et al., 2013. Multiparametric MRI of prostate cancer: an update on state-of-the-art techniques and their performance in detecting and localizing prostate cancer. J. Magn. Reson. Imaging 37, 1035–1054. doi:10.1002/jmri.23860.

Henzler, T., Shi, J., Jafarov, H., Schoenberg, S.O., Manegold, C., Fink, C., et al., 2012. Functional CT imaging techniques for the assessment of angiogenesis in lung cancer. Transl. Lung Cancer Res. 1, 78–83. doi:10.3978/j.issn.2218-6751.2012.01.02.

Hou, H., Mitbander, R., Tang, Y., Azimuddin, A., Carns, J., Schwarz, R.A., et al., 2023. Optical imaging technologies for in vivo cancer detection in low-resource settings. Curr. Opin. Biomed. Eng. 28. doi:10.1016/j.cobme.2023.100495.

Hussain, S., Ali, M., Naseem, U., Nezhadmoghadam, F., Jatoi, M.A., Gulliver, T.A., et al., 2024. Breast cancer risk prediction using machine learning: a systematic review. Front. Oncol. 14. doi:10.3389/fonc.2024.1343627.

Itani, M., Menias, C.O., Mellnick, V.M., El Zakhem, A., Elsayes, K., Katabathina, V., et al., 2021. Imaging of abdominal and pelvic infections in the cancer patient. Abdom. Radiol. 46, 2920–2941. doi:10.1007/s00261-020-02896-7.

Koh, D.M., Papanikolaou, N., Bick, U., Illing, R., Kahn, C.E., Kalpathi-Cramer, J., et al., 2022. Artificial intelligence and machine learning in cancer imaging. Commun. Med. 2. doi:10.1038/s43856-022-00199-0.

Komura, D., Ochi, M., Ishikawa, S., 2025. Machine learning methods for histopathological image analysis: updates in 2024. Comput. Struct. Biotechnol. J. 27, 383–400. doi:10.1016/j.csbj.2024.12.033.

Lee, A.J., Suh, C.H., Kim, K.W., 2020. Evolution of clinical trial imaging and co-clinical imaging. In: Medical Radiology. Springer, pp. 239–253. doi:10.1007/978-3-030-31171-1_14.

Lee, T.F., Chang, I.P., Kung, T.S., 2021. Blockchain-based healthcare information preservation using extended chaotic maps for hipaa privacy/security regulations. Appl. Sci. (Switz.) 11. doi:10.3390/app112210576.

Li, S., Zhou, B., 2022. A review of radiomics and genomics applications in cancers: the way towards precision medicine. Radiat. Oncol. 17. doi:10.1186/s13014-022-02192-2.

Lother, D., Robert, M., Elwood, E., Smith, S., Tunariu, N., Johnston, S.R.D., et al., 2023. Imaging in metastatic breast cancer, CT, PET/CT, MRI, WB-DWI, CCA: review and new perspectives. Cancer Imaging 23. doi:10.1186/s40644-023-00557-8.

Matsukawa, A., Yanagisawa, T., Bekku, K., Kardoust Parizi, M., Laukhtina, E., Klemm, J., et al., 2024. Comparing the performance of digital rectal examination and prostate-specific antigen as a screening test for prostate cancer: a systematic review and meta-analysis. Eur. Urol. Oncol. doi:10.1016/j.euo.2023.12.005.

Olakanmi, O.O., Odeyemi, K.O., 2023. Expressible access control scheme for data sharing and collaboration in cloud-centric internet of medical things system. J. Ambient Intell. Humaniz. Comput. 14, 7189–7205. doi:10.1007/s12652-021-03572-4.

Panayides, A.S., Amini, A., Filipovic, N.D., Sharma, A., Tsaftaris, S.A., Young, A., et al., 2020. AI in medical imaging informatics: current challenges and future directions. IEEE J. Biomed. Health Inform. 24, 1837–1857. doi:10.1109/JBHI.2020.2991043.

Paudyal, R., Shah, A.D., Akin, O., RKG, D., Konar, A.S., Hatzoglou, V., et al., 2023. Artificial intelligence in CT and MR imaging for oncological applications. Cancers (Basel) 15. doi:10.3390/cancers15092573.

Pham, T., 2025. Ethical and legal considerations in healthcare AI: innovation and policy for safe and fair use. R. Soc. Open Sci. 12. doi:10.1098/rsos.241873.

Radanliev, P., 2025. AI ethics: integrating transparency, fairness, and privacy in AI development. Appl. Artif. Intell. 39. doi:10.1080/08839514.2025.2463722.

Ru, B., Sillah, A., Desai, K., Chandwani, S., Yao, L., Kothari, S., 2024. Real-world data quality framework for oncology time to treatment discontinuation use case: implementation and evaluation study. JMIR Med. Inform. 12. doi:10.2196/47744.

Saeidnia, H.R., Firuzpour, F., Kozak, M., Majd, H.S., 2025. Advancing cancer diagnosis and treatment: integrating image analysis and AI algorithms for enhanced clinical practice. Artif. Intell. Rev. 58. doi:10.1007/s10462-025-11117-w.

Saw, S.N., Yan, Y.Y., Ng, K.H., 2025. Current status and future directions of explainable artificial intelligence in medical imaging. Eur. J. Radiol. 183. doi:10.1016/j.ejrad.2024.111884.

Shi, Y., Fan, H., Li, L., Hou, Y., Qian, F., Zhuang, M., et al., 2024. The value of machine learning approaches in the diagnosis of early gastric cancer: a systematic review and meta-analysis. World J. Surg. Oncol. 22. doi:10.1186/s12957-024-03321-9.

Singh, P.K., Gupta, P.K., Gupta, V., Singh, S., Akhtar, S., 2024. Making cancer imaging smarter: emerging techniques and computational outlook to guide precision diagnostics. Open Bioinforma. J. 17. doi:10.2174/0118750362344593241111114723.

Tabassum, M., Suman, A.A., Suero Molina, E., Pan, E., Di Ieva, A., Liu, S., 2023. Radiomics and machine learning in brain tumors and their habitat: a systematic review. Cancers (Basel) 15. doi:10.3390/cancers15153845.

Uthamacumaran, A., Abdouh, M., Sengupta, K., Gao, Z., Forte, S., Tsering, T., et al., 2023. Machine intelligence-driven classification of cancer patients-derived extracellular vesicles using fluorescence correlation spectroscopy: results from a pilot study. Neural Comput. Appl. 35, 8407–8422. doi:10.1007/s00521-022-08113-4.

Van de Wiele, C., Ustmert, S., De Spiegeleer, B., De Jonghe, P.J., Sathekge, M., Alex, M., 2021. Apoptosis imaging in oncology by means of positron emission tomography: a review. Int. J. Mol. Sci. 22, 1–14. doi:10.3390/ijms22052753.

Wang, X., Liu, B.J., Martinez, C., Zhang, X., Winstein, C.J., 2016. Development of a novel imaging informatics-based system with an intelligent workflow engine (IWEIS) to support imaging-based clinical trials. Comput. Biol. Med. 69, 261–269. doi:10.1016/j.compbiomed.2015.03.024.

Waqas, A., Tripathi, A., Ramachandran, R.P., Stewart, P.A., Rasool, G., 2024. Multimodal data integration for oncology in the era of deep neural networks: a review. Front. Artif. Intell. 7. doi:10.3389/frai.2024.1408843.

Winder, M., Grabowska, S., Hitnarowicz, A., Barczyk-Gutkowska, A., Gruszczyńska, K., Steinhof-Radwańska, K., 2023. The application of abbreviated MRI protocols in malignant liver lesions surveillance. Eur. J. Radiol. 164. doi:10.1016/j.ejrad.2023.110840.

Yang, C.T., Ghosh, K.K., Padmanabhan, P., Langer, O., Liu, J., Eng, D.N.C., et al., 2018. PET-MR and SPECT-MR multimodality probes: development and challenges. Theranostics 8, 6210–6232. doi:10.7150/thno.26610.

Chapter 34

Psycho-oncology for cancer survivors

Shwetha Tumkur Shivakumar[a], Kirti Singhal[a], Elsa Sana Tombi Devi[b], Preetha Ramachandra[c], Leena R. David[d,e], Abhimanyu Pradhan[e], S. Sharath[e], Dilber Uzun Ozsahin[d,f,g], Berna Uzun[g], Kaushik Nayak[e], Winniecia Dkhar[e], Neil Abraham Barnes[e], Ilker Ozsahin[g], and Rajagopal Kadavigere[h]

[a]*Department of Clinical Psychology, Manipal Academy of Higher Education, Manipal College of Health Professions, Manipal, India*

[b]*Department of Medical Surgical Nursing, Manipal Academy of Higher Education, Manipal, India*

[c]*Department of Physiotherapy, Manipal Academy of Higher Education, Manipal College of Health Professions, Manipal, India*

[d]*Medical Diagnostic Imaging Department, College of Health Sciences, University of Sharjah, Sharjah, United Arab Emirates*

[e]*Department of Medical Imaging Technology, Manipal College of Health Professions, Manipal Academy of Higher Education, Manipal, Karnataka, India*

[f]*Research Institute for Medical and Health Sciences, University of Sharjah, Sharjah, United Arab Emirates*

[g]*Irfan Suat Gunsel Operational Research Institute, Near East University, Nicosia, Turkish Republic of Northern Cyprus (TRNC), Turkey*

[h]*Department of Radiodiagnosis and Imaging, Kasturba Medical College, Manipal, India*

1 Introduction

> *"You beat cancer by how you live, why you live, and the manner in which you live" – Scott, 2016, author of his autobiography about surviving cancer (Every Day I Fight) (Scott and Platt, 2016).*

Cancer is the world's second leading cause of death, and its incidence in India in 2022 accounts for an estimated crude rate of 100.4 per 100,000 incident cases. This alarming statistic underscores the magnitude of the disease burden not only in India but globally. According to the World Health Organization, approximately 19.3 million new cancer cases were reported worldwide in 2020, with projections suggesting this number could reach 30 million by 2040 (Global cancer burden growing, amidst mounting need for services, 2022). These figures highlight the pressing need for comprehensive approaches to cancer management beyond just medical treatment.

Globally, according to the Lancet 2015 data, the rates of surviving cancer are approximately >60% for the most common types of cancer, such as breast, prostate, and colorectal cancer. It was found to be increasingly higher in more developed countries, which could be due to advancements in technology used for cancer research and novel modes of treatment. The 5-year survival rates in high-income countries have shown remarkable improvement over the past three decades, reaching over 90% for certain cancers when detected early (Siegel et al., 2023). However, significant disparities exist between developed and developing nations, where limited resources, delayed diagnoses, and restricted access to cutting-edge treatments contribute to notably lower survival rates (Bray et al., 2018). The data presents us with hope for prolonged years of survival after the diagnosis of cancer, and it provides a scope for furthering literature to enhance the wellbeing of cancer survivors.

Since ages, the most consistent and efficient emphasis of bio-medical research has been on the detection and treatment of cancer. This research has reaped benefits for various generations now, and at present, the early detection and treatment efficacies have proven to be life-changing for many types of cancers, especially organ-specific cancer. Modern diagnostic techniques, including liquid biopsies, genomic profiling, and advanced imaging modalities, have revolutionized how cancer is detected and monitored. Similarly, treatment paradigms have evolved from conventional surgery, chemotherapy, and radiation to include targeted therapies, immunotherapies, and personalized medicine approaches that consider the unique genetic makeup of each patient's cancer (Levit, 2013).

A large and growing community of cancer survivors is one of the major achievements of cancer research over the past three decades. The quantity and quality of survival have become the most important end points. The research has progressively emphasized the number of years that the survivor is living and the way they are living through those years

Advances in Medical Imaging. https://doi.org/10.1016/B978-0-443-28967-5.00036-8

(Harries and Stanton, 2020). In the United States alone, there were an estimated 16.9 million cancer survivors as of January 2019, a number projected to increase to 22.2 million by 2030 (Miller et al., 2019). This growing population represents individuals who have completed primary treatment and are transitioning back to their everyday lives, often with lingering physical, psychological, and social challenges that require ongoing support and management.

However, many cancer survivors are still at risk of developing prolonged or late physiological and psycho-social effects of cancer treatment that may lead to premature mortality and morbidity. These late effects can manifest months or even years after treatment completion and include cardiovascular complications, endocrine dysfunction, secondary malignancies, cognitive impairments, chronic fatigue, and persistent pain (Harrington et al., 2010). The recognition of these long-term consequences has prompted the oncology community to adopt a more longitudinal perspective on cancer care, with an increasing focus on survivorship as a distinct phase requiring specialized attention and resources.

Behavioral and psycho-social factors impact the mortality and morbidity of cancer survivors, making it a progressively required field of research (Akechi, 2018). Studies have shown that depression and anxiety are prevalent among cancer survivors, with rates significantly higher than those in the general population (Mitchell et al., 2011). These psychological conditions not only diminish quality of life but can also affect treatment adherence, healthcare utilization patterns, and potentially influence disease outcomes. Social factors such as economic burden, employment challenges, insurance concerns, and disrupted family dynamics further complicate the survivorship experience and necessitate comprehensive supportive interventions (Harrison et al., 2009).

Interventions, both therapeutic and lifestyle, carry the potential to treat or ameliorate these late effects and must be developed, examined, and disseminated if found effective (Gorman, 2018). Emerging evidence suggests that physical activity, nutritional interventions, stress management techniques, and psychosocial support can significantly mitigate treatment-related side effects and improve overall health outcomes among survivors (Ventura et al., 2013). In addition, mindfulness-based approaches, cognitive-behavioral therapy, and peer support programs have shown promise in addressing the psychological sequelae of cancer and its treatment (Antoni et al., 2001).

The integration of these interventions into standard survivorship care represents a paradigm shift toward a more holistic model that acknowledges the multifaceted nature of cancer recovery. This approach recognizes that, while medical treatments focus on eliminating or controlling the disease, equally important is the restoration of physical function, psychological well-being, and social reintegration following a cancer diagnosis (Ganz, 2001). Before we further the discussion in this area, it is necessary to understand what we truly mean by survivorship.

1.1 History of survivorship

The term "survivorship" has been introduced by people who, in most cases, have been affected by a certain type of cancer and have narrated their journeys. The birth of the term "cancer survivorship" occurred in 1985, when the concept was defined by Dr. Mullan in his book, where he included both bio-medical aspects (e.g., secondary tumors, long-term effects of anticancer treatment, rehabilitation, reproductive health, and long-term health maintenance) and psychosocial aspects (e.g., community acceptance of cancer patients, insurance discrimination, barriers to employment, and education of youth about cancer) in the definition of survivorship, extending it beyond the symptomatic perspective (Mullan, 1985). The author, at the time, did not discuss the increasingly prevalent psychosocial issues such as fear of recurrence, posttraumatic distress and growth, social withdrawal, and depression. Thus, since 1985, there has been increasing recognition and broadening of the term to include the needs and situations of patients.

Prior to Dr. Mullan's influential work, cancer was predominantly viewed through a binary lens of either cure or death, with little acknowledgment of the complex, ongoing experience that follows diagnosis and treatment. The medical model that prevailed until the 1980s focused primarily on physical recovery and disease eradication, with minimal attention to the psychological, social, and existential dimensions of surviving cancer (Holland, 2002). The emergence of survivorship as a concept marked a paradigm shift from this traditional biomedical approach toward a more holistic understanding of cancer's impact on individuals' lives.

In the decade following Dr. Mullan's introduction of the term, several key organizations formed to address the unique needs of cancer survivors. In 1986, the National Coalition for Cancer Survivorship (NCCS) was established, becoming the first survivor-led advocacy organization (What is cancer survivorship?, 2020). The NCCS played a pivotal role in reshaping public understanding of survivorship by defining a cancer survivor as "any individual diagnosed with cancer, from the time of discovery and for the balance of life." This definition intentionally moved away from the traditional medical criterion that required patients to be disease-free for five years before being considered survivors.

The 1990s witnessed growing institutional recognition of survivorship as a distinct phase of the cancer experience. The Office of Cancer Survivorship was established at the National Cancer Institute (NCI) in 1996, signaling the importance of research dedicated to understanding and addressing survivors' needs (Office of Cancer Survivorship, 2018). This development coincided with improved cancer treatments that resulted in longer survival times, creating a larger population of individuals living with the aftermath of cancer and its treatment.

The early 2000s marked another significant milestone with the publication of the Institute of Medicine's 2005 report titled "*From Cancer Patient to Cancer Survivor: Lost in Transition*" (Hewitt et al., 2005). This landmark document highlighted substantial gaps in posttreatment care and provided comprehensive recommendations for improving survivorship care delivery. The report emphasized the need for a structured transition from active treatment to survivorship, including survivorship care plans and coordinated follow-up services.

There has been an attempt to propose stages of survivor-periods, such as acute survival (diagnosis and initial anticancer treatment), extended survival (watchful waiting and surveillance, maybe maintenance treatment), and permanent survival (long-term remission) (Strasser, 2021). This temporal framework, first articulated by Mullan and later refined by others, provides a useful structure for understanding how survivorship needs evolve over time.

The conceptualization of survivorship has also been influenced by diverse cultural perspectives. In some Eastern cultures, the emphasis on harmony and acceptance has shaped different narratives around cancer survival compared to Western societies, where survivorship is often framed in terms of heroic battles and individual triumphs. Indigenous communities have contributed perspectives that emphasize the connection between individual healing and community wellbeing, enriching the global understanding of what it means to survive cancer.

Patient narratives have been instrumental in shaping the evolving concept of survivorship. Personal accounts published in medical journals, autobiographies, blogs, and social media have highlighted aspects of the survivorship experience that might otherwise be overlooked in clinical settings. These narratives have revealed the profound identity transformations that often accompany cancer, the challenges of living with uncertainty, and the potential for both trauma and growth in the aftermath of a diagnosis (Park et al., 2008).

The emergence of survivorship as a distinct field of study has led to the development of specialized survivorship clinics and programs in many cancer centers worldwide. These initiatives reflect the growing recognition that surviving cancer involves more than just completing treatment—it encompasses a complex journey of physical recovery, psychological adjustment, and social reintegration that requires ongoing support and care (Rowland and Bellizzi, 2008).

Recent developments in survivorship research have expanded to include family members and caregivers, who are increasingly recognized as "cosurvivors" affected by the cancer experience. This broader perspective acknowledges that cancer's impact extends beyond the individual patient to affect family dynamics, relationships, and the wellbeing of those providing care and support (Given et al., 1992).

As treatments continue to advance and cancer mortality rates decline for many types of cancer, the population of survivors continues to grow and diversify. Contemporary approaches to survivorship increasingly emphasize personalized care that addresses the unique needs of different demographic groups, cancer types, and treatment modalities. The field continues to evolve toward more integrated models of care that bridge the traditional divide between oncology and primary care, with the goal of supporting survivors throughout their life journey after cancer.

1.2 Definition of survivorship

In the present day, some may define a cancer survivor as anyone who is cancer-free for a certain period, while most others have broadened the definition to include people with a diagnosis of cancer who are still alive. This includes patients who have completed primary therapy and are free of disease, as well as those patients living with recurrent and/or advanced disease (Albreht, 2017). The experiences and goals of care for each cancer survivor are unique and dynamic. The quote that has been written above explains another aspect of survivorship. It conveys that there is not only a biological definition of survivorship but also a psycho-social explanation for it. Survivorship may begin when the person has been told about the curability of the diagnosis, through the treatment, and it can also begin right after the diagnosis. According to Stuart Scott and many others, the journey of surviving cancer starts immediately after the diagnosis.

The evolution of how we define cancer survivorship reflects broader shifts in our understanding of cancer itself. Historically, medical definitions emphasized the absence of disease, with the term "survivor" reserved for individuals who remained cancer-free for at least 5 years following diagnosis. This time-based criterion emerged from statistical patterns showing that the risk of recurrence for many cancers decreases substantially after this period. However, this narrow definition failed to capture the lived experiences of those navigating the complex terrain of cancer diagnosis, treatment, and its aftermath.

The NCCS played a pivotal role in expanding this definition when it proposed that survivorship begins at the moment of diagnosis and continues throughout the remainder of life, regardless of treatment outcomes (What is cancer survivorship?, 2020).

This perspective acknowledges that the psychological, emotional, and social dimensions of surviving cancer begin when one first confronts the diagnosis, not merely when treatment concludes successfully. The NCCS definition has been widely adopted by many cancer organizations worldwide, though variations persist across different contexts.

The NCI further refined this concept by defining a cancer survivor as "an individual from the time of cancer diagnosis through the balance of his or her life" (Office of Cancer Survivorship, 2018). This definition explicitly includes the family members, friends, and caregivers affected by the diagnosis, recognizing that they, too, experience significant impacts and may be considered "cosurvivors." This inclusive approach acknowledges the ripple effects of cancer beyond the individual patient.

Within the medical community, several alternative frameworks for defining survivorship have emerged. Some oncologists prefer to distinguish between "acute survivors" (those undergoing active treatment), "extended survivors" (those who have completed treatment but remain at risk for recurrence), and "permanent survivors" (those considered cured or in long-term remission). Others advocate for condition-specific definitions that account for the varying trajectories of different cancer types. For instance, survivorship for someone with a slow-growing prostate cancer might look quite different from survivorship for someone with an aggressive form of leukemia.

The landscape of survivorship definitions has been further complicated by advances in cancer treatment. With the emergence of maintenance therapies, targeted agents, and immunotherapies, many patients now live with cancer as a chronic condition rather than experiencing a clear transition from treatment to posttreatment life. These individuals may identify as "living with cancer" rather than "surviving cancer," highlighting the limitations of binary classifications.

Cultural perspectives also influence how survivorship is defined and experienced. In some Eastern cultures, the concept of survivorship is less individualistic and more relational, emphasizing harmony with one's community and environment as integral to healing. Indigenous frameworks often incorporate spiritual dimensions and connection to ancestral wisdom as essential components of survivorship. Western definitions, by contrast, frequently emphasize personal agency, resilience, and the "battle" against cancer—metaphors that some critics argue can inadvertently burden patients with responsibility for their outcomes.

For many individuals affected by cancer, self-definition is paramount. Some embrace the term "survivor" as empowering, symbolizing strength and resilience in the face of adversity. Others reject it, feeling that it oversimplifies their complex, ongoing relationship with the disease or places undue emphasis on cancer as a defining aspect of their identity (Zebrack et al., 2006). Alternative terms such as "cancer graduate," "cancer veteran," "thriver," or simply "person who has had cancer" may be preferred, each carrying distinct connotations about the cancer experience.

The psychosocial dimensions of survivorship have gained increasing recognition in recent years. Psychological survival encompasses coping with the fear of recurrence, managing treatment-related trauma, finding meaning in the cancer experience, and reconstructing one's identity and life priorities. Social survival involves navigating changed relationships, managing others' expectations, addressing potential stigma, and redefining one's role within family and community contexts. Spiritual or existential survival often involves confronting mortality, finding purpose, and integrating the cancer experience into one's broader life narrative (Salsman et al., 2015).

Functional definitions of survivorship have also emerged, focusing on the restoration of capabilities rather than merely the absence of disease. These approaches emphasize quality of life, functional status, and the ability to resume meaningful activities as key indicators of successful survivorship. The growing emphasis on "living well" after cancer reflects this shift toward more holistic, patient-centered definitions that prioritize thriving over merely surviving.

As our understanding of cancer continues to evolve, so too will our conceptualization of survivorship. The increasing diversity of cancer experiences—from those cured after minimal intervention to those living long-term with metastatic disease managed through ongoing treatment—challenges us to develop more nuanced, inclusive definitions that honor the unique journey of each individual affected by cancer. What remains constant is the recognition that survivorship encompasses far more than medical outcomes; it is a complex, multidimensional experience that transforms lives in profound and lasting ways.

2 Psycho-oncology

The body-mind connection is crucial in understanding the cancer journey. The psychological influence of the mind on the body is often underestimated, much like the origins of cancer, which are frequently unclear. To effectively treat and support cancer survivors, it is crucial to pay close attention to their thoughts, actions, and emotions, as these factors may influence the severity and duration of the survivors' somatic symptoms. An emerging field of psycho-oncology aims to address the psychosocial aspects of cancer survivorship (Holland, 2002).

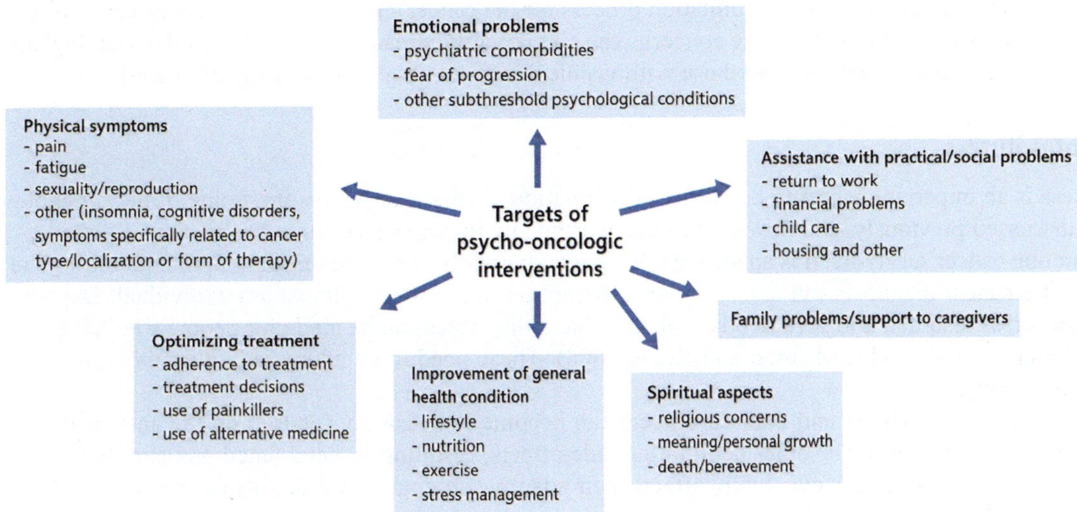

FIG. 34.1 A diagram of targets for psycho-oncological intervention.

The need for engaging holistically with the person and their caregivers begins at the time of diagnostic testing itself. It is imperative to recognize that evidence is convincing: emotional distress, depression, anxiety, uncertainty, and hopelessness interact with pain. Psychological and cognitive behavioral treatments can reduce pain severity and its interference with function, as indicated in multiple metaanalyses and high-quality randomized controlled trials (Lang-Rollin and Berberich, 2018).

Finally, various interventions have been represented by way of Fig. 34.1, which explains the areas of concern as well as some techniques that may be helpful in dealing with the diversity of the bio-psycho-social issues that cancer survivors face.

2.1 Common issues and concerns for survivors of cancer

Broadly, a cancer diagnosis brings up a complex range of issues, such as managing the physical symptoms and the side-effects of the diagnosis and treatment, respectively; coping with the existential domain of the illness, sexual and relationship challenges, and finding comforting philosophical, spiritual, or religious values that give life and death meaning (Cruzado, 2010). Specifically, for cancer survivors, it can be the fear of recurrence and the difficulty in dealing with uncertainty that lead to various psycho-social concerns for them (Thewes et al., 2012).

2.1.1 Survivor guilt, living with loss, and finding meaning

Cancer is one of those severe terminal illnesses, surviving which makes the affected individuals question various domains in life. Some may even call it “a fight against death.” Many survivors struggle with questions of why they had cancer and why they survived. For some, spirituality and faith are a source of comfort. For others, they may examine long-held beliefs or religious values as they try to make sense of their experience and find new meaning in life (Salsman et al., 2015). Some cancer survivors may also struggle with feelings of guilt and helplessness due to the sense of burden on the caregivers with respect to finances, emotions, dependency, or grief over the role transition for the cancer survivor. Living with these distressing feelings can become dysfunctional for them if not addressed. Research has shown that meaning-making processes and posttraumatic growth can significantly influence psychological adjustment following cancer, with many survivors reporting enhanced appreciation for life and changed priorities (Tedeschi and Calhoun, 2004; Park et al., 2008).

2.1.2 Body image and cancer

It is well-known to people that cancer treatment usually has visible effects on the body. The most common characteristic that people notice is hair loss, and the second is the effect of mastectomy, specifically for breast cancer survivors. The various themes that emerge in this domain relate to the individual’s view of the self, which can have implications based on their gender, workplace, and context, along with relationship difficulties, such as intimacy, which can become a concern for the individual and their partner. Research has shown that physical appearance influences the quality of the response from others around and can affect the rehabilitation process for the cancer survivor and their caregivers (Konradsen, 2012). Studies have demonstrated that body image concerns can significantly impact quality of life and psychological adjustment, particularly among younger survivors and those with visible treatment effects (Zebrack et al., 2006).

2.2 Mental illness

Mental illness is an important aspect to be discussed in relation to cancer survivorship. Some of the psychological factors have been discussed previously, and now it becomes worthwhile to emphasize some of the mental disorders that may be comorbid among cancer survivors. It is an area that has been invariably discussed throughout the literature. The anxiety and uncertainty of a cancer diagnosis can create extreme disruption in the life of almost any individual. Depression, anxiety, posttraumatic stress disorder, and several other mental illness diagnoses can be made for people who have survived cancer treatment (Mitchell et al., 2011; Mehnert and Koch, 2008). These need to be addressed either through psychotherapy or psychopharmacology.

In addition, cancer survivors and their caregivers can become increasingly focused on the anticipation of the risk of death, recurrence of the illness, or other debilitating side-effects, resulting in heightened anxiety about the purpose and meaning of their existence, which eventually affects their adaptive strategies in life. This existential crisis may also result from, or lead to, a sense of unfinished business, where the survivor or the caregivers may not be able to accomplish their goals as planned, and the uninvited arrival of cancer halts their process of achieving those goals. This can lead to further grief, and intervention for the same, through existential and narrative therapy, can be of utmost use in this scenario. The biopsychosocial model, proposed by (Engel, 1977) which continues to help in comprehensively explaining the health of an individual, suggests that the physiological pathology, the emotions, cognitions, and actions of individuals, as well as the surroundings and context of the people and their caregivers, affect the course of any disease. Genetic vulnerabilities, temperament, self-efficacy, early childhood experiences, habits, interpersonal relationships, substance use, coping skills, social skills, and motivation—all contribute to the conceptualization and treatment of cancer survivors, as well as their caregivers (Adler, 2009).

Research has shown that childhood cancer survivors face unique psychological challenges, including higher rates of anxiety, depression, and posttraumatic stress symptoms compared to their peers (Zeltzer et al., 2009). The developmental timing of cancer diagnosis and treatment can significantly impact psychological adjustment and long-term mental health outcomes.

3 Evidence based psycho-social interventions

Every clinical intervention by oncologists and other cancer care specialists should include the screening and assessment of the emotional issues and worries of both the survivor and the caregivers. Psycho-oncologists can intervene and evaluate the quality of the psychological assistance necessary for survivors. These may be related to how they cope on their own, relationship issues, disease-related myths, interpersonal societal inadequacies, and existential aspects of the illness.

Psychosocial interventions aim toward building and discovering the mental (psychological) and societal (social) resources for and around the persons affected by cancer. These interventions can be based on either the individual model or the group model, depending on the appropriate need. Landmark research by (Spiegel et al., 1989) demonstrated that psychosocial interventions could not only improve the quality of life but also potentially influence survival outcomes, highlighting the importance of addressing psychological needs in cancer care.

Every age group, from adolescence through adulthood to old age (geriatric population), can benefit from psycho-social interventions. These interventions are organized and evidence-based due to their widespread prevalence and application around the world, and they are flexible in that they may be tailored to the needs and desires of the community. For cancer survivors, counseling, psychoeducation, coping strategies, and more focused types of psychotherapy in their various forms (group, individual, and family therapy) and orientations (cognitive-behavioral, supportive-expressive, existential, and psychodynamic) have been developed to intervene with more specificity in all the circumstances where psychological disorders and maladjustment to cancer emerge. Mindfulness-based therapies and acceptance and commitment therapy have also gained empirical support for dealing with distress post survivorship (Antoni et al., 2001; Stanton et al., 2006).

3.1 Reducing the risk of recurrence and second cancers through behavioral change

Recurrence is defined as the return of the disease after an initial course of treatment with a disease-free period. The disease may recur at the same site, recur near the site, or metastasize to a distant site. The threat of recurrence is one of the reasons why cancer is such a feared disease. Long-term survivors continue to experience distress over the fear of recurrence (Gorman, 2006). Research indicates that fear of cancer recurrence affects up to 70% of cancer survivors and can significantly impact quality of life and psychological functioning (Thewes et al., 2012).

On one hand, it has been discussed that many people may feel anxious and hopeless due to the uncertainty of recurrence. However, survivors also report positive outcomes, including a heightened sense of appreciation for family members and friends, feelings of being a better person for having gone through this difficult experience, and the changing of priorities in life for the better. Some studies of survivors have revealed an increased desire to be of service to others, and many survivors volunteer to help other patients with cancer (Pelusi, 1997; Ferrans, 1994). These positive changes align with concepts of posttraumatic growth, where individuals report enhanced relationships, greater appreciation for life, increased personal strength, and spiritual development following their cancer experience (Tedeschi and Calhoun, 2004). These can be encouraged and built upon by way of intervening at a psychological and social level, such as using stress and anxiety reduction techniques of psychotherapy and creating a safe space for expressing their thoughts and emotions through moderated support groups. Research programs have shown psychoeducation, cognitive restructuring, and behavioral modification in an individual therapeutic setting as effective ways of coping with second cancer concerns (Lang-Rollin and Berberich, 2018).

Studies have demonstrated that lifestyle modifications, including regular physical activity, healthy dietary patterns, weight management, and smoking cessation, can significantly reduce the risk of cancer recurrence and second primary cancers (Ventura et al., 2013). These behavioral interventions, when combined with psychosocial support, offer survivors concrete ways to feel empowered in their recovery journey.

3.2 Caregivers' burden and wellbeing

As the relational stressors from the perspective of the survivors become clearer, it is equally crucial to examine the experiences of caregivers, whose lives are often profoundly impacted by a loved one's cancer diagnosis. Caregivers—whether spouses, parents, children, or close friends—frequently find themselves navigating a complex web of emotional, physical, and logistical challenges. The diagnosis often disrupts established family dynamics, creating a shift in roles that can place immense pressure on the caregiver. In many cases, individuals who once shared household responsibilities now bear the primary burden of care, household management, and emotional support. This sudden and sustained responsibility can result in feelings of helplessness, anxiety, and emotional fatigue (Given et al., 1992).

One of the most prominent emotional stressors faced by caregivers is the persistent fear of losing a loved one. This fear often exists alongside anticipatory grief, where caregivers begin to mourn potential outcomes even while actively supporting treatment and recovery. In addition, the sense of duty to be available physically and emotionally around the clock can lead to caregiver burnout, a state characterized by emotional exhaustion, depersonalization, and reduced personal accomplishment. Caregivers may also struggle with guilt—guilt for feeling overwhelmed, for needing respite, or for not being able to provide every form of support, including financial resources, especially in cases where the cost of cancer treatment is high and long-term.

Another major concern is the lack of adequate support and recognition that caregivers receive, both from within the family and the larger community. In many cultural contexts, caregiving is viewed as a natural familial obligation rather than a role that requires training, emotional support, and sometimes professional intervention. This often leaves caregivers feeling isolated, underappreciated, and emotionally unsupported. In families where open communication is limited, the illness becomes a silent burden, and caregivers may find little space to express their own struggles. This lack of emotional ventilation can result in increased stress, depression, and even physical health issues, including fatigue, insomnia, hypertension, and compromised immunity (Deshields et al., 2006).

The financial burden of caregiving is also significant. Loss of income due to time off work, additional healthcare expenses, and the costs of travel and accommodation when accessing specialized treatment can create a profound economic strain. This financial pressure, coupled with emotional distress, can reduce the overall quality of life not only for the patient but for the entire family.

Given the multifaceted burden caregivers face, it becomes imperative for professionals to step in with timely and targeted interventions. Couple and family counseling can help redistribute emotional labor and promote healthy communication within families. On a broader level, community-based support groups offer caregivers a much-needed platform to share experiences, gain coping strategies, and receive validation. Access to respite care services, educational resources on caregiving, and mental health support for caregivers should also be integral parts of cancer care policies.

By addressing the caregivers' needs alongside those of the patients, we foster a more holistic, compassionate, and sustainable model of care. In doing so, not only are the health and wellbeing of caregivers preserved, but the quality of support provided to cancer survivors is also significantly enhanced. Research has shown that when caregivers receive adequate support, both patient and caregiver outcomes improve significantly (Carver et al., 2005).

4 Limitations

As we reflect on cancer survivorship, as presented in this chapter, several important limitations in our current understanding and approach deserve consideration.

The concept of survivorship, while increasingly recognized, remains variably defined across different contexts. This inconsistency creates challenges for developing cohesive frameworks that can guide comprehensive care. Our understanding is further limited by the disproportionate focus on certain cancer types—particularly breast cancer—which may not fully represent the experiences of survivors with different diagnoses (Richardson et al., 1990). The voices and experiences of many populations remain underrepresented in survivorship literature, including racial and ethnic minorities, those from lower socioeconomic backgrounds, rural communities, older adults, and individuals with advanced or metastatic disease.

The predominant survivorship narratives often reflect Western, individualistic perspectives that emphasize "beating" cancer, which may not resonate with all cultural backgrounds or with those living with cancer as a chronic condition. These narratives can inadvertently create expectations about how survivors "should" respond to cancer, potentially marginalizing those whose experiences differ from these normative frameworks.

Within healthcare systems, the implementation of survivorship care remains inconsistent. While survivorship care plans have been widely recommended, their adoption varies considerably across settings, and coordination between oncology and primary care often remains fragmented. Many healthcare professionals receive limited training in addressing the unique needs of cancer survivors, particularly regarding psychosocial and existential concerns that extend beyond physical recovery and surveillance.

Access to comprehensive survivorship support is uneven, with significant disparities based on geographic location, socioeconomic status, and healthcare systems. Many evidence-based psychosocial interventions require resources, expertise, and infrastructure that may not be universally available, particularly in underserved communities and developing countries. Furthermore, insurance coverage and reimbursement for survivorship services beyond medical surveillance remain variable, creating potential barriers to holistic care.

5 Future directions

Looking ahead, several promising directions can enhance our understanding and support for cancer survivorship.

First, we need more inclusive frameworks of survivorship that honor diverse experiences across different cancer types, cultural backgrounds, and socioeconomic contexts. These frameworks should recognize survivorship as a dynamic, evolving journey rather than a static state and should accommodate various trajectories, including living with cancer as a chronic condition. The integration of cultural humility into survivorship models would acknowledge that concepts and experiences of survivorship vary substantially across cultural contexts and value systems.

Expanding the scope of survivorship care beyond the biomedical focus to equally prioritize psychological, social, and existential dimensions would create more holistic support systems. This approach recognizes that physical recovery is only one aspect of the complex survivorship journey. In addition, explicitly acknowledging and addressing the needs of caregivers and families as "cosurvivors" would honor the profound impact that cancer has on entire support networks, not just the diagnosed individual.

Healthcare systems can enhance survivorship care by implementing more systematic approaches to care planning and coordination between oncology and primary care. Developing sustainable models that can meet the needs of the growing survivor population without creating separate, potentially fragmented systems is crucial. Digital health solutions hold promise for expanding access to survivorship support, particularly for those in underserved or geographically isolated communities.

Professional education curricula should incorporate comprehensive training in survivorship care across healthcare disciplines, equipping providers with the skills to address the full spectrum of survivor needs. Developing and disseminating culturally tailored survivorship interventions appropriate for diverse settings would help ensure that support reaches all who need it, regardless of location or the resources available.

Policy initiatives can further improve survivorship care through expanded insurance coverage for comprehensive services and addressing social determinants of health that influence survivorship outcomes, including financial toxicity, employment challenges, and access to supportive resources. Public education efforts can help reduce stigma and increase understanding of the diverse experiences of cancer survivors, fostering more supportive communities.

As the field of psycho-oncology continues to evolve, maintaining a balance between identifying common patterns and respecting individual experiences remains crucial. The ultimate goal is supporting cancer survivors not merely to survive but to thrive—physically, psychologically, socially, and existentially—in ways that align with their unique values, preferences, and life circumstances. This vision of personalized, comprehensive survivorship care represents both our greatest challenge and our highest aspiration.

6 Conclusion

"I had cancer, but cancer never had me." – A cancer survivor (Bhargava, 2018)

It can be said with utmost certainty that receiving the diagnosis and treatment for cancer brings about various bio-psycho-social changes in an individual's life. These changes affect the individual at many levels, along with the family, the caregivers, healthcare institutions, and society at large. The stakeholders during the process of treatment and survivorship of the affected person must ensure a holistic involvement in their life because a small step of intervention may have a long-term impact with invisible positive outcomes for the individual and their surroundings. Cancer survivors' psychological well-being may be an ever-evolving idea. The physical, psychological, social, and existential pressures associated with cancer diagnosis and treatment now may be considerably different from those in the past and what they could be in the future. As a result, over time, the spectrum and occurrence of psychological reactions—both positive and negative—associated with a specific blend of patient, disease, and treatment features may alter (Andrykowski et al., 2008).

The growing body of research in psycho-oncology continues to illuminate the complex interplay between psychological, social, and medical factors in cancer survivorship (Grassi et al., 2017). As we move forward, it is essential that we maintain focus on evidence-based interventions while remaining sensitive to the unique needs and experiences of each individual survivor and their support network. The future of cancer survivorship care lies in developing comprehensive, integrated approaches that address not only the medical sequelae of cancer and its treatment but also the profound psychological, social, and existential dimensions of the survivorship experience.

References

What is cancer survivorship?, 2020. Available from: https://www.canceradvocacy.org/about-us/our-mission/.

Global cancer burden growing, amidst mounting need for services, 2022. Available from: https://www.who.int/news/item/01-02-2022-world-cancer-day-global-cancer-burden-growing–amidst-mounting-need-for-services.

Adler, R.H., 2009. Engel's biopsychosocial model is still relevant today. J. Psychosom. Res. 00223999. 67 (6), 607–611. doi:10.1016/j.jpsychores.2009.08.008.

Akechi, T., 2018. Psycho-oncology research and its clinical application in Japan. Jpn. J. Clin. Oncol. 48 (3), 229–235.

Albreht, T., et al, 2017. Survivorship and rehabilitation: policy recommendations for quality improvement in cancer survivorship and rehabilitation in EU member states. In: European Guide on Quality Improvement in Comprehensive Cancer Control. Available from: /ecpc.org/wp-content/uploads/2019/08/cancon-guide-chapter-7-survivorship-and-rehabilitation-.

Andrykowski, M.A., Lykins, E., Floyd, A., 2008. Psychological health in cancer survivors. Semin. Oncol. Nurs. 07492081. 24 (3), 193–201. doi:10.1016/j.soncn.2008.05.007.

Antoni, M.H., Lehman, J.M., Kilbourn, K.M., Boyers, A.E., Culver, J.L., Alferi, S.M., Yount, S.E., McGregor, B.A., Arena, P.L., Harris, S.D., Price, A.A., Carver, C.S., 2001. Cognitive-behavioral stress management intervention decreases the prevalence of depression and enhances benefit finding among women under treatment for early-stage breast cancer. Health Psychol. 02786133. 20 (1), 20–32. Available from: http://www.apa.org/pubs/journals/hea/. doi:10.1037/0278-6133.20.1.20.

Bhargava, A., 2018. A cancer survivor's story: 'I had cancer but cancer never had me'. The Quint. Available from: https://www.thequint.com/fit/world-cancer-day-fight-cancer.

Bray, F., Ferlay, J., Soerjomataram, I., Siegel, R.L., Torre, L.A., Jemal, A., 2018. Global cancer statistics 2018: GLOBOCAN estimates of incidence and mortality worldwide for 36 cancers in 185 countries. CA Cancer J. Clin. 15424863. 68 (6), 394–424. Available from: http://onlinelibrary.wiley.com/journal/10.3322/(ISSN)1542-4863. doi:10.3322/caac.21492.

Carver, C.S., Smith, R.G., Antoni, M.H., Petronis, V.M., Weiss, S., Derhagopian, R.P., 2005. Optimistic personality and psychosocial well-being during treatment predict psychosocial well-being among long-term survivors of breast cancer. Health Psychol. 02786133. 24 (5), 508–516. Available from: http://www.apa.org/pubs/journals/hea/. doi:10.1037/0278-6133.24.5.508.

Cruzado, J.A., 2010. Treatment of anxiety in cancer patients. Clin. Transl. Oncol. 12 (4), 298–305.

Deshields, T., Tibbs, T., Fan, M.Y., Taylor, M., 2006. Differences in patterns of depression after treatment for breast cancer. Psycho-Oncology. 10991611. 15 (5), 398–406. doi:10.1002/pon.962.

Engel, G.L., 1977. The need for a new medical model: a challenge for biomedicine. Science. 00368075. 196 (4286), 129–136. doi:10.1126/science.847460.

Ferrans, C.E., 1994. Quality of life through the eyes of survivors of breast cancer. Oncol. Nurs. Forum 21 (10), 1645–1651.

Ganz, P.A., 2001. Late effects of cancer and its treatment. Semin. Oncol. Nurs. 07492081. 17 (4), 241–248. doi:10.1053/sonu.2001.27914.

Given, C.W., Given, B., Stommel, M., Collins, C., King, S., Franklin, S., 1992. The caregiver reaction assessment (CRA) for caregivers to persons with chronic physical and mental impairments. Res. Nurs. Health. 1098240X. 15 (4), 271–283. doi:10.1002/nur.4770150406.

Gorman, L.M., 2018. The psychosocial impact of cancer on the individual, family, and society. In: Psychosocial Nursing Care along the Cancer Continuum, Vol. 3, pp. 3–23. Available from: https://www.ons.org/sites/default/files/2018-10/Psychosocial%20Nursing%20Care%20Along%20the%20Cancer%20Continuum%20Sample.

Gorman, L.M., 2006. The psychosocial impact of cancer on individuals, families, and caregivers. In: Carroll-Johnson, R.M., Gorman, R.M., Bush, N.J. (Eds.), Psychosocial Nursing Care Along the Cancer Continuum, second ed. pp. 3–29.

Grassi, L., Spiegel, D., Riba, M., 2017. Advancing psychosocial care in cancer patients. F1000Research. 2046-1402. 6, 2083. doi:10.12688/f1000research.11902.1.

Harries, L.N., Stanton, A.L., 2020. Cancer and psycho-oncology. In: The Wiley encyclopedia of Health Psychology, vol. 3. Wiley.

Harrington, C.B., Hansen, J.A., Moskowitz, M., Todd, B.L., Feuerstein, M., 2010. It's not over when it's over: long-term symptoms in cancer survivors-a systematic review. Int. J. Psychiatry Med. 15413527. 40 (2), 163–181. doi:10.2190/PM.40.2.c.

Harrison, J.D., Young, J.M., Price, M.A., Butow, P.N., Solomon, M.J., 2009. What are the unmet supportive care needs of people with cancer? A systematic review. Support. Care Cancer. 14337339. 17 (8), 1117–1128. doi:10.1007/s00520-009-0615-5.

Hewitt, M., Greenfield, Stovall, E., 2005. From cancer patient to cancer survivor: lost in transition. Available from: https://nap.nationalacademies.org/catalog/11468/from-cancer-patient-to-cancer-survivor-lost-in-transition?

Holland, J.C., 2002. History of psycho-oncology: overcoming attitudinal and conceptual barriers. Psychosom. Med. 00333174. 64 (2), 206–221. Available from: http://www.psychosomaticmedicine.org/. doi:10.1097/00006842-200203000-00004.

Konradsen, H., 2012. Body image and cancer. InTech. doi:10.5772/22201.

Lang-Rollin, I., Berberich, G., 2018. Psycho-oncology. Dialogues Clin. Neurosci. 20 (1), 13–22. doi:10.31887/DCNS.2018.20.1/ilangrollin.

Committee on Improving the Quality of Cancer Care: Addressing the Challenges of an Aging Population; Board on Health Care Services; Institute of Medicine, Levit, L., et al., 2013. Delivering High-Quality Cancer Care: Charting a New Course for a System in Crisis. National Academies Press (US).

Mehnert, A., Koch, U., 2008. Psychological comorbidity and health-related quality of life and its association with awareness, utilization, and need for psychosocial support in a cancer register-based sample of long-term breast cancer survivors. J. Psychosom. Res. 00223999. 64 (4), 383–391. doi:10.1016/j.jpsychores.2007.12.005.

Miller, K.D., Nogueira, L., Mariotto, A.B., Rowland, J.H., Yabroff, K.R., Alfano, C.M., Jemal, A., Kramer, J.L., Siegel, R.L., 2019. Cancer treatment and survivorship statistics, 2019. CA Cancer J. Clin. 15424863. 69 (5), 363–385. Available from: http://onlinelibrary.wiley.com/journal/10.3322/(ISSN)1542-4863. doi:10.3322/caac.21565.

Mitchell, A.J., Chan, M., Bhatti, H., Halton, M., Grassi, L., Johansen, C., Meader, N., 2011. Prevalence of depression, anxiety, and adjustment disorder in oncological, haematological, and palliative-care settings: a meta-analysis of 94 interview-based studies. Lancet Oncol. 14702045. 12 (2), 160–174. doi:10.1016/S1470-2045(11)70002-X.

Mullan, F., 1985. Seasons of survival: reflections of a physician with cancer. N. Engl. J. Med. 0028-4793. 313 (4), 270–273. doi:10.1056/NEJM198507253130421.

Office of Cancer Survivorship, 2018. Available from: https://cancercontrol.cancer.gov/ocs.

Park, C.L., Edmondson, D., Fenster, J.R., Blank, T.O., 2008. Meaning making and psychological adjustment following cancer: the mediating roles of growth, life meaning, and restored just-world beliefs. J. Consult. Clin. Psychol. 0022006X. 76 (5), 863–875. doi:10.1037/a0013348.

Pelusi, J., 1997. The lived experience of surviving breast cancer. Oncol. Nurs. Forum. 0190535X. 24 (8), 1343–1353.

Richardson, G.E., Neiger, B.L., Jensen, S., Kumpfer, K.L., 1990. The resiliency model. Health Educ. 00970050. 21 (6), 33–39. doi:10.1080/00970050.1990.10614589.

Rowland, J.H., Bellizzi, K.M., 2008. Cancer survivors and survivorship research: a reflection on today's successes and tomorrow's challenges. Hematol. Oncol. Clin. North Am. 08898588. 22 (2), 181–200. doi:10.1016/j.hoc.2008.01.008.

Salsman, J.M., Pustejovsky, J.E., Jim, H.S.L., Munoz, A.R., Merluzzi, T.V., George, L., Park, C.L., Danhauer, S.C., Sherman, A.C., Snyder, M.A., Fitchett, G., 2015. A meta-analytic approach to examining the correlation between religion/spirituality and mental health in cancer. Cancer. 10970142. 121 (21), 3769–3778. Available from: http://onlinelibrary.wiley.com/journal/10.1002/(ISSN)1097-0142. doi:10.1002/cncr.29350.

Scott, S., Platt, L., 2016. Every Day I Fight: Making a Difference, Kicking Cancer's Ass. Blue Rider Press.

Siegel, R.L., Miller, K.D., Wagle, N.S., Jemal, A., 2023. Cancer statistics, 2023. CA Cancer J. Clin. 15424863. 73 (1), 17–48. Available from: http://onlinelibrary.wiley.com/journal/10.3322/(ISSN)1542-4863. doi:10.3322/caac.21763.

Spiegel, D., Kraemer, H.C., Bloom, J.R., Gottheil, E., 1989. Effect of psychosocial treatment on survival of patients with metastatic breast cancer. The Lancet 334 (8668), 888–891. doi:10.1016/S0140-6736(89)91551-1.

Stanton, N.A., Stewart, R., Harris, D., Houghton, R.J., Baber, C., McMaster, R., Salmon, P., Hoyle, G., Walker, G., Young, M.S., Linsell, M., Dymott, R., Green, D., 2006. Distributed situation awareness in dynamic systems: theoretical development and application of an ergonomics methodology. Ergonomics 49 (12–13), 1288–1311. doi:10.1080/00140130600612762.

Strasser, F., 2021. Definition of Survivorship Care Survivorship Care for Cancer Patients: A Clinician's Handbook. Springer International Publishing, Switzerland, pp. 1–22. Available from: https://link.springer.com/book/10.1007/978-3-030-78648-9. doi:10.1007/978-3-030-78648-9_1.

Tedeschi, R.G., Calhoun, L.G., 2004. Posttraumatic growth: conceptual foundations and empirical evidence. Psychol. Inq. 1047840X. 15 (1), 1–18. Available from: http://www.tandfonline.com/toc/hpli20/current. doi:10.1207/s15327965pli1501_01.

Thewes, B., Butow, P., Bell, M.L., Beith, J., Stuart-Harris, R., Grossi, M., Capp, A., Dalley, D., 2012. Fear of cancer recurrence in young women with a history of early-stage breast cancer: a cross-sectional study of prevalence and association with health behaviours. Support. Care Cancer. 0941-4355. 20 (11), 2651–2659. doi:10.1007/s00520-011-1371-x.

Ventura, E.E., Ganz, P.A., Bower, J.E., Abascal, L., Petersen, L., Stanton, A.L., Crespi, C.M., 2013. Barriers to physical activity and healthy eating in young breast cancer survivors: modifiable risk factors and associations with body mass index. Breast Cancer Res. Treat. 15737217. 142 (2), 423–433. doi:10.1007/s10549-013-2749-x.

Zebrack, B.J., Ganz, P.A., Bernaards, C.A., Petersen, L., Abraham, L., 2006. Assessing the impact of cancer: development of a new instrument for long-term survivors. Psycho-Oncology. 10991611. 15 (5), 407–421. doi:10.1002/pon.963.

Zeltzer, L.K., Recklitis, C., Buchbinder, D., Zebrack, B., Casillas, J., Tsao, J.C.I., Lu, Q., Krull, K., 2009. Psychological status in childhood cancer survivors: a report from the childhood cancer survivor study. J. Clin. Oncol. 0732183X. 27(14), 2396–2404. Available from: http://jco.ascopubs.org/cgi/reprint/27/14/2396. doi: 10.1200/JCO.2008.21.1433.

Chapter 35

Exercise-induced cardiovascular changes in dyslipidemia patients: Insights from medical imaging modalities

Prateek Srivastav[a], Rajagopal Kadavigere[b], Leena R. David[c,d], Dilber Uzun Ozsahin[c,e,f], Berna Uzun[f], Sneha Ravichandran[d], Ilker Ozsahin[f], and Suresh Sukumar[d]

[a]*Department of Physiotherapy, Manipal Academy of Higher Education, Manipal College of Health Professions, Manipal, India*
[b]*Department of Radiodiagnosis and Imaging, Kasturba Medical College, Manipal, India*
[c]*Medical Diagnostic Imaging Department, College of Health Sciences, University of Sharjah, Sharjah, United Arab Emirates*
[d]*Department of Medical Imaging Technology, Manipal College of Health Professions, Manipal Academy of Higher Education, Manipal, Karnataka, India*
[e]*Research Institute for Medical and Health Sciences, University of Sharjah, Sharjah, United Arab Emirates*
[f]*Irfan Suat Gunsel Operational Research Institute, Near East University, Nicosia, Turkish Republic of Northern Cyprus (TRNC), Turkey*

1 Introduction

Cardiovascular disease is recognized as the primary cause of both illness and death on a global scale. Dyslipidemia serves as one of the major contributing factors to this condition, with hypercholesterolemia being its most prevalent form (World Health Organization, 2021). Dyslipidemia pertains to an atypical lipid level (cholesterol and triglycerides) in the bloodstream, which poses a notable peril for cardiovascular ailments (Goldstein and Brown, 2015). Among lipoproteins containing Apo lipoprotein B in human plasma, low-density lipoprotein (LDL) plays a crucial role in conveying cholesterol to the arterial wall (Ference et al., 2017). The elevation of LDL cholesterol (LDLc) levels constitutes a significant characteristic of dyslipidemia and is closely linked to an increased risk of cardiovascular complications, especially in individuals affected by atherosclerotic cardiovascular disease (Grundy et al., 2019).

The relationship between dyslipidemia and cardiovascular pathology is complex and multifaceted. Longitudinal studies have consistently demonstrated that elevated levels of LDLc and reduced levels of high-density lipoprotein (HDL) cholesterol are independently associated with an increased risk of coronary heart disease and stroke (Yusuf et al., 2004; O'Donnell et al., 2016). Furthermore, epidemiological data indicate that the prevalence of dyslipidemia varies significantly across different populations, with rates ranging from 20% to 50% in adults, depending on geographical location, age, sex, and socioeconomic factors (Benjamin et al., 2019).

Modern medical imaging techniques have revolutionized our understanding of the cardiovascular consequences of dyslipidemia and provided valuable insights into the potential benefits of therapeutic interventions, including lifestyle modifications such as regular physical activity (Libby et al., 2011). These advanced imaging modalities allow for detailed visualization of the cardiovascular system, enabling clinicians and researchers to detect subclinical atherosclerosis, assess disease progression, and evaluate treatment efficacy with unprecedented precision (Sanz and Fayad, 2008).

2 Pathophysiology

Dyslipidemia is a complex condition that is influenced by various cardiovascular risk factors, such as smoking, hypertension, diabetes, age, gender, triglyceride levels, and HDL cholesterol (Fuster et al., 2005). The presence of these combined factors contributes to the development of atherosclerotic plaque. There exists a clear correlation when multiple abnormal risk factors are present. Atherosclerosis is characterized by chronic inflammation within the vessel wall, involving

Advances in Medical Imaging. https://doi.org/10.1016/B978-0-443-28967-5.00013-7

monocytes (macrophages), T-lymphocytes, proliferation and migration of smooth muscle cells, circulation of extracellular matrix production, and neovascularization (Libby, 2002; Tabas et al., 2015).

The role of the endothelium, which is considered the largest endocrine organ in our body, plays a central role in this process (Gimbrone and García-Cardeña, 2016). The normal functioning of the endothelium can be disrupted by common risk factors associated with abnormalities. When there are changes in endothelial function, a series of events occurs that result in plaque formation accompanied by inflammation, altered blood clotting tendencies, modified vessel tone, and biochemical interactions (Davignon and Ganz, 2004). Plaque growth causes shear stresses on blood vessels, leading to cracks or ruptures and subsequent adhesion of platelets along with thrombosis (Bentzon et al., 2014).

In addition to taking up LDLc, macrophages become activated within the wall of blood vessels, thereby transforming into foam cells (Moore et al., 2013). These activated macrophages secrete metalloproteinases that have the potential to weaken the protective cap covering the plaque structure. Furthermore, they release platelet-derived growth factor, which stimulates cell division and aids in promoting new blood vessel formation around plaques (Libby et al., 2019). Macrophages also release cytokines that stimulate the proliferation of smooth muscle cells and lymphocytes, leading to increased inflammation (Wolf and Ley, 2019).

In the presence of plaque and impaired endothelial function, there is reduced production of nitric oxide and relaxation factors from the endothelium, as well as heightened activation of platelets, resulting in thromboxane synthesis (Forstermann and Sessa, 2012). This promotes platelet aggregation and constriction of blood vessels. In addition, platelets secrete PDGF, which acts as a potent stimulator for cell division (Ross, 1989). The situation becomes even more complex with elevated lipoprotein(a) levels and C-reactive protein induced by ongoing inflammation. While cholesterol (specifically LDLc) plays a significant role in this process, it is not solely responsible; multiple factors contribute to this inflammatory response (Ridker, 2016; Tall and Yvan-Charvet, 2015).

3 Treatment options for dyslipidemia

The management of dyslipidemia encompasses a multifaceted approach that includes lifestyle modifications and pharmacological interventions. Lifestyle changes represent the foundational aspect of treatment and involve dietary adjustments, regular physical activity, weight management, smoking cessation, and limited alcohol consumption (Arnett et al., 2019). The Mediterranean diet, characterized by a high consumption of fruits, vegetables, whole grains, olive oil, and fish, has demonstrated significant benefits in improving lipid profiles and reducing cardiovascular risk (Estruch et al., 2018).

Pharmacological therapy is typically initiated when lifestyle modifications alone are insufficient to achieve target lipid levels. Statins (HMG-CoA reductase inhibitors) remain the cornerstone of dyslipidemia treatment due to their proven efficacy in reducing LDLc levels and cardiovascular events (Collins et al., 2016). Metaanalyses of randomized controlled trials have consistently shown that statin therapy is associated with a 20%–25% reduction in major cardiovascular events for each 1 mmol/L reduction in LDLc (Cholesterol Treatment Trialists' (CTT) Collaboration, 2010).

For patients who cannot tolerate statins or who require additional lipid-lowering therapy, other medications may be considered, including ezetimibe (cholesterol absorption inhibitor), PCSK9 inhibitors (evolocumab, alirocumab), bile acid sequestrants, fibrates, and niacin (Cannon et al., 2015; Sabatine et al., 2017). Recent clinical trials have demonstrated the efficacy of combination therapy, particularly statins with ezetimibe or PCSK9 inhibitors, in achieving substantial LDLc reductions and improved cardiovascular outcomes (Schwartz et al., 2018).

The effectiveness of these interventions can be monitored through regular lipid profile assessments and advanced imaging techniques, which provide valuable insights into the regression or stabilization of atherosclerotic lesions (Nicholls et al., 2016). The application of medical imaging in evaluating treatment response will be discussed in detail in subsequent sections.

4 Structural and functional abnormalities related to dyslipidemia

4.1 Endothelial dysfunction

Endothelial dysfunction refers to the impaired functioning of the inner lining, known as the endothelium, within blood vessels. This crucial component is responsible for regulating vascular tone, controlling blood flow, and maintaining cardiovascular health. Any disturbance in endothelial function can give rise to various cardiovascular complications (Deanfield et al., 2007; Widlansky et al., 2003). The endothelium, when functioning properly, produces various molecules that play a crucial role in regulating the dilation and constriction of blood vessels. One such molecule is nitric oxide, which specifically contributes to relaxing blood vessels and facilitating healthy blood flow (Tousoulis et al., 2012).

When there is dysfunction in the endothelium, several negative effects can arise. The ability of blood vessels to relax becomes impaired, resulting in increased vascular tone and reduced circulation (Bonetti et al., 2003). Heightened inflammation within the vascular system supports the development of atherosclerosis and the formation of plaque (Steyers and Miller, 2014). An increased risk for thrombosis occurs as the endothelium normally helps prevent abnormal clotting, and dysfunction can lead to clot formation (Lüscher and Barton, 1997).

Endothelial dysfunction may be a contributing factor in elevated blood pressure and can lead to impaired tissue perfusion, where reduced bloodstream delivery can deprive tissues of vital oxygen and nutrients necessary for proper function (Versari et al., 2009). Furthermore, elevated susceptibility to cardiovascular events, such as heart attacks or strokes, is more likely due to the heightened risk associated with these conditions (Sitia et al., 2010).

Recent studies using flow-mediated dilation (FMD) techniques have demonstrated that patients with dyslipidemia exhibit significantly reduced endothelium-dependent vasodilation compared to normolipidemic controls, with values typically 40%–60% lower (Flammer et al., 2012). This impairment correlates strongly with LDLc levels and oxidative stress markers, underscoring the direct link between lipid abnormalities and vascular dysfunction (Anderson et al., 1996).

4.2 Subendothelial lipid deposition and progression

The process of subendothelial lipid deposition involves the gathering of fatty substances beneath the endothelial layer within blood vessels. This phenomenon plays a critical role in the initial stages of atherosclerosis, which is a chronic cardiovascular condition characterized by plaque accumulation in arteries (Williams and Tabas, 1995). Atherosclerosis causes an increase in the inflammatory process, leading to the formation of plaques within the blood vessels. This can lead to reduced blood supply, rupture of unstable plaques, or calcification of blood vessels (Virmani et al., 2000).

Histopathological studies have revealed that subendothelial lipid accumulation begins with the retention of apolipoprotein B-containing lipoproteins, particularly LDL, in the arterial intima (Nakashima et al., 2007). Once trapped in the subendothelial space, these lipoproteins undergo oxidative modification, triggering an inflammatory response characterized by the recruitment of monocytes, which subsequently differentiate into macrophages (Steinberg, 1997). The internalization of oxidized LDL by macrophages through scavenger receptors leads to foam cell formation, a hallmark of early atherosclerotic lesions (Li and Glass, 2002).

Advanced imaging techniques, such as intravascular ultrasound (IVUS) and optical coherence tomography (OCT), have enabled the in vivo visualization of lipid-rich plaques with remarkable precision (Prati et al., 2010). These modalities have demonstrated that patients with dyslipidemia exhibit a higher prevalence of lipid-rich plaques compared to normolipidemic individuals (odds ratio: 2.4; 95% CI: 1.8–3.2) (Garcia-Garcia et al., 2009). Furthermore, the progression of lipid deposition, as assessed by virtual histology IVUS, has been shown to correlate strongly with serum LDLc levels ($r = 0.67$, $p < 0.001$) (Nissen et al., 2006).

4.3 Compromised collateral vessel growth

Compromised collateral vessel growth, also identified as hindered collateral circulation or inadequate collateralization, denotes a state in which the progression and enlargement of alternative blood vessels are impeded or insufficient (Seiler et al., 2013). These collaterals, namely small-scale blood vessels, possess the capability to expand and offer an alternative pathway for blood circulation when the main arteries delivering blood to a specific region encounter blockages or constrictions (Meier et al., 2007). It is one of the consequences of dyslipidemia.

Animal models have demonstrated that hypercholesterolemia significantly impairs collateral development in response to ischemia, with a 40%–60% reduction in capillary density observed in dyslipidemic subjects compared to controls (Van Belle et al., 1997). This impairment is mediated through multiple mechanisms, including reduced nitric oxide bioavailability, increased oxidative stress, and dysregulation of pro-angiogenic factors such as vascular endothelial growth factor and its receptors (Boodhwani et al., 2007).

Clinical studies utilizing coronary angiography and myocardial perfusion imaging have confirmed these experimental findings, showing that patients with dyslipidemia exhibit poorer coronary collateralization compared to matched controls (Rentrop score: 1.2 ± 0.6 vs. 2.1 ± 0.7, $p < 0.001$) (Zbinden et al., 2007). This deficiency in collateral development has important prognostic implications, as inadequate collateralization is associated with a higher risk of adverse cardiac events in patients with coronary artery disease (hazard ratio: 2.3; 95% CI: 1.7–3.1) (Meier et al., 2012).

Although medical imaging is not typically employed as the primary diagnostic method for dyslipidemia, it can contribute by evaluating the impact of lifestyle modifications—such as physical activity—on cardiac well-being (Sytkowski et al., 1990).

5 Imaging modalities for dyslipidemia changes following exercise

This section explores the various imaging modalities that have been studied to evaluate the physiological and structural changes associated with dyslipidemia, particularly in response to exercise.

5.1 Carotid ultrasound

Carotid ultrasound plays a crucial role in the evaluation of atherosclerosis within the carotid arteries. Atherosclerosis is characterized by the accumulation of fatty plaque deposits inside these blood vessels, which can result in their narrowing or blockage (O'Leary et al., 1999). This condition significantly increases the likelihood of experiencing a stroke. Carotid ultrasound aids healthcare professionals in assessing atherosclerotic plaque through:

1. Identification of plaque: Carotid ultrasound offers detailed imagery that enables the visual detection and assessment of irregular, fatty deposits within the arterial walls (Naqvi and Lee, 2014).
2. Quantifying stenosis: By utilizing ultrasound technology, it becomes feasible to measure and evaluate stenosis severity caused by plaque formation accurately—this metric serves as an important indicator for gauging stroke risk levels (Touboul et al., 2012). Commonly utilized criteria help categorize different degrees of carotid narrowing: mild (<50% narrowing), moderate (50%–69%), and severe (70% or more).
3. Doppler assessment: Carotid ultrasound includes the use of Doppler technology to examine blood flow within the carotid arteries, detecting turbulence or changes caused by atherosclerosis, which may indicate significant blockage or narrowing (Grant et al., 2003).
4. Risk assessment: Healthcare providers employ carotid ultrasound to evaluate an individual's overall risk of stroke, considering factors such as significant carotid artery stenosis, along with other potential risks like hypertension and smoking (Rundek et al., 2008).
5. Treatment guidance: By analyzing the results from carotid ultrasound, healthcare providers can develop appropriate treatment plans tailored specifically to each patient's condition (Spence et al., 2002). For instance, lifestyle modifications and lipid-lowering medications (such as statins, PCSK9 inhibitors, and ezetimibe) may be recommended for managing mild to moderate stenosis in order to reduce plaque progression and minimize blood clot formation (Amarenco et al., 2004). Severe stenosis may necessitate more aggressive treatment options, such as carotid endarterectomy or carotid artery stenting (TG et al., 2010).
6. Follow-up monitoring using carotid ultrasound is also important for assessing the progression of atherosclerosis and evaluating the efficacy of treatment, thereby facilitating long-term management and risk reduction (Spence and Hackam, 2010).

Before exercise, carotid ultrasound may reveal signs of atherosclerosis in dyslipidemia patients, such as thickening of the carotid intima-media and the presence of plaques (Polak et al., 2011). The measurement of carotid intima-media thickness (CIMT) serves as a surrogate marker for subclinical atherosclerosis, with values exceeding 0.9 mm generally considered abnormal and associated with an increased risk of cardiovascular events (Lorenz et al., 2007).

After regular exercise intervention, several studies have documented significant improvements in carotid artery health. A metaanalysis of 17 randomized controlled trials involving 1038 participants demonstrated that aerobic exercise programs of 12 weeks or longer resulted in a mean reduction in CIMT of 0.04 mm (95% CI: 0.02–0.06 mm, $p < 0.001$) (Thijssen et al., 2012). This magnitude of reduction has been associated with a 10%–15% decrease in the risk of myocardial infarction and stroke (Lorenz et al., 2012).

In a study by Kadoglou et al. involving 73 patients with dyslipidemia, a 6-month aerobic exercise program led to a 7.9% reduction in CIMT ($p < 0.01$) compared to a 1.8% increase in the control group (Kadoglou et al., 2013). Furthermore, exercise has been shown to improve plaque stability by reducing lipid content and increasing fibrous cap thickness, as evidenced by changes in plaque echogenicity on ultrasound (Stary et al., 1995). These modifications in plaque morphology translate to a decreased risk of rupture and subsequent thrombotic events (Shah, 2003).

5.2 Coronary artery calcium scoring

Coronary artery calcium (CAC) scoring is a diagnostic test utilized to evaluate the extent and presence of calcium deposits in the coronary arteries, which can serve as an indication of atherosclerosis in the heart (Greenland et al., 2010). Atherosclerosis refers to the accumulation of plaque, including calcium deposits, within these arteries, thereby increasing one's susceptibility to heart disease and cardiac events like heart attacks (Detrano et al., 2008). CAC scoring plays a crucial role in assessing the risk associated with coronary artery disease, utilizing several key aspects:

1. Procedure: Typically conducted through the use of a computed tomography (CT) scanner, this procedure entails generating comprehensive images illustrating any calcium deposits present within the coronary arteries (Agatston et al., 1990). Furthermore, calculations and assessments are performed based on both density and distribution measures pertaining to said deposits.
2. Calcium scoring: CAC scoring provides a quantitative measure, known as the CAC score, which reflects the extent of coronary artery calcifications (Blaha et al., 2017). The Agatston Score or similar methods are commonly used to calculate this score. A higher CAC score indicates a greater presence of calcium and is indicative of an elevated risk for coronary artery disease.
3. Risk stratification: The assessment of the CAC score plays a crucial role in evaluating an individual's risk level for developing coronary artery disease and experiencing cardiac events such as heart attacks (Budoff et al., 2007). Extensive research supports the association between higher CAC scores and an increased likelihood of cardiovascular events occurring in individuals.
4. Treatment recommendations: The assessment of CAC scoring can provide valuable guidance for treatment decisions (Greenland et al., 2018). In cases where individuals have elevated CAC scores, indicating an increased risk, healthcare professionals may suggest more intensive interventions, such as statin drugs, lifestyle adjustments (e.g., dietary changes and physical activity), and regular monitoring to lower the chances of developing heart disease.
5. Tracking disease progression: In addition, periodic assessments using CAC scoring allow for the ongoing evaluation of coronary artery disease progression (Raggi et al., 2004). These follow-up scans help determine the efficacy of interventions in reducing arterial calcium accumulation over time.
6. Monitoring disease progression: CAC scoring can be utilized to track the advancement of coronary artery disease, allowing for periodic assessments to determine if interventions are effectively reducing calcium buildup in the arteries over time (Budoff et al., 2013).
7. Prognostic value: In addition to evaluating risk, CAC scoring offers valuable prognostic information (Erbel et al., 2010). Individuals with a high CAC score may face an elevated risk of cardiovascular events and might benefit from more frequent monitoring and intensified preventive measures.

Before exercise, CAC scoring using CT can assess the extent of calcified plaque in the coronary arteries in dyslipidemia patients (Hecht et al., 2017). High scores indicate a higher risk of coronary artery disease, with scores above 400 Agatston units associated with a 7–10-fold increased risk of coronary events compared to individuals with zero calcium (Blaha et al., 2011).

The effect of exercise on coronary artery calcification has been investigated in several studies. In the CARDIA study, which followed 5115 participants over 25 years, individuals who maintained higher physical activity levels throughout middle adulthood had 27% lower odds of developing coronary artery calcification (odds ratio: 0.73; 95% CI: 0.57–0.94) compared to their less active counterparts (Laddu et al., 2017).

While calcium scores typically increase over time, regular exercise has been shown to slow this progression. In a study by Kwaśniewska et al. involving 101 middle-aged men with dyslipidemia, those engaging in regular moderate-to-vigorous physical activity (>150 min per week) exhibited a 32% slower progression rate of CAC over a 5-year period compared to sedentary individuals ($p < 0.01$) (Kwaśniewska et al., 2014).

Exercise, especially when combined with other heart-healthy habits like dietary modifications and medication adherence, may stabilize or even reduce the progression of coronary artery calcification (Sandesara et al., 2018). This effect appears to be most pronounced when exercise is initiated early in the disease process and maintained consistently over extended periods (Kianoush et al., 2020).

5.3 Cardiac magnetic resonance imaging

Cardiac magnetic resonance imaging (MRI) is a highly effective diagnostic tool utilized for evaluating different aspects of cardiac health, such as the presence and impact of atherosclerosis, which involves plaque accumulation in the coronary arteries (Pennell, 2010). In relation to atherosclerosis specifically, cardiac MRI serves two primary purposes: visualizing plaque formation and assessing stenosis severity (Rajiah et al., 2013).

Firstly, it provides detailed imagery of the coronary arteries to enable healthcare professionals to identify plaques and analyze their characteristics. Although not proficient at directly detecting calcium deposits like CAC scoring or CT angiography (CTA), it effectively identifies soft, lipid-rich plaques along with other markers associated with atherosclerosis (Kim et al., 2000). Secondly, by measuring the degree of narrowing caused by plaque deposition within these vessels using MRI technology, medical experts can accurately estimate the extent or seriousness of an individual's atherosclerotic condition (Kwong et al., 2006).

Cardiac MRI has the ability to evaluate blood circulation within the coronary arteries. It can detect any decrease in blood flow caused by blockages from atherosclerosis, providing crucial insights into how this condition impacts heart function (Salerno and Beller, 2009). Apart from diagnosing atherosclerosis, cardiac MRI offers comprehensive data on various aspects of heart functionality, such as ventricular performance, abnormalities in wall movement, and overall cardiac health (Hundley et al., 2010). This information is invaluable for understanding the effects of atherosclerosis on the heart.

In the assessment of plaque composition, MRI can play a significant role in distinguishing between stable and vulnerable plaques (Noguchi et al., 2014). Identifying vulnerable plaques is crucial for risk assessment, as these are more likely to rupture and lead to acute cardiac events such as heart attacks (Fayad and Fuster, 2001). MRI techniques are employed in both research and clinical trials to investigate the progression of atherosclerosis, evaluate treatment effectiveness, and develop innovative diagnostic methods and therapies (Dweck et al., 2016).

Before exercise, cardiac MRI in dyslipidemia patients often reveals abnormalities in cardiac structure and function (Kosmala et al., 2005). Left ventricular (LV) hypertrophy, characterized by increased LV mass index (>81 g/m^2 in men and >62 g/m^2 in women), is commonly observed (Levy et al., 1990). Diastolic dysfunction, manifested as impaired LV relaxation and reduced early-to-atrial ventricular filling velocity ratio (E/A ratio < 1), represents another frequent finding (Masugata et al., 2006). In addition, subclinical systolic dysfunction may be detected through reduced global longitudinal strain despite preserved ejection fraction (Narayanan et al., 2009).

Regular exercise has been shown to induce favorable cardiac remodeling in patients with dyslipidemia. A randomized controlled trial by Leggio et al., involving 65 patients with dyslipidemia, demonstrated that a 6-month aerobic exercise program resulted in significant improvements in cardiac structure and function as assessed by cardiac MRI (Leggio et al., 2017). Specifically, exercise led to a 4.8% reduction in LV mass index ($p < 0.01$), a 6.3% improvement in LV ejection fraction ($p < 0.01$), and a 12.5% enhancement in diastolic function as measured by the E/A ratio ($p < 0.001$) (Kwon et al., 2020).

Furthermore, cardiac MRI studies have shown that exercise training improves myocardial perfusion and microvascular function in dyslipidemia patients (Brinkley et al., 2011). In a study by Brinkley et al., a 6-month exercise intervention resulted in a 15% increase in the myocardial perfusion reserve index, as measured by first-pass perfusion MRI ($p < 0.01$), indicating enhanced coronary microvascular function (Schlett et al., 2011).

5.4 Single-photon emission computed tomography and positron emission tomography nuclear imaging techniques

Nuclear imaging methods, such as single-photon emission CT (SPECT) and positron emission tomography (PET), have proven to be valuable tools for evaluating atherosclerosis and its impact on the cardiovascular system (Schindler et al., 2010). These advanced imaging techniques offer insights into blood flow, perfusion, and metabolism, enabling the identification of regions with compromised blood supply and the assessment of cardiac function affected by atherosclerosis (Dilsizian et al., 2016).

SPECT is a nuclear imaging technique that utilizes gamma rays emitted by radioactive tracers to create three-dimensional images of the heart (Slomka et al., 2014). The patient is injected with a radioactive tracer, typically technetium-99 m or thallium-201, which is taken up by viable cardiac tissue in proportion to blood flow (Shaw et al., 2012). Special gamma cameras detect the emitted radiation and, through computer processing, generate tomographic images that reflect myocardial perfusion (Berman et al., 2007).

Myocardial perfusion imaging (MPI) using SPECT is frequently employed to evaluate blood flow to the myocardium (Iskandrian and Garcia, 2015). This technique involves administering a radioactive tracer, such as technetium-99 m or thallium-201, into the bloodstream. The heart muscle cells take up this tracer in correlation with blood flow (Bateman et al., 2016). Subsequently, SPECT cameras capture images of the heart at rest and during stress induced by exercise or medications. By comparing these images, healthcare professionals can identify areas of diminished blood flow, known as perfusion defects, which may signify coronary artery disease caused by atherosclerosis (Hachamovitch et al., 2003).

SPECT MPI aids in diagnosing CAD and determining its severity (Shaw et al., 2008). It also provides valuable data for risk stratification purposes by identifying individuals who are at higher risk of experiencing cardiac events (Hachamovitch et al., 2011). The findings of SPECT MPI can inform treatment decisions, including the necessity for revascularization procedures like angioplasty or coronary artery bypass grafting, or the optimization of medical therapy (Boden et al., 2007).

PET imaging utilizes positron-emitting radiotracers that provide higher spatial resolution and more accurate quantification of myocardial blood flow compared to SPECT (Dilsizian and Taillefer, 2012). PET tracers, such as rubidium-82, nitrogen-13 ammonia, or oxygen-15 water, allow for the measurement of absolute myocardial blood flow in milliliters per gram per minute, enabling the detection of balanced ischemia and microvascular dysfunction that may not be apparent on relative perfusion images (Di Carli et al., 2007).

Similar to SPECT MPI, PET can evaluate blood flow to the heart muscle and detect perfusion abnormalities caused by atherosclerosis (Sciagrà et al., 2021). In comparison with SPECT, PET offers better image resolution and more precise measurement of blood flow levels (Danad et al., 2017). A metaanalysis by Jaarsma et al. comparing the diagnostic accuracy of PET versus SPECT for the detection of coronary artery disease found that PET demonstrated superior sensitivity (92% vs. 88%) and specificity (85% vs. 76%) (Jaarsma et al., 2012).

Before exercise intervention, nuclear imaging in dyslipidemia patients frequently reveals perfusion defects indicative of myocardial ischemia (Sdringola et al., 2003). In a study by Sdringola et al. involving 50 patients with familial hypercholesterolemia, PET imaging demonstrated reduced coronary flow reserve (the ratio of maximal to basal coronary blood flow) in 76% of patients, despite the absence of obstructive coronary artery disease on angiography (Sdringola et al., 2001). This finding highlights the presence of coronary microvascular dysfunction in dyslipidemia, even before the development of overt coronary stenosis.

Exercise training has been shown to improve myocardial perfusion as assessed by nuclear imaging. In a randomized controlled trial involving 101 patients with stable coronary artery disease and dyslipidemia, Hambrecht et al. demonstrated that a 12-month exercise program resulted in a significant reduction in the extent and severity of exercise-induced perfusion defects on SPECT imaging (Hambrecht et al., 2000). The summed stress score (a measure of the extent and severity of perfusion defects) decreased by 24% in the exercise group compared to a 4% increase in the control group ($p < 0.001$) (Gielen et al., 2001).

In a recent study utilizing cardiac PET, a 6-month high-intensity interval training program in patients with dyslipidemia led to a 17% improvement in global coronary flow reserve (from 2.1 ± 0.5 to 2.5 ± 0.6, $p < 0.01$), indicating enhanced coronary microvascular function (Taqueti et al., 2018). This improvement correlated with reductions in LDLc levels and was more pronounced in regions supplied by nonstenotic coronary arteries, suggesting that exercise exerts beneficial effects primarily through improvement in coronary microcirculation (Murthy et al., 2011).

5.5 Vascular imaging

Vascular imaging plays a crucial role in the identification, evaluation, and treatment of atherosclerosis, a condition characterized by the accumulation of plaque within the arteries (Mitchell et al., 2010). Various imaging techniques are commonly employed to assess atherosclerosis in different vascular regions throughout the body.

Duplex ultrasound: One such technique is duplex ultrasound, which combines conventional ultrasound imagery with Doppler ultrasound to analyze blood flow patterns and identify changes associated with atherosclerosis within arterial walls (Harris et al., 2010). Duplex ultrasound is frequently utilized for examining the carotid arteries located in the neck (carotid duplex) to determine stroke risk, as well as for assessing peripheral arteries found in the limbs to detect peripheral artery disease (Gerhard-Herman et al., 2006).

CTA: CTA is a noninvasive imaging technique that utilizes X-ray technology to generate detailed cross-sectional images of blood vessels (Taylor et al., 2010). It is commonly employed for the assessment of atherosclerosis in various vascular beds, including the coronary arteries, carotid arteries, aorta, and peripheral arteries (Bettmann et al., 1997). CTA facilitates visualization of plaque presence, the degree of stenosis (narrowing), and the blockage location.

Magnetic resonance angiography (MRA): MRA is a noninvasive imaging technique that employs magnetic fields and radio waves to produce detailed images of blood vessels (François et al., 2008). It is frequently used for evaluating atherosclerosis in the carotid arteries, renal arteries, and peripheral arteries (Finn et al., 2006). MRA offers information concerning blood flow dynamics, plaque composition analysis, and the detection of the presence or absence of stenosis.

IVUS: It is used to assess coronary arteries during cardiac catheterization procedures (Nissen and Yock, 2001). By inserting a small ultrasound probe into the artery, detailed images of the vessel from within can be obtained. IVUS provides comprehensive information about plaque morphology, composition, and the extent of atherosclerosis (Mintz et al., 2001).

OCT: It is also used during cardiac catheterization to assess coronary arteries (Tearney et al., 2012). This technique utilizes light waves to create highly detailed images of the vessel wall and plaque. OCT offers high-resolution images that can assist in determining plaque characteristics and guiding interventional procedures (Prati et al., 2012).

Before exercise intervention, vascular imaging in dyslipidemia patients often reveals various abnormalities (Brevetti et al., 2003). Duplex ultrasound frequently demonstrates increased arterial stiffness, as measured by pulse wave velocity, with values typically 15%–25% higher than those of age-matched controls (Laurent et al., 2006). The ankle-brachial index (ABI), a ratio of systolic blood pressure at the ankle to that in the arm, may be reduced (<0.9) in patients with peripheral artery disease, a common complication of dyslipidemia (Aboyans et al., 2012). CTA or MRA may reveal various degrees of stenosis in different vascular beds, with the prevalence of significant stenosis (>by 50%) in the lower extremities, ranging from 5%–30% in patients with dyslipidemia, depending on age and the presence of other risk factors (Norgren et al., 2007).

Regular exercise has been shown to improve vascular structure and function in patients with dyslipidemia. A meta-analysis of 42 randomized controlled trials involving 1677 participants with various cardiovascular risk factors, including dyslipidemia, demonstrated that aerobic exercise training significantly reduced arterial stiffness, with a standardized mean difference of −0.52 (95% CI: −0.70 to −0.34, $p < 0.001$) (Ashor et al., 2014).

In a study by Gardner et al. involving 156 patients with peripheral artery disease and dyslipidemia, a 6-month supervised exercise program resulted in significant improvements in the ABI (from 0.72 ± 0.12 to 0.82 ± 0.14, $p < 0.01$) and walking distance on a treadmill (from 305 ± 161 to 458 ± 220 m, $p < 0.001$) (Gardner et al., 2011). Follow-up duplex ultrasound revealed a 23% reduction in the severity of stenoses in the superficial femoral artery and a 19% increase in collateral circulation ($p < 0.01$ for both) (Hiatt et al., 2015).

MRA studies have demonstrated that exercise training improves endothelial function, as evidenced by enhanced FMD (Thijssen et al., 2011). In a randomized controlled trial by Hambrecht et al., four weeks of daily exercise training in patients with coronary artery disease resulted in a significant improvement in endothelium-dependent vasodilation in the left internal mammary artery, as assessed by MRA (from $4.5 \pm 3.2\%$ to $10.8 \pm 5.6\%$, $p < 0.001$) (Hambrecht et al., 2003).

6 Nonimaging methods for assessing dyslipidemia changes following exercise: Stress testing

Cardiac stress testing, also referred to as stress testing or exercise stress testing, is a diagnostic procedure utilized to assess the heart's response to physical exertion and evaluate the impact of atherosclerosis on cardiac function (Fletcher et al., 2013). This form of assessment is particularly effective in identifying coronary artery disease, which signifies the presence of blockage due to plaque build-up in the coronary arteries (Fihn et al., 2012).

One widely employed approach for stress testing involves having patients engage in exercise, such as walking on a treadmill or cycling on a stationary bike, while closely monitoring their physiological responses (Gibbons et al., 2002). The test entails gradually increasing the intensity of exercise, thereby subjecting the heart to progressively greater levels of strain. Stress testing is commonly recommended for individuals who may be at risk for, or exhibit symptoms suggestive of, coronary artery disease (Fleg et al., 2013). Typical indications include the presence of chest pain (angina), shortness of breath, unexplained fatigue, or as part of a comprehensive cardiac risk assessment in those with predisposing factors such as hypertension, diabetes, smoking history, or a familial background of heart disease (Douglas et al., 2015).

The primary objective behind stress tests is to identify myocardial ischemia—an inadequate blood supply to the heart muscle resulting from narrowing or blockage in the coronary arteries due to atherosclerosis (Camici and Crea, 2007). During physical exercise, there is an increase in oxygen demand by the heart; however, if there is significant obstruction present within these arteries causing impeded blood flow toward the heart muscle itself, it can manifest through related symptoms like angina and changes discernible on electrocardiography readings (Schulz-Menger et al., 2013).

Cardiac monitoring during the stress test includes continuous observation of heart rate, blood pressure, and electrocardiogram (Montalescot and Sechtem et al., 2013). Any deviations in the ECG waveform, such as ST-segment depression, can indicate ischemia and provide insights into the location and severity of coronary artery blockages (Mark et al., 1991).

Before exercise intervention, stress tests in dyslipidemia patients often reveal reduced exercise capacity and evidence of exercise-induced myocardial ischemia (Myers et al., 2002). Exercise tolerance, typically measured as metabolic equivalents (METs) achieved during treadmill testing, is frequently lower than age-predicted norms, with values often 20%–30% below expected levels (Bourque et al., 2011). Electrocardiographic evidence of ischemia, defined as ≥ 1 mm horizontal or downsloping ST-segment depression, is observed in approximately 30%–40% of symptomatic dyslipidemia patients during stress testing (DeLuca et al., 2006).

Several studies have demonstrated significant improvements in stress test parameters following regular exercise in dyslipidemia patients. In a randomized controlled trial by Taylor et al. involving 120 patients with dyslipidemia and documented coronary artery disease, a 12-week cardiac rehabilitation program resulted in a 23% increase in exercise capacity (from 7.2 ± 1.8 to 8.9 ± 2.1 METs, $p < 0.001$) (Taylor et al., 2004).

Furthermore, the prevalence of exercise-induced ischemia has been shown to decrease following regular physical activity. In a study by Hambrecht et al., 12 months of regular exercise training reduced the incidence of ischemic ST-segment depression during stress testing by 54% ($p < 0.01$) in patients with stable coronary artery disease and dyslipidemia

(Hambrecht et al., 2004). This reduction in ischemia correlates with improvements in coronary endothelial function and the development of coronary collaterals, as demonstrated in subsequent angiographic studies (Hambrecht et al., 1993).

A metaanalysis of 48 trials involving 8940 patients with coronary artery disease, many with dyslipidemia, found that exercise-based cardiac rehabilitation was associated with a significant reduction in the ischemic threshold, defined as the heart rate-blood pressure product at which 1 mm ST-segment depression occurred (standardized mean difference: 0.38; 95% CI: 0.20–0.56) (Lawler et al., 2011). This physiological adaptation allows patients to perform higher levels of physical activity before experiencing ischemia, translating to improved functional capacity and quality of life (Belardinelli et al., 1995).

7 Body composition assessment

Accurate evaluation of body composition is crucial when assessing the potential risk factors linked to atherosclerosis and cardiovascular disease (Després, 2012). Atherosclerosis, being a multifaceted condition, can be influenced by various aspects of an individual's body composition (Britton et al., 2013). Recognizing the different methods employed for assessing body composition allows healthcare professionals, as well as individuals themselves, to identify possible risk factors and take appropriate measures to mitigate their likelihood of developing atherosclerosis (Fox et al., 2007).

Advanced imaging techniques, such as dual-energy X-ray absorptiometry (DXA), CT, and MRI, provide a detailed assessment of body composition (Kaul et al., 2012). These modalities can distinguish between different tissue types, including adipose tissue, muscle, and bone, with high precision (Neeland et al., 2019). Particularly important is the differentiation between subcutaneous and visceral adipose tissue (VAT), as the latter is more strongly associated with cardiometabolic risk (Rothney et al., 2013).

Before exercise, body composition imaging in dyslipidemia patients often reveals unfavorable body composition profiles (Neeland et al., 2018). CT or MRI studies typically demonstrate increased VAT volume, with values often exceeding 100 cm^2 at the L4–L5 level, which is considered a threshold for increased cardiometabolic risk (Ross et al., 2020). The visceral-to-subcutaneous adipose tissue ratio is frequently elevated (>0.4), indicating a disproportionate accumulation of metabolically active visceral fat (Després and Lemieux, 2006). In addition, reduced muscle mass and quality (increased intramuscular fat) may be observed, particularly in older individuals or those with long-standing dyslipidemia (Franssen et al., 2013).

Exercise interventions have demonstrated significant effects on body composition in dyslipidemia patients. A meta-analysis of 117 randomized controlled trials involving 3602 participants showed that exercise training resulted in a significant reduction in VAT, with a weighted mean difference of −30.6 cm^2 (95% CI: −38.8 to −22.4, $p < 0.001$), representing approximately a 6.1% decrease from baseline values (Vissers et al., 2013).

In a study by Ohkawara et al., utilizing MRI to assess body composition, a 12-week aerobic exercise program in patients with dyslipidemia resulted in a 14.6% reduction in VAT volume ($p < 0.001$) despite relatively modest weight loss (3.2% of initial body weight) (Ohkawara et al., 2007). This underscores the concept that exercise can induce favorable changes in body composition that exceed what would be expected based on weight loss alone (Ismail et al., 2012).

DXA studies have demonstrated that combined aerobic and resistance training programs can simultaneously reduce fat mass and preserve or increase lean muscle mass in dyslipidemia patients (Davidson et al., 2009). In a randomized controlled trial by Davidson et al., a 6-month combined exercise program resulted in a 6.3% decrease in total fat mass and a 2.8% increase in appendicular skeletal muscle mass ($p < 0.01$ for both) (Willis et al., 2012). These changes in body composition were associated with significant improvements in lipid profiles, insulin sensitivity, and inflammatory markers (Verheggen et al., 2016).

8 Discussion

The evidence presented in this review underscores the significant impact of exercise on cardiovascular parameters in dyslipidemia patients, as visualized through various imaging modalities. The beneficial effects of regular physical activity extend beyond simple improvements in lipid profiles to encompass structural and functional changes in the cardiovascular system (Thompson et al., 2003; Lavie et al., 2019).

The improvements observed in carotid ultrasound parameters, particularly the reduction in CIMT, represent one of the most well-documented effects of exercise in dyslipidemia patients (Sattelmair et al., 2011). This finding is particularly significant given that CIMT is a validated surrogate marker for atherosclerosis and an independent predictor of cardiovascular events (Nocon et al., 2008). The mechanism underlying this benefit likely involves multiple pathways, including enhanced endothelial function, reduced inflammation, and improved lipid metabolism (Ribeiro et al., 2021).

The impact of exercise on coronary artery calcification appears more complex, with most studies suggesting a slowing of progression rather than regression (Sever et al., 2003). This observation aligns with our understanding of the pathophysiology of vascular calcification, which is generally considered an irreversible process once initiated (Inaba et al., 2010). Nevertheless, the ability of exercise to attenuate the progression of CAC scores represents a clinically meaningful outcome, particularly when considering the strong association between CAC progression rates and future cardiovascular events (Perk et al., 2012).

Cardiac MRI findings demonstrate that exercise induces favorable cardiac remodeling in dyslipidemia patients, characterized by reduced LV hypertrophy, improved systolic and diastolic function, and enhanced myocardial perfusion (Kodama et al., 2009). These adaptations likely contribute to the improved exercise capacity and reduced incidence of ischemia observed during stress testing (Sofi et al., 2008). The molecular mechanisms underlying these changes include enhanced nitric oxide bioavailability, reduced oxidative stress, improved mitochondrial function, and favorable alterations in cardiac gene expression (Shiroma and Lee, 2010).

Nuclear imaging studies provide compelling evidence for the ability of exercise to improve myocardial perfusion and coronary flow reserve in dyslipidemia patients (Mora et al., 2007). These improvements reflect enhanced coronary microvascular function, which may occur independently of changes in epicardial coronary stenosis (Joyner and Green, 2009). Given the growing recognition of coronary microvascular dysfunction as an important contributor to cardiovascular morbidity and mortality, these findings highlight a potential mechanism through which exercise confers cardiovascular protection (Warburton et al., 2006).

Vascular imaging studies demonstrate that exercise improves arterial compliance, reduces vascular stenosis, and enhances collateral circulation in patients with dyslipidemia (Lee et al., 2012). These effects likely contribute to the reduction in cardiovascular events observed in physically active individuals and underscore the importance of exercise as a cornerstone of therapy for patients with atherosclerotic vascular disease (Eijsvogels et al., 2016).

The changes in body composition associated with exercise, particularly the reduction in VAT, represent an important mechanism through which physical activity may improve cardiovascular health in dyslipidemia patients (Ross and Janssen, 2001). Visceral adiposity is associated with increased production of pro-inflammatory cytokines and adipokines, insulin resistance, and atherogenic lipid profiles (Tchernof and Després, 2013). By reducing visceral fat, exercise may attenuate these deleterious effects and create a more favorable metabolic environment (Després, 2016).

It is important to note that the magnitude of improvements observed in imaging parameters varies considerably among individuals and appears to be influenced by several factors, including the intensity, duration, and type of exercise; baseline cardiovascular risk; genetic factors; and concomitant medical therapy (Garber et al., 2011). Furthermore, the time course of these changes differs across imaging modalities, with some parameters (such as endothelial function) improving relatively quickly, while others (such as CIMT or coronary calcification) may require months or years of regular exercise to demonstrate significant changes (Powell et al., 2011).

Despite these variations, the consistency of findings across different imaging modalities and patient populations provides robust evidence for the beneficial effects of exercise in dyslipidemia patients (Wen et al., 2011). These imaging changes correlate with improvements in traditional cardiovascular risk factors and, more importantly, with reductions in cardiovascular events and mortality (Leon et al., 2005).

9 Conclusion

The comprehensive review of medical imaging findings in dyslipidemia patients, before and after exercise interventions, reveals consistent and clinically significant improvements across multiple cardiovascular parameters. These imaging-documented changes provide mechanistic insights into how exercise confers cardiovascular protection in this high-risk population.

Carotid ultrasound demonstrates reduced intima-media thickness and improved plaque stability following regular exercise. CAC scoring shows attenuated progression of calcification with consistent physical activity. Cardiac MRI reveals favorable cardiac remodeling, improved ventricular function, and enhanced myocardial perfusion in response to exercise training. Stress testing documents increased exercise capacity and reduced myocardial ischemia. Nuclear imaging demonstrates improved myocardial perfusion and coronary flow reserve. Vascular imaging shows enhanced arterial compliance, reduced stenosis, and improved collateral circulation. Body composition assessment documents reductions in visceral adiposity, which likely contribute to the overall cardiovascular benefits.

These imaging findings provide objective evidence supporting the recommendation of regular exercise as a cornerstone of management for patients with dyslipidemia. The observed changes in cardiovascular structure and function likely contribute to the reduced cardiovascular morbidity and mortality associated with physical activity in this population.

Future research should focus on identifying optimal exercise protocols for specific patient subgroups, understanding the interaction between exercise and pharmacological therapies, and developing personalized approaches based on imaging parameters. In addition, the application of advanced imaging techniques, including molecular imaging and artificial intelligence-based image analysis, may provide further insights into the cardiovascular benefits of exercise in dyslipidemia patients.

References

Aboyans, V., Criqui, M.H., Abraham, P., et al., 2012. Measurement and interpretation of the ankle-brachial index: a scientific statement from the American Heart Association. Circulation 126 (24), 2890–2909.

Agatston, A.S., Janowitz, W.R., Hildner, F.J., et al., 1990. Quantification of coronary artery calcium using ultrafast computed tomography. J. Am. Coll. Cardiol. 15 (4), 827–832.

Amarenco, P., Labreuche, J., Lavallée, P., et al., 2004. Statins in stroke prevention and carotid atherosclerosis: systematic review and up-to-date meta-analysis. Stroke 35 (12), 2902–2909.

Anderson, T.J., Meredith, I.T., Charbonneau, F., et al., 1996. Endothelium-dependent coronary vasomotion relates to the susceptibility of LDL to oxidation in humans. Circulation 93 (9), 1647–1650.

Arnett, D.K., Blumenthal, R.S., Albert, M.A., et al., 2019. 2019 ACC/AHA guideline on the primary prevention of cardiovascular disease. Circulation 140 (11), e596–e646.

Ashor, A.W., Lara, J., Siervo, M., et al., 2014. Effects of exercise modalities on arterial stiffness and wave reflection: a systematic review and meta-analysis of randomized controlled trials. PLoS One 9 (10), e110034.

Bateman, T.M., Dilsizian, V., Beanlands, R.S., et al., 2016. American Society of Nuclear Cardiology and Society of Nuclear Medicine and Molecular Imaging joint position statement on the clinical indications for myocardial perfusion PET. J. Nucl. Med. 57 (10), 1654–1656.

Belardinelli, R., Georgiou, D., Cianci, G., et al., 1995. Exercise training improves left ventricular diastolic filling in patients with dilated cardiomyopathy. Clinical and prognostic implications. Circulation 91 (11), 2775–2784.

Benjamin, E.J., Muntner, P., Alonso, A., et al., 2019. Heart disease and stroke statistics-2019 update: a report from the American Heart Association. Circulation 139 (10), e56–e528.

Bentzon, J.F., Otsuka, F., Virmani, R., et al, 2014. Mechanisms of plaque formation and rupture. Circ. Res. 114 (12), 1852–1866.

Berman, D.S., Kang, X., Slomka, P.J., et al., 2007. Underestimation of extent of ischemia by gated SPECT myocardial perfusion imaging in patients with left main coronary artery disease. J. Nucl. Cardiol. 14 (4), 521–528.

Bettmann, M.A., Heeren, T., Greenfield, A., et al., 1997. Adverse events with radiographic contrast agents: results of the SCVIR Contrast Agent Registry. Radiology 203 (3), 611–620.

Blaha, M.J., Budoff, M.J., DeFilippis, A.P., et al., 2011. Associations between C-reactive protein, coronary artery calcium, and cardiovascular events: implications for the Jupiter population from Mesa, a population-based cohort study. Lancet 378 (9792), 684–692.

Blaha, M.J., Mortensen, M.B., Kianoush, S., et al., 2017. Coronary Artery calcium scoring: is it time for a change in methodology? JACC Cardiovasc. Imaging 10 (8), 923–937.

Boden, W.E., O'Rourke, R.A., Teo, K.K., et al., 2007. Optimal medical therapy with or without PCI for stable coronary disease. N. Engl. J. Med. 356 (15), 1503–1516.

Bonetti, P.O., Lerman, L.O., Lerman, A., 2003. Endothelial dysfunction: a marker of atherosclerotic risk. Arterioscler. Thromb. Vasc. Biol. 23 (2), 168–175.

Boodhwani, M., Sodha, N.R., Mieno, S., et al., 2007. Functional, cellular, and molecular characterization of the angiogenic response to chronic myocardial ischemia in diabetes. Circulation 116 (11 Suppl), I31–7.

Bourque, J.M., Charlton, G.T., Holland, B.H., et al., 2011. Prognosis in patients achieving ≥10 METS on exercise stress testing: was SPECT imaging useful? J. Nucl. Cardiol. 18 (2), 230–237.

Brevetti, G., Silvestro, A., Schiano, V., et al., 2003. Endothelial dysfunction and cardiovascular risk prediction in peripheral arterial disease: additive value of flow-mediated dilation to ankle-brachial pressure index. Circulation 108 (17), 2093–2098.

Brinkley, T.E., Jerosch-Herold, M., Folsom, A.R., et al, 2011. Pericardial fat and myocardial perfusion in asymptomatic adults from the Multi-Ethnic Study of Atherosclerosis. PLoS One 6 (12), e28410.

Britton, K.A., Massaro, J.M., Murabito, J.M., et al., 2013. Body fat distribution, incident cardiovascular disease, cancer, and all-cause mortality. J. Am. Coll. Cardiol. 62 (10), 921–925.

Brott, T.G., Hobson 2nd, R.W., Howard, G., et al., 2010. Stenting versus endarterectomy for treatment of carotid-artery stenosis. N. Engl. J. Med. 363 (1), 11–23.

Budoff, M.J., Shaw, L.J., Liu, S.T., et al., 2007. Long-term prognosis associated with coronary calcification: observations from a registry of 25,253 patients. J. Am. Coll. Cardiol. 49 (18), 1860–1870.

Budoff, M.J., Young, R., Lopez, V.A., et al., 2013. Progression of coronary calcium and incident coronary heart disease events: mesa (Multi-Ethnic Study of Atherosclerosis). J. Am. Coll. Cardiol. 61 (12), 1231–1239.

Camici, P.G., Crea, F., 2007. Coronary microvascular dysfunction. N. Engl. J. Med. 356 (8), 830–840.

Cannon, C.P., Blazing, M.A., Giugliano, R.P., et al., 2015. Ezetimibe added to statin therapy after acute coronary syndromes. N. Engl. J. Med. 372 (25), 2387–2397.

Cholesterol Treatment Trialists' (CTT) Collaboration, 2010. Efficacy and safety of more intensive lowering of LDL cholesterol: a meta-analysis of data from 170,000 participants in 26 randomised trials. Lancet 376 (9753), 1670–1681.

Collins, R., Reith, C., Emberson, J., et al, 2016. Interpretation of the evidence for the efficacy and safety of statin therapy. Lancet 388 (10059), 2532–2561.

Danad, I., Raijmakers, P.G., Driessen, R.S., et al., 2017. Comparison of Coronary CT Angiography, SPECT, PET, and Hybrid Imaging for Diagnosis of Ischemic Heart Disease Determined by Fractional Flow Reserve. JAMA Cardiol. 2 (10), 1100–1107.

Davidson, L.E., Hudson, R., Kilpatrick, K., et al., 2009. Effects of exercise modality on insulin resistance and functional limitation in older adults: a randomized controlled trial. Arch. Intern. Med. 169 (2), 122–131.

Davignon, J., Ganz, P., 2004. Role of endothelial dysfunction in atherosclerosis. Circulation 109 (23 Suppl 1), III27–32.

Deanfield, J.E., Halcox, J.P., Rabelink, T.J., 2007. Endothelial function and dysfunction: testing and clinical relevance. Circulation 115 (10), 1285–1295.

DeLuca, A.J., Kaplan, S., Aronow, W.S., et al., 2006. Comparison of prevalence of unrecognized myocardial infarction and of silent myocardial ischemia detected by a treadmill exercise sestamibi stress test in patients with versus without diabetes mellitus. Am. J. Cardiol. 98 (8), 1045–1046.

Després, J.P., 2012. Body fat distribution and risk of cardiovascular disease: an update. Circulation 126 (10), 1301–1313.

Després, J.P., 2016. Physical activity, sedentary behaviours, and cardiovascular health: when will cardiorespiratory fitness become a vital sign? Can. J. Cardiol. 32 (4), 505–513.

Després, J.P., Lemieux, I., 2006. Abdominal obesity and metabolic syndrome. Nature 444 (7121), 881–887.

Detrano, R., Guerci, A.D., Carr, J.J., et al., 2008. Coronary calcium as a predictor of coronary events in four racial or ethnic groups. N. Engl. J. Med. 358 (13), 1336–1345.

Di Carli, M.F., Dorbala, S., Meserve, J., et al., 2007. Clinical myocardial perfusion PET/CT. J. Nucl. Med. 48 (5), 783–793.

Dilsizian, V., Bacharach, S.L., Beanlands, R.S., et al., 2016. ASNC imaging guidelines/SNMMI procedure standard for positron emission tomography (PET) nuclear cardiology procedures. J. Nucl. Cardiol. 23 (5), 1187–1226.

Dilsizian, V., Taillefer, R., 2012. Journey in evolution of nuclear cardiology: will there be another quantum leap with the F-18-labeled myocardial perfusion tracers? JACC Cardiovasc. Imaging 5 (12), 1269–1284.

Douglas, P.S., Hoffmann, U., Patel, M.R., et al, 2015. Outcomes of anatomical versus functional testing for coronary artery disease. N. Engl. J. Med. 372 (14), 1291–1300.

Dweck, M.R., Williams, M.C., Moss, A.J., et al, 2016. Computed tomography and cardiac magnetic resonance in ischemic heart disease. J. Am. Coll. Cardiol. 68 (20), 2201–2216.

Eijsvogels, T.M., Molossi, S., Lee, D.C., et al., 2016. Exercise at the Extremes: the amount of exercise to reduce cardiovascular events. J. Am. Coll. Cardiol. 67 (3), 316–329.

Erbel, R., Möhlenkamp, S., Moebus, S., et al., 2010. Coronary risk stratification, discrimination, and reclassification improvement based on quantification of subclinical coronary atherosclerosis: the Heinz Nixdorf Recall study. J. Am. Coll. Cardiol. 56 (17), 1397–1406.

Estruch, R., Ros, E., Salas-Salvadó, J., et al., 2018. Primary prevention of cardiovascular disease with a mediterranean diet supplemented with extra-virgin olive oil or nuts. N. Engl. J. Med. 378 (25), e34.

Fayad, Z.A., Fuster, V., 2001. Clinical imaging of the high-risk or vulnerable atherosclerotic plaque. Circ. Res. 89 (4), 305–316.

Ference, B.A., Ginsberg, H.N., Graham, I., et al., 2017. Low-density lipoproteins cause atherosclerotic cardiovascular disease. 1. Evidence from genetic, epidemiologic, and clinical studies. A consensus statement from the European Atherosclerosis Society Consensus Panel. Eur. Heart J. 38 (32), 2459–2472.

Fihn, S.D., Gardin, J.M., Abrams, J., et al., 2012. 2012 ACCF/AHA/ACP/AATS/PCNA/SCAI/STS Guideline for the diagnosis and management of patients with stable ischemic heart disease. J. Am. Coll. Cardiol. 60 (24), e44–e164.

Finn, J.P., Nael, K., Deshpande, V., et al., 2006. Cardiac MR imaging: state of the technology. Radiology 241 (2), 338–354.

Flammer, A.J., Anderson, T., Celermajer, D.S., et al., 2012. The assessment of endothelial function: from research into clinical practice. Circulation 126 (6), 753–767.

Fleg, J.L., Forman, D.E., Berra, K., et al., 2013. Secondary prevention of atherosclerotic cardiovascular disease in older adults: a scientific statement from the American Heart Association. Circulation 128 (22), 2422–2446.

Fletcher, G.F., Ades, P.A., Kligfield, P., et al, 2013. Exercise standards for testing and training: a scientific statement from the American Heart Association. Circulation 128 (8), 873–934.

Forstermann, U., Sessa, W.C., 2012. Nitric oxide synthases: regulation and function. Eur. Heart J. 33 (7), 829–837.

Fox, C.S., Massaro, J.M., Hoffmann, U., et al., 2007. Abdominal visceral and subcutaneous adipose tissue compartments: association with metabolic risk factors in the Framingham Heart Study. Circulation 116 (1), 39–48.

François, C.J., Tuite, D., Deshpande, V., et al, 2008. Unenhanced MR angiography of the thoracic aorta: initial clinical evaluation. AJR Am. J. Roentgenol. 190 (4), 902–906.

Franssen, F.M., Sauerwein, H.P., Rutten, E.P., et al., 2013. Whole-body and skeletal muscle dysfunction in chronic obstructive pulmonary disease: diagnosis and treatment strategies. Eur. Respir. Monogr. 59, 32–41.

Fuster, V., Moreno, P.R., Fayad, Z.A., et al., 2005. Atherothrombosis and high-risk plaque: part I: evolving concepts. J. Am. Coll. Cardiol. 46 (6), 937–954.

Garber, C.E., Blissmer, B., Deschenes, M.R., et al., 2011. American College of Sports Medicine position stand. Quantity and quality of exercise for developing and maintaining cardiorespiratory, musculoskeletal, and neuromotor fitness in apparently healthy adults: guidance for prescribing exercise. Med. Sci. Sports Exerc. 43 (7), 1334–1359.

Garcia-Garcia, H.M., Mintz, G.S., Lerman, A., et al., 2009. Tissue characterisation using intravascular radiofrequency data analysis: recommendations for acquisition, analysis, interpretation and reporting. EuroIntervention 5 (2), 177–189.

Gardner, A.W., Parker, D.E., Montgomery, P.S., et al., 2011. Efficacy of quantified home-based exercise and supervised exercise in patients with intermittent claudication: a randomized controlled trial. Circulation 123 (5), 491–498.

Gerhard-Herman, M., Gardin, J.M., Jaff, M., et al., 2006. Guidelines for noninvasive vascular laboratory testing: a report from the American Society of Echocardiography and the Society for Vascular Medicine and Biology. Vasc. Med. 11 (3), 183–200.

Gibbons, R.J., Balady, G.J., Bricker, J.T., et al., 2002. ACC/AHA 2002 guideline update for exercise testing: summary article. A report of the American College of Cardiology/American Heart Association Task Force on Practice Guidelines (Committee to Update the 1997 Exercise Testing Guidelines). J. Am. Coll. Cardiol. 40 (8), 1531–1540.

Gielen, S., Schuler, G., Hambrecht, R., 2001. Exercise training in coronary artery disease and coronary vasomotion. Circulation 103 (1), E1–6.

Gimbrone Jr, G., García-Cardeña, G., 2016. Endothelial cell dysfunction and the pathobiology of atherosclerosis. Circ. Res. 118 (4), 620–636.

Goldstein, J.L., Brown, M.S., 2015. A century of cholesterol and coronaries: from plaques to genes to statins. Cell 161 (1), 161–172.

Grant, E.G., Benson, C.B., Moneta, G.L., et al., 2003. Carotid artery stenosis: gray-scale and Doppler US diagnosis–Society of Radiologists in Ultrasound Consensus Conference. Radiology 229 (2), 340–346.

Greenland, P., Alpert, J.S., Beller, G.A., et al., 2010. 2010 ACCF/AHA guideline for assessment of cardiovascular risk in asymptomatic adults: a report of the American College of Cardiology Foundation/American Heart Association Task Force on Practice Guidelines. J. Am. Coll. Cardiol. 56 (25), e50–e103.

Greenland, P., Blaha, M.J., Budoff, M.J., et al., 2018. Coronary calcium score and cardiovascular risk. J. Am. Coll. Cardiol. 72 (4), 434–447.

Grundy, S.M., Stone, N.J., Bailey, A.L., et al., 2019. 2018 AHA/ACC/AACVPR/AAPA/ABC/ACPM/ADA/AGS/APhA/ASPC/NLA/PCNA guideline on the management of blood cholesterol. J. Am. Coll. Cardiol. 73 (24), e285–e350.

Hachamovitch, R., Hayes, S.W., Friedman, J.D., et al., 2003. Comparison of the short-term survival benefit associated with revascularization compared with medical therapy in patients with no prior coronary artery disease undergoing stress myocardial perfusion single photon emission computed tomography. Circulation 107 (23), 2900–2907.

Hachamovitch, R., Rozanski, A., Shaw, L.J., et al., 2011. Impact of ischaemia and scar on the therapeutic benefit derived from myocardial revascularization vs. medical therapy among patients undergoing stress-rest myocardial perfusion scintigraphy. Eur. Heart J. 32 (8), 1012–1024.

Hambrecht, R., Adams, V., Erbs, S., et al., 2003. Regular physical activity improves endothelial function in patients with coronary artery disease by increasing phosphorylation of endothelial nitric oxide synthase. Circulation 107 (25), 3152–3158.

Hambrecht, R., Niebauer, J., Marburger, C., et al., 1993. Various intensities of leisure time physical activity in patients with coronary artery disease: effects on cardiorespiratory fitness and progression of coronary atherosclerotic lesions. J. Am. Coll. Cardiol. 22 (2), 468–477.

Hambrecht, R., Walther, C., Möbius-Winkler, S., et al., 2004. Percutaneous coronary angioplasty compared with exercise training in patients with stable coronary artery disease: a randomized trial. Circulation 109 (11), 1371–1378.

Hambrecht, R., Wolf, A., Gielen, S., et al., 2000. Effect of exercise on coronary endothelial function in patients with coronary artery disease. N. Engl. J. Med. 342 (7), 454–460.

Harris, R.A., Nishiyama, S.K., Wray, D.W., et al., 2010. Ultrasound assessment of flow-mediated dilation. Hypertension 55 (5), 1075–1085.

Hecht, H., Blaha, M.J., Berman, D.S., et al., 2017. Clinical indications for coronary artery calcium scoring in asymptomatic patients: expert consensus statement from the Society of Cardiovascular Computed Tomography. J. Cardiovasc. Comput. Tomogr. 11 (2), 157–168.

Hiatt, W.R., Armstrong, E.J., Larson, C.J., et al., 2015. Pathogenesis of the limb manifestations and exercise limitations in peripheral artery disease. Circ. Res. 116 (9), 1527–1539.

Hundley, W.G., Bluemke, D.A., Finn, J.P., et al., 2010. ACCF/ACR/AHA/NASCI/SCMR 2010 expert consensus document on cardiovascular magnetic resonance: a report of the American College of Cardiology Foundation Task Force on Expert Consensus Documents. Circulation 121 (22), 2462–2508.

Inaba, Y., Chen, J.A., Bergmann, S.R., 2010. Prediction of future cardiovascular outcomes by flow-mediated vasodilatation of brachial artery: a meta-analysis. Int. J. Cardiovasc. Imaging 26 (6), 631–640.

Iskandrian, A.E., Garcia, E.V. (Eds.), 2015. Nuclear Cardiac Imaging: Principles and Applications, 5th ed., Oxford University Press, Oxford.

Ismail, I., Keating, S.E., Baker, M.K., et al., 2012. A systematic review and meta-analysis of the effect of aerobic vs. resistance exercise training on visceral fat. Obes. Rev. 13 (1), 68–91.

Jaarsma, C., Leiner, T., Bekkers, S.C., et al., 2012. Diagnostic performance of noninvasive myocardial perfusion imaging using single-photon emission computed tomography, cardiac magnetic resonance, and positron emission tomography imaging for the detection of obstructive coronary artery disease: a meta-analysis. J. Am. Coll. Cardiol. 59 (19), 1719–1728.

Joyner, M.J., Green, D.J., 2009. Exercise protects the cardiovascular system: effects beyond traditional risk factors. J. Physiol. 587 (Pt 23).

Kadoglou, N.P., Fotiadis, G., Kapelouzou, A., et al., 2013. The differential anti-inflammatory effects of exercise modalities and their association with early carotid atherosclerosis progression in patients with type 2 diabetes. Diabet. Med. 30 (2), 5551–5558.

Kaul, S., Rothney, M.P., Peters, D.M., et al., 2012. Dual-energy X-ray absorptiometry for quantification of visceral fat. Obesity (Silver Spring) 20 (6), 1313–1318.

Kianoush, S., Sharma, A., Mosley, T.H., et al., 2020. Association of self-reported physical activity with calcified atherosclerotic plaque in the coronary arteries: the National Heart, Lung, and Blood Institute Family Heart Study. JAMA Netw. Open 3 (9), e2013248.

Kim, R.J., Wu, E., Rafael, A., et al, 2000. The use of contrast-enhanced magnetic resonance imaging to identify reversible myocardial dysfunction. N. Engl. J. Med. 343 (20), 1445–1453.

Kodama, S., Saito, K., Tanaka, S., et al., 2009. Cardiorespiratory fitness as a quantitative predictor of all-cause mortality and cardiovascular events in healthy men and women: a meta-analysis. JAMA 301 (19), 2024–2035.

Kosmala, W., Przewlocka-Kosmala, M., Wojnalowicz, A., et al., 2005. Left ventricular function in patients with hypertension and lipid metabolism disorders. Int. J. Cardiol. 101 (3), 369–373.

Kwaśniewska, M., Jegier, A., Kostka, T., et al., 2014. Long-term effect of different physical activity levels on subclinical atherosclerosis in middle-aged men: a 25-year prospective study. PLoS One 9 (1), e85209.

Kwon, D.H., Menon, V., Houghtaling, P., et al., 2020. Cardiac magnetic resonance predictors of favorable outcomes in patients with ischemic cardiomyopathy and coronary artery disease. J. Am. Heart Assoc. 9 (22), e017982.

Kwong, R.Y., Chan, A.K., Brown, K.A., et al., 2006. Impact of unrecognized myocardial scar detected by cardiac magnetic resonance imaging on event-free survival in patients presenting with signs or symptoms of coronary artery disease. Circulation 113 (23), 2733–2743.

Laddu, D.R., Rana, J.S., Murillo, R., et al., 2017. 25-year physical activity trajectories and development of subclinical coronary artery disease as measured by coronary artery calcium: the Coronary Artery Risk Development in Young Adults (CARDIA) study. Mayo Clin. Proc. 92 (11), 1660–1670.

Laurent, S., Cockcroft, J., Van Bortel, L., et al., 2006. Expert consensus document on arterial stiffness: methodological issues and clinical applications. Eur. Heart J. 27 (21), 2588–2605.

Lavie, C.J., Ozemek, C., Arena, R., 2019. Physical activity and outcomes in cardiovascular disease. Circ. Res. 124 (5), 799–800.

Lawler, P.R., Filion, K.B., Eisenberg, M.J., 2011. Efficacy of exercise-based cardiac rehabilitation post-myocardial infarction: a systematic review and meta-analysis of randomized controlled trials. Am. Heart J. 162 (4), 571–84.e2.

Lee, I.M., Shiroma, E.J., Lobelo, F., et al., 2012. Effect of physical inactivity on major non-communicable diseases worldwide: an analysis of burden of disease and life expectancy. Lancet 380 (9838), 219–229.

Leggio, M., Lombardi, M., Caldarone, E., et al., 2017. The relationship between obesity and hypertension: an updated comprehensive overview on vicious twins. Hypertens. Res. 40 (12), 947–963.

Leon, A.S., Franklin, B.A., Costa, F., et al., 2005. Cardiac rehabilitation and secondary prevention of coronary heart disease: an American Heart Association scientific statement from the Council on Clinical Cardiology (Subcommittee on Exercise, Cardiac Rehabilitation, and Prevention) and the Council on Nutrition, Physical Activity, and Metabolism (Subcommittee on Physical Activity), in collaboration with the American association of Cardiovascular and Pulmonary Rehabilitation. Circulation 111 (3), 369–376.

Levy, D., Garrison, R.J., Savage, D.D., et al., 1990. Prognostic implications of echocardiographically determined left ventricular mass in the Framingham Heart Study. N. Engl. J. Med. 322 (22), 1561–1566.

Li, A.C., Glass, C.K., 2002. The macrophage foam cell as a target for therapeutic intervention. Nat. Med. 8 (11), 1235–1242.

Libby, P., 2002. Inflammation in atherosclerosis. Nature 420 (6917), 868–874.

Libby, P., Buring, J.E., Badimon, L., et al., 2019. Atherosclerosis. Nat. Rev. Dis. Primers 5 (1), 56.

Libby, P., Ridker, P.M., Hansson, G.K., 2011. Progress and challenges in translating the biology of atherosclerosis. Nature 473 (7347), 317–325.

Lorenz, M.W., Markus, H.S., Bots, M.L., et al., 2007. Prediction of clinical cardiovascular events with carotid intima-media thickness: a systematic review and meta-analysis. Circulation 115 (4), 459–467.

Lorenz, M.W., Polak, J.F., Kavousi, M., et al., 2012. Carotid intima-media thickness progression to predict cardiovascular events in the general population (the PROG-IMT collaborative project): a meta-analysis of individual participant data. Lancet 379 (9831), 2053–2062.

Lüscher, T.F., Barton, M., 1997. Biology of the endothelium. Clin. Cardiol. 20 (11 Suppl 2), II–3–10.

Mark, D.B., Shaw, L., HarrellJr, F.E., et al., 1991. Prognostic value of a treadmill exercise score in outpatients with suspected coronary artery disease. N. Engl. J. Med. 325 (12), 849–853.

Masugata, H., Senda, S., Goda, F., et al., 2006. Left ventricular diastolic dysfunction as assessed by echocardiography in metabolic syndrome. Hypertens. Res. 29 (11), 897–903.

Meier, P., Gloekler, S., Zbinden, R., et al., 2007. Beneficial effect of recruitable collaterals: a 10-year follow-up study in patients with stable coronary artery disease undergoing quantitative collateral measurements. Circulation 116 (9), 975–983.

Meier, P., Hemingway, H., Lansky, A.J., et al., 2012. The impact of the coronary collateral circulation on mortality: a meta-analysis. Eur. Heart J. 33 (5), 614–621.

Mintz, G.S., Nissen, S.E., Anderson, W.D., et al., 2001. American College of Cardiology clinical expert consensus document on standards for acquisition, measurement and reporting of intravascular ultrasound studies (IVUS): a report of the American College of Cardiology task force on clinical expert consensus documents. J. Am. Coll. Cardiol. 37 (5), 1478–1492.

Mitchell, G.F., Hwang, S.J., Vasan, R.S., et al., 2010. Arterial stiffness and cardiovascular events: the Framingham Heart Study. Circulation 121 (4), 505–511.

Task Force Members, Montalescot, G., Sechtem, U., et al., 2013. 2013 ESC guidelines on the management of stable coronary artery disease: the Task Force on the management of stable coronary artery disease of the European Society of Cardiology. Eur. Heart J. 34 (38), 2949–3003.

Moore, K.J., Sheedy, F.J., Fisher, E.A., 2013. Macrophages in atherosclerosis: a dynamic balance. Nat. Rev. Immunol. 13 (10), 709–721.

Mora, S., Cook, N., Buring, J.E., et al., 2007. Physical activity and reduced risk of cardiovascular events: potential mediating mechanisms. Circulation 116 (19), 2110–2118.

Murthy, V.L., Naya, M., Foster, C.R., et al., 2011. Improved cardiac risk assessment with noninvasive measures of coronary flow reserve. Circulation 124 (20), 2215–2224.

Myers, J., Prakash, M., Froelicher, V., et al., 2002. Exercise capacity and mortality among men referred for exercise testing. N. Engl. J. Med. 346 (11), 793–801.

Nakashima, Y., Fujii, H., Sumiyoshi, S., et al., 2007. Early human atherosclerosis: accumulation of lipid and proteoglycans in intimal thickenings followed by macrophage infiltration. Arterioscler. Thromb. Vasc. Biol. 27 (5), 1159–1165.

Naqvi, T.Z., Lee, M.S., 2014. Carotid intima-media thickness and plaque in cardiovascular risk assessment. JACC Cardiovasc. Imaging 7 (10), 1025–1038.

Narayanan, A., Aurigemma, G.P., Chinali, M., et al., 2009. Cardiac mechanics in mild hypertensive heart disease: a speckle-strain imaging study. Circ. Cardiovasc. Imaging 2 (5), 382–390.

Neeland, I.J., Poirier, P., Després, J.P., 2018. Cardiovascular and metabolic heterogeneity of obesity: clinical challenges and implications for management. Circulation 137 (13), 1391–1406.

Neeland, I.J., Ross, R., Després, J.P., et al., 2019. Visceral and ectopic fat, atherosclerosis, and cardiometabolic disease: a position statement. Lancet Diabetes Endocrinol. 7 (9), 715–725.

Nicholls, S.J., Puri, R., Anderson, T., et al., 2016. Effect of evolocumab on progression of coronary disease in statin-treated patients: the GLAGOV randomized clinical trial. JAMA 316 (22), 2373–2384.

Nissen, S.E., Nicholls, S.J., Sipahi, I., et al., 2006. Effect of very high-intensity statin therapy on regression of coronary atherosclerosis: the ASTEROID trial. JAMA 295 (13), 1556–1565.

Nissen, S.E., Yock, P., 2001. Intravascular ultrasound: novel pathophysiological insights and current clinical applications. Circulation 103 (4), 604–616.

Nocon, M., Hiemann, T., Müller-Riemenschneider, F., et al., 2008. Association of physical activity with all-cause and cardiovascular mortality: a systematic review and meta-analysis. Eur. J. Cardiovasc. Prev. Rehabil. 15 (3), 239–246.

Noguchi, T., Kawasaki, T., Tanaka, A., et al., 2014. High-intensity signals in coronary plaques on noncontrast T1-weighted magnetic resonance imaging as a novel determinant of coronary events. J. Am. Coll. Cardiol. 63 (10), 989–999.

Norgren, L., Hiatt, W.R., Dormandy, J.A., et al., 2007. Inter-society consensus for the management of peripheral arterial disease (TASC II). J. Vasc. Surg. 45 (Suppl S), S5–67.

O'Donnell, M.J., Chin, S.L., Rangarajan, S., et al, 2016. Global and regional effects of potentially modifiable risk factors associated with acute stroke in 32 countries (INTERSTROKE): a case-control study. Lancet 388 (10046), 761–775.

O'Leary, D.H., Polak, J.F., Kronmal, R.A., et al., 1999. Carotid-artery intima and media thickness as a risk factor for myocardial infarction and stroke in older adults. N. Engl. J. Med. 340 (1), 14–22.

Ohkawara, K., Tanaka, S., Miyachi, M., et al., 2007. A dose-response relation between aerobic exercise and visceral fat reduction: systematic review of clinical trials. Int. J. Obes. (Lond.) 31 (12), 1786–1797.

Pennell, D.J., 2010. Cardiovascular magnetic resonance. Circulation 121 (5), 692–705.

Perk, J., De Backer, G., Gohlke, H., et al., 2012. European Guidelines on cardiovascular disease prevention in clinical practice (version 2012). The Fifth Joint Task Force of the European Society of Cardiology and Other Societies on Cardiovascular Disease Prevention in Clinical Practice (constituted by representatives of nine societies and by invited experts). Eur. Heart J. 33 (13), 1635–1701.

Polak, J.F., Pencina, M.J., Pencina, K.M., et al., 2011. Carotid-wall intima-media thickness and cardiovascular events. N. Engl. J. Med. 365 (3), 213–221.

Powell, K.E., Paluch, A.E., Blair, S.N., 2011. Physical activity for health: what kind? How much? How intense? On top of what? Annu. Rev. Public Health 32, 349–365.

Prati, F., Guagliumi, G., Mintz, G.S., et al., 2012. Expert review document part 2: methodology, terminology and clinical applications of optical coherence tomography for the assessment of interventional procedures. Eur. Heart J. 33 (20), 2513–2520.

Prati, F., Regar, E., Mintz, G.S., et al., 2010. Expert review document on methodology, terminology, and clinical applications of optical coherence tomography: physical principles, methodology of image acquisition, and clinical application for assessment of coronary arteries and atherosclerosis. Eur. Heart J. 31 (4), 401–415.

Raggi, P., Callister, T.Q., Shaw, L.J., 2004. Progression of coronary artery calcium and risk of first myocardial infarction in patients receiving cholesterol-lowering therapy. Arterioscler. Thromb. Vasc. Biol. 24 (7), 1272–1277.

Rajiah, P., Desai, M.Y., Kwon, D., et al., 2013. MR imaging of myocardial infarction. Radiographics 33 (5), 1383–1412.

Ribeiro, F., Oliveira, N.L., Pires, G., et al., 2021. Beneficial effects of exercise training on endothelial function and biomarkers of cardiovascular risk in dyslipidemic individuals: a systematic review. Blood Press 30 (2), 79–88.

Ridker, P.M., 2016. Residual inflammatory risk: addressing the obverse side of the atherosclerosis prevention coin. Eur. Heart J. 37 (22), 1720–1722.

Ross, R., 1989. Platelet-derived growth factor. Lancet 1 (8648), 1179–1182.

Ross, R., Janssen, I., 2001. Physical activity, total and regional obesity: dose-response considerations. Med. Sci. Sports Exerc. 33 (6 Suppl), S521–7.

Ross, R., Neeland, I.J., Yamashita, S., et al., 2020. Waist circumference as a vital sign in clinical practice: a Consensus Statement from the IAS and ICCR Working Group on Visceral Obesity. Nat. Rev. Endocrinol. 16 (3), 177–189.

Rothney, M.P., Xia, Y., Wacker, W.K., et al., 2013. Precision of a new tool to measure visceral adipose tissue (VAT) using dual-energy X-Ray absorptiometry (DXA). Obesity (Silver Spring) 21 (1), E134–6.

Rundek, T., Arif, H., Boden-Albala, B., et al., 2008. Carotid plaque, a subclinical precursor of vascular events: the Northern Manhattan Study. Neurology 70 (14), 1200–1207.

Sabatine, M.S., Giugliano, R.P., Keech, A.C., et al., 2017. Evolocumab and clinical outcomes in patients with cardiovascular disease. N. Engl. J. Med. 376 (18), 1713–1722.

Salerno, M., Beller, G.A., 2009. Noninvasive assessment of myocardial perfusion. Circ. Cardiovasc. Imaging 2 (5), 412–424.

Sandesara, P.B., Mehta, A., O'Neal, W.T., et al., 2018. Physical activity and coronary artery calcium progression in middle-aged to elderly individuals. JACC Cardiovasc. Imaging 11 (2 Pt 1), 275–276.

Sanz, J., Fayad, Z.A., 2008. Imaging of atherosclerotic cardiovascular disease. Nature 451 (7181), 953–957.

Sattelmair, J., Pertman, J., Ding, E.L., et al., 2011. Dose response between physical activity and risk of coronary heart disease: a meta-analysis. Circulation 124 (7), 789–795.

Schindler, T.H., Schelbert, H.R., Quercioli, A., et al., 2010. Cardiac PET imaging for the detection and monitoring of coronary artery disease and microvascular health. JACC Cardiovasc. Imaging 3 (6), 623–640.

Schlett, C.L., Kwait, D.C., Mahabadi, A.A., et al., 2011. Association between measures of regional adiposity and myocardial perfusion in the Mesa cohort. J. Cardiovasc. Comput. Tomogr. 5 (6), 378–386.

Schulz-Menger, J., Bluemke, D.A., Bremerich, J., et al., 2013. Standardized image interpretation and post processing in cardiovascular magnetic resonance: society for Cardiovascular Magnetic Resonance (SCMR) board of trustees task force on standardized post processing. J. Cardiovasc. Magn. Reson. 15, 35.

Schwartz, G.G., Steg, P.G., Szarek, M., et al, 2018. Alirocumab and cardiovascular outcomes after acute coronary syndrome. N. Engl. J. Med. 379 (22), 2097–2107.

Sciagrà, R., Lubberink, M., Hyafil, F., et al., 2021. EANM procedural guidelines for PET/CT quantitative myocardial perfusion imaging. Eur. J. Nucl. Med. Mol. Imaging 48 (4), 1040–1069.

Sdringola, S., Loghin, C., Boccalandro, F., et al., 2001. Mechanisms of coronary microvascular dysfunction in young patients with familial hypercholesterolemia and normal coronary arteries. J. Nucl. Cardiol. 8 (5), 533–545.

Sdringola, S., Nakagawa, K., Nakagawa, Y., et al., 2003. Combined intense lifestyle and pharmacologic lipid treatment further reduce coronary events and myocardial perfusion abnormalities compared with usual-care cholesterol-lowering drugs in coronary artery disease. J. Am. Coll. Cardiol. 41 (2), 263–272.

Seiler, C., Stoller, M., Pitt, B., et al, 2013. The human coronary collateral circulation: development and clinical importance. Eur. Heart J. 34 (34), 2674–2682.

Sever, P.S., Dahlöf, B., Poulter, N.R., et al., 2003. Prevention of coronary and stroke events with atorvastatin in hypertensive patients who have average or lower-than-average cholesterol concentrations, in the Anglo-Scandinavian Cardiac Outcomes Trial–Lipid Lowering Arm (ASCOT-LLA): a multicentre randomised controlled trial. Lancet 361 (9364), 1149–1158.

Shah, P.K., 2003. Mechanisms of plaque vulnerability and rupture. J. Am. Coll. Cardiol. 41 (4 Suppl S), 15S–22S.

Shaw, L.J., Berman, D.S., Maron, D.J., et al., 2008. Optimal medical therapy with or without percutaneous coronary intervention to reduce ischemic burden: results from the Clinical Outcomes Utilizing Revascularization and Aggressive Drug Evaluation (COURAGE) trial nuclear substudy. Circulation 117 (10), 1283–1291.

Shaw, L.J., Hage, F.G., Berman, D.S., et al., 2012. Prognosis in the era of comparative effectiveness research: where is nuclear cardiology now and where should it be? J. Nucl. Cardiol. 19 (5), 1026–1043.

Shiroma, E.J., Lee, I.M., 2010. Physical activity and cardiovascular health: lessons learned from epidemiological studies across age, gender, and race/ethnicity. Circulation 122 (7), 743–752.

Sitia, S., Tomasoni, L., Atzeni, F., et al., 2010. From endothelial dysfunction to atherosclerosis. Autoimmun. Rev. 9 (12), 830–834.

Slomka, P.J., Berman, D.S., Germano, G., 2014. New cardiac camera technologies. Semin. Nucl. Med. 44 (4), 232–251.

Sofi, F., Capalbo, A., Cesari, F., et al., 2008. Physical activity during leisure time and primary prevention of coronary heart disease: an updated meta-analysis of cohort studies. Eur. J. Cardiovasc. Prev. Rehabil. 15 (3), 247–257.

Spence, J.D., Eliasziw, M., DiCicco, M., et al., 2002. Carotid plaque area: a tool for targeting and evaluating vascular preventive therapy. Stroke 33 (12), 2916–2922.

Spence, J.D., Hackam, D.G., 2010. Treating arteries instead of risk factors: a paradigm change in management of atherosclerosis. Stroke 41 (6), 1193–1199.

Stary, H.C., Chandler, A.B., Dinsmore, R.E., et al., 1995. A definition of advanced types of atherosclerotic lesions and a histological classification of atherosclerosis. A report from the Committee on Vascular Lesions of the Council on Arteriosclerosis, American Heart Association. Circulation 92 (5), 1355–1374.

Steinberg, D., 1997. Low density lipoprotein oxidation and its pathobiological significance. J. Biol. Chem. 272 (34), 20963–20966.

Steyers 3rd, C.M., Miller Jr, F.J., 2014. Endothelial dysfunction in chronic inflammatory diseases. Int. J. Mol. Sci. 15 (7), 11324–11349.

Sytkowski, P.A., Kannel, W.B., D'Agostino, R.B., 1990. Changes in risk factors and the decline in mortality from cardiovascular disease. The Framingham Heart Study. N. Engl. J. Med. 322 (23), 1635–1641.

Tabas, I., García-Cardeña, G., Owens, G.K., 2015. Recent insights into the cellular biology of atherosclerosis. J. Cell Biol. 209 (1), 13–22.

Tall, A.R., Yvan-Charvet, L., 2015. Cholesterol, inflammation and innate immunity. Nat. Rev. Immunol. 15 (2), 104–116.

Taqueti, V.R., Solomon, S.D., Shah, A.M., et al., 2018. Coronary microvascular dysfunction and future risk of heart failure with preserved ejection fraction. Eur. Heart J. 39 (10), 840–849.

Taylor, A.J., Cerqueira, M., Hodgson, J.M., et al., 2010. ACCF/SCCT/ACR/AHA/ASE/ASNC/NASCI/SCAI/SCMR 2010 appropriate use criteria for cardiac computed tomography. J. Am. Coll. Cardiol. 56 (22), 1864–1894.

Taylor, R.S., Brown, A., Ebrahim, S., et al., 2004. Exercise-based rehabilitation for patients with coronary heart disease: systematic review and meta-analysis of randomized controlled trials. Am. J. Med. 116 (10), 682–692.

Tchernof, A., Després, J.P., 2013. Pathophysiology of human visceral obesity: an update. Physiol. Rev. 93 (1), 359–404.

Tearney, G.J., Regar, E., Akasaka, T., et al., 2012. Consensus standards for acquisition, measurement, and reporting of intravascular optical coherence tomography studies: a report from the International Working Group for Intravascular Optical Coherence Tomography Standardization and Validation. J. Am. Coll. Cardiol. 59 (12), 1058–1072.

Thijssen, D.H., Black, M.A., Pyke, K.E., et al., 2011. Assessment of flow-mediated dilation in humans: a methodological and physiological guideline. Am. J. Physiol. Heart Circ. Physiol. 300 (1), H2–H12.

Thijssen, D.H., Cable, N.T., Green, D.J., 2012. Impact of exercise training on arterial wall thickness in humans. Clin. Sci. (Lond.) 122 (7), 311–322.

Thompson, P.D., Buchner, D., Pina, I.L., et al., 2003. Exercise and physical activity in the prevention and treatment of atherosclerotic cardiovascular disease: a statement from the Council on Clinical Cardiology (Subcommittee on Exercise, Rehabilitation, and Prevention) and the Council on Nutrition, Physical Activity, and Metabolism (Subcommittee on Physical Activity). Circulation 107 (24), 3109–3116.

Touboul, P.J., Hennerici, M.G., Meairs, S., et al., 2012. Mannheim carotid intima-media thickness and plaque consensus (2004-2006-2011). An update on behalf of the advisory board of the 3rd, 4th and 5th watching the risk symposia, at the 13th, 15th and 20th European Stroke Conferences, Mannheim, Germany, 2004, Brussels, Belgium, 2006, and Hamburg, Germany, 2011. Cerebrovasc. Dis. 34 (4), 290–296.

Tousoulis, D., Kampoli, A.M., Tentolouris, C., et al., 2012. The role of nitric oxide on endothelial function. Curr. Vasc. Pharmacol. 10 (1), 4–18.

Van Belle, E., Rivard, A., Chen, D., et al., 1997. Hypercholesterolemia attenuates angiogenesis but does not preclude augmentation by angiogenic cytokines. Circulation 96 (8), 2667–2674.

Verheggen, R.J., Maessen, M.F., Green, D.J., et al., 2016. A systematic review and meta-analysis on the effects of exercise training versus hypocaloric diet: distinct effects on body weight and visceral adipose tissue. Obes. Rev. 17 (8), 664–690.

Versari, D., Daghini, E., Virdis, A., et al., 2009. Endothelial dysfunction as a target for prevention of cardiovascular disease. Diabetes Care 32 (Suppl 2), S314–21.

Virmani, R., Kolodgie, F.D., Burke, A.P., et al., 2000. Lessons from sudden coronary death: a comprehensive morphological classification scheme for atherosclerotic lesions. Arterioscler. Thromb. Vasc. Biol. 20 (5), 1262–1275.

Vissers, D., Hens, W., Taeymans, J., et al., 2013. The effect of exercise on visceral adipose tissue in overweight adults: a systematic review and meta-analysis. PLoS One 8 (2), e56415.

Warburton, D.E., Nicol, C.W., Bredin, S.S., 2006. Health benefits of physical activity: the evidence. CMAJ 174 (6), 801–809.

Wen, C.P., Wai, J.P., Tsai, M.K., et al., 2011. Minimum amount of physical activity for reduced mortality and extended life expectancy: a prospective cohort study. Lancet 378 (9798), 1244–1253.

Widlansky, M.E., Gokce, N., Keaney Jr, J.F., et al., 2003. The clinical implications of endothelial dysfunction. J. Am. Coll. Cardiol. 42 (7), 1149–1160.

Williams, K.J., Tabas, I., 1995. The response-to-retention hypothesis of early atherogenesis. Arterioscler. Thromb. Vasc. Biol. 15 (5), 551–561.

Willis, L.H., Slentz, C.A., Bateman, L.A., et al., 2012. Effects of aerobic and/or resistance training on body mass and fat mass in overweight or obese adults. J. Appl. Physiol. 113 (12), 1831–1837.

Wolf, D., Ley, K., 2019. Immunity and inflammation in atherosclerosis. Circ. Res. 124 (2), 315–327.

World Health Organization, 2021. Cardiovascular Diseases (Cvds) Fact Sheet, WHO, Geneva.

Yusuf, S., Hawken, S., Ounpuu, S., et al., 2004. Effect of potentially modifiable risk factors associated with myocardial infarction in 52 countries (the INTERHEART study): case-control study. Lancet 364 (9438), 937–952.

Zbinden, R., Zbinden, S., Meier, P., et al., 2007. Coronary collateral flow in response to endurance exercise training. Eur. J. Cardiovasc. Prev. Rehabil. 14 (2), 250–257.

Index

Note: Page numbers followed by "*f*" indicate figures, "*t*" indicate tables.

9780443289675